MATHEMATICAL PAPERS.

London: C. J. CLAY & SONS,
CAMBRIDGE UNIVERSITY PRESS WAREHOUSE,
AVE MARIA LANE.

Cambridge: DEIGHTON, BELL AND CO.
Leipzig: F. A. BROCKHAUS.

THE COLLECTED

MATHEMATICAL PAPERS

OF

ARTHUR CAYLEY, Sc.D., F.R.S.,

SADLERIAN PROFESSOR OF PURE MATHEMATICS IN THE UNIVERSITY OF CAMBRIDGE.

VOL. I.

CAMBRIDGE:

AT THE UNIVERSITY PRESS.

1889

CAMBRIDGE:
PRINTED BY C. J. CLAY, M.A. AND SONS,
AT THE UNIVERSITY PRESS.

PREFACE.

RATHER more than a year ago I was requested by the Syndics of the University Press to allow my mathematical papers to be reprinted in a collected form: I had great pleasure in acceding to a request so complimentary to myself, and I willingly undertook the work of superintending the impression of them, and of adding such notes and references as might appear to me desirable.

The present volume contains one hundred papers (numbered 1, 2, 3,...,100) originally published in the years 1841 to 1853. They are here reproduced nearly, but not exactly, in chronological order: and as nearly as may be in their original forms; but in a few cases where the paper is controversial, or where it is a translation (into French or English) of an English or French paper, only the title is given: there are in some few cases omissions which are indicated where they occur. The number is printed in the upper inside corner of the page; it is intended that the numbers shall run consecutively through all the volumes; and thus a paper can be referred to simply by its number.

I have of course corrected obvious typographical errors, and in particular have freely altered punctuation, but I have not attempted to verify formulæ. Additions are made in square brackets [ ]; to avoid confusion with these, square brackets occurring in the original papers have in general been changed into twisted ones { }, but where they occur in a formula it was not always possible to make the alteration. The addition in a square bracket is very frequently that of a date: it appears to me that the proper reference to a serial work is by the number of the volume, accompanied by the date on the title page: the date is always useful, and, in the case of two or more series of a Journal or set of Transactions, we avoid the necessity of a reference to the series; Liouville t. I. (1850) is better than Liouville, Série 2, t. I. I regret that this rule has not been strictly followed as regards the titles of some of the earlier papers, see the remark at the end of the Contents.

A. CAYLEY.

January 23, 1889.

CONTENTS.

[An Asterisk denotes that the Paper is not printed in full.]

Volumes II, III, and IV of the *Cambridge Mathematical Journal* have on the title pages the dates 1841, 1843, 1845 respectively, and volume VIII of the *Cambridge Philosophical Transactions* has the date 1849. As each of these volumes extends over more than a single year, I have added, see p. vii, the year of publication for the papers 1, 2, ... 12. In all other cases, the year of publication is shown by the date on the title page of the volume.

CLASSIFICATION.

GEOMETRY

ANALYSIS

1.

ON A THEOREM IN THE GEOMETRY OF POSITION.

[From the *Cambridge Mathematical Journal*, vol. II. (1841), pp. 267—271.]

WE propose to apply the following (new ?) theorem to the solution of two problems in Analytical Geometry.

Let the symbols

$$|\,\alpha\,|, \qquad \begin{vmatrix} \alpha, & \beta \\ \alpha', & \beta' \end{vmatrix}, \qquad \begin{vmatrix} \alpha\,, & \beta\,, & \gamma \\ \alpha', & \beta', & \gamma' \\ \alpha'', & \beta'', & \gamma'' \end{vmatrix}, \text{ \&c.}$$

denote the quantities

$$\alpha,\ \alpha\beta' \quad \alpha'\beta,\ \alpha\beta'\gamma'' - \alpha\beta''\gamma' + \alpha'\beta''\gamma - \alpha'\beta\gamma'' + \alpha''\beta\gamma' - \alpha''\beta'\gamma, \text{ \&c.}$$

(the law of whose formation is tolerably well known, but may be thus expressed,

$$|\,\alpha\,| = \alpha, \qquad \begin{vmatrix} \alpha, & \beta \\ \alpha', & \beta' \end{vmatrix} = \alpha\,|\,\beta'\,| - \alpha'\,|\,\beta\,|,$$

$$\begin{vmatrix} \alpha\,, & \beta\,, & \gamma \\ \alpha', & \beta', & \gamma' \\ \alpha'', & \beta'', & \gamma'' \end{vmatrix} = \alpha \begin{vmatrix} \beta', & \gamma' \\ \beta'', & \gamma'' \end{vmatrix} + \alpha' \begin{vmatrix} \beta'', & \gamma'' \\ \beta\,, & \gamma \end{vmatrix} + \alpha'' \begin{vmatrix} \beta, & \gamma \\ \beta', & \gamma' \end{vmatrix}, \text{ \&c.}$$

the signs + being used when the number of terms in the side of the square is odd, and + and − alternately when it is even.)

Then the theorem in question is

$$\begin{vmatrix} \rho\,\alpha + \sigma\,\beta + \tau\,\gamma.., & \rho\,\alpha' + \sigma\,\beta' + \tau\,\gamma'.., & \rho\,\alpha'' + \sigma\,\beta'' + \tau\,\gamma''.. \\ \rho'\alpha + \sigma'\beta + \tau'\gamma.., & \rho'\alpha' + \sigma'\beta' + \tau'\gamma'.., & \rho'\alpha'' + \sigma'\beta'' + \tau'\gamma''.. \\ \rho''\alpha + \sigma''\beta + \tau''\gamma.., & \rho''\alpha' + \sigma''\beta' + \tau''\gamma'.., & \rho''\alpha'' + \sigma''\beta'' + \tau''\gamma''.. \\ \vdots & \vdots & \vdots \end{vmatrix} = \begin{vmatrix} \rho\,, & \sigma\,, & \tau\,.. \\ \rho', & \sigma', & \tau'.. \\ \rho'', & \sigma'', & \tau''.. \\ \vdots & \vdots & \vdots \end{vmatrix} \begin{vmatrix} \alpha'', & \beta'', & \gamma''.. \\ \alpha', & \beta', & \gamma'.. \\ \alpha\,, & \beta\,, & \gamma\,.. \\ \vdots & \vdots & \vdots \end{vmatrix}$$

(This theorem admits of a generalisation which we shall not have occasion to make use of, and which therefore we may notice at another opportunity.)

To find the relation that exists between the distances of five points in space.

We have, in general, whatever x_1, y_1, z_1, w_1, &c. denote

$$\begin{vmatrix} x_1^2+y_1^2+z_1^2+w_1^2, & -2x_1, & -2y_1, & -2z_1, & -2w_1, & 1 \\ x_2^2+y_2^2+z_2^2+w_2^2, & -2x_2, & -2y_2, & -2z_2, & -2w_2, & 1 \\ \cdot & \cdot & \cdot & \cdot & \cdot & \cdot \\ \cdot & \cdot & \cdot & \cdot & \cdot & \cdot \\ x_5^2+y_5^2+z_5^2+w_5^2, & -2x_5, & -2y_5, & -2z_5, & -2w_5, & 1 \\ 1 \quad , & 0\,, & 0\,, & 0\,, & 0\,, & 0 \end{vmatrix}$$

multiplied into

$$\begin{vmatrix} 1, & x_1, & y_1, & z_1, & w_1, & x_1^2+y_1^2+z_1^2+w_1^2 \\ 1, & x_2, & y_2, & z_2, & w_2, & x_2^2+y_2^2+z_2^2+w_2^2 \\ \cdot & \cdot & \cdot & \cdot & \cdot & \cdot \\ \cdot & \cdot & \cdot & \cdot & \cdot & \cdot \\ 1, & x_5, & y_5, & z_5, & w_5, & x_5^2+y_5^2+z_5^2+w_5^2 \\ 0, & 0, & 0, & 0, & 0, & 1 \end{vmatrix}$$

$$= \begin{vmatrix} \overline{x_1-x_1}^2+\overline{y_1-y_1}^2+\overline{z_1-z_1}^2+\overline{w_1-w_1}^2, & \overline{x_1-x_2}^2+\dots, & \overline{x_1-x_3}^2+\dots, & \overline{x_1-x_4}^2+\dots, & \overline{x_1-x_5}^2+\dots, & 1 \\ \overline{x_2-x_1}^2+\dots\quad, & \overline{x_2-x_2}^2+\dots, & \overline{x_2-x_3}^2+\dots, & \overline{x_2-x_4}^2+\dots, & \overline{x_2-x_5}^2+\dots, & 1 \\ \cdot & \cdot & \cdot & \cdot & \cdot & \cdot \\ \cdot & \cdot & \cdot & \cdot & \cdot & \cdot \\ \overline{x_5-x_1}^2+\dots\quad, & \overline{x_5-x_2}^2+\dots, & \overline{x_5-x_3}^2+\dots, & \overline{x_5-x_4}^2+\dots, & \overline{x_5-x_5}^2+\dots, & 1 \\ 1 \quad , & 1 \quad , & 1 \quad , & 1 \quad , & 1 \quad , & 0 \end{vmatrix}$$

Putting the w's equal to 0, each factor of the first side of the equation vanishes, and therefore in this case the second side of the equation becomes equal to zero. Hence x_1, y_1, z_1, x_2, y_2, z_2, &c. being the coordinates of the points 1, 2, &c. situated arbitrarily in space, and $\overline{12}^2$, $\overline{13}^2$, &c. denoting the squares of the distances between these points, we have immediately the required relation

$$\begin{vmatrix} 0, & \overline{12}^2, & \overline{13}^2, & \overline{14}^2, & \overline{15}^2, & 1 \\ \overline{21}^2, & 0, & \overline{23}^2, & \overline{24}^2, & \overline{25}^2, & 1 \\ \overline{31}^2, & \overline{32}^2, & 0, & \overline{34}^2, & \overline{35}^2, & 1 \\ \overline{41}^2, & \overline{42}^2, & \overline{43}^2, & 0, & \overline{45}^2, & 1 \\ \overline{51}^2, & \overline{52}^2, & \overline{53}^2, & \overline{54}^2, & 0, & 1 \\ 1, & 1, & 1, & 1, & 1, & 0 \end{vmatrix} = 0,$$

which is easily expanded, though from the mere number of terms the process is somewhat long.

Precisely the same investigation is applicable to the case of four points in a plane, or three points in a straight line. Thus the former gives

$$\begin{vmatrix} 0, & \overline{12}^2, & \overline{13}^2, & \overline{14}^2, & 1 \\ \overline{21}^2, & 0, & \overline{23}^2, & \overline{24}^2, & 1 \\ \overline{31}^2, & \overline{32}^2, & 0, & \overline{34}^2, & 1 \\ \overline{41}^2, & \overline{42}^2, & \overline{43}^2, & 0, & 1 \\ 1, & 1, & 1, & 1, & 0 \end{vmatrix} = 0.$$

The latter gives

$$\begin{vmatrix} 0, & \overline{12}^2, & \overline{13}^2, & 1 \\ \overline{21}^2, & 0, & \overline{23}^2, & 1 \\ \overline{31}^2, & \overline{32}^2, & 0, & 1 \\ 1, & 1, & 1, & 0 \end{vmatrix} = 0;$$

or expanding,

$$\overline{12}^4 + \overline{13}^4 + \overline{23}^4 - 2 \,.\, \overline{12}^2\,\overline{13}^2 - 2 \,.\, \overline{13}^2\,\overline{23}^2 - 2 \,.\, \overline{12}^2\,\overline{23}^2 = 0;$$

which may be derived immediately from the equation

$$\pm\,\overline{12} \pm \overline{13} = \pm\,\overline{23},$$

and is the simplest form under which this equation, cleared of the ambiguous signs, can be put.

(The above result may be deduced so elegantly from the general theory of elimination, that notwithstanding its simplicity it is perhaps worth mentioning.)

Let $$x_{,,} - x_{,,,} = \alpha, \quad x_{,,,} - x_{,} = \beta, \quad \overline{x_{,} - x_{,,}} = \gamma;$$

then $$\overline{12}^2 = \gamma^2, \quad \overline{23}^2 = \alpha^2, \quad \overline{31}^2 = \beta^2, \quad \text{and} \quad \alpha + \beta + \gamma = 0;$$

from which α, β, γ are to be eliminated. Multiplying the last equation by $\beta\gamma$, $\gamma\alpha$, $\alpha\beta$, and reducing by the three first,

$$\begin{aligned} 0 \,.\, \alpha + \overline{12}^2 \,.\, \beta + \overline{31}^2 \,.\, \gamma + \quad \alpha\beta\gamma &= 0, \\ \overline{12}^2 \,.\, \alpha + \ 0 \,.\, \beta + \overline{23}^2 \,.\, \gamma + \quad \alpha\beta\gamma &= 0, \\ \overline{31}^2 \,.\, \alpha + \overline{23}^2 \,.\, \beta + \ 0 \,.\, \gamma + \quad \alpha\beta\gamma &= 0, \\ \alpha + \quad \beta + \quad \gamma + 0 \,.\, \alpha\beta\gamma &= 0; \end{aligned}$$

from which, eliminating α, β, γ, $\alpha\beta\gamma$ by the general theory of simple equations,

$$\begin{vmatrix} 0, & \overline{12}^2, & \overline{13}^2, & 1 \\ \overline{21}^2, & 0, & \overline{23}^2, & 1 \\ \overline{31}^2, & \overline{32}^2, & 0, & 1 \\ 1, & 1, & 1, & 0 \end{vmatrix} = 0.$$

The (additional) equation that exists between the distances of five points on a sphere or four points in a circle, has such a remarkable analogy with the preceding, that they almost require to be noticed at the same time.

If α, β, γ, r be the coordinates of the centre, and the radius of the sphere, and $\delta = \alpha^2 + \beta^2 + \gamma^2 - r^2$, we have immediately

$$\begin{array}{c} x_1^2 + y_1^2 + z_1^2 - 2\alpha x_1 - 2\beta y_1 - 2\gamma z_1 + \delta = 0, \\ \vdots \\ x_5^2 + y_5^2 + z_5^2 - 2\alpha x_5 - 2\beta y_5 - 2\gamma z_5 + \delta = 0; \end{array}$$

whence eliminating α, β, γ, δ,

$$\begin{vmatrix} x_1^2 + y_1^2 + z_1^2, & -2x_1, & -2y_1, & -2z_1, & 1 \\ \vdots & \vdots & \vdots & \vdots & \\ x_5^2 + y_5^2 + z_5^2, & -2x_5, & -2y_5, & -2z_5, & 1 \end{vmatrix} = 0;$$

whence, multiplying by

$$\begin{vmatrix} 1, & x_1, & y_1, & z_1, & x_1^2 + y_1^2 + z_1^2 \\ \vdots & \vdots & \vdots & \vdots & \vdots \\ 1, & x_5, & y_5, & z_5, & x_5^2 + y_5^2 + z_5^2 \end{vmatrix}$$

we have immediately

$$\begin{vmatrix} 0, & \overline{12}^2, & \overline{13}^2, & \overline{14}^2, & \overline{15}^2 \\ \overline{21}^2, & 0, & \overline{23}^2, & \overline{24}^2, & \overline{25}^2 \\ \overline{31}^2, & \overline{32}^2, & 0, & \overline{34}^2, & \overline{35}^2 \\ \overline{41}^2, & \overline{42}^2, & \overline{43}^2, & 0, & \overline{45}^2 \\ \overline{51}^2, & \overline{52}^2, & \overline{53}^2, & \overline{54}^2, & 0 \end{vmatrix} = 0.$$

Forming the analogous equation for four points in a circle, and expanding, we readily deduce

$$\overline{14}^4\,\overline{23}^4 + \overline{12}^4\,\overline{34}^4 + \overline{13}^4\,\overline{24}^4 - 2\,.\,\overline{12}^2\,\overline{34}^2\,\overline{13}^2\,\overline{24}^2 - 2\,.\,\overline{14}^2\,\overline{23}^2\,\overline{13}^2\,\overline{24}^2 - 2\,.\,\overline{14}^2\,\overline{23}^2\,\overline{12}^2\,\overline{34}^2 = 0,$$

which is the rational, and therefore analytically the most simple form of

$$\overline{12}\;\overline{34} + \overline{14}\;\overline{23} = \overline{13}\;\overline{24}.$$

Euclid, B. vi., last proposition.

(It may be remarked that the two factors we have employed in the preceding eliminations, only differ by a numerical factor.)

2.

ON THE PROPERTIES OF A CERTAIN SYMBOLICAL EXPRESSION.

[From the *Cambridge Mathematical Journal*, vol. III. (1841), pp. 62—71.]

THE series

$$\mathfrak{S}_p{}_0^\infty \cdot \zeta_p (a^2 + b^2 \ldots \mathfrak{n} \text{ terms})^{p+i} \left(\frac{l}{1+l} \cdot \frac{d^2}{da^2} + \frac{m}{1+m} \cdot \frac{d^2}{db^2} \cdots\right)^p \frac{1}{\{(1+l)\, a^2 + (1+m)\, b^2 \ldots\}^i}$$

$$\left(\zeta_p = \frac{1}{2^{2p+1}\, 1 \,.\, 2 \ldots p \,.\, i\, (i+1) \ldots (i+p)}\right) \ldots\ldots (\psi),$$

possesses some remarkable properties, which it is the object of the present paper to investigate. We shall prove that the symbolical expression (ψ) is independent of a, b, &c., and equivalent to the definite integral

$$\int_0^1 \frac{x^{2i-1}\, dx}{\{(1+lx^2)(1+mx^2) \ldots\}^{\frac{1}{2}}},$$

a property which we shall afterwards apply to the investigation of the attractions of an ellipsoid upon an external point, and to some other analogous integrals. The demonstration of this, which is one of considerable complexity, may be effected as follows:

Writing the symbol $\frac{l}{1+l} \cdot \frac{d^2}{da^2} + \frac{m}{1+m} \cdot \frac{d^2}{db^2} \cdots$ under the form

$$\left(\frac{d^2}{da^2} + \frac{d^2}{db^2} \cdots\right) - \left(\frac{1}{1+l} \cdot \frac{d^2}{da^2} + \frac{1}{1+m} \cdot \frac{d^2}{db^2} \cdots\right) = \Delta - \left(\frac{1}{1+l} \cdot \frac{d^2}{da^2} + \frac{1}{1+m} \cdot \frac{d^2}{db^2} \cdots\right) \text{ suppose,}$$

let the pth power of this quantity be expanded in powers of Δ. The general term is

$$(-1)^q \cdot \frac{p\,(p-1) \ldots (p-q+1)}{1 \,.\, 2 \ldots q} \cdot \Delta^{p-q} \left(\frac{1}{1+l} \cdot \frac{d^2}{da^2} \cdots\right)^q,$$

which is to be applied to $\qquad \dfrac{1}{\{(1+l)\, a^2 \ldots\}^i}.$

Considering the expression

$$\left(\frac{1}{1+l}\cdot\frac{d^2}{da^2}\ldots\right)^q \frac{1}{\{(1+l)\,a^2\ldots\}^i};$$

if for a moment we write

$$(1+l)\,a^2 = a_1{}^2,\ \&\text{c.};\qquad \Delta_1 = \frac{d^2}{da_1{}^2}+\frac{d^2}{db_1{}^2}\ldots;\qquad \rho_1 = a_1{}^2+b_1{}^2\ldots,$$

this becomes

$$\Delta_1{}^q\,\frac{1}{\rho_1{}^i}.$$

Now it is immediately seen that $\Delta_1\dfrac{1}{\rho_1{}^{i'}} = \dfrac{2i'(2i'+2-\mathfrak{n})}{\rho_1{}^{i'+1}}$;

from which we may deduce

$$\Delta_1{}^q\,\frac{1}{\rho_1{}^i} = \frac{2i\,(2i+2)\ldots(2i+2q-2)\,(2i+2-\mathfrak{n})\ldots(2i+2q-\mathfrak{n})}{\rho_1{}^{i+q}},$$

or, restoring the value of ρ_1, and forming the expression for the general term of (ψ), this is

$$\zeta_p\,.\,\rho^{p+1}\left\{\begin{aligned} &\Delta^p\,\frac{1}{(a^2+b^2\ldots+la^2+mb^2+\&\text{c.})^i}\\ &-\frac{p}{1}\,2i\,(2i+2-\mathfrak{n})\,\Delta^{p-1}\,\frac{1}{(a^2+b^2+\ldots+la^2+mb^2+\ldots)^i}\\ &+\&\text{c.}\end{aligned}\right.$$

ρ representing the quantity $a^2+b^2+\&\text{c.}$

Hence, selecting the terms of the sth order in l, m, &c. the expression for the part of (ψ) which is of the sth order in l, m &c. may be written under the form

$$\mathop{\mathbf{S}}\nolimits_{p\,0}^{\,s}\frac{(-1)^s\,\rho^{p+1}\,\zeta_p}{1\,.\,2\ldots s}$$

multiplied by

$$\left\{\begin{aligned} &i\,(i+1)\ldots(i+s-1)\,\Delta^p\,\frac{U}{\rho^{i+s}}\\ &-\frac{p}{1}\,2i\,(2i+2-\mathfrak{n})\,(i+1)\ldots\quad(i+s)\,\Delta^{p-1}\,\frac{U}{\rho^{i+s+1}}\\ &+\frac{p\,(p-1)}{1\,.\,2}\,2i\,(2i+2)\,(2i+2-\mathfrak{n})\,(2i+4-\mathfrak{n})\,(i+2)\ldots(i+s+1)\,\Delta^{p-2}\,\frac{U}{\rho^{i+s+2}}\\ &-\&\text{c.}\qquad [la^2+mb^2\ldots = U\ \text{suppose}]\end{aligned}\right.$$

which for conciseness we shall represent by

$$\frac{(-1)^s}{1\,.\,2\ldots s}\mathop{\mathbf{S}}\nolimits_{p\,0}^{\,s}\,\rho^{p+1}\,.\,\zeta_p\left\{\begin{aligned} &\alpha_s\,\Delta^p\,\frac{U}{\rho^{i+s}}\\ &-\frac{p}{1}\,\beta_s\,\Delta^{p-1}\,\frac{U}{\rho^{i+s+1}}\\ &+\frac{p\,(p-1)}{1\,.\,2}\,\gamma_s\,\Delta^{p-2}\,\frac{U}{\rho^{i+s+2}}\\ &-\&\text{c.}\end{aligned}\right.$$

$= S$ suppose.

Now U representing any homogeneous function of the order $2s$, it is easily seen that

$$\Delta \frac{U}{\rho^i} = \frac{\Delta U}{\rho^i} + 2i\,(2i+2-4s-\mathfrak{n})\,\frac{U}{\rho^{i+1}};$$

and repeating continually the operation Δ, observing that ΔU, $\Delta^2 U$, &c. are of the orders $2(s-1)$, $2(s-2)$, &c. we at length arrive at

$$\begin{aligned}\Delta^q . \frac{U}{\rho^i} = {} & \Delta^q\, U . \frac{1}{\rho^i} \\ & + \frac{q}{1}\, 2i\,(2i+2q-4s-\mathfrak{n})\,\Delta^{q-1}\, U . \frac{1}{\rho^{i+1}} \\ & + \frac{q(q-1)}{1.2}\, 2i\,(2i+2)\,(2i+2q-4s-\mathfrak{n})\,(2i+2q-4s-\mathfrak{n}-2)\,\Delta^{q-2}\, U . \frac{1}{\rho^{i+2}} \\ & \vdots \\ & + 2i\,(2i+2)\ldots(2i+2q)\,(2i+2q-4s-\mathfrak{n})\ldots\quad(2i+2-4s-\mathfrak{n})\, U . \frac{1}{\rho^{i+q}}.\end{aligned}$$

Changing i into $s+i+i'$, we have an equation which we may represent by

$$\Delta^q \frac{U}{\rho^{s+i+i'}} = A_{q,\,i'} \frac{\Delta^q\, U}{\rho^{s+i+i'}} + {}^1A_{q,\,i'} \frac{\Delta^{q-1}\, U}{\rho^{s+i+i'+1}} \ldots + {}^qA_{q,\,i'} \frac{U}{\rho^{s+i+i'+q}} \ldots \qquad (\alpha),$$

where in general

$$\begin{aligned}{}^rA_{q,\,i'} = {} & \frac{q\,(q-1)\ldots(q-r+1)}{1.2\ldots r} \\ & \times (2s+2i+2i')\,(2s+2i'+2i+2)\ldots(2s+2i'+2i+2r-2) \\ & \times (2i+2i'+2q-2s-\mathfrak{n})\ldots(2i+2i'+2q-2s-\mathfrak{n}-2r+2).\end{aligned}$$

Now the value of S, written at full length, is

$$\frac{(-1)^s}{1.2\ldots s}\left\{\begin{aligned} & \zeta_s \rho^{s+1}\left(\alpha_s \Delta^s \frac{U}{\rho^{s+1}} - \frac{s}{1}\beta_s \Delta^{s-1}\frac{U}{\rho^{s+i+1}}\ldots\right. \\ & + \zeta_{s-1}\,\rho^{s+i-1}\left(\alpha_s \Delta^{s-1}\frac{U}{\rho^{s+1}} - \frac{s-1}{1}\beta_s \Delta^{s-2}\frac{U}{\rho^{s+i+1}} + \ldots\right. \\ & + \&\text{c.}\end{aligned}\right.$$

and substituting for the several terms of this expansion the values given by the equation (α), we have

$$S = \frac{(-1)^s}{1.2\ldots\ldots s}\left(k_0 \Delta^s U + k_1 \frac{1}{\rho}\Delta^{s-1} U \ldots\ldots + k_s \frac{1}{\rho^s}\, U\right)$$

where in general

$$\begin{aligned} k_{\text{x}} = {} & \alpha_s\,({}^{\text{x}}A_{s,\,0}\,\zeta_s + {}^{\text{x}-1}A_{s-1,\,0}\,\zeta_{s-1} \ldots + A_{s-\text{x},\,0}\,\zeta_{s-\text{x}}) \\ & - \beta_s\left(\frac{s}{1}\,{}^{\text{x}-1}A_{s-1,\,1}\,\zeta_s \ldots + \frac{(s-\text{x}+1)}{1} A_{s-\text{x},\,1}\,\zeta_{s-\text{x}+1}\right) \\ & \vdots \\ & \pm \lambda_s\left(\frac{s\,(s-1)\ldots(s-\text{x}+1)}{1.2\ldots\text{x}} A_{s-\text{x},\,\text{x}}\,\zeta_s\right),\end{aligned}$$

λ_s being the $(\text{x}+1)^{\text{th}}$ of the series $\alpha_s, \beta_s \ldots$

Substituting for the quantities involved in this expression, and putting, for simplicity $2i+2-\mathfrak{n}=2\gamma$, we have, without any further reduction, except that of arranging the factors of the different terms, and cancelling those which appear in the numerator and denominator of the same term,

$$\frac{(-1)^s k_{\text{x}}}{1\,.\,2\ldots\ldots s}=\frac{(-1)^{s-\text{x}}(1-\gamma)(2-\gamma)\ldots(\text{x}-\gamma)}{2^{2s+1}\,.\,1\,.\,2\ldots s\,.\,1\,.\,2\ldots(s-\text{x})\,.\,1\,.\,2\ldots\text{x}}$$

multiplied by the series

$$(i+s+1)\ldots(i+s+\text{x}-1)\ \text{into}$$

$$\left\{1+\frac{\gamma}{1}\frac{\text{x}}{\text{x}-\gamma}+\frac{\gamma(\gamma+1)}{1\,.\,2}\frac{\text{x}(\text{x}-1)}{(\text{x}-\gamma)(\text{x}-1-\gamma)}+\ldots\qquad(\text{x}+1)\ \text{terms}\right\}$$

$$-\frac{(i+s)\ldots(i+s+\text{x}-2)}{1-\gamma}\ \text{into}$$

$$\left\{\text{x}+\frac{\gamma}{1}\frac{\text{x}(\text{x}-1)}{\text{x}-\gamma}+\frac{\gamma(\gamma+1)}{1\,.\,2}\frac{\text{x}(\text{x}-1)(\text{x}-2)}{(\text{x}-\gamma)(\text{x}-1-\gamma)}+\ldots\qquad\text{x terms}\right\}$$

$$\vdots$$

$$+(-1)^r\frac{(i+s-r+1)\ldots(i+s+\text{x}-r-1)}{(1-\gamma)(2-\gamma)\ldots(r-\gamma)}\ \text{into}$$

$$\left\{\text{x}(\text{x}-1)\ldots(\text{x}-r+1)+\frac{\gamma}{1}\frac{\text{x}(\text{x}-1)\ldots(\text{x}-r)}{\text{x}-\gamma}+\ldots(\text{x}+r-1)\ \text{terms}\right\}$$

to $r=\text{x}$.

Now it may be shown that

$$\frac{1}{(1-\gamma)(2-\gamma)\ldots(r-\gamma)}$$

$$\left\{\text{x}(\text{x}-1)\ldots(\text{x}-r+1)+\frac{\gamma}{1}\frac{\text{x}(\text{x}-1)\ldots(\text{x}-r)}{\text{x}-\gamma}+\&\text{c.}\ldots(\text{x}+1-r)\ \text{terms}\right\}$$

$$=\frac{\text{x}(\text{x}-1)\ldots(r+1)\,.\,\text{x}(\text{x}-1)\ldots(\text{x}-r+1)}{(1-\gamma)(2-\gamma)\ldots(\text{x}-\gamma)},$$

which reduces the expression for k_{x} to the form

$$\frac{(-1)^s k_{\text{x}}}{1\,.\,2\ldots s}=\frac{(-1)^{s+\text{x}}}{2^{2s+1}\,.\,1\,.\,2\ldots s\,.\,1\,.\,2\ldots(s-\text{x})}\left\{\begin{aligned}&(i+s+1)\ldots(i+s+\text{x}-1)\\&-\frac{\text{x}}{1}(i+s)\ldots(i+s+\text{x}-2)\\&+\frac{\text{x}(\text{x}-1)}{1\,.\,2}(i+s-1)\ldots(i+s+\text{x}-3)\\&\pm\&\text{c.}\ (\text{x}+1)\ \text{terms};\end{aligned}\right.$$

from which it may be shown, that except for $\text{x}=0$, $k_{\text{x}}=0$.

The value $\text{x}=0$, observing that the expression

$$(i+s+1)(i+s+2)\ldots(i+s-1)$$

represents $\dfrac{1}{i+s}$, gives

$$\frac{(-1)^s k_0}{1.2\ldots s} = \frac{(-1)^s}{2^{2s}(1.2\ldots s)^2.(2i+2s)};$$

or we have simply

$$S = \frac{(-1)^s}{2^{2s}(1.2\ldots s)^2.(2i+2s)} \Delta^s U,$$

where $$\Delta = \frac{d^2}{da^2} + \frac{d^2}{db^2} + \ldots, \quad U = (la^2 + mb^2 \ldots)^s.$$

Consider the term $$\frac{1.2\ldots s}{1.2\ldots\lambda.1.2\ldots\mu.\&c.} a^{2\lambda} b^{2\mu} \ldots l^\lambda m^\mu \ldots;$$

with respect to this, Δ^s reduces itself to

$$\frac{1.2\ldots s}{1.2\ldots\lambda.1.2\ldots\mu.\&c.} \left(\frac{d}{da}\right)^{2\lambda} \ldots$$

and the corresponding term of S is

$$\frac{(-1)^s}{2^{2s}(2i+2s)(1.2\ldots\lambda.1.2\ldots\mu.\&c.)^2} 1.2\ldots 2\lambda.1.2\ldots 2\mu.\&c.\ l^\lambda m^\mu \ldots$$

$$= \frac{(-1)^s.1.3\ldots(2\lambda-1).1.3\ldots(2\mu-1).\&c.}{(2i+2s)\,2.4\ldots 2\lambda.2.4\ldots 2\mu.\&c.} l^\lambda m^\mu \ldots$$

which, omitting the factor $\frac{1}{2i+2s}$, and multiplying by x^{2s}, is the general term of the sth order in $l, m, \ldots$ of

$$\frac{1}{\sqrt{\{(1+lx^2)(1+mx^2)\ldots\}}}.$$

The term itself is therefore the general term of

$$\int_0^1 \frac{x^{2i-1}\,dx}{\sqrt{\{(1+lx^2)(1+mx^2)\ldots\}}};$$

or taking the sum of all such terms for the complete value of S, and the sum of the different values of S for the values 0, 1, 2... of the variable s, we have the required equation

$$\psi = \int_0^1 \frac{x^{2i-1}\,dx}{\sqrt{\{(1+lx^2)(1+mx^2)\ldots\}}}.$$

Another and perhaps more remarkable form of this equation may be deduced by writing $\frac{a^2}{1+l}, \frac{b^2}{1+m}$, &c. for a^2, b^2, &c., and putting $\frac{a^2}{1+l} + \frac{b^2}{1+m} + \&c. = \eta^2$, $l\eta^2 = \alpha^2$, $m\eta^2 = \beta^2$, &c.: we readily deduce

$$\eta^{n-2i} \int_0^1 \frac{x^{2i-1}\,dx}{\sqrt{\{(\eta^2+\alpha^2x^2)(\eta^2+\beta^2x^2)\ldots\}}}$$

$$= \mathbf{S}_p{}_0^\infty \frac{1}{2^{2p+1}.1.2\ldots p.i(i+1)\ldots(i+p)} \left(\alpha^2\frac{d^2}{da^2} + \beta^2\frac{d^2}{db^2}\ldots\right)^p \frac{1}{(a^2+b^2\ldots)^i},$$

η being determined by the equation

$$\frac{a^2}{\eta^2+\alpha^2}+\frac{b^2}{\eta^2+\beta^2}\ldots=1;$$

or, as it may otherwise be written,

$$\eta^2=a^2+b^2+\ldots-\frac{a^2\alpha^2}{\eta^2+\alpha^2}-\frac{b^2\beta^2}{\eta^2+\beta^2}-\&c.$$

$\mathfrak{n}$, it will be recollected, denotes the number of the quantities a, b, &c.

Now suppose

$$V=\iint\ldots\phi(a-x,\ b-y,\ldots)\,dx\,dy\ldots$$

(the integral sign being repeated $\mathfrak{n}$ times) where the limits of the integral are given by the equation

$$\frac{x^2}{h^2}+\frac{y^2}{h_{,}^{2}}+\&c.=1;$$

and that it is permitted, throughout the integral to expand the function $\phi(a-x,\ldots)$ in ascending powers of x, y, &c. (the condition for which is apparently that of ϕ not becoming infinite for any values of x, y, &c., included within the limits of the integration): then observing that any integral of the form $\iint\ldots x^p y^q\ldots dx\,dy$ &c.... where any one of the exponents p, q, &c.... is odd, when taken between the required limits contains equal positive and negative elements and therefore vanishes, the general term of V assumes the form

$$\frac{1}{1.2\ldots2r.1.2\ldots2s\ldots}\left(\frac{d}{da}\right)^{2r}\left(\frac{d}{db}\right)^{2s}\ldots\phi(a,\ b\ldots)\iint\ldots x^{2r}y^{2s}\ldots dx\,dy\ldots$$

Also, by a formula quoted in the eleventh No. of the *Mathematical Journal*, the value of the definite integral $\iint\ldots x^{2r}y^{2s}\ldots dx\,dy\ldots$ is

$$h^{2r+1}h_{,}^{2s+1}\ldots\frac{\Gamma(r+\frac{1}{2})\,\Gamma(s+\frac{1}{2})\ldots}{\Gamma(r+s+\ldots+\frac{1}{2}\mathfrak{n}+1)},$$

(observing that the value there given referring to positive values only of the variables, must be multiplied by $2^{\mathfrak{n}}$): or, as it may be written

$$h^{2r+1}h_{,}^{2s+1}\ldots\pi^{\frac{1}{2}\mathfrak{n}}.\frac{1}{2^{r+s\ldots}}\frac{1.3\ldots(2r-1).1.3\ldots(2s-1)\ldots}{\frac{1}{2}\mathfrak{n}(\frac{1}{2}\mathfrak{n}+1)\ldots(\frac{1}{2}\mathfrak{n}+r+s\ldots)\,\Gamma(\frac{1}{2}\mathfrak{n})}:$$

hence the general term of V takes the form

$$\frac{hh_{,}\ldots\pi^{\frac{1}{2}\mathfrak{n}}}{\Gamma(\frac{1}{2}\mathfrak{n})}\frac{1}{\frac{1}{2}\mathfrak{n}(\frac{1}{2}\mathfrak{n}+1)\ldots(\frac{1}{2}\mathfrak{n}+r+s\ldots)}.\frac{1}{2^{2r+2s\ldots}}\frac{1}{1.2.3\ldots r.1.2\ldots s\ldots}$$

$$\times\left(h^2\frac{d^2}{da^2}\right)^r\left(h_{,}^2\frac{d^2}{db^2}\right)^s\ldots\phi(a,\ b,\ \ldots);$$

and putting $r+s+\&c.=p$, and taking the sum of the terms that answer to the same value of p, it is immediately seen that this sum is

$$=\frac{hh_{,}\ldots\pi^{\frac{1}{2}\mathfrak{n}}}{\Gamma(\frac{1}{2}\mathfrak{n})}.\frac{1}{2^{2p}.1.2\ldots p.\frac{1}{2}\mathfrak{n}(\frac{1}{2}\mathfrak{n}+1)\ldots(\frac{1}{2}\mathfrak{n}+p)}\left(h^2\frac{d^2}{da^2}+h_{,}^2\frac{d^2}{db^2}\ldots\right)^p.\phi(a,\ b\ldots).$$

Or the function $\phi(a-x,\ b-y\ldots)$ not becoming infinite within the limits of the integration, we have

$$\iint\ldots\phi(a-x,\ b-y\ldots)\,dx\,dy\ldots$$

$$=\frac{2hh_{\prime}\ldots\pi^{\frac{1}{2}\mathfrak{n}}}{\Gamma(\frac{1}{2}\mathfrak{n})}\mathbf{S}_{p\,0}^{\ \infty}\frac{1}{2^{2p+1}.1.2\ldots p.\frac{1}{2}\mathfrak{n}(\frac{1}{2}\mathfrak{n}+1)\ldots(\frac{1}{2}\mathfrak{n}+p)}\left(h^2\frac{d^2}{da^2}+h_{\prime}^{\,2}\frac{d^2}{db^2}\ldots\right)^p\phi(a,\ b\ldots),$$

the integral on the first side of the equation extending to all real values of x, y, &c., subject to $\dfrac{x^2}{h^2}+\dfrac{y^2}{h^2}+\ldots<1$.

Suppose in the first place $\qquad \phi(a,\ b\ldots)=\dfrac{1}{(a^2+b^2\ldots)^{\frac{1}{2}\mathfrak{n}}}$.

By a preceding formula the second side of the equation reduces itself to

$$\frac{2hh_{\prime}\ldots\pi^{\frac{1}{2}\mathfrak{n}}}{\Gamma(\frac{1}{2}\mathfrak{n})}\int_0^1\frac{x^{\mathfrak{n}-1}\,dx}{\sqrt{\{(\eta^2+h^2x^2)(\eta^2+h_{\prime}^{\,2}x^2)\ldots(\mathfrak{n}\text{ factors})\}}},$$

η being given by $\qquad \dfrac{a^2}{\eta^2+h^2}+\dfrac{b^2}{\eta^2+h_{\prime}^{\,2}}\ldots=1.$

Hence the formula

$$\iint\ldots\mathfrak{n}\text{ times }\frac{dx\,dy\ldots}{\{(a-x)^2+(b-y)^2\ldots\}^{\frac{1}{2}\mathfrak{n}}}$$

$$=\frac{2hh_{\prime}\ldots\pi^{\frac{1}{2}\mathfrak{n}}}{\Gamma(\frac{1}{2}\mathfrak{n})}\int_0^1\frac{x^{\mathfrak{n}-1}.\,dx}{\sqrt{\{(\eta^2+h^2x^2)(\eta^2+h_{\prime}^{\,2}x^2)\ldots(\mathfrak{n}\text{ factors})\}}};$$

where the integral on the first side of the equation extends to all real values of x, y, &c. satisfying $\dfrac{x^2}{h^2}+\dfrac{y^2}{h_{\prime}^{\,2}}+\text{\&c.}\ldots<1$; η^2, as we have seen, is determined by

$$\frac{a^2}{\eta^2+h^2}+\frac{b^2}{\eta^2+h_{\prime}^{\,2}}+\text{\&c.}=1;$$

and finally, the condition of $\phi(a-x,\ b-y\ldots)$ not becoming infinite within the limits of the integration, reduces itself to $\dfrac{a^2}{h^2}+\dfrac{b^2}{h_{\prime}^{\,2}}+\ldots>1$, which must be satisfied by these quantities.

Suppose in the next place that the function $\phi(a,\ b\ldots)$ satisfies $\dfrac{d^2\phi}{da^2}+\dfrac{d^2\phi}{db^2}+\text{\&c.}=0$. The factor $\left(h^2\dfrac{d^2}{da^2}+\text{\&c.}\right)$ may be written under the form

$$(h_{\prime}^{\,2}-h^2)\frac{d^2}{db^2}+(h_{\prime\prime}^{\,2}-h^2)\frac{d^2}{dc^2}+\text{\&c.}+h^2\left(\frac{d^2}{da^2}+\frac{d^2}{db^2}\ldots\right)=(h_{\prime}^{\,2}-h^2)\frac{d^2}{db^2}+\text{\&c.}$$

since, as applied to the function ϕ, $\frac{d^2}{da^2}+\frac{d^2}{db^2}+\&c.$ is equivalent to 0; we have in this case

$$\iint\ldots\phi(a-x,\ b-y\ldots)\,dx\,dy\ldots$$

$$=\frac{2hh_,\ldots\pi^{\frac{1}{2}n}}{\Gamma(\frac{1}{2}n)}\,\mathbf{S}_{p}{}_0^\infty\,\frac{1}{2^{2p+1}.1.2\ldots p.\frac{1}{2}n\ldots(\frac{1}{2}n+p)}\left\{(h_,^2-h^2)\frac{d^2}{db^2}+\ldots\right\}^p\phi(a,\ b\ldots);$$

or the first side divided by $hh_,\ldots$ has the remarkable property of depending on the differences $h_,^2-h^2$, &c. only; this is the generalisation of a well-known property of the function V, in the theory of the attraction of a spheroid upon an external point.

If in this equation we put $\phi(a,\ b\ldots)=\frac{a}{(a^2+b^2\ldots)^{\frac{1}{2}n}}$, which satisfies the required condition $\frac{d^2\phi}{da^2}+\&c.=0$, then transferring the factor a to the left-hand side of the sign S, and putting in a preceding formula, $\alpha^2=0$, $\beta^2=h_,^2-h^2$, &c. and η^2+h^2 for η^2, we obtain

$$\iint\ldots(n\text{ times})\ \frac{(a-x)\,dx\,dy\ldots}{\{(a-x)^2+(b-y)^2\ldots\}^{\frac{1}{2}n}}$$

$$=\frac{2hh_,\ldots\pi^{\frac{1}{2}n}\,a}{\sqrt{(\eta^2+h^2)}\,.\,\Gamma(\frac{1}{2}n)}\int_0^1\frac{x^{n-1}\,dx}{\sqrt{[\{\eta^2+h^2+(h_,^2-h^2)\,x^2\}\,\{\eta^2+h^2+(h_{,,}^2-h^2)\,x^2\}\ldots(n-1)\text{ factors}]}},$$

where, as before, the integrations on the first side extend to all real values of x, y, &c., satisfying $\frac{x^2}{h^2}+\frac{y^2}{h_,^2}\ldots<1$; η^2 is determined by $\frac{a^2}{\eta^2+h^2}+\&c.=1$; and $a,\ b,\ldots h,\ h_,$, &c. are subject to $\frac{a^2}{h^2}+\frac{b^2}{h_,^2}+\&c.\ldots>1$.

For $n=3$, this becomes,

$$\iiint\frac{(a-x)\,dx\,dy\,dz}{\{(a-x)^2+(b-y)^2+(c-z)^2\}^{\frac{3}{2}}}$$

$$=\frac{4\pi hh_,h_{,,}a}{\sqrt{(h^2+\eta^2)}}\int_0^1\frac{x^2dx}{\sqrt{[\{\eta^2+h^2+(h_,^2-h^2)\,x^2\}\,\{\eta^2+h^2+(h_{,,}^2-h^2)\,x^2\}]}}:$$

the integrations on the first side extending over the ellipsoid whose semiaxes are $h,\ h_,,\ h_{,,}$, and the point whose coordinates are a, b, c, being exterior to this ellipsoid; also $\frac{a^2}{\eta^2+h^2}+\frac{b^2}{\eta^2+h_,^2}+\frac{c^2}{\eta^2+h_{,,}^2}=1$: a known theorem.

3.

ON CERTAIN DEFINITE INTEGRALS.

[From the *Cambridge Mathematical Journal*, vol. III. (1841), pp. 138—144.]

IN the first place, we shall consider the integral

$$V = \iint \ldots (\mathfrak{n} \text{ times}) \frac{dx\, dy \ldots}{\{(a-x)^2 + (b-y)^2 \ldots\}^{\frac{1}{2}\mathfrak{n}-1}},$$

the integration extending to all real values of the variables, subject to the condition

$$\frac{x^2}{h^2} + \frac{y^2}{h_,^2} + \ldots < \text{ or } = 1,$$

and the constants a, b, &c. satisfying the condition

$$\frac{a^2}{h^2} + \frac{b^2}{h_,^2} \ldots > 1.$$

We have

$$\frac{dV}{da} = -(\mathfrak{n}-2) \iint \ldots (\mathfrak{n} \text{ times}) \frac{(a-x)\, dx\, dy \ldots}{\{(a-x)^2 + (b-y)^2 \ldots\}^{\frac{1}{2}\mathfrak{n}}},$$

$$= -(\mathfrak{n}-2) \frac{2hh_, \ldots \pi^{\frac{1}{2}\mathfrak{n}} a}{\sqrt{(\xi + h^2)} \,.\, \Gamma(\frac{1}{2}\mathfrak{n})} \int_0^1 \frac{x^{\mathfrak{n}-1}\, dx}{\sqrt{[\{\xi + h^2 + (h_,^2 - h^2)\, x^2\} \{\xi + h^2 + (h_{,,}^2 - h^2)\, x^2\} \ldots]}},$$

ξ being determined by the equation

$$\frac{a^2}{\xi + h^2} + \frac{b^2}{\xi + h_,^2} \ldots = 1,$$

by a formula [see p. 12] in a paper, [2], "On the Properties of a Certain Symbolical Expression," in the preceding No. of this Journal: ξ having been substituted for the η^2 of the formula.

Let the variable x, on the second side of the equation, be replaced by ϕ, where

$$x^2 = \frac{\xi + h^2}{\xi + h^2 + \phi};$$

we have without difficulty

$$\frac{dV}{da} = -(\mathfrak{n}-2)\frac{hh_{,}\dots\pi^{\frac{1}{2}\mathfrak{n}}a}{\Gamma(\frac{1}{2}\mathfrak{n})}\int_0^\infty \frac{d\phi}{(\xi + h^2 + \phi)\sqrt{\Phi}},$$

where $\Phi = (\xi + h^2 + \phi)(\xi + h_{,}^2 + \phi)\dots$

and similarly

$$\frac{dV}{db} = -(\mathfrak{n}-2)\frac{hh_{,}\dots\pi^{\frac{1}{2}\mathfrak{n}}b}{\Gamma(\frac{1}{2}\mathfrak{n})}\int_0^\infty \frac{d\phi}{(\xi + h_{,}^2 + \phi)\sqrt{\Phi}},$$

&c. ......

From these values it is easy to verify the equation

$$V = \frac{(\mathfrak{n}-2)hh_{,}\dots\pi^{\frac{1}{2}\mathfrak{n}}}{2\Gamma(\frac{1}{2}\mathfrak{n})}\int_0^\infty \left(1 - \frac{a^2}{\xi + h^2 + \phi} - \frac{b^2}{\xi + h_{,}^2 + \phi}\dots\right)\frac{d\phi}{\sqrt{\Phi}}.$$

For this evidently verifies the above values of $\frac{dV}{da}$, $\frac{dV}{db}$, &c. if only the term $\frac{dV}{d\xi}d\xi$ vanishes; and we have

$$\frac{dV}{d\xi} = \frac{(\mathfrak{n}-2)hh_{,}\dots\pi^{\frac{1}{2}\mathfrak{n}}}{2\Gamma(\frac{1}{2}\mathfrak{n})}\int_0^\infty d\phi\,.\,\frac{d}{d\xi}\left(1 - \frac{a^2}{\xi + h^2 + \phi}\dots\right)\frac{1}{\sqrt{\Phi}};$$

or, observing that

$$\frac{d}{d\xi}\left(1 - \frac{a^2}{\xi + h^2 + \phi} - \dots\right)\frac{1}{\sqrt{(\Phi)}} = \frac{d}{d\phi}\left(1 - \frac{a^2}{\xi + h^2 + \phi}\dots\right)\frac{1}{\sqrt{(\Phi)}},$$

and taking the integral from 0 to ∞,

$$\frac{dV}{d\xi} = -\frac{(\mathfrak{n}-2)hh_{,}\dots\pi^{\frac{1}{2}\mathfrak{n}}}{2\Gamma(\frac{1}{2}\mathfrak{n})}\left(1 - \frac{a^2}{\xi + h^2} - \frac{b^2}{\xi + h_{,}^2}\dots\right)\frac{1}{\sqrt{\{(\xi + h^2)(\xi + h_{,}^2)\dots\}}}, = 0,$$

in virtue of the equation which determines ξ.

No constant has been added to the value of V, since the two sides of the equation vanish as they should do for $a, b\dots$ infinite, for which values ξ is also infinite and the quantity

$$\left(1 - \frac{a^2}{\xi + h^2 + \phi}\dots\right)\frac{1}{\sqrt{(\Phi)}},$$

which is always less than $\frac{1}{\sqrt{(\Phi)}}$, vanishes.

Hence, restoring the values of V and Φ,

$$\iint\dots(\mathfrak{n}\text{ times})\frac{dx\,dy\dots}{\{(a-x)^2 + (b-y)^2\dots\}^{\frac{1}{2}\mathfrak{n}-1}}$$

$$= \frac{(\mathfrak{n}-2)hh_{,}\dots\pi^{\frac{1}{2}\mathfrak{n}}}{2\Gamma(\frac{1}{2}\mathfrak{n})}\int_0^\infty \left(1 - \frac{a^2}{\xi + h^2 + \phi} - \frac{b^2}{\xi + h_{,}^2 + \phi}\dots\right)\frac{d\phi}{\sqrt{\{(\xi + h^2 + \phi)(\xi + h_{,}^2 + \phi)\dots\}}}$$

the limits of the first side of the equation, and the condition to be satisfied by a, b, &c., also the equation for the determination of ξ, being as above.

The integral

$$V' = \iint \dots (\mathfrak{n} \text{ times}) \frac{dx\, dy \dots}{\{(a-x)^2 + (b-y)^2 \dots\}^{\frac{1}{2}\mathfrak{n}}},$$

between the same limits, and with the same condition to be satisfied by the constants, has been obtained [see p. 11] in the paper already quoted. Writing ξ instead of η^2, and $x^2 = \dfrac{\xi}{\xi+\phi}$, we have

$$V' = \frac{hh_{\prime} \dots \pi^{\frac{1}{2}\mathfrak{n}}}{\Gamma(\frac{1}{2}\mathfrak{n})} \int_0^\infty \frac{d\phi}{(\xi+\phi)\sqrt{\{(\xi+h^2+\phi)(\xi+h_{\prime}^2+\phi)\dots\}}},$$

where

$$\frac{a^2}{\xi+h^2} + \frac{b^2}{\xi^2+h_{\prime}^2} \dots = 1.$$

Let $\nabla = \dfrac{d^2}{da^2} + \dfrac{d^2}{db^2} + \dots$ Then by the assistance of a formula,

$$\nabla^q \frac{1}{(a^2+b^2\dots)^i} = 2i(2i+2)\dots(2i+2q-2)(2i+2-\mathfrak{n})\dots(2i+2q-\mathfrak{n}) \cdot \frac{1}{(a^2+b^2\dots)^{i+q}}$$

given in the same paper [see p. 6], in which it is obvious that a, $b \dots$ may be changed into $a-x$, $b-y$, &c. ...; also putting $i = \frac{1}{2}\mathfrak{n}$; we have

$$\iint \dots (\mathfrak{n} \text{ times}) \frac{dx\, dy \dots}{\{(a-x)^2 + \dots\}^{\frac{1}{2}\mathfrak{n}+q}} = \frac{hh_{\prime} \dots\dots \pi^{\frac{1}{2}\mathfrak{n}}}{2^{2q}.1.2\dots q.\Gamma(\frac{1}{2}\mathfrak{n}+q)} \int_0^\infty d\phi \,.\, \nabla^q \frac{1}{(\xi+\phi)\sqrt{\{(\xi+h^2+\phi)\dots\}}}.$$

Now in general, if $\chi\xi$ be any function of ξ,

$$\nabla\chi\xi = \chi'\xi\left(\frac{d^2\xi}{da^2} + \frac{d^2\xi}{db^2}\dots\right) + \chi''\xi\left\{\left(\frac{d\xi}{da}\right)^2 + \left(\frac{d\xi}{db}\right)^2\dots\right\} = \chi'\xi\,\Sigma\left(\frac{d^2\xi}{da^2}\right) + \chi''\xi\,\Sigma\left(\frac{d\xi}{da}\right)^2, \text{ suppose.}$$

But from the equation

$$\Sigma\frac{a^2}{(\xi+h^2)} = 1,$$

we obtain

$$\frac{2a}{\xi+h^2} - \left\{\Sigma\frac{a^2}{(\xi+h^2)^2}\right\}\frac{d\xi}{da} = 0,$$

whence

$$\Sigma\left(\frac{d\xi}{da}\right)^2 = \frac{4}{\Sigma\dfrac{a^2}{(\xi+h^2)^2}}.$$

Also

$$\frac{2}{\xi+h^2} - 4\frac{a}{(\xi+h^2)^2}\frac{d\xi}{da} + 2\left\{\Sigma\frac{a^2}{(\xi+h^2)^3}\right\}\left(\frac{d\xi}{da}\right)^2 - \left\{\Sigma\frac{a^2}{(\xi+h^2)^2}\right\}\frac{d^2\xi}{da^2} = 0;$$

whence taking the sum Σ, and observing that

$$-4\Sigma\frac{a}{(\xi+h^2)^2}\frac{d\xi}{da} = -8\frac{\Sigma\dfrac{a^2}{(\xi+h^2)^3}}{\Sigma\dfrac{a^2}{(\xi+h^2)^2}} = -2\Sigma\frac{a^2}{(\xi+h^2)^3}\,.\,\Sigma\left(\frac{d\xi}{da}\right)^2,$$

$$2\Sigma\frac{1}{\xi+h^2} - \left\{\Sigma\frac{a^2}{(\xi+h^2)^2}\right\}\Sigma\left(\frac{d^2\xi}{da^2}\right) = 0;$$

we find

$$\Sigma\left(\frac{d^2\xi}{da^2}\right)=\frac{2\Sigma\,\dfrac{1}{\xi+h^2}}{\Sigma\,\dfrac{a^2}{(\xi+h^2)^2}};$$

and we hence obtain

$$\nabla\chi\xi=\frac{2\chi'\xi\,\Sigma\,\dfrac{1}{\xi+h^2}+4\,\chi''\,\xi}{\Sigma\,\dfrac{a^2}{(\xi+h^2)^2}}.$$

Hence the function

$$\int_0^\infty d\phi\,.\,\nabla\,\frac{1}{(\xi+\phi)\sqrt{\{(\xi+h^2+\phi)\ldots\}}}$$

(observing that differentiation with respect to ξ is the same as differentiation with respect to ϕ) becomes integrable, and taking the integral between the proper limits, its value is

$$-\frac{2\chi_0\xi\,\Sigma\,\dfrac{1}{\xi+h^2}+4\chi_0{}'\,\xi}{\Sigma\,\dfrac{a^2}{(\xi+h^2)^2}};$$

where

$$\chi_0\xi=\frac{1}{\xi\sqrt{\{(\xi+h^2)\,(\xi+h_{,}^2)\ldots\}}}.$$

We have immediately

$$\frac{\chi_0{}'\xi}{\chi_0\xi}=-\tfrac{1}{2}\left(\frac{2}{\xi}+\Sigma\,\frac{1}{\xi+h^2}\right);$$

or

$$2\chi_0\xi\,\Sigma\left(\frac{1}{\xi+h^2}\right)+4\chi_0{}'\xi=-\,4\,\frac{\chi_0\xi}{\xi};$$

whence

$$\int_0^\infty d\phi\,.\,\nabla\,\frac{1}{(\xi+\phi)\sqrt{\{(\xi+h^2+\phi)\ldots\}}}=\frac{4}{\xi^2\sqrt{\{(\xi+h^2)\,(\xi+h_{,}^2)\ldots\}}\left\{\dfrac{a^2}{(\xi+h^2)^2}+\dfrac{b^2}{(\xi+h_{,}^2)^2}+\ldots\right\}}.$$

Hence restoring the value of ∇, and of the first side of the equation,

$$\iint\ldots\,(\mathfrak{n}\text{ times})\,\frac{dx\,dy\ldots}{\{(a-x)^2+(b-y)^2\ldots\}^{\frac{1}{2}\mathfrak{n}+q}}$$

$$=\frac{hh_{,}\ldots\,\pi^{\frac{1}{2}\mathfrak{n}}}{2^{2q-2}\,.\,1\,.\,2\ldots q\,.\,\Gamma(\frac{1}{2}\mathfrak{n}+q)}\left(\frac{d^2}{da^2}+\frac{d^2}{db^2}\ldots\right)^{q-1}\frac{1}{\xi^2\sqrt{\{(\xi+h^2)\,(\xi+h_{,}^2)\ldots\}}\left\{\dfrac{a^2}{(\xi+h^2)^2}+\dfrac{b^2}{(\xi+h_{,}^2)^2}+\ldots\right\}},$$

with the condition

$$\frac{a^2}{\xi+h^2}+\frac{b^2}{\xi+h_{,}^2}\ldots=1\,;$$

from which equation the differential coefficients of ξ, which enter into the preceding result, are to be determined.

In general if u be any function of ξ, a, $b\dots$

$$\left(\frac{d^2}{da^2}+\frac{d^2}{db^2}\dots\right)u=\frac{4\frac{d^2u}{d\xi^2}+2\frac{du}{d\xi}\Sigma\frac{1}{\xi+h^2}+4\frac{d^2u}{d\xi da}\Sigma\frac{a}{\xi+h^2}}{\Sigma\frac{a^2}{(\xi+h^2)^2}}+\Sigma\frac{d^2u}{da^2},$$

from which the values of the second side for $q=1$, $q=2$, &c. may be successively calculated.

The performance of the operation $\left(\frac{d}{da}\right)^p\left(\frac{d}{db}\right)^q\left(\frac{d}{dc}\right)^r\dots$, upon the integral V', leads in like manner to a very great number of integrals, all of them expressible algebraically, for a single differentiation renders the integration with respect to ϕ possible. But this is a subject which need not be further considered at present.

We shall consider, lastly, the definite integral

$$U=\iint\dots(\mathfrak{n}\text{ times})\frac{(a-x)f\left(\frac{x^2}{h^2}+\frac{y^2}{h_{\prime}^{2}}+\dots\right)dxdy\dots}{\{(a-x)^2+(b-y)^2\dots\}^{\frac{1}{2}\mathfrak{n}}},$$

limits, &c. as before. This is readily deduced from the less general one

$$\iint\dots(\mathfrak{n}\text{ times})\frac{(a-x)\,dxdy\dots}{\{(a-x)^2+(b-y)^2\dots\}^{\frac{1}{2}\mathfrak{n}}}.$$

For representing this quantity by $F(h,\ h_{\prime}\dots)$, it may be seen that

$$U=\int_0^1 f(m^2)\frac{d}{dm}F(mh,\ mh_{\prime}\dots)\,dm;$$

but in the value of $F(h,\ h_{\prime}\dots)$, changing $h,\ h_{\prime}\dots$ into $mh,\ mh_{\prime}\dots$ also writing $m^2\phi$ instead of ϕ, and $m^2\xi'$ for ξ, we have

$$F(mh,\ mh_{\prime}\dots)=\frac{hh_{\prime}\dots\pi^{\frac{1}{2}\mathfrak{n}}}{\Gamma(\frac{1}{2}\mathfrak{n})}a\int_0^\infty\frac{d\phi}{(\xi'+h^2+\phi)\sqrt{(\Phi')}};$$

where

$$\Phi'=(\xi'+h^2+\phi)(\xi'+h_1^2+\phi)\dots$$

and

$$\frac{a^2}{\xi'+h^2}+\frac{b^2}{\xi'+h_{\prime}^{2}}+\dots=m^2.$$

Hence

$$\frac{d}{dm}F(mh,\ mh_{\prime}\dots)=\frac{d\xi'}{dm}\frac{d}{d\xi'}F(mh,\ mh_{\prime}\dots),$$

$$=\frac{hh_{\prime}\dots\pi^{\frac{1}{2}\mathfrak{n}}}{\Gamma(\frac{1}{2}\mathfrak{n})}a\frac{d\xi'}{dm}\int_0^\infty d\phi\frac{d}{d\xi'}\frac{1}{(\xi'+h^2+\phi)\sqrt{(\Phi')}},$$

or, observing that $\frac{d}{d\xi'}$ is equivalent to $\frac{d}{d\phi}$, and effecting the integration between the proper limits,

$$\frac{d}{dm}F(mh,\ mh_{\prime}\dots)=-\frac{hh_{\prime}\dots\pi^{\frac{1}{2}\mathfrak{n}}}{\Gamma(\frac{1}{2}\mathfrak{n})}a\frac{1}{(\xi'+h^2)\sqrt{\{(\xi'+h^2)(\xi'+h_{\prime}^{2})\dots\}}}.$$

Substituting this value, also $f\left\{\frac{a^2}{\xi'+h^2}+\frac{b^2}{\xi'+h_,^2}+\ldots\right\}$ for $f(m^2)$, in the value of U, and observing that $m=0$ gives $\xi'=\infty$, $m=1$ gives $\xi'=\xi$, where ξ is a quantity determined as before by the equation

$$\frac{a^2}{\xi+h^2}+\frac{b^2}{\xi+h_,^2}+\ldots=1,$$

we have
$$U=-\frac{hh_,\ldots\pi^{\frac{1}{2}\mathfrak{n}}a}{\Gamma(\frac{1}{2}\mathfrak{n})}\int_\infty^\xi\frac{f\left\{\frac{a^2}{\xi'+h^2}+\ldots\right\}d\xi'}{(\xi'+h^2)\sqrt{\{(\xi'+h^2)(\xi'+h_,^2)\ldots\}}},$$

or writing $\phi+\xi$ for ξ', $d\xi'=d\phi$, the limits of ϕ are 0, ∞; or, inverting the limits and omitting the negative sign,

$$U=\frac{hh_,\ldots\pi^{\frac{1}{2}\mathfrak{n}}a}{\Gamma(\frac{1}{2}\mathfrak{n})}\int_0^\infty\frac{f\left\{\frac{a^2}{\xi+h^2+\phi}+\frac{b^2}{\xi+h_,^2+\phi}+\ldots\right\}d\phi}{(\xi+h^2+\phi)\sqrt{\{(\xi+h^2+\phi)(\xi+h_,^2+\phi)\ldots\}}};$$

which, in the particular case of $\mathfrak{n}=3$, may easily be made to coincide with known results. The analogous integral

$$\iint\ldots(\mathfrak{n}\text{ times})\frac{f\left\{\frac{x^2}{h^2}+\frac{y^2}{h_,^2}+\ldots\right\}dx\,dy\ldots}{\{(a-x)^2+(b-y)^2\ldots\}^{\frac{1}{2}\mathfrak{n}}}$$

is apparently not reducible to a single integral.

4.

ON CERTAIN EXPANSIONS, IN SERIES OF MULTIPLE SINES AND COSINES.

[From the *Cambridge Mathematical Journal*, vol. III. (1842), pp. 162—167.]

In the following paper we shall suppose ϵ the base of the hyperbolic system of logarithms; e a constant, such that its modulus, and also the modulus of $\frac{1}{e}\{1-\sqrt{1-e^2}\}$, are each of them less than unity; $\chi\{\epsilon^{u\surd(-1)}\}$ a function of u, which, as u increases from 0 to π, passes continuously from the former of these values to the latter, without becoming a maximum in the interval, $f\{\epsilon^{u\surd(-1)}\}$ any function of u which remains finite and continuous for values of u included between the above limits. Hence, writing

$$\chi\{\epsilon^{u\surd(-1)}\}=m \quad\ldots\ldots\ldots\ldots(1),$$

and considering the quantity

$$\frac{\sqrt{1-e^2}\,f\{\epsilon^{u\surd(-1)}\}}{\sqrt{-1}\,\epsilon^{u\surd(-1)}\chi'\{\epsilon^{u\surd(-1)}\}\,(1-e\cos u)} \quad\ldots\ldots\ldots\ldots(2),$$

as a function of m, for values of m or u included between the limits 0 and π, we have

$$\frac{\sqrt{1-e^2}\,f\{\epsilon^{u\surd(-1)}\}}{\sqrt{-1}\,\epsilon^{u\surd(-1)}\,\chi'\{\epsilon^{u\surd(-1)}\}\,(1-e\cos u)}=\frac{2}{\pi}\Sigma_{-\infty}^{\infty}\cos rm\int_0^\pi\frac{\sqrt{1-e^2}\,f\{\epsilon^{u\surd(-1)}\}\cos rm\,dm}{\sqrt{-1}\,\epsilon^{u\surd(-1)}\,\chi'\{\epsilon^{u\surd(-1)}\}\,(1-e\cos u)}\ldots(3),$$

(Poisson, *Mec.* tom. I. p. 650); which may also be written

$$\frac{\sqrt{1-e^2}\,f\{\epsilon^{u\surd(-1)}\}}{\sqrt{-1}\,\epsilon^{u\surd(-1)}\,\chi'\{\epsilon^{u\surd(-1)}\}\,(1-e\cos u)}=\frac{2}{\pi}\Sigma_{-\infty}^{\infty}\cos rm\int_0^\pi\frac{\sqrt{1-e^2}\,f\{\epsilon^{u\surd(-1)}\}\cos r\chi\{\epsilon^{u\surd(-1)}\}\,du}{1-e\cos u}\ldots(4);$$

and if the first side of the equation be generally expansible in a series of multiple cosines of m, instead of being so in particular cases only, its expanded value will always be the one given by the second side of the preceding equation.

Now, between the limits 0 and π, the function

$$f\{\epsilon^{u\sqrt{(-1)}}\}\cos r\chi\{\epsilon^{u\sqrt{(-1)}}\}$$

will always be expansible in a series of multiple cosines of u; and if by any algebraical process the function $f\rho\cos r\chi\rho$ can be expanded in the form

$$f\rho\cos r\chi\rho = \Sigma_{-\infty}^{\infty}\alpha_s\rho^s, \quad (\alpha_s = \alpha_{-s}) \quad\ldots\ldots(5);$$

we have, in a convergent series,

$$f\{\epsilon^{u\sqrt{(-1)}}\}\cos r\chi\{\epsilon^{u\sqrt{(-1)}}\} = \alpha_0 + 2\Sigma_1^{\infty}\alpha_s\cos su \quad\ldots\ldots(6).$$

Again, putting $$\frac{1}{e}\{1-\sqrt{1-e^2}\} = \lambda \quad\ldots\ldots(7),$$

we have $$\frac{\sqrt{1-e^2}}{1-e\cos u} = 1 + 2\Sigma_1^{\infty}\lambda^p\cos pu \quad\ldots\ldots(8).$$

Multiplying these two series, and effecting the integration, we obtain

$$\frac{1}{\pi}\int_0^{\pi}\frac{\sqrt{1-e^2}f\{\epsilon^{u\sqrt{(-1)}}\}\cos r\chi\{\epsilon^{u\sqrt{(-1)}}\}du}{1-e\cos u} = 2\{\tfrac{1}{2}\alpha_0 + \Sigma_1^{\infty}(\alpha_s\lambda^s)\} \quad\ldots\ldots(9),$$

and the second side of this equation being obviously derived from the expansion of $f\lambda\cos r\chi\lambda$ by rejecting negative powers of λ and dividing by 2, the term independent of λ may conveniently be represented by the notation

$$\overbrace{\underbrace{2f\lambda\cos r\chi\lambda}} \quad\ldots\ldots(10);$$

where in general, if $\Gamma\lambda$ can be expanded in the form

$$\Gamma\lambda = \Sigma_{-\infty}^{\infty}(A_s\lambda^s), \qquad [A_{-s} = A_s] \quad\ldots\ldots(11),$$

we have $$\overbrace{\underbrace{\Gamma\lambda}} = \tfrac{1}{2}A_0 + \Sigma_1^{\infty}A_s\lambda^s \quad\ldots\ldots(12).$$

(By what has preceded, the expansion of $\Gamma\lambda$ in the above form is always possible in a certain sense; however, in the remainder of the present paper, $\Gamma\lambda$ will always be of a form to satisfy the equation $\Gamma\left(\frac{1}{\lambda}\right) = \Gamma\lambda$, except in cases which will afterwards be considered, where the condition $A_{-s} = A_s$ is unnecessary.)

Hence, observing the equations (4), (9), (10),

$$\frac{\sqrt{1-e^2}f\{\epsilon^{u\sqrt{(-1)}}\}}{\sqrt{-1}\,\epsilon^{u\sqrt{(-1)}}\chi'\{\epsilon^{u\sqrt{(-1)}}\}(1-e\cos u)} = \Sigma_{-\infty}^{\infty}\cos rm\,\overbrace{\underbrace{2\cos r\chi\lambda f\lambda}} \quad\ldots\ldots(13);$$

from which, assuming a system of equations analogous to (1), and representing by $\Pi(\Phi)$ the product $\Phi_1\Phi_2\ldots$, it is easy to deduce

$$\Pi\left\{\frac{\sqrt{1-e^2}}{\sqrt{-1}\,\epsilon^{u\sqrt{(-1)}}\chi'\{\epsilon^{u\sqrt{(-1)}}\}(1-e\cos u)}\right\}f\{\epsilon^{u_1\sqrt{(-1)}},\ \epsilon^{u_2\sqrt{(-1)}}\ldots\}$$

$$= \Sigma_{-\infty}^{\infty}\Sigma_{-\infty}^{\infty}\ldots\Pi\cos rm\,\overbrace{\underbrace{\Pi(2\cos r\chi\lambda)f(\lambda_1,\ \lambda_2\ldots)}} \quad\ldots\ldots(14),$$

where $\Gamma(\lambda_1, \lambda_2 \ldots)$ being expansible in the form

$$\Gamma(\lambda_1, \lambda_2 \ldots) = \Sigma_{-\infty}^{\infty} \Sigma_{-\infty}^{\infty} \ldots A_{s_1, s_2 \ldots} \lambda_1^{s_1} \lambda_2^{s_2} \ldots [A_{s_1, s_2 \ldots} = A_{-s_1, -s_2 \ldots}] \ldots\ldots (15),$$

$$\Gamma(\lambda_1, \lambda_2 \ldots) = \Sigma_0^{\infty} \Sigma_0^{\infty} \ldots \frac{1}{2^N} A_{s_1, s_2 \ldots} \lambda_1^{s_1} \lambda_2^{s_2} \ldots, \quad \ldots\ldots\ldots\ldots (16),$$

N being the number of exponents which vanish.

The equations (13) and (14) may also be written in the forms

$$f\{\epsilon^{u\sqrt{(-1)}}\} = \Sigma_{-\infty}^{\infty} \cos rm \, \overbrace{2 \cos r\chi\lambda \frac{\sqrt{-1}\, \chi'\lambda \{1 - \frac{1}{2}e(\lambda + \lambda^{-1})\}}{\sqrt{1-e^2}} f\lambda} \ldots\ldots (17),$$

$$f\{\epsilon^{u_1\sqrt{(-1)}}, \epsilon^{u_2\sqrt{(-1)}} \ldots\}$$

$$= \Sigma_{-\infty}^{\infty} \Sigma_{-\infty}^{\infty} \ldots \Pi(\cos rm)\, \overbrace{\Pi \left\{2 \cos r\chi\lambda \frac{\sqrt{-1}\, \chi'\lambda \{1 - \frac{1}{2}e(\lambda + \lambda^{-1})\}}{\sqrt{1-e^2}}\right\} f(\lambda_1, \lambda_2 \ldots)} \ldots (18).$$

As examples of these formulæ, we may assume

$$\chi\{\epsilon^{u\sqrt{(-1)}}\} = m = u - e \sin u \ldots\ldots\ldots\ldots\ldots\ldots\ldots\ldots (19).$$

Hence, putting

$$\lambda^r \epsilon^{-\frac{re}{2}(\lambda - \lambda^{-1})} + \lambda^{-r} \epsilon^{\frac{re}{2}(\lambda - \lambda^{-1})} = \Lambda_r \ldots\ldots\ldots\ldots\ldots\ldots (20),$$

and observing the equation

$$\sqrt{-1}\, \epsilon^{u\sqrt{(-1)}} \chi'\{\epsilon^{u\sqrt{(-1)}}\} = 1 - e \cos u \ldots\ldots\ldots\ldots\ldots\ldots (21),$$

the equation (17) becomes

$$f\{\epsilon^{u\sqrt{(-1)}}\} = \Sigma_{-\infty}^{\infty} \cos rm \Lambda_r \overbrace{\frac{\{1 - \frac{1}{2}e(\lambda + \lambda^{-1})\}^2}{\sqrt{(1-e^2)}} f\lambda} \ldots\ldots\ldots\ldots (22).$$

Thus, if
$$\theta - \varpi = \cos^{-1} \frac{\cos u - e}{1 - e \cos u} \ldots\ldots\ldots\ldots\ldots\ldots\ldots\ldots (23),$$

assuming
$$f\{\epsilon^{u\sqrt{(-1)}}\} = \frac{\cos u - e}{1 - e \cos u} \ldots\ldots\ldots\ldots\ldots\ldots\ldots\ldots (24),$$

$$\cos(\theta - \varpi) = \Sigma_{-\infty}^{\infty} \frac{1}{\sqrt{1-e^2}} \cos rm \, \overbrace{\{1 - \tfrac{1}{2}e(\lambda + \lambda^{-1})\}\{\tfrac{1}{2}(\lambda + \lambda^{-1}) - e\} \Lambda_r} \ldots (25),$$

the term corresponding to $r = 0$ being

$$\frac{1}{2\sqrt{1-e^2}} \{2\lambda - 2e - e(\lambda^2 + 1) + 2e^2\lambda\}, \; = -e \ldots\ldots\ldots\ldots\ldots (26).$$

Again, assuming

$$f\{\epsilon^{u\sqrt{(-1)}}\} = \frac{d\theta}{dm} = \frac{\sqrt{1-e^2}}{(1 - e \cos u)^2} \ldots\ldots\ldots\ldots\ldots\ldots (27),$$

and integrating the resulting equation with respect to m,

$$\theta - \varpi = \Sigma_{-\infty}^{\infty} \frac{\sin rm}{r} \underbrace{\overbrace{\Lambda_r}} = m + 2\Sigma_1^{\infty} \frac{\sin rm}{r} \underbrace{\overbrace{\Lambda_r}} \quad \ldots\ldots\ldots\ldots\ldots (28),$$

a formula given in the fifth No. of the *Mathematical Journal*, and which suggested the present paper.

As another example, let

$$f\{\epsilon^{u\sqrt{(-1)}}\} = \cos(\theta - \varpi) \frac{d\theta}{dm} = \frac{\sqrt{1-e^2}\,(\cos u - e)}{(1 - e\cos u)^3} \quad \ldots\ldots\ldots\ldots\ldots (29).$$

Then integrating with respect to m, there is a term

$$2m \underbrace{\overbrace{\frac{\frac{1}{2}(\lambda + \lambda^{-1}) - e}{1 - \frac{1}{2}e(\lambda + \lambda^{-1})}}} \quad \ldots\ldots\ldots\ldots\ldots (30),$$

which it is evident, *à priori*, must vanish. Equating it to zero, and reducing, we obtain

$$\frac{e}{1-e^2} = \underbrace{\overbrace{\frac{\lambda + \lambda^{-1}}{1 - \frac{1}{2}e(\lambda + \lambda^{-1})}}} \quad \ldots\ldots\ldots\ldots\ldots (31),$$

that is

$$\frac{e}{1-e^2} = \lambda + \frac{e}{2}(\lambda^2 + 1) + \frac{e^2}{4}(\lambda^3 + 3\lambda) + \frac{e^3}{8}(\lambda^4 + 4\lambda^2 + 3) + \ldots \quad \ldots\ldots (32),$$

a singular formula, which may be verified by substituting for λ its value: we then obtain

$$\sin(\theta - \varpi) = 2\Sigma_1^{\infty} \frac{\sin rm}{r} \underbrace{\overbrace{\Lambda_r \frac{\frac{1}{2}(\lambda + \lambda^{-1}) - e}{1 - \frac{1}{2}(\lambda + \lambda^{-1})}}} \quad \ldots\ldots\ldots\ldots\ldots (33).$$

The expansions of $\sin k(\theta - \varpi)$, $\cos k(\theta - \varpi)$, are in like manner given by the formulæ

$$\cos k(\theta - \varpi) = \Sigma_{-\infty}^{\infty} \underbrace{\overbrace{\Lambda_r L' \cos kL}} \cos rm \quad \ldots\ldots\ldots\ldots\ldots (34),$$

$$\sin k(\theta - \varpi) = \Sigma_{-\infty}^{\infty} \underbrace{\overbrace{\Lambda_r \frac{1}{kr} \cos kL}} \frac{\sin rm}{r} \quad \ldots\ldots\ldots\ldots\ldots (35),$$

where, to abbreviate, we have written

$$\cos^{-1}\left\{\frac{\frac{1}{2}(\lambda + \lambda^{-1}) - e}{1 - \frac{1}{2}e(\lambda + \lambda^{-1})}\right\} = L \quad \ldots\ldots\ldots\ldots\ldots (36),$$

$$\frac{\{1 - \frac{1}{2}e(\lambda + \lambda^{-1})\}^2}{\sqrt{1-e^2}} = L' \quad \ldots\ldots\ldots\ldots\ldots (37).$$

Forming the analogous expressions for

$$\cos k(\theta' - \varpi'), \qquad \sin k(\theta' - \varpi'),$$

substituting in

$$\cos k(\theta-\theta') = \cos k(\varpi-\varpi')\{\cos k(\theta-\varpi)\cos k(\theta'-\varpi') + \sin k(\theta-\varpi)\sin k(\theta'-\varpi')\}$$
$$-\sin k(\varpi-\varpi')\{\sin k(\theta-\varpi)\cos k(\theta'-\varpi') - \sin k(\theta'-\varpi')\cos k(\theta-\varpi)\},$$

and reducing the whole to multiple cosines, the final result takes the very simple form

$$\cos k(\theta-\theta') = \Sigma_{-\infty}^{\infty} \cos\{r'm' - rm + k(\varpi-\varpi')\}\, \overbrace{\Lambda_r \Lambda'_{r'} \cos kL \cos kL' \left(L - \frac{1}{kr}\right)\left(L' - \frac{1}{kr'}\right)} \quad \ldots(38).$$

Again, formulæ analogous to (14), (18), may be deduced from the equation

$$\Gamma(m_1, m_2 \ldots)$$
$$= \Sigma_{-\infty}^{\infty} \Sigma_{-\infty}^{\infty} \ldots \left\{ \begin{aligned} &\cos(r_1 m_1 + r_2 m_2 \ldots) \int_0^{2\pi} \frac{dm_1}{2\pi} \int_0^{2\pi} \frac{dm_2}{2\pi} \ldots \cos(r_1 m_1 + r_2 m_2 \ldots)\, \Gamma(m_1, m_2 \ldots) \\ &+ \sin(r_1 m_1 + r_2 m_2 \ldots) \int_0^{2\pi} \frac{dm_1}{2\pi} \int_0^{2\pi} \frac{dm_2}{2\pi} \ldots \sin(r_1 m_1 + r_2 m_2 \ldots)\, \Gamma(m_1, m_2 \ldots) \end{aligned} \right. \quad (39),$$

which holds from $m_1 = 0$ to $m_1 = 2\pi$, &c., but in many cases universally. In this case, writing for $\Gamma(m_1, m_2 \ldots)$ the function

$$\Pi \left\{ \frac{1}{\sqrt{-1}\, \epsilon^{u\sqrt{(-1)}}\, \chi'\{\epsilon^{u\sqrt{(-1)}}\}} \; \frac{\sqrt{1-e^2} - e \sin u \sqrt{-1}}{1 - e\cos u} \right\} f\{\epsilon^{u_1\sqrt{(-1)}}, \epsilon^{u_2\sqrt{(-1)}} \ldots\} \ldots(40);$$

and observing

$$\frac{\sqrt{1-e^2} - e\sin u\sqrt{-1}}{1 - e\cos u} = \frac{1 + \lambda\epsilon^{-u\sqrt{(-1)}}}{1 - \lambda\epsilon^{-u\sqrt{(-1)}}} = 1 + 2\Sigma_1^{\infty}\{\cos su - \sqrt{-1}\sin su\}\lambda^s \ldots\ldots(41),$$

an exactly similar analysis, (except that in the expansion

$$\Gamma(\lambda_1, \lambda_2 \ldots) = \Sigma_{-\infty}^{\infty} \Sigma_{-\infty}^{\infty} \ldots A_{s_1, s_2} \ldots \lambda_1^{s_1} \lambda_2^{s_2} \ldots,$$

the supposition is not made that $A_{s_1, s_2} \ldots = A_{-s_1, -s_2} \ldots$), leads to the result

$$f\{\epsilon^{u_1\sqrt{(-1)}}, \epsilon^{u_2\sqrt{(-1)}} \ldots\}\, \Pi \left\{ \frac{\sqrt{1-e^2} - e\sin u\sqrt{-1}}{\sqrt{-1}\, \epsilon^{u\sqrt{(-1)}}\, \chi'\{\epsilon^{u\sqrt{(-1)}}\}(1 - e\cos u)} \right\}$$
$$= \Sigma_{-\infty}^{\infty} \Sigma_{-\infty}^{\infty} \ldots \left\{ \begin{aligned} &\cos(r_1 m_1 + r_2 m_2 \ldots)\, \overbrace{2^{\mathfrak{n}} \cos(r_1\chi_1\lambda_1 + r_2\chi_2\lambda_2 \ldots) f(\lambda_1, \lambda_2 \ldots)} \\ &+ \sin(r_1 m_1 + r_2 m_2 \ldots)\, \overbrace{2^{\mathfrak{n}} \sin(r_1\chi_1\lambda_1 + r_2\chi_2\lambda_2 \ldots) f(\lambda_1, \lambda_2 \ldots)} \end{aligned} \right. \ldots(42),$$

$(\mathfrak{n})$ being the number of variables $u_1, u_2 \ldots$. Hence also $f\{\epsilon^{u_1\sqrt{(-1)}}, \epsilon^{u_2\sqrt{(-1)}} \ldots\}$

$$= \Sigma_{-\infty}^{\infty} \Sigma_{-\infty}^{\infty} \ldots \left\{ \begin{aligned} &\cos(r_1 m_1 + \ldots)\, \overbrace{\cos(r_1\chi_1\lambda_1 + \ldots)\, \Pi \left\{ \frac{2\sqrt{-1}\,\chi'\lambda\{1 - \tfrac{1}{2}e(\lambda + \lambda^{-1})\}}{\sqrt{1-e^2} - \tfrac{1}{2}e(\lambda - \lambda^{-1})} \right\} f(\lambda_1, \lambda_2 \ldots)} \\ &+ \sin(r_1 m_1 + \ldots)\, \overbrace{\sin(r_1\chi_1\lambda_1 + \ldots)\, \Pi \left\{ \frac{2\sqrt{-1}\,\chi'\lambda\{1 - \tfrac{1}{2}e(\lambda + \lambda^{-1})\}}{\sqrt{1-e^2} - \tfrac{1}{2}e(\lambda - \lambda^{-1})} \right\} f(\lambda_1, \lambda_2 \ldots)} \end{aligned} \right.$$

$$\ldots\ldots\ldots(43).$$

By choosing for $f\{\epsilon^{u_1\sqrt{(-1)}},\ \epsilon^{u_2\sqrt{(-1)}}\ldots\}$, functions expansible without sines, or without cosines, a variety of formulæ may be obtained: we may instance

$$\underparen{\overparen{\frac{(\lambda-\lambda^{-1})\,\{1-\tfrac{1}{2}e\,(\lambda+\lambda^{-1})\}\,\Lambda_r}{\sqrt{1-e^2}-\tfrac{1}{2}e\,(\lambda-\lambda^{-1})}}}=0 \quad\ldots\ldots\ldots\ldots\ldots\ldots(44),$$

Λ_r having the same meaning as before.

Also,

$$\underparen{\overparen{\frac{\{\tfrac{1}{2}\,(\lambda+\lambda^{-1})-e\}\,\{1-\tfrac{1}{2}e\,(\lambda+\lambda^{-1})\}\,\Lambda'_r}{\sqrt{1-e^2}-\tfrac{1}{2}e\,(\lambda-\lambda^{-1})}}}=0 \quad\ldots\ldots\ldots\ldots\ldots\ldots(45),$$

where

$$\Lambda'_r=\lambda^r\epsilon^{-\frac{re}{2}(\lambda-\lambda^{-1})}-\lambda^{-r}\,\epsilon^{\frac{re}{2}(\lambda-\lambda^{-1})} \quad\ldots\ldots\ldots\ldots\ldots\ldots(46).$$

Again,

$$\underparen{\overparen{\frac{\{1-\tfrac{1}{2}e\,(\lambda+\lambda^{-1})\}\,(\lambda-\lambda^{-1})\,\Lambda'_r}{1-\tfrac{1}{2}\dfrac{e}{\sqrt{1-e^2}}\,(\lambda-\lambda^{-1})}}}+\frac{2}{r}\underparen{\overparen{\Lambda_r}}=0 \quad\ldots\ldots\ldots\ldots\ldots\ldots(47),$$

and

$$\underparen{\overparen{\frac{\{1-\tfrac{1}{2}e\,(\lambda+\lambda^{-1})\}\,\{(\lambda+\lambda^{-1})-\tfrac{1}{2}e\}\,\Lambda_r}{1-\tfrac{1}{2}\dfrac{e}{\sqrt{1-e^2}}\,(\lambda-\lambda^{-1})}}}=\underparen{\overparen{\{1-\tfrac{1}{2}e\,(\lambda+\lambda^{-1})\}\,\{(\lambda+\lambda^{-1})-\tfrac{1}{2}e\}\,\Lambda_r}}, \ldots(48);$$

or, what is the same thing,

$$\underparen{\overparen{\frac{(\lambda-\lambda^{-1})\,\{1-\tfrac{1}{2}e\,(\lambda+\lambda^{-1})\}\,\{(\lambda+\lambda^{-1})-\tfrac{1}{2}e\}\,\Lambda_r}{1-\tfrac{1}{2}\dfrac{e}{\sqrt{1-e^2}}\,(\lambda-\lambda^{-1})}}}=0 \quad\ldots\ldots\ldots\ldots(49);$$

or, comparing with (44),

$$\underparen{\overparen{\frac{(\lambda^2-\lambda^{-2})\,\{1-\tfrac{1}{2}e\,(\lambda+\lambda^{-1})\}\,\Lambda_r}{1-\tfrac{1}{2}\dfrac{e}{\sqrt{1-e^2}}\,(\lambda-\lambda^{-1})}}}=0 \quad\ldots\ldots\ldots\ldots\ldots\ldots(50),$$

which are all obtained by applying the formula (43) to the expansion of $\begin{matrix}\sin\\ \cos\end{matrix}(\theta-\varpi)$, and comparing with the equations (25), (33).

5.

ON THE INTERSECTION OF CURVES.

[From the *Cambridge Mathematical Journal*, vol. III. (1843), pp. 211—213.]

THE following theorem is quoted in a note of Chasles' Aperçu Historique &c., *Memoires de Bruxelles*, tom. XI. p. 149, where M. Chasles employs it in the demonstration of Pascal's theorem: "If a curve of the third order pass through eight of the points of intersection of two curves of the third order, it passes through the ninth point of intersection." The application in question is so elegant, that it deserves to be generally known. Consider a hexagon inscribed in a conic section. The aggregate of three alternate sides may be looked upon as forming a curve of the third order, and that of the remaining sides, a second curve of the same order. These two intersect in nine points, viz. the six angular points of the hexagon, and the three points which are the intersections of pairs of opposite sides. Suppose a curve of the third order passing through eight of these points, viz. the aggregate of the conic section passing through the angular points of the hexagon, and of the line joining two of the three intersections of pairs of opposite sides. This passes through the ninth point, by the theorem of Chasles, i.e. the three intersections of pairs of opposite sides lie in the same straight line, (since obviously the third intersection does *not* lie in the conic section); which is Pascal's theorem.

The demonstration of the above property of curves of the third order is one of extreme simplicity. Let $U = 0$, $V = 0$, be the equations of two curves of the third order, the curve of the same order which passes through eight of their points of intersection (which may be considered as eight perfectly arbitrary points), and a ninth arbitrary point, will be perfectly determinate. Let U_0, V_0, be the values of U, V, when the coordinates of this last point are written in place of x, y. Then $UV_0 - U_0V = 0$, satisfies the above conditions, or it is the equation to the curve required; but it is an equation which is satisfied by all the nine points of intersection of the two curves, i.e. any curve that passes through eight of these points of intersection, passes also through the ninth.

Consider generally two curves, $U_m = 0$, $V_n = 0$, of the orders m and n respectively, and a curve of the rth order (r not less than m or n) passing through the mn points of intersection. The equation to such a curve will be of the form

$$U = u_{r-m} U_m + v_{r-n} V_n = 0,$$

u_{r-m}, v_{r-n}, denoting two polynomes of the orders $r - m$, $r - n$, with all their coefficients complete. It would at first sight appear that the curve $U = 0$ might be made to pass through as many as $\{1 + 2 \ldots + (r - m + 1)\} + \{1 + 2 \ldots (r - n + 1)\} - 1$, arbitrary points, i.e.

$$\tfrac{1}{2}(r - m + 1)(r - m + 2) + \tfrac{1}{2}(r - n + 1)(r - n + 2) - 1\,;$$

or, what is the same thing,

$$\tfrac{1}{2}r(r + 3) - mn + \tfrac{1}{2}(r - m - n + 1)(r - m - n + 2)$$

arbitrary points, such being apparently the number of disposable constants. This is in fact the case as long as r is not greater than $m + n - 1$; but when r exceeds this, there arise, between the polynomes which multiply the disposable coefficients, certain linear relations which cause them to group themselves into a smaller number of disposable quantities. Thus, if r be not less than $m + n$, forming different polynomes of the form $x^\alpha y^\beta\, V_n$ $[\alpha + \beta = \text{or} < m]$, and multiplying by the coefficients of $x^\alpha y^\beta$ in U_m and adding, we obtain a sum $U_m V_n$, which might have been obtained by taking the different polynomes of the form $x^\gamma y^\delta\, U_m$ $[\gamma + \delta = \text{or} < n]$, multiplying by the coefficients of $x^\gamma y^\delta$ in V_n, and adding: or we have a linear relation between the different polynomes of the forms $x^\alpha y^\beta V_n$, and $x^\gamma y^\delta U_m$. In the case where r is not less than $m + n + 1$, there are two more such relations, viz. those obtained in the same way from the different polynomes $x^\alpha y^\beta \,.\, x V_n$, $x^\gamma y^\delta \,.\, x U_m$, and $x^\alpha y^\beta \,.\, y V_n$, $x^\gamma y^\delta \,.\, y U_m$, &c.; and in general, whatever be the excess of r above $m + n - 1$, the number of these linear relations is

$$1 + 2 \ldots (r - m - n + 1) = \tfrac{1}{2}(r - m - n + 1)(r - m - n + 2).$$

Hence, if r be not less than $m + n$, the number of points through which a curve of the rth order may be made to pass, in addition to the mn points which are the intersections of $U_m = 0$, $V_n = 0$, is simply $\tfrac{1}{2}r(r + 3) - mn$. In the case of $r = m + n - 1$, or $r = m + n - 2$, the two formulæ coincide. Hence we may enunciate the theorem

"A curve of the rth order, passing through the mn points of intersection of two curves of the mth and n^{th} orders respectively, may be made to pass through $\tfrac{1}{2}r(r + 3) - mn + \tfrac{1}{2}(m + n - r - 1)(m + n - r - 2)$ arbitrary points, if r be not greater than $m + n - 3$: if r be greater than this value, it may be made to pass through $\tfrac{1}{2}r(r + 3) - mn$ points only."

Suppose r not greater than $m + n - 3$, and a curve of the rth order made to pass through

$$\tfrac{1}{2}r(r + 3) - mn + \tfrac{1}{2}(m + n - r - 1)(m + n - r - 2)$$

arbitrary points, and

$$mn - \tfrac{1}{2}(m + n - r - 1)(m + n - r - 2)$$

of the mn points of intersection above. Such a curve passes through $\frac{1}{2}r(r+3)$ given points, and though the $mn-\frac{1}{2}(m+n-r-1)(m+n-r-2)$ latter points are not perfectly arbitrary, there appears to be no reason why the relation between the positions of these points should be such as to prevent the curve from being *completely determined* by these conditions. But if it be so, then the curve must pass through the remaining $\frac{1}{2}(m+n-r-1)(m+n-r-2)$ points of intersection, or we have the theorem

"If a curve of the rth order (r not less than m or n, not greater than $m+n-3$) pass through

$$mn-\tfrac{1}{2}(m+n-r-1)(m+n-r-2)$$

of the points of intersection of two curves of the mth and n^{th} orders respectively, it passes through the remaining

$$\tfrac{1}{2}(m+n-r-1)(m+n-r-2)$$

points of intersection."

6.

ON THE MOTION OF ROTATION OF A SOLID BODY.

[From the *Cambridge Mathematical Journal*, vol. III. (1843), pp. 224—232.]

IN the fifth volume of Liouville's Journal, in a paper "Des lois géometriques qui régissent les déplacemens d'un système solide," M. Olinde Rodrigues has given some very elegant formulæ for determining the position of two sets of rectangular axes with respect to each other, employing rational functions of three quantities only. The principal object of the present paper is to apply these to the problem of the rotation of a solid body; but I shall first demonstrate the formulæ in question, and some others connected with the same subject which may be useful on other occasions.

Let Ax, Ay, Az; $Ax_{,}$, $Ay_{,}$, $Az_{,}$, be any two sets of rectangular axes passing through the point A: x, y, z, $x_{,}$, $y_{,}$, $z_{,}$, being taken for the points where these lines intersect the spherical surface described round the centre A with radius unity. Join $xx_{,}$, $yy_{,}$, $zz_{,}$, by arcs of great circles, and through the central points of these describe great circles cutting them at right angles: these are easily seen to intersect in a certain point P. Let $Px=f$, $Py=g$, $Pz=h$; then also $Px_{,}=f$, $Py_{,}=g$, $Pz_{,}=h$: and $\angle xPx_{,}=\angle yPy_{,}=\angle zPz_{,}, =\theta$ suppose, θ being measured from xP towards yP, yP towards zP, or zP towards xP. The cosines of f, g, h, are of course connected by the equation

$$\cos^2 f+\cos^2 g+\cos^2 h=1.$$

Let α, β, γ; α', β', γ'; α'', β'', γ'', represent the cosines of $x_{,}x$, $y_{,}x$, $z_{,}x$; $x_{,}y$, $y_{,}y$, $z_{,}y$; $x_{,}z$, $y_{,}z$, $z_{,}z$: these quantities are to be determined as functions of f, g, h, θ.

Suppose for a moment,

$$\angle yPz=\mathrm{x},\qquad \angle zPx=\mathrm{y},\qquad \angle xPy=\mathrm{z};$$

then

$$\begin{aligned}
\alpha\ &= \cos^2 f \qquad\quad + \sin^2 f \cos\theta,\\
\alpha'\ &= \cos f \cos g + \sin f \sin g \cos(\mathrm{z}-\theta),\\
\alpha'' &= \cos f \cos h + \sin f \sin h \cos(\mathrm{y}+\theta),\\
\beta\ &= \cos g \cos f + \sin g \sin f \cos(\mathrm{z}+\theta),\\
\beta'\ &= \cos^2 g \qquad\quad + \sin^2 g \cos\theta,\\
\beta'' &= \cos g \cos h + \sin g \sin h \cos(\mathrm{x}-\theta),\\
\gamma\ &= \cos h \cos f + \sin h \sin f \cos(\mathrm{y}-\theta),\\
\gamma'\ &= \cos h \cos g + \sin h \sin g \cos(\mathrm{x}+\theta),\\
\gamma'' &= \cos^2 h \qquad\quad + \sin^2 h \cos\theta.
\end{aligned}$$

Also

$$\begin{aligned}
\sin g \sin h \cos \mathrm{x} &= -\cos g \cos h,\\
\sin h \sin f \cos \mathrm{y} &= -\cos h \cos f,\\
\sin f \sin g \cos \mathrm{z} &= -\cos f \cos g,
\end{aligned}$$

and

$$\begin{aligned}
\sin g \sin h \sin \mathrm{x} &= \cos f,\\
\sin h \sin f \sin \mathrm{y} &= \cos g,\\
\sin f \sin g \sin \mathrm{z} &= \cos h.
\end{aligned}$$

Substituting,

$$\begin{aligned}
\alpha\ &= \cos^2 f \qquad\qquad\qquad\quad + \sin^2 f \cos\theta,\\
\alpha'\ &= \cos f \cos g\,(1-\cos\theta) + \cos h \sin\theta,\\
\alpha'' &= \cos f \cos h\,(1-\cos\theta) - \cos g \sin\theta,\\
\beta\ &= \cos g \cos f\,(1-\cos\theta) - \cos h \sin\theta,\\
\beta'\ &= \cos^2 g \qquad\qquad\qquad\quad + \sin^2 g \cos\theta,\\
\beta'' &= \cos g \cos h\,(1-\cos\theta) + \cos f \sin\theta,\\
\gamma\ &= \cos h \cos f\,(1-\cos\theta) + \cos g \sin\theta,\\
\gamma'\ &= \cos h \cos g\,(1-\cos\theta) - \cos f \sin\theta,\\
\gamma'' &= \cos^2 h \qquad\qquad\qquad\quad + \sin^2 h \cos\theta.
\end{aligned}$$

Assume $\lambda = \tan\frac{1}{2}\theta \cos f$, $\mu = \tan\frac{1}{2}\theta\cos g$, $\nu = \tan\frac{1}{2}\theta \cos h$, and $\sec^2\frac{1}{2}\theta = 1 + \lambda^2 + \mu^2 + \nu^2 = \kappa$;

then

$$\begin{aligned}
\kappa\alpha &= 1 + \lambda^2 - \mu^2 - \nu^2, & \kappa\alpha' &= 2(\lambda\mu + \nu), & \kappa\alpha'' &= 2(\nu\lambda - \mu),\\
\kappa\beta &= 2(\lambda\mu - \nu), & \kappa\beta' &= 1 + \mu^2 - \nu^2 - \lambda^2, & \kappa\beta'' &= 2(\mu\nu + \lambda),\\
\kappa\gamma &= 2(\nu\lambda + \mu), & \kappa\gamma' &= 2(\mu\nu - \lambda), & \kappa\gamma'' &= 1 + \nu^2 - \lambda^2 - \mu^2;
\end{aligned}$$

which are the formulæ required, differing only from those in Liouville, by having λ, μ, ν, instead of $\frac{1}{2}m, \frac{1}{2}n, \frac{1}{2}p$; and $\alpha, \alpha', \alpha''$; β, β', β''; $\gamma, \gamma', \gamma''$, instead of a, b, c; a', b', c'; a'', b'', c''. It is to be remarked, that β', β'', β; $\gamma'', \gamma, \gamma'$, are deduced from $\alpha, \alpha', \alpha''$, by writing μ, ν, λ; ν, λ, μ, for λ, μ, ν.

Let $1 + \alpha + \beta' + \gamma'' = v$; then $\kappa v = 4$, and we have

$$\begin{aligned}
\lambda v &= \beta'' - \gamma', & \mu v &= \gamma - \alpha'', & \nu v &= \alpha' - \beta,\\
\lambda^2 v &= 1 + \alpha - \beta' - \gamma'', & \mu^2 v &= 1 - \alpha + \beta' - \gamma'', & \nu^2 v &= 1 - \alpha - \beta' - \gamma''.
\end{aligned}$$

Suppose that Ax, Ay, Az, are referred to axes $A\mathrm{X}$, $A\mathrm{Y}$, $A\mathrm{Z}$, by the quantities l, m, n, k, analogous to λ, μ, ν, κ, these latter axes being referred to $Ax_{,}$, $Ay_{,}$, $Az_{,}$ by the quantities $l_{,}$, $m_{,}$, $n_{,}$, $k_{,}$.

Let a, b, c; a', b', c'; a'', b'', c''; $a_{,}$, $b_{,}$, $c_{,}$; $a_{,}'$, $b_{,}'$, $c_{,}'$; $a_{,}''$, $b_{,}''$, $c_{,}''$, denote the quantities analogous to α, β, γ; α', β', γ'; α'', β'', γ''. Then we have, by spherical trigonometry, the formulæ

$$\begin{array}{lll} \alpha\ = a\,a_{,} + b\,a_{,}' + c\,a_{,}'', & \beta\ = a\,b_{,} + b\,b_{,}' + c\,b_{,}'', & \gamma\ = a\,c_{,} + b\,c_{,}' + c\,c_{,}''; \\ \alpha' = a'a_{,} + b'a_{,}' + c'a_{,}'', & \beta' = a'b_{,} + b'b_{,}' + c'b_{,}'', & \gamma' = a'c_{,} + b'c_{,}' + c'c_{,}''; \\ \alpha'' = a''a_{,} + b''a_{,}' + c''a_{,}'', & \beta'' = a''b_{,} + b''b_{,}' + c''b_{,}'', & \gamma'' = a''c_{,} + b''c_{,}' + c''c_{,}''. \end{array}$$

Then expressing a, b, c; a', b', c'; a'', b'', c''; $a_{,}$, $b_{,}$, $c_{,}$; $a_{,}'$, $b_{,}'$, $c_{,}'$; $a_{,}''$, $b_{,}''$, $c_{,}''$, in terms of l, m, n; $l_{,}$, $m_{,}$, $n_{,}$, after some reductions we arrive at

$$\begin{aligned} kk_{,}\nu &= 4(1 - ll_{,} - mm_{,} - nn_{,})^2, \ = 4\Pi^2 \text{ suppose}, \\ kk_{,}(\beta'' - \gamma) &= 4(l + l_{,} + n_{,}m - nm_{,})\,\Pi, \\ kk_{,}(\gamma - \alpha') &= 4(m + m_{,} + l_{,}m - lm_{,})\,\Pi, \\ kk_{,}(\alpha' - \beta'') &= 4(n + n_{,} + m_{,}n - mn_{,})\,\Pi; \end{aligned}$$

and hence

$$\begin{array}{ll} \Pi\ = 1 - ll_{,} - mm_{,} - nn_{,}, & \Pi\lambda = l + l_{,} + n_{,}m - nm_{,}, \\ \Pi\mu = m + m_{,} + l_{,}m - lm_{,}, & \Pi\nu = n + n_{,} + m_{,}n - mn_{,}, \end{array}$$

which are formulæ of considerable elegance for exhibiting the combined effect of successive displacements of the axes. The following analogous ones are readily obtained:

$$\begin{array}{ll} P\ = 1 + \lambda l + \mu m + \nu n, & Pl_{,} = \lambda - l - \nu m + \mu n, \\ Pm_{,} = \mu - m - \lambda n + \nu l, & Pn_{,} = \nu - n - \mu l + \lambda m: \end{array}$$

and again,

$$\begin{array}{ll} P_{,}\ = 1 + \lambda l_{,} + \mu m_{,} + \nu n_{,}, & P_{,}l = \lambda - l_{,} + \nu m_{,} - \mu n_{,}, \\ P_{,}m = \mu - m_{,} + \lambda n_{,} - \nu l_{,}, & P_{,}n = \nu - n_{,} + \mu l_{,} - \lambda m_{,}. \end{array}$$

These formulæ will be found useful in the integration of the equations of rotation of a solid body.

Next it is required to express the quantities p, q, r, in terms of λ, μ, ν, where as usual

$$p = \gamma\frac{d\beta}{dt} + \gamma'\frac{d\beta'}{dt} + \gamma''\frac{d\beta''}{dt},$$

$$q = \alpha\frac{d\gamma}{dt} + \alpha'\frac{d\gamma'}{dt} + \alpha''\frac{d\gamma''}{dt},$$

$$r = \beta\frac{d\alpha}{dt} + \beta'\frac{d\alpha'}{dt} + \beta''\frac{d\alpha''}{dt}.$$

Differentiating the values of $\beta\kappa$, $\beta'\kappa$, $\beta''\kappa$, multiplying by γ, γ', γ'', and adding,

$$\kappa p = 2\lambda'(\gamma\mu - \gamma'\lambda + \gamma'') + 2\mu'(\gamma\lambda - \gamma'\mu + \gamma''\nu) + 2\nu'(-\gamma - \gamma'\nu + \gamma''\mu),$$

where λ', μ', ν', denote $\frac{d\lambda}{dt}$, $\frac{d\mu}{dt}$, $\frac{d\nu}{dt}$. Reducing, we have

$$\kappa p = 2(\lambda' + \nu\mu' - \nu'\mu):$$

from which it is easy to derive the system

$$\begin{aligned}\kappa p &= 2(\quad \lambda' + \nu\mu' - \nu'\mu),\\ \kappa q &= 2(-\nu\lambda' + \mu' + \nu'\lambda),\\ \kappa r &= 2(\quad \mu\lambda' - \lambda\mu' + \nu'\quad);\end{aligned}$$

or, determining λ', μ', ν', from these equations, the equivalent system

$$\begin{aligned}2\lambda' &= (1+\lambda^2)\,p + (\lambda\mu - \nu)\,q + (\nu\lambda + \mu)\,r,\\ 2\mu' &= (\lambda\mu + \nu)\,p + (1+\mu^2)\,q + (\mu\nu - \lambda)\,r,\\ 2\nu' &= (\nu\lambda - \mu)\,p + (\mu\nu + \lambda)\,q + (1+\nu^2)\,r.\end{aligned}$$

The following equation also is immediately obtained,

$$\kappa' = \kappa(\lambda p + \mu q + \nu r).$$

The subsequent part of the problem requires the knowledge of the differential coefficients of p, q, r, with respect to λ, μ, ν; λ', μ', ν'. It will be sufficient to write down the six

$$\begin{aligned}\kappa\frac{dp}{d\lambda'} &= 2, & \kappa\frac{dp}{d\lambda} + 2p\lambda &= 0,\\ \kappa\frac{dq}{d\lambda'} &= -2\nu, & \kappa\frac{dq}{d\lambda} + 2q\lambda &= 2\nu',\\ \kappa\frac{dr}{d\lambda'} &= 2\mu, & \kappa\frac{dr}{d\lambda} + 2r\lambda &= -2\mu',\end{aligned}$$

from which the others are immediately obtained.

Suppose now a solid body acted on by any forces, and revolving round a fixed point. The equations of motion are

$$\begin{aligned}\frac{d}{dt}\frac{dT}{d\lambda'} - \frac{dT}{d\lambda} &= \frac{dV}{d\lambda},\\ \frac{d}{dt}\frac{dT}{d\mu'} - \frac{dT}{d\mu} &= \frac{dV}{d\mu},\\ \frac{d}{dt}\frac{dT}{d\nu'} - \frac{dT}{d\nu} &= \frac{dV}{d\nu};\end{aligned}$$

where $\qquad T = \tfrac{1}{2}(Ap^2 + Bq^2 + Cr^2)$; $\;V = \Sigma\,[\int(Xdx + Ydy + Zdz)]\,dm$;

or if $Xdx + Ydy + Zdz$ is not an exact differential, $\frac{dV}{d\lambda}$, $\frac{dV}{d\mu}$, $\frac{dV}{d\nu}$, are independent symbols standing for

$$\Sigma\left(X\frac{dx}{d\lambda} + Y\frac{dy}{d\lambda} + Z\frac{dz}{d\lambda}\right)dm,\ \ldots\ldots$$

see *Mécanique Analytique*, Avertissement, t. I. p. V. [Ed. 3, p. VII.]: only in this latter case V stands for the disturbing function, the principal forces vanishing.

Now, considering the first of the above equations

$$\frac{dT}{d\lambda'} = Ap\frac{dp}{d\lambda} + Bq\frac{dq}{d\lambda} + Cr\frac{dr}{d\lambda}, \quad = \frac{2}{\kappa}(Ap - \nu Bq + \mu Cr);$$

whence, writing $\qquad p', q', r', \kappa'$, for $\frac{dp}{dt}, \frac{dq}{dt}, \frac{dr}{dt}, \frac{d\kappa}{dt}$,

$$\frac{d}{dt}\frac{dT}{d\lambda'} = \frac{2}{\kappa}(Ap' - \nu Bq' + \mu Cr') - \frac{2}{\kappa}Bq\nu' + \frac{2}{\kappa}Cr\mu' - \frac{2\kappa'}{\kappa^2}(Ap - \nu Bq + \mu Cr).$$

Also $\quad \frac{dT}{d\lambda} = Ap\frac{dp}{d\lambda} + Bq\frac{dq}{d\lambda} + Cr\frac{dr}{d\lambda}, \quad = -\frac{2\lambda}{\kappa}(Ap^2 + Bq^2 + Cr^2) + \frac{2}{\kappa}Bq\nu' - \frac{2}{\kappa}Cr\mu'$;

and hence $\frac{1}{2}\left(\frac{d}{dt}\frac{dT}{d\lambda'} - \frac{dT}{d\lambda}\right)$

$$= \frac{1}{\kappa}(Ap' - \nu Bq' + \mu Cr') - \frac{2}{\kappa}Bq\nu' + \frac{2}{\kappa}Cr\mu' + \frac{\lambda}{\kappa}(Ap^2 + Bq^2 + Cr^2) - \frac{\kappa'}{\kappa^2}(Ap - \nu Bq + \mu Cr).$$

Substituting for $\lambda', \mu', \nu', \kappa'$, after all reductions,

$$\frac{1}{2}\left(\frac{d}{dt}\frac{dT}{d\lambda'} - \frac{dT}{d\lambda}\right) = \frac{1}{\kappa}[\{Ap' + (C-B)qr\} - \nu\{Bq' + (A-C)rp\} + \mu\{Cr + (B-A)pq\}];$$

and, forming the analogous quantities in μ, ν, and substituting in the equations of motion, these become

$$\{Ap' + (C-B)qr\} - \nu\{Bq' + (A-C)rp\} + \mu\{Cr' + (B-A)pq\} = \tfrac{1}{2}\kappa\frac{dV}{d\mu},$$

$$\nu\{Ap' + (C-B)qr\} + \{Bq' + (A-C)rp\} - \lambda\{Cr' + (B-A)pq\} = \tfrac{1}{2}\kappa\frac{dV}{d\mu},$$

$$\mu\{Ap' + (C-B)qr\} + \lambda\{Bq' + (A-C)rp\} + \{Cr' + (B-A)pq\} = \tfrac{1}{2}\kappa\frac{dV}{d\nu};$$

or eliminating, and replacing p', q', r', by $\frac{dp}{dt}, \frac{dq}{dt}, \frac{dr}{dt}$, we obtain

$$A\frac{dp}{dt} + (C-B)qr = \tfrac{1}{2}\left\{(1+\lambda^2)\frac{dV}{d\lambda} + (\lambda\mu+\nu)\frac{dV}{d\mu} + (\nu\lambda-\mu)\frac{dV}{d\nu}\right\},$$

$$B\frac{dq}{dt} + (A-C)rp = \tfrac{1}{2}\left\{(\lambda\mu-\nu)\frac{dV}{d\lambda} + (1+\mu^2)\frac{dV}{d\mu} + (\mu\nu+\lambda)\frac{dV}{d\nu}\right\},$$

$$C\frac{dr}{dt} + (B-A)pq = \tfrac{1}{2}\left\{(\nu\lambda+\mu)\frac{dV}{d\lambda} + (\mu\nu-\lambda)\frac{dV}{d\mu} + (1+\nu^2)\frac{dV}{d\nu}\right\};$$

to which are to be joined

$$\kappa p = 2\left(\quad \frac{d\lambda}{dt} + \nu \frac{d\mu}{dt} - \mu \frac{d\nu}{dt}\right),$$

$$\kappa q = 2\left(-\nu \frac{d\lambda}{dt} + \quad \frac{d\mu}{dt} + \lambda \frac{d\nu}{dt}\right),$$

$$\kappa r = 2\left(\quad \mu \frac{d\lambda}{dt} - \lambda \frac{d\mu}{dt} + \quad \frac{d\nu}{dt}\right);$$

where it will be recollected

$$\kappa = 1 + \lambda^2 + \mu^2 + \nu^2;$$

and on the integration of these six equations depends the complete determination of the motion.

If we neglect the terms depending on V, the first three equations may be integrated in the form

$$p^2 = p_1^2 - \frac{C-B}{A}\phi, \qquad q^2 = q_1^2 - \frac{A-C}{B}\phi, \qquad r_2 + r_1^2 - \frac{B-A}{C}\phi,$$

$$2t = \int \frac{d\phi}{\sqrt{\left\{\left(p_1^2 - \frac{C-B}{A}\phi\right)\left(q_1^2 - \frac{A-C}{B}\phi\right)\left(r_1^2 - \frac{B-A}{C}\phi\right)\right\}}};$$

and considering p, q, r as functions of ϕ, given by these equations, the three latter ones take the form

$$\frac{\kappa}{4qr} = \quad \frac{d\lambda}{d\phi} + \nu \frac{d\mu}{d\phi} - \mu \frac{d\nu}{d\phi},$$

$$\frac{\kappa}{4rp} = -\nu \frac{d\lambda}{d\phi} + \quad \frac{d\mu}{d\phi} + \lambda \frac{d\nu}{d\phi},$$

$$\frac{\kappa}{4pq} = \quad \mu \frac{d\lambda}{d\phi} - \lambda \frac{d\mu}{d\phi} + \quad \frac{d\nu}{d\phi};$$

of which, as is well known, the equations following, equivalent to two independent equations, are integrals,

$$\begin{aligned}
\kappa g &= \quad Ap\,(1 + \lambda^2 - \mu^2 - \nu^2) + 2Bq\,(\lambda\mu - \nu) \qquad + 2Cr\,(\nu\lambda + \mu),\\
\kappa g' &= 2Ap\,(\lambda\mu + \nu) \qquad + \quad Bq\,(1 + \mu^2 - \lambda^2 - \nu^2) + 2Cr\,(\mu\nu - \lambda),\\
\kappa g'' &= 2Ap\,(\nu\lambda - \mu) \qquad + 2Bq\,(\mu\nu + \lambda) \qquad + \quad Cr\,(1 + \nu^2 - \lambda^2 - \mu^2);
\end{aligned}$$

where g, g', g'', are arbitrary constants satisfying

$$g^2 + g'^2 + g''^2 = A^2 p_1^2 + B^2 q_1^2 + C^2 r_1^2.$$

To obtain another integral, it is apparently necessary, as in the ordinary theory, to revert to the consideration of the invariable plane. Suppose $g' = 0$, $g'' = 0$,

then $$g'' = \sqrt{(A^2 p_1^2 + B^2 q_1^2 + C^2 r_1^2)}, \; = k \text{ suppose.}$$

We easily obtain, where λ_0, μ_0, ν_0, κ_0 are written for λ, μ, ν, κ, to denote this particular supposition,

$$\kappa_0 Ap = 2(\nu_0\lambda_0 - \mu_0)k,$$

$$\kappa_0 Bq = 2(\mu_0\nu_0 + \lambda_0)k,$$

$$\kappa_0 Cr = (1 + \nu_0^2 - \lambda_0^2 - \mu_0^2)k;$$

whence, and from $\kappa_0 = 1 + \lambda_0^2 + \mu_0^2 + \nu_0^2$, $\kappa_0 Cr = (2 + 2\nu_0^2 - \kappa_0)k$, we obtain

$$\kappa_0 = \frac{(2 + 2\nu_0^2)k}{k + Cr}, \qquad \nu_0\lambda_0 - \mu_0 = \frac{(1 + \nu_0^2)Ap}{k + Cr}, \qquad \mu_0\nu_0 + \lambda_0 = \frac{(1 + \nu_0^2)Bq}{k + Cr}.$$

Hence, writing $h = Ap_1^2 + Bq_1^2 + Cr_1^2$, the equation

$$\frac{d\nu_0}{d\phi} = \frac{1}{4pqr}\{(\nu_0\lambda_0 - \mu_0)p + (\mu_0\nu_0 + \lambda_0)q + (1 + \nu^2)r\}$$

reduces itself to

$$\frac{4}{1 + \nu_0^2}\frac{d\nu_0}{d\phi} = \frac{h + kr}{(k + Cr)pqr},$$

or, integrating,

$$4\tan^{-1}\nu_0 = \int\frac{(h + kr)\,d\phi}{(k + Cr)pqr}.$$

The integral takes rather a simpler form if p, q, ϕ be considered functions of r, and becomes

$$2\tan^{-1}\nu_0 = \int\frac{h + kr}{k + Cr}\,\frac{C\sqrt{(AB)}\,dr}{\sqrt{[\{k^2 - Bh + (B - C)Cr^2\}\{Ah - k^2 + (C - A)Cr^2\}]}};$$

and then, ν_0 being determined, λ_0, μ_0 are given by the equations

$$\lambda_0 = \frac{\nu_0 Ap + Bq}{k + Cr}, \qquad \mu_0 = \frac{\nu_0 Bq - Ap}{k + Cr}.$$

Hence l, m, n, denoting arbitrary constants, the general values of λ, μ, ν, are given by the equations

$$P_0 = 1 - l\lambda_0 - m\mu_0 - n\nu_0,$$

$$P_0\lambda = l + \lambda_0 + m\nu_0 - n\mu_0,$$

$$P_0\mu = m + \mu_0 + n\lambda_0 - l\nu_0,$$

$$P_0\nu = n + \nu_0 + l\mu_0 - m\lambda_0.$$

In a following paper I propose to develope the formulæ for the variations of the arbitrary constants p_1, q_1, r_1, l, m, n, when the terms involving V are taken into account.

Note. It may be as well to verify independently the analytical conclusion immediately deducible from the preceding formulæ, viz. if λ, μ, ν, be given by the differential equations,

$$\kappa p = \frac{d\lambda}{dt} + \nu \frac{d\mu}{dt} - \mu \frac{d\nu}{dt},$$

$$\kappa q = -\nu \frac{d\lambda}{dt} + \frac{d\mu}{dt} + \lambda \frac{d\nu}{dt},$$

$$\kappa r = \mu \frac{d\lambda}{dt} - \lambda \frac{d\mu}{dt} + \frac{d\nu}{dt},$$

where $\kappa = 1 + \lambda^2 + \mu^2 + \nu^2$, and p, q, r, are any functions of t. Then if λ_0, μ_0, ν_0, be particular values of λ, μ, ν, and l, m, n, arbitrary constants, the general integrals are given by the system

$$\begin{aligned} P_0 &= 1 - l\lambda_0 - m\mu_0 - n\nu_0, \\ P_0\lambda &= l + \lambda_0 + m\nu_0 - n\mu_0, \\ P_0\mu &= m + \mu_0 + n\lambda_0 - l\nu_0, \\ P_0\nu &= n + \nu_0 + l\mu_0 - m\lambda_0. \end{aligned}$$

Assuming these equations, we deduce the equivalent system,

$$\begin{aligned} (1 + \lambda\lambda_0 + \mu\mu_0 + \nu\nu_0)\, l &= \lambda - \lambda_0 + \nu_0\mu - \nu\mu_0, \\ (1 + \lambda\lambda_0 + \mu\mu_0 + \nu\nu_0)\, m &= \mu - \mu_0 + \lambda_0\nu - \lambda\nu_0, \\ (1 + \lambda\lambda_0 + \mu\mu_0 + \nu\nu_0)\, n &= \nu - \nu_0 + \mu_0\lambda - \mu\lambda_0. \end{aligned}$$

Differentiate the first of these and eliminate l, the result takes the form $0 =$

$$-(\mu_0^2 + \nu_0^2)(\lambda' + \nu\mu' - \nu'\mu) - (\nu_0 - \lambda_0\mu_0)(-\nu\lambda' + \mu' + \lambda\nu') + (\mu_0 + \lambda_0\nu_0)(\mu\lambda' - \lambda\mu' + \nu') + \kappa_0\lambda',$$

$$+ (\mu^2 + \nu^2)(\lambda_0' + \nu_0\mu_0' - \nu_0'\mu_0) + (\nu - \lambda\mu)(-\nu_0\lambda_0' + \mu_0' + \lambda_0\nu_0') - (\mu + \lambda\nu)(\mu_0\lambda_0' - \lambda_0\mu_0' + \nu_0') - \kappa\lambda_0',$$

where λ', &c. denote $\frac{d\lambda}{dt}$, &c. and $\kappa_0 = 1 + \lambda_0^2 + \mu_0^2 + \nu_0^2$.

Reducing by the differential equations in λ, μ, ν; λ_0, μ_0, ν_0, this becomes

$$\kappa_0 \{\lambda' + \tfrac{1}{2}p(\mu^2 + \nu^2) + \tfrac{1}{2}q(\nu - \lambda\mu) - \tfrac{1}{2}r(\mu + \lambda\nu)\}$$

$$- \kappa \{\lambda_0' + \tfrac{1}{2}p(\mu_0^2 + \nu_0^2) + \tfrac{1}{2}q(\nu_0 - \lambda_0\mu_0) - \tfrac{1}{2}r(\mu_0 + \lambda\nu_0)\} = 0;$$

or substituting for λ', λ_0', we have the identical equation

$$\tfrac{1}{2}p(\kappa_0\kappa - \kappa\kappa_0) = 0:$$

and similarly may the remaining equations be verified.

7.

ON A CLASS OF DIFFERENTIAL EQUATIONS, AND ON THE LINES OF CURVATURE OF AN ELLIPSOID.

[From the *Cambridge Mathematical Journal*, vol. III. (1843), pp. 264—267.]

CONSIDER the primitive equation

$$fx + gy + hz + \ldots\ldots = 0 \quad\ldots\ldots\ldots\ldots(1),$$

between n variables x, y, z, the constants f, g, h being connected by the equation

$$H(f,\ g,\ h\ldots\ldots) = 0 \quad\ldots\ldots\ldots\ldots(2),$$

H denoting a homogeneous function. Suppose that f, g, h...... are determined by the conditions

$$\begin{aligned} &fx_1 \;\; + gy_1 \;\; + hz_1 \;\;\ldots\ldots = 0 \quad\ldots\ldots\ldots\ldots(3),\\ &\;\vdots \qquad\quad \vdots \qquad\quad \vdots \\ &fx_{n-2} + gy_{n-2} + hz_{n-2}\ldots\ldots = 0. \end{aligned}$$

Then writing

$$X = \begin{vmatrix} y & , z & , \ldots \\ y_1 & , z_1 & , \ldots \\ \vdots & \vdots & \\ y_{n-2}, & z_{n-2}, & \ldots \end{vmatrix} \quad\ldots\ldots\ldots\ldots (4),$$

with analogous expressions for y, z......; the equations (3) give f, g, h,...... proportional to x, y, z,...... or eliminating f, g, h...... by the equation (2),

$$H(X,\ Y,\ Z\ldots\ldots) = 0\ldots\ldots\ldots\ldots(5).$$

Conversely the equation (5), which contains, in appearance, $n(n-2)$ arbitrary constants, is equivalent to the system (1), (2). And if H be a rational integral function of the order r, the first side of the equation (5) is the product of r factors, each of them of the form given by the system (1), (2).

Now from the equation (1), we have the system

$$\begin{aligned} &fx \quad + gy \quad + hz \quad \ldots\ldots = 0 \ldots\ldots\ldots\ldots\ldots\ldots\ldots\ldots\ldots (6), \\ &fdx \quad + gdy \quad + hdx \quad \ldots\ldots = 0, \\ &\vdots \qquad\quad \vdots \qquad\quad \vdots \\ &fd^{n-2}x + gd^{n-2}y + hd^{n-2}z \ldots\ldots = 0, \end{aligned}$$

or writing

$$X = \begin{vmatrix} y & , z & , \ldots \\ dy & , dz & , \ldots \\ \vdots & \vdots & \\ d^{n-2}y, & d^{n-2}z, & \ldots \end{vmatrix} \ldots\ldots\ldots\ldots\ldots\ldots\ldots\ldots\ldots\ldots (7),$$

with analogous expressions for Y, Z......; then from the equations (6), f, g, h...... are proportional to X, Y, Z......: or, eliminating by (2),

$$H(X, Y, Z\ldots\ldots) = 0 \ldots\ldots\ldots\ldots\ldots\ldots\ldots\ldots\ldots\ldots (8).$$

Conversely the integral of the equation (8) of the order $(n-2)$ is given either by the system of equations (1), (2), in which it is evident that the number of arbitrary constants is reduced to $(n-2)$; or, by the equation (5), which contains in appearance $n(n-2)$ arbitrary constants, but which we have seen is equivalent in reality to the system (1), (2).

Thus, with three variables, the integral of

$$H(ydz - zdy, zdx - xdz, xdy - ydx) = 0 \ldots\ldots\ldots\ldots\ldots\ldots (9)$$

may be expressed in the form

$$H(yz_1 - y_1z, \quad zx_1 - z_1x, \quad xy_1 - x_1y) = 0 \ldots\ldots\ldots\ldots\ldots\ldots (10),$$

and also in the form

$$fx + gy + hz = 0 \ldots\ldots\ldots\ldots\ldots\ldots\ldots\ldots\ldots\ldots (11),$$

where

$$H(f, g, h) = 0 \ldots\ldots\ldots\ldots\ldots\ldots\ldots\ldots\ldots\ldots (12).$$

The above principles afford an elegant mode of integrating the differential equation for the lines of curvature of an ellipsoid. The equation in question is

$$(b^2 - c^2)\, xdydz + (c^2 - a^2)\, ydzdx + (a^2 - b^2)\, zdxdy = 0 \ldots\ldots\ldots\ldots\ldots (13),$$

where x, y, z are connected by

$$\frac{x^2}{a^2} + \frac{y^2}{b^2} + \frac{z^2}{c^2} = 1 \ldots\ldots\ldots\ldots\ldots\ldots\ldots\ldots\ldots\ldots (14);$$

writing

$$\frac{x^2}{a^2} = u, \quad \frac{y^2}{b^2} = v, \quad \frac{z^2}{c^2} = w \ldots\ldots\ldots\ldots\ldots\ldots\ldots\ldots (15),$$

these become

$$(b^2 - c^2)\, udvdw + (c^2 - a^2)\, vdwdu + (a^2 - b^2)\, wdudv = 0 \ldots\ldots\ldots\ldots\ldots (16),$$

$$u + v + w = 1 \ldots\ldots\ldots\ldots\ldots\ldots\ldots\ldots\ldots\ldots (17).$$

Multiplying by

$$-\{(vdu-udv)(wdv-vdw)(udw-wdu)\}^{-1},$$

the first of these becomes

$$\frac{-a^2du}{(vdu-udv)(udw-wdu)}+\frac{-b^2dv}{(wdv-vdw)(vdu-udv)}+\frac{-c^2dw}{(udw-wdu)(wdv-vdw)}=0 \ldots \text{ (18)};$$

but writing (17) and its derived equations under the form

$$u+\ (v+\ w)=1 \ldots\ldots \text{(19)},$$

$$du+(dv+dw)=0,$$

we deduce

$$-du\,(v+w)+u\,(dv+dw)=-du \ldots\ldots \text{(20)},$$

i.e.

$$-du=-(vdu-udv)+(udw-wdu) \ldots\ldots \text{(21)},$$

and similarly

$$-dv=-(wdv-vdw)+(vdu-udv),$$

$$-dw=-(udw-wdu)+wdv-vdw).$$

Substituting,

$$\frac{b^2-c^2}{wdv-vdw}+\frac{c^2-a^2}{udw-wdu}+\frac{a^2-b^2}{vdu-udv}=0 \ldots\ldots \text{(22)};$$

the integral of which may be written in the form

$$\frac{b^2-c^2}{wv_1-vw_1}+\frac{c^2-a^2}{uw_1-wu_1}+\frac{a^2-b^2}{vu_1-v_1u}=0 \ldots\ldots \text{(23)},$$

where, on account of (17),

$$u_1+v_1+w_1=1 \ldots\ldots \text{(24)};$$

and also in the form

$$fu+gv+hw=0 \ldots\ldots \text{(25)},$$

where f, g, h are connected by

$$\frac{b^2-c^2}{f}+\frac{c^2-a^2}{g}+\frac{a^2-b^2}{g}=0 \ldots\ldots \text{(26)};$$

this last equation is satisfied identically by

$$f=\frac{b^2-c^2}{B^2-C^2},\qquad g=\frac{c^2-a^2}{C^2-A^2},\qquad h=\frac{a^2-b^2}{A^2-B^2} \ldots\ldots \text{(27)}.$$

Restoring x, y, z, x_1, y_1, z_1 for u, v, w, u_1, v_1, w_1, the equations to a line of curvature passing through a given point x_1, y_1, z_1, on the ellipsoid, are the equation (14) and

$$\frac{(b^2-c^2)}{a^2\,(y_1{}^2z^2-y^2z_1{}^2)}+\frac{(c^2-a^2)}{b^2\,(z_1{}^2x^2-z^2x_1{}^2)}+\frac{(a^2-b^2)}{c^2\,(x_1{}^2y^2-x^2y_1{}^2)}=0 \ldots\ldots \text{(28)},$$

or again, under a known form, they are the equation (14) and

$$\frac{(b^2-c^2)}{B^2-C^2}\frac{x^2}{a^2}+\frac{c^2-a^2}{C^2-A^2}\frac{y^2}{b^2}+\frac{a^2-b^2}{A^2-B^2}\frac{z^2}{c^2}=0 \ldots\ldots \text{(29)}.$$

From the equations (14), (29) it is easy to prove the well-known form

$$\frac{x^2}{a^2+\theta}+\frac{y^2}{b^2+\theta}+\frac{z^2}{c^2+\theta}=1 \quad\text{..............................} (30);$$

in fact, multiplying (29) by m, and adding to (14), we have the equation (30), if the equations

$$\frac{1}{a^2}+m\frac{b^2-c^2}{B^2-C^2}\frac{1}{a^2}=\frac{1}{a^2+\theta}, \quad\text{..........................} (31),$$

$$\frac{1}{b^2}+m\frac{c^2-a^2}{C^2-A^2}\frac{1}{b^2}=\frac{1}{b^2+\theta},$$

$$\frac{1}{c^2}+m\frac{a^2-b^2}{A^2-B^2}\frac{1}{c^2}=\frac{1}{(c^2+\theta)},$$

are satisfied.

But on reduction, these take the form

$$(B^2-C^2)\,\theta+(b^2-c^2)\,m\theta+ma^2\,(b^2-c^2)=0, \quad\text{.................} (32),$$

$$(C^2-A^2)\,\theta+(c^2-a^2)\,m\theta+mb^2\,(c^2-a^2)=0,$$

$$(A^2-B^2)\,\theta+(a^2-b^2)\,m\theta+mc^2\,(a^2-b^2)=0,$$

and since, by adding, an identical equation is obtained, m and θ may be determined to satisfy these equations. The values of θ, m are

$$\theta=\frac{(a^2-b^2)\,(b^2-c^2)\,(c^2-a^2)}{a^2\,(B^2-C^2)+b^2\,(C^2-A^2)+c^2\,(A^2-B^2)} \quad\text{.......................} (33),$$

$$m=\frac{b^2c^2\,(B^2-C^2)+c^2a^2\,(C^2-A^2)+a^2b^2\,(A^2-B^2)}{(a^2-b^2)\,(b^2-c^2)\,(c^2-a^2)} \quad\text{.................} (34).$$

8.

ON LAGRANGE'S THEOREM.

[From the *Cambridge Mathematical Journal*, vol. III. (1843), pp. 283—286.]

THE value given by Lagrange's theorem for the expansion of any function of the quantity x, determined by the equation

$$x = u + hfx \qquad (1),$$

admits of being expressed in rather a remarkable symbolical form. The *à priori* deduction of this, independently of any expansion, presents some difficulties; I shall therefore content myself with showing that the form in question satisfies the equations

$$\frac{d}{du} . \int F'x\, fx\, dx = \frac{d}{dh} . \int F'x\, dx \qquad (2),$$

$$Fx = Fu \text{ for } h = 0 \qquad (3),$$

deduced from the equation (1), and which are sufficient to determine the expansion of Fx, considered as a function of u and h in powers of h.

Consider generally the symbolical expression

$$\phi \left(h \frac{d}{dh}\right) \Xi h \qquad (4),$$

$\phi \left(h \frac{d}{dh}\right)$ involving in general symbols of operation relative to any of the other variables entering into Ξh. Then, if Ξh be expansible in the form

$$\Xi h = \Sigma \,(Ah^m) \qquad (5),$$

it is obvious that

$$\phi \left(h \frac{d}{dh}\right) \Xi h = \Sigma \,\{\phi m \,.\, (Ah^m)\} = \Sigma \,\{(\phi m \,.\, A)\, h^m\} \qquad (6).$$

For instance, u representing a variable contained in the function Ξh, and taking a particular form of $\phi\left(h\frac{d}{dh}\right)$,

$$\left(\frac{d}{du}\right)^{h\frac{d}{dh}}\Xi h=\Sigma\left(\frac{d^mA}{du^m}h^m\right) \quad\text{.......... (7).}$$

From this it is easy to demonstrate

$$\frac{d}{du}\left\{\left(\frac{d}{du}\right)^{h\frac{d}{dh}}\Xi h\right\}=\frac{1}{h}\left(\frac{d}{du}\right)^{h\frac{d}{dh}}\{h\Xi h\} \quad\text{.......... (8),}$$

$$\frac{d}{dh}\left\{\left(\frac{d}{du}\right)^{h\frac{d}{dh}}\Xi h\right\}=\frac{1}{h}\left(\frac{d}{du}\right)^{h\frac{d}{dh}}\{h\Xi' h\} \quad\text{.......... (9),}$$

where $\Xi'h$ denotes $\frac{d}{dh}\Xi h$, as usual. Hence also

$$\frac{d}{du}\left\{\left(\frac{d}{du}\right)^{h\frac{d}{dh}}\Xi' h\right\}=\frac{d}{dh}\left\{\left(\frac{d}{du}\right)^{h\frac{d}{dh}}\Xi h\right\} \quad\text{.......... (10),}$$

of which a particular case is

$$\frac{d}{du}\left\{\left(\frac{d}{du}\right)^{h\frac{d}{dh}-1}F'u\,fu\,e^{hfu}\right\}=\frac{d}{dh}\left\{\left(\frac{d}{du}\right)^{h\frac{d}{dh}-1}F'u\,e^{hfu}\right\} \quad\text{.......... (11).}$$

Also,
$$\left(\frac{d}{du}\right)^{h\frac{d}{dh}-1}(F'u\,e^{hfu})=Fu \text{ for } h=0 \quad\text{.......... (12).}$$

Hence the form in question for Fx is

$$Fx=\left(\frac{d}{du}\right)^{h\frac{d}{dh}-1}(F'u\,e^{hfu}) \quad\text{.......... (13);}$$

from which, differentiating with respect to u, and writing F instead of F',

$$\frac{Fx}{1-hf'x}=\left(\frac{d}{du}\right)^{h\frac{d}{dh}}(Fu\,e^{hfu}) \quad\text{.......... (14),}$$

a well-known form of Lagrange's theorem, almost equally important with the more usual one. It is easy to deduce (13) from (14). To do this, we have only to form the equation

$$\frac{-hFx\,f'x}{1-hf'x}=-h\left(\frac{d}{du}\right)^{h\frac{d}{dh}}(Fu\,f'u\,e^{hfu}) \quad\text{.......... (15),}$$

deduced from (14) by writing $Fxf'x$ for fx, and adding this to (14),

$$\begin{aligned}Fx&=\left(\frac{d}{du}\right)^{h\frac{d}{dh}}(Fu\,e^{hfu})-h\left(\frac{d}{du}\right)^{h\frac{d}{dh}}(Fu\,f'u\,e^{hfu})\\&=\left(\frac{d}{du}\right)^{h\frac{d}{dh}-1}\left\{\frac{d}{du}(Fu\,e^{hfu})-hf'u\,Fu\,e^{hfu}\right\}\\&=\left(\frac{d}{du}\right)^{h\frac{d}{dh}-1}(F'u\,e^{hfu}) \quad\text{.......... (16).}\end{aligned}$$

In the case of several variables, if

$$x = u + hf(x, x_1 \ldots), \quad x_1 = u_1 + h_1 f_1(x, x_1 \ldots), \text{ \&c.} \qquad \ldots\ldots\ldots\ldots\ldots (17),$$

writing for shortness

$$F, f, f_1 \ldots \text{ for } F(u, u_1 \ldots), \quad f(u, u_1 \ldots), \quad f_1(u, u_1 \ldots), \ldots$$

then the formula is

$$\frac{F(x, x_1 \ldots)}{\{1 - hf'(x)\}\{1 - h_1 f_1'(x_1) \ldots\}} = \left(\frac{d}{du}\right)^{h\frac{d}{dh}} \left(\frac{d}{du_1}\right)^{h_1\frac{d}{dh_1}} \ldots (Fe^{hf + h_1 f_1 + \cdots}) \ldots\ldots (18),$$

$$\left\{\text{where } f'(x) \text{ is written to denote } \frac{d}{dx} f(x, x_1 \ldots), \text{ \&c.}\right\}$$

or the coefficient of $h^n h_1^{n_1} \ldots\ldots$ in the expansion of

$$\frac{F(x, x_1 \ldots\ldots)}{\{1 - hf'(x)\}\{1 - h_1 f_1'(x_1)\} \ldots\ldots} \qquad \ldots\ldots\ldots\ldots\ldots\ldots\ldots (19)$$

is

$$= \frac{1}{1 . 2 \ldots n . 1 . 2 \ldots n_1} \left(\frac{d}{du}\right)^n \left(\frac{d}{du_1}\right)^{n_1} \ldots F f^n f_1^{n_1} \ldots\ldots\ldots\ldots\ldots\ldots\ldots (20).$$

From the formula (18), a formula may be deduced for the expansion of $F(x, x_1 \ldots\ldots)$, in the same way as (13) was deduced from (14), but the result is not expressible in a simple form by this method. An apparently simple form has indeed been given for this expansion by Laplace, *Mécanique Celeste*, [Ed. 1, 1798] tom. I. p. 176; but the expression there given for the general term, requires first that certain differentiations should be performed, and then that certain changes should be made in the result, quantities $z, z' \ldots\ldots$, are to be changed into $z^n, z_1^{n_1} \ldots\ldots$; in other words, the general term is not really expressed by known symbols of operation only. The formula (18) is probably known, but I have not met with it anywhere.

9.

DEMONSTRATION OF PASCAL'S THEOREM.

[From the *Cambridge Mathematical Journal*, vol. IV. (1843), pp. 18—20.]

LEMMA 1. Let $U = Ax + By + Cz = 0$ be the equation to a plane passing through a given point taken for the origin, and consider the planes

$$U_1 = 0,\quad U_2 = 0,\quad U_3 = 0,\quad U_4 = 0,\quad U_5 = 0,\quad U_6 = 0\ ;$$

the condition which expresses that the intersections of the planes (1) and (2), (3) and (4), (5) and (6) lie in the same plane, may be written down under the form

$$\left|\begin{matrix} A_1, & A_2, & A_3, & A_4 & . & . \\ B_1, & B_2, & B_3, & B_4 & . & . \\ C_1, & C_2, & C_3, & C_4 & . & . \\ . & . & A_3, & A_4, & A_5, & A_6 \\ . & . & B_3, & B_4, & B_5, & B_6 \\ . & . & C_3, & C_4, & C_5, & C_6 \end{matrix}\right| = 0.$$

LEMMA 2. Representing the determinants

$$\left|\begin{matrix} x_1, & y_1, & z_1 \\ x_2, & y_2, & z_2 \\ x_3, & y_3, & z_3 \end{matrix}\right| \ \&\text{c.}$$

by the abbreviated notation $\overline{123}$, &c.; the following equation is identically true:

$$\overline{345}\,.\,\overline{126} - \overline{346}\,.\,\overline{125} + \overline{356}\,.\,\overline{124} - \overline{456}\,.\,\overline{123} = 0.$$

This is an immediate consequence of the equations

$$\begin{vmatrix} . & . & x_3, & x_4, & x_5, & x_6 \\ . & . & y_3, & y_4, & y_5, & y_6 \\ . & . & z_3, & z_4, & z_5, & z_6 \\ x_1, & x_2, & x_3, & x_4, & x_5, & x_6 \\ y_1, & y_2, & y_3, & y_4, & y_5, & y_6 \\ z_1, & z_2, & z_3, & z_4, & z_5, & z_6 \end{vmatrix} = \begin{vmatrix} . & . & x_3, & x_4, & x_5, & x_6 \\ . & . & y_3, & y_4, & y_5, & y_6 \\ . & . & z_3, & z_4, & z_5, & z_6 \\ x_1, & x_2, & . & . & . & . \\ y_1, & y_2, & . & . & . & . \\ z_1, & z_2, & . & . & . & . \end{vmatrix} = 0.$$

Consider now the points 1, 2, 3, 4, 5, 6, the coordinates of these being respectively $x_1, y_1, z_1 \ldots\ldots x_6, y_6, z_6$. I represent, for shortness, the equation to the plane passing through the origin and the points 1, 2, which may be called the plane $\overline{12}$, in the form

$$x\,\overline{12}_x + y\,\overline{12}_y + z\,\overline{12}_z = 0\,;$$

consequently the symbols $\overline{12}_x$, $\overline{12}_y$, $\overline{12}_z$ denote respectively $y_1z_2 - y_2z_1$, $z_1x_2 - z_2x_1$, $x_1y_2 - x_2y_1$, and similarly for the planes $\overline{13}$, &c. If now the intersections of $\overline{12}$ and $\overline{45}$, $\overline{23}$ and $\overline{56}$, $\overline{34}$ and $\overline{61}$ lie in the same plane, we must have, by Lemma (1), the equation

$$\begin{vmatrix} 12_x, & 45_x, & 23_x, & 56_x, & . & . \\ 12_y, & 45_y, & 23_y, & 56_y, & . & . \\ 12_z, & 45_z, & 23_z, & 56_z, & . & . \\ . & . & 23_x, & 56_x, & 34_x, & 61_x \\ . & . & 23_y, & 56_y, & 34_y, & 61_y \\ . & . & 23_z, & 56_z, & 34_z, & 61_z \end{vmatrix} = 0.$$

Multiplying the two sides of this equation by the two sides respectively of the equation

$$\begin{vmatrix} x_6, & x_1, & x_2, & . & . & . \\ y_6, & y_1, & y_2, & . & . & . \\ z_6, & z_1, & z_2, & . & . & . \\ . & . & . & x_3, & x_4, & x_5 \\ . & . & . & y_3, & y_4, & y_5 \\ . & . & . & z_3, & z_4, & z_5 \end{vmatrix} = \overline{612}\,.\,\overline{345},$$

and observing the equations

$$x_6\,\overline{12}_x + y_6\,\overline{12}_y + z_6\,\overline{12}_z = \overline{612}, \quad \overline{112} = 0, \text{ \&c.}$$

this becomes

$$\left|\begin{matrix} 612 & . & . & . & . & . \\ 645, & 145, & 245, & . & . & . \\ 623, & 123, & . & . & 423, & 523 \\ . & 156, & 256, & 356, & 456, & . \\ . & . & . & . & . & 534 \\ . & . & . & 361, & 461, & 561 \end{matrix}\right| = 0,$$

reducible to

$$\overline{612}\ \overline{534}\left|\begin{matrix} \overline{145}, & \overline{245}, & . & . \\ 123, & . & . & 423 \\ 156, & 256, & 356, & 456 \\ . & . & 361, & 461 \end{matrix}\right| = 0;$$

or, omitting the factor $\overline{612}\,.\,\overline{534}$ and expanding,

$$\overline{145}\,.\,\overline{256}\,.\,\overline{423}\,.\,\overline{361} + \overline{245}\,.\,\overline{123}\,.\,\overline{456}\,.\,\overline{361} - \overline{245}\,.\,\overline{123}\,.\,\overline{356}\,.\,\overline{461} - \overline{245}\,.\,\overline{156}\,.\,\overline{423}\,.\,\overline{361} = 0.$$

Considering for instance x_6, y_6, z_6 as variable, this equation expresses evidently that the point 6 lies in a cone of the second order having the origin for its vertex, and the equation is evidently satisfied by writing $x_6, y_6, z_6 = x_1, y_1, z_1$, or x_3, y_3, z_3, or x_4, y_4, z_4, or x_5, y_5, z_5, and thus the cone passes through the points 1, 3, 4, 5. For $x_6, y_6, z_6 = x_2, y_2, z_2$, the equation becomes, reducing and dividing by $\overline{245}\,.\,\overline{123}$,

$$\overline{452}\,.\,\overline{321} - \overline{352}\,.\,\overline{421} + \overline{152}\,.\,\overline{423} = 0,$$

which is deducible from Lemma (2), by writing $x_6, y_6, z_6 = x_2, y_2, z_2$, and is therefore identically true. Hence the cone passes through the point (2), and therefore the points 1, 2, 3, 4, 5, 6 lie in the same cone of the second order, which is Pascal's Theorem. I have demonstrated it in the cone, for the sake of symmetry; but by writing throughout unity instead of z, the above applies directly to the case of the theorem in the plane.

The demonstration of Chasles' form of Pascal's Theorem (viz. that the anharmonic relation of the planes $\overline{61}$, $\overline{62}$, $\overline{63}$, $\overline{64}$ is the same with that of $\overline{51}$, $\overline{52}$, $\overline{53}$, $\overline{54}$), is very much simpler; but as it would require some preparatory information with reference to the analytical definition of the similarity of anharmonic relation, I must defer it to another opportunity.

10.

ON THE THEORY OF ALGEBRAIC CURVES.

[From the *Cambridge Mathematical Journal*, vol. IV. (1843), pp. 102—112.]

SUPPOSE a curve defined by the equation $U=0$, U being a rational and integral function of the mth order of the coordinates x, y. It may always be assumed, without loss of generality, that the terms involving x^m, y^m, both of them appear in U; and also that the coefficient of y^m is equal to unity: for in any particular curve where this was not the case, by transforming the axes, and dividing the new equation by the coefficient of y^m, the conditions in question would become satisfied. Let H_m denote the terms of U, which are of the order m, and let $y-\alpha x$, $y-\beta x \ldots y-\lambda x$ be the factors of H_m. If the quantities α, $\beta \ldots \lambda$ are all of them different, the curve is said to have a number of asymptotic directions equal to the degree of its equation. Such curves only will be considered in the present paper, the consideration of the far more complicated theory of those curves, the number of whose asymptotic directions is less than the degree of their equation, being entirely rejected. Assuming, then, that the factors of H_m are all of them different, we may deduce from the equation $U=0$, by known methods, the series

$$y = \alpha x + \alpha' + \frac{\alpha''}{x} + \ldots\ldots, \qquad \ldots\ldots\ldots\ldots\ldots\ldots\ldots\ldots \quad (1).$$

$$y = \beta x + \beta' + \frac{\beta''}{x} + \ldots\ldots,$$

$$\vdots$$

$$y = \lambda x + \lambda' + \frac{\lambda''}{x} + \ldots\ldots,$$

and these being obtained, we have, identically,

$$U = (y - \alpha x - \alpha' - \ldots)(y - \beta x - \beta' - \ldots)\ldots(y - \lambda x - \lambda' - \ldots) \ldots\ldots\ldots \;(2),$$

the negative powers of x on the second side, in point of fact, destroying each other. Supposing in general that fx containing positive and negative powers of x, Efx denotes the function which is obtained by the rejection of the negative powers, we may write

$$U = E\left(y - \alpha x \ldots - \frac{\alpha^{(m)}}{x^{m-1}}\right)\left(y - \beta x \ldots - \frac{\beta^{(m)}}{x^{m-1}}\right)\ldots\left(y - \lambda x \ldots - \frac{\lambda^{(m)}}{x^{m-1}}\right) \ldots\ldots(3),$$

the symbol E being necessary in the present case, because, when the series are continued only to the power x^{-m+1}, the negative powers no longer destroy each other.

We may henceforward consider U as originally given by the equation (3), the $m(m+1)$ quantities $\alpha, \alpha' \ldots \alpha^{(m)}, \beta, \beta' \ldots \beta^{(m)}, \ldots \lambda, \lambda' \ldots \lambda^{(m)}$ satisfying the equations obtained from the supposition that it is possible to determine the following terms $\alpha^{(m+1)}, \beta^{(m+1)} \ldots \lambda^{(m+1)}, \ldots$ so that the terms containing negative powers of x, on the second side of equation (2), vanish. It is easily seen that $\alpha, \beta \ldots \lambda, \alpha', \beta' \ldots \lambda'$ are entirely arbitrary, $\alpha'', \beta'' \ldots \lambda''$ satisfy a single equation involving only the preceding quantities, $\alpha''', \beta''' \ldots \lambda'''$ two equations involving the quantities which precede them, and so on, until $\alpha^{(m)}, \beta^{(m)} \ldots \lambda^{(m)}$, which satisfy $(m-1)$ relations involving the preceding quantities. Thus the $m(m+1)$ quantities in question satisfy $\frac{1}{2}m(m-1)$ equations, or they may be considered as functions of $m(m+1) - \frac{1}{2}m(m-1) = \frac{1}{2}m(m+3)$ arbitrary constants. Hence the value of U, given by the equation (3), is the most general expression for a function of the mth order. It is to be remarked also that the quantities $\alpha^{(m+1)}, \beta^{(m+1)} \ldots \lambda^{(m+1)}, \ldots\ldots$ are all of them completely determinable as functions of $\alpha, \beta \ldots \lambda, \ldots \alpha^{(m)}, \beta^{(m)} \ldots \lambda^{(m)}$.

The advantage of the above mode of expressing the function U, is the facility obtained by means of it for the elimination of the variable y from the equation $U = 0$, and any other analogous one $V = 0$. In fact, suppose V expressed in the same manner as U, or by the equation

$$V = E\left(y - Ax \ldots \quad \frac{A^{(n)}}{x^{n-1}}\right)\left(y - Bx \ldots - \frac{B^{(n)}}{x^{n-1}}\right)\ldots\left(y - Kx \ldots - \frac{K^{(n)}}{x^{n-1}}\right) \ldots\ldots\ldots(4),$$

n being the degree of the function V. It is almost unnecessary to remark, that $A, B \ldots K, \ldots A^{(n)}, B^{(n)} \ldots K^{(n)}$ are to be considered as functions of $\frac{1}{2}n(n+3)$ arbitrary constants, and that the subsequent $A^{(n+1)}, B^{(n+1)} \ldots K^{(n+1)} \ldots$ can be completely determined as functions of these. Determining the values of y from the equation (3), viz. the values given by the equations (2); substituting these successively in the equation

$$V = (y - Ax - \ldots)(y - Bx - \ldots)\ldots(y - Kx - \ldots) = 0 \;\ldots\ldots\ldots\ldots\; (5),$$

analogous to (2), and taking the product of the quantities so obtained, also observing that this product must be independent of negative powers of x, the result of the elimination may be written down under the form

$$E\left(\begin{array}{l} \left\{(\alpha - A)x \ldots + \dfrac{\alpha^{(mn)} - A^{(mn)}}{x^{mn-1}}\right\} \ldots \left\{(\alpha - Kx) \ldots + \dfrac{\alpha^{(mn)} - K^{(mn)}}{x^{mn-1}}\right\} \\ \quad\vdots \qquad\qquad\qquad\qquad\qquad\qquad\quad \vdots \\ \times\left\{(\lambda - A)x \ldots + \dfrac{\lambda^{(mn)} - A^{(mn)}}{x^{mn-1}}\right\} \ldots \left\{(\lambda - Kx) \ldots + \dfrac{\lambda^{(mn)} - K^{(mn)}}{x^{mn-1}}\right\} \end{array}\right) \ldots (6),$$

the series in { } being continued only to x^{-mn+1}, because the terms after this point produce in the whole product only terms involving negative powers of x. It is for the same reason that the series in ( ), in the equations (3) and (4), are only continued to the terms involving x^{-m+1}, x^{-n+1} respectively.

The first side of the equation (6) is of the order mn, in x, as it ought to be. But it is easy to see, from the form of the expression, in what case the order of the first side reduces itself to a number less than mn. Thus, if n be not greater than m, and the following equations be satisfied,

$$A=\alpha,\quad A^{(1)}=\alpha^{(1)}\,\ldots\,A^{(r-1)}=\alpha^{(r-1)},\quad r \ngtr n \ldots\ldots\ldots\ldots\ldots\ldots(7),$$
$$B=\beta,\quad B^{(1)}=\beta^{(1)}\,\ldots\,B^{(s-1)}=\beta^{(s-1)},\quad s \ngtr r,$$
$$\vdots$$
$$K=\kappa,\quad K^{(1)}=\kappa^{(1)}\,\ldots\,K^{(v-1)}=\kappa^{(v-1)},\quad v \ngtr u,$$

the degree of the equation (6) is evidently $mn-r-s\ldots-v$, or the curves $U=0$, $V=0$ intersect in this number only of points. If $mn-r-s\ldots-v=0$, the curves $U=0$ and $V=0$ do not intersect at all, and if $mn-r-s-v$ be negative, $=-\omega$ suppose, the equation (6) is satisfied identically; or the functions U, V have a common factor, the number ω expressing the degree of this factor in x, y.

Supposing the function V given arbitrarily, it may be required to determine U, so that the curves $U=0$, $V=0$ intersect in a number $mn-k$ points. This may in general be done, and done in a variety of ways, for any value of k from unity to $\tfrac{1}{2}m\,(m+3)$. I shall not discuss the question generally at present, nor examine into the meaning of the quantity $mn-\tfrac{1}{2}m\,(m+3)\,\{=\tfrac{1}{2}m\,(2n-m-3)\}$ becoming negative, but confine myself to the simple case of U and V, both of them functions of the second order. It is required, then, to find the equations of all those curves of the second order which intersect a given curve of the second order in a number of points less than four.

Assume in general

$$V=E\left(y-Ax-A'-\frac{A''}{x}\right)\left(y-Bx-B'-\frac{B''}{x}\right),$$

then A'', B'' satisfy $A''+B''=0$, and putting $B''=\dfrac{K}{A-B}$, and therefore $A''=-\dfrac{K}{A-B}$, and reducing, we obtain

$$V=(y-Ax-A')(y-Bx-B')+K.$$

Similarly assume $$U=E\left(y-\alpha x-\alpha'-\frac{\alpha''}{x}\right)\left(y-\beta x-\beta'-\frac{\beta''}{x}\right),$$

then α'', β'' satisfy $\alpha''+\beta''=0$, and putting $\beta''=\dfrac{k}{\alpha-\beta}$, and therefore $\alpha''=-\dfrac{k}{\alpha-\beta}$, and reducing, we obtain

$$U=(y-\alpha x-\alpha')(y-\beta x-\beta')+k.$$

Suppose

(1) $U=0$, $V=0$ intersect in three points, we must have $\alpha=A$, or the curve $U=0$ must have one of its asymptotes parallel to one of the asymptotes of $V=0$.

(2) The curves intersect in two points. We must have $\alpha=A$, $\alpha'=A'$, or else $\alpha=A$, $\beta=B$; i.e. $U=0$ must have one of its asymptotes coincident with one of the asymptotes of the curve $V=0$, or else it must have its two asymptotes parallel to those of $V=0$. The latter case is that of similar and similarly situated curves.

(3) Suppose the curves intersect in a single point only. Then either $\alpha=A$, $\alpha'=A'$, $\alpha''=A''$, which it is easy to see gives

$$U=(y-Ax-A')(y-\beta x-\beta')+K\frac{A-\beta}{A-B},$$

or else $\alpha=A$, $\alpha'=A'$, $\beta=B$, which is the case of one of the asymptotes of the curve $U=0$, coinciding with one of those of the curve $V=0$, and the remaining asymptotes parallel. As for the first case, if a, a_1 are the transverse axes, θ, θ_1 the inclinations of the two asymptotes to each other, these four quantities are connected by the equation

$$\frac{a^2}{a_1^2}=\frac{\tan\theta\cos^2\frac{1}{2}\theta}{\tan\theta_1\cos^2\frac{1}{2}\theta_1};$$

and besides, one of the asymptotes of the first curve is coincident with one of the asymptotes of the second. This is a remarkable case; it may be as well to verify that $U=0$, $V=0$ do intersect in a single point only. Multiplying the first equation by $y-Bx-B'$, the second by $y-\beta x-\beta'$, and subtracting, the result is

$$(A-\beta)(y-Bx-B')-(A-B)(y-\beta x-\beta')=0,$$

reducible to

$$y-Ax=\frac{A(B'-\beta')+B\beta'-B'\beta}{B-\beta},\quad \text{i.e. } y-Ax-C=0.$$

Combining this with $V=0$, we have an equation of the form $y-Bx-D=0$. And from this and $y-Ax-C=0$, x, y are determined by means of a simple equation.

(4) Lastly, when the curves do not intersect at all. Here $\alpha=A$, $\alpha'=A'$, $\beta=B$, $\beta'=B'$, or the asymptotes of $U=0$ coincide with those of $V=0$; i.e. the curves are similar, similarly situated, and concentric: or else $\alpha=A$, $\alpha'=A'$, $\alpha''=A''$, $\beta=B$; here

$$U=(y-Ax-A')(y-Bx-\beta')+K,$$

or the required curve has one of its asymptotes coincident with one of those of the proposed curve; the remaining two asymptotes are parallel, and the magnitudes of the curves are equal.

In general, if two curves of the orders m and n, respectively, are such that r asymptotes of the first are parallel to as many of the second, s out of these asymptotes being coincident in the two curves, the number of points of intersection is $mn - r - s$; but the converse of this theorem is not true.

In a former paper, *On the Intersection of Curves*, [5], I investigated the number of arbitrary constants in the equation of a curve of a given order ρ subjected to pass through the mn points of intersection of two curves of the orders m and n respectively. The reasoning there employed is not applicable to the case where the two curves intersect in a number of points less than mn. In fact, it was assumed that, $W=0$ being the equation of the required curve, W was of the form $uU+vV$; u, v being polynomials of the degrees $\rho - m$, $\rho - n$ respectively. This is, in point of fact, true in the case there considered, viz. that in which the two curves intersect in mn points; but where the number of points of intersection is less than this, u, v may be assumed polynomials of an order *higher* than $\rho - m$, $\rho - n$, and yet $uU+vV$ reduce itself to the order ρ. The preceding investigations enable us to resolve the question for every possible case.

Considering then the functions U, V determined as before by the equations (3), (4), suppose, in the first place, we have a system of equations

$$\alpha = A, \quad \beta = B \ldots\ldots \theta = H \ (t \text{ equations}) \ldots\ldots\ldots\ldots\ldots\ldots (8).$$

Assume
$$P = (y - \alpha x - \ldots)(y - \beta x - \ldots) \ldots (y - \theta x - \ldots),$$
$$Q = (y - Ax - \ldots)(y - Bx - \ldots) \ldots (y - Hx - \ldots);$$

$$\Upsilon = (y - \iota x - \ldots) \ldots (y - \kappa x - \ldots),$$
$$\Psi = (y - \mathrm{I}x - \ldots) \ldots (y - Kx - \ldots);$$

whence
$$U = P\Upsilon, \qquad V = Q\Psi.$$

Suppose
$$\Upsilon = E\Upsilon + \Delta\Upsilon, \qquad \Psi = E\Psi + \Delta\Psi,$$

$$\begin{aligned} E\Psi \,.\, U - E\Upsilon \,.\, V &= E\Psi \,.\, P\Upsilon - E\Upsilon \,.\, Q\Psi, \\ &= E\Psi \,.\, P(E\Upsilon + \Delta\Upsilon) - E\Upsilon \,.\, Q(E\Psi + \Delta\Psi), \\ &= E\Upsilon \,.\, E\Psi \,.\, (P - Q) + E\Psi \,.\, P \,.\, \Delta\Upsilon - E\Upsilon \,.\, Q \,.\, \Delta\Psi, \\ &= E\{E\Upsilon \,.\, E\Psi \,.\, (P - Q) + E\Psi \,.\, P\Delta\Upsilon - E\Upsilon \,.\, Q \,.\, \Delta\Psi\}, \\ &= \Pi \text{ suppose.} \end{aligned}$$

In this expression $E\Upsilon$, $E\Psi$ are of the degrees $m-t$, $n-t$, $\Delta\Upsilon$, $\Delta\Psi$ of the degree -1, and P, Q, $P-Q$ of the degrees t, t, $t-1$ respectively. The terms of Π are therefore of the degrees $m+n-t-1$, $m-1$, $n-1$ respectively, and the largest of these is in general $m+n-t-1$. Suppose, however, that $m+n-t-1$ is equal to $m-1$ (it cannot be inferior to it), then $t=n$; Ψ becomes equal to unity, or $\Delta\Psi$ vanishes. The remaining two terms of Π are $E\Upsilon(P-Q)$, $P\Delta\Upsilon$, which are of the degrees $m-1$, $n-1$ respectively. Π is still of the degree $m-1$, supposing $m > n$. If $m=n$, the term $P\Delta\Upsilon$ vanishes. Π is still of the degree $m-1$. Hence in every case the degree of Π is $m+n-t-1$: assuming always that $P-Q$ does not reduce itself to a degree lower than $t-1$, (which is always the case as long as the equations

$\alpha' = A'$, $\beta' = B'$ ...... $\theta' = H'$ are not all of them satisfied simultaneously). It will be seen presently that we shall gain in symmetry by wording the theorem thus: the degree of Π is equal to the greatest of the two quantities $m+n-t-1$, $m-1$.

Suppose next, in addition to the equations (8), we have

$$\alpha' = A', \quad \beta' = B' \;......\; \zeta' = F', \quad t' \text{ equations, } t' \not> t \;...\; (8').$$

Then, taking Υ', Ψ', P', Q' the analogous quantities to Υ, Ψ, P, Q, and putting

$$E\Psi' . U - E\Upsilon' . V = \Pi',$$

we have, as before,

$$\Pi' = E\,\{E\Upsilon' . E\Psi' . (P' - Q') + E\Psi' . P'\Delta\Upsilon' - E\Upsilon' . Q'\Delta\Psi'\}.$$

The degree of $P' - Q'$ is $t' - 2$ (unless simultaneously $\alpha'' = A''$, $\beta'' = B''$ ...... $\zeta'' = F''$, in which case the degree may be lower). The degrees, therefore, of the terms of Π' are $m+n-t'-2$, $n-1$, $m-1$. Or we may say that the degree of Π' is equal to the greatest of the quantities $m+n-t'-2$, $m-1$; though to establish this proposition in the case where $t' = n-1$ would require some additional considerations.

Continuing in this manner until we come to the quantity $\Pi^{(k-1)}$, the degree of this quantity is the greatest of the two numbers $m+n-t^{(k-1)}-k$, $m-1$. And we may suppose that none of the equations $\alpha^{(k)} = A^{(k)}$ ...... are satisfied, so that the series Π, Π' ...... $\Pi^{(k-1)}$ is not to be continued beyond this point.

Considering now the equation of the curve passing through the $mn - t - t' \ldots - t^{(k-1)}$ points of intersection of $U=0$, $V=0$, we may write

$$W = uU + vV + p\Pi + p'\Pi' \;......\; + p^{(k-1)}\,\Pi^{(k-1)} = 0 \;.................\; (9),$$

for the required equation; the dimensions of u, v, p, p' ... $p^{(k-1)}$ being respectively

$$\rho - m, \quad \rho - n; \quad \rho - m - n + t + 1 \text{ or } \rho - m + 1;$$
$$\rho - m - n + t' + 2 \text{ or } \rho - m + 1; \;......\; \rho - m - n + t^{(k-1)} + k \text{ or } \rho - m + 1,$$

the lowest of the two numbers being taken for the dimensions of p, p' ... $p^{(k-1)}$. Also, if any of these numbers become negative, the corresponding term is to be rejected. In saying that the degrees of p, p' ...... $p^{(k-1)}$ have these actual values, it is supposed that the degrees of Π, Π' ...... Π^{k-1} actually ascend to the greatest of the values

$$\rho - m - n + t + 1, \text{ or } m - 1; \quad m + n - t' - 2, \text{ or } m - 1; \quad - m + n - t^{(k-1)} + k, \text{ or } m - 1.$$

The cases of exception to this are when several of the consecutive numbers t, t' ... $t^{(k-1)}$ are equal. In this case the corresponding terms of the series Π, Π' ...... $\Pi^{(k-1)}$, are also equal. Suppose for instance t, t' were equal, Π, Π' would also be equal. A term of p of an *order higher* by unity than $\rho - m - n + t + 1$, or $\rho - m + 1$, which is the highest term admissible, produces in $p\Pi$ a term identical with one of the terms of $p'\Pi$; so that nothing is gained in generality by admitting such terms into p. The equation (9), with the preceding values for the dimensions of p, p' ...... $p^{(k-1)}$, may be employed, therefore, even when several consecutive terms of the series $t, t' \ldots\ldots t^{(k-1)}$ are equal. It will be convenient also to assume that $\rho - m$ is not negative, or at least for greater simplicity to examine this case in the first place.

u, U, and v, V, contain terms of the form $x^\alpha y^\beta U$, $x^\gamma y^\delta V$, $\alpha + \beta \not> \rho - m$, $\gamma + \delta \not> \rho - n$; $p\Pi$ contains terms of this form, and in addition terms for which $\alpha + \beta = \rho + 1 - m$, $\gamma + \delta = \rho + 1 - n$. It is useless to repeat the former terms, so that we may assume for p, a *homogeneous* function of the order $\rho - m - n + t + 1$, or $\rho - m + 1$; in which case $p\Pi$ consists only of terms for which $\alpha + \beta = \rho + 1 - m$, $\gamma + \delta = \rho + 1 - n$. And the general expression of p contains $\rho - m - n + t + 2$, or $\rho - m + 2$, arbitrary constants. Similarly $p'\Pi'$ contains terms of the form of those in uU, vV, $p\Pi$, and also terms for which $\alpha + \beta = \rho + 2 - m$, $\gamma + \delta = \rho + 2 - n$; the latter terms only need be considered, or p' may be assumed to be a homogeneous function of the order $\rho - m - n + t' + 2$, or $\rho - m + 1$, containing therefore $\rho - m - n - t' + 3$, or $\rho - m + 2$ arbitrary constants.

Similarly $p^{(k-1)}$ contains $\rho - m - n + t^{(k-1)} + k + 1$ or $\rho - m + 2$ arbitrary constants. Assume

$$\nabla = \begin{pmatrix}\rho - m - n + t + 2 \\ \rho - m + 2\end{pmatrix} + \begin{pmatrix}\rho - m - n + t' + 3 \\ \rho - m + 2\end{pmatrix} \ldots + \begin{pmatrix}\rho - m - n + t^{(k-1)} + k + 1 \\ \rho - m + 2\end{pmatrix} \ldots (10),$$

where, in forming the value of ∇ the least of the two quantities in ( ) is to be taken; this value also, if negative, being replaced by zero. The number of arbitrary constants in $p, p' \ldots p^{(k-1)}$ is consequently equal to ∇.

The numbers of arbitrary constants in u, v, are respectively

$$\{1 + 2 \ldots\ldots + (\rho - m + 1)\} \text{ and } \{1 + 2 \ldots\ldots + (\rho - n + 1)\}$$

i.e.

$$\tfrac{1}{2}(\rho - m + 1)(\rho - m + 2), \text{ and } \tfrac{1}{2}(\rho - n + 1)(\rho - n + 2);$$

thus the whole number of arbitrary constants in W, diminished by unity (since nothing is gained in generality, by leaving the coefficient (for instance of y^ρ) indeterminate, instead of supposing it equal to unity) becomes

$$\tfrac{1}{2}(\rho - m + 1)(\rho - m + 2) + \tfrac{1}{2}(\rho - n + 1)(\rho - n + 2) + \nabla - 1,$$

reducible to

$$\tfrac{1}{2}\rho(\rho + 3) + \tfrac{1}{2}(\rho - m - n + 1)(\rho - m - n + 2) - mn + \nabla.$$

By the reasonings contained in the paper already referred to, if $\rho + k - m - n + 1$ be positive, to find the number of really disposable constants in W, we must subtract from this number a number $\tfrac{1}{2}(\rho + k - m - n + 1)(\rho + k - m - n + 2)$. Hence, calling ϕ the number of disposable constants in W, we have

$$\phi = \tfrac{1}{2}\rho(\rho + 3) + \tfrac{1}{2}(\rho - m - n + 1)(\rho - m - n + 2) - mn + \nabla - \Lambda \ldots\ldots (11),$$

where

$$\Lambda = 0, \text{ if } \rho + k - m - n + 1 \text{ be negative or zero} \ldots\ldots\ldots\ldots\ldots\ldots (12),$$

$$\Lambda = \tfrac{1}{2}(\rho + k - m - n + 1)(\rho + k - m - n + 2),$$

if $\rho + k - m - n + 1$ be positive; and ∇ is given by the equation (9).

Also, if θ be the number of points through which the curve $W=0$ can be drawn, *including* the points of intersection of the curves $U=0$, $V=0$, then

$$\theta=\phi+(mn-t-t'\ldots\ldots-t^{(k-1)})\text{ or}$$

$$\theta=\tfrac{1}{2}\rho\,(\rho+3)+\tfrac{1}{2}(\rho-m-n+1)(\rho-m-n+2)+\nabla-\Lambda-t-t'\ldots-t^{(k-1)}\ \ldots\ldots(13).$$

Any particular cases may be deduced with the greatest facility from these general formulæ. Thus, supposing the curves to intersect in the complete number of points mn, we have

$$\phi=\tfrac{1}{2}\rho\,(\rho+3)+\tfrac{1}{2}\,(1-\aleph)\,(\rho-m-n+1)\,(\rho-m-n+2)-mn,$$

$\aleph$ being zero or unity according as $\rho<m+n-1$ or $\rho>m+n-1$. Reducing, we have, for $\rho \not> m+n-3$,

$$\phi=\tfrac{1}{2}\rho\,(\rho+3)+\tfrac{1}{2}(\rho-m-n+1)\,(\rho-m-n+2)-mn,$$
$$\theta=\tfrac{1}{2}\rho\,(\rho+3)+\tfrac{1}{2}\,(\rho-m-n+1)\,(\rho-m-n+2)\,;$$

and for $\rho>m+n-3$,

$$\phi=\tfrac{1}{2}\rho\,(\rho+3)-mn,$$
$$\theta=\tfrac{1}{2}\rho\,(\rho+3).$$

Suppose, in the next place, the curves have t parallel pairs of asymptotes, none of these pairs being coincident. Then

$$\rho \not> m+n-t-2,$$
$$\phi=\tfrac{1}{2}\rho\,(\rho+3)+\tfrac{1}{2}\,(\rho-m-n+1)\,(\rho-m-n+2)-mn,$$
$$\theta=\tfrac{1}{2}\rho\,(\rho+3)+\tfrac{1}{2}\,(\rho-m-n+1)\,(\rho-m-n+2)-t\,;$$

$$\rho>m+n-t-2,\quad \rho \not> m+n-2,$$
$$\phi=\tfrac{1}{2}\rho\,(\rho+3)+\tfrac{1}{2}\,(\rho-m-n+2)\,(\rho-m-n+3)-mn+t,$$
$$\theta=\tfrac{1}{2}\rho\,(\rho+3)+\tfrac{1}{2}(\rho-m-n+2)\,(\rho-m-n+3),$$

$$\rho>m+n-2,$$
$$\phi=\tfrac{1}{2}\rho\,(\rho+3)-mn+t,$$
$$\theta=\tfrac{1}{2}\rho\,(\rho+3).$$

In these formulæ, if t be equal to 2 or greater than 2, the limiting conditions are more conveniently written

$$\rho \not> m+n-t-2;\quad \rho \not> m+n-t-2>m+n-4;\quad \rho>m+n-4.$$

Similarly may the solution of the question be explicitly obtained when the curves have t asymptotes parallel, and t' out of these coincident, but the number of separate formulæ will be greater.

In conclusion, I may add the following references to two memoirs on the present subject: the conclusions in one point of view are considerably less general even than those of my former paper, though much more so in another. Jacobi, Theoremata de punctis intersectionis duarum curvarum algebraicarum; *Crelle's Journal,* vol. XV. [1836, pp. 285—308]; Plücker, Théorèmes généraux concernant les equations à plusieurs variables, d'un degré quelconque entre un nombre quelconque d'inconnues. D° vol. XVI. [1837, pp. 47—57].

Addition.

As an exemplification of the preceding formulæ, and besides as a question interesting in itself, it may be proposed to determine the asymptotic curves of the rth order of a given curve having all its asymptotic directions distinct,—r being any number less than the degree of the equation of the given curve.

DEFINITION. A curve of the rth order, which intersects a given curve of the mth order in a number of points, $= mr - \frac{1}{2}r(r+3)$, is said to be an asymptotic curve of the rth order to the curve in question. Suppose, as before, $U=0$ being the equation to the given curve,

$$U = E\left(y - \alpha x - \ldots - \frac{\alpha^{(m)}}{x^{m-1}}\right) \ldots \left(y - \lambda x \ldots - \frac{\lambda^{(m)}}{x^{m-1}}\right);$$

and let $\theta, \phi \ldots \omega$ denote any combination of r terms out of the series $\alpha \ldots \lambda$, and $\theta', \phi' \ldots \omega'$, &c. ... the corresponding terms out of $\alpha' \ldots \lambda'$, &c. Then, writing

$$V = E\left\{\left(y - \theta x \ldots - \frac{\theta^{(m)}}{x^{m-1}}\right)\left(y - \phi x \ldots - \frac{\phi^{(m-1)}}{x^{m-2}} - \frac{\Phi^{(m)}}{x^{m-1}}\right) \times \ldots\ldots\right.$$

$$\left.\left(y - \psi x \ldots - \frac{\psi^{(m-2)}}{x^{m-3}} - \frac{\Psi^{(m-1)}}{x^{m-2}} - \frac{\Psi^{(m)}}{x^{m-1}}\right) \ldots \left(y - \omega x - \Omega' \ldots - \frac{\Omega^{(m)}}{x^{m-1}}\right)\right\},$$

(where the quantities $\Phi^{(m)}$, $\Psi^{(m-1)}$, $\Psi^{(m)}$, $\Omega' \ldots \Omega^{(m)}$ are entirely determinate, since, by what has preceded, $\theta', \phi' \ldots \Omega'$ satisfy a certain equation, $\theta'', \phi'', \ldots \Omega''$ two equations...... $\theta^{(m)}, \Phi^{(m)}, \ldots \Omega^{(m)}$ $(m-1)$ equations), we have $V=0$ for the required equation of the asymptotic curve. It is obvious that the whole number of asymptotic curves of the order r, is $n(n-1) \ldots (n-r+1)$, viz. $1.2 \ldots r$ curves for each combination of $\dfrac{n(n-1)\ldots(n-r+1)}{1.2\ldots r}$ asymptotes. Some particular instances of asymptotic curves will be found in a memoir by M. Plücker, *Liouville's Journal*, vol. I. [1836, pp. 229—252], Enumération des courbes du quatrième ordre, &c. The general theory does not seem to be one to which much attention has been paid.

11.

CHAPTERS IN THE ANALYTICAL GEOMETRY OF (n) DIMENSIONS.

[From the *Cambridge Mathematical Journal*, vol. IV. (1843), pp. 119—127.]

CHAP. 1. *On some preliminary formulæ.*

I TAKE for granted all the ordinary formulæ relating to determinants. It will be convenient, however, to write down a few, relating to a certain system of determinants, which are of considerable importance in that which follows: they are all of them either known, or immediately deducible from known formulæ.

Consider the series of terms

$$\begin{matrix} x_1\,, & x_2 \ldots\ldots x_n \\ A_1, & A_2 \ldots\ldots A_n \\ \vdots & \qquad\quad \vdots \\ K_1, & K_2 \ldots\ldots K_n \end{matrix} \qquad \ldots\ldots\ldots\ldots(1).$$

the number of the quantities $A \ldots\ldots K$ being equal to q $(q < n)$. Suppose $q+1$ vertical rows selected, and the quantities contained in them formed into a determinant, this may be done in $\dfrac{n\,(n-1)\,\ldots\ldots\,(q+2)}{1\,.\,2\,\ldots\ldots\,n-q-1}$ different ways. The system of determinants so obtained will be represented by the notation

$$\left\|\begin{matrix} x_1\,, & x_2 \ldots\ldots x_n \\ A_1, & A_2 \ldots\ldots A_n \\ \vdots & \qquad\quad \vdots \\ K_1, & K_2 \ldots\ldots K_n \end{matrix}\right\| \qquad \ldots\ldots\ldots\ldots (2),$$

and the system of equations, obtained by equating each of these determinants to zero, by the notation

$$\left\|\begin{matrix} x_1\,, & x_2 \ldots\ldots x_n \\ A_1, & A_2 \ldots\ldots A_n \\ \vdots & \qquad\quad \vdots \\ K_1 & K_2 \ldots\ldots K_n \end{matrix}\right\| = 0 \ldots\ldots\ldots\ldots (3).$$

The $\frac{n(n-1)\ldots\ldots(q+2)}{1.2\ldots\ldots(n-q+1)}$ equations represented by this formula reduce themselves to $(n-q)$ independent equations. Imagine these expressed by

$$(1)=0,\quad (2)=0\ldots\ldots(n-q)=0 \quad\ldots\ldots\ldots\ldots (4),$$

any one of the determinants of (2) is reducible to the form

$$\Theta_1(1)+\Theta_2(2)\ldots+\Theta_{n-q}(n-q) \quad\ldots\ldots\ldots\ldots (5),$$

where $\Theta_1, \Theta_2\ldots\Theta_{n-q}$ are coefficients independent of $x_1, x_2\ldots x_n$. The equations (3) may be replaced by

$$\left\|\begin{matrix} \lambda_1x_1+\lambda_2x_2+\ldots\lambda_nx_n, & \mu_1x_1+\ldots, & \ldots\tau_1x_1+\ldots \\ \lambda_1A_1+\lambda_2A_2+\ldots\lambda_nx_n, & \mu_1A_1+\ldots, & \ldots\tau_1A_1+\ldots \\ \vdots & & \\ \lambda_1K_1+\lambda_2K_2+\ldots\lambda_nK_n, & \mu_1K_1+\ldots, & \tau_1K_1+\ldots \end{matrix}\right\|=0 \quad\ldots\ldots\ldots\ldots (6),$$

and conversely from (6) we may deduce (3), unless

$$\left|\begin{matrix} \lambda_1, & \lambda_2,\ldots\lambda_n \\ \mu_1, & \mu_2,\ldots\mu_n \\ \vdots & \\ \tau_1, & \tau_2,\ldots\tau_n \end{matrix}\right|=0 \quad\ldots\ldots\ldots\ldots (7).$$

(The number of the quantities $\lambda, \mu\ldots\tau$ is of course equal to n.) The equations (3) may also be expressed in the form

$$\left\|\begin{matrix} x_1 & , & x_2, & \ldots & x_n \\ \lambda_1A_1+\ldots\omega_1K_1, & & \lambda_1A_2+\ldots\omega_1K_2, & \ldots & \lambda_1A_n\ldots+\omega_1K_n \\ \vdots & & \vdots & & \\ \lambda_qA_1+\ldots\omega_qK_1, & & \lambda_qA_2+\ldots\omega_qK_2, & \ldots & \lambda_qA_n\ldots+\omega_qK_n \end{matrix}\right\| \quad\ldots\ldots\ldots\ldots (8),$$

the number of the quantities $\lambda, \mu\ldots\omega$ being q.

And conversely (3) is deducible from (8), unless

$$\left|\begin{matrix} \lambda_1,\ldots\omega_1 \\ \vdots \\ \lambda_q,\ldots\omega_q \end{matrix}\right|=0 \quad\ldots\ldots\ldots\ldots (9).$$

CHAP. 2. *On the determination of linear equations in* $x_1, x_2,\ldots x_n$ *which are satisfied by the values of these quantities derived from given systems of linear equations.*

It is required to find linear equations in $x_1,\ldots x_n$ which are satisfied by the values of these quantities derived—1. from the equations $\mathfrak{A}'=0, \mathfrak{B}'=0\ldots\mathfrak{G}'=0$; 2. from the equations $\mathfrak{A}''=0, \mathfrak{B}''=0\ldots\mathfrak{N}''=0$; 3. from $\mathfrak{A}'''=0, \mathfrak{B}'''=0\ldots\mathfrak{K}'''=0$, &c. &c., where

$$\mathfrak{A}'=A_1x_1+A_2x_2\ldots+A_nx_n, \quad\ldots\ldots\ldots\ldots (1),$$

$$\mathfrak{B}'=B_1x_1+B_2x_2\ldots+B_nx_n,$$

$$\vdots$$

and similarly $\mathfrak{A}'', \mathfrak{B}'',\ldots, \mathfrak{A}''', \mathfrak{B}''',\ldots$, &c. are linear functions of the coordinates $x_1, x_2,\ldots x_n$.

Also $r', r''\ldots$ representing the number of equations in the systems (1), (2) ... and k the number of these given systems,

$$(n-r')+(n-r'')+\ldots \not> n-1 \text{ or } (k-1)\,n+1 \not> r'+r''+\ldots$$

Assume

$$0=\lambda'\mathfrak{A}'+\mu'\mathfrak{B}'+\ldots,$$
$$\lambda'\mathfrak{A}'+\mu'\mathfrak{B}'+\ldots=\lambda''\mathfrak{A}''+\mu''\mathfrak{B}''+\ldots=\lambda'''\mathfrak{A}'''+\mu'''\mathfrak{B}'''+\ldots=\&\text{c.} \;\ldots\ldots\; (2),$$

the latter equations denoting the equations obtained by equating to zero the terms involving x_1, those involving x_2, &c.... separately. Suppose, in addition to these, a set of linear equations in $\lambda', \mu' \ldots \lambda'', \mu''\ldots$ so that, with the preceding ones, there is a sufficient number of equations for the elimination of these quantities. Then, performing the elimination, we thus obtain equations $\Psi=0$, where Ψ is a function of $x_1, x_2\ldots$ which vanishes for the values of these quantities derived from the equations (1) or (2) ... &c. The series of equations $\Psi=0$ may be expressed in the form

$$\left\|\begin{matrix} \mathfrak{A}', & \mathfrak{B}', \ldots \mathfrak{C}', & & & & & & \\ A_1', & B_1', \ldots G_1', & A_1'', \ldots O''_1, & & & & & \\ \vdots & \vdots \quad \vdots & \vdots \quad \vdots & & & & & \\ A_n', & B_n', \ldots G_n', & A_n'', \ldots O_n'', & & & & & \\ & & A_1'', \ldots O_1'', & A_1''', \ldots R_1''', & & & & \\ & & \vdots \quad \vdots & \vdots \quad \vdots & & & & \\ & & A_n'', \ldots O_n'', & A_n''', \ldots R_n''', & & & & \\ & & \vdots \quad \vdots & \vdots \quad \vdots & \vdots & \vdots & & \end{matrix}\right\| = 0 \;\ldots\ldots\; (3).$$

CHAP. 3. *On reciprocal equations.*

Consider a system of equations

$$\begin{aligned} A_1x_1+A_2x_2\ldots+A_nx_n&=0, \quad \ldots\ldots\ldots\ldots\ldots\ldots\ldots\ldots(1),\\ &\vdots\\ K_1x_1+K_2x_2\ldots+K_nx_n&=0, \end{aligned}$$

(r in number).

The reciprocal system with respect to a given function (U) of the second order in $x_1, x_2\ldots x_n$, is said to be

$$\left\|\begin{matrix} d_{x_1}U, & d_{x_2}U, \ldots & d_{x_n}U \\ A_1, & A_2, \ldots & A_n \\ \vdots & \vdots & \vdots \\ K_1, & K_2, \ldots & K_n \end{matrix}\right\| = 0 \;\ldots\ldots\ldots\ldots\ldots\ldots\ldots\ldots(2),$$

$(n-r$ in number).

It must first be shown that the reciprocal system to (2) is the system (1), or that the systems (1), (2) are reciprocals of each other.

Consider, in general, the system of equations

$$\alpha_1 d_{x_1} U + \alpha_2 d_{x_2} U \ldots + \alpha_n d_{x_n} U = 0 \quad \ldots\ldots (3).$$
$$\vdots$$
$$\lambda_1 d_{x_1} U + \lambda_2 d_{x_2} U \ldots + \lambda_n d_{x_n} U = 0.$$

Suppose $\quad 2U = \Sigma (\alpha^2) x_\alpha^2 + 2\Sigma (\alpha\beta)\, x_\alpha x_\beta$, so that $d_{x_s} U = \Sigma (s\alpha)\, x_\alpha \ldots\ldots (4), (5).$

The equations (3) may be written

$$x_1 \{\alpha_1 (1^2) + \alpha_2 (12) \ldots + \alpha_n (1n)\} + \ldots + x_n \{\alpha_1 (n1) + \alpha_2 (n2) \ldots + \alpha_n (n^2)\} = 0 \ \ldots\ldots (6),$$
&c.

and forming the reciprocals of these, also replacing $d_{x_1} U$, $d_{x_2} U \ldots$ by their values, we have

$$\left\| \begin{matrix} x_1 (1^2) + x_2 (12) + \ldots x_n (1n), \ldots x_1 (n1) + x_2 (n2) \ldots + x_n (n^2) \\ \alpha_1 (1^2) + \alpha_2 (12) + \ldots \alpha_n (1n), \ldots \alpha_1 (n1) + \alpha_2 (n2) \ldots + \alpha_n (n^2) \\ \vdots \\ \lambda_1 (1^2) + \lambda_2 (12) + \ldots \lambda_n (1n), \ldots \lambda_1 (n1) + \lambda_2 (n2) \ldots + \lambda_n (n^2) \end{matrix} \right\| = 0 \ldots\ldots (7).$$

From these, assuming

$$\left| \begin{matrix} (1^2), & (12), \ldots & (1n) \\ (21), & (2^2), \ldots & (2n) \\ \vdots & & \\ (n1), & (n2), \ldots & (n^2) \end{matrix} \right| \neq 0 \ \ldots\ldots (8)$$

we obtain, for the reciprocal system of (3),

$$\left\| \begin{matrix} x_1, & x_2, \ldots x_n \\ \alpha_1, & \alpha_2, \ldots \alpha_n \\ \vdots & \\ \lambda_1, & \lambda_2, \ldots \lambda_n \end{matrix} \right\| = 0 \ \ldots\ldots (9).$$

Now, suppose the equations (3) represent the system (2); their number in this case must be $n - r$. Also if θ represent any one of the quantities $\alpha, \beta \ldots \lambda$, we have

$$A_1\theta_1 + A_2\theta_2 \ldots + A_n\theta_n = 0 \ \ldots\ldots (10),$$
$$\vdots$$
$$K_1\theta_1 + K_2\theta_2 \ldots + K_n\theta_n = 0.$$

By means of these equations, the system (9) may be reduced to the form

$$\left\| \begin{matrix} A_1 x_1 + A_2 x_2 \ldots + A_n x_n, \ldots & K_1 x_1 + K_2 x_2 \ldots + K_n x_n, & x_{r+1}, & x_{r+2}, \ldots x_n \\ 0 \qquad , \ldots & 0 \qquad , & \alpha_{r+1}, & \alpha_{r+2}, \ldots \alpha_n \\ \vdots & & \vdots & \\ 0 \qquad , \ldots & 0 \qquad , & \lambda_{r+1}, & \lambda_{r+2}, \ldots \lambda_n \end{matrix} \right\| = 0 \ldots (11),$$

which are satisfied by the equations (1). Hence the reciprocal system to (2) is (1), or (1), (2) are reciprocals to each other.

THEOREM. Consider the equations

$$(\mathfrak{A}' = 0,\ \mathfrak{B}' = 0 \ldots \mathfrak{G}' = 0) \ldots\ldots (12),$$
$$(\mathfrak{A}'' = 0,\ \mathfrak{B}'' = 0 \ldots \mathfrak{O}'' = 0),$$
$$(\mathfrak{A}''' = 0,\ \mathfrak{B}''' = 0 \ldots \mathfrak{K}''' = 0),$$
&c.

of Chap. 2. The equations

$$\left\| \begin{matrix} d_{x_1}U, & d_{x_2}U, \ldots & d_{x_n}U \\ A_1', & A_2', \ldots & A_n' \\ \vdots & & \vdots \\ G_1', & G_2', \ldots & G_n' \end{matrix} \right\| = 0, \quad \left\| \begin{matrix} d_{x_1}U, & d_{x_2}U, \ldots & d_{x_n}U \\ A_1'', & A_2'', \ldots & A_n'' \\ \vdots & & \\ O_1'', & O_2'', \ldots & O_n'' \end{matrix} \right\| = 0, \ldots (13),$$
&c.

which are the reciprocals of these systems, represent taken conjointly the reciprocal of the system of equations (3) of the same chapter.

Let this system, which contains $n - \{(n-r) + (n-r') + \ldots\}$ equations, be represented by

$$\alpha_1 x_1 + \alpha_2 x_2 \ldots + \alpha_n x_n = 0 \ldots\ldots (14),$$
$$\beta_1 x_1 + \beta_2 x_2 \ldots + \beta_n x_n = 0.$$
$$\vdots$$
$$\zeta_1 x_1 + \zeta_2 x_2 \ldots + \zeta_n x_n = 0.$$

The reciprocal system is

$$\left\| \begin{matrix} d_{x_1}U, & d_{x_2}U, \ldots & d_{x_n}U \\ \alpha_1, & \alpha_2, \ldots & \alpha_n \\ \vdots & & \\ \zeta_1, & \zeta_2, \ldots & \zeta_n \end{matrix} \right\| = 0 \ldots\ldots (15),$$

containing $(n-r) + (n-r') +$ &c. ... equations.

Also, by the formulæ in Chap. 2,

$$\alpha_1 x_1 + \ldots + \alpha_n x_n = \lambda_1'\mathfrak{A}' + \mu_1'\mathfrak{B}' + \ldots \sigma_1'\mathfrak{G}' \quad (\lambda,\ \mu \ldots \sigma,\ r' \text{ in number}).$$
$$\beta_1 x_1 + \ldots + \beta_n x_n = \lambda_2'\mathfrak{A}' + \mu_2'\mathfrak{B}' + \ldots \sigma_2'\mathfrak{G}'$$
$$\vdots$$
$$\zeta_1 x_1 \quad \ldots + \zeta_n x_n = \lambda_\theta'\mathfrak{A}' + \mu_\theta'\mathfrak{B}' + \ldots \sigma_\theta'\mathfrak{G} \ldots\ldots (16),$$

writing $\theta = n - \{(n-r) + (n-r') + \ldots\}$.

Also, assuming any arbitrary quantities $\eta_1, \eta_2 \ldots \eta_n \ldots \phi_1, \phi_2 \ldots \phi_n$ (the number of sets being $(r' - \theta)$,) such that

$$\eta_1 x_1 \ldots + \eta_n x_n = \lambda_{\theta+1}'\mathfrak{A}' + \mu_{\theta+1}'\mathfrak{B}' + \ldots \sigma_{\theta+1}\mathfrak{G}' \ldots\ldots (17),$$
$$\vdots$$
$$\phi_1 x_1 \ldots + \phi_n x_n = \lambda_{r'}'\ \mathfrak{A}' + \mu_{r'}'\ \mathfrak{B}' + \ldots \sigma_{r'}'\ \mathfrak{G}'.$$

From the equations (15) we deduce the $(n-r)$ equations

$$\left\| \begin{matrix} d_{x_1}U, & d_{x_2}U, & \dots & d_{x_n}U \\ \eta_1, & \eta_2, & \dots & \eta_n \\ \vdots & & & \\ \phi_1, & \phi_2, & \dots & \phi_n \end{matrix} \right\| = 0 \quad \text{(18)}.$$

Hence, writing

$$\eta = \lambda_1' A + \mu_1' B + \dots \sigma_1' G \quad \text{(19)},$$
$$\vdots$$
$$\phi = \lambda_r' A + \mu_r' B + \dots \sigma_r' G,$$

and reducing, by the formula (8) of Chap. 1, we have

$$\left\| \begin{matrix} d_{x_1}U, & d_{x_2}U, & \dots & d_{x_n}U \\ A_1', & A_2', & \dots & A_n' \\ \vdots & & & \\ G_1', & G_2', & \dots & G_n' \end{matrix} \right\| = 0 \quad \text{(20)};$$

and similarly may the remaining formulæ of (13) be deduced from the equation (15). Hence the required theorem is demonstrated, a theorem which may be more clearly stated as follows:—

The reciprocals of several systems of equations form together the reciprocal of the equation which is satisfied by the values of the variables which satisfy each of the original systems of equations. (The theorem requires that the number of all the reciprocal equations shall be less than the number of variables.)

Conversely, consider several systems of equations, the whole number of the equations being less than the number of variables. These systems, taken conjointly, have for their reciprocal, the equation which is satisfied by the values satisfying the reciprocal system of each of the given systems.

CHAP. 4. *On some properties of functions of the second order.*

Suppose, as before, U denotes the general function of the second order, or

$$2U = \Sigma\,(\alpha^2)\, x_\alpha^2 + 2\Sigma\,(\alpha\beta)\, x_\alpha x_\beta \quad \text{(21)}.$$

Also let V denote a function of the second order of the form

$$V = H\left(\left\| \begin{matrix} x_1, & x_2, & \dots & x_n \\ \alpha_1, & \alpha_2, & \dots & \alpha_n \\ \vdots & & & \\ \rho_1, & \rho_2, & \dots & \rho_n \end{matrix} \right\|\right) \quad \text{(22)},$$

(H being the symbol of a homogeneous function of the second order, and the number r of the quantities $\alpha, \beta \dots \rho$, being less than $n-1$). [Observe that $\alpha_1, \beta_1, \dots \rho_1, \dots \alpha_n, \beta_n, \dots \rho_n$, or say the suffixed quantities $\alpha, \beta, \dots \rho$ (r in number) are used to denote coefficients: α, β, without suffixes, are any two numbers in the series of suffixes $1, 2, 3, \dots n$.] Then $2U - 2kV$, k arbitrary, is of the form

$$\Sigma\,[\alpha^2]\, x_\alpha^2 + 2\Sigma\,[\alpha\beta]\, x_\alpha x_\beta \quad \text{(23)}.$$

Suppose $X_1, X_2, \dots X_n$ determined by the equations

$$[1^2]X_1+[12]X_2\dots+[1n]X_n=0 \quad\dots\dots\dots\dots (24),$$
$$[21]X_1+[2^2]X_2\dots+[2n]X_n=0,$$
$$\vdots$$
$$[n1]X_1+[n2]X_2\dots+[n^2]X_n=0;$$

equations that involve the condition that k satisfies an equation of the order $n-r$, as will be presently proved.

Then shall $x_1=X_1\dots x_n=X_n$ satisfy the system of equations, which is the reciprocal of

$$\left\|\begin{matrix} x_1, & x_2, \dots x_n \\ \alpha_1, & \alpha_2, \dots \alpha_n \\ \vdots & \\ \rho_1, & \rho_2, \dots \rho_n \end{matrix}\right\| = 0 \quad\dots\dots\dots\dots (25).$$

To prove these properties, in the first place we must find the form of V. Consider the quantities $\xi_A, \xi_B, \dots \xi_L$, $(n-r)$ in number, of the form

$$\xi_A=A_1x_1+A_2x_2\dots+A_nx_n, \quad\dots\dots\dots\dots (26),$$
$$\xi_B=B_1x_1+B_2x_2\dots+B_nx_n,$$
$$\vdots$$
$$\xi_L=L_1x_1+L_2x_2\dots+L_nx_n;$$

where, if Θ represent any of the quantities $A, B\dots L$,

$$\alpha_1\Theta_1+\alpha_2\Theta_2\dots+\alpha_n\Theta_n=0, \quad\dots\dots\dots\dots (27),$$
$$\beta_1\Theta_1+\beta_2\Theta_2\dots+\beta_n\Theta_n=0,$$
$$\vdots$$
$$\rho_1\Theta_1+\rho_2\Theta_2\dots+\rho_n\Theta_n=0,$$

$$2V=(A^2)\xi_A^2+(B^2)\xi_B^2+\dots+2(AB)\xi_A\xi_B+\dots=\Sigma(A^2)\xi_A^2+2\Sigma(AB)\xi_A\xi_B.$$

Hence, if $$2V=\Sigma\{\alpha^2\}x_\alpha^2+2\Sigma\{\alpha\beta\}x_\alpha x_\beta \quad\dots\dots\dots\dots (28),$$

we have for the coefficients of this form

$$\{1^2\}=\Sigma(A^2)A_1^2+2\Sigma(AB)A_1B_1, \quad \{12\}=\Sigma(A^2)A_1A_2+\Sigma(AB)(A_1B_2+A_2B_1),$$
$$\vdots \qquad\qquad \vdots$$

and consequently the coefficients of $2U-2kV$ are

$$[1^2]=(1^2)-k\{1^2\}, \qquad [12]=(12)-k\{12\}.$$
$$\vdots \qquad\qquad \vdots$$

Hence, θ representing any of the quantities $\alpha, \beta\dots\rho$,

$$\theta_1\{1^2\}+\theta_2\{12\}\dots+\theta_n\{1n\}=0 \quad\dots\dots\dots\dots (29),$$
$$\vdots$$
$$\theta_1\{n1\}+\theta_2\{n2\}\dots+\theta_n\{n^2\}=0;$$

whence also
$$\theta_1[1^2] + \dots \theta_n[1n] = \theta_1(1^2) + \dots \theta_n(1n),$$
$$\vdots$$
$$\theta_1[n1] + \dots \theta_n[n^2] = \theta_1(n1) + \dots \theta_n(n^2).$$

Hence, the equations for determining $X_1, \dots X_n$ may be reduced to

$$X_1[\alpha_1(1^2) + \dots \alpha_n(1n)] + X_2[\alpha_1(21) \dots + \alpha_n(2n)] \dots + X_n[\alpha_1(n1) \dots + \alpha_n(n^2)] = 0 \dots (30),$$
$$X_1[\beta_1(1^2) + \dots \beta_n(1n)] + X_2[\beta_1(21) \dots + \beta_n(2n)] \dots + X_n[\beta_1(n1) \dots + \beta_n(n^2)] = 0,$$
$$\vdots$$
$$X_1[\rho_1(1^2) + \dots \rho_n(1n)] + X_2[\rho_1(21) \dots + \rho_n(2n)] \dots + X_n[\rho_n(n1) \dots + \rho_n(n^2)] = 0.$$
$$X_1[r+1, 1] + \quad X_2[r+1, 2] \dots + \quad X_n[r+1, n] \quad = 0,$$
$$\vdots$$
$$X_1[n, 1] + \quad X_2[n, 2] \dots + \quad X_n[n^2] \quad = 0.$$

Eliminating $X_1 \dots X_n$, since the first r equations do not contain k, the equation in this quantity is of the order $n - r$.

Next form the reciprocals of the equations (25). These are

$$\left\| \begin{matrix} d_{x_1}U, & d_{x_2}U, \dots d_{x_n}U \\ A_1, & A_2, \dots A_n \\ \vdots & \\ L_1, & L_2, \dots L_n \end{matrix} \right\| = 0 \dots\dots\dots\dots (31).$$

From which we may deduce

$$\left\| \begin{matrix} \alpha_1 d_{x_1}U \dots + \alpha_n d_{x_n}U, & \beta_1 d_{x_1}U \dots + \beta_n d_x U, \dots & \rho_1 d_{x_1}U \dots \rho_n d_{x_n}U, & d_{x_{r+1}}U, \dots d_{x_n}U \\ 0, & 0, \dots & 0, & A_{r+1}, \dots A_n \\ \vdots & & & \\ 0, & 0, \dots & 0, & L_{r+1}, \dots L_n \end{matrix} \right\| = 0 \dots (32),$$

which are evidently satisfied by $x_1 = X_1$, $x_2 = X_2 \dots x_n = X_n$.

In the case of four variables, the above investigation demonstrates the following properties of surfaces of the second order.

I. If a cone intersect a surface of the second order, three different cones may be drawn through the curve of intersection, and the vertices of these lie in the plane which is the polar reciprocal of the vertex of the intersecting cone.

II. If two planes intersect a surface of the second order through the curve of intersection, two cones may be drawn, and the vertices of these lie in the line which is the polar reciprocal of the line of intersection of the two planes.

Both these theorems are undoubtedly known, though I am not able to refer for them to any given place.

12.

ON THE THEORY OF DETERMINANTS.

[From the *Transactions of the Cambridge Philosophical Society*, vol. VIII. (1843), pp. 1—16.]

THE following Memoir is composed of two separate investigations, each of them having a general reference to the Theory of Determinants, but otherwise perfectly unconnected. The name of "Determinants" or "Resultants" has been given, as is well known, to the functions which equated to zero express the result of the elimination of any number of variables from as many linear equations, without constant terms. But the same functions occur in the resolution of a system of linear equations, in the general problem of elimination between algebraic equations, and particular cases of them in algebraic geometry, in the theory of numbers, and, in short, in almost every part of mathematics. They have accordingly been a subject of very considerable attention with analysts. Occurring, apparently for the first time, in Cramer's *Introduction à l'Analyse des Lignes Courbes*, 1750: they are afterwards met with in a Memoir *On Elimination*, by Bezout, *Mémoires de l'Académie*, 1764; in two Memoirs by Laplace and Vandermonde in the same collection, 1774; in Bezout's *Théorie générale des Equations algebriques* [1779]; in Memoirs by Binet, *Journal de l'École Polytechnique*, vol. IX. [1813]; by Cauchy, *ditto*, vol. X. [1815]; by Jacobi, *Crelle's Journal*, vol. XXII. [1841]; Lebesgue, *Liouville*, [vol. II. 1837], &c. The Memoirs of Cauchy and Jacobi contain the greatest part of their known properties, and may be considered as constituting the general theory of the subject. In the first part of the present paper, I consider the properties of certain derivational functions of a quantity U, linear in two separate sets of variables (by the term "Derivational Function," I would propose to denote those functions, the nature of which depends upon the form of the quantity to which they refer, with respect to the variables entering into it, e.g. the differential coefficient of any quantity is a derivational function. The theory of derivational functions is apparently one that would admit of interesting developments). The particular functions of this class which are here considered, are closely connected with the theory of the reciprocal polars of surfaces of the second order, which latter is indeed a particular case of the theory of these functions.

In the second part, I consider the notation and properties of certain functions resolvable into a series of determinants, but the nature of which can hardly be explained independently of the notation.

In the first section I have denoted a determinant, by simply writing down in the form of a square the different quantities of which it is made up. This is not concise, but it is clearer than any abridged notation. The ordinary properties of determinants, I have throughout taken for granted; these may easily be learnt by referring to the Memoirs of Cauchy and Jacobi, quoted above. It may however be convenient to write down the following fundamental property, demonstrated by these authors, and by Binet.

$$\left|\begin{matrix} \alpha, & \beta, \dots \\ \alpha', & \beta', \\ \vdots & \end{matrix}\right| \left|\begin{matrix} \rho, & \sigma, \dots \\ \rho', & \sigma', \\ \vdots & \end{matrix}\right| = \left|\begin{matrix} \rho\alpha+\sigma\beta\dots, & \rho\alpha'+\sigma\beta'\dots, & \dots \\ \rho'\alpha+\sigma'\beta\dots, & \rho'\alpha'+\rho'\sigma'\dots, & \\ \vdots & & \end{matrix}\right| \quad \dots\dots (\odot),$$

an equation, particular cases of which are of very frequent occurrence, e.g. in the investigations on the forms of numbers in Gauss' *Disquisitiones Arithmetica* [1801], in Lagrange's *Determination of the Elements of a Comet's Orbit* [1780], &c. I have applied it in the *Cambridge Mathematical Journal* [1] to Carnot's problem, of finding the relation between the distances of five points in space, and to another geometrical problem. With respect to the notation of the second section, this is so fully explained there, as to render it unnecessary to say anything further about it at present.

§ 1. On the properties of certain determinants, considered as Derivational Functions.

Consider the function

$$\begin{aligned} U = {} & x(\alpha\xi+\beta\eta+\dots)+ \\ & x'(\alpha'\xi+\beta'\eta+\dots)+ \\ & \vdots \end{aligned} \quad \dots\dots\dots\dots\dots\dots\dots\dots\dots (1),$$

(n lines, and n terms in each line);

and suppose

$$KU = \left|\begin{matrix} \alpha, & \beta, \dots \\ \alpha', & \beta', \dots \\ \vdots & \vdots \end{matrix}\right| \quad \dots\dots\dots\dots\dots\dots\dots\dots\dots\dots (2).$$

(The single letter κ being employed instead of KU, in cases where the quantity KU, rather than the functional symbol K, is being considered.) And write

$$FU = -\left|\begin{matrix} & \mathrm{A}x+\mathrm{A}'x'+\dots, & \mathrm{B}x+\mathrm{B}'x'+\dots, & \dots \\ \mathrm{R}\xi+\mathrm{S}\eta+\dots, & \alpha \quad , & \beta \quad , & \dots \\ \mathrm{R}'\xi+\mathrm{S}'\eta+\dots, & \alpha' \quad , & \beta' \quad , & \dots \\ \vdots & & & \end{matrix}\right| \quad \dots\dots\dots (3).$$

$$\daleth U = -\left|\begin{matrix} & \mathrm{R}x+\mathrm{R}'x'+\dots, & \mathrm{S}x+\mathrm{S}'x'+\dots, & \dots \\ \mathrm{A}\xi+\mathrm{B}\eta+\dots, & \alpha \quad , & \beta \quad , & \dots \\ \mathrm{A}'\xi+\mathrm{B}'\eta+\dots, & \alpha' \quad , & \beta' \quad , & \dots \\ \vdots & & & \end{matrix}\right| \quad \dots\dots\dots (4).$$

The symbols K, F, $\daleth$ possess properties which it is the object of this section to investigate.

Let $A, B, \ldots, A', B', \ldots, \ldots$ be given by the equations:

$$A = \begin{vmatrix} \beta', & \gamma', \ldots \\ \beta'', & \gamma'', \\ \vdots & \end{vmatrix}, \quad B = \pm \begin{vmatrix} \gamma', & \delta', \ldots \\ \gamma'', & \delta'', \\ \vdots & \end{vmatrix} \quad \ldots\ldots\ldots\ldots(5).$$

$$A' = \pm \begin{vmatrix} \beta'', & \gamma'', \ldots \\ \beta''', & \gamma''', \\ \vdots & \end{vmatrix}, \quad B' = \begin{vmatrix} \gamma'', & \delta'', \ldots \\ \gamma''', & \delta''', \\ \vdots & \end{vmatrix}$$

(The upper or lower signs according as n is odd or even.)

These quantities satisfy the double series of equations,

$$\begin{aligned} &A\alpha + B\beta + \ldots = \kappa, \quad \ldots\ldots\ldots\ldots(6). \\ &A\alpha' + B\beta' + \ldots = 0, \\ &\vdots \\ &A'\alpha + B'\beta + \ldots = 0, \\ &A'\alpha' + B'\beta' + \ldots = \kappa, \\ &\vdots \\ &\&c. \end{aligned}$$

$$\begin{aligned} &A\alpha + A'\alpha' + \ldots = \kappa, \quad \ldots\ldots\ldots\ldots(7), \\ &A\beta + A'\beta' + \ldots = 0, \\ &\vdots \\ &B\alpha + B'\alpha' + \ldots = 0, \\ &B\beta + B'\beta' + \ldots = \kappa, \\ &\vdots \\ &\&c. \end{aligned}$$

the second side of each equation being 0, except for the rth equation of the rth set of equations in the systems.

Let $\lambda, \mu, \ldots$ represent the rth, $\overline{r+1}^{\text{th}}$, ... terms of the series $\alpha, \beta, \ldots$; $L, M, \ldots$ the corresponding terms of the series $A, B \ldots$, where r is any number less than n, and consider the determinant

$$\begin{vmatrix} A & , \ldots L \\ \vdots & \\ A^{(r-1)}, & \ldots L^{(r-1)} \end{vmatrix} \quad \ldots\ldots\ldots\ldots(8),$$

which may be expressed as a determinant of the nth order, in the form

$$\begin{vmatrix} A & , \ldots L & , & 0, & 0, \ldots \\ \vdots & & & & \\ A^{(r-1)}, & \ldots L^{(r-1)}, & & 0, & 0, \\ 0 & , \quad 0 & , & 1, & 0, \\ 0 & , \quad 0 & , & 0, & 1, \\ \vdots & & & & \end{vmatrix} \quad \ldots\ldots\ldots\ldots(9).$$

Multiplying this by the two sides of the equation

$$\kappa = \begin{vmatrix} \alpha, & \beta, & \dots \\ \alpha', & \beta', & \\ \vdots & & \end{vmatrix} \quad \dots\dots (10),$$

and reducing the result by the equation (⊙), and the equations (6), the second side becomes

$$\begin{vmatrix} \kappa, & 0, \dots & & & \\ 0, & \kappa, & & & \\ \vdots & & \ddots & & \\ & & \kappa, & 0 \quad , & 0 \quad , \dots \\ & & 0, & \mu^{(r)} \quad , & \nu^{(r)} \quad , \\ & & 0, & \mu^{(r+1)}, & \nu^{(r+1)}, \\ & & \vdots & & \end{vmatrix} \quad \dots\dots (11),$$

which is equivalent to

$$\kappa^r \begin{vmatrix} \mu^{(r)} \quad , & \nu^{(r)} \quad , \dots \\ \mu^{(r+1)}, & \nu^{(r+1)}, \\ \vdots & \end{vmatrix} \quad \dots\dots (12),$$

or we have the equation

$$\begin{vmatrix} A & , \dots L \\ \vdots & \\ A^{(r-1)}, & \dots L^{(r-1)} \end{vmatrix} = \kappa^{r-1} \begin{vmatrix} \mu^{(r)} \quad , & \nu^{(r)} \quad , \dots \\ \mu^{(r+1)}, & \nu^{(r+1)}, \\ \vdots & \end{vmatrix} \quad \dots\dots (13),$$

which in the particular case of $r = n$, becomes

$$\begin{vmatrix} A, & B, \dots \\ A', & B', \\ \vdots & \end{vmatrix} = \kappa^{r-1} \quad \dots\dots (14),$$

which latter equation is given by M. Cauchy in the memoirs already quoted; the proof in the "*Exercises*," being nearly the same with the above one of the more general equation (13). The equation (13) itself has been demonstrated by Jacobi somewhat less directly. Consider now the function FU, given by the equation (3). This may be expanded in the form

$$\begin{aligned} FU = (\mathrm{R}\,\xi + \mathrm{S}\,\eta + \dots)\,[A\;(\mathrm{A}x + \mathrm{A}'x' + \dots) + B\;(\mathrm{B}x + \mathrm{B}'x' + \dots) + \dots] + &\quad \dots\dots (15), \\ (\mathrm{R}'\xi + \mathrm{S}'\eta + \dots)\,[A'\,(\mathrm{A}x + \mathrm{A}'x' + \dots) + B'\,(\mathrm{B}x + \mathrm{B}'x' + \dots) + \dots] + & \\ \vdots \qquad & \end{aligned}$$

which may be written

$$\begin{aligned} FU = x\,(\mathrm{A}\,\xi + \mathrm{B}\,\eta + \dots) + &\quad \dots\dots (16), \\ x'\,(\mathrm{A}'\xi + \mathrm{B}'\eta + \dots) + & \\ \vdots \qquad & \end{aligned}$$

by putting

$$\begin{aligned}
\mathrm{A} &= \textsc{a}\,(\textsc{r}A + \textsc{r}'A' + \ldots) + \textsc{b}\,(\textsc{r}B + \textsc{r}'B' + \ldots) + \ldots\,, \qquad \ldots\ldots\ldots (17).\\
\mathrm{B} &= \textsc{a}\,(\textsc{s}A + \textsc{s}'A' + \ldots) + \textsc{b}\,(\textsc{s}B + \textsc{s}'B' + \ldots) + \ldots\,,\\
&\vdots\\
\mathrm{A}' &= \textsc{a}'(\textsc{r}A + \textsc{r}'A' + \ldots) + \textsc{b}'(\textsc{r}B + \textsc{r}'B' + \ldots) + \ldots\,,\\
\mathrm{B}' &= \textsc{a}'(\textsc{s}A + \textsc{s}'A' + \ldots) + \textsc{b}'(\textsc{s}B + \textsc{s}'B' + \ldots) + \ldots\,,\\
&\vdots
\end{aligned}$$

Hence,

$$KFU = \left|\begin{matrix} \mathrm{A}, & \mathrm{B}, \ldots \\ \mathrm{A}', & \mathrm{B}', \\ \vdots & \end{matrix}\right| \quad \ldots\ldots\ldots\ldots (18),$$

$$= \left|\begin{matrix} \textsc{a}, & \textsc{b}, \ldots \\ \textsc{a}', & \textsc{b}', \\ \vdots & \end{matrix}\right| \left|\begin{matrix} \textsc{r}, & \textsc{s}, \ldots \\ \textsc{r}', & \textsc{s}', \\ \vdots & \end{matrix}\right| \left|\begin{matrix} A, & B, \ldots \\ A', & B', \\ \vdots & \end{matrix}\right| \quad \ldots\ldots\ldots\ldots (19),$$

or observing the equation (14), and writing

$$\left|\begin{matrix} \textsc{a}, & \textsc{b}, \ldots \\ \textsc{a}', & \textsc{b}', \\ \vdots & \end{matrix}\right| = J \quad \ldots\ldots\ldots\ldots (20),$$

$$\left|\begin{matrix} \textsc{r}, & \textsc{s}, \ldots \\ \textsc{r}', & \textsc{s}', \\ \vdots & \end{matrix}\right| = \Gamma \quad \ldots\ldots\ldots\ldots (21),$$

this becomes

$$KFU = J\Gamma\,.\,(KU)^{n-1} \quad \ldots\ldots\ldots\ldots (22).$$

Hence likewise

$$K\text{ꟻ}U = J\Gamma\,.\,(KU)^{n-1} \quad \ldots\ldots\ldots\ldots (23).$$

Consider next the equation

$$\text{ꟻ}FU = -\left|\begin{matrix} & \textsc{r}x + \textsc{r}'x' + \ldots, & \textsc{s}x + \textsc{s}'x' + \ldots, \ldots \\ \textsc{a}\,\xi + \textsc{b}\,\eta + \ldots, & \mathrm{A}, & \mathrm{B}, \\ \textsc{a}'\xi + \textsc{b}'\eta + \ldots, & \mathrm{A}', & \mathrm{B}', \\ \vdots & & \end{matrix}\right| \quad \ldots\ldots (24).$$

$$= -\left|\begin{matrix} 1, & & \\ & \textsc{a}, & \textsc{b}, \ldots \\ & \textsc{a}', & \textsc{b}', \end{matrix}\right| \left|\begin{matrix} 1, & & \\ & \textsc{r}, & \textsc{s}, \ldots \\ & \textsc{r}', & \textsc{s}', \end{matrix}\right| \left|\begin{matrix} & \xi, & \eta, \ldots \\ x, & A, & B, \\ x', & A', & B', \\ \vdots & & \end{matrix}\right| \quad \ldots\ldots (25).$$

$$= -J\Gamma \left|\begin{matrix} & \xi, & \eta, \ldots \\ x, & A, & B, \\ x', & A', & B', \\ \vdots & & \end{matrix}\right| \quad \ldots\ldots\ldots\ldots (26).$$

If the two sides of this equation are multiplied by the two sides of the equation (2), written under the form

$$\kappa = \begin{vmatrix} 1, & & \\ & \alpha, & \beta, \ldots \\ & \alpha', & \beta', \\ & \vdots & \end{vmatrix} \quad \ldots\ldots (27),$$

the second side is reduced to

$$-J\Gamma \begin{vmatrix} & \alpha\xi+\beta\eta \ldots, & \alpha'\xi+\beta'\eta \ldots, \ldots \\ x, & \kappa \quad , & . \quad , \\ x', & . \quad , & \kappa \quad , \\ \vdots & & \end{vmatrix} \quad \ldots\ldots (28),$$

$$= -J\Gamma \,.\, \kappa^{n-1} \,.\, U \quad \ldots\ldots (29),$$

and hence

$$\daleth FU = J\Gamma \,.\, (KU)^{n-2} \,.\, U \quad \ldots\ldots (30).$$

Similarly

$$F\daleth U = J\Gamma \,.\, (KU)^{n-2} \,.\, U \quad \ldots\ldots (31);$$

also combining these with the equations (22), (23),

$$\frac{\daleth FU}{KFU} = \frac{F\daleth U}{K\daleth U} = \frac{U}{KU} \quad \ldots\ldots (32).$$

It may be remarked here that if U, V are functions connected by the equation

$$FU = cFV, \quad \text{or } \daleth U = c\daleth V, \quad \ldots\ldots (33),$$

then in general

$$U = c^{\frac{1}{n-1}} V \quad \ldots\ldots (34).$$

To prove this, observing that the first of the equations (33) may be written

$$FU = F(c^{\frac{1}{n-1}} V) \quad \ldots\ldots (35),$$

we have

$$\daleth \,.\, FU = \daleth \,.\, F(c^{\frac{1}{n-1}} V) \quad \ldots\ldots (36),$$

or

$$J\Gamma \,.\, (KU)^{n-2}\, U = J\Gamma\, [K(c^{\frac{1}{n-1}} V)]^{n-2}\, c^{\frac{1}{n-1}}\, V \quad \ldots\ldots (37).$$

If neither J, Γ nor (KU) vanish, this equation is of the form

$$U = kV \quad \ldots\ldots (38),$$

whence substituting in (33),

$$k^{n-1} = c \quad \ldots\ldots (39),$$

which demonstrates the equation (34); and this equation might be proved in like manner from the second of the equations (33). If however, $J=0$, or $\digamma=0$, the above proof fails, and if $KU=0$, the proof also fails, unless at the same line $n=2$. In all these cases probably, certainly in the case of $KU=0$, $n \neq 2$, the equation (34) is not a necessary consequence of (33). In fact FU, or $\daleth U$ may be given, and yet U remain indeterminate.

Let $U_{,}$, $\alpha_{,}$, $\beta_{,}$, ... $A_{,}$, $B_{,}$, &c. ... be analogous to U, α, β ..., $A_{,}$, B, &c. ... and consider the equation

$$K(KU_{,}\,.\,FU+gKU\,.\,FU_{,}) \quad \text{.......... (40)},$$

$$= \begin{vmatrix} \kappa_{,}A+g\kappa A_{,}, & \kappa_{,}B+g\kappa B_{,}, \ldots \\ \kappa_{,}A'+g\kappa A_{,}', & \kappa_{,}B+g\kappa B_{,}', \\ \vdots & \end{vmatrix}$$

Multiply the two sides by the two sides of the equation (2), the second side becomes, after reduction,

$$\begin{vmatrix} \kappa_{,}\kappa+g\kappa\,(A_{,}\alpha+B_{,}\beta+\ldots), & g\kappa\,(A_{,}'\alpha+B_{,}'\beta+\ldots), \ldots \\ g\kappa\,(A_{,}\alpha'+B_{,}\beta'+\ldots), & \kappa_{,}\kappa+g\kappa\,(A_{,}'\alpha'+B_{,}'\beta'+\ldots), \\ \vdots & \end{vmatrix} \quad \text{...... (41)}.$$

Multiplying by the two sides of the analogous equation

$$\kappa_{,}= \begin{vmatrix} \alpha_{,}, & \alpha_{,}', \ldots \\ \beta_{,}, & \beta_{,}, \\ \vdots & \end{vmatrix} \quad \text{.......... (42)},$$

and reducing, the second side becomes

$$\begin{vmatrix} \kappa\kappa_{,}(\alpha_{,}+g\alpha), & \kappa\kappa_{,}(\beta_{,}+g\beta), \ldots \\ \kappa\kappa_{,}(\alpha_{,}'+g\alpha'), & \kappa\kappa_{,}(\beta_{,}'+g\beta'), \\ \vdots & \end{vmatrix} \quad \text{.......... (43)},$$

$$=\kappa^{n}\,.\,\kappa_{,}^{n}\,.\,K(U_{,}+gU) \quad \text{.......... (44)},$$

whence $$K(KU_{,}\,.\,FU+gKU\,.\,FU_{,})=(KU)^{n-1}(KU_{,})^{n-1}K(U_{,}+gU) \quad \text{...... (45)},$$

and similarly

$$K(KU_{,}\,.\,\daleth U+gKU\,.\,\daleth U_{,})=(KU)^{n-1}(KU_{,})^{n-1}K(U_{,}+gU) \quad \text{...... (46)}.$$

In a similar manner is the following equation to be demonstrated,

$$\daleth(KU_{,}\,.\,FU+gKU\,.\,FU_{,})=F(KU_{,}\,.\,\daleth U+gKU\,.\,\daleth U_{,})= \quad \text{.......... (47)},$$

$$-J\digamma\,.\,(KU)^{n-2}(KU_{,})^{n-2}\times \begin{vmatrix} & \alpha_{,}x+\alpha_{,}'x'\ldots, & \beta_{,}x+\beta_{,}'x'\ldots, \ldots \\ \alpha\xi+\beta\eta\ldots, & \alpha_{,}+g\alpha\ldots, & \beta_{,}+g\beta\ldots, \\ \alpha'\xi+\beta'\eta\ldots, & \alpha_{,}'+g\alpha'\ldots, & \beta_{,}'+g\beta_{,}'\ldots, \\ \vdots & & \end{vmatrix}$$

Suppose $$\overline{U} = \Sigma\,(\rho\xi + \sigma\eta + \ldots)\,(ax + a'x' + \ldots) \qquad\ldots\ldots\ldots\ldots (48),$$

this expression being the abbreviation of

$$\begin{aligned}\overline{U} = {} & (\rho\,\xi + \sigma\,\eta + \ldots)\,(a\,x + a'x' + \ldots) + \qquad\ldots\ldots\ldots\ldots (49),\\ & (\rho_,\xi + \sigma_,\eta + \ldots)\,(a_,x + a_,'x' + \ldots) + \\ & + \\ & \vdots \\ & [(n-1) \text{ lines, or a smaller number}].\end{aligned}$$

then $$K\overline{U} = \begin{vmatrix} \Sigma a\,\rho, & \Sigma a\,\sigma, \ldots \\ \Sigma a'\rho, & \Sigma a'\sigma, \\ \vdots & \end{vmatrix} \text{ is } = 0 \qquad\ldots\ldots\ldots\ldots (50),$$

which follows from the equation ($\odot$).

Conversely, whenever $K\overline{U} = 0$, $\overline{U}$ is of the above form.

Also $$F\overline{U} = -\begin{vmatrix} & \text{A}x + \text{A}'x' + \ldots, & \text{B}x + \text{B}'x' + \ldots, \ldots \\ \text{R}\,\xi + \text{S}\,\eta + \ldots, & \Sigma a\,\rho \quad , & \Sigma a\,\sigma \quad , \\ \text{R}'\xi + \text{S}'\eta + \ldots, & \Sigma a'\rho \quad , & \Sigma a'\sigma \quad , \\ \vdots & & \end{vmatrix} \qquad\ldots\ldots\ldots\ldots (51),$$

which may be transformed into

$$F\overline{U} = \begin{vmatrix} \text{A}x + \text{A}'x' \ldots, & \text{B}x + \text{B}'x' \ldots, \ldots \\ \rho \quad , & \sigma \quad , \\ \vdots & \end{vmatrix}\;\begin{vmatrix} \text{R}\xi + \text{S}\eta \ldots, & \text{R}'\xi + \text{S}'\eta \ldots, \ldots \\ a \quad , & a' \quad , \\ \vdots & \end{vmatrix} \ldots (52),$$

(for shortness, I omit the demonstration of this equation).

And similarly,

$$\daleth\overline{U} = \begin{vmatrix} \text{R}x + \text{R}'x' + \ldots, & \text{S}x + \text{S}'x' + \ldots, \ldots \\ \rho \quad , & \sigma \quad , \\ \vdots & \end{vmatrix}\;\begin{vmatrix} \text{A}\xi + \text{B}\eta + \ldots, & \text{A}'\xi + \text{B}'\eta + \ldots, \ldots \\ a \quad , & a' \quad , \\ \vdots & \end{vmatrix} \ldots (53),$$

where it is obvious that if the sum Σ contain fewer than $(n-1)$ terms, $FU = 0$, $\daleth U = 0$.

The equations (52), (53) express the theorem, that whenever $K\overline{U} = 0$, the functions $F\overline{U}$, $\daleth\overline{U}$ are each of them the product of two determinants.

If next $$U_, = U + \overline{U},$$

then in (45) taking $g=-1$ [the Numbers (56) &c. which follow are as in the original memoir]

$$K\{K(U+\overline{U}).FU-KU.F(U+\overline{U})\}=K\{K(U+\overline{U}).\mathfrak{F}U-KU.\mathfrak{F}(U+\overline{U})\} \quad\ldots\ldots\ldots(56),$$
$$=(KU)^{n-1}.(K(U+\overline{U}))^{n-1}.K\overline{U},$$

or observing the equation (50),

$$K\{K(U+\overline{U}).FU-KU.F(U+\overline{U})\}=K\{K(U+\overline{U}).\mathfrak{F}U-KU.\mathfrak{F}(U+\overline{U})\}=0 \quad\ldots\ldots(57).$$

Hence $F\{(K(U+\overline{U}).\mathfrak{F}U-KU.\mathfrak{F}(U+\overline{U})\}=\mathfrak{F}\{K(U+\overline{U}).FU-KU.F(U+\overline{U})\}$ are each of them the product of two determinants. But this result admits of a further reduction: we have

$$F\{K(U+\overline{U}).\mathfrak{F}U-KU.\mathfrak{F}(U+\overline{U})\}=\mathfrak{F}\{K(U+\overline{U}).FU-KU.F(U+\overline{U})\}\ldots\ldots\ldots(58)$$

$$=-J\Gamma(KU)^{n-2}.(K(U+\overline{U}))^{n-2}\begin{vmatrix} & \alpha_{,}x+\alpha_{,}'x'+\ldots, & \beta_{,}x+\beta_{,}'x'+\ldots,\ \ldots \\ \alpha\xi+\beta\eta+\ldots, & \alpha_{,}\ -\alpha\quad , & \beta_{,}\ -\beta\quad , \\ \alpha'\xi+\beta'\eta+\ldots, & \alpha_{,}'\ -\alpha'\quad , & \beta_{,}'\ +\beta'\quad , \\ \vdots & & \end{vmatrix},$$

substituting $\alpha_{,}=\alpha+\Sigma\rho a$, &c...., also observing that if the second line be multiplied by x, the third by x', ... and the sum subtracted from the first line, the value of the determinant is not altered, and that the effect of this is simply to change $\alpha_{,}$, $\alpha_{,}'$ ... into α, α' ... in the first line, and introduce into the corner place a quantity $-U$, which in the expansion of the determinant is multiplied by zero: this may be written in the form

$$-J\Gamma(KU)^{n-2}(K(U+\overline{U}))^{n-2}\begin{vmatrix} & \alpha x+\alpha'x'+\ldots, & \beta x+\beta'x'+\ldots,\ \ldots \\ \alpha\xi+\beta\eta+\ldots, & \Sigma\rho a\quad , & \Sigma\sigma a\quad , \\ \alpha'\xi+\beta'\eta+\ldots, & \Sigma\rho a'\quad , & \Sigma\sigma a'\quad , \\ \vdots & & \end{vmatrix}\ \ldots\ (59),$$

which may be reduced to

$$J\Gamma.(KU)^{n-2}.(K(U+\overline{U}))^{n-2}\times \quad\ldots\ldots\ldots\ldots\ldots\ldots\ldots\ldots\ldots(60),$$

$$\begin{vmatrix} \alpha x+\alpha'x'+\ldots, & \beta x+\beta'x'+\ldots,\ \ldots \\ \rho\quad , & \sigma\quad , \\ \vdots & \end{vmatrix}\quad\begin{vmatrix} \alpha\xi+\beta\eta+\ldots, & \alpha'\xi+\beta'\eta\ \ldots,\ \ldots \\ a\quad , & a'\quad , \\ \vdots & \end{vmatrix}$$

If each of these determinants are multiplied by the quantity $(KU)^{n-1}$, expressed under the two forms

$$\begin{vmatrix} A, & B,\ldots \\ A', & B', \\ \vdots & \end{vmatrix},\quad\begin{vmatrix} A, & A',\ldots \\ B, & B', \\ \vdots & \end{vmatrix}\quad\ldots\ldots\ldots\ldots\ldots\ldots\ldots\ldots\ldots\ldots(61),$$

they would become respectively

$$KU.\left|\begin{array}{lll} x & , \quad x' & ,\ldots \\ A\rho+B\sigma+\ldots, & A'\rho+B'\sigma+\ldots, & \\ \vdots & & \end{array}\right|,\quad KU.\left|\begin{array}{lll} \xi & , \quad \eta & ,\ldots \\ Aa+A'a'+\ldots, & Ba+B'a'+\ldots, & \\ \vdots & & \end{array}\right| \ldots (62),$$

so that finally

$$F\{K(U+\overline{U}).\,\mathfrak{F}U-KU.\,\mathfrak{F}(U+\overline{U})\}=\mathfrak{F}\{K(U+\overline{U})FU-KU.\,F(U+\overline{U})\}\ \ldots\ldots\ldots (63).$$

$$=J\Gamma\cdot\left(\frac{K(U+\overline{U})}{KU}\right)^{n-2}\times\left|\begin{array}{lll} x & , \quad x' & ,\ldots \\ A\rho+B\sigma+\ldots, & A'\rho+B'\sigma+\ldots, & \\ \vdots & & \end{array}\right|\ \left|\begin{array}{lll} \xi & , \quad \eta & ,\ldots \\ Aa+A'a'+\ldots, & Ba+B'a'+\ldots, & \\ \vdots & & \end{array}\right|$$

The second side of this may be written under the forms

$$\left(\frac{K(U+\overline{U})}{KU}\right)^{n-2}\left|\begin{array}{lll} \mathrm{A}x+\mathrm{A}'x'+\ldots & , \quad \mathrm{B}x+\mathrm{B}'x'+\ldots & ,\ldots \\ \mathrm{A}(A\rho+B\sigma..)+\mathrm{A}'(A'\rho+B'\sigma..)+\ldots, & \mathrm{B}(A\rho+B\sigma..)+\mathrm{B}'(A'\rho+B'\sigma..)+\ldots, & \\ \vdots & & \end{array}\right|$$

multiplied into

$$\left|\begin{array}{lll} R\xi+S\eta+\ldots & , \quad R'\xi+S'\eta+\ldots & ,\ldots \\ R(Aa+A'a'..)+S(Ba+B'a'..)+\ldots, & R'(Aa+A'a'..)+S'(Ba+B'a'..)+\ldots, & \\ \vdots & & \end{array}\right| \ldots (64).$$

And

$$\left(\frac{K(U+\overline{U})}{KU}\right)^{n-2}\left|\begin{array}{lll} \mathrm{R}x+\mathrm{R}'x'+\ldots & , \quad \mathrm{S}x+\mathrm{S}'x'+\ldots & ,\ldots \\ \mathrm{R}(A\rho+B\sigma..)+\mathrm{R}'(A'\rho+R'\sigma..)+\ldots, & \mathrm{S}(A\rho+B\sigma..)+\mathrm{S}'(A'\rho+B'\sigma..)+\ldots, & \\ \vdots & & \end{array}\right|$$

multiplied into

$$\left|\begin{array}{lll} A\xi+B\eta+\ldots & , \quad A'\xi+B'\eta+\ldots & ,\ldots \\ A(Aa+A'a'..)+B(Ba+B'a'..)+\ldots, & A'(Aa+A'a'..)+B'(Ba+B'a'..)+\ldots, & \\ \vdots & & \end{array}\right| \ldots (65).$$

And again, by the equations (52), (53), in the new forms

$$\left(\frac{K(U+\overline{U})}{KU}\right)^{n-2} F.\,\Sigma\{[(A\rho+B\sigma\ \ldots)(\mathrm{A}\xi+\mathrm{B}\eta\ \ldots)+(A'\rho+B'\sigma\ldots)(\mathrm{A}'\xi+\mathrm{B}'\eta\ldots)\ldots]$$
$$\times[(Aa+A'a'\ldots)(\mathrm{R}x+\mathrm{R}'x'\ldots)+(Ba+B'a'\ \ldots)(\mathrm{S}x+\mathrm{S}'x'\ldots)\ldots]\}\ldots\ldots(66),$$

$$\left(\frac{K(U+\overline{U})}{(KU)}\right)^{n-2} \mathfrak{F}.\,\Sigma\{[(A\rho+B\sigma\ \ldots)(\mathrm{R}\xi+\mathrm{S}\eta\ \ldots)+(A'\rho'+B'\sigma\ldots)(\mathrm{R}'\xi+\mathrm{S}'\eta\ldots)\ldots]$$
$$\times[(Aa+A'a'\ldots)(\mathrm{A}x+\mathrm{A}'x'\ldots)+(Ba+B'a'\ \ldots)(\mathrm{B}x+\mathrm{B}'x'\ldots)\ldots]\}\ \ldots (67).$$

Comparing these latter forms with the two equivalent quantities forming the first side of (53), and observing (33), (34), it would appear at first sight that

$$K(U+\bar{U}) \,.\, \mathfrak{F}U - KU \,.\, \mathfrak{F}(U+\bar{U})$$
$$= \left(\frac{K(U+\bar{U})}{KU}\right)^{\frac{n-2}{n-1}} \{\Sigma\, [(A\rho + B\sigma \ldots)(\mathrm{A}\xi + \mathrm{B}\eta \ldots) + (A'\rho + B'\sigma \ldots)(\mathrm{A}'\xi + \mathrm{B}'\eta \ldots)\ldots]$$
$$\times [(Aa + A'a' \ldots)(\mathrm{R}x + \mathrm{R}'x' \ldots) + (Ba + B'a' \ldots)(\mathrm{S}x + \mathrm{S}'x' \ldots)\ldots]\},$$

$$K(U+\bar{U}) \,.\, FU - KU \,.\, F(U+\bar{U})$$
$$= \left(\frac{K(U+\bar{U})}{KU}\right)^{\frac{n-2}{n-1}} \Sigma\, \{[(A\rho + B\sigma \ldots)(\mathrm{R}\xi + \mathrm{S}\eta \ldots) + (A'\rho + B'\sigma \ldots)(\mathrm{R}'\xi + \mathrm{S}'\eta \ldots)\ldots]$$
$$\times [(Aa + A'a' \ldots)(\mathrm{A}x + \mathrm{A}'x' \ldots) + (Ba + B'a' \ldots)(\mathrm{B}x + \mathrm{B}'x' \ldots)\ldots]\},$$

which however are not true, except for $n = 2$, on account of the equation (57). In the case of $n = 2$, these equations become

$$K(U+\bar{U}) \,.\, \mathfrak{F}U - KU \,.\, \mathfrak{F}(U+\bar{U})$$
$$= [(A\rho + B\sigma \ldots)(\mathrm{A}\xi + \mathrm{B}\eta \ldots) + (A'\rho + B'\sigma \ldots)(\mathrm{A}'\xi + \mathrm{B}'\eta \ldots) + \ldots]$$
$$\times [(Aa + A'a' \ldots)(\mathrm{R}x + \mathrm{R}'x' \ldots) + (Ba + B'a' \ldots)(\mathrm{S}x + \mathrm{S}'x' \ldots)\ldots] \quad \ldots\ldots\ldots (68),$$

$$K(U+\bar{U})\, FU - KU \,.\, F(U+\bar{U})$$
$$= [(A\rho + B\sigma \ldots)(\mathrm{R}\xi + \mathrm{S}\eta \ldots) + (A'\rho + B'\sigma \ldots)(\mathrm{R}'\xi + \mathrm{S}'\eta \ldots)\ldots]$$
$$\times [(Aa + A'a' \ldots)(\mathrm{A}x + \mathrm{A}'x' \ldots) + (Ba + B'a' \ldots)(\mathrm{B}x + \mathrm{B}'x' + \ldots)\ldots] \quad \ldots\ldots (69),$$

and it is remarkable that these equations ((68), (69)) are true whatever be the value of n, provided Σ contains a single term only. The demonstration of this theorem is somewhat tedious, but it may perhaps be as well to give it at full length. It is obvious that the equation (69) alone need be proved, (68) following immediately when this is done.

I premise by noticing the following general property of determinants. The function

$$\begin{vmatrix} \alpha + \Sigma\rho a, & \beta + \Sigma\sigma a, \ldots \\ \alpha' + \Sigma\rho a', & \beta' + \Sigma\sigma' a, \\ \vdots & \end{vmatrix} \quad \ldots\ldots\ldots\ldots\ldots\ldots\ldots\ldots (70),$$

(where $\Sigma\rho a = \rho_1 a_1 + \rho_2 a_2 \ldots + \rho_s a_s$), contains no term whose dimension in the quantities $a, a' \ldots$, or in the other quantities $\rho, \sigma \ldots$, is higher than s. (Of course if the order of the determinant be less than s or equal to it, this number becomes the limit of the dimension of any term in $a, a' \ldots$ or $\rho, \sigma \ldots$, and the theorem is useless.) This is easily proved by means of a well-known theorem,

$$\begin{vmatrix} \Sigma\rho a, & \Sigma\sigma a, \ldots \\ \Sigma\rho' a, & \Sigma\sigma' a, \\ \vdots & \end{vmatrix} = 0 \quad \ldots\ldots\ldots\ldots\ldots\ldots\ldots\ldots (71),$$

whenever s is less than the number expressing the order of the determinant. Hence in the formula (70), if Σ contain a single term only, the first side of the equation is linear in $\rho, \sigma, \ldots$ and also in $a, a', \ldots$, i.e. it consists of a term independent of all these quantities, and a second term linear in the products $\rho a, \rho a', \ldots \sigma a, \sigma a', \ldots$ This is therefore the form of $K(U+\overline{U})$.

Consider the several equations

$$\begin{aligned} \kappa = KU &= A\alpha + B\beta + \ldots \\ &= A'\alpha' + B'\beta' + \ldots \\ &= \&c. \end{aligned} \qquad \text{(72)},$$

it is easy to deduce

$$\begin{aligned} \kappa_{,} = K(U+\overline{U}) = KU &+ A\rho a + B\sigma a + \\ &+ A'\rho a' + B'\sigma a' + \\ &\vdots \end{aligned} \qquad \text{(73)}.$$

To find the values of A, B, &c. corresponding to $U+\overline{U}$, we must write

$$\begin{aligned} A &= \mathrm{M}'\beta + \mathrm{N}'\gamma' + \\ &= \mathrm{M}''\beta + \mathrm{N}''\gamma'' + \\ &= \&c. \end{aligned} \qquad \text{(74)},$$

where

$$\mathrm{M}' = \begin{vmatrix} \gamma'', & \delta'', \ldots \\ \gamma''', & \delta''', \\ \vdots & \end{vmatrix}, \quad \mathrm{N}' = \pm \begin{vmatrix} \delta'', & \epsilon'', \ldots \\ \delta''', & \epsilon''', \\ \vdots & \end{vmatrix} \qquad \text{(75)},$$

$$\mathrm{M}'' = \pm \begin{vmatrix} \gamma''', & \delta''', \ldots \\ \gamma'''', & \delta'''', \\ \vdots & \end{vmatrix}, \quad \mathrm{N}'' = \begin{vmatrix} \delta''', & \epsilon''', \ldots \\ \delta'''', & \epsilon'''', \\ \vdots & \end{vmatrix}, \ \&c.$$

the order of each of these determinants being $\overline{n-2}$, and the upper or lower signs being used according as $\overline{n-1}$ is odd or even, i.e. as n is even or odd. Hence

$$\begin{aligned} A_{,} = A &+ \mathrm{M}'\sigma a' + \mathrm{N}'\tau a' + \ldots \\ &+ \mathrm{M}''\sigma a'' + \mathrm{N}''\tau a'' + \ldots \\ &\vdots \end{aligned} \qquad \text{(76)},$$

and therefore

$$\begin{aligned} \kappa_{,}A - \kappa A_{,} = \quad & A^2\rho a + (AB \qquad)\sigma a + (AC \qquad)\tau a + \ldots \\ &+ AA'\rho a' + (AB' - \kappa\mathrm{M}')\sigma a' + (AC' - \kappa\mathrm{N}')\tau a' + \ldots \\ &+ AA''\rho a'' + (AB'' - \kappa\mathrm{M}'')\sigma a'' + (AC'' - \kappa\mathrm{N}'')\tau a'' + \\ &\vdots \end{aligned} \qquad \text{(77)},$$

the additional quantities C, τ having been introduced for greater clearness. Now the equations

$$\begin{aligned} &AB' - \kappa M' = A'B, \quad AC' - \kappa N' = A'C, \ldots \\ &AB'' - \kappa M'' = A''B, \quad AC'' - \kappa N'' = A''C, \\ &\vdots \end{aligned} \qquad \text{(78)},$$

written under the form

$$\begin{array}{ll} AB' - A'B = \kappa M', & AC' - A'C = \kappa N', \ldots \\ AB'' - A''B = \kappa M'', & AC'' - A''C = \kappa N'', \\ \vdots & \end{array} \quad \ldots\ldots\ldots\ldots\ldots\ldots(79),$$

are particular cases of the equation (13), and are therefore identically true. Hence, substituting in (77),

$$\begin{aligned} \kappa_{,}A - \kappa A_{,} = \quad & A^2\rho a + A\,B\sigma a + A\,C\tau a \;\ldots + \quad \ldots\ldots\ldots\ldots\ldots\ldots (80), \\ & + A\,A'\rho a' + A'\,B\sigma a' + A'\,C\tau a' \;\ldots + \\ & + A''A\rho a'' + A''B\sigma a'' + A''C\tau a'' \ldots + \\ & \vdots \\ & = (\rho A + \sigma B + \ldots)(Aa + A'a' + \ldots). \end{aligned}$$

Forming in a similar manner, the combinations $\kappa_{,}B - \kappa B_{,}, \ldots \kappa_{,}A' - \kappa A_{,}', \; \kappa_{,}B - \kappa B_{,}', \ldots$, multiplying by the products of the different quantities $Ax + A'x' \ldots$, $Bx + B'x' \ldots, \ldots$ $R\xi + S\eta \ldots$, $R'\xi + S'\eta \ldots, \ldots$ and adding so as to form the function $K(U + \overline{U}) . FU - KU . F(U + \overline{U})$, we obtain the required formula, viz. that the value of this quantity is

$$\begin{aligned} &= [(\rho A + \sigma B \;\ldots)(\mathrm{R}\xi + \mathrm{S}\eta \;\ldots) + (A'\rho + B'\sigma \ldots)(\mathrm{R}'\xi + \mathrm{S}'\eta \ldots) + \ldots] \ldots\ldots (81); \\ &\times [(Aa + A'a' \ldots)(\mathrm{A}x + \mathrm{A}'x' \ldots) + (Ba + B'a' \ldots)(\mathrm{B}x + \mathrm{B}'x' \ldots) + \ldots] \end{aligned}$$

with this theorem, I conclude the present section,—noticing only, as a problem worthy of investigation, the discovery of the forms of the second sides of the equations (68), (69), in the case of Σ containing more than a single term.

§ 2. On the notation and properties of certain functions resolvable into a series of determinants.

Let the letters $\qquad r_1, \; r_2, \ldots r_k \; \ldots\ldots\ldots\ldots\ldots\ldots\ldots\ldots(1)$,

represent a permutation of the numbers

$$1, \; 2, \ldots k \quad \ldots\ldots\ldots\ldots\ldots\ldots\ldots\ldots(2).$$

Then in the series (1), if one of the letters succeeds mediately or immediately a letter representing a higher number than its own, for each time that this happens there is said to be a "derangement" or "inversion." It is to be remarked that if any letter succeed s letters representing higher numbers, this is reckoned for the same number s of inversions.

Suppose next the symbol

$$\pm_r \quad \ldots\ldots\ldots\ldots\ldots\ldots\ldots\ldots\ldots\ldots (3),$$

denotes the sign + or −, according as the number of inversions in the series (1) is even or odd.

This being premised, consider the symbol

$$\left\{\begin{matrix} A\rho_1\sigma_1 \ldots (n) \\ \vdots \\ \rho_k\sigma_k \ldots \end{matrix}\right\} \quad \ldots\ldots (4),$$

denoting the sum of all the different terms of the form

$$\pm_r \pm_s \ldots A\rho_{r_1}\sigma_{s_1} \ldots \quad \ldots A\rho_{r_k}\sigma_{s_k} \ldots \quad \ldots\ldots (5),$$

the letters $r_1, \ r_2 \ldots r_k; \ s_1, \ s_2 \ldots s_k$; &c. ...... (6),

denoting any permutations whatever, the same or different, of the series of numbers (2) [and the several combinations of $\rho\sigma \ldots$ being understood as denoting suffixes of the A's]. The number of terms represented by the symbol (5) is evidently

$$(1 . 2 \ldots k)^n \ldots\ldots (7).$$

In some cases it will be necessary to leave a certain number of the vertical rows $\rho, \sigma \ldots$ unpermuted. This will be represented by writing the mark (†) immediately above the rows in question. So that for instance

$$\left\{\begin{matrix} A\rho_1\sigma_1 \ldots \overset{\dagger}{\theta_1}\overset{\dagger}{\phi_1} \ldots (n) \\ \vdots \\ \rho_k\sigma_k \ldots \theta_k\phi_k \end{matrix}\right\} \ldots\ldots (8),$$

the number of rows with the (†) being x, denotes the sum of the

$$(1 . 2 \ldots k)^{n-x} \ldots\ldots (9)$$

terms, of the form

$$\pm_r \pm_s \ldots A\rho_{r_1}\sigma_{s_1} \ldots \theta_1\phi_1 \ldots A\rho_{r_k}\sigma_{s_k} \ldots \theta_k\phi_k \ldots \quad \ldots\ldots (10).$$

Then it is obvious, that if all the rows have the mark (†) the notation (8) denotes a single product only, and if the mark (†) be placed over all but one of the rows the notation (8) belongs to a determinant. It is obvious also that we may write

$$\left\{\begin{matrix} A\rho_1\sigma_1 \ldots \theta_1\phi_1 \ldots (n) \\ \vdots \\ \rho_k\sigma_k \ldots \theta_k\phi_k \end{matrix}\right\} = \Sigma \pm_u \pm_v \ldots \left\{\begin{matrix} A\rho_1\sigma_1 \ldots \overset{\dagger}{\theta}_{u_1}\overset{\dagger}{\phi}_{v_1} \ldots (n) \\ \vdots \\ \rho_k\sigma_k \ldots \theta_{u_k}\phi_{v_k} \end{matrix}\right\} \ldots\ldots (11),$$

where Σ refers to the different permutations,

$$u_1, \ u_2, \ldots u_k; \ v_1, \ v_2, \ldots v_k; \ \&c. \ldots\ldots (12),$$

which can be formed out of the numbers (2). The equation (11) would still be true, if the mark (†) were placed over any number of the columns $\rho, \sigma \ldots$

Suppose in this equation a single column only is left without the mark (†) on the second side of the equation; the first side is then expressed as the sum of a number

$$(1 . 2 \ldots k)^{n-1}, \text{ or generally } (1 . 2 \ldots k)^{n-x-1} \ldots\ldots (13),$$

of determinants, according as we consider the symbol (4) or the more general one (8). And this may be done in n or $(n-x)$ different ways respectively.

It may be remarked, that the symbol (8) is the same in form as if a single column only had the mark (†) over it; the number n being at the same time reduced from n to $(n-x+1)$: for the marked columns of symbols may be replaced by a single marked column of new symbols. Hence, without loss of generality, the theorems which follow may be stated with reference to a single marked column only.

Suppose the letters

$$\rho_1, \quad \rho_2, \ldots \rho_k; \quad \sigma_1, \quad \sigma_2, \ldots \sigma_k; \text{ \&c.} \qquad \ldots\ldots\ldots\ldots (14)$$

denote certain permutations of

$$\alpha_1, \quad \alpha_2, \ldots \alpha_k; \quad \beta_1, \quad \beta_2, \ldots \beta_k; \text{ \&c.} \qquad \ldots\ldots\ldots\ldots (15),$$

in such a manner that

$$\rho_1 = \alpha_{g_1}, \quad \rho_2 = \alpha_{g_2}, \ldots \rho_k = \alpha_{g_k}; \quad \sigma_1 = \beta_{h_1}, \quad \sigma_2 = \beta_{h_2}, \ldots \sigma_k = \beta_{h_k} \ldots \qquad \ldots\ldots(16).$$

Then the two following theorems may be proved:

$$\left\{\begin{matrix} A\overset{\dagger}{\rho_1}\sigma_1 \ldots (n) \\ \vdots \\ \rho_k\sigma_k \end{matrix}\right\} = \pm_g \pm_h \ldots \left\{\begin{matrix} A\overset{\dagger}{\alpha_1}\beta_1 \ldots (n) \\ \vdots \\ \alpha_k\beta_k \end{matrix}\right\} \ldots\ldots\ldots\ldots (17),$$

if n be even: but in the contrary case

$$\left\{\begin{matrix} A\overset{\dagger}{\rho_1}\sigma_1 \ldots (n) \\ \vdots \\ \rho_k\sigma_k \end{matrix}\right\} = + \pm_g \ldots \left\{\begin{matrix} A\overset{\dagger}{\alpha_1}\beta_1 \ldots (n) \\ \vdots \\ \alpha_k\beta_k \end{matrix}\right\} \ldots\ldots\ldots\ldots (18).$$

By means of these, and the equation (11), a fundamental property of the symbol (3) may be demonstrated. We have

$$\left\{\begin{matrix} A\alpha_1\beta_1 \ldots (n) \\ \vdots \\ \alpha_k\beta_k \end{matrix}\right\} = \Sigma \pm_g \left\{\begin{matrix} A\overset{\dagger}{\rho_1}\beta_1 \ldots (n) \\ \vdots \\ \rho_k\beta_k \end{matrix}\right\} \ldots\ldots\ldots\ldots (19),$$

which when n is even, reduces itself by (17) to

$$\left\{\begin{matrix} A\alpha_1\beta_1 \ldots (n) \\ \vdots \\ \alpha_k\beta_k \end{matrix}\right\} = \left\{\begin{matrix} A\overset{\dagger}{\alpha_1}\beta_1 \ldots (n) \\ \vdots \\ \alpha_k\beta_k \end{matrix}\right\} \Sigma (\pm_g \pm_g . 1) \ldots\ldots\ldots\ldots (20)$$

$$= 1 . 2 \ldots k \left\{\begin{matrix} A\overset{\dagger}{\alpha_1}\beta_1 \ldots (n) \\ \vdots \\ \alpha_k\beta_k \end{matrix}\right\}.$$

But when n is odd, from the equation (18),

$$\left\{\begin{matrix} A\overset{+}{\alpha}_1, & \beta_1 \ldots (n) \\ \vdots & \\ \alpha_k\beta_k & \end{matrix}\right\} = \left\{\begin{matrix} A\alpha_1\overset{+}{\beta}_1 \ldots (n) \\ \vdots \\ \alpha_k, \beta_k \end{matrix}\right\} \Sigma(\pm_g 1) = 0 \quad \ldots\ldots (21),$$

since the number of negative and positive values of $\pm_g$ are equal.

From the equation (20), it follows that when n is even, the value of a symbol of the form

$$\left\{\begin{matrix} A\overset{+}{\alpha}_1\beta_1, & (n) \\ \vdots & \\ \alpha_k\beta_k & \end{matrix}\right\} \quad \ldots\ldots (22)$$

is the same, over whichever of the columns α, β... the mark ($\dagger$) is placed. To denote this indifference, the preceding quantity is better represented by

$$\left\{\begin{matrix} \overset{+}{A}\alpha_1, & \beta_1 \ldots (n) \\ \vdots & \\ \alpha_k\beta_k & \end{matrix}\right\} \quad \ldots\ldots (23),$$

this last form being never employed when n is odd, in which case the same property does not hold. Hence also an ordinary determinant is represented by

$$\left\{\begin{matrix} \overset{+}{A}\alpha_1\beta_1 \\ \vdots \\ \alpha_k\beta_k \end{matrix}\right\}, \quad \left\{\begin{matrix} \overset{+}{A}\,1\,1 \\ \vdots\;\vdots \\ k\;k \end{matrix}\right\} \quad \ldots\ldots (24),$$

the latter form being obviously equally general with the former one.

It is obvious from the equations (17), (18), that the expression (22) vanishes, in the case of n even whenever any two of the symbols α are equivalent, or any two of the symbols β, &c.; but if n be odd, this property holds for the symbols β, &c., but not for the marked ones α. In fact, the interchange of the two equal symbols, in each case, changes the sign of the expression (22), but they evidently leave it unaltered, i.e. the quantity in question must be zero.

Consider now the symbol

$$\left\{\begin{matrix} \overset{+}{A}\,1\,1 \ldots (2p) \\ \vdots\;\vdots \\ k\;k \end{matrix}\right\} \quad \ldots\ldots (25),$$

which, for shortness, may be denoted by

$$\{\overset{+}{A}\,.\,k\,.\,2p\} \quad \ldots\ldots (26).$$

I proceed to prove a theorem, which may be expressed as follows:

$$\{\overset{\dagger}{A}\,.\,k\,.\,2p\}\,.\,\{\overset{\dagger}{B}\,.\,k\,.\,2q\}=\{\overline{AB}|\,k\,.\,2p+2q-2\}\;\;\ldots\ldots\ldots\ldots\ldots\ldots(27),$$

where
$$\overline{AB}|_{rs\ldots xy\ldots}=S\,.\,A_{rs\ldots l}\,B_{xy\ldots l}\;\;\ldots\ldots\ldots\ldots\ldots\ldots\ldots\ldots\;(28),$$

the number of the symbols r, s,... being obviously $2p-1$, and that of x, y,... being $2q-1$. The summatory sign S refers to l, and denotes the sum of the several terms corresponding to values of l from $l=1$ to $l=k$. Also the theorem would be equally true if l had been placed in any position whatever in the series r, $s\ldots l$; and again, in any position whatever in the series x, $y\ldots l$, instead of at the end of each of these. With a very slight modification this may be made to suit the case of an odd number instead of one of the two even numbers $2p$, $2q$; (in fact, it is only necessary to place the mark (†) in $\{\overline{AB}|\ldots\}$ over the column corresponding to the marked column in $\{A\ldots\}$, $\{A\ldots\}$ being the symbol for which the number of columns is odd), but it is inapplicable where the two numbers are odd. Consider the second side of (27); this may be expanded in the form

$$\Sigma+\pm_s\ldots\pm_x\pm_y\ldots\overline{AB}|_{1\,s_1\ldots x_1y_1\ldots}\,.\,\overline{AB}|_{2\,s_2\ldots x_2y_2\ldots}\ldots\overline{AB}|_{k\,s_k\ldots x_ky_k\ldots}\;\;\ldots\ldots(29),$$

where Σ refers to the different quantities $s,\ldots,x,y,\ldots$ as in (11).

Substituting from (28), this becomes

$$\Sigma\,.\,S_{l_1}\ldots S_{l_k}\,(+\pm_s\ldots\pm_x\pm_y\,A_{1\,s_1\ldots l_1}\ldots A_{k\,s_k\ldots l_k}\ldots B_{x_1\,y_1\ldots l_1}\ldots B_{x_k\,y_k\ldots l_k})\ldots(30).$$

Effecting the summation with respect to x, $y\ldots$ this becomes

$$\Sigma\,.\,S_{l_1}\ldots S_{l_k}+\pm_s\ldots A_{1s_1\ldots l_1}\ldots A_{ks_k\ldots l_k}\begin{Bmatrix}\overset{\dagger}{B}1\;1\ldots l_1\\ \vdots\\ k\,k\ldots l_k\end{Bmatrix}\ldots\ldots\ldots\ldots(31),$$

Σ now referring to $s,\ldots$ only. The quantity under the sign Σ vanishes if any two of the quantities l are equal, and in the contrary case, we have

$$\begin{Bmatrix}\overset{\dagger}{B}1\;1\ldots l_1\\ \vdots\\ k\,k\ldots l_k\end{Bmatrix}=\pm_l\,\{\overset{\dagger}{B}\,.\,k\,.\,2q\}\;\;\ldots\ldots\ldots\ldots\ldots\ldots\ldots\ldots(32),$$

which reduces the above to

$$\{\overset{\dagger}{B}\,.\,k\,.\,2q\}\,.\,\Sigma+\pm_s\ldots\pm_l A_{l.\,s_1\ldots l_1}\ldots A_{k.\,s_k\ldots l_k}\;\;\ldots\ldots\ldots\ldots\ldots\ldots(33),$$

Σ referring to the quantities $s\ldots$, and also to the quantities l. And this is evidently equivalent to

$$\{\overset{\dagger}{A}\,.\,k\,.\,2p\}\,\{\overset{\dagger}{B}\,.\,k\,.\,2q\}\;\;\ldots\ldots\ldots\ldots\ldots\ldots\ldots\ldots\ldots(34),$$

the theorem to be proved. It is obvious that when $p=1$, $q=1$, the equation (27), coincides with the theorem (⊙), quoted in the introduction to this paper.

13.

ON THE THEORY OF LINEAR TRANSFORMATIONS.

[From the *Cambridge Mathematical Journal*, vol. IV. (1845), pp. 193—209.]

THE following investigations were suggested to me by a very elegant paper on the same subject, published in the *Journal* by Mr Boole. The following remarkable theorem is there arrived at. If a rational homogeneous function U, of the nth order, with the m variables $x, y \ldots$, be transformed by linear substitutions into a function V of the new variables, $\xi, \eta \ldots$; if, moreover, θU expresses the function of the coefficients of U, which, equated to zero, is the result of the elimination of the variables from the series of equations $d_x U=0$, $d_y U=0$, &c., and of course θV the analogous function of the coefficients of V: then $\theta V = E^{na} . \theta U$, where E is the determinant formed by the coefficients of the equations which connect $x, y \ldots$ with $\xi, \eta \ldots$,[1] and $a = (n-1)^{m-1}$. In attempting to demonstrate this very beautiful property, it occurred to me that it might be generalised by considering for the function U, not a homogeneous function of the nth order between m variables, but one of the same order, containing n sets of m variables, and the variables of each set entering linearly. The form which Mr Boole's theorem thus assumes is $\theta V = E_1{}^a . E_2{}^a \ldots E_n{}^a . \theta U$. This it was easy to demonstrate would be true, if θU satisfied a certain system of partial differential equations. I imagined at first that these would determine the function θU, (supposed, in analogy with Mr Boole's function, to represent the result of the elimination of the variables from $d_{x_1} U = 0$, $d_{y_1} U = 0, \ldots d_{x_2} U = 0$, &c.): this I afterwards found was not the case; and thus I was led to a class of functions, including as a particular case the function θU, all of them possessed of the same characteristic property. The system of partial differential equations was without difficulty replaced by a more fundamental system of equations, upon which, assumed as definitions, the theory appears to me naturally to depend; and it is this view of it which I intend partially to develope in the present paper.

[1] The value of a was left undetermined, but Mr Boole has since informed me, he was acquainted with it at the time his paper was written; and has given it in a subsequent paper.

I have already employed the notation

$$\left\|\begin{matrix}\alpha, & \beta, & \gamma, & \delta, \ldots \\ \alpha', & \beta', & \gamma', & \delta', \\ \alpha'', & \beta'', & \gamma'', & \delta'', \\ \vdots & & & \end{matrix}\right\| \quad \ldots\ldots\ldots\ldots (1)$$

(where the number of horizontal rows is less than that of vertical ones) to denote the series of determinants,

$$\left|\begin{matrix}\alpha, & \beta, & \gamma, \ldots \\ \alpha', & \beta', & \gamma', \\ \alpha'', & \beta'', & \gamma'', \\ \vdots & & \end{matrix}\right| \quad \ldots\ldots\ldots\ldots (2),$$

which can be formed out of the above quantities by selecting any system of vertical rows; these different determinants not being connected together by the sign +, or in any other manner, but being looked upon as perfectly separate.

The fundamental theorem for the multiplication of determinants gives, applied to these, the formula

$$\left\|\begin{matrix}A, & B, & C, & D, \ldots \\ A', & B', & C', & D', \\ A'', & B'', & C'', & D'', \\ \vdots & & & \end{matrix}\right\| = E \left\|\begin{matrix}\alpha, & \beta, & \gamma, & \delta, \ldots \\ \alpha', & \beta', & \gamma', & \delta', \\ \alpha'', & \beta'', & \gamma'', & \delta'', \\ \vdots & & & \end{matrix}\right\| \quad \ldots\ldots\ldots\ldots (3),$$

where

$$\left.\begin{aligned} A &= \lambda\alpha + \lambda'\alpha' + \lambda''\alpha'' + \ldots \\ B &= \lambda\beta + \lambda'\beta' + \lambda''\beta'' + \ldots \\ &\vdots \\ A' &= \mu\alpha + \mu'\alpha' + \mu''\alpha'' + \ldots \\ B' &= \mu\beta + \mu'\beta' + \mu''\beta'' + \ldots \\ &\vdots \\ &\&c. \end{aligned}\right\} \quad \ldots\ldots\ldots\ldots (4),$$

$$E = \left|\begin{matrix}\lambda, & \mu, \ldots \\ \lambda', & \mu', \\ \vdots & \end{matrix}\right| \quad \ldots\ldots\ldots\ldots (5),$$

and the meaning of the equation is, that the terms on the first side are equal, each to each, to the terms on the second side.

This preliminary theorem being explained, consider a set of arbitrary coefficients, represented by the general formula

$$rst \ldots \quad \ldots\ldots\ldots\ldots (6),$$

in which the number of symbolical letters $r, s, \ldots$ is n, and where each of these is supposed to assume all integer values, from 1 to m inclusively.

Let $$\alpha\, s_{\prime}'t_{\prime}' \dots, \quad \alpha\, s_{\prime}''t_{\prime}'' \dots, \qquad \dots\dots\dots\dots (7)$$

represent the whole series, taken in any order, in which the first symbolical letter is α. Similarly,

$$r_{\prime\prime}'\alpha\, t_{\prime\prime}' \dots, \quad r_{\prime\prime}''\alpha\, t_{\prime\prime}'', \dots \qquad \dots\dots\dots\dots (8),$$

the whole series of those in which the second symbolical letter is α, and so on.

Imagine a function u of the coefficients, which is simultaneously of the forms

$$u = H_p \left\| \begin{array}{lll} 1\, s_{\prime}'t_{\prime}' \dots, & 1\, s_{\prime}''t_{\prime}'' \dots, & \dots \\ 2\, s_{\prime}'t_{\prime}' \dots, & 2\, s_{\prime}''t_{\prime}'' \dots, & \\ \vdots & & \end{array} \right\| \qquad \dots\dots\dots\dots (A),$$

$$u = H_p \left\| \begin{array}{lll} r_{\prime\prime}'1\, t_{\prime\prime}' \dots, & r_{\prime\prime}''1\, t_{\prime\prime}'' \dots, & \dots \\ r_{\prime\prime}'2\, t_{\prime\prime}' \dots, & r_{\prime\prime}''2\, t_{\prime\prime}'' \dots, & \\ \vdots & & \end{array} \right\|$$

&c.; in which H_p denotes a rational homogeneous function of the order p. The function H is not necessarily supposed to be the same in the above equations, and in point of fact it will not in general be so. The number of equations is of course $= n$.

The function u, whose properties we proceed to investigate, may conveniently be named a "Hyperdeterminant." Any function satisfying any of the equations (A), without satisfying all of them, will be an "Incomplete Hyperdeterminant." But, considering in the first place such as are complete—

Let $\dot{r}st \dots$ be a new set of coefficients connected with the former ones by a system of equations of the form

$$\dot{r}st \dots = \lambda_1{}^r\, 1st \dots + \lambda_2{}^r\, 2st \dots + \lambda_m{}^r\, mst \dots \qquad \dots\dots\dots\dots (9),$$

(where the r in $\lambda_1{}^r \dots$ is not an exponent, but an affix).

Suppose $\dot{u}$ is the same function of these new coefficients that u was of the former ones. Then consider the first of the equations (A) and the equation (3), and writing

$$L = \left| \begin{array}{ll} \lambda_1{}^1, & \lambda_1{}^2, \dots \\ \lambda_2{}^1, & \lambda_2{}^2, \\ \vdots & \end{array} \right| \qquad \dots\dots\dots\dots (10),$$

we have immediately the equation

$$\dot{u} = L^p \,.\, u \qquad \dots\dots\dots\dots (11).$$

Consider the new set of coefficients

$$\dot{r}\dot{s}t \dots = \mu_1{}^s\, \dot{r}1t \dots + \mu_2{}^s\, \dot{r}2t \dots + \dots + \mu_m{}^s\, \dot{r}mt \dots \qquad \dots\dots\dots\dots (12),$$

and $\ddot{u}$ the analogous function of these; then, from the second of the equations (A) and the equation (3), and writing

$$M = \begin{vmatrix} \mu_1^1, & \mu_1^2, \dots \\ \mu_2^1, & \mu_2^2, \\ \vdots & \end{vmatrix} \qquad \text{.......... (13)},$$

$$\ddot{u} = M^p \,.\, \dot{u} = L^p M^p \,.\, u \qquad \text{.......... (14)}.$$

In like manner, considering the new coefficients $\dot{r}\dot{s}\dot{t} \dots$, where

$$\dot{r}\dot{s}\dot{t} \dots = \nu_1^t \, \dot{r}\dot{s}1 \dots + \nu_2^t \, \dot{r}\dot{s}2 \dots \quad \dots + \nu_n^t \, \dot{r}\dot{s}m \dots \qquad \text{.......... (15)},$$

the new function $\dddot{u}$, and the quantity N given by

$$N = \begin{vmatrix} \nu_1^1, & \nu_1^2, \dots \\ \nu_2^1, & \nu_2^2, \\ \vdots & \end{vmatrix} \qquad \text{.......... (16)},$$

we have, as before,

$$\dddot{u} = N^p \, \ddot{u} = L^p M^p N^p \, u \qquad \text{.......... (17)},$$

or

$$\dddot{u} = L^p M^p N^p u \qquad \text{.......... (18)};$$

whence, generally, denoting the last result by u',

$$u' = L^p M^p N^p \dots u \qquad \text{.......... (B, 1)}.$$

Consider now the function

$$\Sigma\Sigma\Sigma \dots (rst \dots x_r y_s z_t \dots) \qquad \text{.......... (19)},$$

where the Σ's refer successively to r, s, t, ..., and denote summations from 1 to m inclusively. If u be looked upon as a derivative from the above function, we may write

$$u = \Theta \,.\, \Sigma\Sigma\Sigma \dots (rst \dots x_r y_s z_t \dots) \qquad \text{.......... (20)}.$$

Assume

$$\left.\begin{aligned} x_r &= \lambda_r^1 \dot{x}_1 + \lambda_r^2 \dot{x}_2 \dots + \lambda_r^m \dot{x}_m \\ y_s &= \mu_s^1 \dot{y}_1 + \mu_s^2 \dot{y}_2 \dots + \mu_s^m \dot{y}_m \\ &\vdots \end{aligned}\right\} \qquad \text{.......... (21)}.$$

It is easy to obtain

$$\Sigma\Sigma\Sigma \dots (rst \dots x_r y_s z_t \dots) = \Sigma\Sigma\Sigma \dots (\dot{r}st \dots \dot{x}_r y_s z_t \dots), \dots = \Sigma\Sigma\Sigma \dots (\dot{r}\dot{s}\dot{t} \dots \dot{x}_r \dot{y}_s \dot{z}_t \dots) \dots \quad \text{...(22)},$$

and the formula for (u) becomes

$$\Theta \, \Sigma\Sigma\Sigma \dots (\dot{r}\dot{s}\dot{t} \dots \dot{x}_r \dot{y}_s \dot{z}_t) \dots = L^p M^p N^p \dots \Theta \, \Sigma\Sigma\Sigma \dots (rst \dots x_r y_s z_t) \qquad \text{.......... (B, 2)}.$$

Proceeding to obtain the expression of the coefficients $\dot{r}\dot{s}\dot{t} \dots$ in terms of the coefficients $rst \dots$, we have

$$\dot{r}\dot{s}\dot{t} \dots = \Sigma\Sigma\Sigma \dots (\lambda_f^r \mu_g^s \nu_h^t \dots fgh \dots) \qquad \text{.......... (C)},$$

where the Σ's refer successively to $f, g, h, \ldots$, denoting summations from 1 to m inclusively. Having this equation, it is perhaps as well to retain

$$u' = L^p M^p N^p \ldots u \quad \ldots\ldots\ldots\ldots \quad (\text{B}, 1),$$

instead of (B, 2), that form being principally useful in showing the relation of the function u to the theory of the transformation of functions.

It may immediately be seen, that in the equations (B), (C) we may, if we please, omit any number of the marks of variation (·), omitting at the same time the corresponding signs Σ, and the corresponding factors of the series $L, M, N \ldots$

Also, if u be such as only to satisfy some of the equations (A); then, if in the same formulæ we omit the corresponding marks (·), summatory signs, and terms of the series $L, M, N \ldots$, the resulting equations are still true.

From the formulæ (A) we may obtain the partial differential equations

$$\Sigma\Sigma \ldots \left(\alpha st \ldots \frac{d}{d\beta st \ldots}\right) u = 0, \text{ or } pu, \quad \ldots\ldots\ldots\ldots \quad (\text{D}),$$

$$\Sigma\Sigma \ldots \left(rat \ldots \frac{d}{dr\beta t \ldots}\right) u = 0, \text{ or } pu,$$

according as α is not equal, or is equal, to β;

and so on: the summatory signs referring in every case to those of the series $r, s, t, \ldots$, which are left variable, and extending from 1 to m inclusively.

To demonstrate this, consider the general form of u, as given by the first of the equations (A). This is evidently composed of a series of terms, each of the form

$$cPQR \ldots (p \text{ factors}),$$

in which

$$P = \left|\begin{array}{lll} 1s_,'t_,' \ldots, & 1s_,''t_,'' \ldots, & \ldots (m \text{ terms}) \\ \vdots & & \\ \alpha s_,'t_,' \ldots, & \alpha s_,''t_,'' \ldots, & \\ \vdots & & \\ \beta s_,'t_,' \ldots, & \beta s_,''t_,'' \ldots, & \\ \vdots & & \end{array}\right|$$

Q, R, &c. being of the same form; and we have

$$\Sigma\Sigma \ldots \left(\alpha st \ldots \frac{d}{d\beta st \ldots}\right) u = cQR \ldots \Sigma\Sigma \ldots \left(\alpha st \ldots \frac{d}{d\beta st \ldots}\right) P + \&\text{c.} + \&\text{c.},$$

and

$$\Sigma\Sigma \ldots \left(\alpha st \ldots \frac{d}{d\beta st \ldots}\right) P = \left|\begin{array}{lll} 1s_,'t_,' \ldots, & 1s_,''t_,'' \ldots, & \ldots (m \text{ terms}) \\ \vdots & & \\ \alpha s_,'t_,' \ldots, & \alpha s_,''t_,'' \ldots, & \ldots \\ \vdots & & \\ \alpha s_,'t_,' \ldots, & \alpha s_,''t_,'' \ldots, & \ldots \\ \vdots & & \end{array}\right| = 0;$$

so that all the terms on the second side of the equation vanish. If, however, $\beta = \alpha$,

$$\Sigma\Sigma \ldots \left(\alpha st \ldots \frac{d}{d\beta st \ldots}\right) P = P\,;$$

whence, on the second side, we have

$$\begin{aligned} cQR \ldots P + cPR \ldots Q + \&\text{c.} &= p \,.\, cPQR \ldots + \&\text{c.} + \&\text{c.} \\ &= pU, \end{aligned}$$

or the theorem in question is proved.

In the case of an incomplete hyperdeterminant, the corresponding systems of equations are of course to be omitted. In every case it is from these equations that the form of the function u is to be investigated; they entirely replace the system (A).

A very important case of the general theory is, when we suppose the coefficients $rst \ldots$ to have the property $r's't' \ldots = r''s''t'' \ldots$, whenever $r's't' \ldots$ and $r''s''t'' \ldots$ denote the same combination of letters; and also that the coefficients λ are equal to the coefficients $\mu, \nu \ldots$, each to each. In this case the coefficients $\dot{r}\dot{s}\dot{t} \ldots$ have likewise the same property, viz. that $\dot{r}''\dot{s}''\dot{t}'' \ldots = \dot{r}'\dot{s}'\dot{t}' \ldots$, whenever $r's't' \ldots$ and $r''s''t'' \ldots$ denote the same combination of letters.

The equations (B, 1), (B, 2), become in this case

$$u' = L^{np} \,.\, u \quad \ldots\ldots\ldots\ldots\ldots\ldots\ldots\ldots\ldots\ldots (\text{B, } 3),$$

$$\Theta\ \Sigma\Sigma\Sigma \ldots \left\{\frac{[n]^n}{[\alpha]^\alpha\,[\beta]^\beta \ldots}\, \dot{r}\dot{s}\dot{t} \ldots \dot{x}_r\dot{x}_s\dot{x}_t \ldots\right\} = L^{np} \,.\, \Theta \left\{\frac{[n]^n}{[\alpha]^\alpha\,[\beta]^\beta \ldots}\, rst \ldots x_r x_s x_t \ldots\right\} \ldots\ldots (\text{B, } 4),$$

where only different combinations of values are to be taken for $r, s, t, \ldots$ and $\alpha, \beta, \ldots$ express how often the same number occurs in the series. In the equation (C), μ, ν must be replaced by λ, the equations (D) are no longer satisfied; the equations (A) reduce themselves to a single one, (so that there can be no question here of incomplete hyperdeterminants): but this is no longer sufficient to determine the function sought after. For this reason, the particular case, treated separately, would be far more difficult than the general one; but the formulæ of the general case being first established, these apply immediately to the particular one[1]. The case in question may be defined as that of symmetrical hyperdeterminants, (a denomination already adopted for ordinary determinants). It would be easily seen what on the same principle would be meant by partially symmetrical hyperdeterminants.

I have not yet succeeded in obtaining the general expression of a hyperdeterminant; the only cases in which I can do so are the following: I. $p=1$, n even, (if n be odd, there only exist incomplete hyperdeterminants). II. $p=2$, $m=2$, n even. III. $p=3$, $m=2$, $n=4$.

I. The first case is, in fact, that of the functions considered at the termination of a paper in the *Cambridge Philosophical Transactions*, vol. VIII. part I. [12]; though at that time I was quite unacquainted with the general theory.

[1] See concluding paragraph of this paper.

Using the notation there employed, we have

$$u = \left\{\begin{matrix} \overset{\dagger}{1}\,1 \ldots (n) \\ 2\,2 \\ \vdots\ \vdots \\ mm \end{matrix}\right\},$$

a complete hyperdeterminant when n is even; and when n is odd the functions

$$\left\{\begin{matrix} \overset{\dagger}{1}\,1 \ldots (n) \\ 2\,2 \\ \vdots\ \vdots \\ mm \end{matrix}\right\}, \qquad \left\{\begin{matrix} \overset{\dagger}{1}\,1 \ldots (n) \\ 2\,2 \\ \vdots\ \vdots \\ mm \end{matrix}\right\}$$

are each of them incomplete hyperdeterminants.

(A) In the case of $n = 2$, the complete hyperdeterminant is simply the ordinary determinant

$$\left|\begin{matrix} 1\,1, & 1\,2, \ldots & 1m \\ 2\,1, & 2\,2, \ldots & 2m \\ \vdots & & \\ m1, & m2, \ldots & mm \end{matrix}\right|.$$

Stating the general conclusion as applied to this case, which is a very well known one,

"If the function

$$U = \Sigma\Sigma\,(rs\,.\,x_r y_s)$$

be transformed into a similar function

$$\Sigma\Sigma\,(\dot{r}\dot{s}\,.\,\dot{x}_r \dot{y}_s),$$

by means of the substitutions

$$x_r = \lambda_r^{\,1}\,\dot{x}_1 + \lambda_r^{\,2}\,\dot{x}_2 \ldots + \lambda_r^{\,m}\,\dot{x}_m,$$

$$y_s = \mu_s^{\,1}\,\dot{y}_1 + \mu_s^{\,2}\,\dot{y}_2 \ldots + \mu_s^{\,m}\,\dot{y}_m;$$

then

$$\left|\begin{matrix} \dot{1}\dot{1}, & \dot{1}\dot{2}, \ldots \\ \dot{2}\dot{1}, & \dot{2}\dot{2}, \\ \vdots & \end{matrix}\right| = \left|\begin{matrix} \lambda_1^{\,1}, & \lambda_2^{\,1}, \ldots \\ \lambda_1^{\,2}, & \lambda_2^{\,2}, \\ \vdots & \end{matrix}\right| \left|\begin{matrix} \mu_1^{\,1}, & \mu_2^{\,1}, \ldots \\ \mu_1^{\,2}, & \mu_2^{\,2}, \\ \vdots & \end{matrix}\right| \left|\begin{matrix} 11, & 12, \ldots \\ 21, & 22, \\ \vdots & \end{matrix}\right|.$$

Also, by what has preceded,

$$\dot{r}\dot{s} = \Sigma\Sigma\,(\lambda_f^{\,r}\,.\,\mu_g^{\,s}\,.fg);$$

so that the theorem is easily seen to amount to the following one—"If the terms of a determinant of the mth order be of the form $\Sigma_r\Sigma_s\,(rs\,.\,x_{r\rho}\,y_{s\sigma})$, r, s extending as before, from 1 to m inclusively, the determinant itself is the product of three determinants; the first formed with the coefficients rs, the second with the quantities x, and the third with the quantities y."

In a following number of the *Journal* I shall prove, and apply to the theories of Maxima and Minima and of Spherical Coordinates, (I may just mention having obtained, in an elegant form, the formulæ for transforming from one oblique set of coordinates to another oblique one) the more general theorem,

"If k be the order of the determinant formed as above, the determinant itself is a quadratic function, its coefficients being determinants formed with the coefficients rs, its variables being determinants formed respectively with the variables x and the variables y; and the number of variables in each set being the number of combinations of k things out of m, ($=1$ if $k=m$; if $k>m$ the determinant vanishes)."

I shall give in the same paper the demonstration of a very beautiful theorem, rather relating, however, to determinants than to quadratic functions, proved by Hesse in a Memoir in *Crelle's Journal*, vol. XX., "De curvis et superficiebus secundi ordinis;" and from which he has deduced the most interesting geometrical results. Another Memoir, by the same author, *Crelle*, vol. XXVIII., "Ueber die Elimination der Variabeln aus drei algebraischen Gleichungen vom zweiter Grade mit zwei Variabeln," though relating in point of fact rather to functions of the third order, contains some most important results. A few theorems on quadratic functions, belonging, however, to a different part of the subject, will be found in my paper already quoted in the *Cambridge Philosophical Transactions* [12]; and likewise in a paper in the *Journal*, Chapters in the Algebraical Geometry of n dimensions [11].

I shall, just before concluding this case, write down the particular formula corresponding to three variables, and for the symmetrical case. It is, as is well known, the theorem,

"If $$U = Ax^2 + By^2 + Cz^2 + 2Fyz + 2Gxz + 2Hxy$$

be transformed into

$$\mathfrak{A}\xi^2 + \mathfrak{B}\eta^2 + \mathfrak{C}\theta^2 + 2\mathfrak{F}\eta\theta + 2\mathfrak{G}\xi\theta + 2\mathfrak{H}\xi\eta$$

by means of
$$\begin{aligned} x &= \alpha\ \xi + \beta\ \eta + \gamma\ \theta, \\ y &= \alpha'\ \xi + \beta'\ \eta + \gamma'\ \theta, \\ z &= \alpha''\xi + \beta''\eta + \gamma''\theta, \end{aligned}$$

then
$$(\mathfrak{ABC} - \mathfrak{AF}^2 - \mathfrak{BG}^2 - \mathfrak{CH}^2 + 2\mathfrak{FGH}) =$$
$$(\alpha\beta'\gamma'' - \alpha\beta''\gamma' + \alpha'\beta''\gamma - \alpha'\beta\gamma'' + \alpha''\beta\gamma' - \alpha''\beta'\gamma)^2 (ABC - AF^2 - BG^2 - CH^2 + 2FGH)."$$

(B) Let $n=3$, and for greater simplicity $m=2$; write

$$\begin{aligned} a &= 111, & e &= 112, \\ b &= 211, & f &= 212, \\ c &= 121, & g &= 122, \\ d &= 221, & h &= 222, \end{aligned}$$

so that $$U = a\,x_1y_1z_1 + b\,x_2y_1z_1 + c\,x_1y_2z_1 + d\,x_2y_2z_1 + e\,x_1y_1z_2 + f\,x_2y_1z_2 + g\,x_1y_2z_2 + h\,x_2y_2z_2.$$

There is no complete hyperdeterminant (i.e. for $p=1$), and the incomplete ones are

$$ah - bg - cf + de = u_{,} \text{ , suppose,}$$
$$ah - de - bg + cf = u_{,,} ,$$
$$ah - cf - de + bg = u_{,,,}.$$

Thus, suppose the transforming equations are

$$x_1 = \lambda_1^1 \dot{x}_1 + \lambda_1^2 \dot{x}_2 ,$$
$$x_2 = \lambda_2^1 \dot{x}_1 + \lambda_2^2 \dot{x}_2 ;$$
$$y_1 = \mu_1^1 \dot{y}_1 + \mu_1^2 \dot{y}_2 ,$$
$$y_2 = \mu_2^1 \dot{y}^1 + \mu_2^2 \dot{y}_2 ;$$
$$z_1 = \nu_1^1 \dot{z}_1 + \nu_1^2 \dot{z}_2 ,$$
$$z_2 = \nu_2^1 \dot{z}_1 + \nu_2^2 \dot{z}_2 ;$$

then

$$\ddot{u}_{,} = (\mu_1^1 \mu_2^2 - \mu_2^1 \mu_1^2)(\nu_1^1 \nu_2^2 - \nu_2^1 \nu_1^2)\, u_1 \text{ , where } y, z \text{ are changed}$$
$$\ddot{u}_{,,} = (\nu_1^1 \nu_2^2 - \nu_2^1 \nu_1^2)(\lambda_1^1 \lambda_2^2 - \lambda_2^1 \lambda_1^2)\, u_{,,} \text{ , ,, } z, x \text{ ,,}$$
$$\ddot{u}_{,,,} = (\lambda_1^1 \lambda_2^2 - \lambda_2^1 \lambda_1^2)(\mu_1^1 \mu_2^2 - \mu_2^1 \mu_1^2)\, u_{,,,} \text{ , ,, } x, y \text{ ,,}$$

We might also have assumed

$$u_{,} = ad - bc, \text{ or } eh - gf,$$
$$u_{,,} = af - be, \text{ or } ch - dg,$$
$$u_{,,,} = ag - ce, \text{ or } bh - df,$$

but these are ordinary determinants.

(C) $n = 4$, $m = 2$.

$$\begin{aligned} a &= 1111, & i &= 1112,\\ b &= 2111, & j &= 2112,\\ c &= 1211, & k &= 1212,\\ d &= 2211, & l &= 2212,\\ e &= 1121, & m &= 1122,\\ f &= 2121, & n &= 2122,\\ g &= 2211, & o &= 2212,\\ h &= 2221, & p &= 2222. \end{aligned}$$

$$\begin{aligned} U = \;& a\, x_1y_1z_1w_1 + b\, x_2y_2z_1w_1 + c\, x_1y_2z_1w_1 + d\, x_2y_2z_1w_1 \\ & + e\, x_1y_1z_2w_1 + f\, x_2y_1z_2w_1 + g\, x_2y_2z_1w_1 + h\, x_2y_2z_2w_1 \\ & + i\, x_1y_1z_1w_2 + j\, x_2y_1z_1w_2 + k\, x_1y_2z_1w_2 + l\, x_2y_2z_1w_2 \\ & + m\, x_1y_1z_2w_2 + n\, x_2y_1z_2w_2 + o\, x_2y_2z_1w_2 + p\, x_2y_2z_2w_2, \end{aligned}$$

we have

$$u = ap - bo - cn + dm - el + fk + gj - hi ;$$

so that, with the same sets of transforming equations as above, and the additional one,

$$w_1 = \rho_1^{\,1}\dot{w}_1 + \rho_1^{\,2}\dot{w}_2,$$
$$w_n = \rho_2^{\,1}\dot{w}_1 + \rho_2^{\,2}\dot{w}_2,$$

we have $$\ddot{u} = (\lambda_1^{\,1}\lambda_2^{\,2} - \lambda_2^{\,1}\lambda_1^{\,2})\,(\mu_1^{\,1}\mu_2^{\,2} - \mu_2^{\,1}\mu_1^{\,2})\,(\nu_1^{\,1}\nu_2^{\,2} - \nu_2^{\,1}\nu_1^{\,2})\,(\rho_1^{\,1}\rho_2^{\,2} - \rho_2^{\,1}\rho_1^{\,2})\,.\,u;$$

this is important when viewed in reference to a result which will presently be obtained.

If we take the symmetrical case, we have

$$U = \alpha x^4 + 4\beta x^3y + 6\gamma x^2y^2 + 4\delta xy^3 + \epsilon y^4;$$

which is transformed into

$$U' = \alpha' x'^4 + 4\beta' x'^3y' + 6\gamma' x'^2y'^2 + 4\delta' x'y'^3 + \epsilon' y'^4,$$

by means of

$$x = \lambda\, x' + \mu\, y',$$
$$y = \lambda_{,} x' + \mu_{,} y';$$

then, if $$u = \alpha\,\epsilon - 4\beta\,\delta + 3\gamma^2,$$
$$u' = \alpha'\epsilon' - 4\beta'\delta' + 3\gamma'^2,$$

we have $$u' = (\lambda\mu_{,} - \lambda_{,}\mu)^4\,.\,u.$$

II. Where $p = 2$, $m = 2$, n is odd.

The expression

$$u = \left|\begin{matrix} \overset{\dagger}{\left\{\begin{matrix}1111 \ldots (n)\\ 1222\end{matrix}\right\}}, & \overset{\dagger}{\left\{\begin{matrix}1111 \ldots (n)\\ 2222\end{matrix}\right\}}, \\ \overset{\dagger}{\left\{\begin{matrix}1111 \ldots (n)\\ 2222\end{matrix}\right\}}, & \overset{\dagger}{\left\{\begin{matrix}2111 \ldots (n)\\ 2222\end{matrix}\right\}}, \end{matrix}\right|$$

is a complete hyperdeterminant; and that over whichever row the mark (†) of nonpermutation is placed. The different expressions so obtained are not, however, all of them independent functions: thus, in the following example, where $n = 3$, the three functions are absolutely identical.

(A). $n = 3$, notation as in I. (B).

$$u = a^2h^2 + b^2g^2 + c^2f^2 + d^2e^2 - 2ahbg - 2ahcf - 2ahde - 2bgcf - 2bgde - 2cfde + 4adfg + 4bech,$$

and then $$\ddot{u} = (\lambda_1^{\,1}\lambda_2^{\,2} - \lambda_2^{\,1}\lambda_1^{\,2})^2\,(\mu_1^{\,1}\mu_2^{\,2} - \mu_2^{\,1}\mu_1^{\,2})^2\,(\nu_1^{\,1}\nu_2^{\,2} - \nu_2^{\,1}\nu_1^{\,2})^2\,u.$$

This is in many respects an interesting example. We see that the function u may be expressed in the three following forms:

$$u = (ah - bg - cf + de)^2 + 4\,(ad - bc)\,(fg - eh) \ldots\ldots\ldots\ldots (1),$$
$$u = (ah - bg - de + cf)^2 + 4\,(af - be)\,(dg - ch) \ldots\ldots\ldots\ldots (2),$$
$$u = (ah - cf - de + bg)^2 + 4\,(ag - ce)\,(df - bh) \ldots\ldots\ldots\ldots (3),$$

which are indeed the direct results of the general form above given, the sign (†) being placed in succession over the different columns: and the three forms, as just remarked, are in this case identical.

We see from the first of these that u is of the second or third, from the second that u is of the first or third, from the third that u is of the first or second of the three following forms:

$$u = H_2 \left\| \begin{matrix} a, & b, & c, & d \\ e, & f, & g, & h \end{matrix} \right\|, \quad u = H_2 \left\| \begin{matrix} a, & b, & e, & f \\ c, & d, & g, & h \end{matrix} \right\|, \quad u = H_2 \left\| \begin{matrix} a, & c, & e, & g \\ b, & d, & f, & h \end{matrix} \right\|,$$

which is as it should be.

The following is a singular property of u.

Let $$a' = \tfrac{1}{2}\frac{du}{da}, \quad b' = \tfrac{1}{2}\frac{du}{db}, \ \ldots\ldots \ h' = \tfrac{1}{2}\frac{du}{dh},$$

then, u' being the same function of these new coefficients that u is of the former ones,

$$u' = u^3.$$

To prove this, write

$$p = ah - bg - cf + de, \quad q = (ad - bc), \quad r = eh - fg;$$

$$\begin{aligned} a_{\prime} &= ap - 2qe, & e_{\prime} &= -2ra + pe, \\ b_{\prime} &= bp - 2qf, & f_{\prime} &= -2rb + pf, \\ c_{\prime} &= cp - 2qg, & g_{\prime} &= -2rc + pg, \\ d_{\prime} &= dp - 2qh, & h_{\prime} &= -2rd + ph; \end{aligned}$$

we have, as a particular case of the general formula just obtained,

$$u_{\prime} = (p^2 - 4qr)^2\, u = u^2 \,.\, u, \ = u^3.$$

Also $$\begin{aligned} a_{\prime} &= h', & e_{\prime} &= d', \\ b_{\prime} &= -g', & f_{\prime} &= -c', \\ c_{\prime} &= -f', & g_{\prime} &= -b', \\ d_{\prime} &= e', & h_{\prime} &= a'; \end{aligned}$$

whence $u_{\prime} = u'$, that is $u' = u^3$.

There is no difficulty in showing also, that if a'', b'', ... h'' are derived from a', b' ... h', as these are from a, b, ... h, then

$$a'' = u^2 a, \quad b'' = u^2 b, \ \ldots \ h'' = u^2 h.$$

The particular case of this theorem, which corresponds to symmetrical values of the coefficients, is given by M. Eisenstein, *Crelle*, vol. XXVII. [1844], as a corollary to his researches on the cubic forms of numbers.

Considering this symmetrical case

$$U = \alpha x^3 + 3\beta x^2 y + 3\gamma x y^2 + \delta y^3,$$
$$u = \alpha^2\delta^2 - 6\alpha\delta\beta\gamma - 3\beta^2\gamma^2 + 4\beta^3\delta + 4\alpha^3\gamma,$$

so that if U be transformed into

$$U' = \alpha' x'^3 + 3\beta' x'^2 y' + 3\gamma' x' y'^2 + \delta' y^3,$$

by means of

$$x = \lambda\, x' + \mu\, y',$$
$$y = \lambda_{,} x' + \mu_{,} y',$$

then if

$$u' = \alpha'^2\delta'^2 - 6\alpha'\delta'\beta'\gamma' - 3\beta'^2\gamma'^2 + 4\beta'^3\delta' + 4\alpha'^3\gamma',$$

we have

$$u' = (\lambda\mu_{,} - \lambda_{,}\mu)^6 . u.$$

III. $\quad p = 3, \quad m = 2, \quad n = 4.$

Notation as in I. (C),

$$u = A\,(\mathfrak{A} + 3\mathfrak{B} + 3\mathfrak{C} + 6\mathfrak{D} + 6\mathfrak{E}) - B\,(\mathfrak{C} + \mathfrak{D} - 3\mathfrak{E} - \mathfrak{F} + 2\mathfrak{G} + 3\mathfrak{H}),$$

where A, B are arbitrary constants, and $\mathfrak{A}$, $\mathfrak{B}$, &c. ... $\mathfrak{H}$ are functions of the coefficients, given as follows:—

$$\mathfrak{A} = a^3p^3 - b^3o^3 - c^3n^3 + d^3m^3 - e^3l^3 + f^3k^3 + g^3j^3 - h^3i^3.$$

$$\begin{aligned}\mathfrak{B} = &- a^2p^2bo + b^2o^2ap + c^2n^2dm - d^2m^2cn + e^2l^2fk - f^2k^2el - g^2j^2hi + h^2i^2gj\\ &- a^2p^2cn + b^2o^2dm + c^2n^2ap - d^2m^2bo + e^2l^2gj - f^2k^2hi - g^2j^2el + h^2i^2fk\\ &- a^2p^2el + b^2o^2fk + c^2n^2gj - d^2m^2hi + e^2l^2ap - f^2k^2bo - g^2j^2cn + h^2i^2dm\\ &- a^2p^2hi + b^2o^2gj + c^2n^2fk - d^2m^2el + e^2l^2dm - f^2k^2cn - g^2j^2bo + h^2i^2ap.\end{aligned}$$

$$\begin{aligned}\mathfrak{C} = &+ a^2p^2dm - b^2o^2cn - c^2n^2bo + d^2m^2ap - e^2l^2hi + f^2k^2gj + g^2j^2fk - h^2i^2el\\ &+ a^2p^2fk - b^2o^2el - c^2n^2hi + d^2m^2gj - e^2l^2bo + f^2k^2ap + g^2j^2dm - h^2i^2cn\\ &+ a^2p^2gj - b^2o^2hi - c^2n^2el + d^2m^2fk - e^2l^2cn + f^2k^2dm + g^2j^2ap - h^2i^2bo.\end{aligned}$$

$$\begin{aligned}\mathfrak{D} = &\; apbocn - apbodm - apcndm + bocndm - elfkgj + elfkhi + elgjhi - higjfk\\ &+ apboel - apbofk - cndmgj + dmcnhi - elfkap + elfkbo + gjhicn - higjdm\\ &- apbogj + apbohi + cndmel - dmcnkf + elfkcn - elfkdm - gjhiap + higjbo\\ &+ apcnel - bodmfk - cnapgj + dmbohi - elgjap + fkhibo + gjelcn - hifkdm\\ &- apcnfk + bodmel + cnaphi - dmboel + elgjbo - fkhiap - gjeldm + hifkcn\\ &- apdmel + bocnkf + cmbogj - dmaphi + elhiap - fkgjbo - gjfkcn + hidmel.\end{aligned}$$

$$\mathfrak{E} = apdmfk - bocnel - bocnih + dmapgj - elhibo + fkapgj + gjfkdm - hielcn.$$

$$\begin{aligned}\mathfrak{F} = &\; a^2phjo - b^2goip - c^2nflm + d^2emkn - e^2dlkn + f^2ckml + g^2bjpi - h^2aijo\\ &- i^2phbg + j^2ogah + k^2nfde - l^2mecf + m^2ldcf - n^2ckde - o^2bjah + p^2aibg\\ &+ a^2phkn - b^2golm - c^2nfip + d^2emjo - e^2dljo + f^2ckip + g^2bjpl - h^2aink\\ &- i^2phcf + j^2ogde + k^2nfah - l^2mebg + m^2ldbg - n^2ckah - o^2bjde + p^2aicf\\ &+ a^2phlm - b^2gokn - c^2nfjo + d^2emip - e^2ldpi + f^2ckjo + g^2bjnk - h^2aiml\\ &- i^2phde + j^2gocf + k^2nfbg - l^2meah + m^2dlah - n^2ckbg - o^2bjcf + p^2aide\\ &+ a^2pdno - b^2cmpo - c^2bpmn + d^2aomn - e^2hjkl + f^2gilk + g^2flij - h^2ekji\\ &- i^2lfgh + j^2kehg + k^2jhef - l^2igfe + m^2pbcd - h^2aodc - o^2ndab + p^2mbca\\ &+ a^2plng - b^2hkmo - c^2ejpn + d^2fiom - e^2cpjl + f^2doik + g^2anlj - h^2bmki\\ &- i^2odfh + j^2pceg + k^2mhbf - l^2nage + m^2khbd - n^2lgac - o^2ifbd + p^2ieca\\ &+ a^2plfo - b^2ekpo - c^2hjmn + d^2gimn - e^2bpkl + f^2aolk + g^2dnij - h^2cmji\\ &- i^2ndgh + j^2mchg + k^2pbef - l^2oafe + m^2jhcd - n^2igcd - o^2lfab + p^2keba.\end{aligned}$$

$$\begin{aligned}\mathfrak{G} = {} & apbgkn - boalhm - cndiep + dmcjfo - elfocj + fkpied + gjhmla - ignbk \\ & - ihjocf + gjidep + kflamh - lekbng + mdnbkg - ncmhal - obpied + paofjc \\ & + apbglm - boahkn - cndejo + dmcfip - elcfip + fkedoj + gjhakn - hibgml \\ & - ihjode + gijpcf + kflmbg - leknah + mdknah - ncmlbg - obpicf + paojde \\ & + apcflg - bojehk - cnjehk + dmilgf - elmpbc + fkadno + gjadno - hipmcb \\ & - ihadno + gjbmcp + kfcbpm - eladno + mdjehk - ncilfg - obilfg + pahejk \\ & + apcflm - bodkne - cnahjo + dmbgip - elbgip + fkahjo + gjednk - hicfml \\ & - ihknde + mdahjo + kfipbg - bocfml + gjcfml - ncpigb - leahjo + paknde \\ & + apidng - bojcmh - cnbkpe + dmaflo - elmhjc - fkidng + gjaflo - ihbkpe \\ & - ihaflo + gjbkpe + fkcjhm - elidng + mdbkpe - cnaflo - boidng + pahmcj \\ & + apidfo - bojcep - nbkhm + cdmalng - elmnbk - fkalng + gjidfo - ihpecj \\ & - ihalng + gjkbmh + fkcjpe - elidfo + mdjcep - ncidfo - obalng + pahmbk.\end{aligned}$$

$$\begin{aligned}\mathfrak{H} = {} & a^2honl - b^2pgmk - c^2pfmj + d^2onie - e^2dpkj + f^2ilco + g^2blni - h^2amkj \\ & - i^2pdfg + j^2oech + k^2nbeh - l^2mafg + m^2blch - n^2kadg - o^2jadf + p^2ibec:\end{aligned}$$

we have, as usual,

$$\overset{....}{u} = (\lambda_1^1\lambda_2^2 - \lambda_2^1\lambda_1^2)^3 (\mu_1^1\mu_2^2 - \mu_2^1\mu_1^2)^3 (\nu_1^1\nu_2^2 - \nu_2^1\nu_1^2)^3 (\rho_1^1\rho_2^2 - \rho_1^2\rho_2^1)^3 \,.\, u.$$

Particular forms of U are

$A = 1,\quad B = 0:$

$u = \mathfrak{A} + 3\mathfrak{B} + 3\mathfrak{C} + 6\mathfrak{D} + 6\mathfrak{E} = (ap - bo - cn + dm - el + fk + gj - hi)^3, \ = \nu$ suppose.

$A = 1,\quad B = 9:$

$u = \mathfrak{A} + 3\mathfrak{B} - 6\mathfrak{C} - 3\mathfrak{D} + 33\mathfrak{E} + 9\mathfrak{F} - 18\mathfrak{G} - 27\mathfrak{H}, \ = \theta U$ suppose,

where $\theta U = 0$ is the result of the elimination of the variables from the equations $d_{x_1}U = 0$, $d_{y_1}U = 0$, $d_{z_1}U = 0$, $d_{w_1}U = 0$, $d_{x_2}U = 0$, $d_{y_2}U = 0$, $d_{z_2}U = 0$, $d_{w_2}U = 0$. In fact, by an investigation similar to Mr Boole's, applied to a function such as U, it is shown that θU has the characteristic property of the function u: also in the present case u is the most general function of its kind, so that θU is obtained from U by properly determining the constant. This has been effected by comparing the value of u, in the symmetrical case, with the value of θU, in the same case, the expanded expression of which is given by Mr Boole in the *Journal*, vol. IV. p. 169. Assuming $A = 1$, the result was $B = 9$. [Incorrect: the result of the elimination is not $\theta U = 0$, but an equation of a higher degree.]

The general form of u now becomes

$$u = \alpha\nu^3 + \beta\theta U,$$

in which α, β, are indeterminate.

We have $$\overset{....}{u} = \alpha\overset{....}{\nu^3} + \beta\theta\overset{....}{U} = M(\alpha\nu^3 + \beta\theta U),$$

where $$M = (\lambda_1^1\lambda_2^2 - \lambda_2^1\lambda_1^2)^3 (\mu_1^1\mu_2^2 - \mu_2^1\mu_1^2)^3 (\nu_1^1\nu_2^2 - \nu_2^1\nu_1^2)^3 (\rho_1^1\rho_2^2 - \rho_2^1\rho_1^2)^3.$$

and thence $$\ddot{\ddot{\nu}}^3 = M\nu^3,$$

which coincides with a previous formula, and

$$\theta \ddot{\ddot{U}} = M\theta U:$$

whence, eliminating M,

$$\frac{\theta \ddot{\ddot{U}}}{\ddot{\ddot{\nu}}^3} = \frac{\theta U}{\nu^3},$$

an equation which is remarkable as containing only the constants of U and $\ddot{\ddot{U}}$: it is an equation of condition which must exist among the constants of $\ddot{\ddot{U}}$ in order that this function may be derivable by linear substitutions from U.

In the symmetrical case, or where

$$U = \alpha x^4 + 4\beta x^3 y + 6\gamma x^2 y^2 + 4\delta x y^3 + \epsilon y^4,$$

it has been already seen that ν is given by

$$\nu = \alpha\epsilon - 4\beta\delta + 3\gamma^2.$$

Proceeding to form θU, we have

$$\mathfrak{A} = \alpha^3\epsilon^3 - 4\beta^3\delta^3 + 3\gamma^6,$$

$$\mathfrak{B} = 4\,(\alpha\epsilon\beta^2\delta^2 - \alpha^2\epsilon^2\beta\delta + 3\gamma^2\beta^2\delta^2 - 3\beta\delta\gamma^4),$$

$$\mathfrak{C} = 3\,(\alpha^2\epsilon^2\gamma^2 + 2\gamma^6 + \alpha\epsilon\gamma^4 - 4\beta^3\delta^3),$$

$$\mathfrak{D} = 6\,(\alpha\epsilon\beta^2\delta^2 - 2\alpha\epsilon\beta\delta\gamma^2 + 3\beta^2\gamma^2\delta^2 - 2\beta\delta\gamma^4),$$

$$\mathfrak{E} = 3\alpha\epsilon\gamma^4 - 4\beta^3\delta^3 + \gamma^4,$$

$$\mathfrak{F} = 6\,(\alpha^2\delta^2\gamma\epsilon + \epsilon^2\beta^2\gamma\alpha - 2\beta^3\epsilon\gamma\delta - 2\delta^3\alpha\beta\gamma - 4\beta^2\gamma^2\delta^2 . + 4 . \beta\delta\gamma^4 + \gamma^3\beta^2\epsilon + \gamma^3\alpha\delta^2),$$

$$\mathfrak{G} = 12\,(\alpha\epsilon\beta\delta\gamma^2 - \alpha\beta\gamma\delta^3 - \epsilon\gamma\delta\beta^3 + \gamma^3\alpha\delta^2 + \gamma^3 c\beta^2 + \beta\delta\gamma^4 - 2\beta^2\gamma^2\delta^2),$$

$$\mathfrak{H} = (\alpha^2\delta^4 + \epsilon^2\beta^4 - 4\beta^2\epsilon\gamma^3 - 4\alpha\gamma^3\delta^2 + 6\beta^2\gamma^2\delta^2),$$

and these values give

$$\theta U = \alpha^3\epsilon^3 - 6\alpha\beta^2\delta^2\epsilon - 12\alpha^2\beta\delta\epsilon^2 - 18\alpha^2\gamma^2\epsilon^2 - 27\alpha^2\delta^4 - 27\beta^4\epsilon^2 + 36\beta^2\gamma^2\delta^2 + 54\,\alpha^2\gamma\delta^2\epsilon + 54\,\alpha\beta^2\gamma\epsilon^2$$
$$- 54\alpha\gamma^3\delta^2 - 54\beta^2\gamma^3\epsilon - 64\beta^3\delta^3 + 81\alpha\gamma^4\epsilon + 108\alpha\beta\gamma\delta^3 + 108\beta^3\gamma\delta\epsilon - 180\alpha\beta\gamma^2\delta\epsilon,$$

so that this function, divided by $(\alpha\epsilon - 4\beta\delta + 3\gamma^2)^3$, is invariable for all functions of the fourth order which can be deduced one from the other by linear substitutions. The function $\alpha\epsilon - 4\beta\delta + 3\gamma^2$ occurs in other investigations: I have met with it in a problem relating to a homogeneous function of two variables, of any order whatever, α, β, γ, δ, ϵ signifying the fourth differential coefficients of the function. But this is only remotely connected with the present subject.

Since writing the above, Mr Boole has pointed out to me that in the transformation of a function of the fourth order of the form $ax^4 + 4bx^3y + 6cx^2y^2 + 4dxy^3 + ey^4$,—besides his function θu, and my quadratic function $ae - 4bd + 3c^2$,—there exists a function of the third order $ace - b^2e - ad^2 - c^3 + 2bdc$, possessing precisely the same characteristic property, and that, moreover, the function θu may be reduced to the form

$$(ae - 4bd + 3c^2)^3 - 27(ace - ad^2 - eb^2 - c^3 + 2bdc)^2;$$

the latter part of which was verified by trial; the former he has demonstrated in a manner which, though very elegant, does not appear to be the most direct which the theorem admits of. In fact, it may be obtained by a method just hinted at by Mr Boole, in his earliest paper on the subject, *Mathematical Journal*, vol. II. p. 70. The equations $d_x^2u = 0$, $d_xd_yu = 0$, $d_y^2u = 0$, imply the corresponding equations for the transformed function: from these equations we might obtain two relations between the coefficients, which, in the case of a function of the fourth order, are of the orders 3 and 4 respectively: these imply the corresponding relations between the coefficients of the transformed function. Let $A = 0$, $B = 0$, $A' = 0$, $B' = 0$, represent these equations; then, since $A = 0$, $B = 0$, imply $A' = 0$, we must have $A' = \Lambda A' + \mathrm{M}B$, Λ, M, being functions of λ, λ', μ, &c. μ': but B being of the fourth order, while A, A' are only of the third order in the coefficients of u, it is evident that the term $\mathrm{M}B$ must disappear, or that the equation is of the form $A' = \Lambda A$. The function A is obviously the function which, equated to zero, would be the result of the elimination of x^2, xy, y^2, considered as independent quantities from the equations $ax^2 + 2bxy + cy^2 = 0$, $bx^2 + 2cxy + dy^2 = 0$, $cx^2 + 2dxy + ey^2 = 0$, viz. the function given above. Hence the two functions on which the linear transformation of functions of the fourth order ultimately depend are the very simple ones

$$ae - 4bd + 3c^2, \quad ace - ad^2 - eb^2 - c^3 + 2bdc,$$

the function of the sixth order being merely a derivative from these. The above method may easily be extended: thus for instance, in the transformation of functions of any even order, I am in possession of several of the transforming functions; that of the fourth order, for functions of the sixth order, I have actually expanded: but it does not appear to contain the complete theory. Again, in the particular case of homogeneous functions of two variables, the transforming functions may be expressed as symmetrical functions of the roots of the equation $u = 0$, which gives rise to an entirely distinct theory. This, however, I have not as yet developed sufficiently for publication. There does not appear to be anything very directly analogous to the subject of this note, in my general theory: if this be so, it proves the absolute necessity of a distinct investigation for the present case, that which I have denominated the symmetrical one.

14.

ON LINEAR TRANSFORMATIONS.

[From the *Cambridge and Dublin Mathematical Journal*, vol. I. (1846), pp. 104—122.]

In continuing my researches on the present subject, I have been led to a new manner of considering the question, which, at the same time that it is much more general, has the advantage of applying directly to the only case which one can possibly hope to develope with any degree of completeness, that of functions of two variables. In fact the question may be proposed, "To find all the derivatives of any number of functions, which have the property of preserving their form unaltered after any linear transformations of the variables." By Derivative I understand a function deduced in any manner whatever from the given functions, and I give the name of Hyperdeterminant Derivative, or simply of Hyperdeterminant, to those derivatives which have the property just enunciated. These derivatives may easily be expressed explicitly, by means of the known method of the separation of symbols. We thus obtain the most general expression of a hyperdeterminant. But there remains a question to be resolved, which appears to present very great difficulties, that of determining the *independent* derivatives, and the relation between these and the remaining ones. I have only succeeded in treating a very particular case of this question, which shows however in what way the general problem is to be attacked.

Imagine p series each of m variables

$$x_1,\quad y_1, \ldots \&c. \qquad x_2,\quad y_2, \ldots \&c. \qquad x_p,\quad y_p, \ldots \&c.,$$

where p is at least as great as m.

Similarly p' series each of m' variables

$$x_1',\quad y_1', \ldots \&c. \qquad x_2',\quad y_2', \ldots, \&c. \qquad \ldots x'_{p'},\quad y'_{p'}, \ldots \&c.,$$

$p^{\backprime}$ at least as great as $m^{\backprime}$, and so on. Let the analogous variables $\dot{x}$, $\dot{y}$ ... be connected with these by the equations

$$\begin{aligned} x &= \lambda\,\dot{x} + \mu\,\dot{y} + \dots, \\ y &= \lambda'\dot{x} + \mu'\dot{y} + \dots, \\ &\vdots \\ x^{\backprime} &= \lambda^{\backprime}\,\dot{x}^{\backprime} + \mu^{\backprime}\,\dot{y}^{\backprime} + \dots, \\ y^{\backprime} &= \lambda'^{\backprime}\dot{x}^{\backprime} + \mu'^{\backprime}\dot{y}^{\backprime} + \dots, \\ &\vdots \end{aligned}$$

where x, y,... stand for x_1, y_1,... or x_2, y_2,... or x_p, y_p,...; $x^{\backprime}$, $y^{\backprime}$,... stand for $x_1^{\backprime}$, $y_1^{\backprime}$,... or $x_2^{\backprime}$, $y_2^{\backprime}$,... or $x^{\backprime}_{p}$, $y^{\backprime}_{p^{\backprime}}$, &c.... The coefficients λ, μ, ..., λ', μ', ... &c.; $\lambda^{\backprime}$, $\mu^{\backprime}$, ..., $\lambda'^{\backprime}$, $\mu'^{\backprime}$, ... remain the same in all these systems. Suppose next,

$$\xi = \delta_x, \quad \eta = \delta_y,$$

i.e.
$$\xi_1 = \delta_{x_1}, \quad \eta_1 = \delta_{y_1}, \dots, \dot{\xi}_1 = \delta_{\dot{x}_1}, \dots$$

(where δ_x, δ_y ... are the symbols of differentiation relative to x, y, &c.). Then evidently

$$\begin{aligned} \dot{\xi} &= \lambda\xi + \lambda'\eta + \dots, \\ \dot{\eta} &= \mu\xi + \mu'\eta + \dots, \\ &\vdots \end{aligned}$$

with similar equations for $\dot{\xi}^{\backprime}$, $\dot{\eta}^{\backprime}$,... Suppose

$$\|\Omega\| = \left\| \begin{matrix} \xi_1, & \xi_2, \dots \xi_p \\ \eta_1, & \eta_2, \dots \eta_p \\ \vdots & \end{matrix} \right\|, \quad \|\Omega^{\backprime}\| = \left\| \begin{matrix} \xi_1^{\backprime}, & \xi_2^{\backprime}, \dots \xi^{\backprime}_{p^{\backprime}} \\ \eta_1^{\backprime}, & \eta_2^{\backprime}, \dots \eta^{\backprime}_{p^{\backprime}} \\ \vdots & \end{matrix} \right\|,$$

that is to say $\|\Omega\|$ is the series of determinants formed by choosing any m vertical columns to compose a determinant, and similarly $\|\Omega^{\backprime}\|$, &c. Suppose, besides,

$$E = \left| \begin{matrix} \lambda, & \mu, \dots \\ \lambda', & \mu', \dots \\ \vdots & \end{matrix} \right|, \quad E^{\backprime} = \left| \begin{matrix} \lambda^{\backprime}, & \mu^{\backprime}, \dots \\ \lambda'^{\backprime}, & \mu'^{\backprime}, \dots \\ \vdots & \end{matrix} \right|.$$

Then, by the known properties of determinants,

$$\|\dot{\Omega}\| = E\,\|\Omega\|, \quad \|\dot{\Omega}^{\backprime}\| = E^{\backprime}\,\|\Omega^{\backprime}\| \text{ \&c.},$$

i.e. the terms on the one side are respectively equal to the terms on the other. Hence if

$$\square = F(\|\Omega\|^{f}, \ \|\Omega^{\backprime}\|^{f^{\backprime}}, \dots),$$

i.e. $\square$ a rational and integral function, homogeneous of the order f in the quantities of the series $\|\Omega\|$, homogeneous of the order $f^{\backprime}$ in the quantities of the series $\|\Omega^{\backprime}\|$, &c., we have immediately

$$\dot{\square} = E^{f} E^{\backprime f^{\backprime}} \dots \square;$$

or if U be any function whatever of the variables $x, y \ldots$ which is transformed by the linear substitutions above into $\dot{U}$, then

$$\dot{\Box}\dot{U} = E^f E'^{f'} \ldots \Box U;$$

or the function $$\Box U$$

is by the above definition a hyperdeterminant derivative. The symbol $\Box$ may be called "symbol of hyperdeterminant derivation," or simply "hyperdeterminant symbol."

Let $A, B, \ldots$ represent the different quantities of the series $\|\Omega\|$,—$A', B', \ldots$ those of the series $\|\Omega'\|$, &c. ..., then $\Box$ may be reduced to a single term, and we may write

$$\Box = A^{\alpha} B^{\beta} \ldots A'^{\alpha'} B'^{\beta'} \ldots$$

Also U may be supposed of the form

$$U = \Theta\Phi \ldots$$

where Θ, Φ are functions of the variables of one of the sets $x, y, \ldots$, of one of the sets $x', y', \ldots$, &c., thus Θ is of the form

$$F(x_1, y_1, \ldots x_1', y_1', \ldots),$$

and so on. The functions $\Theta, \Phi \ldots$ may be the same or different. It may be supposed after the differentiations that several of the sets $x, y, \ldots$ or of the sets $x', y', \ldots$ become identical: in such cases it will always be assumed that the functions $\Theta, \ldots$ into which these sets of variables enter, are similar; so that they become absolutely identical, when the variables they contain are made so. Thus the general expression of a hyperdeterminant is

$$\Box U = A^{\alpha} B^{\beta} \ldots A'^{\alpha'} B'^{\beta'} \ldots \Theta\Phi \ldots$$

in which, after the differentiations, any number of the sets of variables are made equal. For instance, if all the sets $x, y \ldots$ and all the sets $x', y' \ldots$ are made equal, the hyperdeterminant refers to a single function $F(x, y \ldots x', y' \ldots)$. In any other case it refers not to a single function but to several.

What precedes, is the general theory: it might perhaps have been made clearer by confining it to a particular case: and by doing this from the beginning, it will be seen that it presents no real difficulties. Passing at present to some developments, to do this, I neglect entirely the sets $x', y' \ldots$ and I assume that the number m of variables in each of the sets $x, y \ldots$ reduces itself to two; so that I consider functions of two variables x, y only. The functions Θ, Φ, &c. reduce themselves to functions $V_1, V_2 \ldots V_p$ of the variables x_1, y_1, or $x_2, y_2 \ldots$ or x_p, y_p. Writing also

$$\xi_1\eta_2 - \xi_2\eta_1 = \overline{12}, \text{ \&c.}$$

the symbols $A, B \ldots$ reduce themselves to $\overline{12}$, $\overline{13} \ldots$. Hence for functions of two variables, there results the following still tolerably general form

$$\Box U = \overline{12}^{\alpha}\ \overline{13}^{\beta}\ \overline{14}^{\gamma} \ldots \overline{23}^{\beta'}\ \overline{24}^{\gamma'} \ldots \overline{34}^{\gamma''} \ldots V_1 V_2 V_3 V_4 \ldots$$

The functions V_1, $V_2 \ldots$ may be the same or different: but they will be supposed the same whenever the corresponding variables are made equal. This equality will be denoted by writing, for instance,

$$\square\, VV'VV \ldots$$

to represent the value assumed by

$$\square\, V_1 V_2 V_3 V_4 \ldots$$

when after the differentiations

$$\begin{aligned} &x_1,\ y_1 = x_3,\ y_3 = x_4,\ y_4 = x,\ y\ ; \\ &x_2,\ y_2 = x',\ y'\ ; \\ &\&c. \end{aligned}$$

It is easy to determine the general term of $\square U$. To do this, writing for shortness

$$\begin{aligned} &\alpha + \beta + \gamma \ \ldots = f_1, \\ &\alpha + \beta' + \gamma' \ \ldots = f_2, \\ &\beta + \beta' + \gamma'' \ldots = f_3, \\ &\&c. \end{aligned}$$

$$N = (-)^{r+s+t \ldots + r'+s' \ldots t' \ldots} \frac{[\alpha]^r}{[r]^r} \frac{[\beta]^s}{[s]^s} \frac{[\gamma]^t}{[t]^t} \ldots \frac{[\beta']^{s'}}{[r']^{r'}} \frac{[\gamma']^{t'}}{[s']^{t'}} \ldots \frac{[\gamma'']^{t''}}{[t'']^{t''}} \ldots$$

$$\xi^{f-r} \eta^r V \text{ or } \delta_x{}^{f-r} \delta_y{}^r V = \overset{f}{V}{}^{,r} \text{ or } V^{,r},$$

the general term is

$$N \overset{f_1}{V_1}{}^{,r+s+t \ldots} \overset{f_2}{V_2}{}^{,\alpha-r+s'+t' \ldots} \overset{f_3}{V_3}{}^{,\beta-s+\beta'-s'+t' \ldots}$$

where r, s, $t, \ldots s'$, $t', \ldots t'', \ldots$ extend from 0 to α, β, $\gamma \ldots \beta'$, $\gamma', \ldots \gamma'' \ldots$ respectively. It would be easy to change this general term in a way similar to that which will be employed presently for the particular case of $\square\, V_1 V_2 V_3$.

If several of the functions become identical, and for these some of the letters f are equivalent, it is clear that the derivative $\square U$ refers to a certain number of functions V_1, $V_2 \ldots$ the same or different, of the variables x, y; x', y'; ... and besides that this derivative is homogeneous, of the degrees θ_1, $\theta_1', \ldots$ with respect to the differential coefficients of the orders f_1, $f_1', \ldots$ &c. of V_1, (consequently homogeneous of the order $\theta_1 + \theta_1' + \ldots$ with respect to these differential coefficients collectively), homogeneous and of the degrees θ_2, $\theta_2', \ldots$ with respect to the differential coefficients of the orders f_2, $f_2' \ldots$ of V_2, (consequently of the order $\theta_2 + \theta_2' \ldots$ with respect to these collectively), and so on. The degree with respect to all the functions is of course $\theta_1 + \theta_1' \ldots + \theta_2 + \theta_2' + \ldots, = p$ suppose. In general, only a single function will be considered, and it will be assumed that $\square U$ only contains the differential coefficients of the fth order. In this case, the derivative is said to be of the degree p and of the order f. The most convenient classification is by degrees, rather than by orders.

Commencing with the simplest case, that of functions of the second order (and writing V, W instead of V_1, V_2), we have

$$\square VW = \overline{12}^{\alpha} VW,$$

(where ξ_1, η_1 apply to V and ξ_2, η_2 to W). This will be constantly represented in the sequel by the notation

$$\overline{12}^{\alpha} VW = B_{\alpha}(V, W).$$

Hence, writing $\quad \delta_x{}^{\alpha} V = V^{,0}, \quad \delta_x{}^{\alpha-1}\delta_y V = V^{,1} \ldots,$

we have $\quad B_{\alpha}(V, W) = V^{,0}W^{,\alpha} - \dfrac{[\alpha]^1}{[1]^1} V^{,1}W^{,\alpha-1} + \ldots;$

and in particular, according as α is odd or even,

$$B_{\alpha}(V, V) = 0,$$

$$\tfrac{1}{2}B_{\alpha}(V, V) = V^{,0}V^{,\alpha} - \frac{\alpha}{1} V^{,1}V^{,\alpha-1} + \ldots,$$

continued to the term which contains $V^{,\frac{1}{2}\alpha}V^{,\frac{1}{2}\alpha}$, the coefficient of this last term being divided by two.

Thus, for the functions $\frac{1}{2}(ax^2 + 2bxy + cy^2)$, $\frac{1}{24}(ax^4 + 4bx^3y + 6cx^2y^2 + 4dxy^3 + ey^4)$, &c., if α be made equal to 2, 4, &c. respectively, we have the *constant* derivatives

$$\begin{aligned}
&ac - b^2,\\
&ae - 4bd + 3c^2,\\
&ag - 6bf + 15ce - 10d^2,\\
&ai - 8bh + 28cg - 56df + 35e^2,\\
&\vdots
\end{aligned}$$

which have all of them the property of remaining unaltered, *à un facteur près*, when the variables are transformed by means of $x = \lambda\dot{x} + \mu\dot{y}$, $y = \lambda'\dot{x} + \mu'\dot{y}$. Thus, for instance, if these equations give

$$ax^2 + 2bxy + cy^2 = \dot{a}\dot{x}^2 + 2\dot{b}\dot{x}\dot{y} + \dot{c}\dot{y}^2,$$

then $\quad \dot{a}\dot{c} - \dot{b}^2 = (\lambda\mu' - \lambda'\mu)^2 \cdot (ac - b^2),$

and so on. This is the general property, which we call to mind for the case of these constant derivatives.

The above functions may be transformed by means of the identical equation

$$B_\alpha(V,\ W) = \overline{12}^{\alpha-k} B_k(V,\ W),$$

to make use of which, it is only necessary to remark the general formula

$$\xi_1{}^\lambda \eta_1{}^\mu \xi_2{}^\rho \eta_2{}^\sigma B_k(V,\ W) = B_k(\xi^\lambda \eta^\mu V,\ \xi^\rho \eta^\sigma W).$$

Thus, if $k = 1$, we obtain for the above series, the new forms

$$\begin{aligned}
&ac - b^2, \\
&(ae - bd) - 3(bd - c^2), \\
&(ag - bf) - 5(bf - ce) + 10(ce - d^2), \\
&(ai - bh) - 7(bh - cg) + 21(cg - df) - 35(df - e^2), \\
&\&c.,
\end{aligned}$$

the law of which is evident. This shows also that these functions may be linearly expressed by means of the series of determinants

$$\begin{Vmatrix} a, & b \\ b, & c \end{Vmatrix} \quad \begin{Vmatrix} a, & b, & c \\ b, & c, & d \end{Vmatrix} \quad \&c.$$

We may also immediately deduce from them the derivatives B which relate to two functions. For example, for functions of the sixth order this is

$$ag' + a'g - 6(bf' + b'f) + 15(ce' + c'e) - 20dd',$$

which has an obvious connection with

$$ag - 6bf + 15ce - 10d^2;$$

and the same is the case for functions of any order.

The following theorem is easily verified; but I am unacquainted with the general theory to which it belongs.

"If U, V are any functions of the second order, and $W = \lambda U + \mu V$; then

$$B_2'[B_2(W,\ W),\ \ B_2(W,\ W)] = 0$$

(where B_2' relates to λ, μ) is the same that would be obtained by the elimination of x, y between $U = 0$, $V = 0$." (See Note[1].)

In fact this becomes

$$4(ac - b^2)(a'c' - b'^2) - (ac' + a'c - 2bb')^2 = 0,$$

which is one of the forms under which the result of the elimination of the variables from two quadratic equations may be written. This is a result for which I am indebted to Mr Boole.

[1] Not given with the present paper.

Passing to the third degree, we may consider in particular the derivatives

$$\square UVW = \overline{23}^{\alpha}\ \overline{31}^{\alpha}\ \overline{12}^{\alpha}\ UVW = C_\alpha(U,\ V,\ W):$$

writing for shortness

$$A_r = \frac{[\alpha]^r}{[r]^r},\quad \delta_x^{2\alpha-r}\delta_y^{\,r}U = U_{\cdot r},$$

we have the general term

$$C_\alpha(U,\ V,\ W) = \Sigma\left\{(-)^{r+s+t} A_r A_s A_t U_{\cdot\alpha+t-s} V_{\cdot\alpha+r-t} W_{\cdot\alpha+s-r}\right\},$$

where r, s, t extend from 0 to α. By changing the suffixes r, s the following more convenient formula

$$C_\alpha(U,\ V,\ W) = \Sigma\Sigma\left\{(-)^{\sigma+\rho} U_{\cdot\rho} V_{\cdot\sigma} W_{\cdot 3\alpha-\rho-\sigma}\Sigma\left[(-)^t A_{\rho-t} A_{\sigma+t-\alpha} A_t\right]\right\},$$

where t extends from 0 to 2α: ρ, σ, and $3\alpha-\rho-\sigma$ must be positive and not greater than 2α.

In particular, according as α is odd or even,

$$C_\alpha(U,\ U,\ U) = 0,$$

$$C_\alpha(U,\ U,\ U) = 6\Sigma\Sigma\left\{(-)^{\rho+\sigma} U_{\cdot\rho} U_{\cdot\sigma} U_{\cdot 3\alpha-\rho-\sigma}\Sigma\left[(-)^t A_{\rho-t} A_{\sigma+t-\alpha} A_t\right]\right\},$$

omitting therein those values of ρ, σ for which $\rho > \sigma$ or $\sigma > 3\alpha-\rho-\sigma$, and dividing by two the terms in which $\rho=\sigma$ or $\sigma = 3\alpha-\rho-\sigma$, and by six the term for which

$$\rho = \sigma = 3\alpha-\rho-\sigma,\ = \alpha.$$

In particular, for functions of the fourth or eighth orders we have the constant derivatives

$$ace - ad^2 - b^2e - c^3 + 2bcd\,;$$

$$aei - 4ibd - 4afh + 3ag^2 + 3ic^2 + 12beh - 8chd - 8bgf - 22ceg + 24cf^2 + 24d^2g - 36def + 15e^3;$$

the first of which is a simple determinant. Thus we have been led to the functions $ae - 4bd + 3c^2$ and $ace - ad^2 - eb^2 - c^3 + 2bcd$, which occur in my "Note sur quelques formules &c." (*Crelle*, vol. XXIX. [1845] [15]), and in the forms which M. Eisenstein has given for the solutions of equations of the first four degrees.

Let U be a function of the order 4α: the derivative C may be expressed by means of the derivatives B.

For, consider the function

$$B_{4\alpha}[U,\ B_{2\alpha}(V,\ W)]\,;$$

paying attention to the signification of B, this may be written

$$\overline{1\theta}^{4\alpha}\ \overline{23}^{2\alpha}\ UVW,$$

where the symbols ξ_θ, η_θ refer to the two systems $x_2, y_2 : x_3, y_3$. Thus it is easily seen that we may write

$$\xi_\theta = \xi_2 + \xi_3, \quad \eta_\theta = \eta_2 + \eta_3, \text{ or } \overline{1\theta} = \overline{12} + \overline{13} = \overline{12} - \overline{31},$$

whence the function becomes

$$(\overline{12} - \overline{31})^{4\alpha}\, \overline{23}^{2\alpha}\, UVW,$$

of which all the terms vanish except

$$\frac{[4\alpha]^{2\alpha}}{[2\alpha]^{2\alpha}}\, \overline{12}^{2\alpha}\, \overline{23}^{2\alpha}\, \overline{31}^{2\alpha}\, UVW.$$

Hence putting

$$K = \frac{[4\alpha]^{2\alpha}}{[2\alpha]^{2\alpha}} = \frac{2^{4\alpha}\, 1\,.\,3\, \ldots\, (4\alpha - 1)}{2\,.\,4\, \ldots\, 4\alpha},$$

we have $$B_{4\alpha}[U,\ B_{2\alpha}(V,\ W)] = KC_\alpha(U,\ V,\ W),$$

or in particular

$$B_{4\alpha}[U,\ B_{2\alpha}(U,\ U)] = KC_\alpha(U,\ U,\ U).$$

Thus for example, neglecting a numerical factor,

$$(ax^2 + 2bxy + cy^2)(cx^2 + 2dxy + ey^2) - (bx^2 + 2cxy + dy^2)^2$$
$$= (ac - b^2)\, x^4 + 2\,(ad - bc)\, x^3y + (ae + 2bd - 3c^2)\, x^2y^2 + 2\,(be - cd)\, xy^3 + (ce - d^2)\, y^4,$$

and then

$$e\,(ac - b^2) - 4d\, \tfrac{2}{4}\,(ad - bc) + 6c\, \tfrac{1}{6}\,(ae + 2bd - 3c^2) - 4b\, \tfrac{2}{4}\,(be - cd) + a\,(ce - d^2)$$
$$= 3\,(ace - ad^2 - b^2e - c^3 + 2bcd).$$

We have likewise the singular equation

$$B_{2\alpha}(V,\ W) = K\left(x^{4\alpha}\frac{d}{da_{4\alpha}} - x^{4\alpha-1}y\,\frac{d}{da_{4\alpha-1}} \ldots + y^{4\alpha}\frac{d}{da_0}\right) C_\alpha(U,\ V,\ W)$$

where $$U = \frac{1}{[4\alpha]^{4\alpha}}\left(a_0x^{4\alpha} - \frac{[4\alpha]^1}{1}\, a_1x^{4\alpha-1}y \ldots + a_{4\alpha}y^{4\alpha}\right), \text{ \&c.}$$

If however $U = V = W$, we must write

$$B_{2\alpha}(U,\ U) = \tfrac{1}{3}K\left(x^{4\alpha}\frac{d}{da_{4\alpha}} - x^{4\alpha-1}y\,\frac{d}{da_{4\alpha-1}} \ldots + y^{4\alpha}\frac{d}{da_0}\right) C_\alpha(U,\ U,\ U),$$

the reason of which is easily seen. This subject will be resumed in the sequel.

The functions C may be transformed in the same way as the functions B have been. In fact

$$C_\alpha(U,\ V,\ W) = \overline{12}^{\alpha-k}\ \overline{23}^{\alpha-k}\ \overline{31}^{\alpha-k}\ C_k(U,\ V,\ W);$$

if in particular $k = 1$, then

$$C_1(U,\ V,\ W) = \begin{vmatrix} U_{\cdot 0} & U_{\cdot 1} & U_{\cdot 2} \\ V_{\cdot 0} & V_{\cdot 1} & V_{\cdot 2} \\ W_{\cdot 0} & W_{\cdot 1} & W_{\cdot 2} \end{vmatrix},\quad U_{\cdot 0} \text{ for } \overset{2}{U}_{\cdot 0},\ \&\text{c.}$$

but in general

$$\xi_1^{\rho'} \eta_1^{\rho} \xi_2^{\sigma'} \eta_2^{\sigma} \xi_3^{\tau'} \eta_3^{\tau} C_1(U,\ V,\ W),\ \text{where } \rho+\rho' = \sigma+\sigma' = \tau+\tau' = 2\alpha-2,$$

$$= C_1(\overset{2\alpha-2}{U}_{\cdot\rho}\ \overset{2\alpha-2}{V}_{\cdot\sigma}\ \overset{2\alpha-2}{W}_{\cdot\tau}) = \begin{vmatrix} U_{\cdot\rho} & U_{\cdot\rho+1} & U_{\cdot\rho+2} \\ V_{\cdot\sigma} & V_{\cdot\sigma+1} & V_{\cdot\rho+2} \\ W_{\cdot\tau} & W_{\cdot\tau+1} & W_{\cdot\rho+2} \end{vmatrix},\quad U_{\cdot\rho} \text{ for } \overset{2\alpha}{U}_{\cdot\rho},\ \&\text{c.}$$

whence $$C_\alpha(U,\ V,\ W) = \Sigma\Sigma\left\{(-)^{\rho+\sigma} \begin{vmatrix} U_{\cdot\rho} & V_{\cdot\sigma-1} & W_{\cdot 3\alpha-\rho-\sigma-2} \\ U_{\cdot\rho+1} & V_{\cdot\sigma} & W_{\cdot 3\alpha-\rho-\sigma-1} \\ U_{\cdot\rho+2} & V_{\cdot\sigma+1} & W_{\cdot 3\alpha-\rho-\sigma} \end{vmatrix}\ \Sigma\left[(-)^t A_t' A'_{\rho-t} A'_{\sigma-\alpha+t}\right]\right\},$$

where $A_t' = \dfrac{[\alpha-1]^t}{[t]^t}$; t extends from 0 to $\overline{\alpha-1}$; ρ, $\sigma-1$, and $3\alpha-\rho-\sigma-2$ may have each of them any positive values not greater than $2\alpha-2$.

In particular

$$C_\alpha(U,\ U,\ U) = 6\Sigma\Sigma\left\{(-)^{\rho+\sigma} \begin{vmatrix} U_{\cdot\rho} & U_{\cdot\sigma-1} & U_{\cdot 3\alpha-\rho-\sigma-2} \\ U_{\cdot\rho+1} & U_{\cdot\sigma} & U_{\cdot 3\alpha-\rho-\sigma-1} \\ U_{\cdot\rho+2} & U_{\cdot\sigma+1} & U_{\cdot 3\alpha-\rho-\sigma} \end{vmatrix}\ \Sigma\left[(-)^t A'_t A'_{\rho-t} A'_{\sigma-\alpha-t}\right]\right\}.$$

where ρ, σ need only have such values that $\rho < \sigma-1$, $\sigma-1 < 3\alpha-\rho-\sigma-2$.

In particular the derivative $aei - \ldots + 15e^3$ may be transformed into

$$\begin{vmatrix} a, & d, & g \\ b, & e, & h \\ c, & f, & i \end{vmatrix} - 3 \begin{vmatrix} a, & e, & f \\ b, & f, & g \\ c, & g, & h \end{vmatrix} - 3 \begin{vmatrix} b, & c, & g \\ c, & d, & h \\ d, & e, & i \end{vmatrix} + 6 \begin{vmatrix} b, & d, & f \\ c, & e, & g \\ d, & f, & h \end{vmatrix} - 15 \begin{vmatrix} c, & d, & e \\ d, & e, & f \\ e, & f, & g \end{vmatrix}$$

in which form it is obviously a linear function of the determinants

$$\left\| \begin{matrix} a, & b, & c, & d, & e, & f, & g \\ b, & c, & d, & e, & f, & g, & h \\ c, & d, & e, & f, & g, & h, & i \end{matrix} \right\|,$$

which is true generally.

Omitting for the present the theory of derivatives of the form

$$\square UVW = \overline{23}^{\alpha}\ \overline{31}^{\beta}\ \overline{12}^{\gamma}\ UVW,$$

we pass on to the derivatives of the fourth degree, considering those forms in which all the differential coefficients are of the same order. We may write

$$\square UVWX = (\overline{12}\,.\,\overline{34})^{\alpha}\,(\overline{13}\,.\,\overline{42})^{\beta}\,(\overline{14}\,.\,\overline{23})^{\gamma}\ UVWX = D_{\alpha,\beta,\gamma}(U,\ V,\ W,\ X) = D_{\alpha,\beta,\gamma};$$

or if for shortness

$$\overline{12}\,.\,\overline{34} = \mathfrak{A},\quad \overline{13}\,.\,\overline{42} = \mathfrak{B},\quad \overline{14}\,.\,\overline{23} = \mathfrak{C},$$

we have

$$D_{\alpha,\beta,\gamma} = \mathfrak{A}^{\alpha}\mathfrak{B}^{\beta}\mathfrak{C}^{\gamma}\,.\,UVWX.$$

Suppose $U = V = W = X$, and consider the derivatives which correspond to the same value f of $\alpha+\beta+\gamma$. The question is to determine how many of these are independent, and to express the remaining ones in terms of these. Since the functions become equal after the differentiations, we are at liberty before the differentiations to interchange the symbolic numbers 1, 2, 3, 4 in any manner whatever. We have thus

$$D_{\alpha,\beta,\gamma} = D_{\beta,\gamma,\alpha} = D_{\gamma,\alpha,\beta} = (-)^f D_{\alpha,\gamma,\beta} = (-)^f D_{\gamma,\beta,\alpha} = (-)^f D_{\beta,\alpha,\gamma};$$

but the identical equation

$$\mathfrak{A} + \mathfrak{B} + \mathfrak{C} = 0,$$

multiplied by $\mathfrak{A}^a\mathfrak{B}^b\mathfrak{C}^c$ and applied to the product $UVWX$, gives

$$D_{a+1,b,c} + D_{a,b+1,c} + D_{a,b,c+1} = 0;$$

whence if $a+b+c = f-1$, we have a set of equations between the derivatives $D_{\alpha,\beta,\gamma}$ for which $\alpha+\beta+\gamma = f$. Reducing these by the conditions first found, suppose Θf is the number of divisions of an integer f into three parts, zero admissible, but permutations of the same three parts rejected. The number of derivatives is Θf, and the number of relations between them is $\Theta(f-1)$. Hence $\Theta f - \Theta(f-1)$ of these derivatives are independent: only when f is even, one of these is $D_{f,0,0}$, i.e. $\overline{12}^f\,\overline{34}^f\,.\,UVWX$, i.e. $\overline{12}^f UV\,.\,\overline{34}^f WX$, or $B_f(U,\ V)\,B_f(X,\ W)$, i.e. $[B_f(U,\ U)]^2$; rejecting this, the number of independent derivatives, when f is even, is $\Theta f - \Theta(f-1) - 1$. Let $E\left(\dfrac{a}{b}\right)$ be the greatest integer contained in the fraction $\dfrac{a}{b}$; the number required may be shown to be

$$E\frac{f}{6}\ \text{or}\ E\frac{f+3}{6},$$

according as f is even or odd. Giving to f the six forms

$$6g,\quad 6g+1,\quad 6g+2,\quad 6g+3,\quad 6g+4,\quad 6g+5,$$

the corresponding numbers of the independent derivatives are

$$g,\quad g,\quad g,\quad g+1,\quad g,\quad g+1;$$

thus there is a single derivative for the orders 3, 5, 6, 7, 8, 10,... two for the orders 9, 11, 12, 13, 14, 16, ... &c.

When f is even, the terms $D_{f-3,3,0}$, $D_{f-6,6,0}\ldots$, and when f is odd, the terms $D_{f-1,1,0}$, $D_{f-4,4,0}$, $D_{f-7,7,0}$, &c. may be taken for independent derivatives: by stopping immediately before that in which the second suffix exceeds the first, the right number of terms is always obtained. Thus, when $f=9$ the independent derivatives are D_{810}, D_{540}, and we have the system of equations

$$\begin{aligned}
&D_{900}+D_{810}+D_{801}=0, &\qquad &D_{621}+D_{531}+D_{522}=0,\\
&D_{810}+D_{720}+D_{711}=0, & &D_{540}+D_{450}+D_{441}=0,\\
&D_{720}+D_{630}+D_{621}=0, & &D_{531}+D_{441}+D_{432}=0,\\
&D_{711}+D_{621}+D_{612}=0, & &D_{522}+D_{432}+D_{423}=0,\\
&D_{630}+D_{531}+D_{540}=0, & &D_{432}+D_{342}+D_{333}=0,
\end{aligned}$$

which are to be reduced by

$$D_{900}=-D_{900}=0,\qquad D_{801}=-D_{810},\ \&\text{c}.$$

It is easy to form the table

$$\begin{aligned}
&D_{200}=\ B_2^2, & &D_{500}=0, & &D_{700}=0,\\
&D_{110}=-\tfrac{1}{2}B_2^2, & &D_{410}, & &D_{610},\\
& & &D_{320}=-D_{410}, & &D_{520}=-D_{610},\\
&D_{300}=0, & &D_{311}=0, & &D_{511}=0,\\
&D_{210}, & &D_{221}=0, & &D_{430}=\ D_{610},\\
&D_{111}=0, & & & &D_{421}=0,\\
& & &D_{600}=\ B_6^2, & &D_{331}=0,\\
&D_{400}=\ B_4^2, & &D_{510}=-\tfrac{1}{2}B_6^2, & &D_{222}=0,\\
&D_{310}=-\tfrac{1}{2}B_4^2, & &D_{420}=-\tfrac{2}{3}D_{330}+\tfrac{1}{6}B_6^2, & &\\
&D_{220}=\ \tfrac{1}{2}B_4^2, & &D_{411}=\ \tfrac{2}{3}D_{330}+\tfrac{1}{3}B_6^2, & &\\
&D_{211}=0, & &D_{330}, & &\\
& & &D_{321}=-\tfrac{1}{3}D_{330}-\tfrac{1}{6}B_6^2, & &\\
& & &D_{222}=\ \tfrac{2}{3}D_{330}+\tfrac{1}{3}B_6^2, & &
\end{aligned}$$

$$\begin{array}{ll}
D_{800}= \qquad\qquad B_8^2, & D_{900}=0, \\
D_{710}= \qquad -\tfrac{1}{2}B_8^2, & D_{810}, \\
D_{620}= -\tfrac{2}{3}D_{530}+\tfrac{1}{6}B_8^2, & D_{720}=-\ D_{810}, \\
D_{611}= \tfrac{2}{3}D_{530}+\tfrac{1}{3}B_8^2, & D_{711}=0, \\
D_{530}, & D_{630}=\tfrac{1}{2}D_{810}-\tfrac{1}{2}D_{540}, \\
D_{521}= -\tfrac{1}{3}D_{530}-\tfrac{1}{12}B_8^2, & D_{621}=\tfrac{1}{2}D_{810}+\tfrac{1}{2}D_{540}, \\
D_{440}= -\tfrac{16}{15}D_{530}-\tfrac{1}{30}B_8^2, & D_{540}, \\
D_{431}= \tfrac{1}{15}D_{530}-\tfrac{1}{30}B_8^2, & D_{531}=-\tfrac{1}{2}D_{810}-\tfrac{1}{2}D_{540}, \\
D_{422}= \tfrac{4}{15}D_{530}+\tfrac{2}{15}B_8^2, & D_{522}=0, \\
D_{332}= -\tfrac{2}{15}D_{530}-\tfrac{1}{15}B_8^2, & D_{441}=0, \\
 & D_{432}=\tfrac{1}{2}D_{810}+\tfrac{1}{2}D_{540}, \\
 & D_{333}=0.
\end{array}$$

Whatever be the value, all the tables except the three first commence thus, according as f is even or odd,

$$\begin{array}{lll}
D_{f,0,0} = B_f^2, & \text{or} & D_{f,0,0}=0, \\
D_{f-1,1,0}=-\tfrac{1}{2}B_f^2, & & D_{f-1,1,0}, \\
D_{f-2,2,0}=-\tfrac{2}{3}D_{f-3,3,0}+\tfrac{1}{6}B_f^2, & & D_{f-2,2,0}=-D_{f-1,1,0}, \\
D_{f-2,1,1}=\tfrac{2}{3}D_{f-3,3,0}+\tfrac{1}{3}B_f^2, & & D_{f-2,1,1}=0, \\
D_{f-3,3,0} & & \vdots \\
\vdots & &
\end{array}$$

but beyond this I am not acquainted with the law.

To give some formulæ for the transformation of these derivatives; we have, for example,

$$D_{f-1,1,0}=(\overline{12}\,.\,\overline{34})^{f-1}\,\overline{13}\,.\,\overline{42}\;UUUU=\overline{13}\,.\,\overline{42}\,B_{f-1}(U,\ U)\,B_{f-1}(U,\ U).$$

But $$\overline{13}\,.\,\overline{42}=\xi_1\eta_2\eta_3\xi_4-\xi_1\xi_2\eta_3\eta_4-\eta_1\eta_2\xi_3\xi_4+\eta_1\xi_2\xi_3\eta_4,$$

and $$\begin{aligned}\xi_1\eta_2\eta_3\xi_4\,B_{f-1}(U,\ U)\,B_{f-1}(U,\ U)&=B_{f-1}(\xi U,\ \eta U)\,B_{f-1}(\eta U,\ \xi U)\\ &=B_{f-1}(U^{\cdot 0}\,U^{\cdot 1})\,B_{f-1}(U^{\cdot 1}\,U^{\cdot 0}),\ \&\text{c.}\end{aligned}$$

(where $U^{\cdot 0}$, $U^{\cdot 1}$ stand for $\overset{1}{U}{}^{\cdot 0}\,\overset{1}{U}{}^{\cdot 1}$, &c.); or

$$D_{f-1,1,0}=-2\,\{B_{f-1}(U^{\cdot 0}\,U^{\cdot 0})\,B_{f-1}(U^{\cdot 1}\,U^{\cdot 1})-B_{f-1}(U^{\cdot 0}\,U^{1})\,B_{f-1}(U^{\cdot 1}\,U^{\cdot 0})\},$$

which reduces itself to

$$D_{f-1,1,0}=-2\,\{B_{f-1}(U^{\cdot 0}\,U^{\cdot 1})\}^2,$$
$$D_{f-1,1,0}=-2\,\{B_{f-1}(U^{\cdot 0}\,U^{\cdot 0})\,B_{f-1}(U^{\cdot 1}\,U^{\cdot 1})-[B_{f-1}(U^{\cdot 0}U^{\cdot 1})]^2\},$$

according as f is even or odd.

For example, for the orders 3, 5, 7, 9, we have

$$D_{210} = -2\,\{4\,(ac - b^2)\,(bd - c^2) - (ad - bc)^2\},$$

$$D_{410} = -2\,\{4\,(ae - 4bd + 3c^2)\,(bf - 4ce + 3d^2) - (af - 3be + 2cd)^2\},$$

$$D_{610} = -2\,\{4\,(ag - 6bf + 15ce - 10d^2)\,(bh - 6cg + 15df - 10e^2) - (ah - 5bg + 9cf - 5de)^2\},$$

$$D_{810} = -2\,\{4\,(ai - 8bh + 28cg - 56df + 35e^2)\,(bj - 8ci + 28dh - 56eg + 35f^2) \\ - (aj - 7bi + 20ch - 28dg + 14ef)^2\}.$$

The derivatives D will be presently calculated in a completely expanded form up to the ninth order. We have, therefore, still to find the derivatives of the sixth and eighth orders, and a second derivative of the ninth order. For the sixth order, the simplest method is to make use of D_{222}, which is easily seen to be equal to

$$24 \begin{vmatrix} a, & b, & c, & d \\ b, & c, & d, & e \\ c, & d, & e, & f \\ d, & e, & f, & g \end{vmatrix}.$$

For the two others we have the general formulæ

$$D_{f-2,\,2,\,0} = 2\,\{B_{f-2}(U^{\cdot 0}\,U^{\cdot 0})\,B_{f-2}(U^{\cdot 2}\,U^{\cdot 2}) - 4B_{f-2}(U^{\cdot 0}\,U^{\cdot 1})\,B_{f-2}(U^{\cdot 2}\,U^{\cdot 1}) \\ + B_{f-2}(U^{\cdot 0}\,U^{\cdot 2})\,B_{f-2}(U^{\ 2}\,U^{\cdot 0}) + 2\,[B_{f-2}(U^{\cdot 1}\,U^{\cdot 1})]^2\},$$

where $U^{\cdot 0}$, $U^{\cdot 1}$, $U^{\cdot 2}$ have been written for $\overset{2}{U}{}^{\cdot 0}$, $\overset{2}{U}{}^{\cdot 1}$, $\overset{2}{U}{}^{\cdot 2}$; a formula which is demonstrated in precisely the same way as that for $D_{f-1,\,1,\,0}$.

$$D_{f-3,\,3,\,0} = -2\,\{\quad B_{f-3}(U^{\cdot 0}\,U^{\cdot 3})\,B_{f-3}(U^{\cdot 3}\,U^{\cdot 0}) - 6B_{f-3}(U^{\cdot 0}\,U^{\cdot 1})\,B_{f-3}(U^{\cdot 3}\,U^{\cdot 2}) \\ + 6B_{f-3}(U^{\cdot 0}\,U^{\cdot 2})\,B_{f-3}(U^{\cdot 3}\,U^{\cdot 1}) + 9B_{f-3}(U^{\cdot 1}\,U^{\cdot 1})\,B_{f-3}(U^{\cdot 2}\,U^{\cdot 2}) \\ - 9B_{f-3}(U^{\cdot 1}\,U^{\cdot 2})\,B_{f-3}(U^{\cdot 2}\,U^{\cdot 1}) - B_{f-3}(U^{\cdot 0}\,U^{\cdot 3})\,B_{f-3}(U^{\cdot 3}\,U^{\cdot 0})\},$$

(in which $U^{\cdot 0}$, &c. stand for $\overset{0}{U}{}^{\cdot 0}$, &c.). In particular

$$D_{620} = 2\,\{4\,(ag - 6bf + 15ce - 10d^2)\,(ci - 6dh + 15eg - 10f^2) - 4\,(ah - 5bg + 9cf - 5de)\,\times \\ (bi + 5ch + 9dg - 5ef) + (ai - 6bh + 16cg - 26df + 15e^2)^2 + 8\,(bh - 6cg + 15df - 10e^2)^2\},$$

$$D_{630} = -2\,\{4\,(ag - 6bf + 15ce - 10d^2)\,(dj - 6ei + 15fh - 10g^2) - 6\,(ah - 5bg + 9cf - 5de)\,\times \\ (cj - 5di + 9eh - 5fg) + 6\,(ai - 6bh + 16cg - 26df + 15e^2)\,(bj - 6ci + 16dh - 26eg + 15f^2) \\ + 36\,(bh - 6cg + 15df - 10e^2)\,(ci - 6dh + 15eg - 10f^2) - 9\,(bi - 5ch + 9dg - 5ef)^2 \\ - (aj - 6bi + 15ch - 19dg + 9ef)^2\}.$$

Hence we have all the elements necessary for the calculation of the following table of the independent constant derivatives of the fourth degree, up to the ninth order. [I have arranged the terms alphabetically and in tabular form as in my Memoirs on Quantics, and have corrected some inaccuracies];

$D_{210} = 2 \times$

a^2d^2	$+1$
$abcd$	-6
ac^3	$+4$
b^3d	$+4$
b^2c^2	-3

± 9

$D_{410} = 2 \times$

a^2f^2	$+1$
$abef$	-10
$acdf$	$+4$
ace^2	$+16$
ad^2e	-12
b^2df	$+16$
b^2e^2	$+9$
bc^2f	-12
$bcde$	-76
bd^3	$+48$
c^3e	$+48$
c^2d^2	-32

± 142

$D_{222} = 24 \times$

$aceg$	$+1$
acf^2	-1
ad^2g	-1
$adef$	$+2$
ae^3	-1
b^2eg	-1
b^2f^2	$+1$
$bcdg$	$+2$
$bcef$	-2
bd^2f	-2
bde^2	$+2$
c^3g	-1
c^2df	$+2$
c^2e^2	$+1$
cd^2e	-3
d^4	$+1$

± 12

$D_{610} = 2 \times$

a^2h^2	$+1$
$abgh$	-14
$acfh$	$+18$
acg^2	$+24$
$adeh$	-10
$adfg$	-60
ae^2g	$+40$
b^2fh	$+24$
b^2g^2	$+25$
$bceh$	-60
$bcfg$	-234
bd^2h	$+40$
$bdeg$	$+50$
bdf^2	$+360$
be^2f	-240
c^2eg	$+360$
c^2f^2	$+81$
cd^2g	-240
$cdef$	-990
ce^3	$+600$
d^3f	$+600$
d^2e^2	$+375$

± 2223

$D_{620} = 2 \times$

a^2i^2	$+1$
$abhi$	-16
$acgi$	$+36$
ach^2	$+20$
$adfi$	-52
$adgh$	-60
ae^2i	$+30$
$aefh$	$+20$
aeg^2	$+60$
af^2g	-40
b^2gi	$+20$
b^2h^2	$+44$
$bcfi$	-60
$bcgh$	-388
$bdei$	$+20$
$bdfh$	$+696$
bdg^2	$+180$
be^2h	-340
$befg$	-460
bf^3	$+240$
c^2ei	$+60$
c^2fh	$+180$
c^2g^2	$+544$
cd^2i	-40
$cdeh$	-460
$cdfg$	-2596
ce^2g	$+2340$
cef^2	-420
d^3h	$+240$
d^2eg	-420
d^2f^2	$+12876$
de^2f	-3280
e^4	$+1025$

± 8632

	$D_{810} = 2 \times$	$D_{621} = 4 \times$
a^2j^2	$+1$	......
$abij$	-18	......
$achj$	$+40$	$+2$
aci^2	$+32$	-2
$adgj$	-56	-7
$adhi$	-112	$+7$
$aefj$	$+28$	$+5$
$aegi$	$+224$	$+22$
aeh^2	......	-27
af^2i	-140	-25
$afgh$	......	$+45$
ag^3	......	-20
b^2hj	$+32$	-2
b^2i^2	$+49$	$+2$
$bcgj$	-112	$+7$
$bchi$	-536	-7
$bdfj$	$+224$	$+22$
$bdgi$	$+392$	-74
bdh^2	$+896$	$+52$
be^2j	-140	-25
$befi$	-196	$+73$
$begh$	-1792	-23
bf^2h	$+1120$	-70
bfg^2	......	$+45$
c^2fj	......	-27
c^2gi	$+896$	$+52$
c^2h^2	$+400$	-25
$cdej$	......	$+45$
$cdfi$	-1792	-23
$cdgh$	-4256	-22
ce^2i	$+1120$	-70
$cefh$	$+560$	$+127$
ceg^2	$+6272$	-32
cf^2g	-3920	-25
d^3j	......	-20
d^2fh	$+6272$	-32
d^2g^2	$+784$	$+47$
d^2ei	......	$+45$
de^2h	-3920	-25
$defg$	-13328	-85
df^3	$+7840$	$+50$
e^3g	$+7840$	$+50$
e^2f^2	-4704	-30
	± 35022	± 698

$$D_{330} = \tfrac{3}{2} D_{222} + \tfrac{1}{2} B_6{}^2,$$
$$D_{530} = -\tfrac{3}{2} D_{620} + \tfrac{1}{4} B_8{}^2,$$
$$D_{540} = 2 D_{621} + D_{810},$$

equations which give the functions D_{330}, D_{530}, D_{540}, by means of which the others have been expressed.

We may now proceed to demonstrate an important property of the derivatives of the fourth degree, analogous to the one which exists for the third degree. Let U, V, W, X be functions of any order f: then, investigating the value of the expression

$$B_{2f-2\alpha}[B_\alpha(U, V), \quad B_\alpha(W, X)],$$

this reduces itself in the first place to

$$\overline{\theta\phi}^{2f-2\alpha}\ \overline{12}^\alpha\ \overline{34}^\alpha\ UVWX,$$

where ξ_θ, η_θ refer to U and V, and ξ_ϕ, η_ϕ to W and X: this comes to writing $\xi_\theta = \xi_1 + \xi_2$, $\eta_\theta = \eta_1 + \eta_2$, and $\xi_\phi = \xi_3 + \xi_4$, $\eta_\phi = \eta_3 + \eta_4$; whence

$$\overline{\theta\phi} = \overline{13} + \overline{14} + \overline{23} + \overline{24},$$

or the function in question is

$$(13 + 14 + 23 + 24)^{2f-2\alpha}\ \overline{12}^\alpha\ \overline{34}^\alpha\ UVWX.$$

But all the terms of this where the sum of the indices of ξ_1, η_1 or ξ_2, η_2 or ξ_3, η_3 or ξ_4, η_4, exceed f, vanish: whence it is only necessary to consider those of the form

$$K_r (\overline{13}\,.\,\overline{42})^r\, (\overline{14}\,.\,\overline{23})^{f-\alpha-r}\, (\overline{12}\,.\,\overline{34})^\alpha\, UVWX,$$

where K_r denotes the numerical coefficient

$$\frac{(-)^r\,[2f-2\alpha]^{2f-2\alpha}}{[r]^r\,[r]^r\,[f-\alpha-r]^{f-\alpha-r}\,[f-\alpha-r]^{f-\alpha-r}},$$

or $$B_{2f-2\alpha}[B_\alpha(U, V), \quad B_\alpha(W, X)] = \Sigma\,\{K_r D_{\alpha,\,r,\,f-\alpha-r}(U, V, W, X)\}.$$

In particular, if $U = V = W = X$, writing also B_α for $B_\alpha(U, U)$,

$$B_{2f-2\alpha}(B_\alpha, B_\alpha) = \Sigma\,(K_r\, D_{\alpha,\,r,\,f-\alpha-r}).$$

If α is odd, this becomes

$$0 = \Sigma\,(K_r\, D_{\alpha,\,r,\,f-\alpha-r}),$$

an equation which must be satisfied identically by the relations that exist between the quantities D: if, on the contrary, α is even, we see that there are as many independent functions of the form

$$B_{2f-2\alpha}(B_\alpha, B_\alpha)$$

as there are of the form D; and that these two systems may be linearly expressed, either by means of the other. Thus, for the orders 3, 5, 7, the derivatives D are respectively equal, neglecting a numerical factor, to

$$B_6(U^2, U^2), \quad B_{10}(U^2, U^2), \quad B_{14}(U^2, U^2);$$

for the sixth order they may be linearly expressed by means of

$$B_{12}(U^2, U^2), \quad B_6^2,$$

and so on. All that remains to complete the theory of the fourth degree is to find the general solution of this system of equations, as also of the system connecting the derivatives D.

Passing on to a more general property; let $U_1, U_2, \dots U_p$ be functions of the orders $f_1, f_2 \dots f_p$; and suppose

$$\Theta(U_2 \dots U_p), \quad = \square U_2 \dots U_p,$$

a function of the degree f_1 in the variables: suppose that $\Theta(U_2 \dots U_p)$ contains the differential coefficients of the order r_2 for U_2, r_3 for U_3, &c., so that $f_1 = (f_2 - r_2) + \dots (f_p - r_p)$. Consider the expression

$$B_{f_1}\{U_1, \Theta(U_2 \dots U_p)\},$$

which reduces itself in the first place to

$$(\overline{12} + \overline{13} \dots + \overline{1p})^{f_1} \square U_1 U_2 \dots U_p,$$

then to

$$K(\overline{12}^{f_2 - r_2} \overline{13}^{f_3 - r_3} \dots \overline{1p}^{f_p - r_p} \square U_1 U_2 \dots U_p;$$

where for shortness

$$K = \frac{[f_1]^{f_1}}{[f_2 - r_2]^{f_2 - r_2} \dots [f_p - r_p]^{f_p - r_p}}.$$

For if one of the indices were smaller another would be greater, for instance that of $\overline{12}$: and the symbols ξ_2, η_2 in $\overline{12}^{f_2 - r_2 - \lambda} \square$ would rise to an order higher than f_2, or the term would vanish. Hence, writing

$$\square' = \overline{12}^{f_2 - r_2} \overline{13}^{f_3 - r_3} \dots \overline{1p}^{f_p - r_p},$$

and

$$\Theta'(U_1, U_2 \dots, U_p) = \square' U_1 U_2 \dots U_p,$$

we have

$$B_{f_1}\{U_1, \Theta(U_2, \dots U_p)\} = K\Theta'(U_1, U_2 \dots U_p);$$

i.e. the first side is a constant derivative of $U_1, U_2 \dots U_p$.

Suppose

$$U_1 = \frac{1}{[f_1]^{f_1}}(a_0 x^{f_1} + \dots),$$

$$\Theta(U_2, \dots U_p) = \frac{1}{[f_1]^{f_1}}(A_0 x^{f_1} + \dots),$$

then

$$K\Theta'(U_1 \dots U_p) = a_0 A_{f_1} - \frac{f_1}{1} a_1 A_{f_1 - 1} + \dots;$$

i.e.

$$A_{f_1} = K \frac{d}{da_0} \Theta'(U_1 \dots U_p), \quad \frac{f_1}{1} A_{f_1 - 1} = K \frac{d}{da_1} \Theta'(U_1, U_2 \dots U_p) \dots,$$

or finally,

$$\Theta(U_2, \dots U_p) = \frac{K}{[f_1]^{f_1}} \left(x^{f_1} \frac{d}{da_{f_1}} - x^{f_1 - 1} y \frac{d}{da_{f_1 - 1}} + \dots \right) \Theta'(U_1, \dots U_p),$$

an equation which holds good (changing, however, the numerical factor,) when several of the functions $U_1 \dots U_p$ become identical. Hence the theorem: if U be a function given by

$$U = \frac{1}{[f]^f}(a_0 x^f + a_1 x^{f-1} y + \dots),$$

and Θ be any constant derivative whatever of U, then

$$\left(x^f \frac{d}{da_f} - x^{f-1}y \frac{d}{da_{f-1}} + \dots\right)\Theta$$

is a derivative of U, and its value, neglecting a numerical factor, may be found by omitting in the symbol $\square$, which corresponds to the derivative Θ, the factors which contain any one, no matter which, of the symbolic numbers.

If, for example,

$$-\tfrac{1}{2} D_{210} = \Theta = 6abcd - 4ac^3 - 4bd^3 + 3b^2c^2 - a^2d^2,$$

or

$$\square = \overline{12}^3 . \overline{34}^2 . \overline{13} . \overline{42};$$

then

$$\left(x^3 \frac{d}{dd} - x^2y \frac{d}{dc} + xy^2 \frac{d}{db} - y^3 \frac{d}{da}\right)\Theta$$

reduces itself, omitting a numerical factor, to

$$\overline{12}^2\, \overline{13}\, UUU = -\tfrac{1}{2}B_1 \{U,\ B_1(U,\ U)\}.$$

This may be compared with some formulæ of M. Eisenstein's (*Crelle*, vol. XXVII. [1844, pp. 105, 106]; adopting his notation, we have

$$\Phi = ax^3 + 3bx^2y + 3cxy^2 + dy^3,$$

$$F = \tfrac{1}{36} B_2(\Phi,\ \Phi) = (ac - b^2)\,x^2 + (ad - bc)\,xy + (bd - c^2)\,y^2,$$

$$\Phi_1 = -\tfrac{1}{2}\left(x^3 \frac{d}{da} - x^2y \frac{d}{dc} + xy^2 \frac{d}{db} - y^3 \frac{d}{da}\right) D,$$

where D is the same as Θ. Hence to the system of formulæ which he has given, we may add the two following:

$$\Phi_1 = \tfrac{1}{3}\left(\frac{d\Phi}{dx}\frac{dF}{dy} - \frac{d\Phi}{dy}\frac{dF}{dx}\right),$$

$$\Phi_1 = -\tfrac{1}{216}\left\{\frac{d^3\Phi}{dx^3}\frac{d^2\Phi}{dx^2}\frac{d\Phi}{dy} - \frac{d^3\phi}{dx^2\,dy}\left(2\frac{d^2\Phi}{dx\,dy}\frac{d\Phi}{dy} + \frac{d^2\Phi}{dy^2}\frac{d\Phi}{dx}\right)\right.$$
$$\left. + \frac{d^3\Phi}{dx\,dy^2}\left(2\frac{d^2\Phi}{dx\,dy}\frac{d\Phi}{dx} + \frac{d^2\Phi}{dx^2}\frac{d\Phi}{dy}\right) - \frac{d^3\Phi}{dy^3}\frac{d^2\Phi}{dx^2}\frac{d\Phi}{dx}\right\},$$

the first of which explains most simply the origin of the function Φ_1.

It will be sufficient to indicate the reductions which may be applied to derivatives of the form

$$C_{\alpha,\beta,\gamma}(U,\ V,\ W) = \overline{23}^\alpha . \overline{31}^\beta . \overline{12}^\gamma\, UVW,$$

where U, V, W are *homogeneous* functions. In fact, if

$$\xi x + \eta y = \Xi,$$

the above becomes, neglecting a numerical factor,

$$(\Xi_1 . \overline{23})^\alpha . (\Xi_2 . \overline{31})^\beta . (\Xi_3 . \overline{12})^\gamma\, UVW,$$

where the symbols ξ, η are supposed not to affect the x, y which enter into the expressions Ξ. But we have identically

$$\Xi_1 \overline{23} + \Xi_2 \overline{31} + \Xi_3 \overline{12} = 0,$$

an equation which gives rise to reductions similar to those which have been found for the derivatives $D_{\alpha, \beta, \gamma}$, but which require to be performed with care, in order to avoid inacccuracies with respect to the numerical factors. It may, however, be at once inferred, that the number of independent derivatives $C_{\alpha, \beta, \gamma}$ is the same with that of the independent derivatives $D_{\alpha, \beta, \gamma}$ for the same value of $\alpha + \beta + \gamma$.

From similar reasonings to those by which $B\{U, B(U, U)\}$ has been found, the following general theorem may be inferred.

"The derivative of any number of the derivatives of one or more functions, or even of any number of functions of these derivatives, is itself a derivative of the original functions."

For the complete reduction of these double derivatives, it would be sufficient, theoretically, to be able to reduce to the smallest number possible, the derivatives of any given degree whatever. This has been done for the derivatives of the third degree $C_{\alpha, \beta, \gamma}$, and for those of the fourth degree, in which all the differentiations rise to the same order $(D_{\alpha, \beta, \gamma})$: it seems, however, very difficult to extend these methods even to the next simplest cases,—extensive researches in the theory of the division of numbers would probably be necessary. Important results might be obtained by connecting the theory of hyperdeterminants with that of elimination, but I have not yet arrived at anything satisfactory upon this subject. I shall conclude with the remark, that it is very easy to find a series, or rather a series of series of hyperdeterminants of all degrees, viz. the determinants

$$\begin{vmatrix} a, & b \\ b, & c \end{vmatrix}, \quad \begin{vmatrix} a, & b, & c \\ b, & c, & d \\ c, & d, & e \end{vmatrix}, \quad \begin{vmatrix} a, & b, & c, & d \\ b, & c, & d, & e \\ c, & d, & e, & f \\ d, & e, & f, & g \end{vmatrix} \text{ \&c.}$$

$$\begin{vmatrix} . & a, & 2b, & c \\ . & b, & 2c, & d \\ a, & 2b, & c, & . \\ b, & 2c, & d, & . \end{vmatrix}, \quad \begin{vmatrix} . & . & a, & 4b, & 6c, & 4d, & e \\ . & . & b, & 4c, & 6d, & 4e, & f \\ . & . & c, & 4d, & 6e, & 4f, & g \\ a, & 4b, & 6c, & 4d, & e, & . & . \\ b, & 4c, & 6d, & 4e, & f, & . & . \\ c, & 4d, & 6e, & 4f, & g, & . & . \end{vmatrix} \text{ \&c.} \quad \begin{vmatrix} . & . & . & a, & 3b, & 3c, & d \\ . & . & . & b, & 3c, & 3d, & e \\ . & . & a, & 3b, & 3c, & d, & . \\ . & . & b, & 3c, & 3d, & e, & . \\ a, & 3b, & 3c, & d, & . & . & . \\ b, & 3c, & 3d, & e, & . & . & . \end{vmatrix} \text{ \&c.}$$

[I have inserted in these determinants the numerical coefficients which were by mistake omitted.]

However, these functions are not all independent; e.g. the last may be linearly expressed by the square of the second and the cube of $(ae - 4bd + 3c^2)$; nor do I know the symbolical form of these hyperdeterminant determinants.

15.

NOTE SUR DEUX FORMULES DONNEES PAR M. M. EISENSTEIN ET HESSE.

[From the *Journal für die reine und angewandte Mathematik*, (Crelle) vol. XXIX. (1845), pp. 54—57.]

MR. EISENSTEIN a donné [*Journal*, t. XXVII. (1844), pp. 105—106] cette formule assez remarquable :

$$(a^2d^2 - 3b^2c^2 + 4ac^3 + 4bd^3 - 6abcd)^3 = A^2D^2 - 3B^2C^2 + 4AC^3 + 4DB^3 - 6ABCD,$$

où A, B, C, D sont des fonctions données de a, b, c, d. Cela peut se généraliser comme suit.

Soit

$$u = a^2h^2 + b^2g^2 + c^2f^2 + d^2e^2 - 2ahbg - 2ahcf - 2ahde - 2bgcf - 2bgde - 2cfde + 4adfg + 4bceh,$$

et de plus

$$A = \tfrac{1}{2}\frac{du}{da}, \qquad B = \tfrac{1}{2}\frac{du}{db}, \qquad C = \tfrac{1}{2}\frac{du}{dc}, \ldots \qquad H = \tfrac{1}{2}\frac{du}{dh}.$$

Représentons par U la même fonction de A, B, C, ... H, que la fonction u l'est des quantités a, b, c, ... h, l'on a l'équation

$$U = u^3.$$

C'est un cas particulier d'une propriété générale de la fonction u, que l'on peut énoncer ainsi. Imaginons la fonction

$$ax_1y_1z_1 + bx_1y_1z_2 + cx_1y_2z_1 + dx_1y_2z_2 + ex_2y_1z_1 + fx_2y_1z_2 + gx_2y_2z_1 + hx_2y_2z_2,$$

transformée en

$$a'x'_1y'_1z'_1 + b'x'_1y'_1z'_2 + \ldots \qquad + h'x'_2y'_2z'_2,$$

par les substitutions

$$x_1 = \lambda_1 x_1' + \lambda_2 x_2', \quad y_1 = \mu_1 y_1' + \mu_2 y_2', \quad z_1 = \nu_1 z_1' + \nu_2 z_2',$$
$$x_2 = \lambda_1' x_1' + \lambda_2' x_2', \quad y_2 = \mu_1' y_1' + \mu_2' y_2', \quad z_2 = \nu_1' z_1' + \nu_2' z_2';$$

et soit u', la fonction analogue à u, des nouveaux coefficients a', b', c', ... h': alors

$$u' = (\lambda_1 \lambda_2' - \lambda_2 \lambda_1')^2 (\mu_1 \eta_2' - \mu_2 \mu_1')^2 (\nu_1 \nu_2' - \nu_2 \nu_1')^2 . u.$$

En échangeant seulement les z, ceci donne

$$u' = (\nu_1\nu_2' - \nu_2\nu_1')^2 . u,$$

ou
$$a' = \nu_1 a + \nu_1' e, \quad b' = \nu_1 b + \nu_1' f, \quad c' = \nu_1 c + \nu_1' g, \quad d' = \nu_1 d + \nu_1' h,$$
$$e' = \nu_2 a + \nu_2' e, \quad f' = \nu_2 b + \nu_2' f, \quad g' = \nu_2 c + \nu_2' g, \quad h' = \nu_2 d + \nu_2' h.$$

Soit
$$\nu_1 = ah - bg - cf - de,$$
$$\nu_1' = -2(ad - bc),$$
$$\nu_2 = -2(eh - fg),$$
$$\nu_2' = ah - bg - cf - de \ (= \nu_1),$$

on trouve d'abord

$$(\nu_1\nu_2' - \nu_2\nu_2') = u^2, \quad \text{ou} \quad u' = u^3,$$

et puis, en ayant égard aux valeurs de A, B, C, ... H:

$$a' = H, \quad b' = -G, \quad c' = -F, \quad d' = E,$$
$$e' = D, \quad f' = -B, \quad g' = -C, \quad h' = A,$$

de manière que $u' = U$, d'où enfin

$$U = u^3.$$

La propriété de la fonction u que je viens d'énoncer, se rapporte à une théorie assez générale, d'une nouvelle espèce de fonctions algébriques dont je m'occupe actuellement, et lesquelles à cause de leur analogie avec les déterminantes, on pourrait comme je crois nommer "Hyperdéterminantes." Je me propose de poser les premiers fondemens de cette théorie dans un mémoire qui va paraître dans le prochain No. du *Cambridge Mathematical Journal* (No. XXIII.) [13].

A présent je passe à une formule de Mr. Hesse [*Journal*, t. XXVIII. (1844), p. 88], qui se rapporte aussi à la même théorie. Soit V, une fonction homogène du troisième ordre, et à trois variables x, y, z. Soit ∇U, la déterminante formée avec les coefficients du second ordre de V, arrangés en cette forme:

$$\begin{matrix} \dfrac{d^2V}{dx^2}, & \dfrac{d^2V}{dxdy}, & \dfrac{d^2V}{dxdz}, \\[2ex] \dfrac{d^2V}{dxdy}, & \dfrac{d^2V}{dy^2}, & \dfrac{d^2V}{dydz}, \\[2ex] \dfrac{d^2V}{dzdx}, & \dfrac{d^2V}{dzdy}, & \dfrac{d^2V}{dz^2}; \end{matrix}$$

soit a une quantité constante quelconque et soient A et B deux autres constantes déterminés: Mr. Hesse à démontré l'équation remarquable

$$\nabla (U + a\nabla U) = AU + B\nabla U;$$

mais sans donner la forme des coefficients A et B, ce qui parait être très difficile à effectuer. En considérant le cas d'une fonction homogène de deux variables seulement, mais d'ailleurs d'un ordre quelconque, on parvient à un théorème analogue qui m'a paru intéressant.

"Soit U une fonction homogène et de l'ordre ν des deux variables x, y, et ∇U la déterminante $\frac{d^2V}{dx^2}\cdot\frac{d^2V}{dy^2}-\left(\frac{d^2V}{dxdy}\right)^2$, l'on a

$$(\nu-2)(\nu-3)^3 . \nabla (U + a\nabla U) = \{-\nu(\nu-1)(\nu-3)^2 Ja + \nu(\nu-1)(2\nu-5)^2 Ia^2\} U$$
$$+ \{(\nu-2)(\nu-3)^3 + (\nu-2)(\nu-3)(2\nu-5) Ja^2\} \nabla U.$$

En représentant par i, j, k, l, m, les coefficients différentiels du quatrième ordre de U, on a

$$I = ikm - il^2 - jm^2 - k^3 + 2jkl,$$
$$J = 4jl - 3k^2 - mi,$$

de manière que I, J sont des fonctions de x, y des ordres $3(\nu-4)$ et $2(\nu-4)$ respectivement."

Ce qu'il y a de remarquable, c'est la forme de ces deux quantités I, J. On voit d'abord que la fonction I est la déterminante formée avec les termes

$$\begin{matrix} i, & j, & k \\ j, & k, & l \\ k, & l, & m. \end{matrix}$$

Mais de plus les deux fonctions I, J ont la propriété suivante. Si l'on imagine une fonction du quatrième ordre

$$l\xi^4 + 4j\xi^3\eta + 6k\xi^2\eta^2 + 4l\xi\eta^3 + m\eta^4,$$

transformée en

$$l'\xi'^4 + 4j'\xi'^3\eta' + 6k'\xi'^2\eta'^2 + 4l'\xi'\eta'^3 + m'\eta'^4,$$

au moyen de

$$\xi = \lambda\xi' + \mu\eta',$$
$$\eta = \lambda'\xi' + \mu'\eta',$$

en représentant par I', J', les mêmes fonctions de i', j', k', l', m', on a

$$J' = (\lambda\mu' - \lambda'\mu)^6 . J, \quad I' = (\lambda\mu' - \lambda'\mu)^4 . I.$$

L'on parviendrait, je crois, à des résultats plus simples en considérant une fonction U, homogène en x, y, et aussi en x_1, y_1, et en posant

$$\nabla U = \frac{d^2V}{dxdx_1}\cdot\frac{d^2V}{dydy_1} - \frac{d^2V}{dxdy_1}\cdot\frac{d^2V}{dydx_1}.$$

Par exemple, si U est du second ordre en x, y et aussi en x_1, y_1, de manière que

$$\begin{aligned} U = \quad & x_1^2 \;(A\,x^2 + 2B\,xy + C\,y^2) \\ + & 2x_1y_1\,(A'\,x^2 + 2B'\,xy + C'\,y^2) \\ + & y_1^2 \;(A''x^2 + 2B''xy + C''y^2);\end{aligned}$$

on a simplement

$$\nabla\nabla U = \mathfrak{A}U,$$

en représentant par $\mathfrak{A}$ la déterminante formée avec les coefficients

$$\begin{matrix} A\;, & B\;, & C; \\ A', & B', & C'; \\ A'', & B'', & C''; \end{matrix}$$

et multipliée par 2^8. Mais il faudrait développer tout cela beaucoup plus loin.

P.S. Je ne sais pas si l'on a jamais discuté la question suivante, qui doit conduire, il me semble, à des résultats importants. Soient L, M, L', M', U, V des fonctions de x. En éliminant cette variable entre les équations $LU + MV = 0$ et $L'U + M'V = 0$, et en représentant l'équation ainsi obtenue par $[LU + MV,\ L'U + M'V] = 0$, et de même par $[U,\ V] = 0$ l'équation obtenue en éliminant x entre $U = 0$, $V = 0$, il est clair que $[LU + MU,\ L'U + M'V]$ contiendra $[U,\ V]$ comme facteur: quelles sont maintenant les propriétés de l'autre facteur qu'on peut écrire sous les trois formes: $[LU + MV,\ L'U + M'V] : [U,\ V]$, $[LU + MV,\ LM' - L'M] : [L,\ M]$ et $[L'U + M'V,\ LM' - L'M] : [L',\ M']$) ?

16.

MÉMOIRE SUR LES HYPERDÉTERMINANTS (TRADUCTION D'UN MÉMOIRE ANGLAIS INSÉRÉ DANS LE "CAMBRIDGE MATHEMATICAL JOURNAL" AVEC QUELQUES ADDITIONS DE L'AUTEUR).

[From the *Journal für die reine und angewandte Mathematik* (Crelle), vol. XXX. (1846), pp. 1—37.]

This is a translation of the foregoing papers, [13] and [14], On Linear Transformations, the additions are inconsiderable: in date of publication it was subsequent to [13] but I think preceded [14].

17.

NOTE ON MR BRONWIN'S PAPER ON ELLIPTIC INTEGRALS.

[From the *Cambridge Mathematical Journal*, vol. III. (1843), pp. 197—198.]

This Note was in answer to objections raised by Mr Bronwin in his paper "On Elliptic Functions" Camb. Math. Jour. vol. III. (1843) pp. 123—131, to some of the formulæ of transformation in the *Fundamenta Nova*. There is in it a serious error which was afterwards pointed out by Mr Bronwin. The Note is omitted, as are also the other papers, [18] and part of [21], which refer to the same subject.

18.

REMARKS ON THE REV. B. BRONWIN'S PAPER ON JACOBI'S THEORY OF ELLIPTIC FUNCTIONS.

[From the *Philosophical Magazine*, vol. XXII. (1843), pp. 358—368.]

This paper is omitted: see [17].

19.

INVESTIGATION OF THE TRANSFORMATION OF CERTAIN ELLIPTIC FUNCTIONS.

[From the *Philosophical Magazine*, vol. XXV. (1844), pp. 352—354.]

THE function sinam u (ϕu for shortness) may be expressed in the form

$$\phi u = u\,\Pi\left(1+\frac{u}{2mK+2m'K'i}\right)\div\Pi\left(1+\frac{u}{2mK+(2m'+1)K'i}\right)\ldots\ldots\ldots(1)$$

where m, m' receive any integer, positive or negative, values whatever, omitting only the combination $m=0$, $m'=0$ in the numerator (Abel, *Œuvres*, t. I. p. 212, [Ed. 2, p. 343] but with modifications to adapt it to Jacobi's notation; also the positive and negative values of m, m' are not collected together as in Abel's formulæ). We deduce from this

$$\frac{\phi(u+\theta)}{\phi\theta} = \Pi\left(1+\frac{u}{2mK+2m'K'i+\theta}\right)\div\Pi\left(1+\frac{u}{2mK+(2m'+1)K'i+\theta}\right)\ldots(2).$$

Suppose now $K=aH+a'H'i$, $K'i=bH+b'H'i$, a, b, a', b' integers, and $ab'-a'b$ a positive number ν. Also let $\theta=fH+f'H'i$; f, f' integers such that $af'-a'f$, $bf'-b'f$, ν, have not all three any common factor. Consider the expression

$$v=\frac{\phi u\,\phi(u+2\omega)\ldots\phi(u+2(\nu-1)\omega)}{\phi(2\omega)\ldots\phi(2(\nu-1)\omega)}\ldots\ldots\ldots\ldots\ldots\ldots\ldots(3),$$

from which

$$v=u\Pi\left(1+\frac{u}{2mK+2m'K'i+2r\theta}\right)\div\Pi\left(1+\frac{u}{2mK+(2m'+1)K'i+2r\theta}\right)\ldots\ldots(4)$$

where r extends from 0 to $\nu-1$ inclusively, the single combination $m=0$, $m'=0$, $r=0$ being omitted in the numerator. We may write

$$mK+m'K'i+r\theta=\mu H+\mu'H'i,$$

μ, μ' denoting any integers whatever. Also to given values of μ, μ' there corresponds only a single system of values of m, m', r. To prove this we must show that the equations

$$ma + m'b + rf = \mu,$$
$$ma' + m'b' + rf' = \mu',$$

can always be satisfied, and satisfied in a single manner only. Observing the value of ν,

$$\nu m + r(b'f - bf') = \mu b' - \mu' b;$$

then if ν and $b'f - bf'$ have no common factor, there is a single value of r less than ν, which gives an integer value for m. This being the case, $m'b$ and $m'b'$ are both integers, and therefore, since b, b' have no common factor (for such a factor would divide ν and $b'f - bf'$), m' is also an integer. If, however, ν and $b'f - bf'$ have a common factor c, so that $\nu = ab' - a'b = c\phi$, $b'f - bf' = c\phi'$; then $(af' - a'f)b' = c(\phi f' - \phi' f)$, or since no factor of c divides $af' - a'f$, c divides b', and consequently b. The equation for ν may therefore be divided by c. Hence, putting $\frac{\nu}{c} = \nu_{,}$, we may find a value of r, say $r_{,}$, less than $\nu_{,}$, which makes m an integer; and the general value of r less than ν which makes m an integer, is $r = r_{,} + s\nu_{,}$, where s is a positive integer less than c. But m being integral, bm', $b'm'$, and consequently cm' are integral; we have also

$$c\nu_1 m' + (r_{,} + s\nu_{,})(af' - a'f) = a\mu' - a'\mu;$$

and there may be found a single value of s less than c, giving an integer value for m'. Hence in every case there is a single system of values of m, m', r, corresponding to any assumed integer values whatever of μ, μ'. Hence

$$U = u\Pi\left(1 + \frac{u}{2\mu H + 2\mu' H'i}\right) \div \Pi\left(1 + \frac{u}{2\mu H + (2\mu'+1)H'i}\right) = \phi_{,}u \quad \ldots\ldots\ldots (5)$$

$\phi_{,}u$ being a function similar to ϕu, or $\sin\text{am}\, u$, but to a different modulus, viz. such that the complete functions are H, H' instead of K, K'. We have therefore

$$\phi_{,}u = \frac{\phi u \phi(u + 2\omega) \ldots \phi(u + 2(\nu-1)\omega)}{\phi(2\omega) \ldots \phi(2(\nu-1)\omega)} \quad \ldots\ldots\ldots\ldots\ldots\ldots\ldots\ldots (6).$$

Expressing ω in terms of K, K', we have $\nu H = b'K - a'K'i$, $-\nu H'i = bK - aK'i$, and therefore $\nu\omega = (b'f - bf')K - (a'f - af')K'i$. Let g, g' be any two integer numbers having no common factor, which is also a factor of ν, we may always determine a, b, a', b', so that $\nu\omega = gK - g'K'i$. This will be the case if $g = b'f - bf'$, $g' = a'f - af'$. One of the quantities f, f' may be assumed equal to 0. Suppose $f' = 0$, then $g = b'f$, $g' = a'f$; whence $ag - bg' = \nu f$. Let k be the greatest common measure of g, g', so that $g = kg_{,}$, $g' = kg'_{,}$; then, since no factor of k divides ν, k must divide f, or $f = kf_{,}$, but $g_{,} = b'f$, $g'_{,} = af'$, and a', b' are integers, or f must divide g, g'; whence $f_{,} = 1$, or $f = k$. Also $ag_{,} - bg'_{,} = \nu$, where $g_{,}$ and $g'_{,}$ are prime to each other, so that integer values may always be found for a and b; so that in the equation (1) we have

$$\omega = \frac{gK - g'K'i}{\nu} \quad \ldots\ldots\ldots\ldots\ldots\ldots\ldots\ldots\ldots\ldots\ldots\ldots (7),$$

g, g' being any integer numbers such that no common factor of g, g' also divides ν.

The above supposition, $f'=0$, is, however, only a particular one; omitting it, the conditions to be satisfied by a, b, a', b', may be written under the form

$$\left.\begin{aligned} ab' - a'b &= \nu, \\ ag - bg' &\equiv 0 \text{ [mod. } \nu], \\ a'g - b'g' &\equiv 0 \text{ [mod. } \nu], \end{aligned}\right\} \quad \text{.......................................(8)}$$

to which we may join the equations before obtained,

$$\left.\begin{aligned} \nu H &= b'K - a'K'i, \\ -\nu H'i &= bK - aKi, \end{aligned}\right\} \quad \text{.......................................(9)}$$

which contain the theory of the modular equation. This, however, involves some further investigations, which are not sufficiently connected with the present subject to be attempted here.

20.

ON CERTAIN RESULTS RELATING TO QUATERNIONS.

[From the *Philosophical Magazine*, vol. XXVI. (1845), pp. 141—145.]

IN his last paper on Quaternions [*Phil. Mag.* vol. XXV. (1844), p. 491] Sir William R. Hamilton has alluded to a paper of mine on the Analytical Geometry of (n) Dimensions, in the *Cambridge Mathematical Journal* [11], as one that might refer to the same subject. It may perhaps be as well to notice that the investigations there contained have no reference whatever to Sir William Hamilton's very beautiful theory; a more correct title for them would have been, a Generalization of the Analysis which occurs in ordinary Analytical Geometry.

I take this opportunity of communicating one or two results relating to quaternions; the first of them does appear to me rather a curious one.

Observing that

$$(A+Bi+Cj+Dk)^{-1}=(A-Bi-Cj-Dk)\div(A^2+B^2+C^2+D^2)\quad\ldots\ldots\ldots(1)$$

it is easy to form the equation

$$\left.\begin{aligned}
&(A+Bi+Cj+Dk)^{-1}(\alpha+\beta i+\gamma j+\delta k)(A+Bi+Cj+Dk)\\
&=\frac{1}{A^2+B^2+C^2+D^2}\,.\\
&\left\{\begin{array}{llll}
& \alpha(A^2+B^2+C^2+D^2) & & \\
+i\,[& \beta\,(A^2+B^2-C^2-D^2)+2\gamma\,(BC+AD) & +2\delta\,(BD-AC) &] \\
+j\,[& 2\beta\,(BC-AD)\quad+\gamma\,(A^2-B^2+C^2-D^2) & +2\delta\,(CD+AB) &] \\
+k\,[& 2\beta\,(BD+AC)\quad+2\gamma\,(CD-AB) & +\delta\,(A^2-B^2-C^2+D^2) &]
\end{array}\right.
\end{aligned}\right\}\ldots(2)$$

which I have given with these letters for the sake of reference; it will be convenient to change the notation and write

$$\left.\begin{array}{l}(1+\lambda i+\mu j+\nu k)^{-1}(ix+jy+kz)(1+\lambda i+\mu j+\nu k)\\ =\dfrac{1}{1+\lambda^2+\mu^2+\nu^2}\\ \left\{\begin{array}{llll} i\,[\;x(1+\lambda^2-\mu^2-\nu^2) & +2y(\lambda\mu+\nu) & +2z(\lambda\nu-\mu) &]\\ +j\,[2x(\lambda\mu-\nu) & +\;y(1-\lambda^2+\mu^2-\nu^2) & +2z(\mu\nu+\lambda) &]\\ +k\,[2x(\lambda\nu+\mu) & +2y(\mu\nu-\lambda) & +\;z(1-\lambda^2-\mu^2+\nu^2) &]\end{array}\right.\end{array}\right\}\ldots(3)$$

$$\left.\begin{array}{l}=\;\;i(\alpha x+\alpha' y+\alpha'' z)\\ \;\;\;+j(\beta x+\beta' y+\beta'' z)\\ \;\;\;+k(\gamma x+\gamma' y+\gamma'' z)\end{array}\right\}\ldots\ldots(4)$$

suppose. The peculiarity of this formula is, that the coefficients α, β, ... are precisely such that a system of formulæ

$$\left.\begin{array}{l}x_{,}=\alpha x+\alpha' y+\alpha'' z\\ y_{,}=\beta' x+\beta' y+\beta'' z\\ z_{,}=\gamma x+\gamma' y+\gamma'' z\end{array}\right\}\ldots\ldots(5)$$

denotes the transformation from one set of rectangular axes to another set, also rectangular. Nor is this all, the quantities λ, μ, ν may be geometrically interpreted. Suppose the axes Ax, Ay, Az could be made to coincide with the axes $Ax_{,}$, $Ay_{,}$, $Az_{,}$ by means of a revolution through an angle θ round an axis AP inclined to Ax, Ay, Az, at angles f, g, h then

$$\lambda=\tan\tfrac{1}{2}\theta\cos f,\qquad \mu=\tan\tfrac{1}{2}\theta\cos g,\qquad \nu=\tan\tfrac{1}{2}\theta\cos h.$$

In fact the formulæ are precisely those given for such a transformation by M. Olinde Rodrigues Liouville, t. v., "Des lois géometriques qui régissent les déplacemens d'un système solide" (or *Camb. Math. Journal*, t. iii. p. 224 [6]). It would be an interesting question to account, *à priori*, for the appearance of these coefficients here.

The ordinary definition of a determinant naturally leads to that of a quaternion determinant. We have, for instance,

$$\begin{vmatrix}\varpi, & \phi\\ \varpi', & \phi'\end{vmatrix}=\varpi\phi'-\varpi'\phi,\ldots\ldots(6),$$

$$\begin{vmatrix}\varpi, & \phi, & \chi\\ \varpi', & \phi', & \chi'\\ \varpi'', & \phi'', & \chi''\end{vmatrix}=\varpi(\phi'\chi''-\phi''\chi')+\varpi'(\phi''\chi-\phi\chi'')+\varpi''(\phi\chi'-\phi'\chi),$$

&c., the same as for common determinants, only here the order of the factors on each term of the second side of the equation is essential, and not, as in the other case, arbitrary. Thus, for instance,

$$\begin{vmatrix}\varpi, & \varpi'\\ \varpi, & \varpi'\end{vmatrix}=\varpi\varpi'-\varpi\varpi',\;=0\;\ldots\ldots(7),$$

but
$$\begin{vmatrix} \varpi, & \varpi \\ \varpi', & \varpi' \end{vmatrix} = \varpi\varpi' - \varpi'\varpi, \ \neq 0 \quad\text{.............................(8)},$$

that is, a quaternion determinant does not vanish when two vertical rows become identical. One is immediately led to inquire what the value of such determinants is. Suppose

$$\varpi = x + iy + jz + kw, \quad \varpi' = x' + iy' + jz' + kw', \text{ \&c.},$$

it is easy to prove

$$\begin{vmatrix} \varpi & \varpi \\ \varpi' & \varpi' \end{vmatrix} = -2 \begin{vmatrix} i, & j, & k \\ x, & y, & z \\ x', & y', & z' \end{vmatrix} \quad\text{............................. (9)},$$

$$\begin{vmatrix} \varpi, & \varpi, & \varpi \\ \varpi', & \varpi', & \varpi' \\ \varpi'', & \varpi'', & \varpi'' \end{vmatrix} = -2 \begin{vmatrix} 3, & i, & j, & k \\ x, & y, & z, & w \\ x', & y', & z', & w' \\ x'', & y'', & z'', & w'' \end{vmatrix} \quad\text{.................. (10)},$$

$$\begin{vmatrix} \varpi, & \varpi, & \varpi, & \varpi \\ \varpi', & \varpi', & \varpi', & \varpi' \\ \varpi'', & \varpi'', & \varpi'', & \varpi'' \\ \varpi''', & \varpi''', & \varpi''', & \varpi''' \end{vmatrix} = 0 \quad\text{.. (11)},$$

or a quaternion determinant vanishes when four or more of its vertical rows become identical.

Again, it is immediately seen that

$$\begin{vmatrix} \varpi, & \phi \\ \varpi', & \phi' \end{vmatrix} + \begin{vmatrix} \phi, & \varpi \\ \phi', & \varpi' \end{vmatrix} = \begin{vmatrix} \varpi, & \varpi \\ \phi', & \phi' \end{vmatrix} - \begin{vmatrix} \varpi', & \varpi' \\ \phi, & \phi \end{vmatrix} \quad\text{........... (12)}$$

&c. for determinants of any order, whence the theorem, if any four (or more) adjacent vertical columns of a quaternion determinant be transposed in every possible manner, the sum of all these determinants vanishes, which is a much less simple property than the one which exists for the horizontal rows, viz. the same that in ordinary determinants exists for the horizontal or vertical rows indifferently. It is important to remark that the equations

$$\begin{vmatrix} \varpi, & \phi \\ \varpi', & \phi' \end{vmatrix} = 0 \text{ or } \begin{vmatrix} \varpi, & \varpi' \\ \phi, & \phi' \end{vmatrix} = 0, \text{ \&c.} \quad\text{..............................(13)}$$

i.e.
$$\varpi\phi' - \varpi'\phi = 0, \text{ or } \varpi\phi' - \phi\varpi' = 0, \text{ \&c.}$$

are none of them the result of the elimination of Π, Φ, from the two equations

$$\begin{aligned} \varpi\Pi + \phi\Phi &= 0, \quad\text{.................................... (14)}. \\ \varpi'\Pi + \phi'\Phi &= 0, \end{aligned}$$

On the contrary, the result of this elimination is the very different equation

$$\varpi^{-1} . \phi - \varpi'^{-1} . \phi' = 0 \quad\text{......(15)},$$

equivalent of course to four independent equations, one of which may evidently be replaced by

$$M\varpi . M\phi' - M\varpi' . M\phi = 0 \quad\text{......(16)},$$

if $M\varpi$, &c. denotes the modulus of ϖ, &c. An equation analogous to this last will undoubtedly hold for any number of equations, but it is difficult to say what is the equation analogous to the one immediately preceding this, in the case of a greater number of equations, or rather, it is difficult to give the result in a symmetrical form independent of extraneous factors.

I may just, in conclusion, mention what appears to me a possible application of Sir William Hamilton's interesting discovery. In the same way that the circular functions depend on infinite products, such as

$$x\Pi\left(1+\frac{x}{m\pi}\right), \text{ \&c.} \quad\text{......(17)},$$

$\{m$ any integer from ∞ to $-\infty$, omitting $m=0\}$

and the inverse elliptic functions on the doubly infinite products

$$x\Pi\left(1+\frac{x}{mw+n\varpi i}\right), \text{ \&c.} \quad\text{......(18)}$$

$\{m$ and n integers from ∞ to $-\infty$, omitting $m=0$, $n=0\}$,

may not the inverse ultra-elliptic functions of the next order of complexity depend on the quadruply infinite products

$$x\Pi\left(1+\frac{x}{mw+n\varpi i+o\phi j+p\psi k}\right) ? \quad\text{......(19)}$$

$\{m, n, o, p$ integers from ∞ to $-\infty$, omitting $m=0$, $n=0$, $o=0$, $p=0\}$.

It seems as if some supposition of this kind would remove a difficulty started by Jacobi (Crelle, t. ix.) with respect to the multiple periodicity of these functions. Of course this must remain a mere suggestion until the theory of quaternions is very much more developed than it is at present; in particular the theory of quaternion exponentials would have to be developed, for even in a product, such as (18), there is a certain singular exponential factor running through the theory, as appears from some formulæ in Jacobi's *Fund. Nova* (relative to his functions Θ, H), the occurrence of which may be accounted for, *à priori*, as I have succeeded in doing in a paper to be published shortly in the *Cambridge Mathematical Journal* [24].

21.

ON JACOBI'S ELLIPTIC FUNCTIONS, IN REPLY TO THE REV. B. BRONWIN; AND ON QUATERNIONS.

[From the *Philosophical Magazine*, vol. XXVI. (1845), pp. 208, 211.]

The first part of this Paper is omitted, see [17]: only the Postscript on Quaternions, pp. 210, 211, is printed.

It is possible to form an analogous theory with seven imaginary roots of (-1) (? with $\nu = 2^n - 1$ roots when ν is a prime number). Thus if these be i_1, i_2, i_3, i_4, i_5, i_6, i_7, which group together according to the types

123, 145, 624, 653, 725, 734, 176,

i.e. the type 123 denotes the system of equations

$$\begin{aligned} i_1 i_2 &= i_3, & i_2 i_3 &= i_1, & i_3 i_1 &= i_2, \\ i_2 i_1 &= -i_3, & i_3 i_2 &= -i_1, & i_1 i_3 &= -i_2, \end{aligned}$$

&c. We have the following expression for the product of two factors:

$$\begin{aligned} &(X_0 + X_1 i_1 + \dots X_7 i_7)(X'_0 + X'_1 i_1 + \dots X'_7 i_7) \\ &= X_0 X'_0 - X_1 X'_1 - X_2 X'_2 \dots - X_7 X'_7 \\ &\quad + [\overline{23} + \overline{45} + \overline{76} + (01)]\, i_1 \\ &\quad + [\overline{31} + \overline{46} + \overline{57} + (02)]\, i_2 \\ &\quad + [\overline{12} + \overline{65} + \overline{47} + (03)]\, i_3 \\ &\quad + [\overline{51} + \overline{62} + \overline{47} + (04)]\, i_4 \\ &\quad + [\overline{14} + \overline{36} + \overline{72} + (05)]\, i_5 \\ &\quad + [\overline{24} + \overline{53} + \overline{17} + (06)]\, i_6 \\ &\quad + [\overline{25} + \overline{34} + \overline{61} + (07)]\, i_7 \end{aligned}$$

where $\quad (01) = X_0 X'_1 + X_1 X'_0 \dots; \ \overline{12} = X_1 X'_2 - X_2 X'_1$ &c.;

and the modulus of this expression is the product of the moduli of the factors. The above system of types requires some care in writing down, and not only with respect to the combinations of the letters, but also to their order; it would be vitiated, e.g. by writing 716 instead of 176. A theorem analogous to that which I gave before, for quaternions, is the following:—If $\Lambda = 1 + \lambda_1 i_1 \dots + \lambda_7 i_7$, $X = x_1 i_1 \dots + x_7 i_7$: it is immediately shown that the possible part of $\Lambda^{-1} X \Lambda$ vanishes, and that the coefficients of $i_1, \dots i_7$ are linear functions of $x_1, \dots x_7$. The modulus of the above expression is evidently the modulus of X; hence "we may determine seven linear functions of $x_1 \dots x_7$, the sum of whose squares is equal to $x_1^2 + \dots + x_7^2$." The number of arbitrary quantities is however only seven, instead of twenty-one, as it should be.

22.

ON ALGEBRAICAL COUPLES.

[From the *Philosophical Magazine*, vol. XXVII. (1845), pp. 38—40.]

It is worth while, in connection with the theory of quaternions and the researches of Mr Graves (*Phil. Mag.* [Vol. XXVI. (1845), pp. 315—320]), to investigate the properties of a couple $ix+jy$ in which i, j are symbols such that

$$
\begin{aligned}
i^2 &= \alpha i + \beta j,\\
ij &= \alpha' i + \beta' j,\\
ji &= \gamma i + \delta j,\\
j^2 &= \gamma' i + \delta' j.
\end{aligned}
$$

If
$$\overline{ix+jy}\;\overline{ix_1+jy_1} = iX + jY,$$

then
$$
\begin{aligned}
X &= \alpha\, xx_1 + \alpha' xy_1 + \gamma x_1 y + \gamma' yy_1,\\
Y &= \beta' xx_1 + \beta' xy_1 + \delta' x_1 y + \delta' yy_1.
\end{aligned}
$$

Imagine the constants α, β... so determined that $ix+jy$ may have a modulus of the form $K(x+\lambda y)(x+\mu y)$; there results one of the four following essentially independent systems

A.
$$
\begin{aligned}
i^2 &= \frac{1}{\lambda\mu}(\delta\lambda\mu + \gamma\,\overline{\lambda+\mu})\, i - \frac{\gamma}{\lambda\mu} j,\\
ij &= ji = \gamma i + \delta j,\\
j^2 &= -\lambda\mu\delta i + (\gamma + \overline{\lambda+\mu}\,\delta)\, j,
\end{aligned}
$$
$$
\begin{cases}
X + \lambda Y = \dfrac{1}{\lambda}(\gamma + \lambda\delta)(x+\lambda y)(x_1 + \lambda y_1),\\[2ex]
X + \mu Y = \dfrac{1}{\mu}(\gamma + \mu\delta)(x+\mu y)(x + \mu y_1).
\end{cases}
$$

The couple may be said to have the *two linear moduli*,

$$\frac{1}{\lambda}(\gamma+\lambda\delta)(x+\lambda y), \quad \frac{1}{\mu}(\gamma+\mu\delta)(x+\mu y);$$

as well as the quadratic one,

$$\frac{1}{\lambda\mu}(\gamma+\lambda\delta)(\gamma+\mu\delta)(x+\lambda y)(x+\mu y),$$

the product of these, which is the modulus, and the only modulus in the remaining systems.

B.

$$i^2 = -\delta i + \frac{1}{\lambda\mu}(\gamma + \delta\overline{\lambda+\mu})j,$$

$$ij = ji = \gamma i + \delta j,$$

$$j^2 = (\overline{\lambda+\mu}\,\gamma + \lambda\mu\delta)\,i - \gamma j,$$

$$\begin{cases} X + \lambda Y = \dfrac{1}{\mu}(\gamma+\lambda\delta)(x+\mu y)(x_1+\mu y_1), \\ X + \mu Y = \dfrac{1}{\lambda}(\gamma+\mu\delta)(x+\lambda y)(x_1+\lambda y_1). \end{cases}$$

C.

$$i^2 = \frac{1}{\lambda\mu}(\delta\lambda\mu + \gamma\,\overline{\lambda+\mu})\,i - \frac{\gamma}{\lambda\mu}j,$$

$$ij = \left(\frac{\mu^2+\mu\lambda+\lambda^2}{\mu\lambda}\gamma + \overline{\mu+\lambda}\,\delta\right)i + \left(-\delta - \frac{\lambda+\mu}{\lambda\mu}\gamma\right)j,$$

$$ji = \quad \gamma i + \delta j,$$

$$j^2 = (\overline{\lambda+\mu}\,\gamma + \lambda\mu\delta)\,i - \gamma j,$$

$$\begin{cases} X + \lambda Y = \dfrac{1}{\lambda}(\gamma+\lambda\delta)(x+\lambda y)(x_1+\mu y_1), \\ X + \mu Y = \dfrac{1}{\mu}(\gamma+\mu\delta)(x+\mu y)(x_1+\lambda y_1). \end{cases}$$

D.

$$i^2 = -\delta i + \frac{1}{\lambda\mu}(\gamma + \delta\,\overline{\lambda+\mu})j,$$

$$ij = (-\gamma - \overline{\lambda+\mu}\,\delta)\,i + \left(\frac{\lambda+\mu}{\lambda\mu}\gamma + \frac{\lambda^2+\lambda\mu+\mu^2}{\lambda\mu}\delta\right)j,$$

$$ji = \quad \gamma i + \delta j,$$

$$j^2 = -\lambda\mu\delta i + (\gamma + \overline{\lambda+\mu}\,\delta)j,$$

$$\begin{cases} X + \lambda Y = \dfrac{1}{\mu}(\gamma+\lambda\delta)(x+\mu y)(x_1+\lambda y_1), \\ X + \mu Y = \dfrac{1}{\lambda}(\gamma+\mu\delta)(x+\lambda y)(x_1+\mu y_1). \end{cases}$$

The formulæ are much simpler and not essentially less general, if $\mu = -\lambda$. They thus become

A′.
$$\begin{aligned} i^2 &= \delta i + \frac{\gamma}{\lambda^2} j, \\ ij = ji &= \gamma i + \delta j, \\ j^2 &= \lambda^2 \delta i + \gamma j, \end{aligned}$$
$$X \pm \lambda Y = \pm \frac{1}{\lambda} (\gamma \pm \lambda\delta)(x \pm \lambda y)(x_1 \pm \lambda y_1).$$

(Two linear moduli.)

B′.
$$\begin{aligned} i^2 &= -\delta i - \frac{\gamma}{\lambda^2} j, \\ ij = ji &= \gamma i + \delta j, \\ j^2 &= -\lambda^2 \delta i - \gamma j, \end{aligned}$$
$$X \pm \lambda Y = \mp \frac{1}{\lambda} (\gamma \pm \lambda\delta)(x \mp \lambda y)(x_1 \mp \lambda y_1).$$

C′.
$$\begin{aligned} i^2 &= -\delta i + \frac{\gamma}{\lambda^2} j, \\ ij &= -\gamma i - \delta j, \\ ji &= \gamma i + \delta j, \\ j^2 &= -\lambda^2 \delta i - \gamma j, \end{aligned}$$
$$X \pm \lambda Y = \pm \frac{1}{\lambda} (\gamma \pm \lambda\delta)(x \pm \lambda y)(x_1 \mp \lambda y_1).$$

D′.
$$\begin{aligned} i^2 &= -\delta i - \frac{\gamma}{\lambda^2} j, \\ ij &= -\gamma i - \delta j, \\ ji &= \gamma i + \delta j, \\ j^2 &= \lambda^2 \delta i + \gamma j, \end{aligned}$$
$$X \pm \lambda Y = \mp \frac{1}{\lambda} (\gamma \pm \lambda\delta)(x \mp \lambda y)(x_1 \pm \lambda y).$$

There is a system more general than (A.) having a *single* linear modulus $q(\theta X + Y)$: this is

E.
$$\begin{aligned} i^2 &= \alpha (i - \theta j) + \theta^2 q j, \\ ij &= \alpha' (i - \theta j) + \theta q j, \\ ji &= \gamma (i - \theta j) + \theta q j, \\ j^2 &= \gamma' (i - \theta j) + q j, \end{aligned}$$
$$\theta X + Y = q(\theta x + y)(\theta x_1 + y_1);$$

or, without real loss of generality,

E′.

$$\begin{aligned} i^2 &= \alpha\, i, \\ ij &= \alpha' i, \\ ji &= \gamma\, i, \\ j^2 &= \gamma' i + qj, \\ Y &= qyy_1. \end{aligned}$$

To complete the theory of this system, one may add the identical equation

$$X + \frac{1}{\theta - qM} Y = \frac{q(\theta^2 - M\alpha)}{\theta - qM} \left(x + \frac{\alpha' - \theta\gamma}{\alpha - \theta\alpha'} y\right) \left(x_1 + \frac{\alpha' - \theta\gamma'}{\alpha - \theta\gamma} y_1\right).$$

where

$$M = \frac{\theta(\gamma - \theta\gamma') - (\alpha - \theta\alpha')}{\alpha'\gamma - \alpha\gamma'}.$$

By determining the constants, so that $\frac{1}{\theta - qM} = \frac{\alpha' - \theta\gamma}{\alpha - \theta\alpha'} = \frac{\alpha' - \theta\gamma'}{\alpha - \theta\gamma}$, the system would reduce itself to the form A.

23.

ON THE TRANSFORMATION OF ELLIPTIC FUNCTIONS.

[From the *Philosophical Magazine and Journal of Science*, vol. XXVII. (1845), pp. 424—427.]

IN a former paper [19] I gave a proof of Jacobi's theorem, which I suggested [21] would lead to the resolution of the very important problem of finding the relation between the complete functions. This is in fact effected by the formulæ there given, but there is an apparent indeterminateness in them, the cause of which it is necessary to explain, and which I shall now show to be inherent in the problem. For the sake of supplying an omission, for the detection of which I am indebted to Mr Bronwin, I will first recapitulate the steps of the demonstration.

If $\frac{1}{2}\omega$, $\frac{1}{2}\upsilon$ [1] be the complete functions corresponding to ϕ, x, then this function is expressible in the form

$$\phi x = x\Pi\left(1+\frac{x}{m\omega+n\upsilon}\right) \div \Pi\left(1+\frac{x}{\overline{m+\tfrac{1}{2}}\,\omega+\overline{n+\tfrac{1}{2}}\,\upsilon}\right).$$

Let p be any prime number, μ, ν integers not divisible by p, and

$$\theta = \frac{\mu\omega+\nu\upsilon}{p}.$$

The function

$$\phi_1 x = \phi x\,\frac{\phi(x+2\theta)}{\phi 2\theta}\,\frac{\phi(x+4\theta)}{\phi 4\theta}\cdots\frac{\phi(x+2(p-1)\theta)}{\phi(2(p-1)\theta)}$$

is always reducible to the form

$$x\Pi\left(1+\frac{x}{m'\omega'+n'\upsilon'}\right) \div \Pi\left(1+\frac{x}{\overline{m'+\tfrac{1}{2}}\,\omega'+\overline{n'+\tfrac{1}{2}}\,\upsilon'}\right);$$

[1] Analogous to the K, $K'i$ of M. Jacobi.

and thus $\phi_1 x$ is an inverse function, the complete functions of which are $\tfrac{1}{2}\omega'$, $\tfrac{1}{2}\upsilon'$: and ω', υ' are connected with ω, υ by the equations

$$\omega' = \frac{1}{p}(\alpha\omega + \beta\upsilon),$$

$$\upsilon' = \frac{1}{p}(\alpha'\omega + \beta'\upsilon),$$

α, β, α', β' being any integers subject to the conditions that α, β' are odd and α', β even; also

$$\begin{aligned} \alpha\beta' - \alpha'\beta &= p, \\ \mu\beta' - \nu\alpha' &= l'p, \\ \mu\beta - \nu\alpha &= l\,p, \end{aligned}$$

l, l' being any integers whatever. In fact, to prove this, we have only to consider the general form of a factor in the numerator of $\phi_1 x$. Omitting a constant factor, this is

$$\left(1 + \frac{x}{m\omega + n\upsilon + 2r\theta}\right) \quad [r < p];$$

and it is to be shown that we can always satisfy the equation

$$m\omega + n\upsilon + 2r\theta = m'\omega' + n'\upsilon',$$

or the equations

$$\begin{aligned} pm + 2r\mu &= m'\alpha + n'\alpha', \\ pn + 2r\nu &= m'\beta + n'\beta'; \end{aligned}$$

and also that to each set of values of m, n, r, there is a unique set of values of m', n', and *vice versâ*. This is done in the paper referred to. Moreover, with the suppositions just made as to the numbers α, β' being odd and α', β even, it is obvious that m' is odd or even, according as m is, and n' according as n is, which shows that we can likewise satisfy

$$\overline{m + \tfrac{1}{2}}\,\omega + \overline{n + \tfrac{1}{2}}\,\upsilon + 2r\theta = \overline{m' + \tfrac{1}{2}}\,\omega' + \overline{n' + \tfrac{1}{2}}\,\upsilon';$$

and thus the denominator of $\phi_1 x$ is also reducible to the required form.

Now proceeding to the immediate object of this paper, α, β, α', β', and consequently ω', υ' are to a certain extent indeterminate. Let A, B, A', B' be a particular set of values of α, β, α', β', and O, P the corresponding values of ω', υ'. We have evidently A, B' odd and A', B even. Also

$$\begin{aligned} AB' - A'B &= p, \\ \mu B' - \nu A' &= L'p, \\ \mu B - \nu A &= L\,p, \end{aligned}$$

$$O = \frac{1}{p}(A\,\omega + B\,\upsilon),$$

$$U = \frac{1}{p}(A'\omega + B'\upsilon).$$

By eliminating ω, υ from these equations and the former system, it is easy to obtain

$$\omega' = a\,O + b\,U,$$
$$\upsilon' = a'O + b'U,$$

where
$$a = \frac{1}{p}(\alpha B' - \beta A'), \quad b = -\frac{1}{p}(\alpha B - \beta A),$$
$$a' = \frac{1}{p}(\alpha' B' - \beta' A'), \quad b' = -\frac{1}{p}(\alpha' B - \beta' A).$$

The coefficients a, b, a', b' are integers, as is obvious from the equation $\mu(\alpha' B - \beta' A) = p(L'\alpha - lA')$, and the others analogous to it; moreover, a, b' are odd and a', b are even, and

$$ab' - a'b = \frac{1}{p^2}(A'B - A'B)(\alpha\beta' - \alpha'\beta);$$

that is
$$ab' - a'b = 1.$$

Hence the theorem,—"The general values ω', υ' of the complete functions are linearly connected with the particular system of values O, U by the equations, $\omega' = aO + bU$, $\upsilon' = a'O + b'U$, in which a, b' are odd integers and a', b even ones, satisfying the condition $ab' - a'b = 1$."

With this relation between O, U and ω', υ', it is easy to show that the function $\phi_1 x$ is precisely the same, whether O, U or ω', υ' be taken for the complete functions. In fact, stating the proposition relatively to ϕx, we have,—"The inverse function ϕx is not altered by the change of ω, υ into ω', υ', where $\omega' = \alpha\omega + \beta\upsilon$, $\upsilon' = \alpha'\omega + \beta'\upsilon$, and α, β, α', β' satisfy the conditions that α, β' are odd, α', β even, and $\alpha\beta' - \alpha'\beta = 1$." This is immediately shown by writing

$$m\omega + n\upsilon = m'\omega' + n'\upsilon',$$

or
$$m = m'\alpha + n'\alpha',$$
$$n = m'\beta + n'\beta'.$$

It is obvious that to each set of values of m, n there is a unique set of values of m', n', and *vice versâ*: also that odd or even values of m, m' or n, n' always correspond to each. It is, in fact, the preceding reasoning applied to the case of $p = 1$.

Hence finally the theorem,—"The only conditions for determining ω', υ' are the equations

$$\omega' = \frac{1}{p}(\alpha\omega + \beta\upsilon), \qquad \upsilon' = \frac{1}{p}(\alpha'\omega + \beta'\upsilon),$$

where α, β' are odd and α', β even, and

$$\alpha\beta' - \alpha'\beta = p, \quad \mu\beta' - \nu\alpha' = l'p, \quad \mu\beta - \nu\alpha = lp,$$

l and l' arbitrary integers: and it is absolutely indifferent what system of values is adopted for ω', υ', the value of $\phi_1 x$ is precisely the same."

We derive from the above the somewhat singular conclusion, that the complete functions are not absolutely determinate functions of the modulus; notwithstanding that they are given by the apparently determinate conditions,

$$\tfrac{1}{2}\,\omega = \int_0^{\frac{1}{c}} \frac{dx}{\sqrt{(1-c^2x^2)(1+e^2x^2)}},$$

$$\tfrac{1}{2}\,\upsilon = \int_0^{\frac{1}{e}} \frac{dx}{\sqrt{(1+c^2x^2)(1-e^2x^2)}}.$$

In fact definite integrals are in many cases really indeterminate, and acquire different values according as we consider the variable to pass through real values, or through imaginary ones. Where the limits are real, it is tacitly supposed that the variable passes through a succession of real values, and thus ω, υ may be considered as completely determined by these equations, but only in consequence of this tacit supposition. If c and e are imaginary, there is absolutely no system of values to be selected for ω, υ in preference to any other system. The only remaining difficulty is to show from the integral itself, independently of the theory of elliptic functions, that such integrals contain an indeterminateness of two arbitrary integers; and this difficulty is equally great in the simplest cases. Why, *à priori*, do the functions

$$\sin^{-1}x = \int_0 \frac{dx}{\sqrt{1-x^2}}, \qquad \text{or } \log x = \int_0 \frac{dx}{x}$$

contain a single indeterminate integer?

Obs. I am of course aware, that in treating of the properties of such products as $\Pi\left(1+\dfrac{x}{m\omega+n\upsilon}\right)$, it is absolutely necessary to pay attention to the relations between the infinite limiting values of m and n; and that this introduces certain exponential factors, to which no allusion has been made. But these factors always disappear from the quotient of two such products, and to have made mention of them would only have been embarrassing the demonstration without necessity.

24.

ON THE INVERSE ELLIPTIC FUNCTIONS.

[From the *Cambridge Mathematical Journal*, t. IV. (1845), pp. 257—277.]

THE properties of the inverse elliptic functions have been the object of the researches of the two illustrious analysts, Abel and Jacobi. Among their most remarkable ones may be reckoned the formulæ given by Abel (*Œuvres*, t. I. p. 212 [Ed. 2, p. 343]), in which the functions $\phi\alpha$, $f\alpha$, $F\alpha$, (corresponding to Jacobi's $\sin\text{am}\,.\,\alpha$, $\cos\text{am}\,.\,\alpha$, $\Delta am\,.\,\alpha$, though not precisely equivalent to these, Abel's radical being $[(1-c^2x^2)(1+e^2x^2)]^{\frac{1}{2}}$, and Jacobi's, like that of Legendre's $[(1-x^2)(1-k^2x^2)]_{\frac{1}{2}}$), are expressed in the form of fractions, having a common denominator; and this, together with the three numerators, resolved into a doubly infinite series of factors; i.e. the general factor contains two independent integers. These formulæ may conveniently be referred to as "Abel's double factorial expressions" for the functions ϕ, f, F. By dividing each of these products into an infinite number of partial products, and expressing these by means of circular or exponential functions, Abel has obtained (pp. 216—218) two other systems of formulæ for the same quantities, which may be referred to as "Abel's first and second single factorial systems." The theory of the functions forming the above numerators and denominator, is mentioned by Abel in a letter to Legendre (*Œuvres*, t. II. p. 259 [Ed. 2, p. 272]), as a subject to which his attention had been directed, but none of his researches upon them have ever been published. Abel's double factorial expressions have nowhere anything analogous to them in Jacobi's *Fund. Nova;* but the system of formulæ analogous to the first single factorial system is given by Jacobi (p. 86), and the second system is implicitly contained in some of the subsequent formulæ. The functions forming the numerator and denominator of $\sin\text{am}\,.\,u$, Jacobi represents, omitting a constant factor, by $\text{H}\,(u)$, $\Theta\,(u)$; and proceeds to investigate the properties of these new functions. This he principally effects by means of a very remarkable equation of the form

$$l\Theta\,(u) = \tfrac{1}{2}\,A u^2 + B\textstyle\int_0 du\,.\int_0 du \sin^2\text{am}\, u,$$

(*Fund. Nova*, pp. 145, 133), by which $\Theta(u)$ is made to depend on the known function $\sin\text{am}.u$. The other two numerators are easily expressed by means of the two functions H, Θ.

From the omission of Abel's double factorial expressions, which are the only ones which display clearly the real nature of the functions in the numerators and denominators; and besides, from the different form of Jacobi's radical, which complicates the transformation from an impossible to a possible argument, it is difficult to trace the connection between Jacobi's formulæ; and in particular to account for the appearance of an exponential factor which runs through them. It would seem therefore natural to make the whole theory depend upon the definitions of the new transcendental functions to which Abel's double factorial expressions lead one, even if these definitions were not of such a nature, that one only wonders they should never have been assumed *à priori* from the analogy of the circular functions sin, cos, and quite independently of the theory of elliptic integrals. This is accordingly what I have done in the present paper, in which therefore I assume no single property of elliptic functions, but demonstrate them all, from my fundamental equations. For the sake however of comparison, I retain entirely the notation of Abel. Several of the formulæ that will be obtained are new.

The infinite product

$$x\Pi\left(1+\frac{x}{m\omega}\right) \quad\ldots\ldots\ldots\ldots\ldots\ldots\ldots\ldots\ldots\ldots\ldots (1),$$

where m receives the integer values $\pm 1, \pm 2, \ldots \pm r$, converges, as is well known, as r becomes indefinitely great to a determinate function $\sin\frac{\pi x}{\omega}$ of x; the theory of which might, if necessary, be investigated from this property assumed as a definition. We are thus naturally led to investigate the properties of the new transcendant

$$u = x\Pi\Pi\left(1+\frac{x}{m\omega+n\upsilon i}\right) \ldots\ldots\ldots\ldots\ldots\ldots\ldots\ldots\ldots\ldots (2):$$

m and n are integer numbers, positive or negative; and it is supposed that whatever positive value is attributed to either of these, the corresponding negative one is also given to it. $i=\sqrt{(-1)}$, ω and υ are real positive quantities. (At least this is the standard case, and the only one we shall explicitly consider. Many of the formulæ obtained are true, with slight modifications, whatever ω and υ represent, provided only $\omega : \upsilon i$ be not a real quantity; for if it were so, $m\omega+n\upsilon i$ for some values of m, n would vanish, or at least become indefinitely small, and u would cease to be a determinate function of x.)[1]

Now the value of the above expression, or, as for the sake of shortness it may be written, of the function

$$u = x\Pi\Pi\left\{1+\frac{x}{(m,\ n)}\right\} \ldots\ldots\ldots\ldots\ldots\ldots\ldots\ldots\ldots\ldots (3),$$

[1] I have examined the case of impossible values of ω and υ in a paper which I am preparing for *Crelle's Journal*. [The paper here referred to is [25], actually published in *Liouville's Journal*].

depends in a remarkable manner on the mode in which the superior limits of m, n are assigned. Imagine m, n to have any positive or negative integer values satisfying the equation

$$\phi(m^2, n^2) < T \quad \ldots\ldots (4).$$

Consider, for greater distinctness, m, n as the coordinates of a point; the equation $\phi(m^2, n^2) = T$ belongs to a certain curve symmetrical with respect to the two axes. I suppose besides that this is a continuous curve without multiple points, and such that the minimum value of a radius vector through the origin continually increases as T increases, and becomes infinite with T. The curve may be *analytically* discontinuous, this is of no importance. The condition with respect to the limits is then that m and n must be integer values denoting the coordinates of a point *within* the above curve, the whole system of such integer values being successively taken for these quantities.

Suppose, next, u' denotes the same function as u, except that the limiting condition is

$$\phi'(m^2, n^2) < T' \quad \ldots\ldots (5).$$

The curve $\phi'(m^2, n^2) = T'$ is supposed to possess the same properties with the other limiting curve, and, for greater distinctness, to lie entirely outside of it; but this last condition is nonessential.

These conditions being satisfied, the ratio $u' : u$ is very easily determined in the limiting case of T and T' infinite. In fact

$$\frac{u'}{u} = \Pi\Pi\left\{1 + \frac{x}{(m, n)}\right\} \quad \ldots\ldots (6),$$

or

$$l\frac{u'}{u} = \Sigma\Sigma l\left\{1 + \frac{x}{(m, n)}\right\} \quad \ldots\ldots (7),$$

the limiting conditions being

$$\phi(m^2, n^2) > T \quad \ldots\ldots (8),$$
$$\phi'(m^2, n^2) < T'.$$

Now

$$l\left\{1 + \frac{x}{(m, n)}\right\} = \frac{x}{(m, n)} - \tfrac{1}{2}\cdot\frac{x^2}{(m, n)^2} + \ldots\ldots (9),$$

$$l\frac{u'}{u} = x\,.\,\Sigma\Sigma\frac{1}{(m, n)} - \tfrac{1}{2}x^2\,.\,\Sigma\Sigma\frac{1}{(m, n)^2} + \ldots\ldots (10),$$

or, the alternate terms vanishing on account of the positive and negative values destroying each other,

$$l\frac{u'}{u} = -\tfrac{1}{2}x^2\,.\,\Sigma\Sigma\frac{1}{(m, n)^2} - \tfrac{1}{4}x^4\,.\,\Sigma\Sigma\frac{1}{(m, n)^4} - \ldots\ldots (11).$$

In general

$$\Sigma\Sigma\psi(m, n) = \iint\psi(m, n)\,dm\,dn + P \quad \ldots\ldots (12),$$

P denoting a series the first term of which is of the form $C\psi(m, n)$, and the remaining ones depending on the differential coefficients of this quantity with respect to m and n. The limits between which the two sides are to be taken, are identical.

In the present case, supposing T and T' indefinitely great, it is easy to see that the first term of the expression for $l\frac{u'}{u}$ is the only one which is not indefinitely small; and we have

$$l\frac{u'}{u} = -\tfrac{1}{2}Ax^2, \text{ or } u' = u\epsilon^{-\frac{1}{2}Ax^2} \ldots\ldots\ldots (13),$$

where

$$A = \iint \frac{dmdn}{(m,\ n)^2} = \iint \frac{dmdn}{(m\omega + nvi)^2} \ldots\ldots\ldots (14);$$

the limits of the integration being given by

$$\phi\,(m^2,\ n^2) > T \ldots\ldots\ldots (15),$$
$$\phi'(m^2,\ n^2) < T'.$$

Some particular cases are important. Suppose the limits of u' are given by

$$m^2\omega^2 < T^2, \quad n^2v^2 < T^2 \ldots\ldots\ldots (16),$$

and those of u, by

$$m^2\omega^2 + n^2v^2 < T^2 \ldots\ldots\ldots (17);$$

we have

$$A = \iint \frac{dmdn}{(m\omega + nvi)^2} \ldots\ldots\ldots (18),$$

$$= -\frac{1}{\omega}\int dn \left\{\frac{1}{T + nvi} - \frac{1}{\sqrt{(T^2 - n^2v^2)} + nvi} - \frac{1}{-\sqrt{(T^2 - n^2v^2)} + nvi} - \frac{1}{-T + nvi}\right\}$$

$$= -\frac{2}{\omega}\int dn \left\{\frac{T}{T^2 + n^2v^2} - \frac{\sqrt{(T^2 - n^2v^2)}}{T^2}\right\} \quad (nv = -T, \quad nv = T)$$

$$= -\frac{2}{\omega v}\int_{-1}^{1} d\theta \left\{\frac{1}{1+\theta^2} - \sqrt{(1-\theta^2)}\right\} = -\frac{2}{\omega v}(\pi - \pi) = 0;$$

or, in this case,

$$u' = u \ldots\ldots\ldots (19).$$

Again, let the limits of u' be

$$m^2\omega^2 < R'^2, \quad n^2v^2 < S'^2 \ldots\ldots\ldots (20),$$

and those of u,

$$m^2\omega^2 < R^2, \quad n^2v^2 < S^2 \ldots\ldots\ldots (21).$$

$$A = \iint \frac{dmdn}{(m\omega + nvi)^2} \ldots\ldots\ldots (22),$$

$$= -\frac{1}{\omega}\int dn \left\{\frac{1}{R' + nvi} - \frac{1}{R + nvi} + \frac{1}{-R + nvi} - \frac{1}{-R' + nvi}\right\},$$

where the limits are $n^2v^2 < S'^2$, for the terms containing R', $n^2v^2 < S^2$, for the terms containing R,

$$= -\frac{2}{\omega vi}\, l\, \frac{R'+S'i}{R'-S'i}\, \frac{R-Si}{R+Si} \quad \dots\dots(23),$$

$$= -\frac{4}{\omega v}(\lambda'-\lambda), \text{ if } \lambda' = \tan^{-1}\frac{S'}{R'}, \quad \lambda = \tan^{-1}\frac{S}{R},$$

the arcs λ, λ' being included between the limits 0, $\frac{1}{2}\pi$. Hence

$$u' = u\epsilon^{2(\lambda'-\lambda)\frac{x^2}{\omega v}} \quad \dots\dots(24).$$

In particular if $\frac{S'}{R'} = \frac{S}{R}$, $u' = u$. If $\frac{S'}{R'} = 0$, $\frac{S}{R} = 1$, $u' = u\epsilon^{-\frac{1}{2}\beta x^2}$; if $\frac{S'}{R'} = \infty$, $\frac{S}{R} = 1$, $u' = u\epsilon^{\frac{1}{2}\beta x^2}$: where $\beta = \frac{\pi}{\omega v}$, for which quantity it will continue to be used.

We may now completely define the functions whose properties are to be investigated. Writing, for shortness,

$$(m, n) = m\omega + nvi \quad \dots\dots(A),$$

$$(\overline{m}, n) = (m+\tfrac{1}{2})\,\omega + nvi,$$

$$(m, \overline{n}) = m\omega + (n+\tfrac{1}{2})\,vi,$$

$$(\overline{m}, \overline{n}) = (m+\tfrac{1}{2})\,\omega + (n+\tfrac{1}{2})\,vi\,;$$

we may put

$$\gamma x = x\Pi\Pi\left\{1 + \frac{x}{(m, n)}\right\} \quad \dots\dots(B),$$

$$gx = \Pi\Pi\left\{1 + \frac{x}{(\overline{m}, n)}\right\},$$

$$Gx = \Pi\Pi\left\{1 + \frac{x}{(m, \overline{n})}\right\},$$

$$\mathfrak{G}x = \Pi\Pi\left\{1 + \frac{x}{(\overline{m}, \overline{n})}\right\};$$

the limits being given respectively by the equations

$$\text{mod.}\,(m, n) < T, \quad \text{mod.}\,(\overline{m}, n) < T, \quad \text{mod.}\,(m, \overline{n}) < T, \quad \text{mod.}\,(\overline{m}, \overline{n}) < T,$$

T being finally infinite. The system of values $m = 0$, $n = 0$, is of course omitted in γx.

The functions γx, gx, Gx, $\mathfrak{G}x$, are all of them real finite functions of x, possessing properties analogous to that of u. Thus, representing any one of them by Jx, we have

$$Jx = \epsilon^{\pm\frac{1}{2}\beta x^2} J_{\pm\beta}\, x \quad \dots\dots(C),$$

where $J_{\pm\beta}x$ is the same as Jx, only for $J_\beta x$ the limits are given by $m^2\omega^2$ or $(m+\frac{1}{2})^2\omega^2 < R^2$, $n^2\upsilon^2$ or $(n+\frac{1}{2})^2\upsilon^2 < S$, ($R$, S, and $\frac{R}{S}$ infinite), and for $J_{-\beta}x$, by the same formulæ, (R, S, and $\frac{S}{R}$ infinite). It is to this equation that the most characteristic properties of the functions Jx are due.

The following equations are deduced immediately from the above definitions:

$$\gamma(-x) = -\gamma x, \quad g(-x) = gx, \quad G(-x) = Gx, \quad \mathfrak{G}(-x) = \mathfrak{G}x \ldots\ldots\ldots (D),$$
$$\gamma(0) = 0, \quad g(0) = 1, \quad G(0) = 1, \quad \mathfrak{G}(0) = 1,$$
$$\gamma'(0) = 1.$$

Suppose $\gamma_1 x$, $g_1 x$, $G_1 x$, $\mathfrak{G}_1 x$, are the values that would have been obtained for γx, gx, Gx, $\mathfrak{G}x$ by interchanging ω and υ,—then changing x into xi, and interchanging m and n, by which means the limiting equations are the same in the two cases, we obtain the following system of equations:

$$\gamma_1(xi) = i\gamma x \ldots\ldots\ldots\ldots\ldots\ldots\ldots\ldots\ldots\ldots\ldots\ldots\ldots (E),$$
$$g_1(xi) = Gx,$$
$$G_1(xi) = gx,$$
$$\mathfrak{G}_1(xi) = \mathfrak{G}x;$$

or otherwise,

$$\gamma(xi) = i\gamma_1 x \ldots\ldots\ldots\ldots\ldots\ldots\ldots\ldots\ldots\ldots\ldots\ldots\ldots (F),$$
$$g(xi) = G_1 x,$$
$$G(xi) = g_1 x,$$
$$\mathfrak{G}(xi) = \mathfrak{G}_1 x,$$

equations which are useful in transforming almost any other property of the functions J.

The functions $J_\beta x$ are changed one into another, except as regards a constant multiplier, by the change of x into $x+\frac{\omega}{2}$. This will be shown in a Note, or it may be seen from some formulæ deduced immediately from the definitions of the functions $J_\beta x$, which will be given in the sequel[1]. Observing the relation between Jx and $J_\beta x$, we have in particular

$$\gamma\left(x+\frac{\omega}{2}\right) = \epsilon^{\frac{1}{2}\beta\omega x} A gx \ldots\ldots\ldots\ldots\ldots\ldots\ldots\ldots\ldots\ldots (G),$$

$$g\left(x+\frac{\omega}{2}\right) = \epsilon^{\frac{1}{2}\omega\beta x} B\gamma x,$$

$$G\left(x+\frac{\omega}{2}\right) = \epsilon^{\frac{1}{2}\beta\omega x} C\mathfrak{G}x,$$

$$\mathfrak{G}\left(x+\frac{\omega}{2}\right) = \epsilon^{\frac{1}{2}\beta\omega x} DGx,$$

[1] Not given in the present paper. [The Note *was* given, see p. 154, and the formulæ referred to must have been the formulæ (M) p. 144.]

where A, B, C, D, are most simply determined by writing $x=0$, $x=-\frac{\omega}{2}$. Putting at the same time $\epsilon^{\beta\omega^2}=\epsilon^{\frac{\pi\omega}{\upsilon}}=q_1^{-1}$,

$$A=\gamma\left(\frac{\omega}{2}\right)\quad\text{.......................}(H),$$

$$B=-q_1^{-\frac{1}{4}}\div\gamma\left(\frac{\omega}{2}\right),$$

$$C=G\left(\frac{\omega}{2}\right)=q_1^{-\frac{1}{4}}\div\mathfrak{G}\left(\frac{\omega}{2}\right),$$

$$D=\mathfrak{G}\left(\frac{\omega}{2}\right)=q_1^{-\frac{1}{4}}\div G\left(\frac{\omega}{2}\right);$$

whence also

$$G\left(\frac{\omega}{2}\right)\mathfrak{G}\left(\frac{\omega}{2}\right)=q_1^{-\frac{1}{4}}\quad\text{.......................}(25).$$

Similarly, the functions $J_{-\beta}x$ are changed one into the other by the change of x into $x+\frac{1}{2}\upsilon i$. We have in the same way

$$\gamma\left(x+\frac{\upsilon i}{2}\right)=\epsilon^{-\frac{1}{2}\beta\upsilon xi}A'Gx\quad\text{.......................}(I),$$

$$g\left(x+\frac{\upsilon i}{2}\right)=\epsilon^{-\frac{1}{2}\beta\upsilon xi}B'\mathfrak{G}x,$$

$$G\left(x+\frac{\upsilon i}{2}\right)=\epsilon^{-\frac{1}{2}\beta\upsilon xi}C'\gamma x,$$

$$\mathfrak{G}\left(x+\frac{\upsilon i}{2}\right)=\epsilon^{-\frac{1}{2}\beta\upsilon xi}D'gx.$$

Whence

$$A'=\gamma\left(\frac{\upsilon i}{2}\right)\quad\text{.......................}(J).$$

$$B'=g\left(\frac{\upsilon i}{2}\right)=q^{-\frac{1}{4}}\div\mathfrak{G}\left(\frac{\upsilon i}{2}\right),$$

$$C'=-q^{-\frac{1}{4}}\div\gamma\left(\frac{\upsilon i}{2}\right),$$

$$D'=\mathfrak{G}\left(\frac{\upsilon i}{2}\right)=q^{-\frac{1}{4}}\div g\left(\frac{\upsilon i}{2}\right);$$

where $\epsilon^{\beta\upsilon^2}=\epsilon^{\frac{\pi\upsilon}{\omega}}=q^{-1}$. It is obvious that the relation between q and q_1 is $lq\,.\,lq_1=-\pi^2$.

We obtain from the above

$$g\left(\frac{\upsilon i}{2}\right)\mathfrak{G}\left(\frac{\upsilon i}{2}\right)=q^{-\frac{1}{4}}\quad\text{.......................}(26).$$

Also, by making $x=\frac{\upsilon i}{2}$ in the expression for $\gamma\left(x+\frac{\omega}{2}\right)$ and $x=\frac{\omega}{2}$ in that for $\gamma\left(x+\frac{\upsilon i}{2}\right)$, we have

$$\gamma\left(\frac{\omega}{2}\right) g\left(\frac{\upsilon i}{2}\right)=-i\gamma\left(\frac{\upsilon i}{2}\right) G\left(\frac{\omega}{2}\right) \quad\quad (27),$$

and the same or an equivalent one would have been obtained from the functions g, G, $\mathfrak{G}$.

By combining the above systems, we deduce one of the form

$$\gamma\left(x+\frac{\omega}{2}+\frac{\upsilon i}{2}\right)=\epsilon^{\frac{1}{2}\beta x(\omega-\upsilon i)} A''\mathfrak{G}x \quad\quad (K),$$

$$g\left(x+\frac{\omega}{2}+\frac{\upsilon i}{2}\right)=\epsilon^{\frac{1}{2}\beta x(\omega-\upsilon i)} B''Gx,$$

$$G\left(x+\frac{\omega}{2}+\frac{\upsilon i}{2}\right)=\epsilon^{\frac{1}{2}\beta x(\omega-\upsilon i)} C''gx,$$

$$\mathfrak{G}\left(x+\frac{\omega}{2}+\frac{\upsilon i}{2}\right)=\epsilon^{\frac{1}{2}\beta x(\omega-\upsilon i)} D''\gamma x;$$

and, observing the equation $\epsilon^{\beta\omega\upsilon i}=\epsilon^{\pi i}=(-1)$, with the following values for the coefficients,

$$A''=(-1)^{\frac{1}{4}}.\gamma\left(\frac{\omega}{2}\right) g\left(\frac{\upsilon i}{2}\right) \quad\quad (L),$$

$$B''=-(-1)^{\frac{1}{4}}.q_1^{-\frac{1}{4}}.\gamma\left(\frac{\upsilon i}{2}\right)\div\gamma\left(\frac{\omega}{2}\right),$$

$$C''=(-1)^{\frac{1}{4}}.\mathfrak{G}\left(\frac{\omega}{2}\right)\mathfrak{G}\left(\frac{\upsilon i}{2}\right),$$

$$D''=-(-1)^{\frac{1}{4}}.q^{-\frac{1}{4}}.\mathfrak{G}\left(\frac{\omega}{2}\right)\div\gamma\left(\frac{\upsilon i}{2}\right).$$

Collecting the formulæ which connect $\gamma\left(\frac{\omega}{2}\right)$, $\gamma\left(\frac{\upsilon i}{2}\right)$, ... these are

$$g\left(\frac{\omega}{2}\right)=0 \quad\quad (L\ bis),$$

$$G\left(\frac{\upsilon i}{2}\right)=0,$$

$$G\left(\frac{\omega}{2}\right)\mathfrak{G}\left(\frac{\omega}{2}\right)=q_1^{-\frac{1}{4}},$$

$$g\left(\frac{\upsilon i}{2}\right)\mathfrak{G}\left(\frac{\upsilon i}{2}\right)=q^{-\frac{1}{4}},$$

$$\gamma\left(\frac{\omega}{2}\right) g\left(\frac{\upsilon i}{2}\right)=-i\gamma\left(\frac{\upsilon i}{2}\right) G\left(\frac{\omega}{2}\right).$$

And by the assistance of these

$$B''C'' \div A''D'' = B'D' \div A'C' = CD \div AB = -1 \quad\ldots\ldots\ldots\ldots\ldots\ldots (28),$$

$$A'B' \div C'D' = -A''B'' \div C''D'' = -\gamma^2\left(\frac{\upsilon i}{2}\right) \div \mathfrak{G}^2\left(\frac{\upsilon i}{2}\right),$$

$$A''C'' \div B''D'' = -A'C' \div B'D' = \gamma^2\left(\frac{\omega}{2}\right) \div \mathfrak{G}^2\left(\frac{\omega}{2}\right),$$

$$A\,D \div B\,C = -A'D' \div B'C' = \gamma^2\left(\frac{\omega}{2}\right) \div G^2\left(\frac{\omega}{2}\right) = -\gamma^2\left(\frac{\upsilon i}{2}\right) \div g^2\left(\frac{\upsilon i}{2}\right),$$

which will be required presently.

It is now easy to proceed to the general systems of formulæ,

$$\Theta = (-1)^{mn}\, \epsilon^{\beta x(m\omega - n\upsilon i)}\, q_1^{-\frac{1}{2}m^2}\, q^{-\frac{1}{2}n^2} \quad\ldots\ldots\ldots\ldots (M),$$

$$\begin{aligned}
\gamma\{x+(m,\ n)\} &= (-1)^{m+n}\,.\,\Theta\gamma x\,,\\
g\{x+(m,\ n)\} &= (-1)^{m}\,.\,\Theta g x\,,\\
G\{x+(m,\ n)\} &= (-1)^{n}\,.\,\Theta G x\,,\\
\mathfrak{G}\{x+(m,\ n)\} &= \Theta\mathfrak{G} x.
\end{aligned}$$

$$\Phi = (-1)^{n(m+\frac{1}{2})}\, \epsilon^{\beta x[(m+\frac{1}{2})\omega - n\upsilon i]}\, q_1^{-\frac{1}{2}m^2 - \frac{1}{2}m}\, q^{-\frac{1}{2}n^2}.$$

$$\begin{aligned}
\gamma\{x+(\overline{m},\ n)\} &= (-1)^{m+n}\,.\,\Phi A g x\,,\\
g\{x+(\overline{m},\ n)\} &= (-1)^{m}\,.\,\Phi B\gamma x\,,\\
G\{x+(\overline{m},\ n)\} &= (-1)^{n}\,.\,\Phi C\mathfrak{G} x,\\
\mathfrak{G}\{x+(\overline{m},\ n)\} &= \Phi D G x.
\end{aligned}$$

$$\Psi = (-1)^{m(n+\frac{1}{2})}\, \epsilon^{\beta x[m\omega - (n+\frac{1}{2})\upsilon i]}\, q_1^{-\frac{1}{2}m^2}\, q^{-\frac{1}{2}n^2 - \frac{1}{2}n}$$

$$\begin{aligned}
\gamma\{x+(m,\ \overline{n})\} &= (-1)^{m+n}.\,\Psi A' G x,\\
g\{x+(m,\ \overline{n})\} &= (-1)^{m}\,.\,\Psi B'\mathfrak{G} x,\\
G\{x+(m,\ \overline{n})\} &= (-1)^{n}\,.\,\Psi C'\gamma x\,,\\
\mathfrak{G}\{x+(m,\ \overline{n})\} &= \Psi D' g x\,.
\end{aligned}$$

$$\Omega = (-1)^{mn+\frac{1}{2}m+\frac{1}{2}n}\, \epsilon^{\beta x[(m+\frac{1}{2})\omega - (n+\frac{1}{2})\upsilon i]}\, q_1^{-\frac{1}{2}m^2 - \frac{1}{2}m}\, q^{-\frac{1}{2}n^2 - \frac{1}{2}n}.$$

$$\begin{aligned}
\gamma\{x+(\overline{m},\ \overline{n})\} &= (-1)^{m+n}.\,\Omega A''\mathfrak{G} x,\\
g\{x+(\overline{m},\ \overline{n})\} &= (-1)^{m}\,.\,\Omega B'' G x\,,\\
G\{x+(\overline{m},\ \overline{n})\} &= (-1)^{n}\,.\,\Omega C'' g x\,,\\
\mathfrak{G}\{x+(\overline{m},\ \overline{n})\} &= .\,\Omega D''\gamma x\,.
\end{aligned}$$

Suppose $x=0$, we have the new systems,

$$\Theta_0=(-1)^{mn}\,q_1^{-\frac{1}{2}m^2}q^{-\frac{1}{2}n^2}\ \dots\dots\dots\dots\ (M\ bis).$$

$$\gamma\,(m,\ n)=0,\qquad\qquad \gamma'\,(m,\ n)=(-1)^{m+n}\,\Theta_0,$$
$$g\,(m,\ n)=(-1)^m\,\Theta_0,$$
$$G\,(m,\ n)=(-1)^n\,\Theta_0,$$
$$\mathfrak{G}\,(m,\ n)=\qquad\Theta_0.$$

$$\Phi_0=(-1)^{n\,(m+\frac{1}{2})}\,q_1^{-\frac{1}{2}m^2-\frac{1}{2}m}\,q^{-\frac{1}{2}n^2}.$$

$$\gamma\,(\overline{m},\ n)=(-1)^{m+n}\,\Phi_0A,$$
$$g\,(\overline{m},\ n)=0,\qquad\qquad g'\,(\overline{m},\ n)=(-1)^m\,\Phi_0B,$$
$$G\,(\overline{m},\ n)=(-1)^n\,\Phi_0C,$$
$$\mathfrak{G}\,(\overline{m},\ n)=\qquad\Phi_0D.$$

$$\Psi_0=(-1)^{m\,(n+\frac{1}{2})}\,q_1^{-\frac{1}{2}m^2}\,q^{-\frac{1}{2}n^2-\frac{1}{2}n}.$$

$$\gamma\,(m,\ \overline{n})=(-1)^{m+n}\,\Psi_0A',$$
$$g\,(m,\ \overline{n})=(-1)^m\quad\Psi_0\,B',$$
$$G\,(m,\ \overline{n})=0,\qquad\qquad G'\,(m,\ \overline{n})=(-1)^n\,\Psi_0C',$$
$$\mathfrak{G}\,(m,\ \overline{n})=\qquad\Psi_0D'.$$

$$\Omega_0=(-1)^{mn+\frac{1}{2}m+\frac{1}{2}n}\,q_1^{-\frac{1}{2}m^2-\frac{1}{2}m}\,q^{-\frac{1}{2}n^2-\frac{1}{2}n}.$$

$$\gamma\,(\overline{m},\ \overline{n})=(-1)^{m+n}\,\Omega_0A'',$$
$$g\,(\overline{m},\ \overline{n})=(-1)^m\quad\Omega_0B'',$$
$$G\,(\overline{m},\ \overline{n})=(-1)^n\quad\Omega_0C'',$$
$$\mathfrak{G}\,(\overline{m},\ \overline{n})=0,\qquad\qquad \mathfrak{G}'\,(\overline{m},\ \overline{n})=\Omega_0D''.$$

We obtain immediately, by taking the logarithmic differentials of the functions γx, gx, Gx, $\mathfrak{G}x$, the equations

$$\gamma'x\div\gamma x\ =\Sigma\Sigma\,\{x-(m,\ n)\}^{-1},\quad m=0,\quad n=0\ \text{admissible},\ \dots\dots\ (N),$$
$$g'x\div gx\ =\Sigma\Sigma\,\{x-(\overline{m},\ n)\}^{-1},$$
$$G'x\div Gx=\Sigma\Sigma\,\{x-(m,\ \overline{n})\}^{-1},$$
$$\mathfrak{G}'x\div\mathfrak{G}x=\Sigma\Sigma\,\{x-(\overline{m},\ \overline{n})\}^{-1},$$

the limits being the same as in the case of the factorial expressions.

Consider an equation

$$gx\,Gx\div\gamma x\,\mathfrak{G}x=\Sigma\Sigma\,[\mathfrak{A}\,\{x-(m,\ n)\}^{-1}+\mathfrak{B}\,\{x-(\overline{m},\ \overline{n})\}^{-1}]\ \dots\dots\dots(29),$$

we have $\mathfrak{A} = g(m, n)\, G(m, n) \div \gamma'(m, n)\, \mathfrak{G}(m, n) = 1$ ....(30),

$$\mathfrak{B} = g(\bar{m}, \bar{n})\, G(\bar{m}, \bar{n}) \div \gamma(\bar{m}, \bar{n})\, \mathfrak{G}'(\bar{m}, \bar{n}) = B''C'' \div A''D'' = -1 \ \dots\ (31).$$

(The application of the ordinary method of decomposition into partial fractions, which is in general exceedingly precarious when applied to transcendental functions, is justified here by a theorem of Cauchy's, which will presently be quoted.) We have thus

$$gxGx \div \gamma x\mathfrak{G}x = (\gamma'x \div \gamma x) - (\mathfrak{G}'x \div \mathfrak{G}x),$$

and similarly

$$gx\,\mathfrak{G}x \div \gamma x\ \ Gx = (\gamma'x \div \gamma x) - (G'x \div Gx), \qquad \dots\dots\dots\dots\dots (O),$$

$$Gx\mathfrak{G}x \div \gamma x\ \ gx = (\gamma'x \div \gamma x) - (g'x \div gx),$$

$$-b^2\gamma x\mathfrak{G}x \div gx\ \ Gx = (g'x \div gx) - (G'x \div Gx),$$

$$e^2\gamma x\ gx \div Gx\,\mathfrak{G}x = (G'x \div Gx) - (\mathfrak{G}'x \div \mathfrak{G}x),$$

$$c^2\gamma xGx \div \mathfrak{G}x\ gx = (\mathfrak{G}'x \div \mathfrak{G}x) - (g'x \div gx);$$

in which we have written

$$\gamma\left(\frac{\upsilon i}{2}\right) \div \mathfrak{G}\left(\frac{\upsilon i}{2}\right) = \frac{i}{e} \ \dots\dots\dots\dots\dots (32),$$

$$\gamma\left(\frac{\omega}{2}\right) \div \mathfrak{G}\left(\frac{\omega}{2}\right) = \frac{1}{c},$$

$$\gamma\left(\frac{\omega}{2}\right) \div G\left(\frac{\omega}{2}\right) = -i\left\{\gamma\left(\frac{\upsilon i}{2}\right) \div g\left(\frac{\upsilon i}{2}\right)\right\} = \frac{1}{b}.$$

Eliminating the derived coefficients,

$$G^2x - \mathfrak{G}^2x = e^2\gamma^2x, \qquad \dots\dots\dots\dots\dots (33),$$

$$g^2x - G^2x = -b^2\gamma^2x,$$

$$\mathfrak{G}^2x - g^2x = c^2\gamma^2x.$$

Adding these equations, $b^2 = e^2 + c^2$, or $b = \sqrt{(e^2 + c^2)}$, in which sense it will continue to be used.

Also,

$$g^2x = \mathfrak{G}^2x - c^2\gamma^2x, \qquad \dots\dots\dots\dots\dots (P).$$

$$G^2x = \mathfrak{G}^2x + e^2\gamma^2x.$$

Suppose

$$\phi x = \gamma x \div \mathfrak{G}x, \qquad \dots\dots\dots\dots\dots (Q),$$

$$fx = gx \div \mathfrak{G}x,$$

$$Fx = Gx \div \mathfrak{G}x,$$

then

$$f^2x = 1 - c^2\phi^2x, \qquad \dots\dots\dots\dots\dots (R),$$

$$F^2x = 1 + e^2\phi^2x,$$

and also

$$\begin{aligned}\phi' x &= \ \ fx\, Fx\ , \ \ldots\ldots\ldots\ldots\ldots\ldots\ldots\ldots\ldots\ldots (S),\\ f' x &= - c^2 \phi x\, Fx,\\ F' x &= \ \ e^2 \phi x f x\ .\end{aligned}$$

Hence, putting for fx, Fx, their values,

$$1 = \frac{\phi' x}{\sqrt{(1 - c^2\phi^2 x)\,(1 + e^2\phi^2 x)}} \ \ldots\ldots\ldots\ldots\ldots\ldots (T);$$

or writing $\phi x = y$, and integrating,

$$x = \int_0^{\phi x} \frac{dy}{\sqrt{(1 - c^2 y^2)\,(1 + e^2 y^2)}} \ \ldots\ldots\ldots\ldots\ldots\ldots (U),$$

or

$$\phi^{-1} y = \int_0^{y} \frac{dy}{\sqrt{(1 - c^2 y^2)\,(1 + e^2 y^2)}},$$

which shows that ϕ is an inverse elliptic function.

The equations which are the foundation of the theory of the functions ϕ, f, F, are deduced immediately from the equations (S). (Abel, *Œuvres*, tom. I. p. 143 [Ed. 2, p. 268.]) These are

$$\phi\,(x + y) = \frac{\phi x\, fy\, Fy + \phi y\, fx\, Fx}{1 + e^2 c^2\, \phi^2 x\, \phi^2 y} \ \ldots\ldots\ldots\ldots\ldots\ldots (V),$$

$$f\,(x + y) = \frac{fx\, fy - c^2\, \phi x\, \phi y\, Fx\, Fy}{1 + e^2 c^2\, \phi^2 x\, \phi^2 y},$$

$$F\,(x + y) = \frac{Fx\, Fy + e^2\, \phi x\, \phi y\, fx\, fy}{1 + e^2 c^2\, \phi^2 x\, \phi^2 y};$$

so that from this point we may take for granted any properties of these functions. We see, for instance, immediately,

$$\phi\left(\frac{vi}{2}\right) = \frac{i}{e},\quad \phi\left(\frac{\omega}{2}\right) = \frac{1}{c};$$

whence

$$\frac{\omega}{2} = \int_0^{\frac{1}{c}} \frac{dy}{\sqrt{(1 - c^2 y^2)\,(1 + e^2 y^2)}} \ldots\ldots\ldots\ldots\ldots\ldots\ldots (W),$$

$$\frac{vi}{2} = \int_0^{\frac{i}{e}} \frac{dy}{\sqrt{(1 - c^2 y^2)\,(1 + e^2 y^2)}},\ \text{ or }\ \frac{v}{2} = \int_0^{\frac{1}{e}} \frac{dy}{\sqrt{(1 + c^2 y^2)\,(1 - e^2 y^2)}} \ \ldots\ldots (X),$$

which give the values of ω, v in terms of c, e; values which may be developed in a variety of ways, in infinite series. We may also express $\gamma\left(\frac{\omega}{2}\right)$, &c., and consequently A, B... &c., by means of the quantities c, e. We have only to combine the equations

$$\gamma\left(\frac{\omega}{2}\right) \div \mathfrak{G}\left(\frac{\omega}{2}\right) = \frac{1}{c},\quad \gamma\left(\frac{vi}{2}\right) \div \mathfrak{G}\,\frac{vi}{2} = \frac{i}{e},\quad G\left(\frac{\omega}{2}\right) \div \mathfrak{G}\left(\frac{\omega}{2}\right) = \frac{b}{c},\quad g\left(\frac{vi}{2}\right) \div \mathfrak{G}\left(\frac{vi}{2}\right) = \frac{b}{e} \ldots \ (34),$$

with the former relations between these quantities, and we have

$$\gamma\left(\frac{\omega}{2}\right) = b^{-\frac{1}{2}} c^{-\frac{1}{2}} q_1^{-\frac{1}{8}}, \qquad \gamma\left(\frac{\upsilon i}{2}\right) = ib^{-\frac{1}{2}} e^{-\frac{1}{2}} q^{-\frac{1}{8}}, \ldots\ldots\ldots\ldots (Y),$$

$$g\left(\frac{\omega}{2}\right) = 0, \qquad g\left(\frac{\upsilon i}{2}\right) = b^{\frac{1}{2}} e^{-\frac{1}{2}} q^{-\frac{1}{8}},$$

$$G\left(\frac{\omega}{2}\right) = b^{\frac{1}{2}} c^{-\frac{1}{2}} q_1^{-\frac{1}{8}}, \qquad \mathfrak{G}\left(\frac{\upsilon i}{2}\right) = 0,$$

$$\mathfrak{G}\left(\frac{\omega}{2}\right) = b^{-\frac{1}{2}} c^{\frac{1}{2}} q_1^{-\frac{1}{8}}, \qquad \mathfrak{G}\left(\frac{\upsilon i}{2}\right) = b^{-\frac{1}{2}} e^{\frac{1}{2}} q^{-\frac{1}{8}}.$$

$$\begin{aligned}
A &= b^{-\frac{1}{2}} c^{-\frac{1}{2}} q_1^{-\frac{1}{8}}, & A' &= ib^{-\frac{1}{2}} e^{-\frac{1}{2}} q^{-\frac{1}{8}}, & A'' &= (-1)^{\frac{1}{4}} c^{-\frac{1}{2}} e^{-\frac{1}{2}} q_1^{-\frac{1}{8}} q^{-\frac{1}{8}}, \quad \ldots (Z).\\
B &= -b^{\frac{1}{2}} c^{\frac{1}{2}} q_1^{-\frac{1}{8}}, & B' &= b^{\frac{1}{2}} e^{-\frac{1}{2}} q^{-\frac{1}{8}}, & B'' &= -(-1)^{\frac{1}{4}} ic^{\frac{1}{2}} e^{-\frac{1}{2}} q_1^{-\frac{1}{8}} q^{-\frac{1}{8}},\\
C &= b^{\frac{1}{2}} c^{-\frac{1}{2}} q_1^{-\frac{1}{8}}, & C' &= -ib^{\frac{1}{2}} e^{\frac{1}{2}} q^{-\frac{1}{8}}, & C'' &= (-1)^{\frac{1}{4}} c^{-\frac{1}{2}} e^{\frac{1}{2}} q_1^{-\frac{1}{8}} q^{-\frac{1}{8}},\\
D &= b^{-\frac{1}{2}} c^{\frac{1}{2}} q_1^{-\frac{1}{8}}, & D' &= b^{-\frac{1}{2}} e^{\frac{1}{2}} q^{-\frac{1}{8}}, & D'' &= -(-1)^{\frac{1}{4}} ic^{\frac{1}{2}} e^{\frac{1}{2}} q_1^{-\frac{1}{8}} q^{-\frac{1}{8}},
\end{aligned}$$

which are to be substituted in any formulæ into which these quantities enter.

The following is Cauchy's Theorem, (*Exercises de Math.* t. II. p. 289).

"If in attributing to the modulus r of the variable

$$z = r\{\cos p + \surd(-1)\sin p\} \ldots\ldots\ldots\ldots (35),$$

infinitely great values, these can be chosen so that the two functions

$$\frac{fz + f(-z)}{2}, \quad \frac{fz - f(-z)}{2z}, \ldots\ldots\ldots\ldots (36),$$

sensibly vanish, whatever be the value of p, or vanish in general, though ceasing to do so and obtaining *finite* values for certain particular values of p; then

$$fx = \mathcal{E}\,\frac{\{(fz)\}}{x - z} \ldots\ldots\ldots\ldots (37),$$

the integral residue being reduced to its principal value."

To understand this, it is only necessary to remark that the integral residue in question is the series of fractions that would be obtained by the ordinary process of decomposition; and by the principal value is meant, that *all* those roots are to be taken, the modulus of which is not greater than a certain limit, this limit being afterwards made infinite.

Suppose now fx is a fraction, the numerator and denominator of which are monomials of the form $(\gamma x)^l (gx)^m \ldots$, l, $m \ldots$ being positive integers, and of course no common factor being left in the numerator and denominator.

Let λ be the excess of the degree of the denominator over that of the numerator. Suppose the modulus r of (z) has any value not the same with any of the moduli of

$$(m,\ n), \quad (\overline{m},\ n), \quad (m,\ \overline{n}), \quad (\overline{m},\ \overline{n}) \ldots\ldots\ldots\ldots (38).$$

Then we have

$$r(\cos p + i \sin p) = m\omega + nvi + \theta \text{ ....(39)},$$

θ being a finite quantity, such that none of the functions $J\theta$ vanish. m and n are the greatest integer values which allow the possible part of θ and the coefficient of its impossible part to remain positive. We have therefore

$$m^2\omega^2 + n^2v^2 = r^2 - M \text{ .... (40)},$$

M being finite; or when r is infinite, at least one of the values m, n is infinite. The function fz reduces itself to the form

$$q_1^{\frac{1}{2}\lambda m^2}\, q^{\frac{1}{2}\lambda n^2}\, \epsilon^{mA+nB} F \text{ .... (41)},$$

where F is finite. Hence q_1 and q being always less than unity, fz, and consequently both $\frac{1}{2}\{fz + f(-z)\}$ and $\frac{1}{2z}\{fz - f(-z)\}$ vanish for $r = \infty$, as long as λ is positive.

In the case of $\lambda = 0$, the conditions are still satisfied, if we suppose fx to denote an uneven function of x: for when $\lambda = 0$, the index of exponential in the above expression vanishes, or fz is constantly finite. But fz being an odd function of z, $fz + f(-z) = 0$. And $\frac{1}{2z}\{fz - f(-z)\}$ vanishes for z infinite, on account of the z in the denominator: hence the expansion is admissible in this case. But it is certainly so also, in a great many cases at least, where fz is an even function of z; for these may be deduced from the others by a simple change in the value of the variable. For instance, from the expansion of $\gamma x \div gx$, which is an odd function, by writing $x + \frac{vi}{2}$ for x, we obtain that of $Gx \div \mathfrak{G}x$, which is even.

A case of some importance is when the function is of the above form, multiplied by an exponential $\epsilon^{\frac{1}{2}ax^2+bx}$. Here writing $z = m\omega + nvi + \theta$, the admissibility of the formula depends on the evanescence of

$$\epsilon^{\frac{1}{2}a(m\omega+nvi)^2}\, q_1^{\frac{1}{2}\lambda m^2}\, q^{\frac{1}{2}\lambda n^2} \text{ .... (42)};$$

or, if $a = h + ki$, this becomes, omitting a finite factor,

$$\epsilon^{-\frac{1}{2}m^2(\lambda\beta-h)\omega^2 - \frac{1}{2}n^2(\lambda\beta+h)v^2 - kmn\,\omega v} \text{ ....(43)},$$

which vanishes if $h^2 + k^2 < \lambda^2\beta^2$, i.e. the modulus of a is less than $\lambda\beta$. The limiting case is admissible when the series is convergent.

We obtain in this way a very great variety of formulæ. For instance,

$$\epsilon^{\frac{1}{2}ax^2+bx} \div \gamma x = \Sigma\Sigma\left[(-1)^{-mn-m-n}\, \epsilon^{\frac{1}{2}a(m,\,n)^2+b(m,\,n)}\, q_1^{\frac{1}{2}m^2}\, q^{\frac{1}{2}n^2}\{x-(m,\,n)\}^{-1}\right] \text{ .... } (A'),$$

$$\epsilon^{\frac{1}{2}ax^2+bx} \div gx = -b^{\frac{1}{2}}c^{-\frac{1}{2}}\,\Sigma\Sigma\left[(-1)^{-mn-m-\frac{1}{2}n}\, \epsilon^{\frac{1}{2}a(\overline{m},\,n)^2+b(\overline{m},\,n)}\, q_1^{\frac{1}{2}(m+\frac{1}{2})^2}\, q^{\frac{1}{2}n^2}\{x-(\overline{m},\,n)\}^{-1}\right],$$

$$\epsilon^{\frac{1}{2}ax^2+bx} \div Gx = ib^{-\frac{1}{2}}e^{-\frac{1}{2}}\,\Sigma\Sigma\left[(-1)^{-mn-\frac{1}{2}m-n}\, \epsilon^{\frac{1}{2}a(m,\,\overline{n})^2+b(m,\,\overline{n})}\, q_1^{\frac{1}{2}m^2}\, q^{\frac{1}{2}(n+\frac{1}{2})^2}\{x-(m,\,\overline{n})\}^{-1}\right],$$

$$\epsilon^{\frac{1}{2}ax^2+bx} \div \mathfrak{G}x = ic^{-\frac{1}{2}}e^{-\frac{1}{2}}\,\Sigma\Sigma\left[(-1)^{-(m+\frac{1}{2})(n+\frac{1}{2})}\, \epsilon^{\frac{1}{2}a(\overline{m},\,\overline{n})^2+b(\overline{m},\,\overline{n})}\, q_1^{\frac{1}{2}(m+\frac{1}{2})^2}\, q^{\frac{1}{2}(n+\frac{1}{2})^2}\{x-(\overline{m},\,\overline{n})\}^{-1}\right],$$

in which the modulus of a must not exceed β: in the limiting cases, for $a=\beta$, b must be entirely impossible, and for $a=-\beta$, b must be entirely real. The formulæ for γx are

$$\epsilon^{\frac{1}{2}\beta x^2+bx} \div \gamma x = \Sigma\Sigma\,(-1)^{-m-n}\, q^{n^2}\, \epsilon^{b\,(m,\,n)}\, \{x-(m,\ n)\}^{-1} \quad\ldots\ldots\ldots\ldots (44),$$

$$\epsilon^{-\frac{1}{2}\beta x^2+bx} \div \gamma x = \Sigma\Sigma\,(-1)^{m-n}\, q_1^{m^2}\, \epsilon^{b\,(m,\,n)}\, \{x-(m,\ n)\}^{-1};$$

and for $b=0$,

$$\epsilon^{\frac{1}{2}\beta x^2} \div \gamma x = \Sigma\Sigma\,(-1)^{-m-n}\, q^{n^2}\, \{x-(m,\ n)\}^{-1} \quad\ldots\ldots\ldots\ldots (45),$$

$$\epsilon^{-\frac{1}{2}\beta x^2} \div \gamma x = \Sigma\Sigma\,(-1)^{-m-n}\, q_1^{m^2}\, \{x-(m,\ n)\}^{-1}.$$

Next the system,

$$\mathfrak{G}x \div \gamma x = \Sigma\Sigma\,(-1)^{m+n}\, \{x-(m,\ n)\}^{-1} \quad\ldots\ldots\ldots\ldots (B').$$

$$gx \div \gamma x = \Sigma\Sigma\,(-1)^{m}\, \{x-(m,\ n)\}^{-1},$$

$$Gx \div \gamma x = \Sigma\Sigma\,(-1)^{n}\, \{x-(m,\ n)\}^{-1};$$

$$\gamma x \div gx = -\,b^{-1}c^{-1}\,\Sigma\Sigma\,(-1)^{n}\, \{x-(\overline{m},\ n)\}^{-1},$$

$$Gx \div gx = -\,c^{-1}\,\Sigma\Sigma\,(-1)^{m+n}\, \{x-(\overline{m},\ n)\}^{-1},$$

$$\mathfrak{G}x \div gx = -\,b^{-1}\,\Sigma\Sigma\,(-1)^{m}\, \{x-(\overline{m},\ n)\}^{-1};$$

$$\gamma x \div Gx = -\,b^{-1}e^{-1}\,\Sigma\Sigma\,(-1)^{m}\, \{x-(m,\ \overline{n})\}^{-1},$$

$$gx \div Gx = ie^{-1}\,\Sigma\Sigma\,(-1)^{m+n}\, \{x-(m,\ \overline{n})\}^{-1},$$

$$\mathfrak{G}x \div Gx = ib^{-1}\,\Sigma\Sigma\,(-1)^{n}\, \{x-(m,\ \overline{n})\}^{-1};$$

$$\gamma x \div \mathfrak{G}x = ic^{-1}e^{-1}\,\Sigma\Sigma\,(-1)^{m+n}\, \{x-(\overline{m},\ \overline{n})\}^{-1},$$

$$gx \div \mathfrak{G}x = e^{-1}\,\Sigma\Sigma\,(-1)^{m}\, \{x-(\overline{m},\ \overline{n})\}^{-1},$$

$$Gx \div \mathfrak{G}x = ic^{-1}\,\Sigma\Sigma\,(-1)^{n}\, \{x-(\overline{m},\ \overline{n})\}^{-1},$$

which is partially given by Abel.

We may obtain, in like manner, expressions for the functions

$$\frac{1}{\gamma x\, gx},\ \frac{1}{\gamma x\, Gx},\ \ldots \text{(six terms of this form)} \quad\ldots\ldots\ldots\ldots (C'),$$

$$\frac{Gx}{\gamma x\, gx},\ \ldots \quad \text{(twelve)} \quad\ldots\ldots\ldots\ldots (D'),$$

$$\frac{\gamma x\, gx}{\mathfrak{G}x\, Gx},\ \ldots \quad \text{(six)} \quad\ldots\ldots\ldots\ldots (E'),$$

$$\frac{1}{\gamma x\, gx\, Gx},\ \ldots \quad \text{(four)} \quad\ldots\ldots\ldots\ldots (F'),$$

$$\frac{\mathfrak{G}x}{\gamma x\, gx\, Gx},\ \ldots \quad \text{(four)} \quad\ldots\ldots\ldots\ldots (G'),$$

$$\frac{1}{\gamma x\, gx\, Gx\, \mathfrak{G}x},\ \ldots \quad \text{(one)} \quad\ldots\ldots\ldots\ldots (H');$$

each of them, except (E'), (the system for which, admitting no exponential, has already been given,) multiplied by an exponential $\epsilon^{\frac{1}{2}ax^2+bx}$, the limits of a being $\pm 2\beta$, $\pm\beta$, $\pm\beta$, $\pm 3\beta$, $\pm 2\beta$, $\pm 4\beta$. For the limiting values, b must be entirely impossible for the superior limit, and entirely possible for the inferior one.

Thus the last case is

$$\frac{1}{\gamma x\, gx\, Gx\, \mathfrak{G}x}\,\epsilon^{\frac{1}{2}ax^2+bx} \quad\ldots\ldots\ldots\ldots\ldots\ldots (H'),$$

$$\begin{aligned}
&=\Sigma\Sigma\left[\epsilon^{\frac{1}{2}a(m,\,n)^2+b(m,\,n)}\, q_1^{2m^2}\, q^{2n^2}\, \{x-(m,\,n)\}^{-1}\right]\\
&-\Sigma\Sigma\left[\epsilon^{\frac{1}{2}a(\overline{m},\,n)^2+b(\overline{m},\,n)}\, q_1^{2(m+\frac{1}{2})^2}\, q^{2n^2}\, \{x-(\overline{m},\,n)\}^{-1}\right]\\
&+\Sigma\Sigma\left[\epsilon^{\frac{1}{2}a(m,\,\overline{n})^2+b(m,\,\overline{n})}\, q_1^{2m^2}\, q^{2(n+\frac{1}{2})^2}\, \{x-(m,\,\overline{n})\}^{-1}\right]\\
&-\Sigma\Sigma\left[\epsilon^{\frac{1}{2}a(\overline{m},\,\overline{n})^2+b(\overline{m},\,\overline{n})}\, q_1^{2(m+\frac{1}{2})^2}\, q^{2(n+\frac{1}{2})^2}\, \{x-(\overline{m},\,\overline{n})\}^{-1}\right];
\end{aligned}$$

in particular

$$\begin{aligned}
\frac{1}{\gamma x\, gx\, Gx\, \mathfrak{G}x}\,\epsilon^{2\beta x^2} &= \Sigma\Sigma q_1^{4m^2}\, \{x-(m,\,n)\}^{-1} \quad\ldots\ldots\ldots\ldots (46),\\
&-\Sigma\Sigma\, q_1^{(2m+1)^2}\, \{x-(\overline{m},\,n)\}^{-1}\\
&+\Sigma\Sigma\, q_1^{4m^2}\, \{x-(m,\,\overline{n})\}^{-1}\\
&+\Sigma\Sigma\, q_1^{(2m+1)^2}\, \{x-(\overline{m},\,\overline{n})\}^{-1},
\end{aligned}$$

or the analogous formula obtained by changing β, q_1, m into $-\beta$, q, n.

The function ϕ^2x, which is even, and for which $\lambda=0$, cannot be expanded entirely in a series of partial fractions: but $(x-a)^{-1}\phi^2x$ may be so expanded. Multiply by $(x-a)$, the second side has for its general term

$$(x-a)(Mx+N)\,\{x-(\overline{m},\,\overline{n})\}^{-2},$$

equivalent to

$$K'+(M'x+N')\,\{x-(\overline{m},\,\overline{n})\}^{-2}.$$

Summing all the K''s, we have an equation of the form

$$\phi^2x = A+\Sigma\Sigma\left[L\,\{x-(\overline{m},\,\overline{n})\}^{-2}+M\,\{x-(\overline{m},\,\overline{n})\}^{-1}\right] \ldots\ldots\ldots\ldots (47).$$

To determine the coefficients as simply as possible, change x into $x+\frac{1}{2}\omega+\frac{1}{2}n\omega i$,

$$-e^{-2}c^{-2}(\phi x)^{-2} = A+\Sigma\Sigma\left[L\,\{x-(m,\,n)\}^{-2}+M\,\{x-(m,\,n)\}^{-1}\right]\ldots\ldots (48),$$

$$L=-e^{-2}c^{-2}\left[\{x-(m,\,n)\}^2(\phi x)^{-2}\right],\ x=(m,\,n) \quad\ldots\ldots\ldots (49),$$

$$M=-e^{-2}c^{-2}\,\partial_x\left[\{x-(m,\,n)\}^2(\phi x)^{-2}\right],$$

or writing $x+(m,\,n)$ for x, and therefore $x=0$ in the values of L and M,

$$L=-e^{-2}c^{-2}\{x^2(\phi x)^{-2}\}=e^{-2}c^{-2} \ldots\ldots\ldots\ldots\ldots\ldots\ldots (50),$$

$$M=-e^{-2}c^{-2}\,\partial_x\{x^2(\phi x)^{-2}\}=0;$$

whence

$$\phi^2x = A-e^{-2}c^{-2}\,\Sigma\Sigma\,\{x-(\overline{m},\,\overline{n})\}^{-2} \ldots\ldots\ldots\ldots\ldots\ldots (51).$$

Integrating this last equation twice,

$$\textstyle\int_0 dx \int_0 dx\, \phi^2 x = \frac{1}{2} A x^2 + e^{-2} c^{-2} \Sigma\Sigma l \{x - (\overline{m}, \overline{n})\} \qquad (52),$$

or

$$\mathfrak{G}x = \epsilon^{-\frac{1}{2} e^2c^2 Ax^2 + e^2c^2 \int_0 dx \int_0 dx\, \phi^2 x} \qquad (53),$$

an equation from which it is easy to determine the coefficient A.

Suppose for a moment $\phi_{,}x = \int_0 \phi^2 x\, dx$, $\phi_{,,}x = \int_0 \phi_{,}x\, dx$; then, since $\phi^2(x+\omega) - \phi^2 x = 0$,

$$\phi_{,}(x+\omega) - \phi x = \phi_{,}\omega, \quad \phi_{,,}(x+\omega) - \phi_{,}x = \phi_{,,}\omega + x\phi_{,}\omega.$$

But similarly $\phi^2 x - \phi^2(\omega - x) = 0$; whence

$$\phi_{,}x + \phi_{,}(\omega - x) = \phi_{,}\omega, \quad \phi_{,,}x - \phi_{,,}(\omega - x) + \phi_{,,}\omega = x\phi_{,}\omega\,;$$

whence, writing $x = \frac{\omega}{2}$,

$$\phi_{,}\omega = 2\phi_{,}\left(\frac{\omega}{2}\right), \quad \phi_{,,}\omega = \omega\phi_{,}\left(\frac{\omega}{2}\right), \quad \text{or } \phi_{,,}(x+\omega) - \phi_{,,}x = \phi_{,}\left(\frac{\omega}{2}\right)(2x+\omega).$$

Hence

$$\mathfrak{G}(x+\omega) = \epsilon^{-\frac{1}{2} e^2c^2 \{A - \frac{1}{2}\omega\phi_{,}(\frac{1}{2}\omega)\}(2\omega x + \omega^2)}\, \mathfrak{G}x \qquad (54).$$

But

$$\mathfrak{G}(x+\omega) = \epsilon^{\beta\omega x} q_1^{-\frac{1}{2}} \mathfrak{G}x = \epsilon^{\frac{1}{2}\beta(2\omega x + \omega^2)}\, \mathfrak{G}x \qquad (55);$$

or, comparing these,

$$-e^2c^2 \left\{A - \frac{2}{\omega}\phi_{,}\left(\frac{\omega}{2}\right)\right\} = \beta \qquad (56),$$

$$-\tfrac{1}{2} e^2c^2 A = \tfrac{1}{2}\beta - \frac{e^2c^2}{\omega}\phi_{,}\left(\frac{\omega}{2}\right) \qquad (57),$$

or writing

$$M = \frac{e^2c^2}{\omega}\int_0^{\frac{1}{2}\omega} \phi^2 x\, dx \qquad (58),$$

then

$$\mathfrak{G}x = \epsilon^{(\frac{1}{2}\beta - M)x^2 + e^2c^2 \int_0 dx \int_0 dx\, \phi^2 x} \qquad (I');$$

which is the formula corresponding to the one of Jacobi's referred to at the beginning of this paper. Analogous formulæ may be deduced from it by writing $x + \frac{\omega}{2}$, or $x + \frac{\upsilon i}{2}$, or $x + \frac{\omega}{2} + \frac{\upsilon i}{2}$, instead of x.

The following formulæ, making the necessary changes of notation, are taken from Jacobi. We have

$$\phi^2(x+a) - \phi^2(x-a) = \frac{4\phi a\, fa\, Fa\, \phi x\, fx\, Fx}{(1 + e^2c^2\, \phi^2 a\, \phi^2 x)^2} \qquad (59),$$

whence $$\int_0 \{\phi^2(x+a) - \phi^2(x-a)\}\,dx = \frac{2\phi a\, fa\, Fa\, \phi^2 x}{1+e^2c^2\phi^2a\phi^2x} \quad \text{.................(60)},$$

the first side of which is

$$\int_{-a} \phi^2(x+a)\,dx - \int_a \phi^2(x-a)\,dx - 2\int_0 \phi^2 a\, da \quad \text{.................(61)}.$$

Hence, multiplying by e^2c^2, and observing the value of $\mathfrak{G}x$,

$$\frac{\mathfrak{G}'(x+a)}{\mathfrak{G}(x+a)} - \frac{\mathfrak{G}'(x-a)}{\mathfrak{G}(x-a)} - 2\frac{\mathfrak{G}'a}{\mathfrak{G}a} = \frac{2e^2c^2 fa\, Fa\, \phi a\, \phi^2 x}{1+e^2c^2\phi^2a\phi^2x} \quad \text{..............(62)}.$$

If in this case we interchange x, a and add,

$$\frac{\mathfrak{G}'x}{\mathfrak{G}x} + \frac{\mathfrak{G}'a}{\mathfrak{G}a} - \frac{\mathfrak{G}'(x+a)}{\mathfrak{G}(x+a)} = e^2c^2\,\phi a\,\phi x\,\phi(a+x) \quad \text{.................(63)}.$$

[By subtracting, we should have obtained an equation only differing from the above in the sign of a.]

Integrating the last equation but one, with respect to a,

$$l\mathfrak{G}(x+a) + l\mathfrak{G}(x-a) - 2l\mathfrak{G}x - 2l\mathfrak{G}a = l(1+e^2c^2\phi^2x\phi^2a),$$

the integral being taken from $a=0$. Hence

$$\mathfrak{G}(x+a)\,\mathfrak{G}(x-a) = \mathfrak{G}^2x\mathfrak{G}^2a\,(1+e^2c^2\phi^2x\phi^2a) \quad \text{...........(64)};$$

or $$\mathfrak{G}(x+a)\,\mathfrak{G}(x-a) = \mathfrak{G}^2x\mathfrak{G}^2a + e^2c^2\gamma^2x\gamma^2a,$$

whence also

$$\left.\begin{aligned} \gamma(x+a)\;\gamma(x-a) &= \gamma^2x\mathfrak{G}^2a - \gamma^2a\,\mathfrak{G}^2x,\\ g(x+a)\;g(x-a) &= g^2x\mathfrak{G}^2a - c^2g^2a\,\mathfrak{G}^2x,\\ G(x+a)\;G(x-a) &= G^2x\mathfrak{G}^2a + e^2G^2a\mathfrak{G}^2x, \end{aligned}\right\} \quad \text{...............} (J),$$

these equations being obtained from the first by the change of x into $x+\frac{\omega}{2}$, $x+\frac{vi}{2}$, $x+\frac{\omega}{2}+\frac{vi}{2}$. They form a most important group of formulæ in the present theory. By integrating the same formulæ with respect to x, and representing by $\Pi(x, a)$ the integral $\int_0 \frac{-e^2c^2\,\phi a\, fa\, Fa\, \phi^2x dx}{1+e^2c^2\phi^2a\phi^2x}$, Jacobi obtains

$$\Pi(x, a) = \tfrac{1}{2} l\frac{\mathfrak{G}(x-a)}{\mathfrak{G}(x+a)} + x\frac{\mathfrak{G}'a}{\mathfrak{G}a}:$$

an equation which conducts him almost immediately to the formulæ for the addition of the argument or of the parameter in the function Π. This, however, is not very closely connected with the present subject. For some formulæ also deduced from (63), by which $\dfrac{\mathfrak{G}(x-a)\,\mathfrak{G}(y-a)\,\mathfrak{G}(x+y+a)}{\mathfrak{G}(x+a)\,\mathfrak{G}(y+a)\,\mathfrak{G}(x+y-a)}$ is expressed in terms of the function ϕ, see Jacobi.

NOTE.—We have

$$\gamma_\beta x = x\Pi\Pi\left(1+\frac{x}{(m,\ n)}\right).$$

$$g_\beta x = \ \Pi\Pi\left(1+\frac{x}{(\overline{m},\ n)}\right),$$

the limits of n being $\pm q$, and those of m being $\pm p$, in the first case, and $p,\ -p-1$, in the second case. Also $\dfrac{p}{q}=\infty$.

We deduce immediately

$$\gamma_\beta\left(x+\frac{\omega}{2}\right)=\left(x+\frac{\omega}{2}\right)\Pi\Pi\left\{1+\frac{\left(x+\frac{\omega}{2}\right)}{(m,\ n)}\right\}=\Pi\Pi\left(1+\frac{x}{(\overline{m},\ n)}\right)\div\frac{\omega}{2}\,\Pi\Pi\,\frac{(m,\ n)}{(\overline{m},\ n)}$$

(paying attention to the omission of $(m=0,\ n=0)$ in $\gamma_\beta x$, and supposing that this value enters into the numerator of the expression just obtained, but not into its denominator). This is of the form

$$\gamma_\beta\left(x+\frac{\omega}{2}\right)=A\Pi\Pi\left(1+\frac{x}{(\overline{m},\ n)}\right);$$

but the limits are not the same in this product and in $g_\beta x$. In the latter m assumes the value $-p-1$, which it does not in the former; hence

$$\gamma_\beta\left(x+\frac{\omega}{2}\right)\div g_\beta\,x=A\div\Pi_n\left(1+\frac{x}{-(p+\frac{1}{2})\,\omega+nvi}\right),$$

and the above product reduces itself to unity in consequence of all the values assumed by n being indefinitely small compared with the quantity $(p+\frac{1}{2})$; we have therefore

$$\gamma_\beta\left(x+\frac{\omega}{2}\right)=Ag_\beta\,x \quad \text{..............................(65)},$$

and similar expressions for the remaining functions. To illustrate this further, suppose we had been considering, instead of $\gamma_\beta x$, the function $\gamma_{-\beta}x$, given by the same formula, but with $\dfrac{p}{q}=0$, instead of $\dfrac{p}{q}=\infty$. We have in this case also

$$\gamma_{-\beta}\left(x+\frac{\omega}{2}\right)\div g_{-\beta}\,x=A'\div\Pi_n\left(1+\frac{x}{(-p+\frac{1}{2})\,\omega+nvi}\right),$$

A' different from A on account of the different limits. The divisor of the second side takes the form

$$\{x-(p+\tfrac{1}{2})\,\omega\}\,\Pi\left(1+\frac{x-(p+\frac{1}{2})\,\omega}{nvi}\right)\div(-p+\tfrac{1}{2})\,\omega\Pi\left(1+\frac{(p+\frac{1}{2})\,\omega}{nvi}\right),$$

where the extreme values of n are infinite as compared with p. This may be reduced to

$$-\sin\frac{\pi}{\upsilon i}\{x-(p+\tfrac{1}{2})\,\omega\}\div\sin(p+\tfrac{1}{2})\,\omega,$$

$$=\epsilon^{\frac{\pi}{\upsilon}[(p+\frac{1}{2})\,\omega-x]}\div\epsilon^{\frac{\pi}{\upsilon}[(p+\frac{1}{2})\,\omega]}=\epsilon^{-\frac{\pi x}{\upsilon}},$$

neglecting the exponentials whose indices are infinitely great and negative. Observing the value of β this becomes $\epsilon^{-\omega\beta x}$, and we have

$$\gamma_{-\beta}\left(x+\frac{\omega}{2}\right)=\epsilon^{\beta\omega x}\,.\,A'g_{-\beta}\,.\,x:$$

a result of the form of that which would be deduced from the equations $\gamma_{-\beta}\,x=\epsilon^{\beta x^2}\gamma_\beta\,x$, $g_{-\beta}x=\epsilon^{\beta x^2}g_\beta x$, $\gamma_\beta\left(x+\dfrac{\omega}{2}\right)=Ag_\beta\,x$. It is scarcely necessary to remark that $\gamma_{-\beta}\,x$ has the same relations to the change of x into $x+\dfrac{\upsilon i}{2}$ as $\gamma_\beta\,x$ has to that of x into $x+\dfrac{\omega}{2}$.

25.

MÉMOIRE SUR LES FONCTIONS DOUBLEMENT PÉRIODIQUES.

[From the *Journal des Mathématiques* (Liouville), tom. x. (1845), pp. 385—420.]

UN des plus beaux résultats des recherches de l'illustre Abel, dans la théorie des fonctions elliptiques, consiste dans les expressions qu'il obtint pour les fonctions inverses $\phi\alpha$, $f\alpha$, $F\alpha$ (équivalentes à-peu-près à $\sin \text{am}\alpha$, $\cos \text{am}\alpha$, $\Delta \text{am}\alpha$) en forme de fractions avec un dénominateur commun: ce dénominateur et les trois numérateurs étant chacun le produit d'une suite infinie double de facteurs. On ne sait pas à quel point Abel avait poussé l'investigation des propriétés de ces nouvelles fonctions; on trouve seulement, dans une Lettre à Legendre, imprimée parmi ses Œuvres [t. II. p. 259, Ed. 2, p. 274], qu'il s'en était occupé. Depuis, les fonctions H, Θ, qui sont essentiellement les mêmes que ces fonctions d'Abel, ont été l'objet des savantes recherches de M. Jacobi, à qui l'on doit, en particulier, la belle formule

$$\log \Theta (\alpha) - \log \Theta (0) = \tfrac{1}{2}\alpha^2 \left(1 - \frac{E_{\prime}}{K}\right) - k^2 \int_0^\alpha d\alpha \int_0^\alpha d\alpha \sin^2 \text{am}\alpha,$$

qui est vraiment fondamentale, et sur laquelle on peut dire que sa théorie est basée. Mais les expressions qu'obtient M. Jacobi pour les fonctions H, Θ, sont sous la forme d'un produit d'une suite infinie simple de facteurs, ce qui ne met pas à beaucoup près si bien en évidence la vraie nature de ces fonctions que les expressions d'Abel; celles-ci sont, en outre, si analogues aux formules en produits infinis des fonctions circulaires, que l'on est seulement étonné que personne ne se soit avisé jusqu'ici de les poser, à priori, comme les définitions les plus simples des fonctions doublement périodiques, pour en déduire la théorie de ces fonctions. C'est de cette manière que je me propose de traiter ici la question. Je prends pour définitions les formules d'Abel, en supposant, pour plus de généralité, que les fonctions complètes Ω, Υ (K, K' de M. Jacobi) sont chacune de la forme $A + B\sqrt{-1}$ (ce qui donne lieu à quelques intégrations assez délicates). Et de ces seules équations, sans me servir en rien de la théorie des fonctions elliptiques, je déduis les propriétés fondamentales des fonctions en question, et de là des fonctions elliptiques. On a ainsi quatre fonctions à considérer au lieu des deux H, Θ, dont l'une est pour ainsi dire analogue à un sinus et les autres à des cosinus. Mais ce qu'il y a de remarquable, c'est l'apparition d'un facteur exponentiel qui entre presque partout. On le prévoyait d'après les formules de M. Jacobi, mais ces formules n'expliquent pas, ce me semble, pourquoi ce facteur s'y rencontre: mon analyse le fait voir de la manière la plus satisfaisante. Ce facteur résulte, en effet, de ce que, pour les produits infinis doubles, il ne suffit pas, pour obtenir un résultat déterminé, d'attribuer aux deux entiers variables des valeurs quelconques depuis $-\infty$ jusqu'à ∞, même en supposant l'égalité des valeurs positives et négatives; il faut, en outre, établir une relation entre les valeurs infinies que reçoivent les deux variables. Mais dans les produits que je considère, on démontre qu'en supposant toujours cette égalité des valeurs positives et négatives, quelque liaison que l'on établisse entre les valeurs infinies, il résulte toujours la même valeur du produit, à un facteur exponentiel près, dont l'indice est le carré de x, multiplié par une constante dont la valeur s'exprime au moyen d'une intégrale définie double, et qui dépend de la liaison établie entre les valeurs infinies des variables. C'est-à-dire qu'en multipliant par un facteur exponentiel de cette forme, convenablement choisi, on peut changer à volonté la relation en question sans affecter la valeur du produit. Voilà l'idée fondamentale du Mémoire qui suit.

En me servant de quelques formules de M. Cauchy, relatives à la décomposition des fonctions en fractions simples, j'établis d'une manière rigoureuse des relations entre les trois quotients de mes quatre fonctions, qui sont les mêmes par lesquelles Abel démontre les propriétés fondamentales des fonctions ϕ, f, F. Ces formules une fois obtenues, on peut supposer connue toute la théorie des fonctions elliptiques. Ces théorèmes de M. Cauchy me conduisent, en outre, à un grand nombre de nouvelles formules qui contiennent des suites infinies doubles. Parmi celles-ci, il y en a une qui me fournit la démonstration du théorème déjà cité de M. Jacobi, théorème duquel il déduit une foule de résultats intéressants. Je finis en citant ceux qui se rapportent de plus près aux fonctions dont je parle. J'espère reprendre une autre fois la considération d'une autre partie de la théorie, dans laquelle j'entrevois des conclusions intéressantes.

Soient Ω, Υ des quantités finies quelconques, assujetties à la seule condition que la fraction $\Omega : \Upsilon$ ne soit pas réelle. En représentant par m, n des entiers positifs ou négatifs quelconques, mettons, pour abréger,

$$m\Omega + n\Upsilon = (m, n) \qquad \text{(1)},$$

et considérons une expression de cette forme

$$u = x\Pi\left\{1 + \frac{x}{(m, n)}\right\} \qquad \text{(2)},$$

où le symbole Π dénote, comme à l'ordinaire, le produit d'un nombre infini de facteurs que l'on obtient en donnant à m, n des valeurs entières quelconques, depuis $-\infty$ jusqu'à

$+\infty$, en excluant seulement la combinaison $(m=0,\ n=0)$. Pour qu'une telle expression soit finie, il faut que pour chaque combinaison de valeurs de m, n, il y en ait une autre des mêmes valeurs avec les signes contraires. Cependant, comme on l'a déjà expliqué, cela ne suffit pas pour rendre déterminée la valeur de u. Soit ϕ une fonction de m, n qui ne change pas en changeant à la fois les signes de ces deux quantités; et imaginons que l'équation

$$\phi = T$$

représente une courbe fermée, dont tous les points s'éloignent, au cas limite de $T=\infty$, d'une distance infinie de l'origine. Cela posé, en donnant à m, n des valeurs entières qui satisfassent à cette condition $\phi \leqq T$, et puis faisant $T=\infty$, on obtient pour u une valeur parfaitement déterminée, qui dépend de la forme de la fonction ϕ. Soit u' ce que devient la fonction u en changeant seulement l'équation aux limites dans l'équation analogue

$$\phi' = T;$$

on peut, pour simplifier, supposer que la courbe représentée par cette équation soit située entièrement en dehors de celle que représente l'équation

$$\phi = T,$$

mais cela n'est pas essentiel. Il est facile de trouver une relation très-simple qui existe entre ces deux expressions u' et u. En effet,

$$u' : u = \Pi \left\{1 + \frac{x}{(m,\ n)}\right\},$$

en donnant à m, n des valeurs qui satisfassent à la fois aux deux conditions $\phi > T$, $\phi' < T$. Donc, en considérant toujours ces valeurs,

$$\log u' - \log u = \log \Pi \left\{1 + \frac{x}{(m,\ n)}\right\} = \Sigma \log \left\{1 + \frac{x}{(m,\ n)}\right\} = x \Sigma \frac{1}{(m,\ n)} - \tfrac{1}{2}x^2 \Sigma \frac{1}{(m,\ n)^2} + \ldots$$

Dans cette expression, les termes qui contiennent les puissances impaires x s'évanouissent, à cause des valeurs égales positives et négatives. Mais puisque, à la limite, m, n ne reçoivent que des valeurs infinies, on peut négliger les termes multipliés par x^4, &c. Donc

$$\log u' - \log u = -\tfrac{1}{2}x^2 \Sigma \frac{1}{(m,\ n)^2},$$

ou enfin,

$$\log u' - \log u = -\tfrac{1}{2}Ax^2, \quad u' = u\epsilon^{-\frac{1}{2}Ax^2} \quad \ldots\ldots\ldots\ldots\ldots\ldots\ (3),$$

où j'ai représenté par ϵ la base du système hyperbolique de logarithmes, et où A est donné par l'équation

$$A = \Sigma \frac{1}{(m,\ n)^2} \quad \ldots\ldots\ldots\ldots\ldots\ldots\ldots\ldots\ldots\ldots\ (4).$$

Il est facile de voir que l'on peut changer la sommation en intégration double, et écrire

$$A = \iint \frac{dm\, dn}{(m,\ n)^2} \quad \text{.......} \quad (5),$$

entre les mêmes limites qu'auparavant, c'est-à-dire que m, n doivent satisfaire aux deux conditions $\phi > T$, $\phi' < T$.

Prenons par exemple, pour limite supérieure,

$$(m^2 = \mathrm{m}^2, \quad n^2 = \mathrm{n}^2),$$

et pour limite inférieure,

$$(m^2 + n^2 = T^2);$$

m, n sont censés contenir T comme facteur, de manière qu'ils deviennent infinis avec cette quantité; on suppose aussi $\mathrm{m} > T$, $\mathrm{n} > T$, mais cela est seulement pour la clarté.

Il devient nécessaire à ce point de définir de plus près les valeurs de Ω, Υ; nous écrirons

$$\Omega = \omega + \omega' i, \quad \Upsilon = \upsilon + \upsilon' i \quad \text{.......} \quad (6),$$

où $i = \sqrt{-1}$; ω, ω', υ, υ' sont des quantités réelles, telles que $\omega\upsilon' - \omega'\upsilon$ ne s'évanouit pas; c'est la condition pour que $\Omega : \Upsilon$ contienne une partie imaginaire.

Avec les coordonnées polaires

$$A = \iint \frac{dr\, d\theta}{r\,(\Omega \cos\theta + \Upsilon \sin\theta)^2} = \int \frac{d\theta\,(\log \mathrm{r} - \log T)}{(\Omega \cos\theta + \Upsilon \sin\theta)^2} \quad \text{.......} \quad (7),$$

en représentant par r ce que devient r à la limite supérieure; l'intégrale doit être prise depuis $\theta = 0$ jusqu'à $\theta = 2\pi$. On voit tout de suite que la partie qui contient $\log T$ s'évanouit; donc

$$A = \int_0^{2\pi} \frac{\log \mathrm{r}\, d\theta}{(\Omega \cos\theta + \Upsilon \sin\theta)^2}.$$

Soit α un angle positif plus petit que $\frac{1}{2}\pi$, tel que $\tan \alpha = \frac{\mathrm{m}}{\mathrm{n}}$, on a évidemment

$$\mathrm{r} = \frac{\pm\, \mathrm{m}}{\cos\theta},$$

depuis $\theta = -\alpha$ jusqu'à $\theta = +\alpha$, ou depuis $\theta = \pi - \alpha$ jusqu'à $\theta = \pi + \alpha$, et

$$\mathrm{r} = \frac{\pm\, \mathrm{n}}{\cos\theta},$$

depuis $\theta = \alpha$ jusqu'à $\theta = \pi - \alpha$, ou depuis $\theta = \pi + \alpha$ jusqu'à $\theta = 2\pi - \alpha$ (le signe ambigu, de manière que r soit toujours positif). En réunissant les parties opposées de l'intégrale, on obtient

$$A = 2\int_{-\alpha}^{\alpha} \frac{(\log \mathrm{m} - \log\cos\theta)\, d\theta}{(\Omega \cos\theta + \Upsilon \sin\theta)^2} + 2\int_{\alpha}^{\pi-\alpha} \frac{(\log \mathrm{n} - \log\sin\theta)\, d\theta}{(\Omega \cos\theta + \Upsilon \sin\theta)^2},$$

ou, en mettant dans la seconde intégrale $\frac{1}{2}\pi - \alpha = \alpha'$, et $\frac{1}{2}\pi - \theta$ au lieu de θ,

$$A = 2\int_{-\alpha}^{\alpha} \frac{(\log \mathrm{m} - \log \cos \theta)\, d\theta}{(\Omega \cos \theta + \Upsilon \sin \theta)^2} + 2\int_{-\alpha'}^{\alpha} \frac{(\log \mathrm{n} - \log \cos \theta)\, d\theta}{(\Upsilon \cos \theta + \Omega \sin \theta)^2}.$$

La première intégrale se réduit à

$$- 2(\log \mathrm{m} - \log \cos \theta) \frac{1}{\Upsilon(\Omega + \Upsilon \tan \theta)} \text{ (entre les limites)}$$

$$+ \frac{2}{\pi}\int_{-\alpha}^{\alpha} \frac{\tan \theta\, d\theta}{(\Omega + \Upsilon \tan \theta)}$$

$$= \frac{4(\log \mathrm{m} - \log \cos \alpha) \tan \alpha}{\Omega^2 - \Upsilon^2 \tan^2 \alpha} - 4\int_0^{\alpha} \frac{\tan^2 \theta\, d\theta}{\Omega^2 - \Upsilon^2 \tan^2 \theta}$$

$$= \frac{4(\log \mathrm{m} - \log \cos \alpha) \tan \alpha}{\Omega^2 - \Upsilon^2 \tan^2 \alpha} + \frac{4}{\Omega^2 + \Upsilon^2}\int_0^{\alpha} d\theta \left[1 - \frac{\Omega^2(1 + \tan^2 \theta)}{\Omega^2 - \Upsilon^2 \tan^2 \theta}\right].$$

L'intégrale, dans cette formule

$$\int_0^{\alpha} \frac{(1 + \tan^2 \theta)\, d\theta}{\Omega^2 - \Upsilon^2 \tan^2 \theta} = \int_0^{\tan \alpha} \frac{dx}{\Omega^2 - \Upsilon^2 x^2},$$

s'exprime tout de suite par les logarithmes, mais il faut apporter une attention particulière à la manière de déterminer quelle valeur doit être attribuée à ce logarithme, dans le cas où la partie réelle en est négative. Je renvoie cette discussion à une Note. Voici le résultat auquel j'arrive.

En représentant par Lu la valeur principale du logarithme de u, quand il y a une valeur principale (c'est-à-dire quand la partie réelle de u est positive), et écrivant

$$L_{\pm\pi i} u = Lu, \text{ ou } L(-u) \pm \pi i,$$

selon qu'il y a une valeur principale de $\log u$, ou de $\log(-u)$, on a

$$\int_0^{\tan \alpha} \frac{dx}{\Omega^2 - \Upsilon^2 x^2} = \frac{1}{2\Omega\Upsilon} L_{\pm\pi i}\left(\frac{\Omega + \Upsilon \tan \alpha}{\Omega - \Upsilon \tan \alpha}\right) \quad \text{(8)},$$

en prenant le signe supérieur pour $\omega\upsilon' - \omega'\upsilon$ positif, et le signe inférieur pour $\omega\upsilon' - \omega'\upsilon$ négatif.

La première partie de A se réduit donc à

$$\frac{4(\log \mathrm{m} - \log \cos \alpha) \tan \alpha}{\Omega^2 - \Upsilon^2 \tan^2 \alpha} + \frac{4}{\Omega^2 + \Upsilon^2}\left(\alpha - \frac{1}{2}\frac{\Omega}{\Upsilon} L_{\pm\pi i} \frac{\Omega + \Upsilon \tan \alpha}{\Omega - \Upsilon \tan \alpha}\right),$$

ou, en rétablissant la valeur de α, à

$$\frac{4\mathrm{mn} \log \sqrt{\mathrm{m}^2 + \mathrm{n}^2}}{\mathrm{m}^2\Omega^2 - \mathrm{n}^2\Upsilon^2} + \frac{4}{\Omega^2 + \Upsilon^2}\left(\arctan \frac{\mathrm{n}}{\mathrm{m}} - \frac{1}{2}\frac{\Omega}{\Upsilon} L_{\pm\pi i} \frac{\mathrm{m}\Omega + \mathrm{n}\Upsilon}{\mathrm{m}\Omega - \mathrm{n}\Upsilon}\right).$$

Pour avoir la seconde partie de A, il faut changer m, n, Ω, Υ en n, m, Υ, Ω. En faisant cela, on change le signe de $\omega v' - \omega' v$; ainsi il faut écrire $L_{\mp\pi i}$ au lieu de $L_{\pm\pi i}$. En réunissant les deux parties, et mettant $\frac{1}{2}\pi$ au lieu de $\text{arc tan}\,\frac{\text{n}}{\text{m}} + \text{arc tan}\,\frac{\text{m}}{\text{n}}$, on obtient enfin

$$A = \frac{2}{\Omega^2 + \Upsilon^2}\left\{\pi + \frac{\Omega}{\Upsilon}L_{\pm\pi i}\left(\frac{\text{m}\Omega + \text{n}\Upsilon}{\text{m}\Omega - \text{n}\Upsilon}\right) - \frac{\Upsilon}{\Omega}L_{\pm\pi i}\left(\frac{\text{n}\Upsilon + \text{m}\Omega}{\text{n}\Upsilon - \text{m}\Omega}\right)\right\} \text{..........(9)},$$

formule que l'on pourrait rendre plus simple en distinguant les deux cas où le premier ou le second logarithme a une valeur principale; mais il vaut mieux la laisser sous cette forme.

On aurait pu croire que la valeur de A pourrait s'obtenir plus facilement en réduisant A à la différence de deux intégrales définies, les conditions pour les limites étant données, dans la première, par $m^2 < \text{m}^2$, $n^2 < \text{n}^2$, et, pour la seconde, par $m^2 + n^2 < T^2$, en désignant généralement les coordonnées par m, n. Cependant, de cette manière, on admet dans les deux intégrales les valeurs $m = 0$, $n = 0$, qui rendent infinie la fonction à intégrer. Ainsi la valeur de l'intégrale dépendrait du choix des variables, et l'on obtiendrait des résultats inexacts. Autrement dit, on obtiendrait de cette façon la valeur de A, à une quantité ∇ près, qui est la différence de deux intégrales de la même forme, entre des limites $m^2 < \mu^2$, $n^2 < \nu^2$ ou $m^2 + n^2 < \tau^2$, μ, ν, τ des quantités infiniment petites. On voit tout de suite que ∇ n'est pas pour cela infiniment petit (en effet, sa valeur dépend du rapport $\mu : \nu$ et nullement des valeurs absolues de ces quantités), et, pour en trouver la valeur, il faudrait se servir de l'analyse précédente.

Soit, comme exemple, $\frac{\text{m}}{\text{n}} = \infty$,

$$A = \frac{2}{\Omega^2 + \Upsilon^2}\left\{\pi - \frac{\Omega}{\Upsilon}L_{\pm\pi i}(1) - \frac{\Upsilon}{\Omega}L_{\mp\pi i}(-1)\right\}$$

$$= \frac{2\pi}{\Omega^2 + \Upsilon^2}\left(1 \pm \frac{\Upsilon i}{\Omega}\right) = \frac{2\pi}{\Omega(\Omega \mp \Upsilon i)} = -\frac{2\pi i}{\Omega\Upsilon}\cdot\frac{\Upsilon}{\Omega \mp \Upsilon i}.$$

De même, pour $\frac{\text{n}}{\text{m}} = \infty$,

$$A = \frac{2}{\Omega^2 + \Upsilon^2}\left\{\pi - \frac{\Omega}{\Upsilon}L_{\pm\pi i}(-1) - \frac{\Pi}{\Upsilon}L_{\mp\pi i}(1)\right\}$$

$$= \frac{2\pi}{\Omega^2 + \Upsilon^2}\left(1 \mp \frac{\Omega i}{\Upsilon}\right) = \pm\frac{2\pi i}{\Upsilon(\Omega \mp \Upsilon)} = \mp\frac{2\pi i}{\Omega\Upsilon}\cdot\frac{\Omega}{\Omega \mp \Upsilon i}.$$

En représentant par A', A'' ces deux valeurs de A, on a

$$A' - A'' = \frac{\mp 2\pi i}{\Omega\Upsilon} = -2\mathfrak{E} \text{ .... (10)},$$

où j'écris

$$\mathfrak{E} = \pm\frac{\pi i}{\Omega\Upsilon} \text{ .... (11)},$$

en prenant toujours le signe supérieur pour $\omega\upsilon' - \omega'\upsilon$ positif, et le signe inférieur pour $\omega\upsilon' - \omega'\upsilon$ négatif.

Soient $u_{\mathfrak{G}}$ ce que devient u en prenant $\frac{\mathrm{m}}{\mathrm{n}} = \infty$, $u_{-\mathfrak{G}}$ ce que devient cette même fonction en prenant $\frac{\mathrm{n}}{\mathrm{m}} = \infty$; on a évidemment

$$u_{-\mathfrak{G}} : u_{\mathfrak{G}} = \epsilon^{-\frac{1}{2}(A'-A'')x^2} = \epsilon^{\mathfrak{G}x^2}.$$

On peut donc s'imaginer une fonction U donnée par les équations

$$U = \epsilon^{-\frac{1}{2}\mathfrak{G}x^2} u_{-\mathfrak{G}} = \epsilon^{\frac{1}{2}\mathfrak{G}x^2} u_{\mathfrak{G}} \quad \ldots\ldots\ldots\ldots \quad (12).$$

On verra dans la suite que $u_{\mathfrak{G}}$ est analogue à la fonction $H(u)$ de M. Jacobi. Il convient cependant, pour la symétrie, de considérer, au lieu de $u_{\mathfrak{G}}$, la nouvelle fonction U qui vient d'être définie. Quoique suffisamment donnée par ces équations, nous allons en chercher encore une autre définition. Pour cela, il faut trouver la valeur de l'intégrale définie qui détermine A, prise depuis la même limite inférieure jusqu'à celle donnée par l'équation

$$\mathrm{mod}\,(m, n) = T.$$

Mettons, comme auparavant,

$$m = r\cos\theta, \qquad n = r\sin\theta;$$

soient aussi

$$\Omega_1 = \omega - \omega' i, \quad \Upsilon_1 = \upsilon - \upsilon' i.$$

L'équation pour la limite supérieure se réduit à

$$r^2(\Omega\cos\theta + \Upsilon\sin\theta)(\Omega_1\cos\theta + \Upsilon_1\sin\theta) = T^2,$$

et celle pour la limite inférieure, à

$$r = T.$$

On trouve tout de suite

$$\left.\begin{aligned}
A &= -\tfrac{1}{2}\int_0^{2\pi} \frac{\log(\Omega\cos\theta + \Upsilon\sin\theta)(\Omega_1\cos\theta + \Upsilon_1\sin\theta)\,d\theta}{(\Omega\cos\theta + \Upsilon\sin\theta)^2} \\
&= \frac{1}{2\Upsilon}\frac{1}{(\Omega + \Upsilon\tan\theta)}\log(\Omega\cos\theta + \Upsilon\sin\theta)(\Omega_1\cos\theta + \Upsilon_1\sin\theta)\ (\text{entre } \theta = 0,\ \theta = 2\pi) \\
&\quad - \frac{1}{2\Upsilon}\int_0^{2\pi} \frac{d\theta}{\Omega + \Upsilon\tan\theta}\left(\frac{\Upsilon - \Omega\tan\theta}{\Omega + \Upsilon\tan\theta} + \frac{\Upsilon_1 - \Omega_1\tan\theta}{\Omega_1 + \Upsilon_1\tan\theta}\right) \\
&= -\frac{1}{\Upsilon}\int_{-\frac{1}{2}\pi}^{\frac{1}{2}\pi} \frac{d\theta}{\Omega + \Upsilon\tan\theta}\left(\frac{\Upsilon - \Omega\tan\theta}{\Omega + \Upsilon\tan\theta} + \frac{\Upsilon_1 - \Omega_1\tan\theta}{\Omega_1 + \Upsilon_1\tan\theta}\right)
\end{aligned}\right\} \quad (13).$$

D'abord

$$\int_{-\frac{1}{2}\pi}^{\frac{1}{2}\pi} \frac{d\theta}{\Omega + \Upsilon \tan\theta} \, \frac{\Upsilon_1 - \Omega_1 \tan\theta}{\Omega_1 + \Upsilon_1 \tan\theta} = \int_{-\frac{1}{2}\pi}^{\frac{1}{2}\pi} \left(\frac{M}{\Omega + \Upsilon \tan\theta} + \frac{M_1}{\Omega_1 + \Upsilon_1 \tan\theta} \right) d\theta,$$

$$\int_{-\frac{1}{2}\pi}^{\frac{1}{2}\pi} \frac{d\theta}{\Omega + \Upsilon \tan\theta} = 2\Omega \int_0^{\frac{1}{2}\pi} \frac{d\theta}{\Omega^2 - \Upsilon^2 \tan^2\theta} = \frac{2\Omega}{\Omega^2 + \Upsilon^2} \int_0^{\frac{1}{2}\pi} \left[1 + \frac{\Upsilon^2 (1 + \tan^2\theta)}{\Omega^2 - \Upsilon^2 \tan^2\theta} \right] d\theta$$

$$= \frac{2\Omega}{\Omega^2 + \Upsilon^2} \left\{ \theta + \frac{\Upsilon}{2\Omega} L_{\pm\pi i} \left(\frac{\Omega + \Upsilon \tan\theta}{\Omega - \Upsilon \tan\theta} \right) \right\} \quad (\text{entre } \theta = 0, \ \ \theta = \tfrac{1}{2}\pi)$$

$$= \frac{2\Omega}{\Omega^2 + \Upsilon^2} \left(\tfrac{1}{2}\pi \pm \frac{\Upsilon}{2\Omega} \pi i \right) = \frac{\pi}{\Omega \mp \Upsilon i}.$$

On a de même, en remarquant qu'en changeant Ω, Υ en Υ, Ω, on change le signe de $\omega\upsilon' - \omega'\upsilon$,

$$\int_{-\frac{1}{2}\pi}^{\frac{1}{2}\pi} \frac{d\theta}{\Omega_1 + \Upsilon_1 \tan\theta} = \frac{\pi}{\Omega_1 \pm \Upsilon_1 i}.$$

D'un autre côté,

$$M = \frac{\Upsilon\Upsilon_1 + \Omega\Omega_1}{\Upsilon\Omega_1 - \Upsilon_1\Omega}, \qquad M_1 = -\frac{\Upsilon_1^2 + \Omega_1^2}{\Upsilon\Omega_1 - \Upsilon_1\Omega};$$

cela donne

$$\int_{-\frac{1}{2}\pi}^{\frac{1}{2}\pi} \frac{d\theta}{\Omega + \Upsilon \tan\theta} \, \frac{\Upsilon_1 - \Omega_1 \tan\theta}{\Omega_1 + \Upsilon_1 \tan\theta} = \frac{\pi}{\Upsilon\Omega_1 - \Upsilon_1\Omega} \left\{ \frac{\Upsilon\Upsilon_1 + \Omega\Omega_1}{\Omega \mp \Upsilon i} - (\Omega \mp \Upsilon i) \right\}$$

$$= \frac{\pi}{\Upsilon\Omega_1 - \Upsilon_1\Omega} \, \frac{\Upsilon\Upsilon_1 + \Omega\Omega_1 - (\Omega + \Upsilon i)(\Omega_1 \pm \Upsilon_1 i \mp 2\Upsilon_1 i)}{\Omega \mp \Upsilon i} = \pm \frac{\pi i}{\Omega \mp \Upsilon i} \pm \frac{2\Upsilon_1 \pi i}{\Upsilon\Omega_1 - \Upsilon_1\Omega},$$

ou enfin

$$\int_{-\frac{1}{2}\pi}^{\frac{1}{2}\pi} \frac{d\theta}{\Omega + \Upsilon \tan\theta} \, \frac{\Upsilon_1 - \Omega_1 \tan\theta}{\Omega_1 + \Upsilon_1 \tan\theta} = \pm \frac{\pi i}{\Omega \mp \Upsilon i} \pm \frac{\pi \Upsilon_1}{\omega\upsilon' - \omega'\upsilon};$$

et de même

$$\int_{-\frac{1}{2}\pi}^{\frac{1}{2}\pi} \frac{d\theta \, (\Upsilon - \Omega \tan\theta)}{(\Omega + \Upsilon \tan\theta)^2} = \frac{\pm \pi i}{\Omega \mp \Upsilon i},$$

en omettant seulement le dernier terme de l'autre intégrale; ce que l'on peut vérifier, au reste, en différentiant par rapport à Ω, Υ l'équation

$$\int_{-\frac{1}{2}\pi}^{\frac{1}{2}\pi} \frac{d\theta}{\Omega + \Upsilon \tan\theta} = \frac{\pi}{\Omega \mp \Upsilon i}.$$

On a donc, en ajoutant les deux parties qui composent l'intégrale,

$$A = \mp \frac{2\pi i}{\Omega\Upsilon} \frac{\Omega}{\Omega \mp \Upsilon i} \mp \frac{\Upsilon_1 \pi}{\Upsilon(\omega v' - \omega' v)} \text{ .... } (14).$$

Il est facile de voir, en écrivant

$$A = \mp \frac{2\pi i}{\Omega\Upsilon} \frac{\Omega}{\Omega \mp \Upsilon i} \pm \frac{\pi \Upsilon_1 i}{\Upsilon(\Upsilon\Omega_1 - \Upsilon_1\Omega)},$$

que cette expression ne change pas de valeur en mettant Ω_1, Υ_1, Ω, Υ au lieu de Ω, Υ, Ω_1, Υ_1, pourvu qu'on change aussi le signe de i; cela doit évidemment être ainsi et peut servir de vérification.

Soit, pour un moment, u_1 ce que devient u en prenant pour limite l'équation

$$m^2 + n^2 = T^2,$$

et supposons que dans la fonction u on ait pour limite l'équation

$$\text{mod}(m, n) = T.$$

En retenant la valeur de A, qui vient d'être trouvée,

$$u = \epsilon^{-\frac{1}{2}Ax^2} u_1;$$

mais aussi

$$U = \epsilon^{\mp\frac{1}{2}\frac{\pi i}{\Omega\Upsilon}x^2} u_{-\mathfrak{e}} = \epsilon^{\left(\mp\frac{1}{2}\frac{\pi i}{\Omega\Upsilon} \pm \frac{\pi i}{\Omega\Upsilon}\frac{\Omega}{\Omega\mp\Upsilon i}\right)x^2} u_1,$$

à cause de l'équation, conséquence facile des résultats précédents,

$$u_{-\mathfrak{e}} = \epsilon^{\pm\frac{\pi i}{\Omega\Upsilon}\frac{\Omega}{\Omega\mp\Upsilon i}x^2} u_1.$$

En éliminant u_1, on obtient, entre u, U, une équation de la forme

$$U = \epsilon^{-\frac{1}{2}Bx^2} u \text{ .... } (15),$$

dans laquelle

$$B = -A \pm \frac{\pi i}{\Omega\Upsilon} \mp \frac{2\pi i}{\Omega \mp \Upsilon i} = \pm \frac{\pi i}{\Omega\Upsilon} \pm \frac{\Upsilon_1 \pi}{\Upsilon(\omega v' - \omega' v)}$$

$$= \frac{\pm\pi}{\Omega\Upsilon(\omega v' - \omega' v)}\left[2(\omega v' - \omega' v) + (\omega + \omega' i)(v - v' i)\right],$$

ou enfin

$$B = \frac{\pm\pi(\omega v + \omega' v')}{\Omega\Upsilon(\omega v' - \omega' v)} = \frac{\pi(\omega v' + \omega' v')}{\Omega\Upsilon \,\text{mod}\,(\omega v' - \omega' v)} \text{ .. .. } (16).$$

En rassemblant tous ces résultats, on a le système de formules

(A)

$$U = \epsilon^{-\frac{1}{2}Bx^2}\, u = \epsilon^{-\frac{1}{2}\mathfrak{G}x^2}\, u_{-\mathfrak{G}} = \epsilon^{\frac{1}{2}\mathfrak{G}x^2}\, u_{\mathfrak{G}},$$

$$u = \epsilon^{\frac{1}{2}(B-\mathfrak{G})\, x^2}\, u_{-\mathfrak{G}} = \epsilon^{\frac{1}{2}(B+\mathfrak{G})\, x^2}\, u_{\mathfrak{G}},$$

$$\mathfrak{G} = \frac{\pi i}{\Omega\Upsilon}\, \frac{(\omega v' - \omega' v)}{\text{mod}\,(\omega v' - \omega' v)},$$

$$B = \frac{\pi\,(\omega v + \omega' v')}{\text{mod}\,(\omega v' - \omega' v)},$$

u, $u_{\mathfrak{G}}$, $u_{-\mathfrak{G}}$ des fonctions de la forme

$$x\Pi\left\{1 + \frac{x}{(m,\ n)}\right\},$$

mais avec des limites différentes, savoir:

mod $(m, n) = T$, $T = \infty$, pour la fonction u;

$m^2 = \mathrm{m}^2$, $n^2 = \mathrm{n}^2$, $\mathrm{m} = \infty$, $\mathrm{n} = \infty$, $\frac{\mathrm{m}}{\mathrm{n}} = \infty$, pour u_β;

$m^2 = \mathrm{m}^2$, $n^2 = \mathrm{n}^2$, $m = \infty$, $\mathrm{n} = \infty$, $\frac{\mathrm{n}}{\mathrm{m}} = \infty$, pour $u_{-\mathfrak{G}}$.

A présent, il importe de remarquer que la fonction $u_{\mathfrak{G}}$ est périodique à l'égard de Ω, de la manière d'un sinus ou d'un cosinus, mais ne l'est pas à l'égard de Υ; de même $u_{-\mathfrak{G}}$ est périodique à l'égard de Υ, mais non à l'égard de Ω. Quant à U, u, elles ne sont simplement périodiques ni à l'égard de Ω, ni à l'égard de Υ. Pour démontrer cela, considérons, par exemple, l'équation

$$u_{\mathfrak{G}} = x\ \Pi\left(1 + \frac{x}{(m,\ n)}\right) \quad \ldots\ldots\ldots\ldots\ldots\ldots\ldots\ldots\ldots\ldots \quad (17)$$

[m depuis $-\mathrm{m}$ jusqu'à m, n depuis $-\mathrm{n}$ jusqu'à n, $\mathrm{m} = \infty$, $\mathrm{n} = \infty$, $\frac{\mathrm{m}}{\mathrm{n}} - \infty$, le système $(m = 0,\ n = 0)$ toujours exclu]; en représentant par $u'_{\mathfrak{G}}$ ce que devient $u_{\mathfrak{G}}$ quand on écrit $x + \Omega$ au lieu de x, on a évidemment

$$u'_{\mathfrak{G}} = (x + \Omega)\ \Pi\left(1 + \frac{x + \Omega}{(m,\ n)}\right)$$

$$= -x\ \Pi\left(1 + \frac{1}{(m+1,\ n)}\right) : \Pi\,\frac{(m+1,\ n)}{(m,\ n)},$$

en admettant dans le premier produit la combinaison $(m = 0,\ n = 0)$, mais excluant $(m + 1 = 0,\ n = 0)$, et excluant l'une et l'autre du second produit. Cette équation est de la forme

$$u'_{\mathfrak{G}} = A'x\Pi\left(1 + \frac{x}{(m+1,\ n)}\right),$$

avec les mêmes limites que dans u; donc

$$u'_{\mathfrak{G}} : u_{\mathfrak{G}} = A'\Pi\left(1 + \frac{x}{(\mathrm{m}+1,\ n)}\right) : \Pi\left(1 + \frac{x}{(-\mathrm{m},\ n)}\right),$$

où Π se rapporte à la seule variable n qui doit s'étendre depuis $-\mathrm{n}$ jusqu'à n. Puisqu'ainsi n ne reçoit jamais des valeurs qui soient comparables à m, chacun de ces produits se réduit à l'unité, et l'on obtient

$$u'_{\mathfrak{G}} : u_{\mathfrak{G}} = A',$$

ou enfin

$$u'_{\mathfrak{G}} = -u_{\mathfrak{G}} \qquad (18),$$

en posant

$$x = -\tfrac{1}{2}\Omega,$$

et remarquant que u est fonction impaire de x, ce qui donne

$$A' = -1.$$

Soit de même $u''_{\mathfrak{G}}$ ce que devient $u_{\mathfrak{G}}$ en changeant x en $x + \Upsilon$. On a pareillement

$$u''_{\mathfrak{G}} : u_{\mathfrak{G}} = A''\Pi\left(1 + \frac{x}{(m,\ \mathrm{n}+1)}\right) : \Pi\left(1 + \frac{x}{(m,\ -\mathrm{n})}\right),$$

où Π se rapporte à la seule variable m. Mais ici les produits ne se réduisent point à l'unité; en effet,

$$\Pi\left(1 + \frac{x}{(m,\ \mathrm{n}+1)}\right) \text{ ou } \Pi\left(1 + \frac{x}{(m,\ \mathrm{n})}\right)$$

$$= (x + \mathrm{n}\Upsilon)\,\Pi\left(1 + \frac{x - \mathrm{n}\Upsilon}{m\Omega}\right) : \mathrm{n}\Upsilon\Pi\left(1 + \frac{\mathrm{n}\Upsilon}{m\Omega}\right)$$

$$= \sin\frac{\pi}{\Omega}(x + \mathrm{n}\Upsilon) : \sin\frac{\pi}{\Omega}\mathrm{n}\Upsilon.$$

Mais quand la partie imaginaire de θ est infinie, on trouve

$$\sin\theta = \frac{1}{2i}(\epsilon^{\theta i} - \epsilon^{-\theta i}) = \pm\frac{1}{2i}\epsilon^{\pm\theta i},$$

selon que la partie réelle de θi est positive ou négative. Or la partie réelle de $\dfrac{i\Upsilon}{\Omega}$ est de signe contraire à $\omega v' - \omega' v$; on a donc

$$\Pi\left(1 + \frac{x}{(m,\ \mathrm{n}+1)}\right) = \epsilon^{\mp\frac{\pi i}{\Omega}x} = \epsilon^{-\mathfrak{G}\Upsilon x};$$

de même,

$$\Pi\left(1 + \frac{x}{(m,\ -\mathrm{n})}\right) = \epsilon^{\mathfrak{G}\Upsilon x},$$

et de là

$$u''_{\mathfrak{G}} = A''\epsilon^{-2\mathfrak{G}\Upsilon x}u_{\mathfrak{G}} \qquad (19),$$

équation de la même forme que celle que l'on obtiendrait en posant

$$u_{\mathfrak{G}} = \epsilon^{-\mathfrak{G}x^2}u_{-\mathfrak{G}},$$

et remarquant que $u_{-\mathfrak{G}}$ est périodique à l'égard de Υ. On voit aussi qu'en sommant par rapport à m, il est permis d'exprimer $u_{\mathfrak{G}}$ par un produit infini simple, dont chaque facteur est de la forme

$$\sin\frac{\pi}{\Omega}(x+n\Upsilon),$$

mais qu'on n'arrive pas à un tel résultat, en sommant par rapport à n. En effet, l'équation

$$u_{\mathfrak{G}} = \epsilon^{-\mathfrak{G}x^2}\, u_{-\mathfrak{G}}$$

fait voir que $u_{\mathfrak{G}}$ s'exprime par un tel produit où chaque terme est de la forme

$$\sin\frac{\pi}{\Upsilon}(x+m\Omega),$$

multiplié par le facteur exponentiel $\epsilon^{-\mathfrak{G}x^2}$. On établit de cette façon l'identité, à un facteur constant près, de $u_{\mathfrak{G}}$ à $H(u)$; mais, à présent, pour éviter les longueurs, je ne me propose de considérer aucun de ces produits infinis simples.

Il faut maintenant, pour le développement de la théorie, introduire les trois fonctions qui correspondent au cosinus. Mettant, pour abréger,

$$\bar{m} = m+\tfrac{1}{2},\quad \bar{n} = n+\tfrac{1}{2}\ \ldots\ldots\ldots\ldots\ldots\ldots\ldots\ldots\ldots(20),$$

j'écris

$$(\mathrm{B})\quad \left\{\begin{array}{lll} \mathrm{y}x = \epsilon^{-\frac{1}{2}Bx^2}\, x\Pi\left\{1+\dfrac{x}{(m,\ n)}\right\}. & & \begin{array}{l}\text{Limites.}\\ \bmod\,(m,\ n)=T,\quad T=\infty.\end{array} \\[2ex] \mathrm{g}x = \epsilon^{-\frac{1}{2}Bx^2}\ \Pi\left\{1+\dfrac{x}{(\bar{m},\ n)}\right\}. & & \bmod\,(\bar{m},\ n)=T. \\[2ex] Gx = \epsilon^{-\frac{1}{2}Bx^2}\ \Pi\left\{1+\dfrac{x}{(m,\ \bar{n})}\right\}. & & \bmod\,(m,\ \bar{n})=T. \\[2ex] Zx = \epsilon^{-\frac{1}{2}Bx^2}\ \Pi\left\{1+\dfrac{x}{(\bar{m},\ \bar{n})}\right\}. & & \bmod\,(\bar{m},\ \bar{n})=T. \end{array}\right.$$

En représentant par $\mathrm{y}_{\mathfrak{G}}x$, $\mathrm{y}_{-\mathfrak{G}}x$, &c., ce que deviennent ces fonctions, et prenant pour limites

$$(m^2=\mathrm{m}^2,\ n^2=\mathrm{n}^2),\quad (\bar{m}^2=\bar{\mathrm{m}}^2,\ n^2=\mathrm{n}^2),$$
$$(m^2=\mathrm{m}^2,\ \bar{n}^2=\bar{\mathrm{n}}^2),\quad (\bar{m}^2=\bar{\mathrm{m}}^2,\ \bar{n}^2=\bar{\mathrm{n}}^2),$$

où $\dfrac{\mathrm{m}}{\mathrm{n}}=\infty$ pour $\mathrm{y}_{\mathfrak{G}}x$, &c., $\dfrac{\mathrm{n}}{\mathrm{m}}=\infty$ pour $\mathrm{y}_{-\mathfrak{G}}x$, &c.,

on a, en général, en mettant J au lieu de l'une quelconque des lettres y, g, G, Z,

$$Jx = \epsilon^{\frac{1}{2}\mathfrak{G}x^2}\, J_{\mathfrak{G}}\,x = \epsilon^{-\frac{1}{2}\mathfrak{G}x^2}\, J_{-\mathfrak{G}}\,x \ldots\ldots\ldots\ldots\ldots\ldots\ldots\ldots(21),$$

où $J_{\mathfrak{G}}x$ est périodique à l'égard de Ω, et $J_{-\mathfrak{G}}x$ à l'égard de Υ. C'est cette équation remarquable qui définit la loi de périodicité des fonctions J, et de laquelle se déduisent presque toutes les propriétés de ces fonctions.

Il est clair qu'en changeant entre elles les quantités Ω, Υ (quantités que nous désignerons comme les *fonctions complètes*), on ne change pas les fonctions $\mathrm{y}x$, Zx, tandis que $\mathrm{g}x$ se change en Gx, et Gx en $\mathrm{g}x$. On change de cette manière $\mathfrak{C}$ en $-\mathfrak{C}$ et B en $-B$ (par exemple, $\mathrm{y}_{\mathfrak{C}}x$ se change en $\mathrm{y}_{-\mathfrak{C}}x$, &c.). Cette considération fait voir que toute propriété des fonctions J est double, et donne le moyen le plus facile pour passer d'une propriété quelconque à la propriété correspondante.

On tire immédiatement des définitions mêmes les équations

$$\text{(C)} \qquad \begin{cases} \mathrm{y}(-x) = -\mathrm{y}x, & \mathrm{g}(-x) = \mathrm{g}x, \quad G(-x) = Gx, \quad Z(-x) = Zx; \\ \mathrm{y}0 = 0, & \mathrm{g}0 = 1, \quad G0 = 1, \quad Z0 = 1; \\ y'0 = 1. \end{cases}$$

Il est facile de démontrer, de la même manière dont nous avons prouvé la périodicité de $J_{\mathfrak{C}}x$ à l'égard de Ω, que ces fonctions se changent l'une en l'autre, à un facteur constant près, en changeant x en $x+\frac{1}{2}\Omega$. Faisant donc attention à l'équation qui lie ensemble Jx et $J_{\mathfrak{C}}x$, on obtient le système de formules

$$\left.\begin{aligned} \mathrm{y}\,(x+\tfrac{1}{2}\,\Omega) &= \epsilon^{\frac{1}{2}\beta\Omega x} A\mathrm{g}x, \\ \mathrm{g}\,(x+\tfrac{1}{2}\,\Omega) &= \epsilon^{\frac{1}{2}\beta\Omega x} B\mathrm{y}x, \\ G\,(x+\tfrac{1}{2}\,\Omega) &= \epsilon^{\frac{1}{2}\beta\Omega x} CZx, \\ Z\,(x+\tfrac{1}{2}\,\Omega) &= \epsilon^{\frac{1}{2}\beta\Omega x} DZx. \end{aligned}\right\} \quad \ldots\ldots\ldots\ldots\ldots\ldots \quad (22)$$

Pour déterminer A, B, C, D, posons

$$x = 0 \quad \text{ou} \quad x = -\tfrac{1}{2}\,\Omega.$$

En écrivant, pour abréger,

$$\text{(D)} \qquad \epsilon^{\mathfrak{C}\Omega^2} = \epsilon^{\pm\frac{\pi\Omega i}{\Upsilon}} = q_1^{-1},$$

on trouve

$$\left.\begin{aligned} A &= \mathrm{y}\,(\tfrac{1}{2}\Omega), \\ B &= -\,q_1^{-\frac{1}{4}} : \mathrm{y}\,(\tfrac{1}{2}\Omega), \\ C &= G\,(\tfrac{1}{2}\,\Omega) = q_1^{-\frac{1}{4}} : Z\,(\tfrac{1}{2}\,\Omega), \\ D &= Z\,(\tfrac{1}{2}\,\Omega) = q_1^{-\frac{1}{4}} : G\,(\tfrac{1}{2}\,\Omega); \end{aligned}\right\} \quad \ldots\ldots\ldots\ldots\ldots\ldots \quad (23)$$

d'où l'on déduit cette équation de condition,

$$G\,(\tfrac{1}{2}\,\Omega)\,Z\,(\tfrac{1}{2}\,\Omega) = q_1^{-\frac{1}{4}} \quad \ldots\ldots\ldots\ldots\ldots\ldots \quad (24).$$

On a de même le système

$$\left.\begin{aligned} \mathrm{y}\,(x+\tfrac{1}{2}\,\Upsilon) &= \epsilon^{-\frac{1}{2}\mathfrak{C}\Upsilon x} A'Gx, \\ \mathrm{g}\,(x+\tfrac{1}{2}\,\Upsilon) &= \epsilon^{-\frac{1}{2}\mathfrak{C}\Upsilon x} B'\,Zx, \\ G\,(x+\tfrac{1}{2}\,\Upsilon) &= \epsilon^{-\frac{1}{2}\mathfrak{C}\Upsilon x} C'\,\mathrm{y}x, \\ Z\,(x+\tfrac{1}{2}\,\Upsilon) &= \epsilon^{-\frac{1}{2}\mathfrak{C}\Upsilon x} D'Gx, \end{aligned}\right\} \quad \ldots\ldots\ldots\ldots\ldots\ldots (25).$$

De là, si

$$\epsilon^{-6\Upsilon^2} = \epsilon^{\pm \frac{\pi \Upsilon i}{\Omega}} = q^{-1},$$

ce qui donne

$$\text{(E)} \qquad \log q \log q_1 = -\pi^2,$$

il résulte

$$\left.\begin{aligned} A' &= \mathrm{y}(\tfrac{1}{2}\Upsilon), \\ B' &= \mathrm{g}(\tfrac{1}{2}\Upsilon) = q^{-\frac{1}{4}} : Z(\tfrac{1}{2}\Upsilon), \\ C' &= -q^{-\frac{1}{4}} : \mathrm{y}(\tfrac{1}{2}\Upsilon), \\ D' &= Z(\tfrac{1}{2}\Upsilon) : \mathrm{g}(\tfrac{1}{2}\Upsilon); \end{aligned}\right\} \quad \ldots\ldots\ldots\ldots\ldots\ldots (26);$$

d'où nous tirons l'équation de condition

$$\mathrm{g}(\tfrac{1}{2}\Upsilon) Z(\tfrac{1}{2}\Upsilon) = q^{-\frac{1}{4}} \ldots\ldots\ldots\ldots\ldots\ldots (27).$$

En posant $x = \frac{1}{2}\Upsilon$ dans l'équation pour $\mathrm{y}(x + \frac{1}{2}\Omega)$ et $x = \frac{1}{2}\Omega$ dans l'équation pour $\mathrm{y}(x + \frac{1}{2}\Upsilon)$, on obtient aussi

$$\mathrm{y}(\tfrac{1}{2}\Omega)\,\mathrm{g}(\tfrac{1}{2}\Upsilon) = -i\mathrm{y}(\tfrac{1}{2}\Upsilon)\,G(\tfrac{1}{2}\Omega) \ldots\ldots\ldots\ldots\ldots\ldots (28),$$

et l'on déduirait cette même équation, ou une équation équivalente, des autres formules, de manière qu'on ne peut pas trouver d'autres équations de condition.

Enfin, en rapprochant ces systèmes, on obtient

$$\left.\begin{aligned} \mathrm{y}(x + \tfrac{1}{2}\Omega + \tfrac{1}{2}\Upsilon) &= \epsilon^{\frac{1}{2}\beta x(\Omega - \Upsilon)} A'' Z x, \\ \mathrm{g}(x + \tfrac{1}{2}\Omega + \tfrac{1}{2}\Upsilon) &= \epsilon^{\frac{1}{2}\beta x(\Omega - \Upsilon)} B'' \mathrm{g} x, \\ G(x + \tfrac{1}{2}\Omega + \tfrac{1}{2}\Upsilon) &= \epsilon^{\frac{1}{2}\beta x(\Omega - \Upsilon)} C'' G x, \\ Z(x + \tfrac{1}{2}\Omega + \tfrac{1}{2}\Upsilon) &= \epsilon^{\frac{1}{2}\beta x(\Omega - \Upsilon)} D'' \mathrm{y} x, \end{aligned}\right\} \quad \ldots\ldots\ldots\ldots\ldots\ldots (29),$$

avec les équations suivantes pour les coefficients,

$$\left.\begin{aligned} A'' &= (-1)^{\frac{1}{4}}\,\mathrm{y}(\tfrac{1}{2}\Omega)\,\mathrm{g}(\tfrac{1}{2}\Upsilon), \\ B'' &= -(-1)^{\frac{1}{4}}\,q_1^{-\frac{1}{4}}\,\mathrm{y}(\tfrac{1}{2}\Upsilon) : \mathrm{y}(\tfrac{1}{2}\Omega), \\ C'' &= (-1)^{\frac{1}{4}}\,R(\tfrac{1}{2}\Omega)\,Z(\tfrac{1}{2}\Upsilon), \\ D'' &= -(-1)^{\frac{1}{4}}\,q^{-\frac{1}{4}}\,Z(\tfrac{1}{2}\Omega) : \mathrm{y}(\tfrac{1}{2}\Upsilon). \end{aligned}\right\} \quad \ldots\ldots\ldots\ldots\ldots\ldots (30).$$

En rassemblant les équations entre $J(\frac{1}{2}\Omega)$, $J(\frac{1}{2}\Upsilon)$, on a

$$\left.\begin{aligned} \mathrm{g}(\tfrac{1}{2}\Omega) &= 0, \\ G(\tfrac{1}{2}\Upsilon) &= 0, \\ G(\tfrac{1}{2}\Omega)\,Z(\tfrac{1}{2}\Omega) &= q_1^{-\frac{1}{4}}, \\ \mathrm{g}(\tfrac{1}{2}\Upsilon)\,Z(\tfrac{1}{2}\Upsilon) &= q^{-\frac{1}{4}}, \\ \mathrm{y}(\tfrac{1}{2}\Omega)\,\mathrm{g}(\tfrac{1}{2}\Upsilon) &= -i\mathrm{y}(\tfrac{1}{2}\Upsilon)\,G(\tfrac{1}{2}\Omega), \end{aligned}\right\} \quad \ldots\ldots\ldots\ldots\ldots\ldots (31),$$

d'où l'on conclut les formules

$$\left.\begin{aligned}
B''C'' : A''D'' = B'D' : A'C' = CD : AB &= -1,\\
A'B' : C'D' = -A''B'' : C''D'' &= -\mathrm{y}^2(\tfrac{1}{2}\Upsilon) : Z^2(\tfrac{1}{2}\Upsilon),\\
A''C'' : B''D'' = -A'C' : B'D' &= \mathrm{y}^2(\tfrac{1}{2}\Omega) : Z^2(\tfrac{1}{2}\Omega),\\
A\,D : B\,C = -A'D' : B'C' &= \mathrm{y}^2(\tfrac{1}{2}\Omega) : G^2(\tfrac{1}{2}\Omega),\\
&= -\mathrm{y}^2(\tfrac{1}{2}\Upsilon) : \mathrm{g}^2(\tfrac{1}{2}\Upsilon),
\end{aligned}\right\}\ldots\ldots\ldots(32),$$

dont nous aurons bientôt besoin.

On peut passer maintenant à ce système général de formules, où j'écris $(-)^m$ au lieu de $(-1)^m$:

$$(\mathrm{F})\quad\left\{\begin{aligned}
&\Theta = (-)^{mn}\,\epsilon^{\mathcal{E}x\,(m,\,-n)}\,q_1^{-\frac{1}{2}m^2}\,q^{-\frac{1}{2}n^2},\\
&\text{là, comme toujours, } (m,\,-n) = m\Omega - n\Upsilon\,;\\
&\mathrm{y}\,\{x+(m,\,n)\} = (-)^{m+n}\,\Theta\mathrm{y}x,\\
&\mathrm{g}\,\{x+(m,\,n)\} = (-)^{m}\,\Theta\mathrm{g}x,\\
&G\,\{x+(m,\,n)\} = (-)^{n}\,\Theta Gx,\\
&Z\,\{x+(m,\,n)\} = \Theta Zx\,;\\
&\Phi = (-)^{nm+\frac{1}{2}n}\,\epsilon^{\mathcal{E}x\,(\overline{m},\,-n)}\,q_1^{-\frac{1}{2}m^2-\frac{1}{2}m}\,q^{-\frac{1}{2}n^2}\,;\\
&\mathrm{y}\,\{x+(\overline{m},\,n)\} = (-)^{m+n}\,\Phi A\mathrm{g}x,\\
&\mathrm{g}\,\{x+(\overline{m},\,n)\} = (-)^{m}\,\Phi B\mathrm{y}x,\\
&G\,\{x+(\overline{m},\,n)\} = (-)^{n}\,\Phi CZx,\\
&Z\,\{x+(\overline{m},\,n)\} = \Phi DGx\,;\\
&\Psi = (-)^{mn+\frac{1}{2}m}\,\epsilon^{\mathcal{E}x\,(m,\,-\overline{n})}\,q_1^{-\frac{1}{2}m^2}\,q^{-\frac{1}{2}n^2-\frac{1}{2}n}\,;\\
&\mathrm{y}\,\{x+(m,\,\overline{n})\} = (-)^{m+n}\,\Psi A'Gx,\\
&\mathrm{g}\,\{x+(m,\,\overline{n})\} = (-)^{m}\,\Psi B'Zx,\\
&G\,\{x+(m,\,\overline{n})\} = (-)^{n}\,\Psi C'\mathrm{y}x,\\
&Z\,\{x+(m,\,\overline{n})\} = \Psi D'\mathrm{g}x\,;\\
&\mathrm{X} = (-)^{mn+\frac{1}{2}m+\frac{1}{2}n}\,\epsilon^{\mathcal{E}x\,(\overline{m},\,-\overline{n})}\,q_1^{-\frac{1}{2}m^2-\frac{1}{2}m}\,q^{-\frac{1}{2}n^2-\frac{1}{2}n}\,;\\
&\mathrm{y}\,\{x+(\overline{m},\,\overline{n})\} = (-)^{m+n}\,\mathrm{X}A''Zx,\\
&\mathrm{g}\,\{x+(\overline{m},\,\overline{n})\} = (-)^{m}\,\mathrm{X}B''Gx,\\
&G\,\{x+(\overline{m},\,\overline{n})\} = (-)^{n}\,\mathrm{X}C''\mathrm{g}x,\\
&Z\,\{x+(\overline{m},\,\overline{n})\} = \mathrm{X}D''\mathrm{y}x.
\end{aligned}\right.$$

Soit $x=0$:

$$
(\mathrm{G})\quad\left\{\begin{array}{ll}
\Theta_0 = (-)^{mn}\, q_1^{-\frac{1}{2}m^2} q^{-\frac{1}{2}n^2}; & \\
\mathrm{y}\,(m,\ n) = \quad 0, & \mathrm{y}'\,(m,\ n) = (-)^{m+n}\,\Theta_0, \\
\mathrm{g}\,(m,\ n) = (-)^m\,\Theta_0, & \\
G\,(m,\ n) = (-)^n\,\Theta_0, & \\
Z\,(m,\ n) = \quad \Theta_0; & \\
\Phi_0 = (-)^{mn+\frac{1}{2}m}\, q_1^{-\frac{1}{2}m^2-\frac{1}{2}m-\frac{1}{2}n^2}; & \\
\mathrm{y}\,(\overline{m},\ n) = (-)^{m+n}\,\Phi_0 A, & \\
\mathrm{g}\,(\overline{m},\ n) = \quad 0, & \mathrm{g}'\,(\overline{m},\ n) = (-)^m\,\Phi_0 B, \\
G\,(\overline{m},\ n) = (-)^n\quad \Phi_0 C, & \\
Z\,(\overline{m},\ n) = \quad \Phi_0 D; & \\
\Psi_0 = (-)^{mn+\frac{1}{2}n}\, q_1^{-\frac{1}{2}m^2} q^{-\frac{1}{2}n^2-\frac{1}{2}n}; & \\
\mathrm{y}\,(m,\ \overline{n}) = (-)^{m+n}\,\Psi_0 A', & \\
\mathrm{g}\,(m,\ \overline{n}) = (-)^m\quad \Psi_0 B', & \\
G\,(m,\ \overline{n}) = \quad 0, & G'\,(m,\ \overline{n}) = (-)^n\,\Psi_0 C', \\
Z\,(m,\ \overline{n}) = \quad \Psi_0 D'; & \\
\mathrm{X}_0 = (-)^{mn+\frac{1}{2}m+\frac{1}{2}n}\, q_1^{-\frac{1}{2}m^2-\frac{1}{2}m} q^{-\frac{1}{2}n^2-\frac{1}{2}n}; & \\
\mathrm{y}\,(\overline{m},\ \overline{n}) = (-)^{m+n}\,\mathrm{X}_0 A'', & \\
\mathrm{g}\,(\overline{m},\ \overline{n}) = (-)^m\quad \mathrm{X}_0 B'', & \\
G\,(\overline{m},\ \overline{n}) = (-)^n\quad X_0 C'', & \\
Z\,(\overline{m},\ \overline{n}) = \quad 0, & Z'\,(\overline{m},\ \overline{n}) = X_0 D''.
\end{array}\right.
$$

On a de suite, en prenant les différentielles des logarithmes des fonctions Jx,

$$
(\mathrm{H})\quad\left\{\begin{array}{l}
\mathrm{y}'\,x\ :\ \mathrm{y}x = -Bx + \sum\{x-(m,\ n)\}^{-1}, \\
\mathrm{g}'\,x\ :\ \mathrm{g}x = -Bx + \sum\{x-(\overline{m},\ n)\}^{-1}, \\
G'x\ :\ Gx = -Bx + \sum\{x-(m,\ \overline{n})\}^{-1}, \\
Z'x\ :\ Zx = -Bx + \sum\{x-(\overline{m},\ \overline{n})\}^{-1}.
\end{array}\right.
$$

(B est le coefficient de x^2, dans l'exponentielle $\epsilon^{-\frac{1}{2}Bx^2}$ des équations (B); il n'y a pas à craindre de le confondre avec le B qui entre dans les équations pour y $(\overline{m},\ n)$, &c.: c'est par hasard que j'ai pris deux fois la même lettre.)

Réduisons en fractions simples la fonction

$$\mathrm{g}x\, Gx \;:\; \mathrm{y}x\, Zx.$$

En écrivant

$$\mathrm{g}x Gx \;:\; \mathrm{y}x Zx = \sum \left[L\{x-(m,\ n)\}^{-1} + M\{x-(\overline{m},\ \overline{n})\}^{-1}\right],$$

on obtient

$$L = \mathrm{g}(m,\ n)\, G(m,\ n) \;:\; \mathrm{y}'(m,\ n)\, Z(m,\ n) = 1,$$

$$M = \mathrm{g}(\overline{m},\ \overline{n})\, G(\overline{m},\ \overline{n}) \;:\; \mathrm{y}(\overline{m},\ \overline{n})\, Z'(\overline{m},\ \overline{n}) = B''C'' - A''D'' = 1.$$

Donc

$$\mathrm{g}x\, Gx \;:\; \mathrm{y}x\, Zx = \sum \{x-(m,\ n)\}^{-1} - \sum \{x-(\overline{m},\ \overline{n})\}^{-1},$$

c'est-à-dire

$$\mathrm{g}x\, Gx \;:\; \mathrm{y}x\, Zx = (\mathrm{y}'x \;:\; \mathrm{y}x) - (Z'x \;:\; Zx).$$

(Nous allons bientôt justifier, au moyen d'un théorème de M. Cauchy, l'emploi de cette méthode de décomposition en fractions simples qui ne s'applique pas toujours aux fonctions transcendantes.)

On obtient de même

$$(\mathrm{I})\qquad \left\{\begin{aligned} \mathrm{g}x\, Zx \;&:\; \mathrm{y}x\, Gx = (\mathrm{y}'x \;:\; \mathrm{y}x) - (G'x \;:\; Gx),\\ Gx Zx \;&:\; \mathrm{y}x\, \mathrm{g}x = (\mathrm{y}'x \;:\; \mathrm{y}x) - (\mathrm{g}'x \;:\; \mathrm{g}x),\\ -b^2\mathrm{y}zZx \;&:\; \mathrm{g}x\, Gx = (\mathrm{g}'x \;:\; \mathrm{g}x) - (G'x \;:\; Gx),\\ e^2\mathrm{y}x\mathrm{g}x \;&:\; Gx\, Zx = (G'x \;:\; Gx) - (Z'x \;:\; Zx),\\ c^2\mathrm{y}xGx \;&:\; Zx\, \mathrm{y}x = (Z'x \;:\; Zx) - (\mathrm{y}'x \;:\; \mathrm{y}x),\end{aligned}\right.$$

en mettant, pour abréger,

$$(\mathrm{J})\qquad \left\{\begin{aligned} &\mathrm{y}(\tfrac{1}{2}\Upsilon) \;:\; Z(\tfrac{1}{2}\Omega) = \frac{i}{e},\ ^{(1)}\\ &\mathrm{y}(\tfrac{1}{2}\Omega) \;:\; Z(\tfrac{1}{2}\Omega) = \frac{1}{c},\\ &\mathrm{y}(\tfrac{1}{2}\Omega) \;:\; G(\tfrac{1}{2}\Omega) = -i\mathrm{y}(\tfrac{1}{2}\Upsilon) \div \mathrm{g}(\tfrac{1}{2}\Upsilon) = \frac{1}{b}.\end{aligned}\right.$$

Donc, en éliminant les fonctions dérivées,

$$\begin{aligned} G^2x - Z^2x &= \quad e^2\mathrm{y}^2x,\\ \mathrm{g}^2x - G^2x &= -\,b^2\mathrm{y}^2x,\\ Z^2x - \mathrm{g}^2x &= \quad c^2\mathrm{y}^2x,\end{aligned}$$

[1] J'ai écrit $\frac{i}{e}$ au lieu de $\frac{1}{e}$ pour me conformer à la notation d'Abel; il est à peine nécessaire de remarquer que c, e sont, en général, l'une et l'autre des quantités imaginaires.

ce qui donne, en ajoutant,

$$\text{(K)} \qquad b^2 = e^2 + c^2, \text{ ou } b = \sqrt{e^2 + c^2},$$

en nous retiendrons désormais la lettre b dans cette signification. Puis

$$\text{(L)} \qquad \begin{cases} \mathrm{g}^2 x = Z^2 x - c^2 \mathrm{y}^2 x, \\ G^2 x = Z^2 x + e^2 \mathrm{y}^2 x. \end{cases}$$

Soient maintenant

$$\text{(M)} \qquad \begin{cases} \phi x = \mathrm{y} x : Zx, \\ fx = \mathrm{g} x : Zx, \\ Fx = Gx : Zx. \end{cases}$$

On obtient

$$\text{(N)} \qquad \begin{cases} f^2 x = 1 - c^2 \phi^2 x, \\ F^2 x = 1 + e^2 \phi^2 x\,; \end{cases}$$

$$\text{(O)} \qquad \begin{cases} \phi' x = fxFx, \\ f'x = - c^2 \phi x F x, \\ F'x = e^2 \phi x f x. \end{cases}$$

Ces équations sont précisément les équations fondamentales d'Abel (*Œuvres*, t. I., p. 143 [Ed. 2, p. 268]), et il en déduit

$$\text{(P)} \qquad \begin{cases} \phi(x+y) = \dfrac{\phi x f y F y + \phi y f x F x}{1 + e^2 c^2 \phi^2 x \phi^2 y}, \\[2ex] f(x+y) = \dfrac{fxfy - c^2 \phi x \phi y FxFy}{1 + e^2 c^2 \phi^2 x \phi^2 y}, \\[2ex] F(x+y) = \dfrac{FxFy + e^2 \phi x \phi y f x f y}{1 + e^2 c^2 \phi^2 x \phi^2 y}, \end{cases}$$

qui sont les formules connues pour l'addition des fonctions elliptiques.

En effet, on peut écrire l'équation

$$\phi' x = fx\, Fx$$

sous la forme

$$1 = \frac{\phi' x}{\sqrt{(1 - c^2 \phi^2 x)(1 + e^2 \phi^2 x)}},$$

ou, en mettant $y = \phi x$,

$$\text{(Q)} \qquad x = \int_0^{\phi x} \frac{dy}{\sqrt{(1 - c^2 y^2)(1 + e^2 y^2)}}\,;$$

on peut donc de ce point supposer connues toutes les propriétés des fonctions elliptiques. On a, par exemple,

$$\phi\left(\tfrac{1}{2}\Upsilon\right)=\frac{i}{e},\quad \phi\left(\tfrac{1}{2}\Omega\right)=\frac{1}{c},$$

et de là

$$(\mathrm{R})\quad \left\{\begin{aligned} \tfrac{1}{2}\Omega &= \int_0^{\frac{1}{c}} \frac{dy}{\sqrt{(1-c^2y^2)(1+e^2y^2)}}, \\ \tfrac{1}{2}\Upsilon &= \int_0^{\frac{i}{e}} \frac{dy}{\sqrt{(1-c^2y^2)(1+e^2y^2)}}, \end{aligned}\right.$$

qui déterminent Ω, Υ en fonction de c, e. Il paraît au premier abord que Υ, Ω soient des fonctions parfaitement déterminées de c, e. J'ai des raisons pour croire que cela n'est pas précisément le cas, et que la question admettrait des développements intéressants; mais je réserve ce sujet pour une autre occasion.

On peut, à l'aide des équations entre $\mathrm{y}\left(\tfrac{1}{2}\Omega\right)$, &c., et les quantités c, e, exprimer cette suite de fonctions en termes de c, e. On déduit:

$$(\mathrm{S})\quad \left\{\begin{aligned} &\mathrm{y}\left(\tfrac{1}{2}\Omega\right)=b^{-\frac{1}{2}}c^{-\frac{1}{2}}q_1^{-\frac{1}{8}}, && \mathrm{y}\left(\tfrac{1}{2}\Upsilon\right)=ib^{-\frac{1}{2}}e^{-\frac{1}{2}}q^{-\frac{1}{8}}; \\ &\mathrm{g}\left(\tfrac{1}{2}\Omega\right)=0, && \mathrm{g}\left(\tfrac{1}{2}\Upsilon\right)=b^{\frac{1}{2}}e^{-\frac{1}{2}}q^{-\frac{1}{8}}; \\ &G\left(\tfrac{1}{2}\Omega\right)=b^{\frac{1}{2}}c^{-\frac{1}{2}}q_1^{-\frac{1}{8}}, && G\left(\tfrac{1}{2}\Upsilon\right)=0; \\ &Z\left(\tfrac{1}{2}\Omega\right)=b^{-\frac{1}{2}}c^{\frac{1}{2}}q_1^{-\frac{1}{8}}, && Z\left(\tfrac{1}{2}\Upsilon\right)=b^{-\frac{1}{2}}e^{\frac{1}{2}}q^{-\frac{1}{8}}; \end{aligned}\right.$$

et de là

$$(\mathrm{T})\quad \left\{\begin{aligned} &A=b^{-\frac{1}{2}}c^{-\frac{1}{2}}q_1^{-\frac{1}{8}}, && A'=ib^{-\frac{1}{2}}e^{-\frac{1}{2}}q^{-\frac{1}{8}}, && A''=(-)^{\frac{1}{4}}c^{-\frac{1}{2}}e^{-\frac{1}{2}}q_1^{-\frac{1}{8}}q^{-\frac{1}{8}}; \\ &B=-b^{\frac{1}{2}}c^{\frac{1}{2}}q_1^{-\frac{1}{8}}, && B'=b^{\frac{1}{2}}e^{-\frac{1}{2}}q^{-\frac{1}{8}}, && B''=-(-)^{\frac{1}{4}}ic^{\frac{1}{2}}e^{-\frac{1}{2}}q_1^{-\frac{1}{8}}q^{-\frac{1}{8}}; \\ &C=b^{\frac{1}{2}}c^{-\frac{1}{2}}q_1^{-\frac{1}{8}}, && C'=-ib^{\frac{1}{2}}e^{\frac{1}{2}}q^{-\frac{1}{8}}, && C''=(-)^{\frac{1}{4}}c^{-\frac{1}{2}}e^{\frac{1}{2}}q_1^{-\frac{1}{8}}q^{-\frac{1}{8}}; \\ &D=b^{-\frac{1}{2}}c^{\frac{1}{2}}q_1^{-\frac{1}{8}}, && D'=b^{-\frac{1}{2}}e^{\frac{1}{2}}q^{-\frac{1}{8}}, && D''=-(-)^{\frac{1}{4}}ic^{\frac{1}{2}}e^{\frac{1}{2}}q_1^{-\frac{1}{8}}q^{-\frac{1}{8}}; \end{aligned}\right.$$

qu'on doit introduire dans toutes les formules où entrent les quantités A, B, &c.

Voici le théorème de M. Cauchy (*Exercices de Mathématiques,* tome II. page 289):

"Si, en attribuant au module r de la variable

$$z=r(\cos p+i\sin p)$$

des valeurs infiniment grandes, on peut les choisir de manière que les deux fonctions

$$\frac{fz+f(-z)}{2},\quad \frac{fz-f(-z)}{2z}$$

deviennent sensiblement nulles, quel que soit l'angle p, ou du moins que chacune de ces fonctions reste toujours finie ou infiniment petite, et ne cesse d'être infiniment petite, en demeurant finie, que dans le voisinage de certaines valeurs particulières de p, on aura

$$fx = \mathcal{E}\,\frac{((fz))}{x-z},$$

pourvu qu'on réduise le résidu intégral à sa valeur principale."

On se rappelle que cela veut dire qu'en supposant ces conditions satisfaites, la fonction fx peut s'exprimer de la manière ordinaire comme la somme d'une suite de fractions simples, mais qu'il faut étendre d'abord la sommation aux racines de l'équation

$$\frac{1}{fx} = 0,$$

dont les modules sont inférieurs à une certaine limite qu'on fait alors infinie.

Soit, par exemple,

$$fx = \epsilon^{\frac{1}{2}ax^2+bx}\frac{T(x)}{T_1(x)} \qquad \text{(33)},$$

où Tx, T_1x ne contiennent que des facteurs qui sont des puissances entières des fonctions yx, gx, Gx, Zx. En supposant que r n'est d'aucune des formes

$$\text{mod}\,(m,\ n),\quad \text{mod}\,(\overline{m},\ n),\quad \text{mod}\,(m,\ \overline{n}),\quad \text{mod}\,(\overline{m},\ \overline{n}),$$

mais, d'ailleurs, une quantité infinie quelconque, on peut toujours écrire

$$z = r(\cos p + i\sin p) = (m,\ n) + \theta \qquad \text{(34)},$$

où θ est une quantité finie telle qu'aucune des fonctions yθ, gθ, $G\theta$, $Z\theta$ ne s'évanouit, m, n sont des entiers dont l'un au moins est infini, mais qui varient depuis $-\infty$ jusqu'à ∞, avec l'angle p. Soit $\aleph$ le degré de Tx, $\aleph_1$ celui de T_1x à l'égard de yx, &c.; soit aussi $\lambda = \aleph_1 - \aleph$; on a, en général, une équation de cette forme,

$$fz = \epsilon^{\frac{1}{2}a\,(m,\,n)^2}\, q_1^{\frac{1}{2}\lambda m^2}\, q^{\frac{1}{2}\lambda n^2}\, \epsilon^{Im+Jn}\, F \qquad \text{(35)},$$

où F est fini. En supposant donc que la partie réelle de

$$a(m,\ n)^2 - \lambda\mathfrak{G}\Omega^2 m^2 + \lambda\mathfrak{G}\Upsilon^2 n^2 \qquad \text{(36)}$$

est négative quelles que soient les valeurs de m, n, on a

$$fz = 0,$$

et de même

$$\frac{fz + f(-z)}{2} = 0,\qquad \frac{fz - f(-z)}{2z} = 0.$$

Si, au contraire, cette partie réelle est positive, ces trois fonctions sont toujours infinies. Il y a cependant un cas particulier à considérer, savoir, celui où cette partie réelle se réduit à Pm^2 ou Qn^2, P, Q étant des quantités positives. Il faut ici que le coefficient de n ou de m dans la partie réelle de $Im+Jn$ s'évanouisse. Enfin, si les parties réelles s'évanouissent entièrement, ce qui ne peut arriver que pour

$$a=0, \quad b=0, \quad \lambda=0,$$

fx est fini. On a donc pour fx fonction impaire,

$$\frac{fz+f(-z)}{2}=0, \quad \frac{fz-f(-z)}{2z}=0$$

(la seconde équation à cause du dénominateur infini z). Il est cependant certain que, dans plusieurs cas pour lesquels fx est fonction paire, on peut réduire fx en une suite de fractions simples: par exemple, $\text{y}x : \text{g}x$ est une fonction impaire que l'on peut, par ce qui précède, développer en suite de fractions simples; en écrivant $x+\frac{1}{2}\Upsilon$ au lieu de x, on déduit un pareil développement pour $Gx : Zx$ qui est fonction paire.

Remarquons que quand la partie réelle de

$$a\,(m,\ n)^2-\lambda\mathfrak{G}\Omega^2m^2+\lambda\mathfrak{G}\Upsilon^2n^2$$

est négative pour toute valeur de m ou n, la suite pour fx est toujours convergente. En effet, les numérateurs des fractions simples contiennent ce facteur de fz,

$$\epsilon^{\frac{1}{2}a\,(m,\ n)^2}\,q_1^{\frac{1}{2}\lambda m^2}\,q^{\frac{1}{2}\lambda n^2},$$

qui s'évanouit pour les valeurs infiniment grandes de m, n. Dans le cas où la partie réelle est positive, le développement ne peut jamais être vrai; je crois qu'en général il est vrai, dans les cas limites, quand la suite que l'on obtient est convergente. Il y a des exceptions cependant; on en verra une en développant ϕ^2x.

Avant de passer aux exemples, développons la condition pour que la partie réelle en question soit toujours négative.

En supposant

$$a=h+ki,$$

on obtient pour cette quantité l'expression

$$\left.\begin{aligned} &m^2\left\{h\,(\omega^2-\omega'^2)-2k\omega\omega'-\frac{\lambda\pi\bmod(\omega\upsilon'-\omega'\upsilon)}{\upsilon^2+\upsilon'^2}\right\}\\ +\ &n^2\left\{h\,(\upsilon^2-\upsilon'^2)-2k\upsilon\upsilon'-\frac{\lambda\pi\bmod(\omega\upsilon'-\omega'\upsilon)}{\omega^2+\omega'^2}\right\}\\ &+2mn\,\{h\,(\omega\upsilon-\omega'\upsilon')-k\,(\omega\upsilon'+\omega'\upsilon)\},\end{aligned}\right\}\ldots\ldots\ldots\ldots\ldots\ldots(37),$$

qui doit toujours rester négative. Cela donne, après quelques réductions, la condition

$$\left.\begin{aligned} &h^2+k^2+\frac{2\lambda\pi\,(\omega\upsilon+\omega'\upsilon')}{(\omega^2+\omega'^2)\,(\varpi^2+\varpi'^2)\bmod(\omega\upsilon'-\omega'\upsilon)}\left[(\omega\upsilon-\omega'\upsilon')\,h+(\omega\upsilon'+\omega'\upsilon)\,k\right]\\ &\quad-\frac{\lambda^2\pi^2}{(\omega^2+\omega'^2)\,(\upsilon^2+\upsilon'^2)}<0.\end{aligned}\right\}\ldots(38).$$

Les valeurs $$h=0,\quad k=0$$
satisfont à cette condition, laquelle du reste (en considérant h, k comme les coordonnées d'un point) est satisfaite pour tout point situé en dedans d'un certain cercle qui inclut l'origine, et dont on aurait l'équation en remplaçant le signe $<$ par le signe $=$ dans la formule (38).

On obtient de cette façon une grande variété de formules particulières. Par exemple celles-ci :

$$(\mathrm{U})\left\{\begin{aligned}
&\epsilon^{\frac{1}{2}ax^2+bx} : \mathrm{y}x = \sum\left[(-)^{-mn-m-n}\,\epsilon^{\frac{1}{2}a\,(m,\,n)^2+b\,(m,\,n)}\,q_1^{\frac{1}{2}m^2}\,q^{\frac{1}{2}n^2}\,\{x-(m,\ n)\}^{-1}\right],\\
&\epsilon^{\frac{1}{2}ax^2+bx} : \mathrm{g}x = -\,b^{-\frac{1}{2}}\,c^{-\frac{1}{2}}\sum\left[(-)^{-mn-m-\frac{1}{2}n}\,\epsilon^{\frac{1}{2}a\,(\bar{m},\,n)^2+b\,(\bar{m},\,n)}\,q_1^{\frac{1}{2}\,(m+\frac{1}{2})^2}\,q^{\frac{1}{2}n^2}\,\{x-(\bar{m},\ n)\}^{-1}\right],\\
&\epsilon^{\frac{1}{2}ax^2+bx} : Gx = ib^{-\frac{1}{2}}\,c^{-\frac{1}{2}}\sum\left[(-)^{-mn-\frac{1}{2}m-n}\,\epsilon^{\frac{1}{2}a\,(m,\,\bar{n})^2+b\,(m,\,\bar{n})}\,q_1^{\frac{1}{2}m^2}\,q^{\frac{1}{2}\,(n+\frac{1}{2})^2}\,\{x-(m,\ \bar{n})\}^{-1}\right],\\
&\epsilon^{\frac{1}{2}ax^2+bx} : Zx = ic^{-\frac{1}{2}}\,e^{-\frac{1}{2}}\sum\left[(-)^{-\,(m+\frac{1}{2})\,(n+\frac{1}{2})}\,\epsilon^{\frac{1}{2}a\,(\bar{m},\,\bar{n})^2+b\,(\bar{m},\,\bar{n})}\,q_1^{\frac{1}{2}\,(m+\frac{1}{2})^2}\,q^{\frac{1}{2}\,(n+\frac{1}{2})^2}\,\{x-(\bar{m},\ \bar{n})\}^{-1}\right],
\end{aligned}\right.$$

dans lesquelles, pour trouver les limites de a, il faut faire $\lambda=1$ dans la formule (38).

On a ensuite ce système de formules sans exponentielles,

$$(\mathrm{V})\left\{\begin{aligned}
&\mathrm{g}x : \mathrm{y}x = \sum\left[(-)^{n}\,\{x-(m,\ n)\}^{-1}\right],\\
&Gx : \mathrm{y}x = \sum\left[(-)^{m}\,\{x-(m,\ n)\}^{-1}\right],\\
&Zx : \mathrm{y}x = \sum\left[(-)^{m+n}\,\{x-(m,\ n)\}^{-1}\right];\\
&\mathrm{y}x : \mathrm{g}x = -\,b^{-1}c^{-1}\sum\left[(-)^{n}\,\{x-(\bar{m},\ n)\}^{-1}\right],\\
&Gx : \mathrm{g}x = -\,c^{-1}\sum\left[(-)^{m+n}\,\{x-(\bar{m},\ n)\}^{-1}\right],\\
&Zx : \mathrm{g}x = -\,b^{-1}\sum\left[(-)^{m}\,\{x-(\bar{m},\ n)\}^{-1}\right];\\
&\mathrm{y}x : Gx = -\,b^{-1}\,e^{-1}\sum\left[(-)^{m}\,\{x-(m,\ \bar{n})\}^{-1}\right],\\
&\mathrm{g}x : Gx = ie^{-1}\sum\left[(-)^{m+n}\,\{x-(m,\ \bar{n})\}^{-1}\right],\\
&Zx : Gx = ib^{-1}\sum\left[(-)^{n}\,\{x-(m,\ \bar{n})\}^{-1}\right];\\
&\mathrm{y}x : Zx = ic^{-1}\,e^{-1}\sum\left[(-)^{m+n}\,\{x-(\bar{m},\ \bar{n})\}^{-1}\right],\\
&\mathrm{g}x : Zx = e^{-1}\sum\left[(-)^{m}\,\{x-(\bar{m},\ \bar{n})\}^{-1}\right],\\
&Gx : Zx = ic^{-1}\sum\left[(-)^{n}\,\{x-(\bar{m},\ \bar{n})\}^{-1}\right],
\end{aligned}\right.$$

dont quelques-unes ont été données par Abel. Il faut remarquer que celles de ces équations où la fonction est impaire sont justifiées par le théorème de M. Cauchy, et que les autres se déduisent de celles-ci. On peut trouver de même le développement des fonctions

$$\frac{1}{\mathrm{y}x\,\mathrm{g}x},\ \ldots,\quad \frac{Gx}{\mathrm{y}x\,\mathrm{g}x},\ \ldots,\quad \frac{\mathrm{y}x\,\mathrm{g}x}{Zx\,Gx}$$

(celles-ci, qui sont toutes impaires, ont été déja considérées ; nous venons de faire voir que le développement est admissible) ;

$$\frac{1}{\text{y}x\,\text{g}x\,Gx},\ldots,\quad \frac{Zx}{\text{y}x\,\text{g}x\,Gx},\ldots,\quad \frac{1}{\text{y}x\,\text{g}x\,Zx},\ldots,$$

chacune, excepté celles de la suite $\dfrac{\text{y}x\,\text{g}x}{Gx\,Zx}$, multipliée par un facteur exponentiel $\epsilon^{\frac{1}{2}ax^2+bx}$. Par exemple,

$$(\text{W})\quad \left\{\begin{aligned} \frac{1}{\text{y}x\,\text{g}x\,Gx\,Zz} &= \sum \left[\epsilon^{\frac{1}{2}a(m,\,n)^2+b(m,\,n)}\, q_1^{\,2m^2} q^{2n^2} \quad \{x-(m,\,n)\}^{-1}\right] \\ &- \sum \left[\epsilon^{\frac{1}{2}a(\overline{m},\,n)^2+b(\overline{m},\,n)}\, q_1^{\,2(m+\frac{1}{2})^2} q^{2n^2} \quad \{x-(\overline{m},\,n)\}^{-1}\right] \\ &+ \sum \left[\epsilon^{\frac{1}{2}a(m,\,\overline{n})^2+b(m,\,\overline{n})}\, q_1^{\,2m^2} q^{2(n+\frac{1}{2})^2} \quad \{x-(m,\,\overline{n})\}^{-1}\right] \\ &- \sum \left[\epsilon^{\frac{1}{2}a(\overline{m},\,\overline{n})^2+b(\overline{m},\,\overline{n})}\, q_1^{\,2(m+\frac{1}{2})^2} q^{2(n+\frac{1}{2})^2} \{x-(\overline{m},\,\overline{n})\}^{-1}\right]. \end{aligned}\right.$$

Mais on est conduit à un résultat beaucoup plus important en considérant, par exemple, le développement de $\phi^2 x$. Cette fonction contient non-seulement une suite de fractions simples, mais aussi un terme constant ; nous écrirons

$$\phi^2 x = J + \sum \left[L\,\{x-(\overline{m},\,\overline{n})\}^{-2} + M\,\{x-(\overline{m},\,\overline{n})\}^{-1}\right] \;\ldots\ldots\ldots\ldots\ldots (39).$$

Pour déterminer L, M de la manière la plus simple, changeons x en $x+\frac{1}{2}\Omega+\frac{1}{2}\Upsilon$. En faisant attention à la valeur de $\phi x = \text{y}x : Zx$, on obtient

$$-e^{-2}c^{-2}(\phi x)^{-2} = J + \sum \left[L\,\{x-(m,\,n)\}^{-2} + M\,\{x-(m,\,n)\}^{-1}\right];$$

de là

$$L = -e^{-2}c^{-2}\{x-(m,\,n)\}^2(\phi x)^{-2},$$

$$M = -e^{-2}c^{-2}\frac{d}{dx}\left[\{x-(m,\,n)\}^2(\phi x)^{-2}\right],\quad x=(m,\,n),$$

ou enfin

$$L = -e^{-2}c^{-2}x^2(\phi x)^{-2},\quad \text{pour } x=0,$$

$$M = -e^2c^2\frac{d}{dx}\left[x^2(\phi x)^{-2}\right],$$

ce qui donne

$$\left.\begin{aligned} L &= -e^{-2}c^{-2},\quad M=0, \\ \phi^2 x &= J - e^{-2}c^{-2}\sum\{x-(\overline{m},\,\overline{n})\}^{-2}. \end{aligned}\right\}\ldots\ldots\ldots\ldots\ldots\ldots (40).$$

En intégrant deux fois

$$\int_0^x dx\int_0^x \phi^2 x\,dx = \tfrac{1}{2}Jx^2 + e^{-2}c^{-2}\sum\log\{x-(\overline{m},\,\overline{n})\} \;\ldots\ldots\ldots\ldots (41),$$

ou, à cause de $\log Zx = -\frac{1}{2}Bx^2 + \sum \log\{x - (\overline{m}, \overline{n})\}$,

$$\int_0^x dx \int_0^x \phi^2 x\, dx = \tfrac{1}{2}Ix^2 + e^{-2}c^{-2}\log Zx \quad \text{(42)},$$

en mettant, pour abréger,

$$I = J - e^{-2}c^{-2}B$$

(on n'a pas ajouté de constante arbitraire, parce que Zx est fonction paire de x qui se réduit à l'unité pour $x = 0$). Puis, on a

$$Zx = \epsilon^{-\frac{1}{2}e^2c^2Ix^2 + e^2c^2\int_0^x dx \int_0^x \phi^2 x\, dx} \quad \text{(43)}.$$

De là il est facile de déterminer la valeur de I. Soit, pour un moment,

$$\phi_, x = \int_0^x \phi^2 x\, dx, \qquad \phi_{,,} x = \int_0^x \phi_, x\, dx.$$

Puisque

$$\phi^2(x + \Omega) - \phi^2 x = 0,$$

on a

$$\phi_, (x + \Omega) = \phi_, x - \phi_, \Omega,$$

$$\phi_{,,} (x + \Omega) - \phi_{,,} x = x\phi_, \Omega.$$

Mais de l'équation

$$\phi^2 x - \phi^2(\Omega - x) = 0$$

on déduit

$$\phi_, x + \phi_, (\Omega - x) = \phi_, \Omega, \quad \phi_{,,} x - \phi_{,,}(\Omega - x) = x\phi_, \Omega,$$

ou, en faisant $x = \frac{1}{2}\Omega$,

$$\phi_, \Omega = 2\phi_, (\tfrac{1}{2}\Omega), \quad \phi_{,,}\Omega = \Omega\phi_, \tfrac{1}{2}\Omega,$$

c'est-à-dire

$$\phi_{,,}(x + \Omega) - \phi_{,,} x = (2x + \Omega)\, \phi_, \tfrac{1}{2}\Omega\, ;$$

et de là

$$Z(x + \Omega) = \epsilon^{-\frac{1}{2}e^2c^2\left[I - \frac{2}{\Omega}\phi_, (\frac{1}{2}\Omega)\right] 2(\Omega x + \Omega^2)} Zx.$$

Mais on a déjà

$$Z(x + \Omega) = \epsilon^{\beta\Omega x} q_1^{-\frac{1}{2}} Zx = \epsilon^{\frac{1}{2}\beta(2\Omega x + \Omega^2)} Zx\, ;$$

donc enfin

$$-e^2c^2\left[I - \frac{2}{\Omega}\phi_, (\tfrac{1}{2}\Omega)\right] = \beta,$$

c'est-à-dire

$$-\tfrac{1}{2}e^2c^2 I = \tfrac{1}{2}\beta - \frac{e^2c^2}{\Omega}\phi_, (\tfrac{1}{2}\Omega) = \tfrac{1}{2}\beta - \frac{e^2c^2}{\Omega}\int_0^{\frac{1}{2}\Omega} \phi^2 x\, dx \quad \text{(44)}.$$

Soit, pour abréger,

$$M = \frac{e^2c^2}{\Omega}\int_0^{\frac{1}{2}\Omega} \phi^2 x\, dx \ldots\ldots\ldots (45);$$

on a cette équation,

$$(\mathrm{X}) \qquad Zx = \varepsilon^{(\frac{1}{2}\beta - M)x^2 + e^2c^2\int_0^x dx \int_0^x \phi^2 x\, dx},$$

qui exprime la fonction Zx au moyen de ϕx. C'est, en effet, la formule remarquable de M. Jacobi, que nous avons citée dans l'introduction de ce Mémoire.

En changeant seulement la notation, on a, d'après M. Jacobi,

$$\phi^2(x+a) - \phi^2(x-a) = \frac{4(\phi a\, fa\, Fa\, \phi x\, fx\, Fx)}{(1+e^2c^2\phi^2 a\, \phi^2 x)^2} \ldots\ldots\ldots (46),$$

$$\int_0^x \{\phi^2(x+a) - \phi^2(x-a)\}\, dx = \frac{2\phi a\, fa\, Fa\, \phi^2 x}{1+e^2c^2\phi^2 a\, \phi^2 x} \ldots\ldots\ldots (47),$$

ou, ce qui est la même chose,

$$\int_{-a}^x \phi^2(x+a)\, dx - \int_a^x \phi^2(x-a)\, dx - 2\int_0^x \phi^2 a da = \frac{2\phi a\, fa\, Fa\, \phi^2 x}{1+e^2c^2\phi^2 a\, \phi^2 x} \ldots\ldots (48).$$

De là, en multipliant par e^2c^2, et faisant attention à la valeur de Zx,

$$\frac{Z'(x+a)}{Z(x+a)} - \frac{Z'(x-a)}{Z(x-a)} - \frac{2Z'a}{Za} = \frac{2e^2c^2\phi a\, fa\, Fa\, \phi^2 x}{1+e^2c^2\phi^2 a\, \phi^2 x} \ldots\ldots\ldots (49).$$

Écrivant dans cette équation a, x au lieu de x, a, et ajoutant, on obtient

$$\frac{Z'x}{Zx} + \frac{Z'a}{Za} - \frac{Z'(x+a)}{Z(x+a)} = e^2c^2\phi a\, \phi x\, \phi(a+x) \ldots\ldots\ldots (50).$$

En intégrant la même équation par rapport à a,

$$\left.\begin{aligned} &\log Z(x+a) + \log Z(x-a) - 2\log Zx - 2\log Za \\ &\quad = \log(1+e^2c^2\phi^2 a\, \phi^2 x) \end{aligned}\right\} \ldots\ldots\ldots (51),$$

c'est-à-dire

$$Z(x+a)\, Z(x-a) = Z^2 x\, Z^2 a\, (1+e^2c^2\phi^2 a\, \phi^2 x) \ldots\ldots\ldots (52),$$

ou, ce qui est la même chose,

$$Z(x+a)\, Z(x-a) = Z^2 x\, Z^2 a + e^2c^2\, \mathrm{y}^2 x\, \mathrm{y}^2 a \ldots\ldots\ldots (53),$$

de laquelle on déduit facilement, en écrivant $x+\frac{1}{2}\Omega$, $x+\frac{1}{2}\Upsilon$, $x+\frac{1}{2}\Omega+\frac{1}{2}\Upsilon$ au lieu de x, les équations complémentaires

$$(\mathrm{Y}) \ldots\ldots\ldots \left\{\begin{aligned} &\mathrm{y}(x+a)\, \mathrm{y}(x-a) = \mathrm{y}^2 x\, Z^2 a - \mathrm{y}^2 a\, Z^2 x, \\ &\mathrm{g}(x+a)\, \mathrm{g}(x-a) = \mathrm{g}^2 x\, Z^2 a - c^2\mathrm{g}^2 a Z^2 x, \\ &G(x+a)\, G(x-a) = G^2 x\, Z^2 a + e^2 G^2 a\, Z^2 x. \end{aligned}\right.$$

Quoiqu'elle ne soit pas liée très-étroitement avec la théorie actuelle, on peut ajouter ici cette autre formule de M. Jacobi, qu'il obtient en intégrant par rapport à x, au lieu de a,

$$\Pi(x, a) = \int_0^x \frac{-e^2c^2\,\phi a\, fa\, Fa\, \phi^2 x\, dx}{1 + e^2c^2\,\phi^2 a\,\phi^2 x} = \tfrac{1}{2}\log\frac{Z(x-a)}{Z(x+a)} + x\frac{Z'a}{Za} \quad\ldots\ldots\ldots\ (54),$$

au moyen de laquelle il déduit des formules pour l'addition des arguments ou des paramètres des fonctions de la troisième espèce. On trouve aussi, dans les *Fund. Nova*, quelques formules déduites de l'équation (49) pour exprimer $\dfrac{Z(x-a)\,Z(y-a)\,Z(x+y+a)}{Z(x+a)\,Z(y+a)\,Z(x+y-a)}$ au moyen de la fonction ϕ; il serait facile de déduire des équations semblables pour les autres fonctions y, g, G.

Note sur une intégrale définie.

Soient k_1, k des quantités réelles dont la première est la plus grande; $\Omega = \omega + \omega' i$, $\Upsilon = v + v'i$ des quantités quelconques, telles que $\omega v' - \omega' v$ ne s'évanouisse pas, et écrivons

$$u = \int_k^{k_1} \frac{dx}{\Omega + \Upsilon x} \quad\ldots\ldots\ldots\ldots\ldots\ldots\ (55).$$

L'intégrale u a toujours une valeur finie et déterminée, puisque le dénominateur ne devient jamais zéro. On a évidemment

$$u = \int_0^\infty dx \left[\frac{1}{\Omega + \Upsilon(x+k)} - \frac{1}{\Omega + \Upsilon(x+k_1)}\right] \quad\ldots\ldots\ldots\ (56),$$

et l'intégrale indéfinie est

$$u = -\frac{1}{\Upsilon}\log\left\{\pm\frac{\Omega + \Upsilon(x+k)}{\Omega + \Upsilon(x+k_1)}\right\} \quad\ldots\ldots\ldots\ (57);$$

il faut passer de là à l'intégrale définie, entre les limites 0, ∞. Soit

$$\frac{\Omega + \Upsilon(x+k)}{\Omega + \Upsilon(x+k_1)} = A_x + iB_x \quad\ldots\ldots\ldots\ldots\ (58).$$

Il est facile de voir qu'en faisant abstraction d'un dénominateur toujours positif, A_x est une fonction du second degré en x, et B_x se réduit à la quantité constante $(\omega v' - \omega' v)(k_1 - k)$. Le signe de B_x est donc toujours le même que celui de $\omega v' - \omega' v$; quant à celui de A_x, puisque évidemment $A_\infty = 1$, il est clair que si A_0 est positif, A_x reste toujours positif, ou change deux fois de signe quand x passe de 0 à ∞. Si, au contraire, A_0 est négatif, A_x est négatif depuis $x = 0$ jusqu'à une certaine valeur positive de x, $x = \alpha$, et positif depuis $x = \alpha$ jusqu'à $x = \infty$. Considérons d'abord ce dernier cas. En représentant par Lx la valeur principale de $\log x$ (cela suppose que la partie réelle de x est positive), on obtient cette valeur de u,

$$u = \frac{1}{\Upsilon}L\left[-\frac{\Omega + \Upsilon k_1}{\Omega + \Upsilon k}\right] - \frac{1}{\Upsilon}L\left[-\frac{\Omega + \Upsilon(\alpha + k_1 - \epsilon)}{\Omega + \Upsilon(\alpha + k - \epsilon)}\right] - \frac{1}{\Upsilon}L\left[\frac{\Omega + \Upsilon(\alpha + k_1 + \epsilon)}{\Omega + \Upsilon(\alpha + k + \epsilon)}\right],$$

où ϵ est supposé une quantité infiniment petite positive. La somme des derniers termes se réduit à

$$\frac{i}{\Upsilon}\left(-\text{arc tan}\,\frac{B_a}{A_{a-\epsilon}}+\text{arc tan}\,\frac{A_a}{B^{a+\epsilon}}\right)\ldots\ldots\ldots\ldots\ldots\ldots\ldots\ldots (59),$$

où, comme à l'ordinaire, $\text{arc tan}\, x$ doit être situé entre les limites $\pm\frac{1}{2}\pi$.

Dans cette formule, $A_{a-\epsilon}$ est une quantité infiniment petite et négative; $A_{a+\epsilon}$ est une quantité infiniment petite et positive; donc, quand B_a est négatif, les arcs se réduisent à $\frac{1}{2}\pi$, $-\frac{1}{2}\pi$, et si B_a est positif, à $-\frac{1}{2}\pi$, $\frac{1}{2}\pi$. On a donc

$$u=\frac{1}{\Upsilon}L\left(-\frac{\Omega+\Upsilon k_1}{\Omega+\Upsilon k}\right)\pm\frac{1}{\Upsilon}\pi i\ \ldots\ldots\ldots\ldots\ldots\ldots\ldots\ldots (60),$$

où il faut se servir du signe supérieur ou inférieur selon que $\omega\upsilon'-\omega'\upsilon$ est positif ou négatif. Dans le cas où A_0 est positif, si A_x reste toujours positif, on obtient tout de suite

$$u=\frac{1}{\Upsilon}L\left(\frac{\Omega+\Upsilon k_1}{\Omega+\Upsilon k}\right)\ \ldots\ldots\ldots\ldots\ldots\ldots\ldots\ldots\ldots\ldots (61).$$

Si A_x change deux fois de signe, par exemple pour $x=\alpha$ et $x=\beta$, il faut introduire les corrections

$$\frac{i}{\Upsilon}\left(-\text{arc tan}\,\frac{B_a}{A_{a-\epsilon}}+\text{arc tan}\,\frac{B_a}{A_{a+\epsilon}}\right)+\frac{i}{\Upsilon}\left(-\text{arc tan}\,\frac{B_\beta}{A_{\beta-\epsilon}}+\text{arc tan}\,\frac{B_\beta}{A_{\beta+\epsilon}}\right),$$

qui se détruisent l'une l'autre, en sorte que l'on a le même résultat que si A_x était toujours positif. En se servant de la notation employée dans le Mémoire, on a dans tous les cas

$$u=L_{\pm\pi i}\left(\frac{\Omega+\Upsilon k_1}{\Omega+\Upsilon k}\right)\ \ldots\ldots\ldots\ldots\ldots\ldots\ldots\ldots\ldots\ldots (62),$$

où le signe se détermine d'après celui de $\omega\upsilon'-\omega'\upsilon$.

En particulier,

$$\int_{-k}^{k}\frac{dx}{\Omega+\Upsilon x}=\Omega\int_0^k\frac{dx}{\Omega^2-\Upsilon^2x^2}=\frac{1}{\Upsilon}L_{\pm\pi i}\left(\frac{\Omega+\Upsilon k}{\Omega-\Upsilon k}\right),$$

ou

$$\int_0^{\tan\alpha}\frac{dx}{\Omega^2-\Upsilon^2x^2}=\frac{1}{2\Omega\Upsilon}L_{\pm\pi i}\left(\frac{\Omega+\Upsilon\tan\alpha}{\Omega-\Upsilon\tan\alpha}\right)\ \ldots\ldots\ldots\ldots\ldots (63).$$

Je ne sais pas si l'on a cherché avant moi la valeur exacte de cette intégrale définie, qui est cependant la plus simple qu'on puisse imaginer, et qui devrait, je crois, trouver place dans les livres élémentaires.

NOTA. Une partie de ces recherches a été déjà imprimée dans le *Cambridge Mathematical Journal* [24]; je me suis borné au cas où Ω, Υ sont de la forme ω, vi, ce qui simplifie beaucoup la détermination des intégrales définies doubles; mais la forme générale des résultats en est très-peu affectée.

26.

MÉMOIRE SUR LES COURBES DU TROISIÈME ORDRE.

[From the *Journal de Mathématiques Pures et Appliquées* (Liouville), tome IX. (1844), pp. 285—293.]

CONSIDÉRONS d'abord la surface du troisième ordre qui passe par les six arêtes d'un tétraèdre quelconque. Cette surface sera touchée selon chaque arête par un seul plan; je dis que: "les plans tangents selon les arêtes opposées se rencontrent en trois droites qui sont dans le même plan, et chacune de ces droites est située entièrement sur la surface."

En effet, si nous représentons par $P=0$, $Q=0$, $R=0$, $S=0$, les équations des quatre plans du tétraèdre, et par α, β, γ, δ des constantes arbitraires, l'équation de la surface ne peut avoir que la forme

$$\alpha QRS+\beta PRS+\gamma PQS+\delta PQR=0.$$

Soient $\mathfrak{P}$, $\mathfrak{Q}$, $\mathfrak{R}$, $\mathfrak{S}$ ce que deviennent les quantités P, Q, R, S, quand on change les coordonnées x, y, z en de nouvelles variables ξ, η, ζ; l'équation du plan tangent se réduit facilement à la forme

$$\begin{aligned}(\mathfrak{P}-P)(\beta RS+\gamma QS+\delta QR)+(\mathfrak{Q}-Q)(\alpha RS+\gamma PS+\delta PR)&\\ +(\mathfrak{R}-R)(\alpha QS+\beta PS+\delta PQ)+(\mathfrak{S}-S)(\alpha QR+\beta PR+\gamma PQ)&=0.\end{aligned}$$

Soient $P=0$, $Q=0$; cette équation devient

$$\beta\mathfrak{P}+\alpha\mathfrak{Q}=0;$$

et de même, si $R=0$, $S=0$, l'équation devient

$$\delta\mathfrak{R}+\gamma\mathfrak{S}=0.$$

De ces équations on déduit les deux suivantes:

$$\alpha\mathfrak{QRS}+\beta\mathfrak{PRS}+\gamma\mathfrak{PQS}+\delta\mathfrak{PQR}=0,$$

$$\frac{1}{\alpha}\mathfrak{P}+\frac{1}{\beta}\mathfrak{Q}+\frac{1}{\gamma}\mathfrak{R}+\frac{1}{\delta}\mathfrak{S}=0.$$

On conclut de la première, que la droite d'intersection des deux plans tangents est située sur la surface; et de la seconde, que cette droite est dans un même plan, quelles que soient les deux arêtes opposées par lesquelles on a mené les deux plans tangents. Ainsi le théorème est démontré.

En considérant une section quelconque de cette surface, ou même en supposant que P, Q, R, S ne contiennent que deux variables x, y, nous avons ce théorème:

"Etant donnée une courbe du troisième ordre qui passe par les six points d'intersection de quatre droites, les tangentes à la courbe en ces six points se rencontrent deux à deux en trois points qui sont les points d'intersection de la courbe par une droite. Les tangentes qui doivent être prises ensemble sont les tangentes en deux points A, A', tels que A est l'intersection de deux des quatre lignes données, et A' l'intersection des deux autres lignes, points que l'on peut nommer *opposés*.

"De même, si une courbe du troisième ordre passe par cinq de ces points, de telle manière que l'intersection des tangentes à la courbe en deux points opposés soit située sur la courbe, la courbe passe par le sixième point, et par les points d'intersection des tangentes menées par les deux autres paires de points opposés."

A présent, posons une courbe quelconque du troisième ordre et deux points A, A' sur la courbe, tels que les tangentes en ces deux points se rencontrent sur la courbe (cela suppose que la courbe est de la sixième ou quatrième classe, et non pas de la troisième). Un autre point B sur la courbe étant pris à volonté, les droites AB, $A'B$ rencontrent la courbe en H et h; et les droites Ah, $A'H$ se rencontrent en un point B' situé sur la courbe. Les tangentes en B, B', et de même les tangentes en H, h, se rencontrent sur la courbe en deux points qui sont en ligne droite avec le point d'intersection des tangentes en A et A'.

On peut dire que les deux points B, B', ou les deux points H, h, sont une paire de points correspondante à la paire A, A'. Les deux paires B, B' et H, h sont évidemment correspondantes l'une à l'autre, et, de plus, A, A' correspond à ces deux paires, de la même manière que H, h correspond à A, A', et B, B'; de sorte que l'on peut dire que les trois paires A, A'; B, B'; H, h sont supplémentaires l'une aux deux autres.

Soit C, C' une autre paire de points correspondante à A, A'; je dis que *B, B' et C, C' sont des paires correspondantes l'une à l'autre;* et, de plus, *si d'un point P quelconque de la courbe, l'on mène des lignes aux points A, A', B, B', C, C', ces lignes forment un faisceau en involution.*

En effet, considérons six points quelconques A, A', B, B', C, C' dans le même plan, et représentons par f, g, h les points d'intersection de BC' et $B'C$, CA' et $C'A$, AB' et $A'B$, et par F, G, H les points d'intersection de CB et $C'B'$, CA et $C'A'$, AB et $A'B'$. Nous allons faire voir que le lieu d'un point P qui se meut de telle manière que les lignes menées aux points A, B, C, A', B', C' forment toujours un faisceau en involution, est une courbe du troisième ordre qui passe par ces six points, et aussi par les points f, g, h, F, G, H.

Soient α, α', β, β', γ, γ' les tangentes trigonométriques des inclinaisons des lignes PA, PA', PB, PB', PC, PC', sur une ligne fixe quelconque que l'on peut prendre pour l'axe des x. Pour que ces lignes forment un faisceau en involution, nous devrions avoir la relation

$$\alpha\alpha'(\beta+\beta'-\gamma-\gamma')+\beta\beta'(\gamma+\gamma'-\alpha-\alpha')+\gamma\gamma'(\alpha+\alpha'-\beta-\beta')=0,$$

ou, ce qui revient à la même chose, l'une quelconque des quatre équations

$$\begin{aligned}
(\alpha-\beta')(\beta-\gamma')(\gamma-\alpha')+(\alpha'-\beta)(\beta'-\gamma)(\gamma'-\alpha)&=0,\\
(\alpha'-\beta')(\beta-\gamma')(\gamma-\alpha)+(\alpha-\beta)(\beta'-\gamma)(\gamma'-\alpha')&=0,\\
(\alpha-\beta)(\beta'-\gamma')(\gamma-\alpha')+(\alpha'-\beta')(\beta-\gamma)(\gamma'-\alpha)&=0,\\
(\alpha'-\beta)(\beta'-\gamma')(\gamma-\alpha)+(\alpha-\beta')(\beta-\gamma)(\gamma'-\alpha')&=0.
\end{aligned}$$

Soient x_P, y_P, x_A, y_A, etc., les coordonnées de P, A, etc.

$$\alpha=\frac{y_P-y_A}{x_P-y_A},$$

$$\alpha-\beta'=\frac{y_P-y_A}{x_P-x_A}-\frac{y_P-y_{B'}}{x_P-x_{B'}}=-\frac{1}{(x_P-x_A)(x_P-x_{B'})}(PAB'),$$

en représentant par (PAB'), etc., les quantités telles que

$$x_P(y_A-y_{B'})+y_P(x_A-x_{B'})+x_Ay_{B'}-x_{B'}y_A.$$

L'équation de la courbe peut donc se mettre sous l'une quelconque des quatre formes

$$\begin{aligned}
PAB'\,.\,PBC'\,.\,PCA'+PA'B\,.\,PB'C\,.\,PC'A&=0,\\
PA'B'\,.\,PBC'\,.\,PCA+PAB\,.\,PB'C\,.\,PC'A'&=0,\\
PAB\,.\,PB'C'\,.\,PCA'+PA'B'\,.\,PBC\,.\,PC'A&=0,\\
PA'B\,.\,PB'C'\,.\,PCA+PAB'\,.\,PBC\,.\,PC'A'&=0,
\end{aligned}$$

qui sont du troisième ordre. La première fait voir que la courbe cherchée passe par les neuf points A, B, C, A', B', C', f, g, h. La troisième ou la quatrième fait voir qu'elle passe de plus par le point F; la quatrième ou la seconde, qu'elle passe par le point G; la seconde ou la troisième, qu'elle passe par le point H. Ainsi la courbe passe par les douze points A, B, C, A', B', C', f, g, h, F, G, H.

On peut remarquer qu'une courbe du troisième ordre qui passe par dix quelconques de ces points, ou même par neuf quelconques, pourvu que nous exceptions les combinaisons de A, B, C, A', B', C', avec f, g, h, ou f, G, H, ou F, g, H, ou F, G, h, ne peut être que la courbe que nous venons de trouver. On peut aussi remarquer en passant que si A, A', B, B', C, C' sont les points d'intersection de quatre droites, l'équation ci-devant trouvée est satisfaite identiquement, de sorte que la position du point P est absolument arbitraire. Cela étant connu, je ne m'arrête pas pour le démontrer.

Revenons au cas d'une courbe donnée, avec trois paires de points A, A', B, B', C, C', comme auparavant, tels que B, B' et C, C' sont des paires correspondantes à A, A'. Les points A, B, C, A', B', C', G, g, H, h sont situés sur la courbe; ainsi F, f sont aussi sur la courbe (c'est-à-dire que B, B' et C, C' sont des paires correspondantes), et les lignes menées par un point P quelconque de la courbe et les points A, B, C, A', B', C' forment un faisceau en involution: théorème ci-devant énoncé.

Considérons une courbe du troisième ordre, de la quatrième classe (c'est-à-dire, telle que d'un point quelconque on ne peut lui mener que quatre tangentes). D'un point sur la courbe, indépendamment de la tangente en ce point, on ne peut mener que deux tangentes. Prenons les deux points K, L sur la courbe, et menons les tangentes KA, KA', LB, LB' touchant la courbe en A, A', B, B'. Il est clair qu'en ce cas A, A' et B, B' sont des paires correspondantes de points. Mais le cas général où la courbe est de la sixième classe est moins simple. Considérons, en effet, pour une telle courbe, les huit points A_1, A_2, A_3, A_4 et B_1, B_2, B_3, B_4 de contact des tangentes menées par les deux points K et L. En choisissant A, A' et B de quelque manière que ce soit, parmi les points A_1, A_2, A_3, A_4 et B_1, B_2, B_3, B_4 respectivement, le point B', qui doit entrer dans cette dernière série, ne peut pas être choisi à volonté, mais est parfaitement déterminé. En désignant convenablement les points B, on peut toujours supposer que les paires correspondantes soient A_1A_2 ou A_3A_4 avec B_1B_2 ou B_3B_4; A_1A_3 ou A_2A_4 avec B_1B_3 ou B_2B_4; A_1A_4 ou A_2A_3 avec B_1B_4 ou B_2B_3. Cela suppose que

$$A_1B_1,\quad A_2B_2,\quad A_3B_3,\quad A_4B_4;\qquad A_1B_2,\quad A_2B_1,\quad A_3B_4,\quad A_4B_3;$$

$$A_1B_3,\quad A_3B_1,\quad A_2B_4,\quad A_4B_2;\qquad A_1B_4,\quad A_4B_1,\quad A_2B_3,\quad A_3B_2,$$

se rencontrent, chaque système, dans le même point. Il faut expliquer avec plus d'evidence la corrélation de ces huit points.

Imaginons les six points A, A'; B, B'; C, C', tels que AA', BB', CC' se rencontrent dans le même point. Formons, comme auparavant, le système de points f, g, h, F, G, H. Les propriétés de ce système de douze points sont très-nombreuses. Non-seulement F, G, H; F, g, h; f, G, h; f, g, H sont en ligne droite; mais, en outre, les trois droites de chacun des onze systèmes que voici se rencontrent dans le même point:

$$\begin{array}{lll} Af,\ Bg,\ Ch; & A'f,\ B'g,\ C'h; & \\ AA',\ Gg,\ Hh; & BB',\ Hh,\ Ff; & CC',\ Ff,\ Gg; \\ A'f,\ BG,\ CH; & B'g,\ CH,\ AF; & C'h,\ AF,\ BG; \\ Af,\ B'G,\ C'H; & Bg,\ C'H,\ A'F; & Ch,\ A'F,\ B'G. \end{array}$$

(Cela est connu, je crois; au reste, pour le démontrer, considérons un parallélipipède, tel que gf, GF, AB, $A'B'$ en soient quatre arêtes parallèles, les autres arêtes parallèles étant gA, $A'G$, fB, $B'F$; gA', $A'G$, Bf, $B'F$. Soient H le point de rencontre, à l'infini, du premier système d'arêtes parallèles; C' et C les points de rencontre, à l'infini, des arêtes des second et troisième systèmes respectivement; h le point de rencontre des quatre diagonales du parallélipipède; faisons la perspective de ce système, désignant chaque point de la projection par la même lettre: l'on voit sans peine que la figure plane, ainsi obtenue, a les propriétés en question.)

A présent, en examinant la figure, on voit qu'il est permis de prendre pour A_1, A_2, A_3, A_4 les points A, A', F, f, et pour B_1, B_2, B_3, B_4 les points B, B', G, g, pour que les systèmes A_1, A_2, A_3, A_4; B_1, B_2, B_3, B_4 aient la corrélation ci-devant trouvée. Le système C, C', H, h est supplémentaire à ces deux-ci.

Toutes les propriétés que je viens de trouver sont telles qu'il y a à chacune une propriété correspondante que l'on peut obtenir par la théorie des polaires réciproques. Nous avons ainsi des propriétés non moins intéressantes des courbes de la troisième classe et du sixième ou quatrième ordre.

En prenant pour la courbe du troisième ordre l'ensemble d'une section conique et d'une droite, pour que les tangentes en A, A' se rencontrent sur la courbe, il faut que A, A' soient situées sur la section conique de telle manière que AA' passe par le pôle de la droite à l'égard de la conique. Alors B, B'; C, C' étant pris de la même manière, BC', $B'C$ et BC, $B'C'$ se rencontrent sur la droite.

Les lignes tirées d'un point P quelconque de la conique, ou de la droite à A, B, C; A', B', C', forment un faisceau en involution.

Le premier théorème et la première partie du second sont très-bien connus; je ne sais s'il en est de même de la dernière partie de ce théorème.

ADDITION.

La droite PP′, *menée par deux points* P, P′ *d'une courbe du troisième ordre, qui correspondent toujours à une paire correspondante donnée de points* A, A′, *est toujours tangente à une certaine courbe de la troisième classe.*

Pour démontrer ce théorème, imaginons une paire fixe B, B' de points qui corresponde à la paire donnée A, A'. Soient $P=0$, $Q=0$, $R=0$, $S=0$ les équations des lignes qui joignent A, A' avec B, B'; l'équation de la courbe donnée du troisième ordre peut se mettre sous la forme

$$\frac{\alpha}{P}+\frac{\beta}{Q}+\frac{\gamma}{R}+\frac{\delta}{S}=0 \quad\quad (1).$$

On peut toujours supposer, sans perte de généralité, que l'équation

$$P+Q+R+S=0 \quad\quad (2)$$

soit identiquement vraie, car chaque équation de la forme $P=0$ peut être censée contenir une constante arbitraire qui en multiplie tous les termes.

Donc, en faisant

$$P+R=\theta, \qquad P=\theta-R,$$

on peut toujours supposer

$$Q+S=-\theta, \qquad Q=-\theta-S,$$

et l'équation de la courbe se transforme en

$$\frac{\alpha}{\theta - R} - \frac{\beta}{\theta + S} + \frac{\gamma}{R} + \frac{\delta}{S} = 0 \quad \ldots\ldots\ldots\ldots\ldots\ldots(3),$$

ce que l'on peut écrire aussi sous la forme

$$\frac{\theta + (\lambda\mu + \lambda + \mu) S}{(\theta + \lambda R)(\theta + \mu R)} + \frac{-\theta + (\nu\rho + \nu + \rho) R}{(-\theta + \nu S)(-\theta + \rho S)} = 0 \quad \ldots\ldots\ldots(4),$$

en déterminant convenablement les constantes λ, μ, ν, ρ. En effet, en réduisant, les deux équations deviennent identiques au moyen des quatre conditions

$$\left.\begin{array}{l} \delta = -k\lambda\mu, \qquad \gamma = -k\nu\rho, \\ \delta - \beta = k\lambda\mu(\nu\rho + \nu + \rho), \\ \gamma - \alpha = k\nu\rho(\lambda\mu + \lambda + \mu), \end{array}\right\} \quad \ldots\ldots\ldots\ldots\ldots\ldots(5),$$

où k est une quantité arbitraire, de manière que des quantités λ, μ, ν, ρ, il y en a une seule qui peut être prise à volonté.

De l'équation (4) l'on déduit tout de suite cette autre forme,

$$\frac{1}{\mu - \lambda}\left[\frac{\mu(\lambda + 1)}{\theta + \lambda R} - \frac{\lambda(\mu + 1)}{\theta + \mu R}\right] + \frac{1}{\rho - \nu}\left[\frac{\rho(\nu + 1)}{-\theta + \nu S} - \frac{\nu(\rho + 1)}{-\theta + \rho S}\right] = 0 \quad \ldots\ldots(6);$$

et de là nous voyons que les points donnés par les deux systèmes d'équations

$$\left.\begin{array}{l} (\theta + \lambda R = 0, \quad -\theta + \nu S = 0), \\ (\theta + \mu R = 0, \quad -\theta + \rho S = 0), \end{array}\right\} \quad \ldots\ldots\ldots\ldots\ldots\ldots (7)$$

correspondent toujours l'un à l'autre et aux points donnés par les deux systèmes

$$\left.\begin{array}{l} (\theta = 0, \quad R = 0), \\ (\theta = 0, \quad S = 0), \end{array}\right\} \quad \ldots\ldots\ldots\ldots\ldots\ldots(8),$$

c'est-à-dire aux points A, A'.

On trouve assez facilement pour l'équation de la droite menée par les points déterminés par les deux systèmes (7), points que l'on peut prendre pour P et P',

$$(\mu\nu - \rho\lambda)\,\theta + \lambda\mu(\nu - \rho) R + \nu\rho(\lambda - \mu) S = 0 \ldots\ldots\ldots\ldots(9),$$

ou

$$C\theta + AR + BS = 0 \quad \ldots\ldots\ldots\ldots\ldots\ldots(10),$$

en faisant

$$\left.\begin{array}{l} pC = \mu\nu - \rho\lambda \quad , \\ pA = \lambda\mu(\nu - \rho), \\ pB = \nu\rho\,(\lambda - \mu). \end{array}\right\} \quad \ldots\ldots\ldots\ldots\ldots\ldots(11).$$

Éliminons des équations (5) et (11) les cinq quantités λ, μ, ν, ρ, p. Pour opérer de la manière la plus élégante, formons d'abord les équations identiques

$$\left.\begin{aligned}&[\mu\nu - \rho\lambda - \nu\rho\ (\lambda - \mu)]\,(\nu - \rho)\,(\lambda\mu + \lambda + \mu)\\&+ [\mu\nu - \rho\lambda + \lambda\mu\,(\nu - \rho)]\,(\lambda - \mu)\,(\nu\rho + \nu + \rho)\\&\qquad = (\nu\lambda - \mu\rho)\,[\lambda\mu\,(\nu - \rho) - \nu\rho\,(\lambda - \mu) + 2\,(\mu\nu - \rho\lambda)],\\&\lambda\mu\,(\nu - \rho)^2 - \nu\rho\,(\lambda - \mu)^2 = (\mu\nu - \rho\lambda)\,(\lambda\nu - \mu\rho),\end{aligned}\right\} \quad \ldots\ldots\ldots\ldots(12),$$

ce qui donne

$$\left.\begin{aligned}&A\,(C - B)\,(\gamma - \alpha) + B\,(C + A)\,(\delta - \beta)\\&\qquad = \frac{k}{p}\,\lambda\mu\nu\rho\,(\nu\lambda - \mu\rho)\,(2C + A - B),\\&A^2\gamma - B^2\delta = -\frac{k}{p}\,\lambda\mu\nu\rho\,(\nu\lambda - \mu\rho)\,C,\end{aligned}\right\} \quad \ldots\ldots\ldots\ldots\ldots\ldots(13),$$

et de là

$$\left.\begin{aligned}&C\,[A\,(C - B)\,(\gamma - \alpha) + B\,(C + A)\,(\delta - \beta)]\\&\quad + (2C + A - B)\,(A^2\gamma - B^2\delta)\end{aligned}\right\} = 0 \quad \ldots\ldots\ldots\ldots\ldots(14),$$

ou, toute réduction faite,

$$\left.\begin{aligned}&A\,(A + C)\,(A + C - B\gamma) + B\,(C - B)\,(A + B - C\delta)\\&\qquad - AC\,(C - B\alpha) - BC\,(C + A\beta)\end{aligned}\right\} = 0 \quad \ldots\ldots\ldots(15);$$

et puisque cette équation est du troisième ordre à l'égard des quantités A, B, C, il est évident que la ligne donnée par l'équation (10) est toujours tangente à une certaine courbe de la troisième classe. En supposant $\gamma = 0$, $\delta = 0$, ce qui réduit la courbe du troisième ordre aux lignes $R = 0$, $S = 0$, $(\beta - \alpha)\,\theta - \beta R - \alpha S = 0$, l'équation (15) se partage dans les deux équations

$$C = 0, \quad A\,(C - B)\,\alpha + B\,(C + A)\,\beta = 0 \ldots\ldots\ldots\ldots\ldots\ldots\ldots\ldots(16),$$

et la courbe de la troisième classe se réduit au point $(R = 0,\ S = 0)$ et à une conique. Cela est un théorème connu, car l'on sait bien que si A, A' sont des points quelconques sur deux droites données α, α', et B, B' deux autres points déterminés, sur ces droites, de manière que l'intersection des lignes AB', $A'B$ soit toujours sur une ligne droite donnée, les deux lignes α, α' sont divisées homographiquement, B, B' étant deux points correspondants; et de plus, que la ligne qui unit les points correspondants de deux lignes divisées homographiquement, est toujours tangente à une certaine conique. Quant au point d'intersection des lignes α, α', que nous venons de trouver comme formant avec la conique une courbe de la troisième classe, on observera que, dans notre théorie, non-seulement les points des lignes α, α' se correspondent, mais que le point d'intersection des lignes α, α' correspond à tous les points de la troisième droite. La conique touche les deux droites α, α'; cela n'a pas, je crois, d'analogie dans la théorie générale.

27.

NOUVELLES REMARQUES SUR LES COURBES DU TROISIÈME ORDRE.

[From the *Journal de Mathématiques Pures et Appliquées* (Liouville), tom. X. (1845), pp. 102—109.]

JE me propose de développer ici quelques conséquences de la théorie que j'ai donnée, il y a quelques mois, dans ce Journal[1], [26], des points *correspondants* des courbes du troisième ordre. Rappelons d'abord la signification de ce terme et ajoutons-y quelques nouvelles définitions.

On dit que les points A, A' sont correspondants quand les tangentes à la courbe en ces points se rencontrent sur la courbe. En considérant les points correspondants A, A' et les deux autres points correspondants B, B', on dit que ces paires correspondent, quand les points d'intersection H, h de AB', $A'B$ ou AB, $A'B'$ sont situés sur la courbe. Les trois paires A, A', B, B', H, h sont nommées paires supplémentaires ou système supplémentaire. On dirait de même que les deux systèmes AA', BB', Hh et $A_1A'_1$, $B_1B'_1$, H_1h_1 sont des systemes supplémentaires correspondants, si, par exemple, A, A' et A_1, A'_1, &c., formaient des paires correspondantes.

On peut nommer quadrilatère inscrit le système de quatre droites qui passent par les points supplémentaires AA', BB', Hh, et conique d'involution chaque conique tangente à ces quatre droites, ou, en d'autres termes, inscrite dans un quadrilatère inscrit. Il est inutile d'expliquer ce que veulent dire coniques d'involution correspondantes, ou quadrilatères correspondants, ou l'expression conique correspondante à une paire donnée de points correspondants, &c.

Démontrons les théorèmes suivants :

THÉORÈME I. "Les tangentes menées par un point P de la courbe à trois coniques d'involution correspondantes forment un faisceau en involution."

[1] Voyez page 285 du tome IX.

Chaque conique peut se réduire à une paire de points qui correspondent aussi aux coniques (ce qui justifie la dénomination que nous avons donnée à ces coniques). Considérons un quadrilatère inscrit AA', BB', Hh, la conique d'involution tangente aux côtés de ce quadrilatère, et deux paires LL', MM' de points correspondants, ces paires étant correspondantes l'une à l'autre et à la conique d'involution. On sait que les lignes menées d'un point quelconque, et ainsi du point P de la courbe, par les points A, A', B, B', forment avec les tangentes à la conique menées par ce même point, un faisceau en involution. Mais PA, PA', PB, PB', PL, PL', et de même PB, PB', PL, PL', PM, PM', forment aussi des faisceaux en involution; donc les deux tangentes forment, avec PL, PL' et PM, PM', un faisceau en involution. De même, avec une conique correspondante à la première, PL, PL', PM, PM' et les deux tangentes forment un faisceau en involution; donc PL, PL' et les quatre tangentes forment un faisceau en involution, et de même en introduisant la troisième conique.

THÉORÈME II. "On peut circonscrire à une conique d'involution donnée une infinité de quadrilatères inscrits correspondants au premier."

Soient A, A', B, B', H, h comme auparavant; et A_1, A'_1 une paire de points correspondants qui correspondent à ceux-ci; en menant par A_1, A'_1 des tangentes à la conique qui se rencontrent en B_1, B'_1, H_1, h_1, on voit d'abord que les lignes PA, PA', PA_1, PA'_1 et les tangentes à la conique forment un faisceau en involution; et reciproquement, chaque point P qui satisfait à cette condition appartient à la courbe. Mais en prenant, par exemple, pour P le point B_1, les lignes PA_1, PA'_1 deviennent identiques avec les deux tangentes, ce qui satisfait à la condition d'involution. Donc B_1, B'_1, H_1, h_1 appartiennent à la courbe, ou A_1, A'_1, B_1, B'_1, H_1, h_1 forment un quadrilatère inscrit correspondant au premier ou à la conique.

THÉORÈME III. "Les centres d'homologie de deux coniques d'involution correspondantes forment un quadrilatère inscrit correspondant aux coniques."

Considérons les deux coniques et une troisième conique quelconque. Chaque point P pour lequel les six tangentes forment un faisceau en involution appartient à la courbe. En prenant pour P un centre d'homologie des deux premières coniques, les deux paires de tangentes deviennent identiques, ce qui satisfait à la condition d'involution; donc les six centres d'homologie sont sur la courbe, ou ces six points sont les sommets d'un quadrilatère inscrit qui correspond aussi aux coniques.

Réciproquement,

THÉORÈME IV. "Le lieu d'un point P qui se meut de manière que les tangentes menées par ce point à trois coniques données quelconques, forment toujours un faisceau en involution, est une courbe du troisième ordre qui passe par les dix-huit centres d'homologie des coniques prises deux à deux, ces centres formant six à six des quadrilatères inscrits correspondants."

La démonstration analytique de la première partie de ce théorème, quoique longue, me paraît assez intéressante pour trouver place ici. Pour plus de symétrie, prenons $\frac{\xi}{\zeta}$, $\frac{\eta}{\zeta}$ pour les coordonnées indéfinies d'un point, et représentons par

$$\Upsilon = 0, \quad \Upsilon' = 0, \quad \Upsilon'' = 0,$$

les équations des trois coniques, Υ, &c. étant des fonctions homogènes de la forme

$$\Upsilon = A\xi^2 + B\eta^2 + C\zeta^2 + 2F\eta\zeta + 2G\zeta\xi + 2H\xi\eta, \text{ \&c.}$$

Soient $\frac{x}{z}$, $\frac{y}{z}$ les coordonnées du point P; mettons, pour abréger,

$$U = Ax^2 + By^2 + Cz^2 + 2Fyz + 2Gzx + 2Hxy,$$
$$W = Ax\xi + By\eta + Cz\zeta + F(y\zeta + \eta z) + G(z\xi + \zeta x) + H(x\eta + y\xi)$$

(de manière que $W = 0$ serait l'équation de la ligne polaire du point P). Nous avons

$$U\Upsilon - W^2 = 0,$$

pour l'équation des deux tangentes menées par le point P à la conique. Cela est probablement connu. Il est clair d'abord que cette équation appartient à une conique qui a un double contact avec la conique donnée, et par la forme à laquelle nous allons réduire cette équation, on voit ensuite qu'elle appartient à un système de deux droites qui passent par le point P. En développant et écrivant

$$BC - F^2 = a, \quad CA - G^2 = b, \quad AB - H^2 = c,$$
$$GH - AF = f, \quad HF - BG = g, \quad FG - CH = h,$$

on trouve, en effet,

$$a(y\zeta - z\eta)^2 + b(z\xi - x\zeta)^2 + c(x\eta - y\xi)^2$$
$$+ 2f(z\xi - x\zeta)(x\eta - y\xi) + 2g(x\eta - y\xi)(y\zeta - z\eta) + 2h(y\zeta - z\eta)(z\xi - x\zeta) = 0,$$

et de là, en posant

$$\mathfrak{A} = bz^2 + cy^2 - 2fyz, \qquad \mathfrak{F} = ayz - gxy - hxz + fx^2,$$
$$\mathfrak{B} = cx^2 + az^2 - 2gxz, \qquad \mathfrak{G} = bzx - hyz - fxy + gy^2,$$
$$\mathfrak{C} = ay^2 + bx^2 - 2hxy, \qquad \mathfrak{H} = cxy - fxz - gyz + hz^2,$$

on obtient

$$\mathfrak{A}\xi^2 + \mathfrak{B}\eta^2 + \mathfrak{C}\zeta^2 - 2\mathfrak{F}\eta\zeta - 2\mathfrak{G}\xi\zeta - 2\mathfrak{H}\xi\eta = 0,$$

ou, en transportant l'origine au point P,

$$\mathfrak{A}\xi^2 + \mathfrak{B}\eta^2 - 2\mathfrak{H}\xi\eta = 0;$$

et de même, pour l'équation des tangentes, par ce même point aux deux autres coniques

$$\mathfrak{A}'\xi^2 + \mathfrak{B}'\eta^2 - 2\mathfrak{H}'\xi\eta = 0,$$
$$\mathfrak{A}''\xi^2 + \mathfrak{B}''\eta^2 - 2\mathfrak{H}''\xi\eta = 0.$$

On obtient donc, pour la condition que ces lignes forment un faisceau en involution,

$$\begin{bmatrix} \mathfrak{A}, & \mathfrak{B}, & \mathfrak{H} \\ \mathfrak{A}', & \mathfrak{B}', & \mathfrak{H}' \\ \mathfrak{A}'', & \mathfrak{B}'', & \mathfrak{H}'' \end{bmatrix} = 0,$$

en représentant de cette manière le déterminant formé avec ces neuf quantités. Formons d'abord la fonction

$$\mathfrak{A}'\mathfrak{B}'' - \mathfrak{A}'\mathfrak{B}''.$$

Cela se reduit à

$$z[-2(fc)x^2y - 2(cg)xy^2 + (ca)y^2z - 2(fa)yz^2 - 2(bg)xz^2 + (bc)x^2z + 4(fg)xyz + (ba)z^3],$$

où $(fc) = (f'c'' - f''c')$, &c.

Multipliant par $\mathfrak{H}$, formant ensuite les quantités analogues et ajoutant; écrivant aussi

$$c(fg) + c'(f'g') + c''(f''g'') = (cfg),$$

c'est-à-dire (cfg) pour le déterminant formé avec les neuf quantités

$$c, f, g, \quad c', f', g', \quad c'', f'', g'',$$

on obtient d'abord les termes

$$-4(cfg)x^2y^2z^2, \quad +2(fcg)x^2y^2z^2, \quad +2(gfc)x^2y^2z^2,$$

qui se détruisent; les autres termes contiennent z^3 comme facteur, et en écartant cette quantité, l'on obtient en dernière analyse l'équation

$$\begin{aligned}(cbf)x^3 + (acg)y^3 + (bah)z^3 + 4(fgh)xyz \\ + [2(agf) + (cah)]y^2z + [2(bhg) + (abf)]z^2x + [2(cfh) + (bcg)]x^2y \\ + [2(afh) + (abg)]yz^2 + [2(bgf) + (bch)]zx^2 + [2(chg) + (caf)]xy^2 = 0\end{aligned}$$

(où l'on peut, si l'on veut, écrire $z = 1$). C'est l'équation d'une courbe du troisième ordre.

Observation. Cette méthode peut être utile dans l'investigation d'autres problèmes relatifs aux coniques. Par exemple, la question de déterminer les centres d'homologie de deux coniques données revient à celle-ci: satisfaire identiquement à l'équation

$$\begin{aligned}\mathfrak{A}\,\xi^2 + \mathfrak{B}\,\eta^2 + \mathfrak{C}\,\zeta^2 + 2\mathfrak{F}\,\eta\zeta + 2\mathfrak{G}\,\xi\zeta + 2\mathfrak{H}\,\xi\eta \\ k(\mathfrak{A}'\xi^2 + \mathfrak{B}'\eta^2 + \mathfrak{C}'\zeta^2 + 2\mathfrak{F}'\eta\zeta + 2\mathfrak{G}'\xi\zeta + 2\mathfrak{H}'\xi\eta) = 0,\end{aligned}$$

parce que, en effectuant cela, il est facile de voir que $\frac{x}{z}$, $\frac{y}{z}$ sont les coordonnées du centre cherché. Écrivons

$$a + ka' = \mathrm{a}, \quad c + kb' = \mathrm{b}, \quad \&c.,$$

l'on obtient les six équations

$$\begin{aligned}&\mathrm{b}z^2 + \mathrm{c}y^2 - 2\mathrm{f}yz = 0, &\qquad &\mathrm{a}yz - \mathrm{g}xy - \mathrm{h}xz + \mathrm{f}x^2 = 0,\\ &\mathrm{c}x^2 + \mathrm{a}z^2 - 2\mathrm{g}zx = 0, & &\mathrm{b}zx - \mathrm{h}yz - \mathrm{f}yx + \mathrm{g}y^2 = 0,\\ &\mathrm{a}y^2 + \mathrm{b}x^2 - 2\mathrm{h}xy = 0, & &\mathrm{c}xy - \mathrm{f}zx - \mathrm{g}zy + \mathrm{h}z^2 = 0,\end{aligned}$$

dont les trois dernières se déduisent des autres. On peut, de ces six équations, éliminer les six quantités x^2, y^2, z^2, yz, zx, xy, considérées comme indépendantes; on obtient ainsi, toute réduction faite,

$$(\mathrm{abc} - \mathrm{af}^2 - \mathrm{bg}^2 - \mathrm{ch}^2 + 2\mathrm{fgh})^2 = 0,$$

équation qui détermine la quantité k; les trois premières équations deviennent alors équivalentes à deux, qui suffisent pour déterminer les rapports $\frac{x}{z}$, $\frac{y}{z}$. Il serait facile de rapprocher cette solution de celle que l'on déduit de la théorie des polaires réciproques. Par exemple, l'équation qui vient d'être obtenue entre les quantités a, b, ... est précisément celle qui exprime que la fonction

$$\mathrm{a}x^2 + \mathrm{b}y^2 + \mathrm{c}z^2 + 2\mathrm{f}yz + 2\mathrm{g}xz + 2\mathrm{h}xy$$

se divise en facteurs linéaires. Remarquons encore que, dans la géométrie solide, l'équation

$$U\Upsilon - W^2 = 0$$

appartient au cône, ayant le point P pour sommet, et circonscrit à une surface du second ordre qui a pour équation $\Upsilon = 0$. C'est un cas particulier d'une autre formule que voici:

"En représentant par $\mathfrak{P}$ une fonction linéaire des trois variables ξ, η, ζ (ou de quatre variables sans termes constants), et par P la même fonction de x, y, z, l'équation

$$\mathfrak{P}^2 U - 2\mathfrak{P} P W + P^2 \Upsilon = 0$$

appartient au cône ayant pour sommet le point dont les coordonnées sont x, y, z, et passant par la courbe d'intersection du plan $P = 0$, et de la surface du second ordre $\Upsilon = 0$. Et de même pour deux variables."

Je finirai en citant un théorème de géométrie dû à M. Hesse[1], qui a quelques rapports avec le sujet que je viens de traiter:

"Le lieu d'un point P, qui se meut de manière que ses polaires par rapport à trois coniques données se rencontrent dans le même point, est une courbe du troisième ordre; et encore cette courbe ne change pas quand on remplace les coniques données

$$U = 0, \quad U' = 0, \quad U'' = 0,$$

par trois nouvelles coniques de la forme

$$\lambda U + \lambda' U' + \lambda'' U'' = 0."$$

Cette courbe du troisième ordre est tout à fait distincte de celle que j'ai considérée.

[1] *Voyez* le Journal de M. Crelle, tome XXVIII.

28.

SUR QUELQUES INTÉGRALES MULTIPLES.

[From the *Journal de Mathématiques Pures et Appliquées* (Liouville), tome X. (1845), pp. 158—168.]

J'AI donné, il y a trois ans, dans le *Cambridge Mathematical Journal*, [2], une formule assez singulière pour l'intégrale multiple

$$\int \dots dx_1\, dx_2 \dots \phi\,(a_1 - x_1, \quad a_2 - x_2 \dots),$$

prise entre les limites données par l'équation

$$\frac{x_1^2}{h_1^2} + \frac{x_2^2}{h_2^2} + \dots = 1,$$

la fonction ϕ étant seulement assujettie à la condition de ne pas devenir infinie entre les limites de l'intégration. L'expression que j'obtiens est une suite infinie, dont le terme général est de cette forme,

$$A_p \left(h_1^2 \frac{d^2}{da_1^2} + h_2^2 \frac{d^2}{da_2^2} + \dots\right)^p \phi\,(a_1, \; a_2 \dots).$$

En appliquant ce résultat à un cas particulier, j'ai obtenu l'intégrale à n variables analogue à celle qui exprime le *potentiel* d'un ellipsoïde homogène, pour un point extérieur. J'ai depuis cherché à étendre ces résultats au cas d'une densité variable et égale à une fonction rationnelle et entière des coordonnées, et d'une loi d'attraction selon une puissance quelconque de la distance (toujours à n variables); mais quoique j'aie réussi à effectuer cette généralisation, mes formules étaient si confuses et si inférieures à celles que donne la belle analyse de M. Lejeune-Dirichlet, que je ne les ai jamais publiées. Cependant, en revenant il y a quelques jours sur ce sujet, en me fondant sur une intégrale plus générale, j'ai trouvé que la question était à peine plus difficile que dans le cas d'une densité constante, et se laissait traiter exactement de la même manière. J'ai réussi de cette façon à exprimer l'intégrale cherchée au moyen d'une seule intégrale définie *abélienne*, et de ses coefficients différentiels relatifs aux constantes qui y entrent; et il m'a paru que les formules que j'ai ainsi obtenues pourraient n'être pas tout à fait indignes de l'attention des géomètres.

Considérons l'intégrale multiple à n variables V, donnée par l'équation

$$V = \int dx_1 \ldots x_1^{2\alpha_1+1} \ldots (f \text{ termes})\; x_{f+1}^{2\alpha_{f+1}} \ldots (n-f \text{ termes})\; \phi\,(a_1 - x_1 t, \ldots),$$

où les variables $x_1, \ldots$ doivent recevoir des valeurs réelles quelconques, positives ou négatives, qui satisfassent à la condition

$$\frac{x_1^2}{h_1^2} + \ldots \leqq 1\,;$$

et l'on suppose de plus qu'il est permis de développer (sous le signe intégral) la fonction ϕ suivant les puissances ascendantes des variables $x_1, \ldots$.

En faisant ce développement, il est clair que les termes qui contiennent des puissances impaires d'une ou de plusieurs des variables se détruisent par l'intégration; donc, en ne faisant attention qu'aux termes qui contiennent seulement des puissances paires, on a ce terme général,

$$\frac{(-)^f\, t^{2p+f}}{[2r_1+1]^{2r_1+1} \ldots [2r_{f+1}]^{2r_{f+1}} \ldots} \left(\frac{d}{da_1}\right)^{2r_1+1} \ldots \left(\frac{d}{da_{f+1}}\right)^{2r_{f+1}} \ldots \phi\,(a_1, \ldots)$$

$$\times \int dx_1 \ldots x_1^{2\alpha_1+2r_1+2} \ldots x_{f+1}^{2\alpha_{f+1}+2r_{f+1}} \ldots,$$

où $$p = r_1 + \ldots\,;$$

il est à peine nécessaire de remarquer que, dans l'expression

$$[2r_1+1]^{2r_1+1} \ldots [2r_{f+1}]^{2r_{f+1}} \ldots,$$

il faut prendre f termes tels que $[2r_1+1]^{2r_1+1}$, et $\overline{n-f}$ termes de l'autre forme $[2r_{f+1}]^{2r_{f+1}}$, et ainsi dans tous les cas semblables.

L'intégrale qui entre dans cette formule a été trouvée, comme on sait, par M. Dirichlet. Sa valeur est

$$\frac{\Gamma\,(\alpha_1 + r_1 + \frac{3}{2}) \ldots \Gamma\,(\alpha_{f+1} + r_{f+1} + \frac{1}{2}) \ldots}{\Gamma\,(p + k + f + \frac{1}{2}n + 1)}\, h_1^{2\alpha_1+2r_1+3} \ldots h_{f+1}^{2\alpha_{f+1}+2r_{f+1}+1},$$

où $$k = \alpha_1 + \ldots.$$

Le terme entier devient, après quelques réductions très-simples,

$$\frac{(-)^f \pi^{\frac{1}{2}n} t^{2p+f}}{2^{2p+k+f}\Gamma(p+k+f+\frac{1}{2}n+1)}$$

$$\times \frac{L_1 \ldots M_{f+1} \ldots}{[r]^{r_1}\ldots} h_1^{2\alpha_1+2r_1+3} \ldots h_{f+1}^{2\alpha_{f+1}+2r_{f+1}+1} \left(\frac{d}{da_1}\right)^{2r_1+1} \ldots \left(\frac{d}{da_{f+1}}\right)^{2r_{f+1}} \ldots \phi(a_1, \ldots),$$

où l'on écrit, pour un moment,

$$\begin{aligned} L_1 &= (2r_1+3)(2r_1+5)\ldots(2r_1+2\alpha_1+1), \\ &\vdots \\ M_{f+1} &= (2r_{f+1}+1)(2r_{f+1}+3)\ldots(2r_{f+1}+2\alpha_{f+1}-1); \\ &\vdots \end{aligned}$$

puis on le transforme en

$$\frac{(-)^f \pi^{\frac{1}{2}n} t^{2p+f}}{2^{2p+k+f}\Gamma(p+k+f+\frac{1}{2}n+1)}$$

$$\times \left(\frac{d}{da_1}\cdots\frac{d}{da_f}\right)\left(h_1^3\frac{d}{dh_1}\right)^{\alpha_1}\ldots h_1^3 \ldots h_{f+1}\ldots \frac{1}{[r_1]^{r_1}}\left(h_1^2\frac{d^2}{da_1^2}\right)^{r_1}\ldots \phi(\alpha_1, \ldots).$$

En effet, les symboles $\dfrac{d}{da_1}$ et $h_1^3\dfrac{d}{dh_1}$ étant conversibles, on a

$$\frac{d}{da_1}\left(h_1^3\frac{d}{dh_1}\right)^{\alpha_1} h_1^{2r_1+3}\left(\frac{d}{da_1}\right)^{2r_1} = L_1 h_1^{2r_1+2\alpha_1+3}\left(\frac{d}{da_1}\right)^{2r_1+1};$$

et ainsi de suite.

En prenant la somme de tous les termes qui correspondent à une même valeur de p, on a

$$\mathbf{S}\,\frac{1}{[r_1]^{r_1}}\left(h_1^2\frac{d^2}{da_1^2}\right)^{2r_1}\ldots = \frac{1}{[p]^p}\nabla^p.$$

En posant

$$\nabla = \left(h_1^2\frac{d^2}{da_1^2}+\ldots\right);$$

puis, en observant que p doit s'étendre depuis 0 jusqu'à ∞, on a, pour l'intégrale cherchée,

$$V = \frac{(-)^f \pi^{\frac{1}{2}n}}{2^{k+f}}\left(\frac{d}{da_1}\cdots\frac{d}{da_f}\right)\left(h_1^3\frac{d}{dh_1}\right)^{\alpha_1}\ldots$$

$$\times\left[h_1^3\ldots h_{f+1}\ldots t^f\,\mathbf{S}_{p=0}^{p=\infty}\left(\frac{t^{2p}}{2^{2p}[p]^p\Gamma(p+k+f+\frac{1}{2}n+1)}\nabla^p\phi(a_1,\ldots)\right)\right],$$

ce qui fait voir que l'intégrale V dépend de cette seule expression,

$$U = \mathbf{S}_0^\infty\left(\frac{t^{2p}}{2^{2p}[p]^p\Gamma(p+k+f+\frac{1}{2}n+1)}\nabla^p\phi(a_1,\ldots)\right).$$

On peut remarquer, en passant, que cette quantité satisfait à cette équation différentielle,

$$\frac{1}{t}\frac{d}{dt}\left(t^{-2k-2f-n+1}\frac{d}{dt}(t^{2k+2f+n}U)\right)-\nabla U=0,$$

ou

$$\frac{d^2U}{dt^2}+(2k+2f+n+1)\frac{1}{t}\frac{dU}{dt}-\nabla U=0.$$

M. G. Boole, de Lincoln, a déduit une équation semblable de mes formules dans le *Mathematical Journal*. C'est à lui qu'on doit l'introduction dans l'intégrale proposée de cette quantité t, ce qui, au reste, n'est pas d'une grande importance ici; mais j'ai cru devoir la conserver, à cause de cette équation même, qui pourrait conduire à des résultats intéressants.

A présent, soit

$$\phi(a_1-x_1t, \ldots)=\frac{1}{\{(a_1-x_1t)^2+\ldots\}^{\frac{1}{2}n-s}},$$

où s est un nombre entier; je pose, pour abréger,

$$\tfrac{1}{2}n-s=i, \quad k+f+s=\sigma,$$

ce qui donne

$$\Gamma(p+k+f+\tfrac{1}{2}n+1)=\Gamma(p+i+\sigma+1)=[p+i+\sigma]^{p+1}\,\Gamma(i+\sigma);$$

et ainsi

$$U=\frac{1}{\Gamma(i+\sigma)}\mathbf{S}_0^\infty\left(\frac{t^{2p}}{2^{2p}[p]^p[p+i+\sigma]^{p+1}}\nabla^p\frac{1}{(a_1^2+\ldots)^i}\right).$$

Le cas de $\sigma=0$, qui est le plus simple, est, en effet, celui du Mémoire cité, et à ce cas on peut réduire celui de σ entier négatif; il y a même deux manières d'effectuer cette réduction. Représentons, en effet, la valeur de U par cette notation plus complète U_i; on obtient tout de suite

$$U_i^{-\sigma}=2^{-2\sigma}t^{2\sigma-2i}\left(\frac{1}{t}\frac{d}{dt}\right)^\sigma(t^{2i}U_i^0),$$

et

$$U_i^{-\sigma}=\frac{1}{2^{2\sigma}[1-i]^\sigma[1-\frac{1}{2}n]^\sigma}\left(\frac{d^2}{da_1^2}+\ldots\right)^\sigma U_{i-\sigma}^0,$$

la seconde desquelles équations se déduit de cette formule facile à démontrer,

$$\left(\frac{d^2}{da_1^2}+\ldots\right)^\sigma\frac{1}{(a_1^2+\ldots)^{i-\sigma}}=2^{2\sigma}[i-1]^\sigma[i-\tfrac{1}{2}n]^\sigma\frac{1}{(a_1^2+\ldots)^i}.$$

Nous pouvons donc, dans la suite, ne considérer que le cas de σ entier positif.

Commençons par opérer la transformation que voici; en déterminant ζ par l'équation

$$\frac{a_1^2}{\zeta+h_1^2}+\ldots=t^2,$$

je pose $$l_1 = \frac{h_1^2}{\zeta}, \ldots, \quad a_1^2 = (1 + l_1)\, t^2 \alpha_1^2, \ldots,$$

ce qui donne
$$\zeta = \alpha_1^2 + \ldots,$$
$$a_1^2 + \ldots = t^2 [(1 + l_1)\, \alpha_1^2 + \ldots],$$
$$\nabla = \frac{\zeta}{t^2} \left(\frac{l_1}{1 + l_1} \right) \frac{d^2}{d\alpha_1^2} + \ldots \Big),$$

où il faut remarquer que ce ζ, contenu en ∇, ne doit pas être affecté des symboles $\dfrac{d}{d\alpha_1}, \ldots$ de ∇, de manière qu'il faut écrire

$$\nabla^p = t^{-2p} \zeta^p \left(\frac{l_1}{1 + l_1} \frac{d^2}{d\alpha_1^2} + \ldots \right)^p = t^{-2p} (\alpha_1^2 + \ldots)^p \left(\frac{l_1}{1 + l_1} \frac{d^2}{d\alpha_1^2} + \ldots \right)^p,$$

$$U = \frac{1}{\Gamma(i)\, t^{2i} (\alpha_1^2 + \ldots)^i}$$
$$\times \mathbf{S}_0^\infty \left(\frac{1}{2^{2p} [p + i + \sigma]^{p+\sigma+1} [p]^p} (\alpha_1^2 + \ldots)^{p+i} \left(\frac{l_1}{1 + l_1} \frac{d^2}{d\alpha_1^2} + \ldots \right)^p \frac{1}{\{\alpha_1^2 (1 + l_1) + \ldots\}^i} \right),$$

formules qui se prêtent mieux aux réductions, quoique plus compliquées en apparence.

Écrivons d'abord

$$\frac{l_1}{1 + l_1} \frac{d^2}{d\alpha_1^2} + \ldots = \left(\frac{d^2}{d\alpha_1^2} + \ldots \right) - \left(\frac{1}{1 + l_1} \frac{d^2}{d\alpha_1^2} + \ldots \right) = \Delta - \left(\frac{1}{1 + l_1} \frac{d^2}{d\alpha_1^2} + \ldots \right).$$

En développant la $p^{\text{ième}}$ puissance de ce symbole selon les puissances de Δ, on a

$$(-)^{p-q} \frac{[p]^p}{[p - q]^{p-q} [q]^q} \Delta^q \left(\frac{1}{1 + l_1} \frac{d^2}{d\alpha_1^2} + \ldots \right)^{p-q},$$

qui doit s'appliquer à $\dfrac{1}{\{\alpha_1^2 (1 + l_1) + \ldots\}^i}$.

Considérons l'expression

$$\left(\frac{l_1}{1 + l_1} \frac{d^2}{d\alpha_1^2} + \ldots \right)^{p-q} \frac{1}{\{\alpha_1^2 (1 + l_1) + \ldots\}^i},$$

et mettons, pour un moment,

$$(1 + l_1)\, \alpha_1^2 = \alpha'^2_1 \ldots;$$

cette expression se trouve réduite à

$$\left(\frac{d^2}{d\alpha'^2} + \ldots \right)^{p-q} \frac{1}{(\alpha'^2_1 + \ldots)^i},$$

laquelle, par une formule déjà citée, devient

$$2^{2p-2q} [i + p - q - 1]^{p-q} [i + p - q - \tfrac{1}{2} n]^{p-q} \frac{1}{(\alpha'^2_1 + \ldots)^{i+p-q}},$$

c'est-à-dire $$2^{2p-2q}[i+p-q-1]^{p-q}\,[i+p-q-\tfrac{1}{2}n]^{p-q}\,\frac{1}{\{\alpha_1^2(1+l_1)+\ldots\}^{i+p-q}}.$$

Le terme général de U, en faisant abstraction du facteur $\frac{1}{\Gamma(i)\,t^{2i}\,(\alpha_1^2+\ldots)^i}$, se réduit à

$$\frac{(-)^{p-q}(\alpha_1^2+\ldots)^{p+i}}{2^{2q}[p+i+\sigma]^{\sigma+q+1}[p-q]^{p-q}[q]^q}[i+p-q-\tfrac{1}{2}n]^{p-q}\,\Delta^q\,\frac{1}{\{\alpha_1^2(1+l_1)+\ldots\}^{i+p-q}};$$

considérons la partie de ce terme qui est de l'ordre θ en $l_1, \ldots$, elle devient

$$\frac{(-)^{p-q+\theta}(\alpha_1^2+\ldots)^{p+1}}{2^{2q}[p-q]^{p-q}[q]^q[s]^s}[i+p-q+\theta-1]^{\theta-\sigma-q-1}$$

$$\times[i+p-q-\tfrac{1}{2}n]^{p-q}\,\Delta^q\,\frac{(l_1\alpha_1^2+\ldots)^\theta}{(\alpha_1^2+\ldots)^{i+p-q+\theta}}.$$

Soit, en général, Θ une fonction homogène de l'ordre 2θ des lettres $\alpha_1, \ldots$; on a

$$\Delta\,\frac{\Theta}{(\alpha_1^2+\ldots)^i}=\frac{\Delta\Theta}{(\alpha_1^2+\ldots)^i}+2^2 i\,(i+1-2\theta-\tfrac{1}{2}n)\,\frac{\Theta}{(\alpha_1^2+\ldots)^{i+1}};$$

et de là, en répétant toujours l'opération Δ, et faisant attention à ce que $\Delta\Theta$, $\Delta^2\Theta, \ldots$ sont des fonctions homogènes des ordres $\overline{2\theta-1}$, $\overline{2\theta-2}$, &c., on obtient

$$\Delta^q\,\frac{\Theta}{(\alpha_1^2+\ldots)^i}=\frac{[q]^q}{[\lambda]^\lambda[q-\lambda]^{q-\lambda}}\,2^{2q-2\lambda}[i+q-\lambda-1]^{q-\lambda}$$

$$\times[i+q-2\theta-\tfrac{1}{2}n]^{q-\lambda}\,\Delta^\lambda\Theta\,.\,\frac{1}{(\alpha_1^2+\ldots)^{i+q-\lambda}},$$

ou
$$\Delta^q\,\frac{(l_1\alpha_1^2+\ldots)^\theta}{(\alpha_1^2+\ldots)^{i+p-q+\theta}}=\frac{[q]^q}{[\lambda]^\lambda[q-\lambda]^{q-\lambda}}\,2^{2q-2\lambda}[i+p+\theta-\lambda-1]^{q-\lambda}$$

$$\times[i+p-\theta-\tfrac{1}{2}n]^{q-\lambda}\,\Delta^\lambda(l_1\alpha_1^2+\ldots)^\theta\,.\,\frac{1}{(\alpha_1^2+\ldots)^{i+p+\theta-\lambda}},$$

depuis $\lambda=0$ jusqu'à $\lambda=q$.

Puis, pour le terme général de U,

$$\frac{(-)^{p-q+\theta}}{2^{2\lambda}[p-q]^{p-q}[\theta]^\theta[\lambda]^\lambda[q-\lambda]^{q-\lambda}}[i+p+\theta-\lambda-1]^{\theta-\sigma-\lambda-1}$$

$$\times[i+p-q-\tfrac{1}{2}n]^{p-q}[i+p-\theta-\tfrac{1}{2}n]^{q-\lambda}\,\Delta^\lambda(l_1\alpha_1^2+\ldots)^\theta\,.\,\frac{1}{(\alpha_1^2+\ldots)^{\theta-\lambda}},$$

ou
$$\frac{(-)^{-q}}{[p-q]^{p-q}[q-\lambda]^{q-\lambda}}[i+p-q-\tfrac{1}{2}n]^{p-q}[i+p-\theta-\tfrac{1}{2}n]^{q-\lambda}$$

$$\times\frac{(-)^{p+\theta}}{2^{2\lambda}[\theta]^\theta[\lambda]^\lambda}[i+p+\theta-\lambda-1]^{\theta-\sigma-\lambda-1}\,\Delta^\lambda(l_1\alpha_1^2+\ldots)^\theta\,.\,\frac{1}{(\alpha_1^2+\ldots)^{\theta-\lambda}};$$

le dernier facteur étant indépendant de q, q doit s'étendre depuis 0 jusqu'à p. Mais à cause de $[q-\lambda]^{q-\lambda}$ qui devient infini pour $q<\lambda$, on peut faire étendre q depuis $q=\lambda$ jusqu'à $q=p$; ou, en écrivant

$$q-\lambda=q', \quad p-\lambda=p',$$

q' s'étend depuis 0 jusqu'à p'. Le facteur à sommer est donc

$$(-)^\lambda \left\{\frac{(-)^{q'}}{[p'-q']^{p'-q'}\,[q']^{q'}}\,[C-q']^{p'-q'}\,[M]^{q'}\right\},$$

si, pour un moment, l'on pose

$$C=i+p'-\tfrac{1}{2}n, \quad M=i+p-\theta-\tfrac{1}{2}n.$$

Cette somme se réduit à

$$(-)^\lambda \frac{[C-M]^{C-M-p'}}{[C-M-p']^{C-M-p'}},$$

c'est-à-dire à

$$(-)^\lambda \frac{[\theta-\lambda]^{\theta-p}}{[\theta-p]^{\theta-p}},$$

et nous avons ainsi le terme général

$$\frac{(-)^{\lambda+p+\theta}\,[\theta-\lambda]^{\theta-p}\,[i+\theta-\lambda-1+p]^{\theta-\sigma-\lambda-1}}{2^{2\lambda}\,[\theta]^\theta\,[\lambda]^\lambda\,[\theta-p]^{\theta-p}}\,\Delta^\lambda\,(l_1\alpha_1^2+\ldots)^\theta\cdot\frac{1}{(\alpha_1^2+\ldots)^{\theta-\lambda}}.$$

Écrivons

$$p=\theta-p' ;$$

cela devient

$$\frac{(-)^{p'}\,[M]^{p'}\,[C-p']^{C-A}}{[p']^{p'}}\;\frac{(-)^\kappa}{2^{2\lambda}\,[\theta]^\theta\,[\lambda]^\lambda}\,\Delta^\lambda\,(l_1\alpha_1^2+\ldots)^\theta\cdot\frac{1}{(\alpha_1^2+\ldots)^{\theta-\lambda}},$$

où $$M=\theta-\lambda, \quad C=i+2\theta-\lambda-1, \quad C-A=\theta-\sigma-\lambda-1 ;$$

p' doit s'étendre depuis $-\infty$ jusqu'à θ. Mais à cause de $[p']^{p'}$, qui devient infini, pour p' négatif, on peut l'étendre seulement depuis 0 jusqu'à θ, ou même seulement depuis 0 jusqu'à M, à cause du facteur $[M]^{p'}$. La somme se réduit à

$$[C-M]^{C-M-A}\,[C-A]^M=[i+\theta-1]^{-\sigma-1}\,[\theta-\sigma-\lambda-1]^{\theta-\lambda},$$

et le terme général devient

$$\frac{(-)^\lambda}{2^{2\lambda}\,[\theta]^\theta\,[\lambda]^\lambda}\,[i+\theta-1]^{-\sigma-1}\cdot[\theta-\sigma-\lambda-1]^{\theta-\lambda}\,\Delta^\lambda\,(l_1\alpha_1^2+\ldots)^\theta\cdot\frac{1}{(\alpha_1^2+\ldots)^{\theta-\lambda}}.$$

Écrivons enfin

$$\theta=\lambda+\kappa,$$

le terme devient

$$\frac{(-)^\lambda}{2^{2\lambda}[\lambda]^\lambda[\lambda+\kappa]^{\lambda+\kappa}[i+\lambda+\kappa+\sigma]^{\sigma+1}}[\kappa-\sigma-1]^\kappa\,\Delta^\lambda(l_1\alpha_1^2+\ldots)^{\lambda+\kappa}\cdot\frac{1}{(\alpha_1^2+\ldots)^\lambda}.$$

Il faut que κ soit toujours positif, car autrement le terme s'évanouit à cause de $\Delta^\lambda(l_1\alpha_1^2+\ldots)^{\lambda+\kappa}$; mais pour κ plus grand que σ, le facteur $[\kappa-\sigma-1]^\kappa$ s'évanouit, donc κ s'étend seulement depuis 0 jusqu'à σ.

Soit $k_1+\ldots=\kappa$, et considérons les termes de $\Delta^\lambda(l_1\alpha_1^2+\ldots)^{\lambda+\kappa}$ qui contiennent $\alpha_1^{2k_1},\ldots$; ces termes seront de la forme

$$\frac{[\lambda]^\lambda}{[q_1]^{q_1}\ldots}\,\frac{[\lambda+\kappa]^{\lambda+\kappa}}{[q_1+k_1]^{q_1+k_1}\ldots}\left(\frac{d^2}{d\alpha_1^2}\right)^{2q_1}\ldots\,l_1^{q_1+k_1}\,\alpha_1^{2q_1+2k_1}\ldots,$$

où

$$q_1+\ldots=\lambda,$$

c'est-à-dire

$$\frac{[\lambda]^\lambda[\lambda+\kappa]^{\lambda+\kappa}[2q_1+2k_1]^{2q_1}\ldots}{[q_1]^{q_1}\ldots[q_1+k_1]^{q_1+k_1}\ldots}\,l_1^{q_1+k_1}\ldots\,\alpha_1^{2k_1}\ldots,$$

ou, en réduisant,

$$2^{2\lambda}[\lambda]^\lambda[\lambda+\kappa]^{\lambda+\kappa}\frac{\alpha_1^{2k_1}\ldots}{[k_1]^{k_1}\ldots}\,\frac{[q_1+k_1-\tfrac{1}{2}]^{q_1}\ldots}{[q_1]^{q_1}\ldots}\,l_1^{q_1+k_1}\ldots;$$

et cela donne pour U,

$$\frac{(-)^\lambda}{[i+\kappa+\lambda+\sigma]^{\sigma+1}}\,\frac{[q_1+k_1-\tfrac{1}{2}]^{q_1}\ldots}{[q_1]^{q_1}\ldots}\,l_1^{q_1+k_1}\ldots\,\frac{[\kappa-\sigma-1]^\kappa.\,\alpha_1^{2k_1}\ldots}{[k_1]^{k_1}}\,\frac{1}{(\alpha_1^2+\ldots)^\kappa}.$$

Soit

$$\frac{1}{\{(1+l_1u)\ldots\}^{\frac{1}{2}}}=(0)+(1)\,u\ldots+(\lambda)\,u^\lambda+\ldots;$$

on a

$$(\lambda)=(-)^\lambda\frac{[q_1-\tfrac{1}{2}]^{q_1}\ldots}{[q_1]^{q_1}\ldots}\,l_1^{q_1}\ldots+\&\text{c.},\quad q_1+\ldots=\lambda;$$

et de là,

$$\frac{1}{[k_1-\tfrac{1}{2}]^{k_1}\ldots}\,\frac{1}{l_1^{\frac{1}{2}}\ldots}\left(l_1^2\frac{d}{dl_1}\right)^{k_1}\ldots l_1^{\frac{1}{2}}(\lambda)=(-)^\lambda\frac{[q_1+k_1-\tfrac{1}{2}]^{q_1}\ldots}{[q_1]^{q_1}\ldots}\,l^{q_1+k_1}\ldots,$$

de manière que le terme de U devient

$$\frac{1}{l_1^{\frac{1}{2}}\ldots}\left(l_1^2\frac{d}{dl_1}\right)^{k_1}\ldots\,l_1^{\frac{1}{2}}\left(\frac{1}{[i+\kappa+\lambda+\sigma]^{\sigma+1}}(\lambda)\right)\frac{[\kappa-\sigma-1]^\kappa\,\alpha^{2k_1}\ldots}{[2k_1]^{2k_1}\ldots}\,\frac{1}{(\alpha_1^2+\ldots)^\kappa}.$$

Prenant la somme pour λ, à l'aide de

$$[\sigma]^\sigma\int_0^1(1-u)^\sigma\,u^{i+\kappa+\lambda-1}\,du=\frac{1}{[i+\kappa+\sigma+\lambda]^{\sigma-1}}$$

(ce qui suppose $l+u>0$), on obtient

$$\frac{2^{2\kappa}\,[\sigma]^\sigma\,[\kappa-\sigma-1]^\kappa\,\alpha_1^{2k_1}\ldots}{[2k_1]^{2k_1}\ldots(\alpha_1^2+\ldots)^\kappa}\;\frac{1}{\sqrt{l_1}\ldots}\left(l_1^2\frac{d}{dl_1}\right)^{k_1}\ldots\sqrt{l_1}\ldots\int_0^1\frac{(1-u)^\sigma\,u^{i+\kappa-1}\,du}{\{(1+l_1u)\ldots\}^{\frac{1}{2}}},$$

ou, mettant $l_1=\dfrac{h_1^2}{\zeta},\ldots,$

$$2^\kappa\,[\sigma]^\sigma\,[\kappa-\sigma-1]^\kappa\,\frac{\alpha_1^{2k_1}\ldots}{[2k_1]^{2k_1}\ldots}\,\zeta^{\frac{1}{2}n-2\kappa}\times\frac{1}{h_1\ldots}\left(h_1^3\frac{d}{dh_1}\right)^{k_1}\ldots h_1\ldots\int_0^1\frac{(1-u)^\sigma\,u^{i+\kappa-1}\,du}{\{(\zeta+h_1^2u)\ldots\}^{\frac{1}{2}}};$$

en rétablissant le facteur constant $\dfrac{1}{\Gamma(i)\,t^{2i}\,\zeta^i}$ de U, le terme général de cette quantité est

$$\frac{2^\kappa\,[\sigma]^\sigma\,[\kappa-\sigma-1]^\kappa}{t^{2i}\,\Gamma(i)}\,\frac{\alpha_1^{2k_1}\ldots}{[2k_1]^{2k_1}\ldots}\,\zeta^{\frac{1}{2}n-2\kappa+i}\times\frac{1}{h_1\ldots}\left(h_1^3\frac{d}{dh_1}\right)^{k_1}\ldots h_1\ldots\int_0^1\frac{(1-u)^\sigma\,u^{i+\kappa-1}\,du}{\{(\zeta+h_1^2u)\ldots\}^{\frac{1}{2}}},$$

où k_1, &c., sont des entiers positifs quelconques qui satisfont à

$$k_1+\ldots=\kappa,$$

et κ peut s'étendre depuis 0 jusqu'à σ. Il faut observer qu'en différentiant par $\dfrac{d}{dh_1}$, &c., on ne doit pas considérer ζ comme fonction de $h_1\ldots$, ce qui est cependant nécessaire dans l'équation entre V et U.

29.

ADDITION A LA NOTE SUR QUELQUES INTÉGRALES MULTIPLES.

[From the *Journal de Mathématiques Pures et Appliquées* (Liouville), tome X. (1845), pp. 242—244.]

On démontre, au moyen des formules que j'ai données sur ce sujet, [28], une propriété remarquable de l'intégrale multiple

$$V = \int dx_1 \dots dx_n x_1^{2\alpha_1+1} \dots x_{f+1}^{2\alpha_{f+1}} \dots \phi(a_1 - x_1 t, \dots),$$

prise entre les limites

$$\frac{x_1^2}{h_1^2} + \dots + \frac{x_n^2}{h_n^2} \overline{\overline{<}} 1.$$

En effet, en supposant toujours que l'on peut développer la fonction ϕ selon les puissances entières et positives de t, l'intégrale V peut s'exprimer sous la forme

$$V = \frac{(-)^f \pi^{\frac{1}{2}n}\, t^f}{2^{k+f}} \left(\frac{d}{da_1} \cdots \frac{d}{da_f}\right) \left(h_1^3 \frac{d}{dh_1}\right)^{\alpha_1} \cdots \left(h_n^3 \frac{d}{dh_n}\right)^{\alpha_n} h_1^3 \dots h_{f+1} \dots U,$$

où

$$k = \alpha_1 + \dots + \alpha_n,$$

$$U = \mathbf{S}_{p=0}^{p=\infty} \left(\frac{t^{2p}}{2^{2p}\,\Gamma(p+1)\,\Gamma(p+k+f+\frac{1}{2}n+1)} \nabla^p \phi(a_1, \dots) \right),$$

$$\nabla = h_1^2 \frac{d^2}{da_1^2} + \dots + h_n^2 \frac{d^2}{da_n^2}.$$

Soit

$$W = \int dx_1 \dots \phi(a_1 - x_1 t T, \dots)$$

entre les mêmes limites que V. On a

$$W = \pi^{\frac{1}{2}n} h_1 \dots h_n \mathbf{S}_{p=0}^{p=\infty} \left(\frac{T^{2p}\, t^{2p}}{2^{2p}\,\Gamma(p+1)\,\Gamma(p+\frac{1}{2}n+1)} \nabla^p \phi(a_1, \dots) \right).$$

En multipliant par

$$\frac{2}{\Gamma(k+f)} T^{n+1} (1-T^2)^{k+f} dT,$$

et intégrant depuis $T=0$ jusqu'à $T=1$, on obtient

$$\frac{2}{\Gamma(k+f)} \int_0^1 T^{n+1} (1-T^2)^{k+f}\, W dT$$

$$= \pi^{\frac{1}{2}n} h_1 \dots h_n \mathbf{S}_{p=0}^{p=\infty} \left(\frac{t^{2p}}{2^{2p}\,\Gamma(p+1)\,\Gamma(p+k+f+\frac{1}{2}n+1)} \nabla^p \phi(a_1, \dots) \right),$$

puisque en général

$$\frac{2}{\Gamma(k+f)} \int_0^1 T^{2p+n+1} (1-T^2)^{k+f} dT = \frac{\Gamma(p+\frac{1}{2}n+1)}{\Gamma(p+k+f+\frac{1}{2}n+1)}.$$

On a donc l'équation

$$\frac{2}{\Gamma(k+f)} \int_0^1 T^{n+1} (1-T^2)^{k+f}\, W dT = \pi^{\frac{1}{2}n} h_1 \dots h_n U.$$

Mettons la valeur de U qui en résulte, dans l'équation entre V et U; faisons aussi $t=1$, ce qui ne nuit pas à la généralité. On a, en rassemblant les formules,

$$V = \int dx_1 \dots dx_n \dots x_1^{2a_1+1} \dots x_{f+1}^{2a_{f+1}} \dots \phi(a_1 - x_1, \dots),$$

$$W = \int dx_1 \dots dx_n \dots \phi(a_1 - x_1 T, \dots),$$

$$V = \frac{(-)^f}{2^{k+f-1}\,\Gamma(k+f)} \left(\frac{d}{da_1} \dots \frac{d}{da_f} \right) \left(h_1^3 \frac{d}{dh_1} \right)^{a_1} \dots \left(h_n^3 \frac{d}{dh_n} \right)^{a_n} h_1^2 \dots h_f^2 \int_0^1 T^{n+1} (1-T^2)^{k+f}\, W dT:$$

ce qui établit ce théorème: La suite des intégrales V s'exprime au moyen des coefficients différentiels par rapport à $h_1 \dots h_n$ et $a_1 \dots a_f$ de la seule intégrale

$$\int_0^1 T^{n+1} (1-T^2)^{k+f} \int dx_1 \dots dx_n \dots \phi(a_1 - x_1 T, \dots).$$

En supposant que les indices dans V soient tous pairs, on a $f=0$. Les symboles $\frac{d}{da_1}, \dots$ n'entrent plus dans l'expression de V. En changeant la fonction ϕ, on a ces formules plus simples,

$$\begin{cases} V = \int dx_1 \dots dx_n \, x_1^{2\alpha_1} \dots x_n^{2\alpha_n} \, \phi(x_1, \dots, x_n), \\ W = \int dx_1 \dots dx_n \, \phi(Tx_1, \dots, Tx_n), \\ V = \dfrac{1}{2^{k-1}\Gamma(k)} \left(h_1^3 \dfrac{d}{dh_1}\right)^{\alpha_1} \dots \left(h_n^3 \dfrac{d}{dh_n}\right)^{\alpha_n} \displaystyle\int_0^1 T^{n+1}(1-T^2)^k \, W dT, \end{cases}$$

où V s'exprime au moyen des coefficients différentiels par rapport à $h_1, \dots, h_n$ de l'intégrale

$$\int_0^1 T^{n+1}(1-T^2)^k \int dx_1 \dots dx_n \, \phi(Tx_1, \dots, Tx_n),$$

de manière que nous avons trouvé des relations entre des intégrales définies qui contiennent une fonction indéterminée, assujettie à la seule condition d'être développable dans les limites de l'intégration selon les puissances entières et positives des variables[1], mais dans lesquelles les limites sont données par

$$\frac{x_1^2}{h_1^2} + \dots + \frac{x_n^2}{h_n^2} \overset{=}{<} 1.$$

[1] Cette condition est supposée dans la démonstration, mais il me paraît, d'après la forme du résultat, qu'elle n'est pas essentielle.

30.

MÉMOIRE SUR LES COURBES À DOUBLE COURBURE ET LES SURFACES DÉVELOPPABLES.

[From the *Journal de Mathématiques Pures et Appliquées* (Liouville), tome x. (1845), pp. 245—250.

On trouve, dans la *Théorie des courbes algébriques* de M. Plücker, de très-belles recherches sur le nombre des différentes singularités (points d'inflexion, tangentes doubles, &c.) des courbes planes. Les mêmes principes peuvent s'appliquer au cas des courbes à trois dimensions. Pour cela, considérons une suite continue de points dans l'espace, les lignes qui passent par deux points consécutifs, et les plans qui passent par trois points consécutifs; ou, en envisageant autrement la même figure, une suite de lignes dont chacune rencontre la ligne consécutive, les points d'intersection de deux lignes consécutives, et les plans qui contiennent ces lignes; ou encore de cette manière: une suite de plans, les lignes d'intersection de deux plans consécutifs, les points d'intersection de trois plans consécutifs. C'est ce que l'on peut nommer *un système simple*. Ce système est évidemment formé d'une courbe à double courbure et d'une surface développable; la courbe est l'arête de rebroussement de la surface, la surface est l'osculatrice développable de la courbe. Les points du système sont des points dans la courbe, les lignes du système sont les tangentes à la courbe, les plans du système sont les plans osculateurs de la courbe. De même, les plans sont les plans tangents de la surface, les lignes sont les génératrices de la surface; pour les points, on peut les nommer les points de rebroussement de la surface. J'entendrai dans la suite par les termes *ligne par deux points*, *ligne dans deux plans*, une ligne menée par deux points quelconques (non consécutifs en général) du système, et la ligne d'intersection de deux plans quelconques (non consécutifs en général) du système. Enfin, chaque ligne du système est rencontrée, en général, par un certain nombre d'autres lignes non consécutives du système. Je nommerai le point de rencontre d'une telle paire de lignes *point dans deux lignes*, et le plan qui contient une telle paire *plan par deux lignes*.

Supposons qu'un plan donné contient, *en général*, m points du système, qu'une ligne donnée rencontre, *en général*, r lignes du système, qu'un point donné est situé, *en général*, dans n plans du système. Le système est dit être de l'ordre m, du rang r, de la classe n. On voit tout de suite que l'ordre de la courbe est égal à l'ordre du système, ou à m; la classe de la courbe au rang du système, ou à r. Et de même, l'ordre de la surface au rang du système, ou à r; la classe de la surface à la classe du système, ou à n.

Cela posé, les singularités proprement dites (ouvrage cité, page 202) sont les deux suivantes, analogues aux points d'inflexion et de rebroussement dans les courbes planes:

1. Quand quatre points consécutifs sont situés dans le même plan, ou, autrement dit, quand trois lignes consécutives sont situées dans le même plan, ou quand deux plans consécutifs deviennent identiques; je dirai qu'il y a alors un *plan stationnaire*, et je représenterai par la lettre α le nombre de ces plans.

2. Quand quatre plans consécutifs se rencontrent dans le même point, ou, autrement dit, quand trois lignes se rencontrent au même point, ou quand deux points consécutifs deviennent identiques; je dirai qu'il y a alors un *point stationnaire*, et je représenterai par β le nombre de ces points.

Dans le premier cas, il y a un point d'inflexion sphérique dans la courbe, et une ligne d'inflexion dans la surface. On n'a pas donné de noms à ce qui arrive dans la courbe et la surface dans le second cas; et puisque la singularité est suffisamment distinguée déjà en la nommant *point stationnaire*, il n'est pas necessaire de suppléer à cette omission. On peut dire que ces deux cas sont les singularités simples d'un système. Il y a des singularités d'un ordre plus élevé dont on n'a pas besoin ici. Ensuite, il y a des singularités d'une autre espèce, en quelque sorte analogues aux points et tangentes doubles, mais qui ont rapport à un point ou plan indéterminé (hors du système); savoir:

3. Un plan donné peut contenir, *en général*, un nombre g de *lignes dans deux plans*.

4. Un point donné peut être situé, *en général*, dans un nombre h de *lignes par deux points*.

5. Un plan donné peut contenir, *en général*, un nombre x de *points dans deux lignes*.

6. Un point donné peut être situé, *en général*, dans un nombre y de *plans par deux lignes*.

Ces quatre cas sont les singularités impropres simples du système.

Il faut maintenant chercher les relations qui ont lieu entre les nombres m, r, n, α, β, g, h, x, y.

Citons d'abord les formules de M. Plücker pour les courbes planes (page 211), en changeant seulement les lettres, pour éviter la confusion. En représentant par μ l'ordre d'une courbe, par ν sa classe, par ξ le nombre de ses points doubles, η de ses points de rebroussement, a de ses tangentes doubles, b de ses points d'inflexion, l'on aura les six équations

$$\nu = \mu \,.\, \overline{\mu - 1} - (2\xi + 3\eta),$$

$$b = 3\mu \,.\, \overline{\mu - 2} - (6\xi + 8\eta),$$

$$a = \tfrac{1}{2}\,\mu \,.\, \overline{\mu - 2}\; \overline{\mu^2 - 9} - (2\xi + 3\eta)(\mu \,.\, \overline{\mu - 1} - 6) + 2\xi \,.\, \overline{\xi - 1} + \tfrac{9}{2}\,\eta \,.\, \overline{\eta - 1} + 6\xi\eta\,;$$

$$\mu = \nu \,.\, \overline{\nu - 1} - (2a + 3b),$$

$$\eta = 3\nu \,.\, \overline{\nu - 2} - (6a + 8b),$$

$$\xi = \tfrac{1}{2}\,\nu \,.\, \overline{\nu - 2}\; \overline{\nu^2 - 9} - (2a + 3b)(\nu \,.\, \overline{\nu - 1} - 6) + 2a \,.\, \overline{a - 1} + \tfrac{9}{2}\,b \,.\, \overline{b - 1} + 6ab,$$

dont les trois dernières se dérivent des trois premières, et *vice versâ*.

Considérons ensuite un plan donné quelconque en conjonction avec le système. Ce plan coupe la surface suivant une courbe plane. Les points de cette courbe sont les points de rencontre du plan avec les lignes du système, les tangentes de cette courbe sont les lignes de rencontre du plan avec les plans du système. Il est clair que la courbe est de l'ordre r et de la classe n. Chaque fois que le plan contient un point en deux lignes, la courbe a un point double; aux points où le plan rencontre la courbe à double courbure, il y a dans la courbe un rebroussement (car, dans ce cas, il y a deux lignes du système qui coupent le plan en un même point, c'est-à-dire qu'il y a dans la courbe d'intersection un point stationnaire ou de rebroussement). Quand le plan contient une ligne en deux plans, il y a dans la courbe une tangente double; et enfin, pour chaque plan stationnaire du système, il y a dans la courbe une tangente stationnaire ou une inflexion. Donc nous avons, dans la courbe, r l'ordre, n la classe, x le nombre de points doubles, m de points de rebroussement, g de tangentes doubles, α de points d'inflexion, ce qui donne les six équations (équivalentes à trois):

$$n = r \,.\, \overline{r - 1} - (2x + 3m),$$

$$\alpha = 3r \,.\, \overline{r - 2} - (6x + 8m),$$

$$g = \tfrac{1}{2}\,r \,.\, \overline{r - 2}\; \overline{r^2 - 9} - (2x + 3m)(r \,.\, \overline{r - 1} - 6) + 2x \,.\, \overline{x - 1} + \tfrac{9}{2}\,m \,.\, \overline{m - 1} + 6xm\,;$$

$$r = n \,.\, \overline{n - 1} - (2g + 3\alpha),$$

$$m = 3n \,.\, \overline{n - 2} - (6g + 8\alpha),$$

$$x = \tfrac{1}{2}\,n \,.\, \overline{n - 2}\; \overline{n^2 - 9} - (2g + 3\alpha)(n \,.\, \overline{n - 1} - 6) + 2g \,.\, \overline{g - 1} + \tfrac{9}{2}\,\alpha \,.\, \overline{\alpha - 1} + 6g\alpha$$

(où l'on peut remarquer la symétrie à l'égard des combinaisons n, α, g, et r, m, x).

De même, en considérant un point donné quelconque en conjonction avec le système, ce point détermine, avec la courbe à double courbure, une surface conique, et, par des raisonnements pareils, on fait voir que, dans cette surface conique, l'ordre est m, la classe r, le nombre des lignes doubles n, des lignes de rebroussement β, des plans tangents doubles y, des lignes d'inflexion h, ce qui donne les équations (dont trois seulement sont indépendantes):

$$r = m\,.\,\overline{m-1} - (2h + 3\beta),$$

$$n = 3m\,.\,\overline{m-2} - (6h + 8\beta),$$

$$y = \tfrac{1}{2}\,m\,.\,\overline{m-2}\,\overline{m^2-9} - (2h + 3\beta)\,(m\,.\,\overline{m-1} - 6) + 2h\,.\,\overline{h-1} + \tfrac{9}{2}\,\beta\,.\,\overline{\beta-1} + 6h\beta\,;$$

$$m = r\,.\,\overline{r-1} - (2y + 3n),$$

$$\beta = 3r\,.\,\overline{r-2} - (6y + 8n),$$

$$h = \tfrac{1}{2}\,r\,.\,\overline{r-2}\,\overline{r^2-9} - (2y + 3n)\,(r\,.\,\overline{r-1} - 6) + 2y\,.\,\overline{y-1} + \tfrac{9}{2}\,n\,.\,\overline{n-1} + 6yn$$

(dans lesquelles on remarque la correspondance r, n, y; m, β, h: et, en les comparant avec les autres six équations, la correspondance m, r, n, α, β, g, h, x, y; n, r, m, β, α, h, g, y, x).

En considérant une courbe à double courbure d'un ordre donné m, on peut attribuer à h, β des valeurs quelconques (entre certaines limites), et l'on a alors, pour déterminer les autres quantités, les équations suivantes:

$$r = m\,.\,\overline{m-1} - (2h + 3\beta),$$

$$n = 3m\,.\,\overline{m-2} - (6h + 8\beta),$$

$$y = \tfrac{1}{2}\,m\,.\,\overline{m-2}\,\overline{m^2-9} - (2h + 3\beta)\,(m\,.\,\overline{m-1} - 6) + 2h\,.\,\overline{h-1} + \tfrac{9}{2}\,\beta\,.\,\overline{\beta-1} + 6h\beta\,;$$

$$n = r\,.\,\overline{r-1} - (2x + 3m),$$

$$\alpha = 3r\,.\,\overline{r-2} - (6x + 8m),$$

$$g = \tfrac{1}{2}\,r\,.\,\overline{r-2}\,\overline{r^2-9} - (2x + 3m)\,(r\,.\,\overline{r-1} - 6) + 2x\,.\,\overline{x-1} + \tfrac{9}{2}\,m\,.\,\overline{m-1} + 6xm.$$

Au cas d'une courbe plane les trois premières équations continuent d'être vraies, mais les trois autres n'ont pas de sens. Le cas le plus simple est celui d'une courbe du troisième ordre dans l'espace. On a, dans ce cas, $m = 3$, $n = 1$, $\beta = 0$; et de là le système

$$m,\ n,\ r,\ \alpha,\ \beta,\ g,\ h,\ x,\ y\,;$$
$$3,\ 3,\ 4,\ 0,\ 0,\ 1,\ 1,\ 0,\ 0\,;$$

c'est-à-dire l'osculatrice développable d'une courbe du troisième ordre est une surface du quatrième ordre, &c. On obtient aussi un système très-simple, en écrivant $m = 4$, $h = 2$, $\beta = 1$, ce qui donne

$$m,\ n,\ r,\ \alpha,\ \beta,\ g,\ h,\ x,\ y\,;$$
$$4,\ 4,\ 5,\ 1,\ 1,\ 2,\ 2,\ 2,\ 2.$$

Mais cela n'appartient pas, à ce que je crois, aux courbes les plus générales du quatrième ordre.

Le problème de classifier les courbes à double courbure au moyen des surfaces que l'on peut faire passer par ces courbes, ou de trouver la nature d'une courbe qui est l'intersection de deux surfaces données, paraît appartenir plutôt à la théorie des surfaces qu'à celle des courbes. Je n'ai rien de complet à offrir sur cela. Seulement je crois pouvoir dire que quand la courbe d'intersection de deux surfaces des ordres μ, ν est de l'ordre $\mu\nu$ (ce qui est le cas général), on a toujours

$$2h = \mu\nu \,.\, \overline{\mu - 1}\; \overline{\nu - 1},$$

de manière que la classe de la courbe est au plus $mn(m+n-2)$. Mais j'espère revenir une autre fois sur cette question. Il est presque inutile de remarquer que pour un système qui est réciproque polaire d'un système donné, il faut seulement changer m, r, n, α, β, g, h, x, y en n, r, m, β, α, h, g, y, x. Par exemple, les deux systèmes qui viennent d'être considérés ont des réciproques de la même forme.

31.

DEMONSTRATION D'UN THÉORÈME DE M. CHASLES.

[From the *Journal de Mathématiques Pures et Appliquées* (Liouville), tome X. (1845), pp. 383—384.]

"Soient P, P' des points correspondants de deux figures homographiques; si la droite PP' passe toujours par un point fixe O, les points P sont situés sur une courbe du troisième degré, qui passe par ce même point."

Soient $\frac{x}{w}, \frac{y}{w}, \frac{z}{w}$ les coordonnées de P; $\frac{x'}{w'}, \frac{y'}{w'}, \frac{z'}{w'}$ celles de P'. En supposant que x', y', z', w' sont des fonctions linéaires (sans terme constant) de x, y, z, w, les deux figures seront homographiques.

Soient, de même, $\frac{\alpha}{\delta}, \frac{\beta}{\delta}, \frac{\gamma}{\delta}$ les coordonnées de O; $\frac{\lambda}{\varpi}, \frac{\mu}{\varpi}, \frac{\nu}{\varpi}$ les coordonnées d'un point quelconque T.

Puisque P, P', O sont sur la même droite, on peut faire passer un plan par les quatre points P, P', O, T. Cela donne tout de suite l'équation

$$\left\{\begin{matrix} \lambda, & \mu, & \nu, & \varpi \\ \alpha, & \beta, & \gamma, & \delta \\ x, & y, & z, & w \\ x', & y', & z', & w' \end{matrix}\right\} = 0$$

(en representant de cette manière le déterminant formé avec les quantités λ, μ, ν, &c.), équation qui doit être satisfaite quels que soient λ, μ, ν, ϖ, et qui équivaut ainsi aux deux conditions

$$\left\{\begin{matrix} \alpha, & \beta, & \gamma \\ x, & y, & z \\ x', & y', & z' \end{matrix}\right\} = 0, \qquad \left\{\begin{matrix} \alpha, & \beta, & \delta \\ x, & y, & w \\ x', & y', & w' \end{matrix}\right\} = 0.$$

Ces deux dernières équations sont du second degré par rapport aux quantités $\frac{x}{w}, \frac{y}{w}, \frac{z}{w}$. Le point P est donc situé à l'intersection de deux surfaces du second ordre. Mais ces surfaces ont en commun la droite représentée par les équations

$$\alpha y - \beta x = 0, \quad \alpha x' - \beta y' = 0.$$

Donc elles se coupent de plus suivant une courbe du troisième degré qui passe évidemment par le point O, parce qu'on satisfait aux équations en écrivant

$$x : \alpha = y : \beta = z : \gamma = w : \delta.$$

32.

ON SOME ANALYTICAL FORMULÆ, AND THEIR APPLICATION TO THE THEORY OF SPHERICAL COORDINATES.

[From the *Cambridge and Dublin Mathematical Journal*, vol. I. (1846), pp. 22—33.]

SECTION 1.

THE formulæ in question are only very particular cases of some relating to the theory of the transformation of functions of the second order, which will be given in a following paper. But the case of three variables, here as elsewhere, admits of a symmetrical notation so much simpler than in any other case (on the principle that with three quantities a, b, c, functions of b, c; of c, a; and of a, b, may symmetrically be denoted by A, B, C, which is not possible with a greater number of variables) that it will be convenient to employ here a notation entirely different from that made use of in the general case, and by means of which the results will be exhibited in a more compact form. There is no difficulty in verifying by actual multiplication, any of the equations here obtained.

It will be expedient to employ the abbreviation of making a single letter stand for a system of quantities. Thus for instance, if $\vartheta=\theta,\ \phi,\ \psi$, this merely means that $\Phi(\vartheta)$ is to stand for $\Phi(\theta,\ \phi,\ \psi)$, $k\vartheta$ for $k\theta,\ k\phi,\ k\psi$, &c.

Suppose then

$$\omega=\xi,\ \eta,\ \zeta, \qquad\qquad (1),$$

$$\omega'=\xi',\ \eta',\ \zeta',$$

$$\vdots$$

$$Q=A,\ B,\ C,\ F,\ G,\ H \qquad\qquad (2),$$

$$W(\omega,\ \omega',\ Q)=A\xi\xi'+B\eta\eta'+C\zeta\zeta'+F(\eta\zeta'+\eta'\zeta)+G(\zeta\xi'+\zeta'\xi)+H(\xi\eta'+\xi'\eta) \ldots (3),$$

the function W satisfies a remarkable equation, as follows:

write

$$\mathfrak{A} = BC - F^2, \quad\dots\dots\dots (4),$$
$$\mathfrak{B} = CA - G^2,$$
$$\mathfrak{C} = AB - H^2.$$
$$\mathfrak{F} = GH - AF,$$
$$\mathfrak{G} = HF - BG,$$
$$\mathfrak{H} = FG - CH.$$

$$\mathfrak{Q} = \mathfrak{A}, \mathfrak{B}, \mathfrak{C}, \mathfrak{F}, \mathfrak{G}, \mathfrak{H} \quad\dots\dots\dots (5),$$

$$\overline{\omega\omega'} = \eta\zeta' - \eta'\zeta, \quad \zeta\xi' - \zeta'\xi, \quad \xi\eta' - \xi'\eta \quad\dots\dots\dots (6),$$

we have

$$W(\omega_1, \omega_2, Q)\, W(\omega_3, \omega_4, Q) - W(\omega_1, \omega_3, Q)\, W(\omega_2, \omega_4, Q) = W(\overline{\omega_1\omega_4}, \overline{\omega_2\omega_3}, \mathfrak{Q}) \;\dots\; (7);$$

of which we may notice also the particular cases

$$W(\omega_1, \omega_2, Q)\, W(\omega_3, \omega_3, Q) - W(\omega_1, \omega_3, Q)\, W(\omega_2, \omega_3, Q) = W(\overline{\omega_1\omega_3}, \overline{\omega_2\omega_3}, \mathfrak{Q}) \;\dots\; (8),$$

$$W(\omega_1, \omega_1, Q)\, W(\omega_2, \omega_2, Q) - \{W(\omega_1, \omega_2, Q)\}^2 = W(\overline{\omega_1\omega_2}, \overline{\omega_1\omega_2}, \mathfrak{Q}) \;\dots\; (9).$$

To these we may join the following formulæ, for the transformation of the function W.

Suppose

$$\begin{aligned} \omega_1 &= \mathrm{a}x_1 + \mathrm{a}'y_1 + \mathrm{a}''z_1, \quad \mathrm{b}x_1 + \mathrm{b}'y_1 + \mathrm{b}''z_1, \quad \mathrm{c}x_1 + \mathrm{c}'y_1 + \mathrm{c}''z_1 \quad\dots\dots (10),\\ \omega_2 &= \mathrm{a}x_2 + \mathrm{a}'y_2 + \mathrm{a}''z_2, \quad \mathrm{b}x_2 + \mathrm{b}'y_2 + \mathrm{b}''z_2, \quad \mathrm{c}x_2 + \mathrm{c}'y_2 + \mathrm{c}''z_2,\\ &\;\vdots \end{aligned}$$

then, writing

$$\begin{aligned} g &= \mathrm{a},\; \mathrm{b},\; \mathrm{c} \quad\dots\dots\dots (11),\\ g' &= \mathrm{a}',\; \mathrm{b}',\; \mathrm{c}',\\ g'' &= \mathrm{a}'',\; \mathrm{b}'',\; \mathrm{c}'', \end{aligned}$$

$$\begin{aligned} p_1 &= x_1,\; y_1,\; z_1 \quad\dots\dots\dots (12),\\ p_2 &= x_2,\; y_2,\; z_2,\\ &\;\vdots \end{aligned}$$

$$\Theta = W(g, g, Q),\; W(g', g', Q),\; W(g'', g'', Q),\; W(g', g'', Q),\; W(g'', g, Q),\; W(g, g', Q) \;(13),$$

we have

$$W(\omega_1, \omega_2, Q) = W(p_1, p_2, \Theta) \quad\dots\dots\dots (14).$$

Similarly, writing

$$\begin{aligned} \Psi = W(\overline{g'g''}, \overline{g'g''}, \mathfrak{Q}),\; & W(\overline{g''g}, \overline{g''g}, \mathfrak{Q}),\; W(\overline{gg'}, \overline{gg'}, \mathfrak{Q})\\ & W(\overline{gg'}, \overline{g''g}, \mathfrak{Q}),\; W(\overline{g'g''}, \overline{gg'}, Q),\; W(\overline{g''g}, \overline{g'g''}, \mathfrak{Q}), \;\dots\dots (15), \end{aligned}$$

we have

$$W(\overline{\omega_1\omega_4}, \overline{\omega_2\omega_3}, \mathfrak{Q}) = W(\overline{p_1p_4}, \overline{p_2p_3}, \Psi) \quad\dots\dots\dots (16),$$

in which equations $\mathfrak{Q}$ may obviously be changed into Q.

SECTION 2. *Geometrical Applications.*

Consider any three axes Ax, Ay, Az, and let λ, μ, ν be the cosines of the inclinations of these lines to each other.

Let Λ, M, N be the inclinations of the coordinate planes to each other; l, m, n the inclination of the axes to the coordinate planes. Suppose, besides,

$$\begin{aligned}
\mathfrak{a} &= 1 - \lambda^2 \quad \ldots\ldots\ldots\ldots (17),\\
\mathfrak{b} &= 1 - \mu^2,\\
\mathfrak{c} &= 1 - \nu^2,\\
\mathfrak{f} &= \mu\nu - \lambda,\\
\mathfrak{g} &= \nu\lambda - \mu,\\
\mathfrak{h} &= \lambda\mu - \nu,
\end{aligned}$$

$$k = 1 - \lambda^2 - \mu^2 - \nu^2 + 2\lambda\mu\nu \quad \ldots\ldots\ldots\ldots (18);$$

we have the following systems of equations:

$$\begin{aligned}
&\sqrt{(\mathfrak{bc})}\cos\Lambda = -\mathfrak{f}, && \sqrt{(\mathfrak{bc})}\sin\Lambda = \sqrt{(k)}, && \sqrt{(\mathfrak{a})}\sin l = \sqrt{(k)} \quad \ldots\ldots (19).\\
&\sqrt{(\mathfrak{ca})}\cos \mathrm{M} = -\mathfrak{g}, && \sqrt{(\mathfrak{ca})}\sin \mathrm{M} = \sqrt{(k)}, && \sqrt{(\mathfrak{b})}\sin m = \sqrt{(k)}\\
&\sqrt{(\mathfrak{ab})}\cos \mathrm{N} = -\mathfrak{h}, && \sqrt{(\mathfrak{ab})}\sin \mathrm{N} = \sqrt{(k)}, && \sqrt{(\mathfrak{c})}\sin n = \sqrt{(k)}.
\end{aligned}$$

$$\begin{aligned}
\mathfrak{a} + \nu\mathfrak{h} + \mu\mathfrak{g} &= k, \quad \ldots\ldots\ldots\ldots (20).\\
\nu\mathfrak{a} + \mathfrak{h} + \lambda\mathfrak{g} &= 0,\\
\mu\mathfrak{a} + \lambda\mathfrak{h} + \mathfrak{g} &= 0.
\end{aligned}$$

$$\begin{aligned}
\mathfrak{h} + \nu\mathfrak{b} + \mu\mathfrak{f} &= 0, \quad \ldots\ldots\ldots\ldots (21).\\
\nu\mathfrak{h} + \mathfrak{b} + \lambda\mathfrak{f} &= k,\\
\mu\mathfrak{h} + \lambda\mathfrak{b} + \mathfrak{f} &= 0.
\end{aligned}$$

$$\begin{aligned}
\mathfrak{g} + \nu\mathfrak{f} + \mu\mathfrak{c} &= 0, \quad \ldots\ldots\ldots\ldots (22).\\
\nu\mathfrak{g} + \mathfrak{f} + \lambda\mathfrak{c} &= 0,\\
\mu\mathfrak{g} + \lambda\mathfrak{f} + \mathfrak{c} &= k.
\end{aligned}$$

$$\begin{aligned}
\mathfrak{bc} - \mathfrak{f}^2 &= ka \quad \ldots\ldots\ldots\ldots (23).\\
\mathfrak{ca} - \mathfrak{g}^2 &= kb,\\
\mathfrak{ab} - \mathfrak{h}^2 &= kc,\\
\mathfrak{gh} - \mathfrak{af} &= kf,\\
\mathfrak{hf} - \mathfrak{bg} &= kg,\\
\mathfrak{fg} - \mathfrak{ch} &= kh,
\end{aligned}$$

$$\mathfrak{abc} - \mathfrak{af}^2 - \mathfrak{bg}^2 - \mathfrak{ch}^2 + 2\mathfrak{fgh} = k^2 \quad \ldots\ldots\ldots\ldots (24).$$

Imagine now a line AO, and let α, β, γ be the cosines of its inclinations to the three axes. Suppose also, θ, ϕ, χ being its inclinations to the coordinate planes, we write

$$a=\frac{\sin\theta}{\sqrt{(\mathfrak{a})}},\qquad b=\frac{\sin\phi}{\sqrt{(\mathfrak{b})}},\qquad c=\frac{\sin\chi}{\sqrt{(\mathfrak{c})}}\ \ldots\ldots\ldots\ldots\ (25).$$

If we consider a point P on the line AO, at a distance unity from the origin, we see immediately, by considering the projections in the directions perpendicular to the coordinate planes, that the coordinates of this point are a, b, c. By projecting on the three axes and on the line AO, we then obtain the equations

$$\alpha=\ \ a+\nu b+\mu c\ \ldots\ldots\ldots\ldots\ (26),$$

$$\beta=\nu a+\ \ b+\lambda c,$$

$$\gamma=\mu a+\lambda b+\ \ c,$$

$$1=\alpha a+\beta b+\gamma c\ \ldots\ldots\ldots\ldots\ (27),$$

from which we obtain

$$ka=\mathfrak{a}\alpha+\mathfrak{h}\beta+\mathfrak{g}\gamma\ \ldots\ldots\ldots\ldots\ (28),$$

$$kb=\mathfrak{h}\alpha+\mathfrak{b}\beta+\mathfrak{f}\gamma,$$

$$kc=\mathfrak{g}\alpha+\mathfrak{f}\beta+\mathfrak{c}\gamma,$$

$$1=\alpha a+\beta b+\gamma c\ \ldots\ldots\ldots\ldots\ (29),$$

and hence

$$1=a^2+b^2+c^2+2\lambda bc+2\mu ac+2\nu ab\ \ldots\ldots\ldots\ldots\ (30),$$

$$k=\mathfrak{a}\alpha^2+\mathfrak{b}\beta^2+\mathfrak{c}\gamma^2+2\mathfrak{f}\beta\gamma+2\mathfrak{g}\alpha\gamma+2\mathfrak{h}\alpha\beta\ \ldots\ldots\ldots\ldots\ (31).$$

Hence writing

$$a,\ b,\ c=t\ \ldots\ldots\ldots\ldots\ (32),$$

$$\alpha,\ \beta,\ \gamma=\tau\ \ldots\ldots\ldots\ldots\ (33),$$

$$1,\ 1,\ 1,\ \lambda,\ \mu,\ \nu=\mathrm{q}\ \ldots\ldots\ldots\ldots\ (34),$$

$$\mathfrak{a},\ \mathfrak{b},\ \mathfrak{c},\ \mathfrak{f},\ \mathfrak{g},\ \mathfrak{h}=\mathfrak{q}\ \ldots\ldots\ldots\ldots\ (35),$$

we have the equations

$$1=W(t,\ t,\ \mathfrak{q})\ \ldots\ldots\ldots\ldots\ (36),$$

$$k=W(\tau,\ \tau,\ \mathfrak{q})\ \ldots\ldots\ldots\ldots\ (37).$$

Let AO' be any other line, and δ its inclination to AO: α', β', γ', a', b', c', the quantities corresponding to α, β, γ, a, b, c, and similarly t', τ' to t, τ. We have of course

$$1=W(t',\ t',\ \mathrm{q})\ \ldots\ldots\ldots\ldots\ (38),$$

$$k=W(\tau',\ \tau',\ \mathfrak{q})\ \ldots\ldots\ldots\ldots\ (39).$$

We have besides, by projecting on the line AO', the equation

$$\cos\delta = \alpha'a + \beta'b + \gamma'c \quad \ldots\ldots\ldots\ldots (40),$$

or the analogous one

$$\cos\delta = \alpha'a + \beta'b + \gamma'c \quad \ldots\ldots\ldots\ldots (41).$$

From either of which we deduce

$$\cos\delta = aa' + bb' + cc' + \lambda\,(bc' + b'c) + \mu\,(ca' + c'a) + \nu\,(ab' + a'b) \quad \ldots\ldots (42),$$

$$k\cos\delta = \mathfrak{a}\alpha\alpha' + \mathfrak{b}\beta\beta' + \mathfrak{c}\gamma\gamma' + \mathfrak{f}\,(\beta\gamma' + \beta'\gamma) + \mathfrak{g}\,(\gamma\alpha' + \gamma'\alpha') + \mathfrak{h}\,(\alpha\beta + \alpha'\beta) \quad \ldots\ldots (43);$$

which may otherwise be written

$$\cos\delta = W\,(t,\ t',\ \mathrm{q}) \quad \ldots\ldots\ldots\ldots (44),$$

$$k\cos\delta = W\,(\tau,\ \tau',\ \mathfrak{q}) \quad \ldots\ldots\ldots\ldots (45);$$

or again, observing the equations which connect the quantities t, τ,

$$\cos\delta = \frac{W\,(t,\ t',\ \mathrm{q})}{\sqrt{\{W\,(t,\ t,\ \mathrm{q})\ W\,(t',\ t',\ \mathrm{q})\}}} \quad \ldots\ldots\ldots\ldots (46),$$

$$\cos\delta = \frac{W\,(\tau,\ \tau',\ \mathfrak{q})}{\sqrt{\{W\,(\tau,\ \tau,\ \mathfrak{q})\ W\,(\tau',\ \tau',\ \mathfrak{q})\}}} \quad \ldots\ldots\ldots\ldots (47),$$

forms which, though more complicated, have certain advantages; for instance, we derive immediately from them the new equations

$$\sin\delta = \frac{\sqrt{\{W\,(\overline{\underline{tt'}},\ \overline{\underline{tt'}},\ \mathfrak{q})\}}}{\sqrt{\{W\,(t,\ t,\ \mathrm{q})\ W\,(t',\ t',\ \mathrm{q})\}}} \quad \ldots\ldots\ldots\ldots (48),$$

$$\sin\delta = \frac{\sqrt{\{k\,W\,(\overline{\underline{\tau\tau'}},\ \overline{\underline{\tau\tau'}},\ \mathrm{q})\}}}{\sqrt{\{W\,(\tau,\ \tau,\ \mathfrak{q})\ W\,(\tau',\ \tau',\ \mathfrak{q})\}}} \quad \ldots\ldots\ldots\ldots (49).$$

written more simply thus

$$\sin\delta = \quad W\,(\overline{\underline{tt'}},\ \overline{\underline{tt'}},\ \mathfrak{q}) \quad \ldots\ldots\ldots\ldots (50),$$

$$\sqrt{k}\sin\delta = \sqrt{\{W\,(\overline{\underline{\tau\tau'}},\ \overline{\underline{\tau\tau'}},\ \mathrm{q})\}} \quad \ldots\ldots\ldots\ldots (51);$$

to these we may join

$$\cot\delta = \frac{W\,(t,\ t',\ \mathrm{q})}{\sqrt{\{W\,(\overline{\underline{tt'}},\ \overline{\underline{tt'}},\ \mathfrak{q})\}}} \quad \ldots\ldots\ldots\ldots (52),$$

$$\sqrt{k}\cot\delta = \frac{W\,(\tau,\ \tau',\ \mathfrak{q})}{\sqrt{\{W\,(\overline{\underline{\tau\tau'}},\ \overline{\underline{\tau\tau'}},\ \mathrm{q})\}}} \quad \ldots\ldots\ldots\ldots (53).$$

SECTION 3. *On Spherical Coordinates.*

Consider the points X, Y, Z, on the surface of a sphere, as the intersections of the three axes of the preceding section, with a sphere having its centre in the origin. It is evident that λ, μ, ν are the cosines of the sides of the spherical triangle XYZ,

Λ, M, N are its sides, l, m, n are the perpendiculars from the angles upon the opposite sides. Let P be the point where the line AO intersects the sphere: the position of the point P may be determined by means of the ratios $\xi : \eta : \zeta$, supposing ξ, η, ζ denote quantities proportional to the α, β, γ of the preceding section, i.e.

$$\xi : \eta : \zeta = \cos PX : \cos PY : \cos PZ \quad\text{..................... (54);}$$

or again, by means of the ratios $x : y : z$, supposing x, y, z denote quantities proportional to the a, b, c of the preceding section, i.e.

$$x : y : z = \frac{\sin Px}{\sin X} : \frac{\sin Py}{\sin Y} : \frac{\sin Pz}{\sin Z} \quad\text{......................(55),}$$

(Px, Py, Pz are the perpendiculars from P on the sides of the spherical triangle XYZ).

These last equations may be otherwise written,

$$\frac{x \sin X}{y \sin Y} = \frac{\sin PZY}{\sin PZX} \quad\text{.................................. (56).}$$

$$\frac{y \sin Y}{z \sin Z} = \frac{\sin PXZ}{\sin PXY},$$

$$\frac{z \sin Z}{x \sin X} = \frac{\sin PYX}{\sin PYZ}.$$

The ratios $\xi : \eta : \zeta$, or $x : y : z$, are termed the spherical coordinate ratios of the point P. The two together may be termed conjoint systems: the first may be termed the cosine system, and the second the sine system. The coordinates of the two systems are evidently connected by

$$\xi : \eta : \zeta = x + \nu y + \mu z : \nu x + y + \lambda z : \mu x + \lambda y + z \quad\text{............(57),}$$

or

$$x : y : z = \mathfrak{a}\xi + \mathfrak{h}\eta + \mathfrak{g}\zeta : \mathfrak{h}\xi + \mathfrak{b}\eta + \mathfrak{f}\zeta : \mathfrak{g}\xi + \mathfrak{f}\eta + \mathfrak{c}\zeta \quad\text{...........(58).}$$

The systems may conveniently be represented by the single letters

$$\omega = \xi,\ \eta,\ \zeta \quad\text{....................................... (59),}$$

$$p = x,\ y,\ z \quad\text{....................................... (60).}$$

Fundamental formula of spherical coordinates; distance of two points.

Let P, P' be the points, δ their distance, ω, p the conjoint coordinate systems of the first point, ω', p' of the second; we have obviously

$$\cot \delta = \frac{W(p, p', \mathrm{q})}{\sqrt{\{W(p, p, \mathrm{q})\, W(p', p', \mathrm{q})\}}} \quad\text{.......................(61),}$$

$$\sin \delta = \frac{\sqrt{\{W(\overline{pp'}, \overline{pp'}, \mathfrak{q})\}}}{\sqrt{\{W(p, p, \mathrm{q})\, W(p', p', \mathrm{q})\}}},$$

$$\cot \delta = \frac{W(p, p', \mathrm{q})}{\sqrt{\{W(\overline{pp'}, \overline{pp'}\ \mathfrak{q})\}}};$$

or
$$\cos\delta = \frac{W(\omega,\ \omega',\ \mathfrak{q})}{\sqrt{\{W(\omega,\ \omega,\ \mathfrak{q})\ W(\omega',\ \omega',\ \mathfrak{q})\}}} \quad\ldots\ldots(62).$$

$$\frac{1}{\sqrt{k}}\sin\delta = \frac{\sqrt{\{W(\underline{\overline{\omega\omega'}},\ \underline{\overline{\omega\omega'}},\ \mathrm{q})\}}}{\sqrt{\{W(\omega,\ \omega,\ \mathfrak{q})\ W(\omega',\ \omega',\ \mathfrak{q})\}}},$$

$$\sqrt{k}\,\{\cot\delta = \frac{W(\omega,\ \omega',\ \mathfrak{q})}{\sqrt{\{W(\underline{\overline{\omega\omega'}},\ \underline{\overline{\omega\omega'}},\ \mathrm{q})\}}}.$$

Equation of a great Circle.

Let the conjoint coordinate systems of the pole be

$$e = a,\ b,\ c \quad\ldots\ldots(63),$$

$$\epsilon = \alpha,\ \beta,\ \gamma \quad\ldots\ldots(64),$$

then, expressing that the distance of any point P in the locus from the pole is equal to 90°, we have immediately the equations

$$W(p,\ e,\ \mathrm{q}) = 0 \quad\ldots\ldots(65),$$

$$W(\omega,\ \epsilon,\ \mathfrak{q}) = 0 \quad\ldots\ldots(66),$$

which may otherwise be written in the forms

$$a\xi + b\eta + c\zeta = 0 \quad\ldots\ldots(67),$$

$$\alpha x + \beta y + \gamma\zeta = 0 \quad\ldots\ldots(68),$$

or the equation of a great circle is linear in either coordinate system. Conversely, any linear equation belongs to a great circle.

Suppose the equation given in the form

$$A\xi + B\eta + C\zeta = 0 \quad\ldots\ldots(69);$$

or by an equation between cosine coordinate ratios:—the sine system for the pole is given by

$$e = A,\ B,\ C \quad\ldots\ldots(70),$$

and the cosine system by

$$\epsilon = A + \nu B + \mu C,\quad \nu A + B + \lambda C,\quad \mu A + \lambda B + C \quad\ldots\ldots(71).$$

Suppose the circle given by an equation between sine coordinates, or in the form

$$\mathrm{A}x + \mathrm{B}y + \mathrm{C}z = 0 \quad\ldots\ldots(72),$$

the cosine system of coordinates for the pole is given by

$$\epsilon = \mathrm{A},\ \mathrm{B},\ \mathrm{C} \quad\ldots\ldots(73),$$

and the sine system by

$$e = \mathfrak{a}\mathrm{A} + \mathfrak{h}\mathrm{B} + \mathfrak{g}\mathrm{C},\quad \mathfrak{h}\mathrm{A} + \mathfrak{b}\mathrm{B} + \mathfrak{f}\mathrm{C},\quad \mathfrak{g}\mathrm{A} + \mathfrak{f}\mathrm{B} + \mathfrak{c}\mathrm{C} \quad\ldots\ldots(74).$$

It is hardly necessary to observe, that if

$$A\xi + B\eta + C\zeta = 0 \qquad (75),$$

$$\mathrm{A}x + \mathrm{B}y + \mathrm{C}z = 0 \qquad (76),$$

represent the same great circle,

$$\mathrm{A} : \mathrm{B} : \mathrm{C} = A + \nu B + \mu C : \nu A + B + \lambda C : \mu A + \lambda B + C \qquad (77),$$

$$A : B : C = \mathfrak{a}\mathrm{A} + \mathfrak{h}\mathrm{B} + \mathfrak{g}\mathrm{C} : \mathfrak{h}\mathrm{A} + \mathfrak{b}\mathrm{B} + \mathfrak{f}\mathrm{C} : \mathfrak{g}\mathrm{A} + \mathfrak{f}\mathrm{B} + \mathfrak{c}\mathrm{C} \qquad (78).$$

Inclination of two great Circles.

Let the equations of these be

$$\begin{cases} A\xi + B\eta + C\zeta = 0 & (79), \\ \text{or } \mathrm{A}x + \mathrm{B}y + \mathrm{C}z = 0 & (80), \end{cases}$$

$$\begin{cases} A'\xi + B'\eta + C'\zeta = 0 & (81), \\ \text{or } \mathrm{A}'x + \mathrm{B}'y + C'z = 0 & (82), \end{cases}$$

and let e, ϵ, have the same values as above, and e', ϵ', corresponding ones. To obtain the inclination of the two circles, we have only, in the formulæ given above for the distance of two points, to change p, p', ω, ω', into e, e', ϵ, ϵ'.

The distance of a point from a given circle may be found with equal facility; for this is evidently the complement of the distance of the point from the pole of the circle. In like manner we may find the condition that two great circles intersect at right angles, &c.

There are evidently a whole class of formulæ, not by any means peculiar to the present system of coordinates, such as

$$Ax + By + Cz - s(A'x + B'y + C'z) \qquad (83),$$

for the equation of a great circle subjected to pass through the points of intersection of

$$Ax + By + Cz = 0, \quad A'x + B'y + C'z = 0.$$

Again,

$$\begin{vmatrix} x, & y, & z \\ a, & b, & c \\ a', & b', & c' \end{vmatrix} = 0 \qquad (84),$$

for the equation of the great circle which passes through the points given by the sine systems $a : b : c$ and $a' : b' : c'$, &c., and which are obtained so easily that it is not worth while writing down any more of them.

Transformation of Coordinates.

Let X_1, Y_1, Z_1, be the new points of reference, and suppose X_1 is given by the conjoint systems $e = a, b, c$, $\epsilon = \alpha, \beta, \gamma$; and similarly Y_1, Z_1, by the analogous systems e', ϵ'; e'', ϵ''.

Suppose, as before, P is given by one of the systems ω, p; and let ω_1, p_1 be the new systems which determine the position of P with reference to X_1, Y_1, Z_1.

In the first place, λ_1, μ_1, ν_1, are given by the formulæ

$$\lambda_1 = \frac{W(e', e'', \mathrm{q})}{\sqrt{\{W(e', e', \mathrm{q})\, W(e'', e'', \mathrm{q})\}}} = \frac{W(\epsilon', \epsilon'', \mathfrak{q})}{\sqrt{\{W(\epsilon', \epsilon', \mathfrak{q})\, W(\epsilon'', \epsilon'', \mathfrak{q})\}}} \quad\ldots\ldots (85),$$

$$\mu_1 = \frac{W(e'', e, \mathrm{q})}{\sqrt{\{W(e'', e'', \mathrm{q})\, W(e, e, \mathrm{q})\}}} = \frac{W(\epsilon'', \epsilon, \mathfrak{q})}{\sqrt{\{W(\epsilon'', \epsilon'', \mathfrak{q})\, W(\epsilon, \epsilon, \mathfrak{q})\}}},$$

$$\nu_1 = \frac{W(e, e', \mathrm{q})}{\sqrt{\{W(e, e, \mathrm{q})\, W(e', e', \mathrm{q})\}}} = \frac{W(\epsilon, \epsilon', \mathfrak{q})}{\sqrt{\{W(\epsilon, \epsilon, \mathfrak{q})\, W(\epsilon', \epsilon', \mathfrak{q})\}}}.$$

The system ω_1 is evidently given immediately by

$$\xi_1 : \eta_1 : \zeta_1 = \frac{W(e, p, \mathrm{q})}{\sqrt{\{W(e, e, \mathrm{q})\}}} : \frac{W(e', p, \mathrm{q})}{\sqrt{\{W(e', e', \mathrm{q})\}}} : \frac{W(e'', p, \mathrm{q})}{\sqrt{\{W(e'', e'', \mathrm{q})\}}} \quad\ldots\ldots\ldots (86)$$

$$= \frac{W(\epsilon, \omega, \mathfrak{q})}{\sqrt{\{W(\epsilon, \epsilon, \mathfrak{q})\}}} : \frac{W(\epsilon', \omega, \mathfrak{q})}{\sqrt{\{W(\epsilon', \epsilon', \mathfrak{q})\}}} : \frac{W(\epsilon'', \omega, \mathfrak{q})}{\sqrt{\{W(\epsilon'', \epsilon'', \mathfrak{q})\}}} \quad\ldots\ldots\ldots (87),$$

and from these we may obtain the system p_1, by means of the formulæ

$$x_1 : y_1 : z_1 = \mathfrak{a}_1\xi_1 + \mathfrak{h}_1\eta_1 + \mathfrak{g}_1\zeta_1 : \mathfrak{h}_1\xi_1 + \mathfrak{b}_1\eta_1 + \mathfrak{f}_1\zeta_1 : \mathfrak{g}_1\xi_1 + \mathfrak{f}_1\eta_1 + \mathfrak{c}_1\zeta_1 \ldots\ldots (88).$$

This requires some further development however. We must in the first place form the system $\mathfrak{a}_1$, $\mathfrak{b}_1$, $\mathfrak{c}_1$, $\mathfrak{f}_1$, $\mathfrak{g}_1$, $\mathfrak{h}_1$: this is done immediately from the formulæ of Sect. 2, and we have

$$\mathfrak{a}_1 = \frac{W(\overline{e'e''}, \overline{e'e''}, \mathfrak{q})}{W(e', e', \mathrm{q})\, W(e'', e'', \mathrm{q})} = \frac{kW(\overline{\epsilon'\epsilon''}, \overline{\epsilon'\epsilon''}, \mathrm{q})}{W(\epsilon', \epsilon', \mathfrak{q})\, W(\epsilon'', \epsilon'', \mathfrak{q})} \quad\ldots\ldots\ldots\ldots\ldots\ldots (89),$$

$$\mathfrak{b}_1 = \frac{W(\overline{e''e}, \overline{e''e}, \mathfrak{q})}{W(e'', e'', \mathrm{q})\, W(e'', e, \mathrm{q})} = \frac{kW(\overline{\epsilon''\epsilon}, \overline{\epsilon''\epsilon}, \mathrm{q})}{W(\epsilon'', \epsilon'', \mathfrak{q})\, W(\epsilon, \epsilon, \mathfrak{q})},$$

$$\mathfrak{c}_1 = \frac{W(\overline{ee'}, \overline{ee'}, \mathfrak{q})}{W(e, e, \mathrm{q})\, W(e', e', \mathrm{q})} = \frac{kW(\overline{\epsilon\epsilon'}, \overline{\epsilon\epsilon'}, \mathrm{q})}{W(\epsilon, \epsilon, \mathfrak{q})\, W(\epsilon', \epsilon', \mathfrak{q})},$$

$$\mathfrak{f}_1 = \frac{W(\overline{e''e}, \overline{ee'}, \mathfrak{q})}{W(e, e, \mathrm{q})\sqrt{\{W(e', e', \mathrm{q})\, W(e'', e'', \mathrm{q})\}}} = \frac{kW(\overline{\epsilon''\epsilon}, \overline{\epsilon\epsilon'}, \mathrm{q})}{W(\epsilon, \epsilon, \mathfrak{q})\sqrt{\{W(\epsilon', \epsilon', \mathfrak{q})\, W(\epsilon'', \epsilon'', \mathfrak{q})\}}},$$

$$\mathfrak{g}_1 = \frac{W(\overline{ee'}, \overline{e'e''}, \mathfrak{q})}{W(e', e', \mathrm{q})\sqrt{\{W(e'', e'', \mathrm{q})\, W(e, e, \mathrm{q})\}}} = \frac{kW(\overline{\epsilon\epsilon'}, \overline{\epsilon'\epsilon''}, \mathrm{q})}{W(\epsilon', \epsilon', \mathfrak{q})\sqrt{\{W(\epsilon'', \epsilon'', \mathfrak{q})\, W(\epsilon, \epsilon, \mathfrak{q})\}}},$$

$$\mathfrak{h}_1 = \frac{W(\overline{e'e''}, \overline{e''e}, \mathfrak{q})}{W(e'', e'', \mathrm{q})\sqrt{\{W(e, e, \mathrm{q})\, W(e', e', \mathrm{q})\}}} = \frac{kW(\overline{\epsilon'\epsilon''}, \overline{\epsilon''\epsilon}, \mathrm{q})}{W(\epsilon'', \epsilon'', \mathfrak{q})\sqrt{\{W(\epsilon, \epsilon, \mathfrak{q})\, W(\epsilon', \epsilon', \mathfrak{q})\}}}.$$

$$x_1 : y_1 : z_1 = \sqrt{\{W(e,\ e,\ \text{q})\}} \times \qquad (90),$$

$$\{W(e,\ p,\ \text{q})\ W(\overline{e'e''},\ \overline{e'e''},\ \mathfrak{q}) + W(e',\ p,\ \text{q})\ W(\overline{e'e''},\ \overline{e''e},\ \mathfrak{q}) + W(e'',\ p,\ \mathfrak{q})\ W(\overline{e'e''},\ \overline{ee'},\ \mathfrak{q})\}$$

$$: \sqrt{\{W(e',\ e',\ \text{q})\}} \times$$

$$\{W(e,\ p,\ \text{q})\ W(\overline{e''e},\ \overline{e'e''},\ \mathfrak{q}) + W(e',\ p,\ \text{q})\ W(\overline{e''e},\ \overline{e''e},\ \mathfrak{q}) + W(e'',\ p,\ \text{q})\ W(\overline{e''e},\ \overline{ee'},\ \text{q})\}$$

$$: \sqrt{\{W(e'',\ e'',\ \text{q})\}} \times$$

$$\{W(e,\ p,\ \text{q})\ W(\overline{ee'},\ \overline{e'e''},\ \mathfrak{q}) + W(e',\ p,\ \text{q})\ W(\overline{ee'},\ \overline{e''e},\ \mathfrak{q}) + W(e'',\ p,\ \text{q})\ W(\overline{ee'},\ \overline{ee'},\ \mathfrak{q})\};$$

these may be reduced to the very simple form

$$x_1 : y_1 : z_1 = \sqrt{\{W(e\ ,\ e\ ,\ \text{q})\}}\ W(\overline{e'e''},\ \omega,\ \mathfrak{q}) \qquad (91),$$

$$: \sqrt{\{W(e',\ e',\ \text{q})\}}\ W(\overline{e''e},\ \omega,\ \mathfrak{q}),$$

$$: \sqrt{\{W(e'',\ e'',\ \text{q})\}}\ W(\overline{ee'},\ \omega,\ \mathfrak{q}).$$

and in like manner we obtain

$$x_1 : y_1 : z_1 = \sqrt{\{W(\epsilon\ ,\ \epsilon\ ,\ \mathfrak{q})\}}\ W(\overline{\epsilon'\epsilon'},\ p, \text{q}) \qquad (92),$$

$$: \sqrt{\{W(\epsilon',\ \epsilon',\ \mathfrak{q})\}}\ W(\overline{\epsilon''\epsilon},\ p,\ \text{q}),$$

$$: \sqrt{\{W(\epsilon'',\ \epsilon'',\ \mathfrak{q})\}}\ W(\overline{\epsilon\epsilon'},\ p,\ \text{q}).$$

It will be as well to indicate the steps of this reduction. Consider the quantity in { } in the first line of the equation which gives the ratios $x_1 : y_1 : z_1$; and suppose for a moment $\overline{e'e''} = l,\ m,\ n$, &c.: then, selecting the portion of the expression which is multiplied by $\mathfrak{a}$, this is

$$= \mathfrak{a}l\ \{l\,(a\xi + b\eta + c\zeta) + l'\,(a'\xi + b'\eta + c'\zeta) + l''\,(a''\xi + b''\eta + c''\zeta)\},$$

or, since

$$la + l'a' + l''a'' = \overline{ee'e''},\quad lb + l'b' + l''b'' = 0,\quad lc + l'c' + l''c'' = 0,$$

this reduces itself to $\overline{ee'e''}\,.\,\mathfrak{a}l\xi$, which is a term of

$$\overline{ee'e''}\ W(\overline{e'e''},\ \omega,\ \mathfrak{q});$$

and by comparing the remaining terms in the same manner, it would be seen that the whole reduces itself to

$$\overline{ee'e''}\ W(\overline{e'e''},\ \omega,\ \mathfrak{q});$$

whence the formulæ in question.

The formulæ (86), (87), (91), (92), completely resolve the problem of the transformation of coordinates; they determine respectively p_1 from p or ω, ω_1 from p or ω.

To complete the present part of the subject we may add the following formulæ.

Suppose
$$x_1 : y_1 : z_1 = \text{a}x_1 + \text{a}'y_1 + \text{a}''z_1 \qquad (93),$$

$$: \text{b}x_1 + \text{b}'y_1 + \text{b}''y_1,$$

$$: \text{c}x_1 + \text{c}'y_1 + \text{c}''z_1,$$

which we see from the preceding formulæ is the form of the relation between the systems p_1 and p. And suppose, as before, λ_1, μ_1, ν_1 are the cosines of the distances of the new points of reference X_1, Y_1, Z_1.

We can immediately determine the relations that must exist between these coefficients, in order that they may actually denote such a transformation. For this purpose write

$$\begin{aligned} &\mathrm{a}\ ,\ \mathrm{b}\ ,\ \mathrm{c}\ = j \quad\text{.................................}\quad (94),\\ &\mathrm{a}',\ \mathrm{b}',\ \mathrm{c}' = j',\\ &\mathrm{a}'',\ \mathrm{b}'',\ \mathrm{c}'' = j''. \end{aligned}$$

Then the distance between the point P and any other point P' is given by the formula

$$\cos\delta = \frac{W(p,\ p',\ \mathrm{q})}{\sqrt{\{W(p,\ p,\ \mathrm{q})\ W(p',\ p',\ \mathrm{q})\}}} = \frac{W(p_1,\ p_1',\ \Theta)}{\sqrt{\{W(p_1,\ p_1,\ \Theta)\ W(p_1,\ p_1',\ \Theta)\}}} \ldots (95),$$

where

$$\Theta = W(j, j, \mathrm{q}),\quad W(j', j', \mathrm{q}),\quad W(j'', j'', \mathrm{q}),\quad W(j', j'', \mathrm{q}),\quad W(j'', j, \mathrm{q}),\quad W(j, j', \mathrm{q}) \ldots (96).$$

But we must evidently have

$$\cos\delta = \frac{W(p_1,\ p_1',\ \mathrm{q}_1)}{\sqrt{\{W(p_1,\ p_1,\ \mathrm{q}_1)\ W(p_1',\ p_1',\ \mathrm{q}_1)\}}} \quad\text{....................}\quad (97),$$

or the quantities Θ must be proportional to the quantities q, i.e.

$$\begin{aligned} &W(j,\ j,\ \mathrm{q}) : W(j',\ j',\ \mathrm{q}) : W(j'',\ j'',\ \mathrm{q}) : W(j',\ j'',\ \mathrm{q}) : W(j'',\ j,\ \mathrm{q}) : W(j'',\ j,\ \mathrm{q})\\ &= 1 \quad : \quad 1 \quad : \quad 1 \quad : \quad \lambda_1 \quad : \quad \mu_1 \quad : \quad \nu_1 \ldots (98). \end{aligned}$$

And in precisely the same manner, if instead of x, y, z, x_1, y_1, z_1, in the above formulæ, we had had ξ, η, ζ : ξ_1, η_1, ζ_1, the result would have been

$$\begin{aligned} &W(j,\ j,\ \mathfrak{q}) : W(j',\ j',\ \mathfrak{q}) : W(j'',\ j'',\ \mathfrak{q}) : W(j',\ j'',\ \mathfrak{q}) : W(j'',\ j,\ \mathfrak{q}) : W(j'',\ j,\ \mathfrak{q})\\ &= \mathfrak{a} \quad : \quad \mathfrak{b} \quad : \quad \mathfrak{c} \quad : \quad \mathfrak{f} \quad : \quad \mathfrak{g} \quad : \quad \mathfrak{h} \ldots (99). \end{aligned}$$

It is hardly necessary to remark, that throughout the preceding formulæ an expression, such as $W(p,\ p',\ \mathrm{q})$, is proportional to either of the quantities

$$x\xi' + y\eta' + z\zeta' \ \text{ or } \ x'\xi + y'\eta + z'\zeta,$$

and may be changed into one of these multiplied by an arbitrary constant, which may be always made to disappear by a corresponding change in another quantity of the same form: thus, for instance,

$$\frac{W(p,\ p',\ \mathrm{q})}{W(p,\ p,\ \mathrm{q})} = \frac{x'\xi + y'\eta + z'\zeta}{x\xi + y\eta + z\zeta} \quad\text{......................}\quad (100);$$

but these forms being unsymmetrical, it is better in general not to introduce them.

All the preceding expressions simplify exceedingly, reducing themselves in fact to the ordinary formulæ for the transformation of rectangular coordinates in Geometry of three dimensions, for the case where the triangle XYZ has its sides and angles right angles. As this presents no difficulty, I shall not enter upon it at present.

33.

ON THE REDUCTION OF $\dfrac{du}{\sqrt{U}}$, WHEN U IS A FUNCTION OF THE FOURTH ORDER.

[From the *Cambridge and Dublin Mathematical Journal*, vol. I. (1846), pp. 70—73.]

It is well known that the transformation of this differential expression into a similar one, in which the function in the denominator contains only even powers of the corresponding variable, is the first step in the process of reducing $\int \dfrac{du}{\sqrt{U}}$ to elliptic integrals. And, accordingly, the different modes of effecting this have been examined, more or less, by most of those who have written on the subject. The simplest supposition, that adopted by Legendre, and likewise discussed in some detail by Gudermann, is that u is a fraction, the numerator and denominator of which are linear functions of the new variable. But the theory of this transformation admits of being developed further than it has yet been done, as regards the equation which determines the modulus of the elliptic function. This may be effected most easily as follows.

Suppose

$$U = a + 4bu + 6cu^2 + 4du^3 + eu^4,$$
$$P = ax^4 + 4bx^3y + 6cx^2y^2 + 4dxy^3 + ey^4.$$

Also let

$$P' = a'x'^4 + 4b'x'^2y' + 6c'x'^2y'^2 + 4d'x'y'^3 + e'y'^4$$

be what P becomes after writing

$$x = \lambda x' + \mu y',$$
$$y = \lambda_{\prime} x' + \mu_{\prime} y':$$

and let

$$U' = a' + 4b'u' + 6c'u'^2 + 4d'u'^3 + e'u'^4.$$

Suppose, moreover,

$$\begin{cases} k = \lambda\mu_{,} - \lambda_{,}\mu, \\ I = ae - 4bd + 3c^2, \\ I' = a'e' - 4b'd' + 3c'^2, \\ J = ace - ad^2 - be^2 - c^3 + 2bcd, \\ J' = a'c'e' - a'd'^2 - b'e'^2 - c'^3 + 2b'c'd'; \end{cases}$$

we have evidently

$$xdy - ydx = k\,(x'dy' - y'dx'),$$

or

$$\frac{xdy - ydx}{\sqrt{P}} = k\,\frac{x'dy' - y'dx'}{\sqrt{P'}}.$$

Hence writing

$$u = \frac{y}{x}, \qquad u' = \frac{y'}{x'};$$

and therefore

$$\frac{xdy - ydx}{P^{\frac{1}{2}}} = \frac{du}{U^{\frac{1}{2}}}, \qquad \frac{x'dy' - y'dx'}{P'^{\frac{1}{2}}} = \frac{du'}{U'^{\frac{1}{2}}},$$

we obtain

$$\frac{du}{\sqrt{U}} = k\,\frac{du'}{\sqrt{U'}},$$

the equation between u and u' being

$$u = \frac{\lambda + \mu u'}{\lambda_{,} + \mu_{,} u'}.$$

Next, to determine the relations between the coefficients of U and U'. Since P, P' are obtained from each other by linear transformations (*Math. Journal*, vol. IV. p. 208), [13, p. 94], we have between the coefficients of these functions and of the transforming equations, the relations

$$I' = k^4 I,$$

$$J' = k^6 J;$$

whence also

$$\frac{J'^2}{I'^3} = \frac{J^2}{I^3}.$$

Suppose now

$$U' = a'\,(1 + pu'^2)\,(1 + qu'^2),$$

or

$$b' = 0, \quad d' = 0, \quad 6c' = a'\,(p + q), \quad e' = a'pq;$$

whence also

$$I' = \tfrac{1}{12}\,a'^2\,(p^2 + q^2 + 14pq),$$

$$J' = \tfrac{1}{216}\,a'^3\,(p + q)\,(34pq - p^2 - q^2);$$

$$p^2 + q^2 + 14pq = 12 \frac{k^4}{a'^2} I,$$

$$(p+q)(34pq - p^2 - q^2) = 216 \frac{k^6}{a'^3} J;$$

therefore

$$\frac{(p+q)^2 (34pq - p^2 - q^2)^2}{(p^2 + q^2 + 14pq)^3} = \frac{27J^2}{I^3},$$

whence also

$$\frac{108pq(p-q)^4}{(p^2 + q^2 + 14pq)^3} = 1 - \frac{27J^2}{I^3},$$

which determines the relation between p and q. Also

$$\frac{k}{\sqrt{a'}} = \left(\frac{p^2 + q^2 + 14pq}{12I}\right)^{\frac{1}{4}},$$

so that

$$\frac{du}{\sqrt{U}} = \left(\frac{p^2 + q^2 + 14pq}{12I}\right)^{\frac{1}{4}} \frac{du'}{\sqrt{\{(1 + pu'^2)(1 + qu'^2)\}}}.$$

If in particular $p = -1$, writing also $-q$ for q,

$$\frac{du}{\sqrt{U}} = \left(\frac{q^2 + 14q + 1}{12I}\right)^{\frac{1}{4}} \frac{du'}{\sqrt{\{(1 - u'^2)(1 - qu'^2)\}}},$$

where

$$\frac{108q(1-q)^4}{(q^2 + 14q + 1)^3} = 1 - \frac{27J^2}{I^3}.$$

Suppose, for shortness,

$$M = \tfrac{27}{4} \cdot \frac{1}{\left(1 - \frac{27J^2}{I^3}\right)}, \quad \text{or } \tfrac{1}{108}\left(1 - \frac{27J^2}{I^3}\right) = \frac{1}{16M},$$

then $$(q^2 + 14q + 1)^3 - 16Mq(q-1)^4 = 0,$$

i.e. $$\left(q + \frac{1}{q} + 14\right)^3 - 16M\left(q^{\frac{1}{2}} - \frac{1}{q^{\frac{1}{2}}}\right)^4 = 0.$$

Let $$q^{\frac{1}{2}} - q^{-\frac{1}{2}} = \frac{4}{(\theta - 1)^{\frac{1}{2}}},$$

then $$\theta^3 - M(\theta - 1) = 0,$$

which determines θ. And then

$$q = \frac{7 + \theta + 4(3 + \theta)^{\frac{1}{2}}}{\theta - 1}.$$

Suppose $q = \alpha$ is one of the values of q; the equation becomes

$$\frac{(q^2+14q+1)^3}{q(q-1)^4} = \frac{(\alpha^2+14\alpha+1)^3}{\alpha(\alpha-1)^4}, \quad = \frac{(\beta^8+14\beta^4+1)^3}{\beta^4(\beta^4-1)^4}, \quad \text{if } \alpha = \beta^4.$$

Now if $$q = \left(\frac{1-\beta}{1+\beta}\right)^4,$$

then $$(q^2+14q+1) = \frac{16(\beta^8+14\beta^4+1)}{(1+\beta)^8}, \qquad q-1 = -\frac{8\beta(1+\beta^2)}{(1+\beta)^4},$$

which values satisfy the above equation: hence also, identically,

$$(q^2+14q+1)^3 - q(q-1)^4\frac{(\beta^8+14\beta^4+1)^3}{\beta^4(\beta^4-1)^4}$$

$$= (q-\beta^4)\left(q-\frac{1}{\beta^4}\right)\left\{q-\left(\frac{1-\beta}{1+\beta}\right)^4\right\}\left\{q-\left(\frac{1+\beta}{1-\beta}\right)^4\right\}\left\{q-\left(\frac{1-\beta i}{1+\beta i}\right)^4\right\}\left\{q-\left(\frac{1+\beta i}{1-\beta i}\right)^4\right\};$$

or the values of q take the form

$$\beta^4, \quad \frac{1}{\beta^4}, \quad \left(\frac{1-\beta}{1+\beta}\right)^4, \quad \left(\frac{1+\beta}{1-\beta}\right)^4, \quad \left(\frac{1-\beta i}{1+\beta i}\right)^4, \quad \left(\frac{1+\beta i}{1-\beta i}\right)^4.$$

(Comp. *Abel. Œuv.* tom. I. p. 310 [Ed. 2, p. 459].)

The equation $$\theta^3 - M\theta + M = 0$$

has its three roots real if $27-4M$ is negative, and only a single real root if $27-4M$ is positive. Writing the equation under the form

$$(\theta+3)^3 - 9(\theta+3)^2 + (27-M)(\theta+3) - (27-4M) = 0,$$

we see that in the former case θ has two values greater than -3, and a single value less than -3. Writing the equation under the form

$$(\theta-1)^3 + 3(\theta-1)^2 + (3-M)(\theta-1) + 1 = 0, \quad (3-M \text{ is negative})$$

the positive roots are both greater than 1. Hence, in this case, q has four positive values and two imaginary ones. In the second case θ has a single real value, which is greater than -3 and less than 1. Hence q has two negative values and four imaginary ones. In the former case, I^3-27J^2 is positive, and the function U has either four imaginary factors or four real ones. In the second case, I^3-27J^2 is negative, or the function U has two real and two imaginary factors.

34.

NOTE ON THE MAXIMA AND MINIMA OF FUNCTIONS OF THREE VARIABLES.

[From the *Cambridge and Dublin Mathematical Journal*, vol. I. (1846), pp. 74, 75.]

IF A, B, C, F, G, H, be any real quantities, such that

$$BC + CA + AB - F^2 - G^2 - H^2,$$

and
$$(A + B + C)(ABC - AF^2 - BG^2 - CH^2 + 2FGH)$$

are positive; the six quantities

$$BC - F^2,\ CA - G^2,\ AB - H^2,\ AK,\ BK,\ CK,$$

(where $K = ABC - AF^2 - BG^2 - CH^2 + 2FGH$) are all of them positive. It is unnecessary to point out the connection of this property with the theory of maxima and minima.

To demonstrate this, writing as usual

$$\begin{aligned} BC - F^2 &= A', & GH - AF &= F', \\ CA - G^2 &= B', & HF - BG &= G', \\ AB - H^2 &= C', & FG - CH &= H', \end{aligned}$$

and K as above: then if A'', B'', C'', F'', G'', H'', K' be formed from A', B', C', F', G', H', as these and K are from A, B, C, F, G, H, we have the well-known formulæ

$$\begin{aligned} A'' &= KA, & F'' &= KF, & K' &= K^2. \\ B'' &= KB, & G'' &= KG, \\ C'' &= KC, & H'' &= KH, \end{aligned}$$

It is required to show that if $A'+B'+C'$ and $A''+B''+C''$ are positive, A', B', C', A'', B'', C'' are so likewise.

Consider the cubic equation

$$(A'-k)(B'-k)(C'-k)-(A'-k)F'^2-(B'-k)G'^2-(C'-k)H'^2+2F'G'H'=0,$$

the roots of which are all real. By the formulæ just given this may be written

$$k^3-k^2(A'+B'+C')+k(A''+B''+C'')-K^2=0;$$

and the terms of this equation are alternately positive and negative; i.e. the roots are all positive. Hence the roots of the limiting equation

$$(B'-k)(C'-k)-F'^2=0$$

are positive, i.e. $B'+C'$ and $B'C'$ are positive: but from the second condition B', C' are of the same sign: consequently they are of the same sign with $B'+C'$, or positive. Also $A''=B'C'-F'^2$ is positive. Similarly, considering the other limiting equations, A', B', C', A'', B'', C'' are all of them positive.

In connection with the above I may notice the following theorem. The roots of the equation

$$(A-ka)(B-kb)(C-ck)-(A-ka)(F-kf)^2-(B-kb)(G-kg)^2-(C-kc)(H-kh)^2 \\ +2(F-kf)(G-kg)(H-kh)=0,$$

are all of them real, if either of the functions

$$Ax^2+By^2+Cz^2+2Fyz+2Gxz+2Hxy,$$
$$ax^2+by^2+cz^2+2fyz+2gxz+2hxy,$$

preserve constantly the same sign. The above form parts of a general system of properties of functions of the second order.

35.

ON HOMOGENEOUS FUNCTIONS OF THE THIRD ORDER WITH THREE VARIABLES.

[From the *Cambridge and Dublin Mathematical Journal*, vol. I. (1846), pp. 97—104.]

THE following problem corresponds to the geometrical question of determining the polar reciprocal of a plane curve of the third order: the solution of it is also important, with reference to the linear transformations of homogeneous functions of three variables of the third order; reasons for which it has appeared to me worth while to obtain the completely developed result.

Let

$$3U = ax^3 + by^3 + cz^3 + 3iy^2z + 3jz^2x + 3kx^2y + 3i_1yz^2 + 3j_1zx^2 + 3k_1xy^2 + 6lxyz \quad \text{.........(1).}$$

It is required to eliminate x, y, z, λ from the equations

$$U = 0 \quad \text{..(2),}$$

$$\left.\begin{aligned} \frac{dU}{dx} + \lambda\xi = 0, \\ \frac{dU}{dy} + \lambda\eta = 0, \\ \frac{dU}{dz} + \lambda\zeta = 0, \end{aligned}\right\} \quad \text{......................................(3).}$$

From the equations (2), (3), we obtain immediately

$$\Theta = \xi x + \eta y + \zeta z = 0 \quad \text{..................................(4);}$$

and thence

$$\Theta x = 0, \quad \Theta y = 0, \quad \Theta z = 0 \quad \text{.............................(5);}$$

so that a single equation more, such as

$$\Phi = 0 \quad \ldots\ldots\ldots\ldots\ldots\ldots\ldots\ldots\ldots\ldots\ldots\ldots (6),$$

where Φ is homogeneous and of the second order in x, y, z, would, in conjunction with the equations (3) and (5), enable us to eliminate linearly the seven quantities x^2, y^2, z^2, yz, zx, xy, λ. Such an equation may be thus obtained.

Let L, M, N, R, S, T, be the second differential coefficients of U, each of them divided by two. The equations (3) may be written

$$\begin{aligned} Lx + Ty + Sz + \lambda\xi &= 0, \quad \ldots\ldots\ldots\ldots\ldots\ldots\ldots\ldots (7). \\ Tx + My + Rz + \lambda\eta &= 0, \\ Sx + Ry + Nz + \lambda\zeta &= 0. \end{aligned}$$

And joining to these the equation (4),

$$\xi x + \eta y + \zeta z \qquad = 0,$$

we have, by the elimination of x, y, z, in so far as they explicitly appear, and λ, an equation $\Phi = 0$ of the required form. Hence we may write

$$\Phi = - \begin{vmatrix} L, & T, & S, & \xi \\ T, & M, & R, & \eta \\ S, & R, & N, & \zeta \\ \xi, & \eta, & \zeta, & \end{vmatrix} \quad \ldots\ldots\ldots\ldots\ldots\ldots\ldots\ldots (8);$$

or substituting for L, M, N, R, S, T, and expanding,

$$\Phi = Ax^2 + By^2 + Cz^2 + 2Fyz + 2Gzx + 2Hxy \quad \ldots\ldots\ldots\ldots\ldots (9),$$

where the values of A, B, C, F, G, H are $\ldots\ldots\ldots\ldots\ldots\ldots\ldots\ldots\ldots\ldots\ldots\ldots (10)$.

[I omit these values (10), and the values in the subsequent equations (13), (14), (20): the values (10) and (13) serve for the calculation of (14), $\mathsf{F}U$, the expression for which with the letters (a, b, c, f, g, h, i, j, k, l) in place of (a, b, c, i, j, k, i_1, j_1, k_1, l) is reproduced in my "Third Memoir on Quantics," *Phil. Trans.* vol. CXLVI. (1856) pp. 627—647: the values (20) serve for finding that of (21), $K(U)$, but the developed expression of this has not been calculated.]

Performing the elimination indicated, the result may be represented by

$$\mathsf{F}U = \begin{vmatrix} a, & k_1, & j, & l, & j_1, & k, & \xi \\ k, & b, & i_1, & i, & l, & k_1, & \eta \\ j_1, & i, & c, & i_1, & j, & l, & \zeta \\ 2\xi, & . & . & . & \zeta, & \eta & . \\ . & 2\eta & . & \zeta & . & \xi & . \\ . & . & 2\zeta & \eta & \xi & . & . \\ A & B & C & F & G & H & . \end{vmatrix} = 0 \quad \ldots\ldots\ldots\ldots\ldots (11).$$

Partially expanding,

$$\mathsf{F}U = A\mathrm{a} + B\mathrm{b} + C\mathrm{c} + 2F\mathrm{f} + 2G\mathrm{g} + 2H\mathrm{h} \quad \ldots\ldots\ldots\ldots\ldots (12).$$

The values of the coefficients a, b, c, f, g, h may be useful on other occasions: they are as follows. ....(13).

Substituting these values the result after all reductions becomes

$$0 = F(U) = \quad \ldots\ldots\ldots\ldots\ldots\ldots\ldots\ldots\ldots\ldots\ldots\ldots \quad (14).$$

It would be desirable, in conjunction with the above, to obtain the equation

$$K(U) = 0,$$

which results from the elimination of x, y, z from the equations

$$\frac{dU}{dx} = 0, \quad \frac{dU}{dy} = 0, \quad \frac{dU}{dz} = 0 \quad \ldots\ldots\ldots\ldots\ldots\ldots\ldots\ldots \quad (15),$$

(i.e. the condition of a curve of the third order having a multiple point), but to effect this would be exceedingly laborious. The following is the process of the elimination as given by Dr Hesse, *Crelle*, t. XXVIII. (and which applies also to the case of any three equations of the second order). Forming the function ∇U, of the third order in x, y, z, by means of the equation

$$\nabla U = \begin{vmatrix} L, & T, & S \\ T, & M, & R \\ S, & R, & N \end{vmatrix} \quad \ldots\ldots\ldots\ldots\ldots\ldots\ldots\ldots\ldots\ldots \quad (16),$$

(L, M, N, R, S, T, the same as before).

Then, in consequence of the equations (15), we have not only

$$\nabla U = 0 \ldots\ldots\ldots\ldots\ldots\ldots\ldots\ldots\ldots\ldots\ldots\ldots\ldots \quad (17),$$

which is very easily proved to be the case, but also

$$\frac{d}{dx}\nabla U = 0, \quad \frac{d}{dy}\nabla U = 0, \quad \frac{d}{dz}\nabla U = 0 \ldots\ldots\ldots\ldots\ldots\ldots\ldots \quad (18),$$

as will be shown in a subsequent paper "On Points of Inflection." [I think never written.]

And from the six equations (15), (18), the six quantities x^2, y^2, z^2, yz, zx, xy, may be linearly eliminated: we have

$$\nabla U = Ax^3 + By^3 + Cz^3 + 3Iy^2z + 3Jz^2x + 3Kx^2y + 3I_1yz^2 + 3J_1zx^2 + 3K_1xy^2 + 6\Lambda xyz \ldots (19),$$

where the values of the coefficients A, B, ... Λ are ....(20),

and the result of the elimination is

$$K(U) = \begin{vmatrix} a\,, & k_1\,, & j\,, & l\,, & j_1\,, & k \\ k\,, & b\,, & i_1\,, & i\,, & l\,, & k_1 \\ j_1\,, & i\,, & c\,, & i_1\,, & j\,, & l \\ A, & K_1, & J\,, & L, & J_1, & K \\ K, & B\,, & I_1, & I\,, & L\,, & K_1 \\ J_1, & I\,, & C, & I_1, & J\,, & L \end{vmatrix} = 0 \quad \ldots\ldots\ldots\ldots\ldots\ldots \quad (21).$$

{$K(U)$ is consequently, as is well known, a function of the twelfth order in a, b, c, i, j, k, i_1, j_1, k_1, l}.

The equation $$\nabla U = 0,$$

combined with that of the curve, determine, as Dr Hesse has demonstrated in the paper quoted, the points of inflection of the curve. It may be inferred from this, that if U reduce itself to the form

$$U = (\alpha x^2 + \beta y^2 + \gamma z^2 + 2\iota yz + 2\kappa xz + 2\lambda xy)\, P = VP \quad\ldots\ldots(22),$$

P a linear function of x, y, z: then ∇U takes the form

$$\nabla U = P\,(\rho V + \sigma P^2) \quad\ldots\ldots(23),$$

where ρ is of the second order in the coefficients of P, and also in the coefficients α, β, γ, ι, κ, λ: and σ is equal to the determinant

$$\begin{vmatrix} \alpha, & \lambda, & \kappa \\ \lambda, & \beta, & \iota \\ \kappa, & \iota, & \gamma \end{vmatrix} \quad\ldots\ldots(24),$$

multiplied by a numerical factor. If U is of the form

$$U = PQR \quad\ldots\ldots(25),$$

then $$\nabla U = \rho PQR = \rho U \quad\ldots\ldots(26),$$

and this equation is consequently the condition of the function U being resolvable into linear factors. The equation in question resolves itself into

$$\frac{A}{a} = \frac{B}{b} = \frac{C}{c} = \frac{I}{i} = \frac{J}{j} = \frac{K}{k} = \frac{I_1}{i_1} = \frac{J_1}{j_1} = \frac{K_1}{k_1} = \frac{\Lambda}{l} \quad\ldots\ldots(27);$$

a system which must contain three independent equations only. It would be interesting to verify this *à posteriori*.

36.

ON THE GEOMETRICAL REPRESENTATION OF THE MOTION OF A SOLID BODY.

[From the *Cambridge and Dublin Mathematical Journal*, vol. I. (1846), pp. 164—167.]

LET P, Q, R, ... be consecutive generating lines of a skew surface, and on these take points p', p; q', q; r', r ...... such that pq', qr' ...... are the shortest distances between P and Q, Q and R, &c. Then for the generating line P, the ratio of the inclination of the lines P, Q to the distance pq' is said to be "the torsion," the angle $q'pq$ is said to be the deviation, and the ratio of the inclination of the planes Qpq' and Qqr' to the inclination of P and Q is said to be the "skew curvature." And similarly for any other generating line; so that the torsion and deviation depend on the position of the consecutive line, and the skew curvature on the position of the two consecutive lines. The curve pqr ...... is said to be the minimum distance curve [or curve of striction]. {When the skew surface degenerates into a developable surface, the torsion is infinite, the deviation a right angle, the skew curvature proportional to the curvature of the principal section, i.e. it is the distance of a point from the edge of regression, multiplied into the reciprocal of the radius of curvature, a product which is evidently constant along a generating line. Also the curve of minimum distance becomes the edge of regression.} A skew surface, considered independently of its position in space, is determined when for each generating line we know the torsion, deviation, and skew curvature. For, assuming arbitrarily the line P and the point p, also the plane in which pq' lies, the position of Q is completely determined from the given torsion and deviation; and then Q being known, the position of R is completely determined from the skew curvature for P, and the torsion and deviation for Q; and similarly the consecutive generating lines are to be determined.

Two skew surfaces are said to be "deformations" of each other, when for corresponding generating lines the torsion is always the same. Thus a surface will be deformed if considering the elements between the successive generating lines P, Q ...... as rigid, these elements be made to revolve round the successive generating lines P, Q ...... and to slide along them. {They are "transformations", when not only the torsions but also the deviations are equal at corresponding generating lines: thus, if the sliding of the elements along P, Q ...... be omitted, the new surface will be, not a deformation, but a transformation of the other.} No two skew surfaces can be made to roll and slide one upon the other, so that their successive generating lines coincide, unless one of them is a deformation of the other: and when this is the case, the rolling and sliding motions are *completely determined.* In fact the angular velocity of the generating line is the angular velocity round this line, into the difference of the skew curvatures of the two surfaces; the velocity of translation of the generating line in its own direction is to the angular velocity of the generating line, as the difference of the deviations is to the torsion. {This includes also the case in which one surface is a transformation of the other, where the motion is evidently a rolling one.} A skew surface moving in this manner upon another of which it is the deformation, may be said to "glide" upon it. We may now state the kinematical theorem:

"Any motion whatever of a solid body in space may be represented as the 'gliding' motion of one skew surface upon another fixed in space, and of which it is the deformation."

a theorem which is to be considered as the generalization of the well-known one—

"Any motion of a solid body round a fixed point may be represented as the rolling motion of a conical surface upon a second conical surface fixed in space."

and of the supplementary theorem—

"The angular velocity round the line of contact (the instantaneous axis) is to the angular velocity of this line as the difference of curvatures of the two cones at any point in the same line, to the reciprocal of the distance of the point from the vertex."

The analytical demonstration of this last theorem is rather interesting: it depends on the following formulæ. Forming two determinants, the first with the angular velocities round three axes fixed in space, and the first and second derived coefficients of these velocities with respect to the time; the other in the same way with the angular velocities round axes fixed in the body; the difference of these determinants is equal to the fourth power of the angular velocity into the square of the angular velocity of the instantaneous axis.

To show this, let p, q, r be the angular velocities round the axes fixed in the body; u, v, w those round axes fixed in space; ω the angular velocity round the instantaneous axis; ∇, Ω the two determinants: the theorem comes to

$$\nabla - \Omega = M,$$

where $$M = \omega^2 (p'^2 + q'^2 + r'^2 - \omega'^2), \text{ or } \omega^2 (u'^2 + v'^2 + w'^2 - \omega'^2).$$

Here

$$u = \alpha p + \beta q + \gamma r,$$
$$v = \alpha' p + \beta' q + \gamma' r,$$
$$w = \alpha'' p + \beta'' q + \gamma'' r;$$

whence
$$u' = \alpha\, p' + \beta\, q' + \gamma\, r',$$
$$v' = \alpha' p' + \beta' q' + \gamma' r',$$
$$w' = \alpha'' p' + \beta'' q' + \gamma'' r',$$

(the remaining terms vanishing as is well known); and therefore

$$vw' - v'w = \alpha\ (qr' - q'r) + \beta\ (rp' - r'p) + \gamma\ (pq' - p'q),$$
$$wu' - w'u = \alpha'\ (qr' - q'r) + \beta'\ (rp' - r'p) + \gamma'\ (pq' - p'q),$$
$$uv' - u'v = \alpha''\ (qr' - q'r) + \beta''\ (rp' - r'p) + \gamma''\ (pq' - p'q).$$

Hence
$$vw'' - v''w = \alpha\ (qr'' - q''r) + \beta\ (rp'' - r''p) + \gamma\ (pq'' - p''q) + u'\omega^2 - u\omega\omega',$$
$$wu'' - w''u = \alpha'\ (qr'' - q''r) + \beta'\ (rp'' - r''p) + \gamma'\ (pq'' - p''q) + v'\omega^2 - v\omega\omega',$$
$$uv'' - u''v = \alpha''\ (qr'' - q''r) + \beta''\ (rp'' - r''p) + \gamma''\ (pq'' - p''q) + w'\omega^2 - w\omega\omega',$$

and multiplying these by u', v', w', and adding, the required equation is immediately obtained.

In fact, if r be the distance of a point in the instantaneous axis from the vertex, and ρ, σ the radii of curvature of the two cones at that point, then

$$\frac{r}{\rho} = \frac{\omega^3}{M^{\frac{3}{2}}}\,\Omega, \quad \frac{r}{\sigma} = \frac{\omega^3}{M^{\frac{3}{2}}}\,\nabla,$$

as may be shown without difficulty: and the angular velocity of the instantaneous axis is given by the equation $\varpi = \dfrac{M^{\frac{1}{2}}}{\omega^2}$; hence the relation between the two angular velocities is

$$\omega : \varpi = \frac{1}{\rho} - \frac{1}{\sigma} : \frac{1}{r}.$$

37.

ON THE ROTATION OF A SOLID BODY ROUND A FIXED POINT.

[From the *Cambridge and Dublin Mathematical Journal*, vol. I. (1846), pp. 167—173 and 264—274.]

THE difficulty of completing elegantly the solution of this problem, in the case where no forces act upon the body, arises from the complexity and want of symmetry of the ordinary formulæ for determining the position of one set of rectangular axes with respect to another set; in consequence of which it has hitherto been considered necessary to make a particular supposition relative to the position of the fixed axes in space, viz. that one of them shall be perpendicular to the "invariable plane" of the rotating body. But some formulæ for the above purpose, given also by Euler, are entirely free from these objections. Imagine two sets of axes Ax, Ay, Az, $Ax_{,}$, $Ay_{,}$, $Az_{,}$. The former set can be made to coincide with the second set, by a rotation θ round a certain axis AR, inclined to Ax, Ay, Az at angles f, g, h. (As usual f, g, h are the angles RAx, RAy, RAz considered as positive, and the rotation is in the same direction as a rotation round Az from x towards y.) This axis may be termed the resultant axis, and the angle θ the resultant rotation. The formulæ of Euler express the coefficients of the transformation in terms of the resultant rotation and of the position of the resultant axis, i.e. in terms of θ and of the angles f, g, h, whose cosines are connected by the equation

$$\cos^2 f + \cos^2 g + \cos^2 h = 1.$$

This idea was improved upon by M. Rodrigues (*Liouv.* tom. V. p. 404), who introduced the quantities

$$\tan \tfrac{1}{2}\theta \cos f, \quad \tan \tfrac{1}{2}\theta \cos g, \quad \tan \tfrac{1}{2}\theta \cos h,$$

(quantities which will be represented by λ, μ, ν) by means of which he expressed the coefficients as fractions, the numerators of which are very simple rational functions of the second order of λ, μ, ν, and which have the common denominator $(1+\lambda^2+\mu^2+\nu^2)$. These quantities may conveniently be termed the "coordinates of the resultant rotation," and the denominator or the square of the secant of the semi-angle of resultant rotation will be the "modulus" of the rotation. The elegance of these results led me to apply them to the mechanical question, and I gave in the *Journal* (vol. III. p. 224), [6], the differential equations of motion obtained in terms of λ, μ, ν: which I integrated as in the common theory, by supposing one of the fixed axes to be perpendicular to the invariable plane. Though my attention was again called to the subject, by the connexion of some of these formulæ with Sir William Hamilton's theory of quaternions, no other way of performing the integration occurred to me. The grand discovery however of Jacobi, of the possibility of reducing to quadratures the two final differential equations of any mechanical problem, when the remaining integrals are known, induced me to resume the problem, and at least attempt to bring it so far as to obtain a differential equation of the first order between two variables only, the multiplier of which could be obtained theoretically by Jacobi's discovery. The choice of two new variables to which the equations of the problem led me, enabled me to effect this with the greatest simplicity; and the differential equation which I finally obtained, turned out to be integrable *per se*, so that the laborious process of finding the multiplier became unnecessary. The new variables Ω, υ have the following geometrical interpretations, $\Omega = k\tan\frac{1}{2}\theta\cos I$, where k is the principal moment, θ as before the angle of resultant rotation, and I is the inclination of the resultant axis to the perpendicular upon the invariable plane, and $\upsilon = k^2\cos^2\frac{1}{2}J$; where, if we imagine a line AQ having the same position relatively to the axes in fixed space that the perpendicular upon the invariable plane has to the principal axes of the rotating body, then J is the inclination of this line to the above perpendicular. To the choice of these variables I was led by the analysis only. It will be seen that p, q, r are functions of υ only, while λ, μ, ν contain besides the variable Ω. In obtaining these relations a singular equation $\Omega^2 = \kappa\upsilon - k^2$ occurs (equation 13), which may also be written $1+\tan^2\frac{1}{2}\theta\cos^2 I = \sec^2\frac{1}{2}\theta\cos^2\frac{1}{2}J$, in which form the interpretation of the quantities I, J has just been given. The equation (17), it may be remarked, is self-evident: it expresses that the inclination of the resultant axis to the normal of the invariable plane, is equal to the inclination of the same axis to the line AQ. Now the resultant axis having the same inclination to the axes fixed in space as it has to the principal axes, and the line AQ the same inclinations to these fixed axes that the normal to the invariable plane has to the principal axes, the truth of the proposition becomes manifest. The correspondence in form between the systems (10) and (14) is also worth remarking. The final results at which I arrive are, that the time and the arc whose tangent is $\Omega \div k$, are each of them expressible as the integrals of certain algebraical functions of υ. The notation throughout is the same as that made use of in the paper already quoted.

The equations of rotatory motion are

$$dt = \frac{dp}{P} = \frac{dq}{Q} = \frac{dr}{R} = \frac{d\lambda}{\Lambda} = \frac{d\mu}{\mathrm{M}} = \frac{d\nu}{\mathrm{N}} \quad\ldots\ldots\ldots\ldots (1),$$

where

$$\left.\begin{aligned}
P&=\frac{1}{A}\left[(B-C)\,qr+\tfrac{1}{2}\left\{(1+\lambda^2)\frac{dV}{d\lambda}+(\lambda\mu+\nu)\frac{dV}{d\mu}+(\lambda\nu-\mu)\frac{dV}{d\nu}\right\}\right],\\
Q&=\frac{1}{B}\left[(C-A)\,rp+\tfrac{1}{2}\left\{(\mu\lambda-\nu)\frac{dV}{d\lambda}+(1+\mu^2)\frac{dV}{d\mu}+(\mu\nu+\lambda)\frac{dV}{d\nu}\right\}\right],\\
R&=\frac{1}{C}\left[(A-B)\,pq+\tfrac{1}{2}\left\{(\nu\lambda+\mu)\frac{dV}{d\lambda}+(\mu\nu-\lambda)\frac{dV}{d\mu}+(1+\nu^2)\frac{dV}{d\nu}\right\}\right],
\end{aligned}\right\}\quad\ldots\ldots(2).$$

$$\left.\begin{aligned}
\Lambda&=\tfrac{1}{2}\{(1+\lambda^2)\,p+(\lambda\mu-\nu)\,q+(\lambda\nu+\mu)\,r\},\\
\mathrm{M}&=\tfrac{1}{2}\{(\mu\lambda+\nu)\,p+(1+\mu^2)\,q+(\mu\nu-\lambda)\,r\},\\
\mathrm{N}&=\tfrac{1}{2}\{(\nu\lambda-\mu)\,p+(\mu\nu+\lambda)\,q+(1+\nu^2)\,r\},
\end{aligned}\right\}\quad\ldots\ldots\ldots\ldots(3).$$

[whence also $\lambda\Lambda+\mu\mathrm{M}+\nu\mathrm{N}=\tfrac{1}{2}\kappa\,(\lambda p+\mu q+\nu r)$ .... (3 *bis*)].

In the case where the forces vanish, the first three equations become simply

$$\left.\begin{aligned}
P&=\frac{1}{A}\,(B-C)\,qr,\\
Q&=\frac{1}{B}\,(C-A)\,rp,\\
R&=\frac{1}{C}\,(A-B)\,pq,
\end{aligned}\right\}\quad\ldots\ldots\ldots\ldots(4),$$

and here the usual four integrals of the system are

$$Ap^2+Bq^2+Cr^2=h \quad\ldots\ldots\ldots\ldots(5),$$

$$\left.\begin{aligned}
Ap\,(1+\lambda^2-\mu^2-\nu^2)+2Bq\,(\lambda\mu-\nu)+2Cr\,(\nu\lambda+\mu)&=\mathrm{a}\,(1+\lambda^2+\mu^2+\nu^2),\\
2Ap\,(\lambda\mu+\nu)+Bq\,(1+\mu^2-\nu^2-\lambda^2)+2Cr\,(\mu\nu-\lambda)&=\mathrm{b}\,(1+\lambda^2+\mu^2+\nu^2),\\
2Ap\,(\nu\lambda-\mu)+2Bq\,(\mu\nu+\lambda)+Cr\,(1+\nu^2-\lambda^2-\mu^2)&=\mathrm{c}\,(1+\lambda^2+\mu^2+\nu^2),
\end{aligned}\right\}\quad\ldots\ldots(6),$$

or as they may also be written,

$$\left.\begin{aligned}
\mathrm{a}\,(1+\lambda^2-\mu^2-\nu^2)+2\mathrm{b}\,(\lambda\mu+\nu)+2\mathrm{c}\,(\nu\lambda-\mu)&=Ap\,(1+\lambda^2+\mu^2+\nu^2),\\
2\mathrm{a}\,(\lambda\mu-\nu)+\mathrm{b}\,(1+\mu^2-\nu^2-\lambda^2)+2\mathrm{c}\,(\mu\nu+\lambda)&=Bq\,(1+\lambda^2+\mu^2+\nu^2),\\
2\mathrm{a}\,(\nu\lambda+\mu)+2\mathrm{b}\,(\mu\nu-\lambda)+\mathrm{c}\,(1+\nu^2-\lambda^2-\mu^2)&=Cr\,(1+\lambda^2+\mu^2+\nu^2),
\end{aligned}\right\}\quad\ldots\ldots(6\ bis);$$

to which we may add,

$$A^2p^2+B^2q^2+C^2r^2=k^2 \quad\ldots\ldots\ldots\ldots(7);$$

where

$$k^2=\mathrm{a}^2+\mathrm{b}^2+\mathrm{c}^2 \quad\ldots\ldots\ldots\ldots(8).$$

Introducing the quantities κ, Ω, (the former of which has been already made use of) given by the equations

$$\left.\begin{aligned}
\kappa&=1+\lambda^2+\mu^2+\nu^2,\\
\Omega&=\lambda Ap+\mu Bq+\nu Cr,
\end{aligned}\right\}\quad\ldots\ldots\ldots\ldots(9).$$

The equations (6) may be written under the form

$$\left.\begin{aligned} 2\lambda\Omega \ + 2\mu Cr \ - 2\nu Bq \ &= \kappa\,(Ap + \mathrm{a}) - 2Ap, \\ -\,2\lambda Cr + 2\mu\Omega \ + 2\nu Ap &= \kappa\,(Bq \ + \mathrm{b}) - 2Bq\,, \\ 2\lambda Bq - 2\mu Ap + 2\nu\Omega \ &= \kappa\,(Cr \ + \mathrm{c}) - 2Cr\,, \end{aligned}\right\} \quad \ldots\ldots\ldots\ldots(10),$$

whence also, multiplying by Ap, Bq, Cr, and adding,

$$2\Omega^2 = \kappa\,\{k^2 + (Ap\mathrm{a} + Bq\mathrm{b} + Cr\mathrm{c})\} - 2k^2 \ldots\ldots\ldots\ldots(11),$$

or writing

$$k^2 + (Ap\mathrm{a} + Bq\mathrm{b} + Cr\mathrm{c}) = 2\upsilon \ \ldots\ldots\ldots\ldots(12),$$

this becomes

$$\Omega^2 = \kappa\upsilon - k^2 \ldots\ldots\ldots\ldots(13);$$

an equation, the geometrical interpretation of which has already been given.

From the equations (10) we deduce the inverse system

$$\left.\begin{aligned} \mathrm{a}\Omega \ - \mathrm{b}Cr \ + \mathrm{c}Bq \ &= 2\lambda\upsilon - \Omega Ap, \\ \mathrm{a}Cr + \mathrm{b}\Omega \ - \mathrm{c}Ap &= 2\mu\upsilon - \Omega Bq\,, \\ -\,\mathrm{a}Bq + \mathrm{b}Ap + \mathrm{c}\Omega \ &= 2\nu\upsilon \ - \Omega Cr\,, \end{aligned}\right\} \ldots\ldots\ldots\ldots(14),$$

which are easily verified by multiplying by Ω, Cr, $-Bq$; or by $-Cr$, Ω, Ap; or Bq, $-Ap$, Ω: adding and reducing, by which means the equations (10) are re-obtained. Hence also if for shortness

$$\left.\begin{aligned} \Phi &= \mathrm{a}p + \mathrm{b}q + \mathrm{c}r, \\ \nabla &= \mathrm{a}qr\,(B - C) + \mathrm{b}rp\,(C - A) + \mathrm{c}pq\,(A - B), \end{aligned}\right\} \ldots\ldots\ldots\ldots(15),$$

we have, multiplying by p, q, r, and adding,

$$\Omega\Phi - \nabla = 2\upsilon\,(\lambda p + \mu q + \nu r) - \Omega h \ldots\ldots\ldots\ldots(16).$$

To these may be added the equation

$$\Omega = \mathrm{a}\lambda + \mathrm{b}\mu + \mathrm{c}\nu \ \ldots\ldots\ldots\ldots(17),$$

which follows immediately from either of the systems (10) or (14).

We may also put the equations (10) under this other form,

$$\left.\begin{aligned} 2\lambda\Omega - 2\mu\mathrm{c} \ + 2\nu\mathrm{b} \ &= \kappa\,(Ap + \mathrm{a}) - 2\mathrm{a}, \\ 2\lambda\mathrm{c} \ + 2\mu\Omega - 2\nu\mathrm{a} \ &= \kappa\,(Bq \ + \mathrm{b}) - 2\mathrm{b}, \\ -\,2\lambda\mathrm{b} \ + 2\mu\mathrm{a} \ + 2\nu\Omega &= \kappa\,(Cr \ + \mathrm{c}) - 2\mathrm{c}, \end{aligned}\right\} \quad \ldots\ldots\ldots\ldots(10\ bis).$$

It may be remarked now, that p, q, r are functions of υ; since we have to determine these quantities, the three equations

$$\left.\begin{aligned} Ap^2 + Bq^2 + Cr^2 &= h, \\ A^2p^2 + B^2q^2 + C^2r^2 &= k^2, \\ Ap\mathrm{a} + Bq\mathrm{b} + Cr\mathrm{c} &= 2\upsilon - k^2, \end{aligned}\right\} \ldots\ldots\ldots\ldots(18).$$

Also λ, μ, ν are given by the equations (14) as functions of p, q, r, Ω, i.e. of v, Ω; so that every thing is prepared for the investigation of the differential equation between v, Ω. To find this we have immediately

$$dv = \tfrac{1}{2}(A\mathrm{a}dp + B\mathrm{b}dq + C\mathrm{c}dr) = \tfrac{1}{2}\nabla dt \quad\text{.......................(19)},$$

from the equations (4) and (15). ∇ is of course to be considered as a given function of v. Again,

$$\Omega d\Omega = \tfrac{1}{2}(\kappa dv + v d\kappa) \quad\text{.................................(20)},$$

where

$$d\kappa = 2(\lambda d\lambda + \mu d\mu + \nu d\nu) \quad\text{............................. (21)};$$

or from the equations (1), (3), [and (3 *bis*)],

$$d\kappa = \kappa(\lambda p + \mu q + \nu r)\,dt \quad\text{...............................(22)}.$$

Hence, from (16),

$$2v d\kappa = \kappa\{\Omega(h + \Phi) - \nabla\}\,dt \quad\text{.............................(23)};$$

or

$$2(v d\kappa + \kappa dv) = \kappa\Omega(h + \Phi)\,dt \quad\text{........................... (24)},$$

whence

$$d\Omega = \tfrac{1}{4}\kappa(h + \Phi)\,dt,$$

$$= \tfrac{1}{4}\frac{\Omega^2 + k^2}{v}(h + \Phi)\,dt \quad\text{..........................(25)},$$

and therefore, from (19),

$$\frac{2d\Omega}{\Omega^2 + k^2} = \frac{h + \Phi}{v\nabla}dv \quad\text{.....................................(26)},$$

the required differential equation, in which Φ, ∇ are given functions of v, i.e. they are functions of p, q, r by the equations (15), and these quantities are functions of v by (18). The variables in (26) are therefore separated, and we have the integral equation

$$2\tan^{-1}\frac{\Omega}{k} = \delta + k\int\frac{(h + \Phi)\,dv}{v\nabla} \quad\text{..........................(27)},$$

where δ is the constant of integration. The equation (19) gives also

$$t - \epsilon = 2\int\frac{dv}{\nabla} \quad\text{...(28)};$$

and thus the solution of the problem is completely effected. The integrals may be taken from any particular value v_0 of v. The variable Ω may be exhibited as the integral of an *explicit* algebraical function, by recurring to the variable ϕ of the paper quoted.

Thus if

$$Ap_0^2 + Bq_0^2 + Cr_0^2 = h,$$

$$A^2p_0^2 + B^2q_0^2 + C^2r_0^2 = k^2,$$

$$Ap_0\mathrm{a} + Bq_0\mathrm{b} + Cr_0\mathrm{c} = 2v_0 - k^2;$$

then the values of p, q, r are respectively

$$\sqrt{\left\{p_0^2 - \frac{1}{A}(C-B)\phi\right\}}, \quad \sqrt{\left\{q_0^2 - \frac{1}{A}(A-C)\phi\right\}}, \quad \sqrt{\left\{r_0^2 - \frac{1}{C}(B-A)\phi\right\}},$$

where

$$dt = \tfrac{1}{2}\frac{d\phi}{pqr} = \frac{2dv}{\nabla}, \quad \text{or } \frac{dv}{\nabla} = \tfrac{1}{4}\frac{d\phi}{pqr};$$

and then

$$4\tan^{-1}\frac{\Omega}{k} = 2\delta + k\int_0 \frac{(h + \mathrm{a}p + \mathrm{b}q + \mathrm{c}r)\,d\phi}{(k^2 + Ap\mathrm{a} + Bq\mathrm{b} + Cr\mathrm{c})pqr},$$

in which form it is exactly analogous to the equation there obtained, p. 230, [6, p. 34]

$$4\tan^{-1}\nu_0 = \int \frac{(h + kr)\,d\phi}{(k + Cr)\,pqr}.$$

On the Variation of the Constants, when the body is acted upon by Forces.

The dynamical equations of a problem being expressed in the form

$$\frac{d}{dt}\frac{dT}{d\lambda'} - \frac{dT}{d\lambda} = \frac{dV}{d\lambda},$$

$$\frac{d}{dt}\frac{dT}{d\mu'} - \frac{dT}{d\mu} = \frac{dV}{d\mu},$$

$$\frac{d}{dt}\frac{dT}{d\nu'} - \frac{dT}{d\nu} = \frac{dV}{d\nu},$$

suppose the equations obtained from these by neglecting the function V, are integrated; each of the six integrals may be expressed in the form

$$a = f(\lambda, \mu, \nu, \lambda', \mu', \nu', t),$$

where a denotes any one of the arbitrary constants. Assume

$$\frac{dT}{d\lambda'} = u, \quad \frac{dT}{d\mu'} = v, \quad \frac{dT}{d\nu} = w;$$

then λ', μ', ν' may be expressed in terms of λ, μ, ν, u, v, w, and the integrals may be reduced to the form

$$a = F(\lambda, \mu, \nu, u, v, w, t).$$

These equations may be considered as the integrals of the proposed system, taking into account the terms involving V, provided the constants [say a, b, c, d, e, f] be supposed to become variable. We have, in this case, by Lagrange's theory of the variation of the arbitrary constants, the formulæ

$$\frac{da}{dt}=(a,\, b)\frac{dV}{db}+(a,\, c)\frac{dV}{dc}+(a,\, d)\frac{dV}{dd}+(a,\, e)\frac{dV}{de}+(a, f)\frac{dV}{df};$$

where

$$(a,\, b)=\left(\frac{da}{du}\frac{db}{d\lambda}-\frac{da}{d\lambda}\frac{db}{du}\right)+\left(\frac{da}{dv}\frac{db}{d\mu}-\frac{da}{d\mu}\frac{db}{dv}\right)+\left(\frac{da}{dw}\frac{db}{d\nu}-\frac{da}{d\nu}\frac{db}{dw}\right),$$

and in which V is supposed to be expressed as a function of a, b, c, d, e, f, t.

Thus the solution of the problem requires the calculation of thirty coefficients (a, b), or rather of fifteen only, since evidently $(a, b) = -(b, a)$. It is known that these coefficients are functions of a, b, c, d, e, f, without t; so that, in calculating them, any assumed arbitrary value, e.g. $t = 0$, may be given to the time.

In practice, it often happens that one of the arbitrary constants, e.g. a, may be expressed in the form

$$a = F(\lambda,\ \mu,\ \nu,\ u,\ v,\ w,\ t,\ b,\ c,\ d,\ e,\ f),$$

where b, c, d, e, f are given functions of λ, μ, ν, u, v, w, t. In this case, it is easily seen that we may write

$$(a,\, b)=\{(a,\, b)\}+(c,\, b)\frac{da}{dc}+(d,\, b)\frac{da}{dd}+(e,\, b)\frac{da}{de}+(f,\, b)\frac{da}{df},$$

where, in the calculation of $\{(a, b)\}$, the differentiations upon a are performed without taking into account the variability of b, c......

In the particular problem in question, the following are the values of the new variables u, v, w (*Math. Journal*, memoir already quoted, [6]),

$$u=\frac{2}{\kappa}(\quad Ap-\nu Bq+\mu Cr), \quad\text{.......................... (29)},$$

$$v=\frac{2}{\kappa}(\ \nu Ap+\ Bq-\lambda Cr),$$

$$w=\frac{2}{\kappa}(-\mu Ap+\lambda Bq+\ Cr);$$

equations which may also be expressed in the form

$$2Ap=(1+\lambda^2)\,u+(\lambda\mu+\nu)\,v+(\nu\lambda-\mu)\,w \quad\text{..................(30)},$$

$$2Bq=(\lambda\mu-\nu)\,u+(1+\mu^2)\,v+(\mu\nu+\lambda)\,w,$$

$$2Cr=(\nu\lambda+\mu)\,u+(\mu\nu-\lambda)\,v+(1+\nu^2)\,w,$$

or putting for shortness

$$\lambda u+\mu v+\nu w=\varpi \quad\text{....................................(31)},$$

these become

$$2Ap = \lambda\varpi + \quad u + \nu v - \mu w, \qquad\qquad (32).$$
$$2Bq = \mu\varpi - \nu u + \quad v + \lambda w,$$
$$2Cr = \nu\varpi + \mu u - \lambda v + \quad w,$$

whence also

$$2\Omega = \kappa\varpi \qquad\qquad (33).$$

Substituting the values of Ap, Bq, Cr, given by (30) in the equations (6), we deduce

$$2\mathrm{a} = \lambda\varpi + \quad u - \nu v + \mu w \qquad\qquad (34),$$
$$2\mathrm{b} = \mu\varpi + \nu u + \quad v - \lambda w,$$
$$2\mathrm{c} = \nu\varpi - \mu u + \lambda v + \quad w,$$

whence also

$$2\,(\mathrm{a}\lambda + \mathrm{b}\mu + \mathrm{c}\nu) = \kappa\varpi \qquad\qquad (35),$$

which in fact follows from (33) and (17). And likewise the inverse system,

$$u = \frac{2}{\kappa}(\quad \mathrm{a} + \nu\mathrm{b} - \mu\mathrm{c}) \qquad\qquad (36).$$
$$v = \frac{2}{\kappa}(-\nu\mathrm{a} + \quad \mathrm{b} + \lambda\mathrm{c}),$$
$$w = \frac{2}{\kappa}(\quad \mu\mathrm{a} - \lambda\mathrm{b} + \quad \mathrm{c}).$$

It is easy to deduce

$$k^2 = \tfrac{1}{4}\,\kappa\,[u^2 + v^2 + w^2 + \varpi^2] \qquad\qquad (37),$$
$$\upsilon = \tfrac{1}{4}\,[(u^2 + v^2 + w^2) + (1 + \kappa)\,\varpi^2] \qquad\qquad (38).$$

Again, from the equations (10 *bis*),

$$\begin{aligned}\kappa\,(\mathrm{b}Cr - \mathrm{c}Bq) &= -\,2\lambda\,(\mathrm{a}^2 + \mathrm{b}^2 + \mathrm{c}^2) + 2\mathrm{a}\,(\lambda\mathrm{a} + \mu\mathrm{b} + \nu\mathrm{c}) + 2\,(\mathrm{b}\nu - \mathrm{c}\mu)\,\Omega \\ &= -\,2\lambda k^2 + 2\,(\mathrm{a} + \mathrm{b}\nu - \mathrm{c}\mu)\,\Omega \\ &= -\,2\lambda k^2 + \quad \kappa u\Omega\,;\end{aligned}$$

and, forming also the similar expressions for $\kappa\,(\mathrm{c}Ap - \mathrm{a}Cr)$, and $\kappa\,(\mathrm{a}Bq - \mathrm{b}Ap)$, we thus obtain

$$\Omega u - \frac{2}{\kappa}\,k^2\lambda = \mathrm{b}Cr - \mathrm{c}Bq \qquad\qquad (39),$$
$$\Omega v - \frac{2}{\kappa}\,k^2\mu = \mathrm{c}Ap - \mathrm{b}Cr,$$
$$\Omega w - \frac{2}{\kappa}\,k^2\nu = \mathrm{a}Bq - \mathrm{c}Ap\,;$$

to which many others might probably be joined.

The constants of the problem are a, b, c, h, ϵ, δ. Of these a, b, c are given as functions of λ, μ, ν, u, v, w, by the equations (34); in which ϖ is to be considered as standing for $\lambda u+\mu v+\nu w$. {These determine k^2, which is however given immediately by (37).} As for h, we have

$$h=\frac{1}{A}(Ap)^2+\frac{1}{B}(Bq)^2+\frac{1}{C}(Cr)^2,$$

where Ap, Bq, Cr are given as functions of λ, μ, ν, u, v, w by (32), in which also ϖ stands for $\lambda u+\mu v+\nu w$. Again,

$$\epsilon=t-\tfrac{1}{2}\int\frac{dv}{\nabla},$$

$$\delta=2\tan^{-1}\frac{\kappa\varpi}{2k}-\tfrac{1}{4}k\int\frac{(h+\Phi)\,dv}{v\nabla};$$

in each of which ∇, Φ are functions of v, and of a, b, c, h, partly as entering explicitly into these functions, partly as contained implicitly in p, q, r, which enter into ∇, Φ, and are functions of v, h, k given by (18). After the integration v is to be considered a function of λ, μ, ν, u, v, w given by (38). Both of the integrals may be supposed taken from a certain value v_0 of v, which may be considered as an absolutely invariable arbitrary constant, since without it we have the right number, six, of arbitrary constants.

First to find (a, b), (b, c), and (c, a). From (34) we have

$$\begin{aligned}(\text{a},\text{b})=\tfrac{1}{4}\{\ &(1+\lambda^2)(\mu u-w)-(\lambda u+\varpi)(\lambda\mu+\nu)\\ &+(\lambda\mu-\nu)(\mu v+\varpi)-(\lambda v+w)(1+\mu^2)\\ &+(\nu\lambda+\mu)(u+\mu w)-(\lambda w-v)(\mu\nu-\lambda)\}\\ =\tfrac{1}{2}&(\mu u-\lambda v-w-\nu\varpi)=-\tfrac{1}{2}\,2\text{c}=-\text{c};\end{aligned}$$

whence the system

$$(\text{b},\text{c})=-\text{a},\quad (\text{c},\text{a})=-\text{b},\quad (\text{a},\text{b})=-\text{c}\ \ldots\ldots\ldots\ldots\ldots\ldots(40).$$

Also we may add $\quad (k,\text{a})=\frac{\text{a}}{k}(\text{a},\text{a})+\frac{\text{b}}{k}(\text{b},\text{a})+\frac{\text{c}}{k}(\text{c},\text{a})=0,$

or

$$(k,\text{a})=0,\quad (k,\text{b})=0,\quad (k,\text{c})=0\ \ldots\ldots\ldots\ldots\ldots\ldots(41),$$

which will be useful in calculating some of the following coefficients.

Proceeding to calculate (a, h), (b, h), (c, h). It is seen immediately that

$$(\text{a},h)=2\{p(\text{a},Ap)+q(\text{a},Bq)+r(\text{a},Cr)\},$$

where Ap, Bq, Cr, are given by the equations (32), so that

$$\begin{aligned}(\text{a},Ap)=\tfrac{1}{4}\{\ &(1+\lambda^2)(\lambda u+\varpi)-(1+\lambda^2)(\lambda u+\varpi)\\ &+(\lambda\mu-\nu)(\lambda v-w)-(\lambda\mu+\nu)(\lambda v+w)\\ &+(\nu\lambda+\mu)(v+\lambda w)-(\nu\lambda-\mu)(-v+\lambda w)\}\end{aligned}$$

i.e. $\quad (\text{a},Ap)=0.\ \ldots\ldots\ldots\ldots\ldots\ldots\ldots\ldots\ldots\ldots(42).$

Similarly $(\text{a}, Bq) = \tfrac{1}{4}\{ \ (1 + \lambda^2)(\mu u + w) - (\lambda u + \varpi)(\lambda\mu - \nu)$
$+ (\lambda\mu - \nu)(\mu v + \varpi) - (\lambda v + w)(1 + \mu^2)$
$+ (\nu\lambda + \mu)(\mu w - u) - (-v + \lambda w)(\mu\nu + \lambda)\}$

i.e. $(\text{a}, Bq) = 0,$ ....(43),

and similarly $(\text{a}, Cr) = 0;$

whence $(\text{a}, h) = 0$, and therefore $(\text{b}, h) = 0$, $(\text{c}, h) = 0$ .... (44);

also $(k, h) = 0,$ .... (45).

Next we have to determine (a, ϵ), (b, ϵ), (c, ϵ). Here ϵ being a function of $u, v, w, \lambda, \mu, \nu$, a, b, c, h, we must write

$$(\text{a}, \epsilon) = \{(\text{a}, \epsilon)\} + (\text{a}, \text{b})\frac{d\epsilon}{d\text{b}} + (\text{a}, \text{c})\frac{d\epsilon}{d\text{c}} + (\text{a}, h)\frac{d\epsilon}{dh},$$

i.e.

$$(\text{a}, \epsilon) = \{(\text{a}, \epsilon)\} + \text{b}\frac{d\epsilon}{d\text{c}} - \text{c}\frac{d\epsilon}{d\text{b}}.$$

But $\epsilon = t - 2\int\frac{dv}{\nabla}$; and thence $\{(\text{a}, \epsilon)\} = -\frac{2}{\nabla}(\text{a}, v)$,

and v is given immediately as a function of $\lambda, \mu, \nu, u, v, w$, by the equation (38). Hence

$$\begin{aligned}(\text{a}, v) = \tfrac{1}{4}[\ & (1 + \lambda^2)\{(1+\kappa)u\varpi + \lambda\varpi^2\} - (\lambda u + \varpi)\{u + \lambda(1+\kappa)\varpi\} \\ & + (\lambda\mu - \nu)\{(1+\kappa)v\varpi + \mu\varpi^2\} - (\lambda v + w)\{v + \mu(1+\kappa)\varpi\} \\ & + (\nu\lambda + \mu)\{(1+\kappa)w\varpi + \nu\varpi^2)\} - (-v + \lambda w)\{w + \nu(1+\kappa)\varpi\}] \\ = \tfrac{1}{4}\{ & (1+\kappa)\varpi u - \lambda(1+\kappa)\varpi^2 + \lambda\kappa - \lambda(u^2+v^2+w^2) - u\varpi\} \\ = \tfrac{1}{4}\{ & \kappa u\varpi - \lambda\varpi^2 - \lambda(u^2+v^2+w^2)\} \\ = \tfrac{1}{4} & \kappa u\varpi - \frac{k^2\lambda}{\kappa} = \tfrac{1}{2}\left(\Omega u - \frac{2k^2\lambda}{\kappa}\right) \text{ \{by (37) and (33)\}}, \\ = \tfrac{1}{2} & (\text{b}Cr - \text{c}Bq) \quad \text{..........} \quad (46);\end{aligned}$$

whence

$$\{(\text{a}, \epsilon)\} = -\frac{1}{\nabla}(\text{b}Cr - \text{c}Bq),$$

$$(\text{a}, \epsilon) = -\frac{1}{\nabla}(\text{b}Cr - \text{c}Bq) + \text{b}\frac{d\epsilon}{d\text{c}} - \text{c}\frac{d\epsilon}{d\text{b}}.$$

The terms $\text{b}\frac{d\epsilon}{d\text{c}} - \text{c}\frac{d\epsilon}{d\text{b}}$ are evidently of the form $F(v) - F(v_0)$.

If therefore we suppose $v = v_0$, we have

$$(\text{a}, \epsilon) = -\frac{1}{\nabla_0}(\text{b}Cr_0 - \text{c}Bq_0) \quad \text{..........} \quad (47),$$

if p_0, q_0, r_0, ∇_0 refer to the value v_0 of v, i.e. if

$$A\,p_0^2 + B\,q_0^2 + C\,r_0^2 = h \quad\ldots\ldots(48),$$
$$A^2p_0^2 + B^2q_0^2 + C^2r_0^2 = k^2,$$
$$Ap_0\mathrm{a} + Bq_0\mathrm{b} + Cr_0\mathrm{c} = 2v_0 - k^2.$$

{This implies evidently

$$\mathrm{b}\frac{d\epsilon}{dc} - \mathrm{c}\frac{d\epsilon}{db} = \frac{1}{\nabla}(\mathrm{b}Cr - \mathrm{c}Bq) - \frac{1}{\nabla_0}(\mathrm{b}Cr_0 - \mathrm{c}Bq_0),$$

an equation which it is interesting to verify. In fact, from the value of ϵ

$$\mathrm{b}\frac{d\epsilon}{dc} - \mathrm{c}\frac{d\epsilon}{d\mathrm{b}} = -2\int dv\left(\mathrm{b}\frac{d}{dc} - \mathrm{c}\frac{d}{d\mathrm{b}}\right)\frac{1}{\nabla} = 2\int dv\frac{1}{\nabla^2}\left(\mathrm{b}\frac{d\nabla}{dc} - \mathrm{c}\frac{d\nabla}{d\mathrm{b}}\right);$$

or we have to show that

$$\frac{d}{dv}\frac{1}{\nabla}(\mathrm{b}Cr - \mathrm{c}Bq) = \frac{2}{\nabla^2}\left(\mathrm{b}\frac{d\nabla}{dc} - \mathrm{c}\frac{d\nabla}{d\mathrm{b}}\right) = \frac{2}{\nabla^2}\delta\nabla;$$

if for shortness

$$\delta = \mathrm{b}\frac{d}{dc} - \mathrm{c}\frac{d}{d\mathrm{b}}.$$

Now ∇ containing a, b, c explicitly, and also as involved in p, q, r, we have

$$\delta\nabla = \mathrm{b}pq(A-B) - \mathrm{c}rp(C-A) + \frac{d\nabla}{dp}\delta p + \frac{d\nabla}{dq}\delta q + \frac{d\nabla}{dr}\delta r = \mathrm{b}pq(A-B) - \mathrm{c}rp(C-A) + \delta'\nabla$$

suppose. The equation to be verified becomes

$$\nabla\left(\mathrm{b}C\frac{dr}{dv} - \mathrm{c}B\frac{dq}{dv}\right) - (\mathrm{b}Cr - \mathrm{c}Bq)\frac{d\nabla}{dv} = 2\{\mathrm{b}pq(A-B) - \mathrm{c}rp(C-A) + \delta'\nabla\}.$$

Now, observing that $\delta k = 0$, we have

$$A\,p\delta p + B\,q\delta q + C\,r\delta r = 0,$$
$$A^2p\delta p + B^2q\delta q + C^2r\delta r = 0,$$
$$A\,\mathrm{a}\delta p + B\,\mathrm{b}\delta q + C\,\mathrm{c}\delta r = -(\mathrm{b}Cr - \mathrm{c}Bq).$$

Also,

$$A\,p\frac{dp}{dv} + B\,q\frac{dq}{dv} + C\,r\frac{dr}{dv} = 0,$$

$$A^2p\frac{dp}{dv} + B^2q\frac{dq}{dv} + C^2r\frac{dr}{dv} = 0,$$

$$A\,\mathrm{a}\frac{dp}{dv} + B\,\mathrm{b}\frac{dq}{dv} + C\,\mathrm{c}\frac{dr}{dv} = 2,$$

whence evidently

$$\frac{dp}{dv}=\frac{-2}{\mathrm{b}Cr-\mathrm{c}Bq}\,\delta p,\quad \frac{dq}{dv}=\frac{-2}{\mathrm{b}Cr-\mathrm{c}Bq}\,\delta q,\quad \frac{dr}{dv}=\frac{-2}{\mathrm{b}Cr-\mathrm{c}Bq}\,\delta r,$$

or

$$\frac{d\nabla}{dv}=\frac{-2}{\mathrm{b}Cr-\mathrm{c}Bq}\,\delta'\nabla\,;$$

or the equation to be verified is simply

$$\nabla\left(\mathrm{b}C\frac{dr}{dv}-\mathrm{c}B\frac{dq}{dv}\right)=2\,\{\mathrm{b}pq\,(A-B)-\mathrm{c}rp\,(C-A)\}\,;$$

which follows immediately from the three equations just given for the determination of $\frac{dp}{dv},\ \frac{dq}{dv},\ \frac{dr}{dv}$ }.

From these values also

$$(k,\ \epsilon)=0 \quad\text{.......................................}\quad (49).$$

Next, to calculate $(h,\ \epsilon)$,

$$(h,\ \epsilon)=\{(h,\ \epsilon)\}+(h,\ \mathrm{a})\frac{d\epsilon}{d\mathrm{a}}+(h,\ \mathrm{b})\frac{d\epsilon}{d\mathrm{b}}+(h,\ \mathrm{c})\frac{d\epsilon}{d\mathrm{c}}\,;$$

but the three last terms being evidently such as to vanish for $v=v_0$, we may neglect them, and consider $(h,\ \epsilon)$ as the value which $\{(h,\ \epsilon)\}$ assumes for this value of v.

Now $$\{(h,\ \epsilon)\}=2p\,\{(Ap,\ \epsilon)\}+2q\,\{(Bq,\ \epsilon)\}+2r\,\{(Cr,\ \epsilon)\},$$

where $$\{(Ap,\ \epsilon)\}=-\frac{2}{\nabla}\,(Ap,\ v),$$

and

$$\begin{aligned}(Ap,\ v)=\tfrac{1}{4}\,[\ &(1+\lambda^2)\,\{(1+\kappa)\,u\varpi+\lambda\varpi^2\}-(\lambda u+\varpi)\,\{u+\lambda\,(1+\kappa)\,\varpi\}\\ &+(\lambda\mu+\nu)\,\{(1+\kappa)\,v\varpi+\mu\varpi^2\}-(\lambda v-w)\,\{v+\mu\,(1+\kappa)\,\varpi\}\\ &+(\nu\lambda-\mu)\,\{(1+\kappa)\,w\varpi+\nu\varpi^2\}-(v+\lambda w)\,\{w+\nu\,(1+\kappa)\,\varpi\}]\end{aligned}$$

$$=\tfrac{1}{4}\,\{(1+\kappa)\,\varpi u+\lambda\kappa\varpi^2-\lambda\,(1+\kappa)\,\varpi^2-\lambda\,(u^2+v^2+w^2)-\varpi u\}=(\mathrm{a},\ v)$$

$$=\tfrac{1}{2}\,(\mathrm{b}Cr-\mathrm{c}Bq),\quad\text{..}(50),$$

whence $$\{(Ap,\ \epsilon)\}=-\frac{1}{\nabla}\,(\mathrm{b}Cr-\mathrm{c}Bq)\ \text{..........\ ....}(51),$$

and therefore $$\{(Bq,\ \epsilon)\}=-\frac{1}{\nabla}\,(\mathrm{c}Ap-\mathrm{a}Cr),$$

$$\{(Cr,\ \epsilon)\}=-\frac{1}{\nabla}\,(\mathrm{a}Bq-\mathrm{b}Ap),$$

whence
$$\{(h,\ \epsilon)\} = -2,$$
and therefore
$$(h,\ \epsilon) = -2 \quad \text{.......................................} (52).$$

Next, to find (a, δ), (b, δ), (c, δ), (h, δ),

$$\delta = 2\tan^{-1}\frac{\kappa\varpi}{2k} - k\int\frac{(h+\Phi)\,dv}{v\nabla},$$
$$= \delta' + \delta'' \text{ suppose},$$
$$(\mathrm{a},\ \delta) = (\mathrm{a},\ \delta') + (\mathrm{a},\ \delta''),$$
$$(\mathrm{a},\ \delta') = \frac{k}{\kappa v}(\mathrm{a},\ \kappa\varpi) + (\mathrm{a},\ k)\frac{d\delta'}{dk},$$
{observing
$$\kappa^2\varpi^2 + 4k^2 = 4(\Omega^2 + k^2) = 4\kappa v\}$$
$$= \frac{k}{\kappa v}(\mathrm{a},\ \kappa\varpi),$$
where
$$\begin{aligned}(\mathrm{a},\ \kappa\varpi) = \tfrac{1}{2}\{ &\ (1+\lambda^2)(\kappa v + 2\lambda\varpi) - (\lambda u + \varpi)\kappa\lambda \\ &+ (\lambda\mu - \nu)(\kappa v + 2\mu\varpi) - (\lambda v + w)\kappa\mu \\ &+ (\nu\lambda + \mu)(\kappa w + 2\nu\varpi) - (-v + \lambda w)\kappa\nu\}\end{aligned}$$
$$= \tfrac{1}{2}\kappa(u + \lambda\varpi) = Ap - \nu Bq + \mu Cr + \lambda\Omega = \tfrac{1}{2}(\mathrm{a} + Ap)\kappa \quad \text{...........} (53),$$

by equations (29), (33), and (10);

that is
$$(\mathrm{a},\ \delta') = \frac{k}{2v}(\mathrm{a} + Ap).$$

Also
$$(\mathrm{a},\ \delta'') = -k\frac{h+\Phi}{v\nabla}(\mathrm{a},\ v) + (\mathrm{a},\ \mathrm{b})\frac{d\delta''}{d\mathrm{b}} + \&\mathrm{c}.$$
$$= -\tfrac{1}{2}k\frac{h+\Phi}{v\nabla}(\mathrm{b}Cr - \mathrm{c}Bq) + Fv - Fv_0,$$
whence
$$(\mathrm{a},\ \delta) = \frac{k}{2v}\left\{\mathrm{a} + Ap - \frac{h+\Phi}{\nabla}(\mathrm{b}Cr - \mathrm{c}Bq)\right\} + Fv - Fv_0,$$
or putting $v = v_0$,
$$(\mathrm{a},\ \delta) = \frac{k}{2v_0}\left\{\mathrm{a} + Ap_0 - \frac{h+\Phi_0}{\nabla_0}(\mathrm{b}Cr_0 - \mathrm{c}Bq_0)\right\} \quad \text{..............} (54),$$
and therefore also
$$(\mathrm{b},\ \delta) = \frac{k}{2v_0}\left\{\mathrm{b} + Bq_0 - \frac{h+\Phi_0}{\nabla_0}(\mathrm{c}Ap_0 - \mathrm{a}Cr_0)\right\},$$
$$(\mathrm{c},\ \delta) = \frac{k}{2v_0}\left\{\mathrm{c} + Cr_0 - \frac{h+\Phi_0}{\nabla_0}(\mathrm{a}Bq_0 - \mathrm{b}Ap_0)\right\}.$$

Again,
$$(h,\ \delta) = 2p\,(Ap,\ \delta) + 2q\,(Bq,\ \delta) + 2r\,(Cr,\ \delta),$$
$$(Ap,\ \delta) = (Ap,\ \delta') + (Ap,\ \delta''),$$
$$(Ap,\ \delta') = \frac{k}{\kappa v}(Ap,\ \kappa\varpi) + (Ap,\ k)\frac{d\delta'}{dk}$$
$$= \frac{k}{\kappa v}(Ap,\ \kappa\varpi),$$

$$\begin{aligned}(Ap,\ \kappa\varpi) = \tfrac{1}{2}\{\ & (1+\lambda^2)(\kappa u + 2\lambda\varpi) - (\lambda u + \varpi)\,\kappa\lambda \\ & + (\lambda\mu + \nu)(\kappa v + 2\mu\varpi) - (\lambda v - w)\,\kappa\mu \\ & + (\nu\lambda + \mu)(\kappa w + 2\nu\varpi) - (v + \lambda w)\,\kappa\nu\ \} \\ = \tfrac{1}{2}\,\kappa\,(u + \lambda\varpi) = \tfrac{1}{2}\,\kappa\,(\mathrm{a} + Ap) & \quad\ldots\ldots (55);\end{aligned}$$

therefore
$$(Ap,\ \delta') = \frac{k}{2u}(\mathrm{a} + Ap) \quad\ldots\ldots (56),$$

$$(Ap,\ \delta'') = -k\frac{h+\Phi}{v\nabla}(Ap,\ v) + \&c.$$
$$= -\tfrac{1}{2}k\,\frac{h+\Phi}{v\nabla}(\mathrm{b}Cr - \mathrm{c}Bq) + Fv - Fv_0,$$
$$(Ap,\ \delta) = \frac{k}{2v}\left\{\mathrm{a} - Ap - \frac{h+\Phi}{\nabla}(\mathrm{b}Cr - \mathrm{c}Bq)\right\} + Fv - Fv_0,$$

and similarly for $(Bq,\ \delta)$, $(Cr,\ \delta)$. Substituting, and neglecting the terms which vanish for $v = v_0$,

$$(h,\ \delta) = \frac{k}{v}\left(\Phi + h - \frac{\Phi + h}{\nabla}\,\nabla\right),$$

i.e.
$$(h,\ \delta) = 0 \quad\ldots\ldots (57).$$

Lastly, to find $(\epsilon,\ \delta)$,

$$(\epsilon,\ \delta) = \{(\epsilon,\ \delta)\} + (\mathrm{a},\ \delta)\frac{d\epsilon}{d\mathrm{a}} + (\mathrm{b},\ \delta)\frac{d\epsilon}{d\mathrm{b}} + (\mathrm{c},\ \delta)\frac{d\epsilon}{d\mathrm{c}},$$

where, in $\{(\epsilon,\ \delta)\}$, the differentiations upon ϵ are supposed not to affect the constants a, b, c. Neglecting the terms which vanish for $v = v_0$,

$$(\epsilon,\ \delta) = \{(\epsilon,\ \delta)\},$$
$$\{(\epsilon,\ \delta)\} = \{(\epsilon,\ \delta')\} + \{(\epsilon,\ \delta'')\},$$
$$\{(\epsilon,\ \delta')\} = [\{(\epsilon,\ \delta')\}] + (\epsilon,\ k)\frac{d\delta'}{dk} = [\{(\epsilon,\ \delta')\}];$$

where, in $[\{(\epsilon,\ \delta')\}]$, the differentiations upon ϵ and δ do not affect the constants.

$$\{(\epsilon,\ \delta'')\} = [\{(\epsilon,\ \delta'')\}] + (\epsilon,\ \mathrm{a})\frac{d\delta''}{d\mathrm{a}} + \&c.$$

i.e.
$$\{(\epsilon,\ \delta'')\} = [\{(\epsilon,\ \delta'')\}]:$$

neglecting the terms which vanish for $\upsilon=\upsilon_0$,

therefore
$$(\epsilon,\ \delta)=[\{(\epsilon,\ \delta')\}]+[\{(\epsilon,\ \delta'')\}]$$
$$=[\{(\epsilon,\ \delta')\}];$$

since
$$[\{(\epsilon,\ \delta'')\}]=(\upsilon,\ \upsilon)\frac{d\epsilon}{d\upsilon}\frac{d\delta''}{d\upsilon}=0.$$

Hence
$$(\epsilon,\ \delta)=-\tfrac{1}{2}\frac{k}{\kappa\nabla\upsilon}(\upsilon,\ \kappa\varpi)\ \ldots\ldots\ldots\ldots(58),$$

$$\begin{aligned}(\upsilon,\ \kappa\varpi)&=\tfrac{1}{2}\left(\begin{array}{l}\{u+(1+\kappa)\lambda\varpi\}(2\lambda\varpi+\kappa u)-\{\lambda\varpi^2+(1+\kappa)\varpi u\}\kappa\lambda\\ +\{v+(1+\kappa)\mu\varpi\}(2\mu\varpi+\kappa v)-\{\mu\varpi^2+(1+\kappa)\varpi v\}\kappa\mu\\ +\{w+(1+\kappa)\nu\varpi\}(2\nu\varpi+\kappa w)-\{\nu\varpi^2+(1+\kappa)\varpi w\}\kappa\nu\end{array}\right)\\ &=\tfrac{1}{2}\{2\varpi^2+\kappa(u^2+v^2+w^2)+2(1+\kappa)(\kappa-1)\varpi^2+\kappa(1+\kappa)\varpi^2\\ &\qquad-\kappa(\kappa-1)\varpi^2-\kappa(\kappa+1)\varpi^2\}\\ &=\tfrac{1}{2}\kappa\{(\kappa+1)\varpi^2+(u^2+v^2+w^2)\}=\tfrac{1}{2}4\kappa\upsilon=2\kappa\upsilon\ \ldots\ldots\ldots\ldots(59),\end{aligned}$$

therefore
$$(\epsilon,\ \delta)=-\frac{k}{\nabla_0}\ \ldots\ldots\ldots\ldots(60).$$

Hence, recapitulating,

$$\left.\begin{array}{l}(\mathrm{b},\ \mathrm{c})=-\mathrm{a},\quad(\mathrm{c},\ \mathrm{a})=-\mathrm{b},\quad(\mathrm{a},\ \mathrm{b})=-\mathrm{c},\\ (\mathrm{a},\ h)=0,\quad(\mathrm{b},\ h)=0,\quad(\mathrm{c},\ h)=0,\\ (\mathrm{a},\ \epsilon)=-\dfrac{1}{\nabla_0}(\mathrm{b}Cr_0-\mathrm{c}Bq_0),\\ (\mathrm{b},\ \epsilon)=-\dfrac{1}{\nabla_0}(\mathrm{c}Ap_0-\mathrm{a}Cr_0),\\ (\mathrm{c},\ \epsilon)=-\dfrac{1}{\nabla_0}(\mathrm{a}Bq_0-\mathrm{b}Ap_0),\\ (h,\ \epsilon)=-2,\\ (\mathrm{a},\ \delta)=\dfrac{k}{2\upsilon_0}\left\{\mathrm{a}+Ap_0-\dfrac{h+\Phi_0}{\nabla_0}(\mathrm{b}Cr_0-\mathrm{c}Bq_0)\right\},\\ (\mathrm{b},\ \delta)=\dfrac{k}{2\upsilon_0}\left\{\mathrm{b}+Bq_0-\dfrac{h+\Phi_0}{\nabla_0}(\mathrm{c}Ap_0-\mathrm{a}Cr_0)\right\},\\ (\mathrm{c},\ \delta)=\dfrac{k}{2\upsilon_0}\left\{\mathrm{c}+Cr_0-\dfrac{h+\Phi_0}{\nabla_0}(\mathrm{a}Bq_0-\mathrm{b}Ap_0)\right\},\\ (h,\ \delta)=0,\\ (\epsilon,\ \delta)=-\dfrac{k}{\nabla_0}\end{array}\right\}\ \ldots\ldots\ldots\ldots(61),$$

and therefore

$$\left.\begin{aligned}
\frac{d\mathrm{a}}{dt} &= -\mathrm{c}\frac{dV}{d\mathrm{b}} + \mathrm{b}\frac{dV}{d\mathrm{c}} - \frac{1}{\nabla_0}(\mathrm{b}Cr_0 - \mathrm{c}Bq_0)\frac{dV}{d\epsilon} + \frac{k}{2v_0}\left\{\mathrm{a} + Ap_0 - \frac{h+\Phi_0}{\nabla_0}(\mathrm{b}Cr_0 - \mathrm{c}Bq_0)\right\}\frac{dV}{d\delta}, \\
\frac{d\mathrm{b}}{dt} &= -\mathrm{a}\frac{dV}{d\mathrm{c}} + \mathrm{c}\frac{dV}{d\mathrm{a}} - \frac{1}{\nabla_0}(\mathrm{c}Ap_0 - \mathrm{a}Cr_0)\frac{dV}{d\epsilon} + \frac{k}{2v_0}\left\{\mathrm{b} + Bq_0 - \frac{h+\Phi_0}{\nabla_0}(\mathrm{c}Ap_0 - \mathrm{a}Cr_0)\right\}\frac{dV}{d\delta}, \\
\frac{d\mathrm{c}}{dt} &= -\mathrm{b}\frac{dV}{d\mathrm{a}} + \mathrm{a}\frac{dV}{d\mathrm{b}} - \frac{1}{\nabla_0}(\mathrm{a}Bq_0 - \mathrm{b}Ap_0)\frac{dV}{d\epsilon} + \frac{k}{2v_0}\left\{\mathrm{c} + Cr_0 - \frac{h+\Phi_0}{\nabla_0}(\mathrm{a}Bq_0 - \mathrm{b}Ap_0)\right\}\frac{dV}{d\delta}, \\
\frac{dh}{dt} &= -2\frac{dV}{d\epsilon}, \\
\frac{d\epsilon}{dt} &= \frac{1}{\nabla_0}\left\{(\mathrm{b}Cr_0 - \mathrm{c}Bq_0)\frac{dV}{d\mathrm{a}} + (\mathrm{c}Ap_0 - \mathrm{a}Bq_0)\frac{dV}{d\mathrm{b}} + (\mathrm{a}Bq_0 - \mathrm{b}Ap_0)\frac{dV}{d\mathrm{c}}\right\} + 2\frac{dV}{dh} - \frac{k}{\nabla_0}\frac{dV}{d\delta}, \\
\frac{d\delta}{dt} &= -\frac{k}{2v_0}\Bigg[\quad\left\{\mathrm{a} + Ap_0 - \frac{h+\Phi_0}{\nabla_0}(\mathrm{b}Cr_0 - \mathrm{c}Bq_0)\right\}\frac{dV}{d\mathrm{a}} \\
&\qquad + \left\{\mathrm{b} + Bq_0 - \frac{h+\Phi_0}{\nabla_0}(\mathrm{c}Ap_0 - \mathrm{a}Cr_0)\right\}\frac{dV}{d\mathrm{b}} \\
&\qquad + \left\{\mathrm{c} + Cr_0 - \frac{h+\Phi_0}{\nabla_0}(\mathrm{a}Bq_0 - \mathrm{b}Ap_0)\right\}\frac{dV}{d\mathrm{c}}\Bigg] + \frac{k}{\nabla_0}\frac{dV}{d\epsilon},
\end{aligned}\right\} \quad (62),$$

to which we may join

$$\frac{dk}{dt} = \frac{dV}{d\delta} \quad \text{.......................................(63).}$$

We have thus the complete system of formulæ.

38.

NOTE ON A GEOMETRICAL THEOREM CONTAINED IN A PAPER BY SIR W. THOMSON.

[From the *Cambridge and Dublin Mathematical Journal*, vol. I. (1846), pp. 207, 208.]

IT is easily shown that if three confocal surfaces of the second order pass through a point P, then the square of the distance of this point from the origin is equal to the sum of the squares of three of the axes, no two of which are parallel or belong to the same surface (the squares of one or two of the axes of the hyperboloids being considered negative); i.e. if

$$\frac{x^2}{a^2+h}+\frac{y^2}{b^2+h}+\frac{z^2}{c^2+h}=1,$$

$$\frac{x^2}{a^2+k}+\frac{y^2}{b^2+k}+\frac{z^2}{c^2+k}=1,$$

$$\frac{x^2}{a^2+l}+\frac{y^2}{b^2+l}+\frac{z^2}{c^2+l}=1;$$

then

$$x^2+y^2+z^2=a^2+b^2+c^2+h+k+l.$$

In fact these equations give

$$x^2=\frac{(a^2+h)(a^2+k)(a^2+l)}{(a^2-b^2)(a^2-c^2)},$$

$$y^2=\frac{(b^2+h)(b^2+k)(b^2+l)}{(b^2-a^2)(b^2-c^2)},$$

$$z^2=\frac{(c^2+h)(c^2+k)(c^2+l)}{(c^2-a^2)(c^2-b^2)},$$

and adding these and reducing, we have the relation in question; which is also immediately obtained by forming the cubic whose roots are h, k, l.

From this property may be deduced the theorem given by Mr Thomson in the preceding memoir [" On the Principal Axes of a Solid Body," pp. 127—133 and 195—206, see p. 205]. In fact, writing

$$r^2 = x^2 + y^2 + z^2,$$

and

$$k = - a^2 - b^2 - c^2 + h,$$

we see that in consequence of these relations the equations

$$\frac{x^2}{a^2} + \frac{y^2}{b^2} + \frac{z^2}{c^2} = 1,$$

$$\frac{x^2}{r^2 + a^2 - h} + \frac{y^2}{r^2 + b^2 - h} + \frac{z^2}{r^2 + c^2 - h} = 1,$$

$$\frac{x^2}{a^2 + k} + \frac{y^2}{b^2 + k} + \frac{z^2}{c^2 + k} = 1,$$

are equivalent to two independent equations, i.e. the third can be deduced from the two first. Now the first equation is that of an ellipsoid (or generally a surface of the second order, since a, b, c are not necessarily real). The second is that of what may be called a conjugate equimomental surface, defining the term as follows: "The conjugate equimomental surfaces of an ellipsoid (or surface of the second order) $\frac{x^2}{a^2} + \frac{y^2}{b^2} + \frac{z^2}{c^2} = 1$, are the equimomental surfaces derived in the usual manner from any surface of the second order $\frac{x^2}{h - a^2} + \frac{y^2}{h - b^2} + \frac{z^2}{h - c^2} = 1$, which is confocal with the *conjugate* surface $\frac{x^2}{a^2} + \frac{y^2}{b^2} + \frac{z^2}{c^2} = - 1$ of the given ellipsoid," viz. by measuring along any line through the centre distances equal to the axes of the section by a plane through the centre perpendicular to this line, and taking the locus of the points so determined for the equimomental surface. The third equation is that of a surface confocal with the given ellipsoid; hence the theorem, "The curves of curvature of a given ellipsoid lie upon a system of conjugate equimomental surfaces."

But since the first and second equations are evidently satisfied by the combination of the first equation with the relation $r^2 = h$, which is that of a sphere, we have also, "The curve of intersection of the ellipsoid with any one of the conjugate equimomental surfaces, is composed of the line of curvature and a spherical conic." And these two curves being each of them of the fourth order make up the complete curve of intersection, which should obviously be of the eighth order.

It would be an interesting question to determine the relations existing between the curve of curvature and the spherical conic, which have been thus brought into connection by means of the conjugate equimomental surfaces; i.e. between the two curves obtained by combining the equation of the ellipsoid with

$$\frac{x^2}{a^2 + k} + \frac{y^2}{b^2 + k} + \frac{z^2}{c^2 + k} = 1,$$

$$r^2 = a^2 + b^2 + c^2 + k,$$

respectively: but it will be sufficient at present to have suggested the problem.

39.

ON THE DIAMETRAL PLANES OF A SURFACE OF THE SECOND ORDER.

[From the *Cambridge and Dublin Mathematical Journal*, vol. I. (1846), pp. 274—278.]

LET $U = Ax^2 + By^2 + Cz^2 + 2Fyz + 2Gxz + 2Hxy = 0$, be the equation of a surface of the second order referred to its centre, and let $\alpha x + \alpha' y + \alpha'' z = 0$ be the equation of one of its diametral planes; then, as usual

$$\begin{aligned}(A-u)\,\alpha + \quad H\alpha' + \quad G\alpha'' &= 0,\\ H\alpha + (B-u)\,\alpha' + \quad F\alpha'' &= 0,\\ G\alpha + \quad F\alpha' + (C-u)\,\alpha'' &= 0,\end{aligned}$$

which are equivalent to two independent equations, and consequently capable of determining the ratios $\alpha : \alpha' : \alpha''$, provided that u satisfy the cubic equation that is obtained by eliminating $\alpha, \alpha', \alpha''$ from the three equations.

We have from the second and third, from the third and first, and from the first and second equations respectively,

$$\alpha : \alpha' : \alpha'' = \mathfrak{A}_, : \mathfrak{H}_, : \mathfrak{G}_, = \mathfrak{H}_, : \mathfrak{B}_, : \mathfrak{F}_, = \mathfrak{G}_, : \mathfrak{F}_, : \mathfrak{C}_, ;$$

where, if

$$\begin{aligned}\mathfrak{A} &= BC - F^2\ ,\\ \mathfrak{B} &= CA - G^2\ ,\\ \mathfrak{C} &= AB - H^2\ ,\end{aligned}$$

$$\begin{aligned}\mathfrak{F} &= GH - AF,\\ \mathfrak{G} &= HF - BG,\\ \mathfrak{H} &= FG - CH,\end{aligned}$$

$\mathfrak{A}_,$, $\mathfrak{B}_,$, $\mathfrak{C}_,$, $\mathfrak{F}_,$, $\mathfrak{G}_,$, $\mathfrak{H}_,$ are what these become when A, B, C are changed into

$A-u,\ B-u,\ C-u,$ so that

$$\begin{aligned}
\mathfrak{A}_{\prime} &= \mathfrak{A} - (B+C)\,u + u^2,\\
\mathfrak{B}_{\prime} &= \mathfrak{B} - (C+A)\,u + u^2,\\
\mathfrak{C}_{\prime} &= \mathfrak{C} - (A+B)\,u + u^2,\\
\mathfrak{F}_{\prime} &= \mathfrak{F} + Fu\,,\\
\mathfrak{G}_{\prime} &= \mathfrak{G} + Gu\,,\\
\mathfrak{H}_{\prime} &= \mathfrak{H} + Hu.
\end{aligned}$$

Hence the equation $\alpha x + \alpha' y + \alpha'' z = 0$ may be written in the three forms

$$\begin{aligned}
\mathfrak{A}_{\prime}x + \mathfrak{H}_{\prime}y + \mathfrak{G}_{\prime}z &= 0,\\
\mathfrak{H}_{\prime}x + \mathfrak{B}_{\prime}y + \mathfrak{F}_{\prime}z &= 0,\\
\mathfrak{G}_{\prime}x + \mathfrak{F}_{\prime}y + \mathfrak{C}_{\prime}z &= 0;
\end{aligned}$$

or, what comes to the same thing, as follows,

$$\begin{aligned}
\mathfrak{A}x + \mathfrak{H}y + \mathfrak{G}z + u\,(Ax + Hy + Gz) + vx &= 0,\\
\mathfrak{H}x + \mathfrak{B}y + \mathfrak{F}z + u\,(Hx + By + Fz) + vy &= 0,\\
\mathfrak{G}x + \mathfrak{F}y + \mathfrak{C}z + u\,(Gx + Fy + Cz) + vz &= 0,
\end{aligned}$$

in which for shortness v has been written instead of

$$u^2 - (A+B+C)\,u.$$

The elimination of $u,\ v$ from these equations gives a result $\Theta = 0$, where Θ is a homogeneous function of the third order in $x,\ y,\ z$; and this equation, it is evident, must belong to the three diametral planes jointly, i.e. Θ must be the product of three linear factors, each of which equated to zero would correspond to a diametral plane. Thus the system of diametral planes is given by

$$\Theta = \begin{vmatrix}
\mathfrak{A}x + \mathfrak{H}y + \mathfrak{G}z, & Ax + Hy + Gz, & x\\
\mathfrak{H}x + \mathfrak{B}y + \mathfrak{F}z, & Hx + By + Fz, & y\\
\mathfrak{G}x + \mathfrak{F}y + \mathfrak{C}z, & Gx + Fy + Cz, & z
\end{vmatrix} = 0,$$

or developing the determinant, as follows,

$$\begin{aligned}
\Theta = {} & (G\mathfrak{H} - H\mathfrak{G})\,x^3 + (H\mathfrak{F} - F\mathfrak{H})\,y^3 + (F\mathfrak{G} - G\mathfrak{F})\,z^3\\
& + \{\ \ G\,(\mathfrak{C} - \mathfrak{B}) - \mathfrak{G}\,(C - B) - (H\mathfrak{F} - F\mathfrak{H})\}\,yz^2\\
& + \{\ \ H\,(\mathfrak{A} - \mathfrak{C}) - \mathfrak{H}\,(A - C) - (F\mathfrak{G} - G\mathfrak{F})\}\,zx^2\\
& + \{\ \ F\,(\mathfrak{B} - \mathfrak{A}) - \mathfrak{F}\,(B - A) - (G\mathfrak{H} - H\mathfrak{G})\}\,xy^2\\
& + \{-H\,(\mathfrak{C} - \mathfrak{B}) + \mathfrak{H}\,(C - B) + (F\mathfrak{G} - G\mathfrak{F})\}\,y^2z\\
& + \{-F\,(\mathfrak{A} - \mathfrak{C}) + \mathfrak{F}\,(A - C) + (G\mathfrak{H} - H\mathfrak{G})\}\,z^2x\\
& + \{-G\,(\mathfrak{B} - \mathfrak{A}) + \mathfrak{C}\,(B - A) + (H\mathfrak{F} - F\mathfrak{H})\}\,x^2y\\
& + (C\mathfrak{B} - B\mathfrak{A} + \mathfrak{A}\mathfrak{C} - C\mathfrak{A} + B\mathfrak{A} - A\mathfrak{B})\,xyz\,;
\end{aligned}$$

or reducing

$$\begin{aligned}\Theta = \quad & \{F(G^2-H^2) - GH(C-B)\}\,x^3 \\ + & \{G(H^2-F^2) - HF(A-C)\}\,y^3 \\ + & \{H(F^2-G^2) - FG(B-A)\}\,z^3 \\ + & \{G(A-B)(B-C) + FH(A+B-2C) + G(F^2+G^2-2H^2)\}\,yz^2 \\ + & \{H(B-C)(C-A) + GF(B+C-2A) + H(G^2+H^2-2F^2)\}\,zx^2 \\ + & \{F(C-A)(A-B) + GH(C+A-2B) + F(H^2+F^2-2G^2)\}\,xy^2 \\ + & \{H(B-C)(C-A) + FG(C+A-2B) + H(H^2+F^2-2G^2)\}\,y^2z \\ + & \{F(C-A)(A-B) + GH(A+B-2C) + F(F^2+G^2-2H^2)\}\,z^2x \\ + & \{G(A-B)(B-C) + HF(B+C-2A) + G(G^2+H^2-2F^2)\}\,x^2y \\ - & \{(A-B)(B-C)(C-A) + (B-C)F^2 + (C-A)G^2 + (A-B)H^2\}\,xyz.\end{aligned}$$

In the case of *curves* of the second order, the result is much more simple; we have

$$\Theta = \begin{vmatrix} Ax+Hy, & x \\ Hx+By, & y \end{vmatrix} = 0,$$

i.e.

$$\Theta = H(y^2-x^2) + (A-B)xy = 0,$$

for the equation of the two diameters.

The above formulæ may be applied to the question of finding the diametral planes of the cone circumscribed about a given surface of the second order, (or of the lines bisecting the angles made by two tangents of a curve of the second order). Considering the latter question first: if

$$\frac{x^2}{a^2} + \frac{y^2}{b^2} - 1 = 0$$

be the equation of the curve, and α, β the coordinates of the point of intersection of the two tangents, the equation of the pair of tangents is

$$\left(\frac{x^2}{a^2} + \frac{y^2}{b^2} - 1\right)\left(\frac{\alpha^2}{a^2} + \frac{\beta^2}{b^2} - 1\right) - \left(\frac{\alpha x}{a^2} + \frac{\beta y}{b^2} - 1\right)^2 = 0;$$

or making the point of intersection the origin,

$$\left(\frac{x^2}{a^2} + \frac{y^2}{b^2}\right)\left(\frac{\alpha^2}{a^2} + \frac{\beta^2}{b^2} - 1\right) - \left(\frac{\alpha x}{a^2} + \frac{\beta y}{b^2}\right)^2 = 0,$$

i.e.

$$(\beta x - \alpha y)^2 - (b^2x^2 + a^2y^2) = 0;$$

whence $A = \beta^2 - b^2$, $B = \alpha^2 - a^2$, $H = -\alpha\beta$, and the equation to the lines bisecting the angles formed by the tangents is

$$\alpha\beta(x^2-y^2) - \{\alpha^2 - \beta^2 - (a^2-b^2)\}\,xy = 0,$$

which is the same for all confocal ellipses; whence the known theorem,

"If there be two confocal ellipses, and tangents be drawn to the second from any point P of the first, the tangent and normal of the first conic at the point P, bisect the angles formed by the two tangents in question."

In the case of surfaces, the equation of the circumscribing cone referred to its vertex as origin, is

$$\left(\frac{x^2}{a^2}+\frac{y^2}{b^2}+\frac{z^2}{c^2}\right)\left(\frac{\alpha^2}{a^2}+\frac{\beta^2}{b^2}+\frac{\gamma^2}{c^2}-1\right)-\left(\frac{\alpha x}{a^2}+\frac{\beta y}{b^2}+\frac{\gamma z}{c^2}\right)^2=0\,;$$

whence

$$\begin{aligned}A &= \beta^2c^2+\gamma^2b^2-b^2c^2,\\ B &= \gamma^2a^2+\alpha^2c^2-a^2c^2,\\ C &= \alpha^2b^2+\beta^2a^2-b^2a^2,\\ F &= -a^2\beta\gamma,\\ G &= -b^2\gamma\alpha\,,\\ H &= -c^2\alpha\beta\,.\end{aligned}$$

Hence, omitting the factor $b^2c^2\alpha^2+c^2a^2\beta^2+a^2b^2\gamma^2-a^2b^2c^2$, we have

$$\begin{aligned}\mathfrak{A} &= \alpha^2-a^2,\\ \mathfrak{B} &= \beta^2-b^2,\\ \mathfrak{C} &= \gamma^2-c^2,\\ \mathfrak{F} &= \beta\gamma,\\ \mathfrak{G} &= \gamma\alpha\,,\\ \mathfrak{H} &= \alpha\beta\,;\end{aligned}$$

and the equation of the system of diametral planes becomes

$$\begin{aligned}\Theta=0= \quad & \alpha^2\beta\gamma\,(c^2-b^2)\,x^3+\beta^2\gamma\alpha\,(a^2-c^2)\,y^3+\gamma^2\alpha\beta\,(b^2-a^2)\,z^3\\
&+\gamma\alpha\,\{\alpha^2(c^2-b^2)+\beta^2(b^2+c^2-2a^2)-\gamma^2(b^2-a^2)+(b^2-a^2)(c^2-b^2)\}\,yz^2\\
&+\alpha\beta\,\{-\alpha^2(c^2-b^2)+\beta^2(a^2-c^2)+\gamma^2(c^2+a^2-2b^2)+(c^2-b^2)(a^2-c^2)\}\,zx^2\\
&+\gamma\alpha\,\{\alpha^2(a^2+b^2-2c^2)-\beta^2(a^2-c^2)+\gamma^2(b^2-a^2)+(a^2-c^2)(b^2-a^2)\}\,xy^2\\
&-\alpha\beta\,\{\alpha^2(c^2-b^2)-\beta^2(a^2-c^2)-\gamma^2(b^2+c^2-2a^2)-(a^2-c^2)(c^2-b^2)\}\,y^2z\\
&-\beta\gamma\,\{-\alpha^2(c^2+a^2-2b^2)+\beta^2(a^2-c^2)-\gamma^2(b^2-a^2)-(b^2-a^2)(a^2-c^2)\}\,z^2x\\
&-\gamma\alpha\,\{-\alpha^2(c^2-b^2)-\beta^2(a^2+b^2-2c^2)+\gamma^2(b^2-a^2)-(c^2-b^2)(b^2-a^2)\}\,x^2y\\
&+\{(a^2-b^2)(b^2-c^2)(c^2-a^2)+\\
&\quad(\alpha^4+\beta^2\gamma^2)(b^2-c^2)-(\beta^4+\gamma^2\alpha^2)(c^2-a^2)-(\gamma^4+\alpha^2\beta^2)(a^2-b^2)+\\
&\quad\alpha^2(b^2-c^2)(2a^2-b^2-c^2)+\beta^2(c^2-a^2)(2b^2-c^2-a^2)+\gamma^2(a^2-b^2)(2c^2-a^2-b^2)\}\,xyz\,;\end{aligned}$$

and since this is a function of a^2-b^2, b^2-c^2, and c^2-a^2, the equation is the same for all confocal ellipsoids; whence the known theorem, "The axes of the circumscribing cone having its vertex in a given point P, are tangents to the curves of intersection of the three surfaces, confocal with the given surface, which pass through the point P."

40.

ON THE THEORY OF INVOLUTION IN GEOMETRY.

[From the *Cambridge and Dublin Mathematical Journal*, vol. II. (1847), pp. 52—61.]

WHEN three conics have the same points of intersection, any transversal intersects the system in six points, which are said to be in involution. It appears natural to apply the term to the conics themselves; and then it is easy to generalize the notion of involution so as to apply it to functions of any number of variables. Thus, if $U, V\ldots$ be homogeneous functions of the same order of any number of variables $x, y\ldots$, a function Θ, which is a linear function of $U, V\ldots$, is said to be in involution with these functions. More generally Θ may be said to be in involution with any system of factors of these functions: or if $U, V\ldots$ be given functions of $x, y, z\ldots$, homogeneous of the degrees $m, n\ldots$, and $u, v,\ldots$ arbitrary homogeneous functions of the degrees $r-m$, $r-n\ldots$; then, if $\Theta = uU+vV+\ldots$, Θ is a function of the degree r, which is in involution with $U, V\ldots$. The question which immediately arises, is to find the degree of generality of Θ, or the number of arbitrary constants which it contains. And this is a question which, from the variety and interest of its geometrical interpretations, has very frequently been treated of by geometers, though never, I believe, in quite so general a form, (the number r has almost always had particular values given to it, except in a short paper of my own, on the particular case of two curves, in the *Journal*, vol. III. p. 211 [5]).[1] There is also an analytical application of

[1] The first suggestion of the problem is contained in a memoir of Euler's—"Sur une contradiction apparente dans la doctrine des lignes courbes," *Mém. de Berlin*, t. IV. [1748] p. 219. It is noticed also in Cramer's *Introduction à l'analyse des lignes courbes* [1750]. The following memoirs also have been published on the subject: Plücker, "Recherches sur les courbes algébriques de tous les degrés," *Gerg. Ann.* t. XIX. [1828—29] p. 97; "Recherches sur les surfaces algébriques de tous les degrés," p. 129; (a great number of memoirs on particular applications of the theory are contained in Gergonne;) Jacobi, "De relationibus quæ locum habere debent inter puncta intersectionis duarum curvarum vel trium superficierum dati ordinis, simul cum enodatione paradoxi algebraici," *Crelle*, t. XV. [1836]; Plücker, "Théorèmes généraux concernant les équations d'un degré quelconque entre un nombre quelconque d'inconnues," *Crelle*, t. XVI. [1837], (but this last must be read with caution, as several of the theorems are incorrect, or at least stated without the proper limitations); and the *Einleitende Betrachtungen*, in Plücker's "Theorie der algebraischen Curven" [1839]. The following memoirs of Hesse, containing developments relative to the case of three surfaces of the second order, may likewise be mentioned, "De curvis et superficiebus secundi gradus," *Crelle*, t. XX. [1840] p. 285; and "Ueber die lineare Construction des achten Schnitt-punctes dreier Oberflächen zweiter Ordnung, wenn sieben Schnitt-puncte derselben gegeben sind," *Crelle*, t. XXVI. [1843] p. 147.

the theory, of considerable interest, to the problem of elimination between any number of equations containing the same number of variables. Suppose, for instance, two equations, $U=0$, $V=0$, when U, V are homogeneous functions of x, y of the degrees m, n respectively. To eliminate the variables it is sufficient to multiply the first equation by x^{n-1}, $x^{n-2}y\ldots$, y^{n-1}, and the second by $x^{m-1}\ldots$, y^{m-1}, and from the equations so obtained to eliminate linearly the $(m+n)$ quantities x^{m+n-1}, $x^{m+n-2}y\ldots$, y^{m+n-1}. But in the case of a greater number of equations it is not at first obvious how many new equations should be obtained; and when a number apparently sufficiently great have been found, it may happen that the equations so obtained are not independent, and that the elimination cannot be performed. But in showing the connexion that exists between these different equations, the theory of involution explains in what manner a system is to be formed, which includes all the really independent equations, and gives the means of detecting the extraneous factors which appear in the result of the linear elimination of the different terms; but I do not see at present any mode of obtaining the final result at once in its reduced form free from any extraneous factors.

Let X, Y, ... be given homogeneous functions of the same degree of any number of variables, and suppose

$$\Theta = \alpha X + \beta Y + \ldots,$$

α, $\beta\ldots$ being constants, and the number of terms in the series being g; Θ contains therefore g arbitrary constants. If however, by giving to α, β ... particular values α_1, $\beta_1 \ldots$, or α_2, $\beta_2 \ldots$, and representing by Θ_1, Θ_2 ... the corresponding values of Θ, we have *identically*

$$\Theta_1 = 0, \quad \Theta_2 = 0, \ \ldots \ (h \text{ equations});$$

then the constants in Θ group themselves together into a smaller number $g-h$ of arbitrary constants. This supposes, however, that the last mentioned equations are linearly independent; if there are a certain number k of equations

$$\Phi_1 = 0, \quad \Phi_2 = 0 \ \ldots,$$

(where Φ_1, Φ_2, ... are linear functions of Θ_1, $\Theta_2\ldots$) which are identically satisfied, independently of the h equations, then the equations in question are equivalent to $h-k$ equations, and the function Θ contains $g-(h-k)$ or $g-h+k$, arbitrary constants. Similarly if the functions Φ are not independent; so that the number of arbitrary constants really contained in Θ is always

$$N = g - h + k - \&\text{c.} \ \ldots$$

Consider now the case of a function Θ, homogeneous of the rth degree in the variables x, $y\ldots\{(\theta+1)$ in number$\}$. Let U, $V\ldots$ be functions of the degrees m, $n\ldots$, and suppose

$$\Theta = uU + vV + \ldots$$

where u, $v\ldots$ are arbitrary functions of the degrees $r-m$, $r-n$, ... $\{r$ is supposed throughout greater than m, $n\ldots\}$. Suppose for shortness that the number of terms in the complete function of θ variables, and of the order ρ, i.e. the quotient $\dfrac{[\rho+\theta]^\theta}{[\theta]^\theta}$, is represented by $[\rho, \theta]$; then the function Θ contains apparently a number

$$([r-m, \theta]+[r-n, \theta]+\ldots)$$

of arbitrary constants.

But since we should have identically $\Theta=0$ by assuming $u=LV$, $v=-LU$, $w=0$, &c.... (L the general function of the order $r-m-n$), or $u=MW$, $v=0$, $w=-MU$ (M the general function of the order $r-m-p$) &c., the number N must be diminished by

$$[r-m-n, \theta]+[r-m-p, \theta]+[r-n-p, \theta]+\ldots;$$

but the equations just obtained are themselves not linearly independent, and in consequence of this the number of arbitrary constants has to be increased by

$$[r-m-n-p, \theta]+\ldots;$$

and so on. Hence finally the whole number of arbitrary constants in the function θ is

$$\begin{aligned} N=&[r-m, \theta]+[r-n, \theta]+[r-p, \theta]+\ldots \\ &-[r-m-n, \theta]-[r-m-p, \theta]-[r-n-p, \theta]-\ldots \\ &+[r-m-n-p, \theta]+\ldots \pm \text{\&c. \&c.} \qquad \text{(A)}. \end{aligned}$$

This however supposes that all the numbers $r-m$, $r-n\ldots$, $r-m-n\ldots$, are positive: whenever this is not the case for any one of them, the corresponding term is obviously to be omitted. With this convention the equation (A) gives always the correct number of arbitrary constants in Θ: it will be convenient to represent it in the abbreviated form

$$N=\{r : m, n, p, \ldots : \theta\}.$$

An expression analogous to this, for the particular case of $r=m$, but incorrect on account of the omission of all the terms after the second line, has been given by M. Plücker (*Crelle*, tom. XVI. p. 55), and even some of his particular formulæ are incorrect. But proceeding to examine some particular cases: if $r>m+n+p+\ldots-\theta-1$, then in the expression (A) either no terms are to be omitted, or else the terms to be omitted reduce themselves to zero, so that N is given by this formula continued to its last term. It will be subsequently shown that in this case

$$\{r : m, n, p \ldots : \theta\}=[r, \theta]-mnp\ldots;$$

or in the case of two or three variables, we have the theorem, "If a curve or surface of the order r be determined to pass through the mn points of intersection of two curves of the orders m and n, or the mnp points of intersection of three surfaces of the orders m, n, p; then if $r>m+n-3$, or $r>m+n+p-4$, the curve or surface contains precisely the same number of arbitrary constants as if the mn or mnp points were perfectly arbitrary."

This is natural enough; the peculiarity is in the case where $r \not> m+n-3$, or $r \not> m+n+p-4$. For instance, for two curves, $r \not> m+n-3$, we have

$$\{r : m, n : 2\}=[r-m, 2]+[r-n, 2]=[r, 2]-mn+[r-m-n, 2],$$

or the new curve contains $\frac{1}{2}[m+n-r-1]^2$ more arbitrary constants than it would do if the mn points, through which it was made to pass, had been perfectly arbitrary; a result given before in the *Journal*, [5].

In the case of surfaces, if $r \not> m+n+p-4$. Then assuming $r > m+n-4$, $m+p-4$, or $n+p-4$, we have

$$\begin{aligned}\{r : m,\ n,\ p : 3\} &= [r-m,\ 3]+[r-n,\ 3]+[r-p,\ 3]\\ &\quad -[r-m-n,\ 3]-[r-m-p,\ 3]-[r-n-p,\ 3]\\ &= [r,\ 3]-mnp-[r-m-n-p,\ 3];\end{aligned}$$

or the surface contains $\frac{1}{6}[m+n+p-r-1]^3$ more arbitrary constants than it would do if the mnp points, through which it was made to pass, had been perfectly arbitrary. Similarly, in the case where r is not greater than one or more of the quantities $m+n-4$, $m+p-4$, $n+p-4$. Thus in particular, if r be not greater than the least of these quantities

$$\begin{aligned}\{r : m,\ n,\ p : 3\} &= [r,\ 3]-mnp+[r-n-p,\ 3]+[r-m-p,\ 3]\\ &\quad +[r-m-n,\ 3]-[r-m-n-p,\ 3];\end{aligned}$$

or the surface contains

$$\tfrac{1}{6}[m+n+p-r-1]^3-\tfrac{1}{6}[n+p-r-1]^3-\tfrac{1}{6}[m+p-r-1]^3-\tfrac{1}{6}[m+n-r-1]^3$$

more arbitrary constants than it would otherwise have done. Again, for a surface of the rth order, subjected to pass through the curve of intersection of two surfaces of the orders m, n,

$$\{r : m,\ n,\ 3\}=[r-m,\ 3]+[r-n,\ 3]-[r-m-n,\ 3];$$

in which the last term, or $\frac{1}{6}[m+n-r-1]^3$, is to be omitted when $r \not> m+n-4$.

The function of the rth order, which is satisfied by the systems of values which satisfy the equations of the orders m, n... contains, we have seen, $[r,\ m,\ n,\ p\ \dots\ \theta]$ arbitrary constants; hence it may be determined so as to pass through this number, diminished by unity, of arbitrary points. But the equation being determined in general by the condition of being satisfied by $[r,\ \theta]-1$ systems of variables, it will be completely determined if, in addition to the above number of arbitrary systems, we suppose it to be satisfied by a number $N=[r,\ \theta]-\{r;\ m,\ n,\ p\dots : \theta\}$ of systems satisfying the equations above. Hence the theorem

"The equation of the rth order which is satisfied by a number

$$N=[r,\ \theta]-\{r;\ m,\ n,\ p\dots : \theta\}$$

of systems satisfying the equations of the orders m, n, p... is satisfied by any systems whatever which satisfy these equations."

In particular—"The surface of the rth order which passes through a number

$$[r,\ \theta]-\{r : m,\ n : \theta\}$$

of points in the curve of intersection of two surfaces of the orders m, n,—or through $[r,\ \theta]-\{r : m,\ n,\ p : \theta\}$ of the mnp points of intersection of three surfaces of the orders m, n, p,—passes through the curve of intersection, or through the mnp points of intersection."

Thus a surface of the second order which passes through eight points of the curve of intersection of two surfaces of the second order passes through this curve; and any surface of the second order which passes through seven of the points of intersection of three surfaces of the second order passes through the eighth point. (The first theorem obviously fails if the eight points have the relation in question, i.e. if they are the eight points of intersection of three surfaces of the second order.)

Again—"The curve of the rth order which passes through $[r,\ \theta]-\{r : m,\ n : \theta\}$ of the points of intersection of two curves of the orders mn, passes through the remaining points of intersection." e.g. "Any curve of the third order which passes through eight of the points of intersection of two curves of the third order, passes also through the ninth point."

Consider next the following question, which [as regards particular cases] has been treated of by Jacobi in the memoir already quoted (*Crelle*, tom. XV.). "To find the number of relations which must exist between $K(\theta+1)$ variables, forming K systems, each of which satisfies simultaneously equations of the orders m, n, p... respectively; the number ϕ of these equations being anything less than θ; or ϕ being equal to θ, provided at the same time $K=mnp$ ...."

Suppose $m \not< n$, $n \not< p$... and write

$$[m,\ \theta]-\{m : m,\ n,\ p\ \dots : \theta\}=N,$$
$$[n,\ \theta]-\{n : n,\ p\ \dots\dots : \theta\}=N',$$
&c.

Imagine the equations of the orders n, p... given. Any function of the mth order which is satisfied by N of the systems of values which satisfy the given equations, and any particular equation of the mth order, is satisfied by the remaining $K-N$ systems of values. Hence assuming N systems, satisfying the equations of the orders n, p ... but otherwise arbitrary, the remaining systems must satisfy these equations, and a completely determinate equation of the mth order; i.e. there must be ϕ relations between the variables of each system, and consequently $\phi(K-N)$ relations in all. Similarly, if the equations of the orders p ... were given, N' systems of variables might be assumed satisfying these equations, but otherwise arbitrary; the remaining $N-N'$ systems satisfy $(\phi-1)$ determinate equations, or the number of relations between the variables is $(\phi-1)(N-N')$...; continuing in the same manner the total number of relations between the variables is

$$\phi(K-N)+(\phi-1)(N-N')+(\phi-2)(N'-N'')+\dots$$

in which however any term $(\phi-1)(N-N')$ or $(\phi-2)(N-N')$ ... &c., which becomes negative, *must be omitted*. It is obvious that we may write more simply

$$N=[m,\ \theta]-1-\{m;\ n,\ p\dots\theta\},$$
$$N'=[n,\ \theta]-1-\{n;\ p\dots\theta\},\ \text{\&c.}$$

In particular, to find the relations which must exist between the coordinates of mn points in order that they may be the points of intersection of two curves of the orders m, n respectively: here $K = mn$, $N = \frac{1}{2}[m+2]^2 - \frac{1}{2}[m-n+2]^2 = \frac{1}{2}(2mn - n^2 + 3n)$, $N' = \frac{1}{2}(n^2 + 3n + 2)$, so that $N - N' = m(m-n) - 1$ which becomes negative when $m = n$; hence in general the required number of conditions is $mn - 3n + 1$, but when $m = n$, the number in question becomes $(n-1)(n-2)$.

Passing to the case of surfaces; to determine the number of relations which must exist between the coordinates of mnp points, in order that they may be the points of intersection of surfaces of the orders m, n, p respectively. The number required is

$$3(mnp - N) + 2(N - N') + (N' - N''),$$

where

$$N = [m, 3] - 1 - [m-n, 3] - [m-p, 3] + [m-n-p, 3]$$

(this last term to be omitted when $m < n+p-3$),

$$N' = [n, 3] - 1 - [n-p, 3],$$
$$N'' = [p, 3] - 1.$$

If, for instance, $m > n+p-3$, so as to retain the term $[m-n-p, 3]$, and $n > p$, so as to retain the term $N' - N''$, the number becomes, after all reductions,

$$2mnp + np^2 - 4np - 2p^2 - \tfrac{1}{3}(p-1)(p-2)(p-3),$$

a formula given by Jacobi. If, however, $n = p$, this number must be augmented by unity. Again, for $m < n+p-3$, the required number is

$$2mnp + np^2 - 4np - 2p^2 - \tfrac{1}{3}(p-1)(p-2)(p-3)$$
$$- \tfrac{1}{6}(n+p-m-1)(n+p-m-2)(n+p-m-3),$$

which however must be augmented by unity if $m = n$ or $n = p$, and by 3 if $m = n = p$. But without entering into further details about this part of the subject, which has been sufficiently illustrated by the examples that have been given, I pass on to notice the application of the above theory to the problem of elimination. Imagine $(\theta + 1)$ equations between the $(\theta + 1)$ variables, the first sides of these being, as before, rational and integral homogeneous functions of the variables of the orders m, n, p ... respectively. Writing $m + n + p \ldots - \theta = r$, and multiplying the first equation by all the terms of the form $x^\alpha y^\beta \ldots$ of the degree $r - m$, the second equation by all the terms of the same form, of the degree $r - n$, and so on, there result a certain number of equations, containing all the terms $x^\alpha y^\beta \ldots$ of the degree r. But these equations are not independent; and the reasoning in the former part of the present paper shows that the number of independent equations is given by the symbol $\{r : m, n, p \ldots : \theta\}$; the number of terms $x^\alpha y^\beta \ldots$ is evidently $[r, \theta]$; and it will be shown immediately that for the actual value of r,

$$[r, \theta] - \{r ; m, n, p \ldots : \theta\} = 0 \quad \text{.........................(B)};$$

so that the number of quantities to be linearly eliminated is precisely equal to the number of equations, or the elimination is always possible. I may mention also that, supposing the coefficients of all the equations to be of the order unity, the order of the result, free from extraneous factors, may be shown to be

$$[r-m,\ \theta]+\ldots-2\{[r-m-n,\ \theta]+\ldots\}+3\{[r-m-n-p,\ \theta]+\ldots\}-\&c.$$
$$=mn\ldots+mp\ldots+np\ldots+\&c.\ \ldots\ldots\ldots\ldots\ldots\ldots\ldots\ldots(\text{C}),$$

(the equality of which will be presently proved) a result which agrees with that deduced from the theory of symmetrical functions; but I am not in possession of any mode of directly obtaining the final result in this its most simplified form. My method, which it is not necessary to explain here more particularly, leads me to the formation of a set of functions

$$P,\ Q,\ \ldots\ldots\ X,\ Y,\ Z,$$

θ in number, such that Z divides Y, this quotient divides X, and so on until we have a certain quotient which divides P, and this quotient equated to zero is the result of the elimination freed from extraneous factors. It only remains to demonstrate the formulæ (A), (B), and (C). Suppose in general that (k) denotes the sum of all the terms of the form $m^a n^b\ldots$, which can be formed with a given combination of k letters out of the ϕ letters $m,\ n,\ p\ldots$; and let $\Sigma(k)$ denote the sum of all the series (k) obtained by taking all the possible different combinations of k letters. It is evident that $\Sigma(k)$ is a multiple of (ϕ), $\{(\phi)$ denoting of course the sum of all the terms $m^a n^b\ldots$, $m,\ n\ldots$ being any letters whatever out of the series $m,\ n,\ p\ldots\}$. Let g be the number of exponents $a,\ b,\ \ldots$, then (ϕ) contains $[\phi]^g$ terms, also (k) contains $[k]^g$ terms, and the number of terms such as (k) in the sum $\Sigma(k)$ is $[\phi]^{\phi-k}\div[\phi-k]^{\phi-k}$. Hence evidently

$$\Sigma(k)=\frac{[\phi-g]^{\phi-k}}{[\phi-k]^{\phi-k}}(\phi),$$

or, what comes to the same thing,

$$\Sigma(\phi-k)=\frac{[\phi-g]^k}{[k]^k}(\phi).$$

Let A be an indeterminate coefficient, σ a summatory sign referring to different systems of exponents; then

$$\Sigma\sigma A(\phi-k)=\sigma\frac{[\phi-g]^k}{[k]^k}A(\phi),$$

or, giving to k the values $1,\ 2\ldots\phi$, multiplying each equation by an arbitrary coefficient, and adding, putting also for shortness $\sigma A(\phi-k)=U_{\phi-k}$, we have

$$\alpha_\phi U_\phi+\alpha_{\phi-1}\Sigma U_{\phi-1}+\ldots=\sigma\left(\alpha_\phi+\alpha_{\phi-1}\frac{[\phi-g]^1}{[1]^1}+\ldots\right)A(\phi);$$

whence in particular,

$$U_\phi-\Sigma U_{\phi-1}+\ldots=\sigma\{0^{\phi-g}A(\phi)\},$$
$$\Sigma U_{\phi-1}-2\Sigma U_{\phi-2}+\ldots=\sigma\{(\phi-g)\,0^{\phi-g-1}A(\phi)\},$$

which are still equations of considerable generality. If now $\phi=\theta$ and U_θ is a function of $m+n+p+\ldots$ of the order θ, the quantity $\sigma\{0^{\theta-g}A(\theta)\}$ reduces itself to the single term of U_θ which contains the product $mnp\ldots$. Hence, if

$$U_\theta=[\alpha+m+n+p\;\ldots,\;\theta]$$

in which afterwards $\alpha=r-m-n-p-\ldots$ we have the formula (A). Again, if $\phi=\theta+1$, and $U_{\theta+1}$ is a function of $m+n+p\ldots$ of the order θ, the sum $\sigma\;\{0^{\theta+1-g}A(\phi)\}$ vanishes; whence writing $U_{\theta+1}=[m+n+p\ldots-\theta,\;\theta]$, we have the formula (B). Similarly, if in the second formula $\phi=\theta+1$, and $U_{\theta+1}$ is a function of $m+n+p\ldots$ of the degree θ, then

$$\sigma\;\{(\theta+1-g)\,0^{\theta-g}A\;(\theta+1)\},$$

reduces itself to the term which contains $mn\ldots+np\ldots+mp\ldots+$ &c.; whence, if

$$U_{\theta+1}=[m+n+p+\;\ldots-\theta,\;\theta],$$

we have the formula (C).

41.

ON CERTAIN FORMULÆ FOR DIFFERENTIATION WITH APPLICATIONS TO THE EVALUATION OF DEFINITE INTEGRALS.

[From the *Cambridge and Dublin Mathematical Journal*, vol. II. (1847), pp. 122—128.]

IN attempting to investigate a formula in the theory of multiple definite integrals (which will be noticed in the sequel), I was led to the question of determining the $(i+1)^{\text{th}}$ differential coefficient of the $2i^{\text{th}}$ power of $\sqrt{(x+\lambda)} - \sqrt{(x+\mu)}$; the only way that occurred for effecting this was to find the successive differential coefficients of this quantity, which may be effected as follows. Assume

$$U_{k,i} = \{(x+\lambda)(x+\mu)\}^{\frac{1}{2}k} \{\sqrt{(x+\lambda)} - \sqrt{(x+\mu)}\}^{2i},$$

then

$$\frac{1}{U_{k,i}} \frac{d}{dx} U_{k,i} = \tfrac{1}{2}k \frac{2x+\lambda+\mu}{(x+\lambda)(x+\mu)} - \frac{i}{\sqrt{\{(x+\lambda)(x+\mu)\}}}$$

$$= \tfrac{1}{2}k \frac{\{\sqrt{(x+\lambda)} + \sqrt{(x+\mu)}\}^2 - 2\sqrt{\{(x+\lambda)(x+\mu)\}}}{(x+\lambda)(x+\mu)} - \frac{i}{\sqrt{\{(x+\lambda)(x+\mu)\}}}$$

$$= \tfrac{1}{2}k \frac{(\lambda-\mu)^2}{\{\sqrt{(x+\lambda)} - \sqrt{(x+\mu)}\}^2 (x+\lambda)(x+\mu)} - \frac{k+i}{\sqrt{\{(x+\lambda)(x+\mu)\}}};$$

or, attending to the signification of $U_{k,i}$,

$$\frac{d}{dx} U_{k,i} = \tfrac{1}{2}k (\lambda-\mu)^2 U_{k-2,i-1} - (k+i) U_{k-1,i}.$$

Hence

$$-\frac{1}{i}\frac{d}{dx} U_{0,i} = U_{-1,i}$$

$$\frac{1}{i}\frac{d^2}{dx^2} U_{0,i} = \tfrac{1}{2}(\lambda-\mu)^2 U_{-3,i-1} + (i-1) U_{-2,i},$$

&c.

from which the law is easily seen to be of the form

$$\frac{(-)^r}{i}\left(\frac{d}{dx}\right)^r U_{0,i} = S_\theta\, K_{r,\theta}\,(\lambda-\mu)^{2r-2-2\theta}\, U_{-2r+1+\theta,\, i-r+1+\theta}$$

(where the extreme values of θ are 0 and $(r-1)$ respectively) and $K_{r,\theta}$ is determined by

$$K_{r+1,\theta+1} = (r-1-\tfrac{1}{2}\theta)\, K_{r,\theta+1} + (i-3r+2+2\theta)\, K_{r,\theta}.$$

This equation is satisfied by

$$K_{r,\theta} = \frac{\Gamma(r-\frac{1}{2}-\theta)\,\Gamma(2r-1-\theta)\,\Gamma(i-r+\theta+1)}{\Gamma\frac{1}{2}\,\Gamma(\theta+1)\,\Gamma(2r-1-2\theta)\,\Gamma(i-r+1)};$$

for in the first place this gives

$$(r-1-\tfrac{1}{2}\theta)\, K_{r,\theta+1} = \frac{(r-1-\frac{1}{2}\theta)}{\Gamma\frac{1}{2}\,\Gamma(\theta+2)}\,\frac{\Gamma(r-\frac{3}{2}-\theta)\,\Gamma(2r-2-\theta)\,\Gamma(i-r+\theta+2)}{\Gamma(2r-3-2\theta)\,\Gamma(i-r+1)}$$

$$= \frac{\Gamma(r-\frac{1}{2}-\theta)\,\Gamma(2r-1-\theta)\,\Gamma(i-r+\theta+2)}{\Gamma(\frac{1}{2})\,\Gamma(\theta+2)\,\Gamma(2r-2-2\theta)\,\Gamma(i-r+1)},$$

and hence the second side of the equation reduces itself to

$$\frac{\Gamma(r-\frac{1}{2}-\theta)\,\Gamma(2r-1-\theta)\,\Gamma(i-r+\theta+1)}{\Gamma(\frac{1}{2})\,\Gamma(\theta+2)\,\Gamma(2r-1-2\theta)\,\Gamma(i-r+1)}\,\{2(r-1-\theta)(i-r+\theta+1)+(\theta+1)(i-3r+2-2\theta)\},$$

where the quantity within brackets reduces itself to $(i-r)(2r-1-\theta)$, so that the above value reduces itself to $K_{r+1,\theta+1}$, which verifies the equation in question. Also by comparing the first few terms, it is immediately seen that the above is the correct value of $K_{r,\theta}$, so that

$$\frac{(-)^r}{i}\left(\frac{d}{dx}\right)^r U_{0,i} = S_\theta \frac{\Gamma(r-\frac{1}{2}-\theta)\,\Gamma(2r-1-\theta)\,\Gamma(i-r+\theta+1)}{\Gamma(\frac{1}{2})\,\Gamma(\theta+1)\,\Gamma(2r-1-\theta)\,\Gamma(i-r+1)}\,(\lambda-\mu)^{2r-2-2\theta}\, U_{-2r+1+\theta,\, i-r+\theta+1} \;\ldots(1),$$

θ extending as before from 0 to $(r-1)$. In particular if i be integer and $r=i+1$,

$$\frac{(-)^{i+1}}{i}\left(\frac{d}{dx}\right)^{i+1}\{\sqrt{(x+\lambda)}-\sqrt{(x+\mu)}\}^{2i} = \frac{\Gamma(i+\frac{1}{2})}{\Gamma(\frac{1}{2})}(\lambda-\mu)^{2i}\frac{1}{\{(x+\lambda)(x+\mu)\}^{i+\frac{1}{2}}} \;\ldots\ldots\ldots(2),$$

(since the factor $\Gamma(i-r+\theta+1)\div\Gamma(i-r+1)$ vanishes except for $\theta=0$ on account of $\Gamma(i-r+1)=\infty$). Thus also, if r be greater than $(i+1)$, $=i+1+s$ suppose, then

$$(-)^s\left(\frac{d}{dx}\right)^s \frac{1}{\{(x+\lambda)(x+\mu)\}^{i+\frac{1}{2}}}$$

$$= S_\theta \frac{\Gamma(i+s+\frac{1}{2}-\theta)\,\Gamma(2i+2s+1-\theta)\,\Gamma(\theta-s)}{\Gamma(i+\frac{1}{2})\,\Gamma(\theta+1)\,\Gamma(2i+2s+1-2\theta)\,\Gamma(-s)}\,(\lambda-\mu)^{2s-2\theta}\, U_{-2i-2s-1+\theta,\,-s+\theta} \;\ldots\ldots(3),$$

where θ extends only from $\theta=0$ to $\theta=s$, on account of the factor $\Gamma(\theta-s)\div\Gamma(-s)$, which vanishes for greater values of θ: a rather better form is obtained by replacing this factor by

$$(-)^\theta \frac{\Gamma(1+s)}{\Gamma(1+s-\theta)}.$$

The above formulæ have been deduced on the supposition of i being an integer; assuming that they hold generally, the equation (2) gives, by writing $(i-\frac{1}{2})$ for i,

$$\frac{(-)^{i+\frac{1}{2}}}{i-\frac{1}{2}}\left(\frac{d}{dx}\right)^{i+\frac{1}{2}}\{\sqrt{(x+\lambda)}-\sqrt{(x+\mu)}\}^{2i-1}=\frac{\Gamma i}{\Gamma(\frac{1}{2})}(\lambda-\mu)^{2i-1}\frac{1}{\{(x+\lambda)(x+\mu)\}^i},$$

or integrating $(i+\frac{1}{2})$ times by means of the formula

$$\int_0^\infty x^{i-\frac{1}{2}} fx\,dx=\frac{\Gamma(i+\frac{1}{2})}{(-)^{i+\frac{1}{2}}}\left(\int_\infty d\alpha\right)^{i+\frac{1}{2}} f\alpha,\quad \alpha=0;$$

this gives

$$\int_0^\infty \frac{x^{i-\frac{1}{2}}dx}{\{(x+\lambda)(x+\mu)\}^i}=\frac{\Gamma\frac{1}{2}\,\Gamma(i-\frac{1}{2})}{\Gamma i}\frac{1}{(\sqrt{\lambda}+\sqrt{\mu})^{2i-1}}\quad\ldots\ldots\ldots\ldots\ldots\ldots(4),$$ [1]

whence also

$$\int_0^\infty \frac{x^{i-\frac{1}{2}}dx}{(x+\lambda)^{i+1}(x+\mu)^i}=\frac{\Gamma\frac{1}{2}\,\Gamma(i+\frac{1}{2})}{\Gamma(i+1)}\frac{1}{(\sqrt{\lambda}+\sqrt{\mu})^{2i}\sqrt{\lambda}}\quad\ldots\ldots\ldots\ldots\ldots(5);$$

and from these, by simple transformations,

$$\int_\beta^\alpha \frac{(\alpha-x)^{i-\frac{1}{2}}(x-\beta)^{i-\frac{1}{2}}\,dx}{\{(\alpha-x)+m(x-\beta)\}^i}=\frac{\Gamma\frac{1}{2}\,\Gamma(i+\frac{1}{2})}{\Gamma(i+1)}\frac{(\alpha-\beta)^i}{(\sqrt{m}+1)^{2i}}\quad\ldots\ldots\ldots\ldots\ldots(6),$$

$$\int_\beta^\alpha \frac{(\alpha-x)^{i-\frac{1}{2}}(x-\beta)^{i-\frac{3}{2}}\,dx}{\{(\alpha-x)+m(x-\beta)\}^i}=\frac{\Gamma\frac{1}{2}\,\Gamma(i-\frac{1}{2})}{\Gamma i}\frac{(\alpha-\beta)^{i-1}}{(\sqrt{m}+1)^{2i-1}}\quad\ldots\ldots\ldots\ldots(7).$$

These last two formulæ are connected also by the following general property:

"If
$$(a,\,b,\,i)=\int_\beta^\alpha \frac{(\alpha-x)^{a-1}(x-\beta)^{b-1}\,dx}{\{(\alpha-x)+m(x-\beta)\}^i},$$

then
$$(a,\,b,\,i)=\frac{\Gamma a\Gamma b}{\Gamma(a+b-i)\,\Gamma i}(\alpha-\beta)^{b-i}(a+b-i,\,i,\,\beta)\text{"}\quad\ldots\ldots\ldots\ldots\ldots(8),$$

which I have proved by means of a multiple[2] integral. From (6) we may obtain for $\gamma<1$,

$$\int_{-1}^1 \frac{(1-x^2)^{i-\frac{1}{2}}\,dx}{(1-2\gamma x+\gamma^2)^i}=\frac{\Gamma(\frac{1}{2})\,\Gamma(i+\frac{1}{2})}{\Gamma(i+1)}\quad\ldots\ldots\ldots\ldots\ldots\ldots\ldots\ldots\ldots(9),$$

[1] This is immediately transformed into

$$\int_0^\infty \frac{x^{i-\frac{1}{2}}\,dx}{(ax^2+bx+c)^i}=\frac{\Gamma\frac{1}{2}\,\Gamma(i-\frac{1}{2})}{\Gamma i}\frac{1}{\{b+2\sqrt{(ac)}\}^{i-\frac{1}{2}}},$$

which is a particular case of a formula which will be demonstrated in a subsequent paper. [I am not sure to what this refers.]

[2] [The triple integral $\iiint u^{i-1}x^{a-1}y^{\beta-1}e^{-(x+my)u-x-y}\,dx\,dy\,du$.]

which however is only a particular case of

$$\int_{-1}^{1} dx\,(1-x^2)^{i-\frac{1}{2}}\,(1-2\,\gamma x+\gamma^2)^{-i}\,\frac{d}{d\beta}\left[\beta^i\left(1-2\frac{\beta}{\gamma}x+\frac{\beta^2}{\gamma^2}\right)^{-i}\right]$$
$$=\frac{\Gamma\frac{1}{2}\,\Gamma(i+\frac{1}{2})}{\Gamma(i+1)}\,\beta^{i-1}\,(1-\beta)^{-2i}\;\ldots\ldots\ldots\ldots(10),$$

which supposes γ and $\frac{\beta}{\gamma}$ each less than unity. This formula was obtained in the case of $(i+\frac{1}{2})$ an integer, from a theorem, *Leg. Cal. Int.*, tom. II. p. 258, but there is no doubt that it is generally true.

From (9), by writing $x=\cos\theta$, we have

$$\int_0^{\pi}\frac{\sin^{2i}\theta\,d\theta}{(1-2\gamma\cos\theta+\gamma^2)^i}=\frac{\Gamma\frac{1}{2}\,\Gamma(i+\frac{1}{2})}{\Gamma(i+1)}\;\ldots\ldots\ldots\ldots(11),$$

which may also be demonstrated by the common equation in the theory of elliptic functions $\sin(\phi-\theta)=\gamma\sin\phi$, as was pointed out to me by Mr [Sir W.] Thomson. It may be compared with the following formula of Jacobi's, *Crelle*, tom. XV. [1836] p. 7,

$$\int_0^{\pi}\frac{\sin^{2i-1}\theta\,d\theta}{(1-2\gamma\cos\theta+\gamma^2)^i}=\frac{1}{\Gamma(i+\frac{1}{2})}\int_0^{\pi}\frac{\cos(i-\frac{1}{2})\,\theta\,d\theta}{\sqrt{(1-2\gamma\cos\theta+\gamma^2)}}\;\ldots\ldots\ldots\ldots(12).$$

Consider the multiple integral

$$W=\int\frac{dx\,dy\,\ldots}{\{(x-a)^2+\ldots u^2\}^i}\;\ldots\ldots\ldots\ldots(13),$$

the number of variables being $(2i+1)$ (not necessarily odd), and the equation of the limits being

$$x^2+y^2\ldots=\xi;$$

then, as will presently be shown, W may be expanded in the form

$$W=\pi^{i+\frac{1}{2}}\,S_\lambda\,\frac{(-)^\lambda A^\lambda}{2^{2\lambda}\,\Gamma(\lambda+1)\,\Gamma(i+\lambda+\frac{1}{2})}\left(\frac{d}{du}\right)^{2\lambda}\int_0\xi^{i-\frac{1}{2}}\,(\xi+u^2)^{-i}d\xi\;\ldots\ldots\ldots(14),$$

where $A=a^2+b^2+\ldots\ldots$ and λ extends from 0 to ∞. Suppose next

$$V=\int\frac{dx\,dy\,\ldots}{\{(x-a)^2\ldots+u^2\}^i\,(x^2+\ldots v^2)^{i+1}}\;\ldots\ldots\ldots\ldots(15):$$

the number of variables as before, and the limits for each variable being $-\infty$, ∞. We have immediately

$$V=\int_0^{\infty}\frac{1}{(\xi+v^2)^{i+1}}\,\frac{dW}{d\xi}\,d\xi;$$

W as before, i.e.

$$V=\pi^{i+\frac{1}{2}}\,S_\lambda\,\frac{(-)^\lambda A^\lambda}{2^{2\lambda}\,\Gamma(\lambda+1)\,\Gamma(i+\lambda+\frac{1}{2})}\left(\frac{d}{du}\right)^{2\lambda}\int_0^{\infty}\frac{\xi^{i-\frac{1}{2}}\,d\xi}{(\xi+u^2)^i\,(\xi+v^2)^{i+1}}.$$

But writing u^2, v^2 for λ, μ in the formula (5) (u and v being supposed positive), the integral in this formula is

$$\frac{\sqrt{\pi}\,\Gamma(i+\tfrac{1}{2})}{\Gamma(i+1)}\frac{1}{v(u+v)^{2i}};$$

hence, after a slight reduction,

$$V=\frac{\pi^{i+1}}{v\Gamma(i+1)}\,S\,\frac{(-)^\lambda\,\Gamma(i+\lambda+1)}{\Gamma(i+1)\,\Gamma(\lambda+1)}\frac{A^\lambda}{\{(u+v)^2\}^\lambda};$$

or finally

$$V=\frac{\pi^{i+1}}{\Gamma(i+1)}\frac{1}{v\{(u+v)^2+A\}^i}\quad\ldots\ldots\ldots\ldots(16),$$

a remarkable formula, the discovery of which is due to Mr Thomson. It only remains to prove the formula for W. Out of the variety of ways in which this may be accomplished, the following is a tolerably simple one. In the first place, by a linear transformation corresponding to that between two sets of rectangular axes, we have

$$W=\int\frac{dx\,dy\ldots}{\{(x-\sqrt{A})^2+y^2\ldots+u^2\}^i};$$

or expanding in powers of A, and putting for shortness $R=x^2+y^2\ldots+u^2$, the general term of W is

$$(-)^\sigma A^\lambda\frac{\Gamma(i+\lambda+\sigma)}{\Gamma i\,\Gamma(\lambda-\sigma+1)\,\Gamma(2\sigma+1)}\,2^{2\sigma}\int x^{2\sigma}R^{-i-\lambda-\sigma}\,dx\,dy\ldots$$

the limits being as before $x^2+y^2+\ldots=\xi$. To effect the integrations, write $\sqrt{\xi}\sqrt{x}$, $\sqrt{\xi}\sqrt{y}$, &c. for x, y ... so that the equation of the limits becomes $x+y+\ldots=1$. Also restricting the integral to positive values, we must multiply it by 2^{2i+1}: the integral thus becomes

$$\xi^{\sigma+i+\frac{1}{2}}\int x^{\sigma-\frac{1}{2}}y^{-\frac{1}{2}}\ldots\{\xi(x+y\ldots)+u^2\}^{-i-\lambda-\sigma}\,dx\,dy\ldots$$

equivalent to

$$\xi^{\sigma+i+\frac{1}{2}}\frac{\Gamma(\sigma+\tfrac{1}{2})\,\pi^i}{\Gamma(i+\sigma+\tfrac{1}{2})}\int_0^1\theta^{i+\sigma-\frac{1}{2}}(\xi\theta+u^2)^{-i-\lambda-\sigma}\,d\theta;$$

i.e. to

$$\frac{\Gamma(\sigma+\tfrac{1}{2})\,\pi^i}{\Gamma(i+\sigma+\tfrac{1}{2})}\int_0\xi^{i+\sigma-\frac{1}{2}}(\xi+u^2)^{-i-\lambda-\sigma}\,d\xi.$$

Hence, after a slight reduction, the general term of W is

$$\frac{\pi^{i+\frac{1}{2}}}{\Gamma i}(-)^{\lambda+\sigma}A^\lambda\frac{\Gamma(i+\lambda+\sigma)}{\Gamma(\sigma+1)\,\Gamma(\lambda-\sigma+1)\,\Gamma(i+\sigma+\tfrac{1}{2})}\int_0\xi^{i+\sigma-\frac{1}{2}}(\xi+u^2)^{-i-\lambda-\sigma}\,d\xi,$$

where σ may be considered as extending from 0 to λ inclusively, and then λ from 0 to ∞. But by a formula easily proved

$$\left(\frac{d}{du}\right)^{2\lambda}(\xi+u^2)^{-1}=\frac{2^{2\lambda}\,\Gamma(\lambda+1)\,\Gamma(i+\lambda+\tfrac{1}{2})}{\Gamma i}\times$$

$$S(-)^\sigma\frac{\Gamma(i+\lambda+\sigma)}{\Gamma(\sigma+1)\,\Gamma(\lambda-\sigma+1)\,\Gamma(i+\sigma+\tfrac{1}{2})}\,\xi^\sigma(\xi+u^2)^{-i-\lambda-\sigma},$$

where σ extends from 0 to λ. Hence, substituting and prefixing the summatory sign,

$$W = \pi^{i+\frac{1}{2}}\, S \frac{(-)^\lambda A^\lambda}{2^{2\lambda}\,\Gamma(\lambda+1)\,\Gamma(i+\lambda+\frac{1}{2})} \left(\frac{d}{du}\right)^{2\lambda} \int_0 \xi^{i-\frac{1}{2}} (\xi + u^2)\, d\xi,$$

where λ extends from 0 to ∞, the formula required.

[I annex the following Note added in MS. in my copy of the *Journal*, and referring to the formula, *ante* p. 267; a is written to denote $\lambda - \mu$.

N.B.—It would be worth while to find the general differential coefficient of $U_{k,\,i}$.

$$\partial_x U_{k,\,i} = -(k+i)\, U_{k-1,\,i} + \tfrac{1}{2} k a^2\, U_{k-2,\,i-1},$$

from which it is easy to see that

$$\begin{aligned} \partial_x^r\, U_{k,\,i} = (-)^r\ & K_{r,\,0}\ U_{k-r,\,i} \\ & \vdots \\ +(-)^{r-\theta}\ & K_{r,\,\theta}\, a^{2\theta}\, U_{k-r-\theta,\,i-\theta} \\ & \vdots \\ +\ & K_{r,\,r}\, a^{2r}\, U_{k-2r,\,i-r}. \end{aligned}$$

The general term of $\partial_x^{r+1}\, U_{k,\,i}$ is

$$\begin{aligned} & (-)^{r-\theta}\ K_{r,\,\theta}\, a^{2\theta}\, [\tfrac{1}{2}(k-r-\theta)\, a^2 U_{k-r-\theta-2,\,i-\theta-1}] \\ +\, & (-)^{r-\theta-1}\, K_{r+\theta+1}\, a^{2\theta+2}\, [-(k+i-2\theta-2-r)\, U_{k-r-\theta-2,\,i-\theta-1}], \end{aligned}$$

which must be equal to

$$(-)^{r-\theta}\ K_{r+1,\,\theta+1}\, a^{2\theta+2}\, U_{k-r-\theta-2,\,i-\theta-1},$$

therefore

$$K_{r+1,\,\theta+1} = (k+i-r-2\theta-2)\, K_{r,\,\theta+1} + \tfrac{1}{2}(k-r-\theta)\, K_{r,\,\theta}.$$

In particular

$$\begin{aligned} & K_{r+1,\,0} \quad -(k+i-r) \quad K_{r,\,0} = 0, \\ & K_{r+1,\,1} \quad -(k+i-r-2)\, K_{r,\,1} = \tfrac{1}{2}(k-r)\, K_{r,\,0}, \\ & \vdots \\ & K_{r+1,\,r+1} \quad (\tfrac{1}{2}k \quad r) \quad K_{r,\,r} = 0, \end{aligned}$$

whence

$$\begin{aligned} K_{r,\,0} &= [k+i]^r \\ K_{r,\,1} &= \tfrac{1}{2} r\, \{k^2 + (i-r)\, k - \tfrac{1}{2}(r-1)\, i\}\, [k+i-2]^{r-2}, \\ & \vdots \\ K_{r,\,r} &= [\tfrac{1}{2}k]^r, \end{aligned}$$

which appears to indicate a complicated general law.

Even the verification of $K_{r,\,1}$ is long, thus the equation becomes

$$\overline{r+1}\, [k+i-2]^{r-1}\, \{k^2 + (i-r-1)\, k - \tfrac{1}{2} r i\} - (k+i-r-2)\, r\, \{k^2 + (i-r)\, k - \tfrac{1}{2}(r-1)\, i\}\, [k+i-2]^{r-2} = (k-r)\, [k+i]^r,$$

or

$$\overline{r+1}\, (k+i-r)\, \{k^2 + (i-r-1)\, k - \tfrac{1}{2} r i\} - r\, (k+i-r-2)\, \{k^2 + (i-r)\, k - \tfrac{1}{2}(r-1)\, i\} = (k-r)\, (k+i)\, (k+i-1),$$

which is identical, as may be most easily seen by taking first the coefficient of k^3, and then writing $k=r$, $k=-i$, $k=-i-1$.]

42.

ON THE CAUSTIC BY REFLECTION AT A CIRCLE.

[From the *Cambridge and Dublin Mathematical Journal*, vol. II. (1847), pp. 128—130.]

THE following solution of the problem is that given by M. de St-Laurent (*Annales de Gergonne*, t. XVII. [1826] pp. 128—134); the process of elimination is somewhat different.

The centre of the circle being taken for the origin, let k be its radius; a, b the coordinates of the luminous point; ξ, η those of the point at which the reflection takes place; x, y those of any point in the reflected ray: we have in the first place

$$\xi^2+\eta^2=k^2 \quad \ldots\ldots\ldots\ldots\ldots\ldots\ldots\ldots\ldots\ldots(1).$$

There is no difficulty in finding the equation of the reflected ray[1]; this is

$$(b\xi-a\eta)(\xi x+\eta y-k^2)+(y\xi-x\eta)(a\xi+b\eta-k^2)=0 \quad \ldots\ldots\ldots\ldots(2),$$

[1] To do this in the simplest way, write

$$\rho^2=(\xi-x)^2+(\eta-y)^2, \quad \sigma^2=(\xi-a)^2+(\eta-b)^2,$$

then, by the condition of reflection,

$$\rho+\sigma=\text{min.},$$

ρ, σ being considered as functions of the variables ξ, η, which are connected by the equation (1). Hence

$$\frac{\xi-x}{\rho}+\frac{\xi-a}{\sigma}+\lambda\xi=0,$$

$$\frac{\eta-y}{\rho}+\frac{\eta-b}{\sigma}+\lambda\eta=0;$$

or, eliminating λ,

$$\frac{\eta x-\xi y}{\rho}+\frac{\eta a-\xi b}{\sigma}=0,$$

whence $$(\eta x-\xi y)^2[(\xi-a)^2+(\eta-b)^2]=(\eta a-\xi b)^2[(\xi-x)^2+(\eta-y)^2].$$

This may be written $$\{(\eta x-\xi y)(\xi-a)-(\eta a-\xi b)(\xi-x)\}[(\eta x-\xi y)(\xi-a)+(\eta a-\xi b)(\xi-x)]$$
$$+\{(\eta x-\xi y)(\eta-b)-(\eta a-\xi b)(\eta-y)\}[(\eta x-\xi y)(\eta-b)+(\eta a-\xi b)(\eta-y)]=0;$$

the factors in { } reduce themselves respectively to ξP and ηP, where $P=\xi(b-y)-\eta(a-x)+ay-bx$; omitting the factor P, (which equated to zero, is the equation of the line through (a, b) and (ξ, η),) and replacing $\xi(\xi-a)+\eta(\eta-b)$ and $\xi(\xi-x)+\eta(\eta-y)$ by $k^2-a\xi-b\eta$ and $k^2-\xi x-\eta y$, respectively, we have the equation given above.

or, arranging the terms in a more convenient order,

$$(bx+ay)(\xi^2-\eta^2)+2(by-ax)\xi\eta-k^2(b+y)\xi+k^2(a+x)\eta=0 \quad\dots\dots\dots\dots(2').$$

Hence, considering ξ, η as indeterminate parameters connected by the equation (1), the locus of the curve generated by the continued intersections of the lines (2) will be found by eliminating ξ, η, λ from these equations and the system

$$\xi[\lambda+2(bx+ay)]+\eta[2(by-ax)]-k^2(b+y)=0 \quad\dots\dots\dots\dots(3),$$

$$\xi[2(by-ax)]+\eta[\lambda-2(bx+ay)]+k^2(a+x)=0 \quad\dots\dots\dots\dots(4),$$

and from these, multiplying by ξ, η, adding and reducing by (2), we have

$$-\xi(b+y)+\eta(a+x)-\lambda=0 \quad\dots\dots\dots\dots(5),$$

which replaces the equation (2) or (2'). Thus the equations from which ξ, η, λ are to be eliminated are (1), (3), (4), (5).

From (3), (4), (5), by the elimination of ξ, η, we have

$$\begin{aligned}-\lambda\{\lambda^2-4(bx+ay)^2\}&-4k^2(by-ax)(a+x)(b+y)\\ &-k^2(a+x)^2[\lambda+2(bx+ay)]\\ &-k^2(b+y)^2[\lambda-2(bx+ay)]\\ &+4\lambda(by-ax)^2=0 \quad\dots\dots\dots\dots(6),\end{aligned}$$

or, reducing,

$$\begin{aligned}-\lambda^3+\lambda\{4(a^2+b^2)(x^2+y^2)&-k^2[(a+x)^2+(b+y)^2]\}\\ &-2k^2(bx-ay)(x^2+y^2-a^2-b^2)=0 \quad\dots\dots\dots\dots(7);\end{aligned}$$

which may be represented by

$$-\lambda^3+\lambda Q-2R=0 \quad\dots\dots\dots\dots(7').$$

Again, from the equations (4), (3), transposing the last terms and adding the squares, also reducing by (1),

$$\begin{aligned}k^4[(a+x)^2+(b+y)^2]&=k^2\lambda^2+4k^2(a^2+b^2)(x^2+y^2)\\ &+4\lambda[(\xi^2-\eta^2)(bx+ay)+2\xi\eta(by-ax)] \quad\dots\dots\dots\dots(8);\end{aligned}$$

but from the same equations, multiplying by ξ, η and adding, also reducing by (1),

$$k^2\lambda+2(bx+ay)(\xi^2-\eta^2)+4\xi\eta(by-ax)+k^2[-\xi(b+y)+\eta(a+x)]=0 \quad\dots\dots\dots\dots(9),$$

or reducing by (5) and dividing by two,

$$k^2\lambda+(bx+ay)(\xi^2-\eta^2)+2\xi\eta(by-ax)=0 \quad\dots\dots\dots\dots(10).$$

Using this to reduce (8),

$$k^2 [(a+x)^2 + (b+y)^2] = 4(a^2+b^2)(x^2+y^2) + 3\lambda^2 \quad \text{.....................(11)},$$

or, from the value of P,

$$-3\lambda^2 + Q = 0 \quad \text{.........................(12)},$$

which singularly enough is the derived equation of (7′) with respect to λ: so that the equation of the curve is obtained by expressing that two of the roots of the equation (7′) are equal. Multiplying (12) by λ and reducing by (7′),

$$-\lambda Q + 3R = 0,$$

or, combining this with (12),

$$27R^2 - Q^3 = 0\,;$$

whence, replacing R, Q by their values, we find

$$27k^4 (bx-ay)^2 (x^2+y^2-a^2-b^2)^2 - \{4(a^2+b^2)(x^2+y^2) - k^2[(a+x)^2+(b+y)^2]\}^3 = 0,$$

the equation of M. de St-Laurent.

43.

ON THE DIFFERENTIAL EQUATIONS WHICH OCCUR IN DYNAMICAL PROBLEMS.

[From the *Cambridge and Dublin Mathematical Journal*, vol. II. (1847), pp. 210—219.]

JACOBI, in a very elaborate memoir, "Theoria novi multiplicatoris systemati æquationum differentialium vulgarium applicandi"([1]), has demonstrated a remarkable property of an extensive class of differential equations, namely, that when all the integrals of the system except a single one are known, the remaining integral can always be determined by a quadrature. Included in the class in question are, as Jacobi proceeds to show, the differential equations corresponding to any dynamical problem in which neither the forces nor the equations of condition involve the velocities; i.e. in all ordinary dynamical problems, when all the integrals but one are known, the remaining integral can be determined by quadratures. In the case where the forces and equations of condition are likewise independent of the time, it is immediately seen that the system may be transformed into a system in which the number of equations is less by unity than in the original one, and which does not involve the time, which may afterwards be determined by a quadrature[2]; and, Jacobi's theorem applying to this new system, he arrives at the proposition "In any dynamical problem where the forces and equations of condition contain only the coordinates of the different points of the system, when all the integrals but two are determined, the remaining integrals may be found by quadratures only." In the following paper, which contains the demonstrations of these propositions, the analysis employed by Jacobi has been considerably varied in the details, but the leading features of it are preserved.

[1] *Crelle*, t. XXVII. [1844], pp. 199—268 and t. XXIX. [1845], pp. 213—279 and 333—376. Compare also the memoir in *Liouville*, t. X. [1845], pp. 337—346.

[2] For, representing the velocities by x', y' ... the dynamical system takes the form

$$dt : dx : dy \ldots : dx' : dy' \ldots = 1 : x' : y' \ldots : X : Y \ldots,$$

and the system in question is simply $dx : dy \ldots : dx' : dy' \ldots = x' : y' \ldots : X : Y \ldots$.

§ 1. Let the variables $x, y, z, \ldots$ &c. be connected with the variables $u, v, w, \ldots$ by the same number of equations, so that the variables of each set may be considered as functions of those of the other set. And assume

$$dx\, dy \ldots = \nabla\, du\, dv \ldots ;$$

if from the functions which equated to zero express the relations between the two sets of variables we form two determinants, the former with the differential coefficients of these functions with respect to $u, v, \ldots$ and the latter with the differential coefficients of the same functions with respect to $x, y, \ldots$ the quotient with its sign changed obtained by dividing the first of these determinants by the second is, as is well known, the value of the function ∇.

Putting for shortness

$$\frac{dx}{du} = \alpha, \quad \frac{dy}{du} = \beta, \ldots ; \quad \frac{dx}{dv} = \alpha', \quad \frac{dy}{dv} = \beta', \ldots \text{ \&c.}$$

and

$$\frac{du}{dx} = A, \quad \frac{du}{dy} = B, \ldots ; \quad \frac{dv}{dx} = A', \quad \frac{dv}{dy} = B', \ldots$$

∇ is the reciprocal of the determinant formed with $A, B, \ldots ; A', B', \ldots$, &c.; or it is the determinant formed with $\alpha, \beta, \ldots \alpha', \beta', \ldots$, &c.

From the first of these forms, i.e. considering ∇ as a function of $A, B, \ldots$

$$\frac{d\nabla}{dA} = -\nabla\alpha, \quad \frac{d\nabla}{dB} = -\nabla\beta, \ldots \quad \frac{d\nabla}{dA'} = -\nabla\alpha', \quad \frac{d\nabla}{dB'} = -\nabla\beta', \ldots$$

where the quantities $\alpha, \beta, \ldots \alpha', \beta', \ldots$ and $A, B, \ldots A', B', \ldots$ may be interchanged provided $-\nabla$ be substituted for ∇. (Demonstrations of these formulæ or of some equivalent to them will be found in Jacobi's memoir "De determinantibus functionalibus," *Crelle*, t. XXII. [(1841) pp. 319—359].)

Hence

$$\frac{1}{\nabla} d\nabla + \alpha dA + \beta dB + \ldots + \alpha' dA' + \beta' dB' + \ldots = 0,$$

or reducing by

$$\frac{dA}{dy} = \frac{dB}{dx}, \ldots ; \quad \frac{dA'}{dy} = \frac{dB'}{dx}, \ldots \text{ \&c.}$$

this becomes

$$\left.\begin{aligned} &\frac{1}{\nabla} d\nabla + \alpha\left(\frac{dA}{dx}\, dx + \frac{dB}{dx}\, dy + \ldots\right) + \beta\left(\frac{dA}{dy}\, dx + \frac{dB}{dy}\, dy + \ldots\right) \ldots \\ &\quad + \alpha\left(\frac{dA'}{dx}\, dx + \frac{dB'}{dx}\, dy + \ldots\right) + \beta'\left(\frac{dA'}{dy}\, dx + \frac{dB}{dy}\, dy + \ldots\right) \ldots \\ &\quad \vdots \end{aligned}\right\} = 0 ;$$

or, reducing,

$$\frac{1}{\nabla}d\nabla + \left(\frac{dA}{du} + \frac{dA'}{dv} + \ldots\right)dx + \left(\frac{dB}{du} + \frac{dB'}{dv} + \ldots\right)dy + \ldots = 0\,;$$

whence separating the differentials and replacing $A, A', \ldots; B, B', \ldots$; by their values

$$\frac{1}{\nabla}\frac{d\nabla}{dx} + \frac{d}{du}\cdot\frac{du}{dx} + \frac{d}{dv}\cdot\frac{dv}{dx} + \ldots = 0,$$

$$\frac{1}{\nabla}\frac{d\nabla}{dy} + \frac{d}{du}\cdot\frac{du}{dy} + \frac{d}{dv}\cdot\frac{dv}{dy} + \ldots = 0,$$

$$\vdots$$

(in which $-\nabla, u, v \ldots; x, y \ldots$ may be substituted for $\nabla, x, y \ldots; u, v \ldots$).

§ 2. Let $X, Y \ldots$ be any functions of the variables $x, y, \ldots$ and assume

$$U = X\frac{du}{dx} + Y\frac{du}{dy} + \ldots\,;$$

$$V = X\frac{dv}{dx} + Y\frac{dv}{dy} + \ldots\,;$$

$U, V, \ldots$ being expressed in terms of $u, v, \ldots$. Then

$$\frac{dU}{du} + \frac{dV}{dv} + \ldots = X\left(\frac{d}{du}\cdot\frac{du}{dx} + \frac{d}{dv}\cdot\frac{dv}{dx} + \ldots\right) + Y\left(\frac{d}{du}\cdot\frac{du}{dy} + \frac{d}{dv}\cdot\frac{dv}{dy} + \ldots\right)\ldots$$

$$+\left(\frac{dX}{du}\cdot\frac{du}{dx} + \frac{dX}{dv}\cdot\frac{dv}{dx} + \ldots\right) + \left(\frac{dY}{du}\cdot\frac{du}{dy} + \frac{dY}{dv}\cdot\frac{dv}{dy} + \ldots\right)\ldots$$

i.e.
$$\nabla\left(\frac{dU}{du} + \frac{dV}{dv} + \ldots\right) = -\left(X\frac{d\nabla}{dx} + Y\frac{d\nabla}{dy} + \ldots\right) + \nabla\left(\frac{dX}{dx} + \frac{dY}{dy} + \ldots\right).$$

Also, whatever be the value of M,

$$U\frac{dM\nabla}{du} + V\frac{dM\nabla}{dv} + \ldots = X\frac{dM\nabla}{dx} + Y\frac{dM\nabla}{dy} + \ldots;$$

and from these two properties,

$$\frac{dM\nabla U}{du} + \frac{dM\nabla V}{dv} + \ldots = \nabla\left(\frac{dMX}{dx} + \frac{dMY}{dy} + \ldots\right).$$

§ 3. Consider the system of differential equations

$$dx : dy : dz \ldots = X : Y : Z \ldots$$

(where, for greater clearness, an additional letter z has been introduced). From these we deduce the equivalent system

$$du : dv : dw \ldots = U : V : W \ldots.$$

Suppose that u and v continue to represent arbitrary functions of $x, y, z, \ldots$ but that the remaining functions $w, \ldots$ are such as to satisfy $W=0, \ldots$ (so that $w, \ldots$ may be considered as the constants introduced by obtaining all the integrals but one of the system of differential equations in $x, y, z, \ldots$), we have

$$\frac{dM\nabla U}{du}+\frac{dM\nabla V}{dv}=\nabla\left(\frac{dMX}{dx}+\frac{dMY}{dy}+\frac{dMZ}{dz}+\ldots\right).$$

Also the only one of the transformed equations which remains to be integrated is

$$du : dv = U : V, \quad \text{or } Vdu - Udv = 0,$$

(in which it is supposed that U and V are expressed by means of the other integrals in terms of u and v).

Suppose M can be so determined that

$$\frac{dMX}{dx}+\frac{dMY}{dy}+\frac{dMZ}{dz}+\ldots=0,$$

(M is what Jacobi terms the multiplier of the proposed system of differential equations): then

$$\frac{dM\nabla U}{du}+\frac{dM\nabla V}{dv}=0,$$

or $M\nabla$ is the multiplier of $Vdu - Udv = 0$, so that

$$\int M\nabla\,(Vdu - Udv) = \text{const.}$$

Hence the theorem:—"Given a multiplier of the system of equations

$$dx : dy : dz, \ldots = X : Y : Z \ldots$$

(the meaning of the term being defined as above), then if all the integrals but one of this system are known, the remaining integral depends upon a quadrature."

Jacobi proceeds to discuss a variety of different systems of equations in which it is possible to determine the multiplier M. Among the most important of these may be considered the system corresponding to the general problem of Dynamics, which may be discussed under three different forms.

§ 4. Lagrange's first form[1].

Let the *whole series* of coordinates, each of them multiplied by the square root of the corresponding mass, be represented by $x, y, \ldots$ and in the same way the whole series of forces, each of them multiplied by the square root of the corresponding mass, by $P, Q, \ldots$; then the equations of motion are

$$\frac{d^2x}{dt^2}=X, \quad \frac{d^2y}{dt^2}=Y, \ldots;$$

[1] I have slightly modified the form so as to avoid the introduction of the masses, and to allow x (for instance) to stand for any one of the coordinates of any of the points, instead of standing for a coordinate parallel to a particular axis.

where

$$X = P + \lambda \frac{d\Theta}{dx} + \mu \frac{d\Phi}{dx} + \ldots,$$

$$Y = Q + \lambda \frac{d\Theta}{dy} + \mu \frac{d\Phi}{dy} + \ldots,$$

$$\vdots$$

where $\Theta = 0$, $\Phi = 0$, ... are the equations of condition connecting the variables, and λ, μ, ... coefficients to be determined by substituting the values of $\frac{d^2x}{dt^2}$, &c. in the equations $\frac{d^2\Theta}{dt^2} = 0$, $\frac{d^2\Phi}{dt^2} = 0$, &c. It is supposed that as well P, Q, ... as Θ, Φ, ... are independent of the velocities.

In order to reduce these to an analogous form to that previously employed, we have only to write

$$\frac{dx}{dt} = x', \quad \frac{dy}{dt} = y', \ \ldots$$

which gives

$$\begin{aligned} dt : dx : dy : dz \ \ldots : dx' : dy' : dz' \ \ldots \\ = 1 : x' : y' : z' \ \ldots : X : Y : Z \ \ldots \end{aligned}$$

Supposing that M is independent of x', y', z', ... the equation on which it depends becomes immediately

$$\delta M + M \left(\frac{dX}{dx'} + \frac{dY}{dy'} + \frac{dZ}{dz'} + \ldots\right) = 0,$$

where for shortness

$$\delta = \frac{d}{dt} + x' \frac{d}{dx} + y' \frac{d}{dy} + z' \frac{d}{dz} + \ldots.$$

To reduce this we must first determine the values of λ, μ ..., and for this we have

$$\frac{d^2\Theta}{dt^2} = \delta^2\Theta + \frac{d\Theta}{dx}\frac{d^2x}{dt^2} + \frac{d\Theta}{dy}\frac{d^2y}{dt^2} + \ldots \qquad = 0, \ \&c.$$

i.e.

$$\delta^2\Theta + P\frac{d\Theta}{dx} + Q\frac{d\Theta}{dy} + \ldots + a\lambda + h\mu + g\nu + \ldots = 0,$$

$$\delta^2\Phi + P\frac{d\Phi}{dx} + Q\frac{d\Phi}{dy} + \ldots + h\lambda + b\mu + f\nu + \ldots = 0,$$

$$\delta^2\Psi + P\frac{d\Psi}{dx} + Q\frac{d\Psi}{dy} + \ldots + g\lambda = f\mu + c\nu + \ldots = 0.$$

$$\vdots$$

where for greater clearness an additional letter of the series Θ, Φ ... has been introduced, and where

$$a = \left(\frac{d\Theta}{dx}\right)^2 + \left(\frac{d\Phi}{dy}\right)^2 + \dots,$$

$$b = \left(\frac{d\Phi}{dx}\right)^2 + \left(\frac{d\Phi}{dy}\right)^2 + \dots,$$

$$\vdots$$

$$h = \left(\frac{d\Theta}{dx}\frac{d\Phi}{dx} + \frac{d\Theta}{dy}\frac{d\Phi}{dy}\right) + \dots.$$

$$\vdots$$

Hence differentiating with respect to x',

$$2\delta\frac{d\Theta}{dx} + a\frac{d\lambda}{dx'} + h\frac{d\mu}{dx'} + g\frac{d\nu}{dx'}\dots = 0,$$

$$2\delta\frac{d\Phi}{dx} + h\frac{d\lambda}{dx'} + b\frac{d\mu}{dx'} + f\frac{d\nu}{dx'}\dots = 0,$$

$$y\delta\frac{d\Psi}{dx} + g\frac{d\lambda}{dx'} + f\frac{d\mu}{dx'} + c\frac{d\nu}{dx'}\dots = 0;$$

$$\vdots$$

or representing by K the determinant formed with the quantities $a, h, g, \dots h, b, f, \dots g, f, c, \dots$ and by $A, H, G, \dots H, B, F, \dots G, F, C, \dots$ the inverse system of coefficients, we have

$$2\left(A\delta\frac{d\Theta}{dx} + H\delta\frac{d\Phi}{dx} + G\delta\frac{d\Psi}{dx}\dots\right) + K\frac{d\lambda}{dx'} = 0,$$

$$2\left(H\delta\frac{d\Theta}{dx} + B\delta\frac{d\Phi}{dx} + F\delta\frac{d\Psi}{dx}\dots\right) + K\frac{d\mu}{dx'} = 0,$$

$$2\left(G\delta\frac{d\Theta}{dx} + F\delta\frac{d\Phi}{dx} + C\delta\frac{d\Psi}{dx}\dots\right) + K\frac{d\nu}{dx'} = 0,$$

$$\vdots$$

whence, multiplying by $\frac{d\Theta}{dx}, \frac{d\Phi}{dx}, \frac{d\Psi}{dx}, \dots$ and adding,

$$A\delta\left(\frac{d\Theta}{dx}\right)^2 + B\delta\left(\frac{d\Phi}{dx}\right)^2 \dots + 2H\delta\frac{d\Theta}{dx}\frac{d\Phi}{dx}\dots + K\frac{dX}{dx'} = 0,$$

and, forming the similar equations with the remaining variables and adding,

$$A\delta a + B\delta b + C\delta c \dots + 2F\delta f + 2G\delta g + 2H\delta h + \dots + K\left(\frac{dX}{dx'} + \frac{dY}{dy'} + \frac{dZ}{dz'}\dots\right) = 0;$$

i.e.

$$\delta K + K\left(\frac{dX}{dx'} + \frac{dY}{dy'} + \frac{dZ}{dz'} + \dots\right) = 0.$$

Thus the equation in M reduces itself to

$$K\delta M - M\delta K = 0,$$

which is satisfied by $M = K$. It may be remarked that K reduces itself to the sum of the squares of the different functional determinants formed with the differential coefficients of Θ, Φ ... with respect to the different combinations of as many variables out of the series x, y ....

§ 5. Lagrange's second form.

Here the equations of motion are assumed to be

$$\frac{d}{dt}\frac{dT}{dx'} - \frac{dT}{dx} - P = 0,$$

$$\frac{d}{dt}\frac{dT}{dy'} - \frac{dT}{dy} - Q = 0,$$

$$\frac{d}{dt}\frac{dT}{dz'} - \frac{dT}{dz} - R = 0,$$

$$\vdots$$

where $2T$ represents the vis viva of the system, x, y, z, ... are the independent variables on which the solution of the problem depends, and x', y', z',... their differential coefficients with respect to the time. It is assumed as before P, Q, R ... do not contain x', y', z', ...

Suppose these equations give

$$\begin{aligned} &dt : dx : dy : dz \ldots : dx' : dy' : dz' \ldots \\ &= 1 : x' : y' : z' \ldots : X : Y : Z \ldots; \end{aligned}$$

then the equation which determines the multiplier M takes as before the form

$$\delta M + M\left(\frac{dX}{dx'} + \frac{dY}{dy'} + \frac{dZ}{dz'} + \ldots\right) = 0.$$

To reduce this equation, substituting for T its value which is of the form

$$T = \tfrac{1}{2}(ax'^2 + by'^2 + cz'^2 \ldots + 2fy'z' + 2gz'x' + 2hx'y' \ldots),$$

and putting for shortness

$$\begin{aligned} L &= ax' + hy' + gz' \ldots, \\ M &= hx' + by' + fz' \ldots, \\ N &= gx' + fy' + cz' \ldots \\ &\vdots \end{aligned}$$

the equations which determine X, Y, Z ... are

$$aX + hY + gZ \dots + \delta L - \frac{dT}{dx} - P = 0,$$

$$hX + bY + fZ \dots + \delta M - \frac{dT}{dy} - Q = 0,$$

$$gX + fY + cZ \dots + \delta N - \frac{dT}{dz} - R = 0.$$

$$\vdots$$

Hence, differentiating with respect to x',

$$a\frac{dX}{dx'} + h\frac{dY}{dx'} + g\frac{dZ}{dx'} \dots + \delta a \qquad\qquad = 0,$$

$$h\frac{dX}{dx'} + b\frac{dY}{dx'} + f\frac{dZ}{dx'} \dots + \delta h + \frac{dM}{dx} - \frac{dL}{dy} = 0,$$

$$g\frac{dX}{dx'} + f\frac{dY}{dx'} + c\frac{dZ}{dx'} \dots + \delta g + \frac{dN}{dx} - \frac{dL}{dz} = 0;$$

$$\vdots$$

or representing by K the determinant formed with $a, h, g, \dots h, b, f, \dots g, f, c, \dots$ and by $A, H, G, \dots H, B, F, \dots G, F, C \dots$ the inverse system of coefficients, we have

$$K\frac{dX}{dx'} + A\delta a + H\delta h + G\delta g \dots + \quad * \quad + H\left(\frac{dM}{dx} - \frac{dL}{dy}\right) + G\left(\frac{dN}{dx} - \frac{dL}{dz}\right) \dots = 0,$$

and similarly

$$K\frac{dY}{dy'} + H\delta h + B\delta b + F\delta f \dots + H\left(\frac{dL}{dy} - \frac{dM}{dx}\right) + \quad * \quad + F\left(\frac{dN}{dy} - \frac{dM}{dz}\right) \dots = 0,$$

$$K\frac{dZ}{dz'} + G\delta g + F\delta f + C\delta c \dots + G\left(\frac{dL}{dz} - \frac{dN}{dx}\right) + F\left(\frac{dM}{dz} - \frac{dN}{dy}\right) + \quad * \quad \dots = 0.$$

$$\vdots$$

Hence, adding,

$$K\left(\frac{dX}{dx'} + \frac{dY}{dy'} + \frac{dZ}{dz'} \dots\right) + A\delta a + B\delta b + C\delta c \dots + 2F\delta f + 2G\delta g + 2H\delta h \dots = 0;$$

and thus we have as before, though with symbols bearing an entirely different signification,

$$K\left(\frac{dX}{dx'} + \frac{dY}{dy'} + \frac{dZ}{dz'} + \dots\right) + \delta K = 0;$$

and thence $K\delta M - M\delta K = 0$, and $M = K$.

(The value of K *in this section* may I think be conveniently termed "the determinant of the vis viva," with respect to the variables $x, y, z, \ldots$. It may be remarked that "the determinant of the vis viva" with respect to any other system of variables $u, v, w, \ldots$ is $=\nabla^2 K$, ∇ as before.)

§ 6. Third form of the equations of motion. [Hamiltonian form.]

Here writing

$$\frac{dT}{dx'} = \xi, \ \frac{dT}{dy'} = \eta, \ldots,$$

and taking $t, x, y, \ldots \xi, \eta, \ldots$ for the variables of the problem [and considering T to be expressed as a function of these variables: to denote this change it would have been proper to use instead of T a new letter H] the equations of motion reduce themselves to

$$\left\{\begin{array}{ll} \dfrac{d\xi}{dt} = -\dfrac{dT}{dx} + P, & \dfrac{dx}{dt} = \dfrac{dT}{d\xi}, \\[2ex] \dfrac{d\eta}{dt} = -\dfrac{dT}{d\eta} + Q, & \dfrac{dy}{dt} = \dfrac{dT}{d\eta}; \\ \vdots & \vdots \end{array}\right.$$

or putting for shortness

$$\left\{\begin{array}{ll} P - \dfrac{dT}{dx} = X, & \dfrac{dT}{d\xi} = \Xi, \\[2ex] Q - \dfrac{dT}{d\eta} = Y, & \dfrac{dT}{d\eta} = \mathrm{H}, \\ \vdots & \vdots \end{array}\right.$$

they become

$$\begin{array}{l} dt : dx : dy : dz \ldots : d\xi : d\eta : d\zeta \ldots \\ = 1 : \Xi : \mathrm{H} : \Omega \ldots : X : Y : Z \ldots. \end{array}$$

Hence writing the equation in M under the form

$$\delta M + M \left(\frac{d\Xi}{dx} + \frac{d\mathrm{H}}{dy} + \ldots + \frac{dX}{d\xi} + \frac{dY}{d\eta} + \ldots \right) = 0;$$

$$\left(\text{where } \delta = \frac{d}{dt} + \Xi \frac{d}{dx} + \mathrm{H} \frac{d}{dy} \ldots + X \frac{d}{d\xi} + Y \frac{d}{dy} + \ldots \right),$$

we see immediately that (P, Q ... being as before independent of the velocities, and consequently of $\xi, \eta, \zeta, \ldots$),

$$\frac{d\Xi}{dx} + \frac{dX}{d\xi} = 0, \ \frac{d\mathrm{H}}{dy} + \frac{dY}{d\eta} = 0, \ \&\text{c.}$$

Hence $\delta M = 0$, which is satisfied by $M = 1$.

44.

ON A MULTIPLE INTEGRAL CONNECTED WITH THE THEORY OF ATTRACTIONS.

[From the *Cambridge and Dublin Mathematical Journal*, vol. II. (1847), pp. 219—223.]

MR BOOLE [in the Memoir "On a Certain Multiple definite Integral" Irish Acad. Trans. vol. XXI. (1848), pp. 40—150] has given for the integral with n variables

$$V = \int \frac{\phi\left(\frac{x^2}{f^2} + \frac{y^2}{g^2} + \ldots\right) dx\, dy \ldots}{[(a-x)^2 + (b-y)^2 \ldots + u^2]^{\frac{1}{2}n+q}} \quad \ldots\ldots(1);$$

limits

$$\frac{x^2}{f^2} + \frac{y^2}{g^2} + \ldots \overline{\overline{<}}\, 1,$$

the following formula, or one which may readily be reduced to that form[1],

$$V = \frac{fg \ldots \pi^{\frac{1}{2}n}}{\Gamma(\frac{1}{2}n + q)} \int_\eta^\infty \frac{Ss^{-q-1}\, ds}{\sqrt{\{(s+f^2)(s+g^2)\ldots\}}} \quad \ldots\ldots(2),$$

where

$$S = \frac{(1-\sigma)^{-q}}{\Gamma(-q)} \int_0^1 t^{-q-1} \phi\{\sigma + t(1-\sigma)\}\, dt \quad \ldots\ldots(3);$$

in which

$$\sigma = \frac{a^2}{f^2+s} + \frac{b^2}{g^2+s} \ldots + \frac{u^2}{s} \quad \ldots\ldots(4)$$

and η is determined by

$$1 = \frac{a^2}{f^2+\eta} + \frac{b^2}{g^2+\eta} \ldots + \frac{u^2}{\eta}.$$

[1] See note at the end of this paper.

Suppose $f = g = \ldots = \infty$; also assume

$$\phi(\lambda) = \frac{1}{(f^2\lambda + v^2)^{\frac{1}{2}n+q'}} \quad \ldots\ldots\ldots\ldots\ldots\ldots\ldots\ldots\ldots\ldots\ldots(5);$$

then the integral becomes

$$U = \int \frac{dx\, dy \ldots}{(x^2 + y^2 \ldots + v^2)^{\frac{1}{2}n+q'} \{(x-a)^2 + (y-b^2) \ldots + u^2\}^{\frac{1}{2}n+q}} \quad \ldots\ldots\ldots\ldots\ldots\ldots(6),$$

the limits for each variable being $-\infty$, ∞.

Now, writing $f^2 s$ for s and $f^2\eta$ for η, the new value of η reduces itself to zero, and

$$U = \frac{f^{-2q}\, \pi^{\frac{1}{2}n}}{\Gamma(\frac{1}{2}n + q)} \int_0^\infty \frac{S ds}{(1+s)^{\frac{1}{2}n}}.$$

Also $\sigma = 0$; but

$$f^2\sigma = \frac{r^2}{1+s} + \frac{u^2}{s},$$

where $r^2 = a^2 + b^2 + \ldots$ whence also putting $\frac{t}{f^2}$ for t, $\phi\{\sigma + t(1-\sigma)\}$ becomes

$$\frac{1}{\{f^2\sigma + t(1-\sigma) + v^2\}^{\frac{1}{2}n+q'}};$$

i.e.

$$\frac{1}{(t+A)^{\frac{1}{2}n+q'}},$$

if for a moment

$$A = \frac{r^2}{1+s} + \frac{u^2}{s} + v^2.$$

Hence

$$S = \frac{f^{2q}}{\Gamma(-q)} \int_0^\infty \frac{t^{-q-1}\, dt}{(t+A)^{\frac{1}{2}n+q'}}$$

$$= \frac{\Gamma(\frac{1}{2}n + q + q')}{\Gamma(\frac{1}{2}n + q')} \frac{f^{2q}}{A^{\frac{1}{2}n+q+q'}};$$

or substituting in U, and replacing A by its value,

$$U = \frac{\pi^{\frac{1}{2}n}\, \Gamma(\frac{1}{2}n + q + q')}{\Gamma(\frac{1}{2}n + q)\, \Gamma(\frac{1}{2}n + q')} \int_0^\infty \frac{s^{-q-1}\, ds}{(1+s)^{\frac{1}{2}n} \left(\frac{r^2}{1+s} + \frac{u^2}{s} + v^2\right)^{\frac{1}{2}n+q+q'}};$$

or, what comes to the same,

$$U = \frac{\pi^{\frac{1}{2}n}\, \Gamma(\frac{1}{2}n + q + q')}{\Gamma(\frac{1}{2}n + q)\, \Gamma(\frac{1}{2}n + q')} \int_0^\infty \frac{s^{\frac{1}{2}n+q'-1} (1+s)^{q+q'}\, ds}{(v^2 s^2 + js + u^2)^{\frac{1}{2}n+q+q'}} \quad \ldots\ldots\ldots\ldots\ldots\ldots(7)$$

where

$$j = u^2 + v^2 + r^2.$$

The only practicable case is that of $q' = -q$, for which

$$U = \frac{\pi^{\frac{1}{2}n}\,\Gamma(\frac{1}{2}n)}{\Gamma(\frac{1}{2}n+q)\,\Gamma(\frac{1}{2}n-q)}\int_0^\infty \frac{s^{\frac{1}{2}n-q-1}\,ds}{(v^2s^2+js+u^2)^{\frac{1}{2}n}} \quad\text{.........................(8).}$$

Consider the more general expression

$$\Theta = \int_0^\infty s^{-q-1}\,\phi\left(\frac{v^2s^2+js+u^2}{s}\right)ds \quad\text{.............................(9);}$$

by writing $2u\sqrt{s} = \sqrt{(s'+4uv)} \pm \sqrt{s'}$,

the upper sign from $s=\infty$ to $s=\frac{u}{v}$, and the lower one from $s=\frac{u}{v}$ to $s=0$, it is easy to derive

$$\Theta = (2v)^{2q}\int_0^\infty \frac{\{\sqrt{(s+4uv)}+\sqrt{s}\}^{-2q}+\{\sqrt{(s+4uv)}-\sqrt{s}\}^{-2q}}{\sqrt{s}\,\sqrt{(s+4uv)}}\,\phi(s+j+2uv)\,ds \quad\text{........(10).}$$

Now, by a formula which will presently be demonstrated,

$$\int_0^\infty \frac{\{\sqrt{(s+4uv)}+\sqrt{s}\}^{-2q}+\{\sqrt{(s+4uv)}-\sqrt{s}\}^{-2q}}{\sqrt{s}\,\sqrt{(s+4uv)}}\,e^{-\theta s}\,ds$$

$$= \frac{2\sqrt{\pi}}{\Gamma(\frac{1}{2}-q)}\,\theta^{-q}\int_0^\infty s^{-\frac{1}{2}-q}(s+4uv)^{-\frac{1}{2}-q}\,e^{-\theta s}\,ds \quad\text{..................(11);}$$

whence

$$\int_0^\infty \frac{\{\sqrt{(s+4uv)}+\sqrt{s}\}^{-2q}+\{\sqrt{(s+4uv)}-\sqrt{s}\}^{-2q}}{\sqrt{s}\,\sqrt{(s+4uv)}}\,Fs\,.\,ds$$

$$= \frac{2\sqrt{\pi}}{\Gamma(\frac{1}{2}-q)}\int_0^\infty s^{-\frac{1}{2}-q}(s+4uv)^{-\frac{1}{2}-q}\left(-\frac{d}{ds}\right)^{-q} Fs\,.\,ds \quad\text{.........(12).}$$

Thus, by merely changing the function,

$$\Theta = \frac{2^{2q+1}\,v^{2q}\sqrt{\pi}}{\Gamma(\frac{1}{2}-q)}\int_0^\infty s^{-\frac{1}{2}-q}(s+4uv)^{-\frac{1}{2}-q}\left(-\frac{d}{ds}\right)^{-q}\phi(s+j+2uv)\,ds \quad\text{.........(13);}$$

and hence in the particular case in question

$$U = \frac{2^{2q+1}\,v^{2q}\,\pi^{\frac{1}{2}(n+1)}}{\Gamma(\frac{1}{2}-q)\,\Gamma(\frac{1}{2}n+q)}\int_0^\infty s^{-\frac{1}{2}-q}(s+4uv)^{-\frac{1}{2}-q}(s+j+2uv)^{-\frac{1}{2}n+q}\,ds \quad\text{.........(14),}$$

by means of the formula

$$\left(-\frac{d}{ds}\right)^{-q}(s+a)^{-\frac{1}{2}n} = \frac{\Gamma(\frac{1}{2}n-q)}{\Gamma(\frac{1}{2}n)}(s+a)^{-\frac{1}{2}n+q}.$$

But as there may be some doubt about this formula, which is not exactly equivalent either to Liouville's or Peacock's expression for the general differential coefficient of a power, it is worth while to remark that, by first transforming the $\frac{1}{2}n^{\text{th}}$ power into an exponential, and then reducing as above (thus avoiding the general differentiation), we should have obtained

$$U = \frac{2^{2q+1} v^{2q} \pi^{\frac{1}{2}(n+1)}}{\Gamma(\frac{1}{2}-q)\,\Gamma(\frac{1}{2}n+q)\,\Gamma(\frac{1}{2}n-q)} \int_0^\infty d\theta \int_0^\infty ds\, \theta^{\frac{1}{2}n-q-1} e^{-\theta(s+j+2uv)} s^{-\frac{1}{2}-q} (s+4uv)^{-\frac{1}{2}-q} e^{-\theta s},$$

which reduces itself to the equation (14) by simply performing the integration with respect to θ; thus establishing the formula beyond doubt[1]. The integral may evidently be effected in finite terms when either q or $q-\frac{1}{2}$ is integral. Thus for instance in the simplest case of all, or when $q=-\frac{1}{2}$,

$$U = \frac{\pi^{\frac{1}{2}(n-1)}}{v\Gamma\,\frac{1}{2}(n+1)} \frac{1}{(j+2uv)^{\frac{1}{2}(n-1)}} = \int_{-\infty}^{\infty} \frac{dx\,dy\ldots}{(x^2+y^2\ldots+v^2)^{\frac{1}{2}(n+1)}\,\{(x-a)^2+\ldots u^2\}^{\frac{1}{2}(n-1)}},$$

a formula of which several demonstrations have already been given in the *Journal.*

The following is a demonstration, though an indirect one, of the formula (11): in the first place

$$\int_0^\infty \frac{\{\sqrt{(s+4uv)}+\sqrt{s}\}^{-2q}+\{\sqrt{(s+4uv)}-\sqrt{s}\}^{-2q}}{\sqrt{s}\,\sqrt{(s+4uv)}} e^{-\theta s}\,ds$$

$$= \frac{2\Gamma(\frac{1}{2}-q)\,\theta^q e^{2uv\theta}}{\sqrt{\pi}\,(4uv)^{2q}} \int_0^\infty (4u^2v^2+x^2)^{q-\frac{1}{2}} e^{i\theta x}\,dx \quad \ldots\ldots\ldots\ldots(16),$$

(where as usual $i=\sqrt{-1}$): to prove this, we have

$$\int_{-\infty}^{\infty} (4u^2v^2+x^2)^{q-\frac{1}{2}} e^{i\theta x}\,dx = \frac{1}{\Gamma(\frac{1}{2}-q)} \int_{-\infty}^{\infty} dx \int_0^\infty dt\, t^{-\frac{1}{2}-q} e^{-t(4u^2v^2+x^2)+i\theta x}$$

$$= \frac{\sqrt{\pi}}{\Gamma(\frac{1}{2}-q)} \int_0^\infty dt\, t^{-1-q} e^{-4u^2v^2t-\frac{\theta^2}{4t}};$$

or, putting $4uv\sqrt{t}=\sqrt{(s+4uv)}\pm\sqrt{s}$ (which is a transformation already employed in the present paper), the formula required follows immediately.

Now, by a formula due to M. Catalan, but first rigorously demonstrated by M. Serret,

$$\int_0^\infty \frac{\cos ax\,dx}{(1+x^2)^n} = \frac{\pi}{(\Gamma n)^2} \int_0^\infty e^{-(a+2z)} (z+a)^{n-1} z^{n-1}\,dz,$$

(*Liouville*, t. VIII. [1843] p. 1), and by a slight modification in the form of this equation

$$\int_{-\infty}^{\infty} (4u^2v^2+x^2)^{q-\frac{1}{2}} e^{i\theta x}\,dx = \frac{\pi e^{-2uv\theta}(4uv)^{2q}}{\theta^{2q}\,\Gamma^2(\frac{1}{2}-q)} \int_0^\infty s^{-q-\frac{1}{2}} (s+4uv)^{-q-\frac{1}{2}} e^{-\theta s}\,ds,$$

which, compared with (16), gives the required equation.

[1] A paper by M. Schlömilch "Note sur la variation des constantes arbitraires d'une Integrale definie," *Crelle*, t. XXXIII. [1846], pp. 268—280, will be found to contain formulæ analogous to some of the preceding ones.

NOTE.—One of the intermediate formulæ of Mr Boole [in the Memoir referred to] may be written as follows:

$$S=\frac{1}{\pi}\int_0^1 d\alpha \int_0^\infty dv\, v^q \cos\left[(\alpha-\sigma)\,v+\tfrac{1}{2}\,q\pi\right]\phi\alpha,$$

or what comes to the same thing, putting $i=\sqrt{-1}$, and rejecting the impossible part of the integral,

$$S=\frac{1}{\pi}\,e^{\frac{1}{2}q\pi i}\int_0^1 d\alpha\int_0^\infty dv\, v^q\, e^{2v\,(\alpha-\sigma)}\,\phi\alpha,$$

i.e.

$$S=\int_0^1 I\phi\alpha\, d\alpha,\quad I=\frac{1}{\pi}\,e^{\frac{1}{2}q\pi i}\int_0^\infty dv\, v^q\, e^{iv\,(\alpha-\sigma)}.$$

Now $(\alpha-\sigma)$ being positive,

$$I=\frac{1}{\pi}\,e^{\frac{1}{2}q\pi i}\,\Gamma(q+1)\,e^{\frac{1}{2}(q+1)\pi i}\,(\alpha-\sigma)^{-q-1};$$

i.e.

$$I=\frac{1}{\pi}\,e^{(q+\frac{1}{2})\pi i}\,\Gamma(q+1)\,(\alpha-\sigma)^{-q-1},$$

or, retaining the real part only,

$$I=-\frac{1}{\pi}\sin q\pi\,\Gamma(q+1)\,(\alpha-\sigma)^{-q-1};$$

i.e.

$$I=\frac{1}{\Gamma(-q)}\,(\alpha-\sigma)^{-q-1}.$$

But $(\alpha-\sigma)$ being negative,

$$I=\frac{1}{\pi}\,e^{\frac{1}{2}q\pi i}\,\Gamma(q+1)\,e^{-\frac{1}{2}(q+1)\pi i}\,(\sigma-\alpha)^{-q-1};$$

i.e.

$$I=\frac{1}{\pi}\,e^{-\frac{1}{2}\pi i}\,\Gamma(q+1)\,(\sigma-\alpha)^{-q-1},$$

or, retaining the real part only, $I=0$.

Hence

$$S=\frac{1}{\Gamma(-q)}\int_\sigma^1(\alpha-\sigma)^{-q-1}\,\phi\alpha\, d\alpha;$$

or putting

$$\alpha=\sigma+t\,(1-\sigma),\ \text{or}\ \alpha-\sigma=t\,(1-\sigma),$$

$$S=\frac{(1-\sigma)^{-q}}{\Gamma(-q)}\int_0^1 t^{-q-1}\,\phi\left[\sigma+t\,(1-\sigma)\right]dt;$$

the expression in the text. Mr Boole's final value is

$$S=\left(-\frac{d}{d\sigma}\right)^q\phi(\sigma),$$

which, though simpler, appears to me to be in some respects less convenient.

45.

ON THE THEORY OF ELLIPTIC FUNCTIONS.

[From the *Cambridge and Dublin Mathematical Journal*, vol. II. (1847), pp. 256—266.]

ADOPTING the notation of the *Fund. Nova*, except that for shortness sn u, cn u, dn u are written instead of sinam u, cosam u, ∇ am u, let the functions $\Theta(u)$, H(u) be defined by the equations

$$\Theta u = \sqrt{\left(\frac{2KK'}{\pi}\right)}\, e^{\frac{1}{2}u^2\left(1-\frac{E}{K}\right)-k^2\int_0 du\int_0 du\,\mathrm{sn}^2 u} \quad \text{.......................(1)},$$

$$\mathrm{H} u = -\,ie^{-\frac{\pi(K'-2iu)}{4K}}\,\Theta(u+iK') \quad \text{...............................(2)},$$

it is required from these equations to express sn u in terms of the functions H(u), $\Theta(u)$. To accomplish this we have

$$\frac{d^2}{du^2}\log \mathrm{sn}\, u = \frac{1}{\mathrm{sn}\, u}\frac{d^2}{du^2}\mathrm{sn}\, u - \frac{1}{\mathrm{sn}^2 u}\left(\frac{d}{du}\mathrm{sn}\, u\right)^2$$

$$= -(1+k^2) + 2k^2\,\mathrm{sn}^2 u - \left\{\frac{1}{\mathrm{sn}^2 u} - (1+k^2) + k^2\,\mathrm{sn}^2 u\right\}$$

$$= k^2\,\mathrm{sn}^2 u - \frac{1}{\mathrm{sn}^2 u};$$

whence also

$$\frac{d^2}{du^2}\log \mathrm{sn}\, u = k^2\,\mathrm{sn}^2 u - k^2\,\mathrm{sn}^2(u+iK').$$

If for a moment

$$\psi_{,}\, u = \int_0 du\,\mathrm{sn}^2 u, \qquad \psi_{,,}\, u = \int_0 du\int_0 du\,\mathrm{sn}^2 u,$$

then $$\log \mathrm{sn}\, u = k^2\psi_{,,}\, u - k^2\,\psi_{,,}(u+iK') + Au + B;$$

or writing $-u$ for u and subtracting, $\psi_{,,}u$ being an even function,

$$2Au = \pi i - k^2\psi_{,,}(iK' - u) + k^2\psi_{,,}(iK' + u),$$

or putting $u = K$,

$$2AK = \pi i - k^2\psi_{,,}(iK' - K) + k^2\psi_{,,}(iK' + K).$$

Now $$\text{sn}^2(u + K) - \text{sn}^2(u - K) = 0,$$

and therefore $$\psi_{,}(u + K) - \psi_{,}(u - K) = 2\psi_{,}K,$$
$$\psi_{,,}(u + K) - \psi_{,,}(u - K) = 2u\psi_{,}K;$$

or $$\psi_{,,}(iK' + K) - \psi_{,,}(iK' - K) = 2iK'\psi_{,}K.$$

Also $$E(u) = u - k^2\psi_{,}u,$$

or $$E = K - k^2\psi_{,}K, \quad \text{i.e.} \quad \psi_{,}K = \frac{K}{k^2}\left(1 - \frac{E}{K}\right).$$

Hence

$$A = iK'\left(1 - \frac{E}{K}\right) + \frac{\pi i}{2K},$$

$$\log \text{sn}\, u = k^2\psi_{,,}u - k^2\psi_{,,}(u + iK') + uiK'\left(1 - \frac{E}{K}\right) + \frac{\pi ui}{2K} + B$$

$$= k^2\psi_{,,}u - k^2\psi_{,,}(u + iK') + \tfrac{1}{2}\left[(u + iK')^2 - u^2\right]\left(1 - \frac{E}{K}\right) + \frac{\pi ui}{2K} + B',$$

i.e. $$\log \text{sn}\, u = \log \Theta(u + iK') - \log \Theta u + \frac{\pi ui}{2K} + B',$$

or, changing the constant,

$$\text{sn}\, u = Ce^{\frac{\pi ui}{2K}}\frac{\Theta(u + iK')}{\Theta u}.$$

Now, to determine C, write $u - iK'$ for u; this gives

$$\frac{1}{k\, \text{sn}\, u} = Ce^{\frac{\pi i}{2K}(u - iK')}\frac{\Theta u}{\Theta(u - iK')}:$$

and again changing u into $-u$,

$$-\text{sn}\, u = Ce^{-\frac{\pi ui}{2K}}\frac{\Theta(u - iK')}{\Theta u}:$$

whence, multiplying these last two equations,

$$C^2 = -\frac{1}{k}e^{-\frac{\pi K'}{2K}},$$

or

$$C = \frac{1}{i\sqrt{k}} e^{-\frac{\pi K'}{4K}};$$

whence

$$\text{sn}\, u = \frac{1}{i\sqrt{k}} e^{-\frac{\pi(K'-2iu)}{4K}} \frac{\Theta(u+iK')}{\Theta u},$$

i.e.

$$\sqrt{k}\, \text{sn}\, u = \frac{\text{H}(u)}{\Theta(u)} \quad \ldots\ldots\ldots (3);$$

and the equations (1), (2) and (3) may be considered as comprehending the theory of the functions $\text{H}(u)$, $\Theta(u)$. The preceding process is, in fact, the converse of that made use of in the *Fund. Nova;* Jacobi having obtained for $\text{sn}\, u$ an expression in the form of a fraction, takes the numerator of it for $\text{H}(u)$ and the denominator for $\Theta(u)$, and thence deduces the equations (1), (2), the intermediate steps of the demonstration being conducted by means of infinite series; the necessity of which is avoided by the preceding investigation.

I proceed to investigate certain results relating to these functions, and to the theory of elliptic functions which have been given by Jacobi in two papers, "Suite des notices sur les fonctions elliptiques," *Crelle,* t. III. [1828] p. 306, and t. IV. [1829] p. 185, but without demonstration.

In the first place, the equation

$$\frac{d^2\Sigma}{du^2} - 2u\left(k'^2 - \frac{E}{K}\right)\frac{d\Sigma}{du} + 2kk'^2\frac{d\Sigma}{dk} = 0 \quad \ldots\ldots\ldots (4)$$

is satisfied by $\Sigma = \Theta(u)$ or $\Sigma = \text{H}(u)$. It will be sufficient to prove this for $\Sigma = \Theta(u)$, since a similar demonstration may easily be found for the other value. The following preliminary formulæ will be required:

$$k\frac{dK}{dk} = \frac{E}{k'^2} - K, \qquad k\frac{dE}{dk} = E - K,$$

$$k\frac{dK'}{dk} = -\frac{E'}{k'^2} + \frac{k^2K}{k'^2}, \qquad KK' - EK' - E'K = -\tfrac{1}{2}\pi,$$

which are all of them known.

Now, writing $\Theta(u)$ under the slightly more convenient form

$$\Theta u = \sqrt{\left(\frac{2Kk'}{\pi}\right)}\, e^{\int_0 du \int_0 du\, \text{dn}^2 u - \frac{1}{2}u^2 \frac{E}{K}},$$

we have

$$\frac{d\,\Theta u}{du} = \left(\int_0 du\, \text{dn}^2 u - \frac{E}{K}u\right)\Theta u = \left\{u\left(k'^2 - \frac{E}{K}\right) + k^2\int_0 du\, \text{cn}^2 u\right\}\Theta u,$$

$$\frac{d^2\,\Theta u}{du^2} = \left[\text{dn}^2 u - \frac{E}{K} + \left\{u\left(k'^2 - \frac{E}{K}\right) + k^2\int_0 du\, \text{cn}^2 u\right\}^2\right]\Theta u,$$

$$\frac{d\,\Theta u}{dk} = \left[\frac{1}{2Kk'}\frac{dKk'}{dk} - \tfrac{1}{2}u^2\frac{d}{dk}\frac{E}{K} + \int_0 du \int_0 du \frac{d}{dk}\text{dn}^2 u\right]\Theta u.$$

The success of the process depends upon a transformation of the double integral

$$\int_0 du \int_0 du \frac{d}{dk} \operatorname{dn}^2 u \,;$$

to effect this we have

$$\frac{d}{dk} \operatorname{dn}^2 u = - 2k \operatorname{sn} u \left(\operatorname{sn} u + k \frac{d}{dk} \operatorname{sn} u\right);$$

but, by a known formula,

$$k'^2 \frac{d}{dk} \operatorname{sn} u = - k \operatorname{cn} u \operatorname{dn} u \int_0 \operatorname{cn}^2 u \, du + k \operatorname{cn}^2 u \operatorname{sn} u \,;$$

whence

$$\operatorname{sn} u + k \frac{d}{dk} \operatorname{sn} u = \frac{1}{k'^2} \operatorname{sn} u \operatorname{dn}^2 u - k^2 \operatorname{cn} u \operatorname{dn} u \int_0 du \operatorname{cn}^2 u,$$

or

$$\frac{d}{dk} \operatorname{dn}^2 u = - \frac{2k}{k'^2} (\operatorname{sn}^2 u \operatorname{dn}^2 u - k^2 \operatorname{sn} u \operatorname{cn} u \operatorname{dn} u \int_0 du \operatorname{cn}^2 u)$$

$$= - \frac{2k}{k'^2} \left\{\operatorname{sn}^2 u \operatorname{dn}^2 u + \tfrac{1}{2} k^2 \left(\frac{d}{du} \operatorname{cn}^2 u\right) \int_0 du \operatorname{cn}^2 u\right\};$$

whence

$$\int_0 du \int_0 du \frac{d}{dk} \operatorname{dn}^2 u = - \frac{2k}{k'^2} \{\int_0 du \int_0 du \operatorname{sn}^2 u \operatorname{dn}^2 v + \tfrac{1}{2} k^2 \int_0 du \,(\operatorname{cn}^2 u \int du \operatorname{cn}^2 u - \int du \operatorname{cn}^4 u)\}$$

$$= - \frac{k}{k'^2} \{\int_0 du \int_0 du \,(2 \operatorname{sn}^2 u \operatorname{dn}^2 u - k^2 \operatorname{cn}^4 u) + \tfrac{1}{2} k^2 (\int_0 du \operatorname{cn}^2 u)^2\}.$$

But

$$\frac{d^2}{du^2} \operatorname{sn}^2 u = 2 (\operatorname{cn}^2 u \operatorname{dn}^2 u - \operatorname{sn}^2 u \operatorname{dn}^2 u - k^2 \operatorname{sn}^2 u \operatorname{cn}^2 u) = 2 (k'^2 - 2 \operatorname{sn}^2 u \operatorname{dn}^2 u + k^2 \operatorname{cn}^4 u)\,;$$

or, integrating,

$$\operatorname{sn}^2 u = k'^2 u^2 - 2 \int_0 du \int_0 du \,(2 \operatorname{sn}^2 u \operatorname{dn}^2 u - k^2 \operatorname{cn}^4 u)\,;$$

whence at length

$$\int_0 du \int_0 du \frac{d}{dk} \operatorname{dn}^2 u = - \tfrac{1}{2} k u^2 + \tfrac{1}{2} \frac{k}{k'^2} \operatorname{sn}^2 u + \frac{k^3}{2k'^2} (\int_0 du \operatorname{cn}^2 u)^2.$$

Also

$$\frac{d}{dk} KK' = \frac{E - K}{kk'}, \quad \frac{d}{dk} \frac{E}{K} = \frac{1}{kk'^2} \left\{ k'^2 \left(\frac{2E}{K} - 1\right) - \frac{E^2}{K^2}\right\},$$

so that

$$\frac{d\,\Theta u}{dk} = \frac{1}{2kk'^2} \left\{\frac{E}{K} - \operatorname{dn}^2 u - u^2 \left(k'^2 - \frac{E}{K}\right)^2 - k^4 (\int du \operatorname{cn}^2 u)^2\right\} \Theta u.$$

Substituting these values of $\frac{d}{du}\Theta u$, $\frac{d^2}{du^2}\Theta u$ and $\frac{d}{dk}\Theta u$ in the equation (4) in the place of the corresponding differential coefficients of Σ, all the terms vanish, or the equation is satisfied by $\Sigma = \Theta(u)$, and similarly it would be satisfied by $\Sigma = \mathrm{H}(u)$.

Assume now

$$\omega = \frac{\pi K'}{K}, \quad \upsilon = \frac{\pi u}{2K};$$

then observing the equation

$$\frac{d}{dk}\frac{K'}{K} = \frac{1}{K^2 kk'^2}(KK' - KE' - K'E) = -\frac{\pi}{2K^2 kk'^2},$$

we have

$$\frac{d\Sigma}{du} = \frac{\pi}{2K}\frac{d\Sigma}{d\upsilon}, \quad \frac{d^2\Sigma}{du^2} = \frac{\pi^2}{4K^2}\frac{d^2\Sigma}{d\upsilon^2},$$

$$\frac{d\Sigma}{dk} = \frac{\upsilon}{kk'^2}\left(k'^2 - \frac{E}{K}\right)\frac{d\Sigma}{d\upsilon} - \frac{\pi^2}{2K^2kk'^2}\frac{d\Sigma}{d\omega};$$

whence, substituting in the equation (4), this becomes

$$\frac{d^2\Sigma}{d\upsilon^2} - 4\frac{d\Sigma}{d\omega} = 0 \quad \text{.........................(5)},$$

which is of course satisfied as before by $\Sigma = \Theta(u)$, or $\Sigma = \mathrm{H}(u)$, an equation demonstrated in a different manner (by means of expansions) by Jacobi in the Memoirs referred to.

Consider next the equation

$$\frac{d^2\Sigma}{du^2} - 2nu\left(k'^2 - \frac{E}{K}\right)\frac{d\Sigma}{du} + 2nkk'^2\frac{d\Sigma}{dk} = 0 \quad \text{...................(6)},$$

(n being any positive integer number). Then, by assuming

$$\omega = n\frac{\pi K'}{K}, \quad \upsilon = \frac{n\pi u}{K},$$

we should be led as before to the equation (5). Hence, considering Θu or $\mathrm{H}u$ as functions of u and $\frac{K'}{K}$, the equation (6) is satisfied by assuming for Σ a corresponding function of nu and $\frac{nK'}{K}$. Let λ be the modulus corresponding to a transformation of the nth order; then Λ, Λ' being the complete functions corresponding to this modulus, $\frac{\Lambda'}{\Lambda} = n\frac{K'}{K}$, so that the equation (6) will be satisfied by assuming $\Sigma = \Theta_{,}(nu)$ or $\Sigma = \mathrm{H}_{,}(nu)$, where $\Theta_{,}$, $\mathrm{H}_{,}$ correspond to the new modulus λ.

Assume now in the equation (6),

$$\Sigma = \left(\frac{\pi}{2}\right)^{\frac{1}{2}(n-1)} (Kk')^{-\frac{1}{2}(n-1)}\, \Theta^n u \,.\, z.$$

Hence, substituting,

$$\frac{d^2}{du^2}(\Theta^n u \,.\, z) - 2nu\left(k'^2 - \frac{E}{K}\right)\frac{d}{du}(\Theta^n u \,.\, z) + 2nkk'^2 (Kk')^{\frac{1}{2}(n-1)} \frac{d}{dk}\left[(Kk')^{-\frac{1}{2}(n-1)}\, \Theta^n u \,.\, z\right] = 0\,;$$

but $$(Kk')^{\frac{1}{2}(n-1)} \frac{d}{dk}\left[(Kk')^{-\frac{1}{2}(n-1)}\, \Theta^n u \,.\, z\right] = \frac{d}{dk}(\Theta^n u \,.\, z) - \frac{n-1}{2Kk'}\frac{dKk'}{dk}\,\Theta^n u \,.\, z,$$

or effecting the differentiation, and eliminating $\dfrac{d\Theta u}{dk}$ by means of the equation obtained from (4) by writing $\Sigma = \Theta u$,

$$(Kk')^{\frac{1}{2}(n-1)} \frac{d}{dk}\left[(Kk')^{-\frac{1}{2}(n-1)}\, \Theta^u u \,.\, z\right]$$

$$= \Theta^n u\left[\frac{dz}{dk} - \frac{nz}{2kk'^2\,\Theta u}\left\{\frac{d^2\,\Theta u}{du^2} - 2\left(k'^2 - \frac{E}{K}\right)\frac{d\,\Theta u}{du}\right\} + \frac{n-1}{2kk'^2}\left(1 - \frac{E}{K}\right)z\right].$$

Substituting in (6) and reducing,

$$\frac{d^2z}{du^2} + 2n\left[\frac{1}{\Theta n}\frac{d\,\Theta u}{du} - u\left(k'^2 - \frac{E}{K}\right)\right]\frac{dz}{du} + 2n\,kk'^2\frac{dz}{dk}$$
$$+ n(n-1)\left\{\left[\frac{1}{\Theta^2 u}\left(\frac{d\,\Theta u}{du}\right)^2 - \frac{1}{\Theta u}\frac{d^2\,\Theta u}{du^2}\right] + \left(1 - \frac{E}{K}\right)\right\} z = 0,$$

i.e. $$\frac{d^2z}{du^2} + 2n\left[\frac{d\log\Theta u}{du} - u\left(k'^2 - \frac{E}{K}\right)\right]\frac{dz}{du} + 2n\,kk'^2\frac{dz}{dk}$$
$$+ n(n-1)\left[-\frac{d^2\log\Theta u}{du^2} + \left(1 - \frac{E}{K}\right)\right] z = 0.$$

But
$$\frac{d\log\Theta u}{du} = u\left(k'^2 - \frac{E}{K}\right) = k^2 \textstyle\int_0 du\,\mathrm{cn}^2\,u,$$
$$\frac{d^2\log\Theta u}{du^2} = 1 - \frac{E}{K} - k^2\,\mathrm{sn}^2\,u\,;$$

whence

$$\frac{d^2z}{du^2} + 2nk^2\left(\textstyle\int_0 du\,\mathrm{cn}^2\,u\right)\frac{dz}{du} + 2nkk'^2\frac{dz}{dk} + n(n-1)\,k^2\,\mathrm{sn}^2\,u\,.\,z = 0 \ldots\ldots\ldots\ldots(7)\,;$$

which is therefore satisfied by

$$z = \left(\frac{2Kk'}{\pi}\right)^{\frac{1}{2}(n-1)}\frac{\Theta_{,}nu}{\Theta^n u}, \quad z = \left(\frac{2Kk'}{\pi}\right)^{\frac{1}{2}(n-1)}\frac{\mathrm{H}_{,}nu}{\Theta^n u}:$$

and each of these values is an algebraical function of $\operatorname{sn} u$, (viz. either a rational function or a rational function multiplied by $\operatorname{cn} u \operatorname{dn} u$). Also, in the transformation of the nth order,

$$\sqrt{\lambda}\, \operatorname{sn}_{,} u = \frac{\mathrm{H}_{,}(nu)}{\Theta_{,}(nu)};$$

so that it is clear that the above values of z may be taken for the denominator and numerator respectively of $\sqrt{\lambda}\, \operatorname{sn}_{,} u$; i.e. these quantities each of them satisfy the equation (7).

By assuming

$$x = \sqrt{k}\, \operatorname{sn} u, \quad \alpha = k + \frac{1}{k},$$

this becomes

$$n(n-1)\, x^2 z + (n-1)(\alpha x - 2x^3)\frac{dz}{dx} + (1 - \alpha x^2 + x^4)\frac{d^2 z}{dx^2} - 2n(\alpha^2 - 4)\frac{dz}{d\alpha} = 0 \quad \ldots\ldots\ldots(8);$$

which is therefore satisfied by assuming for z either the numerator or the denominator of $\sqrt{\lambda}\, \operatorname{sn}_{,} u$ (the transformation of the nth order), which is the form in which the property is given by Jacobi.

In the case where n is odd, the denominator is of the form

$$B_0 + B_{,} x^2 \ldots + B_{\frac{1}{2}(n-1)}\, x^{n-1},$$

and then the numerator is

$$x\,(B_{\frac{1}{2}(n-1)} \ldots + B_{,} x^{n-3} + B_0 x^{n-1}),$$

where

$$B_0 = \sqrt{\left(\frac{\lambda'}{k' M}\right)}, \quad B_{\frac{1}{2}(n-1)} = \sqrt{\left(\frac{\lambda\lambda'}{kk' M^3}\right)};$$

and all the remaining coefficients may be determined from these, the modular equation being supposed known. But the principal use of the formula is for the multiplication of elliptic functions, which it is well known corresponds to the case where n is a square number. Writing $n = \nu^2$, when ν is odd, the denominator is

$$1 + B_2 x^4 \ldots + B_{\frac{1}{2}(\nu^2-3)}\, x^{\nu^2-3} \pm \nu x^{\nu^2-1},$$

(the $\pm$ sign according as $\nu = (4p+1)$ or $(4p-1)$); and the numerator is obtained from this by multiplying by x and reversing the order of the coefficients. When ν is even the denominator is

$$1 + B_2 x^4 \ldots \pm B_2 x^{\nu^2-4} \pm x^{\nu^2},$$

(+ or −, according as $\nu = 4p$ or $\nu = 4p+2$), so that there are only half as many coefficients to be determined; but then the numerator must be separately investigated.

In general, by leaving n indeterminate, and integrating in the form of a series arranged according to ascending powers of x^2; then, whenever n is a square number, the series terminates and gives the denominator of the corresponding formula of multiplication; but the general form of the coefficients has not hitherto been discovered.

By writing $\frac{x}{\sqrt{n}}$ instead of x, and then making n infinite, the equation (8) takes the form

$$x^2 z + \alpha x \frac{dz}{dx} + \frac{d^2 z}{dx^2} - 2(\alpha^2 - 4)\frac{dz}{d\alpha} = 0 \quad \ldots\ldots\ldots\ldots(9):$$

and it is worth while, before attempting the solution of the general case, to discuss this more simple one[1].

Assume

$$z = 1 + C_1 \frac{x^2}{1.2} \ldots + C_r \frac{x^{2r}}{1.2 \ldots 2r} + \ldots;$$

then it is easy to obtain

$$C_{r+2} = -(2r+1)(2r+2)C_r - (2r+2)\alpha C_{r+1} + 2(\alpha^2 - 4)\frac{dC_{r+1}}{d\alpha}.$$

The general form may be seen to be

$$C_r = (-)^{r+1}\{2^{2r-3} C_r^1 \alpha^{r-2} + 2^{2r-6} C_r^2 \alpha^{r-4} + \ldots\},$$

and then

$$C_{r+1}^p - pC_r^p = -r(2r-1)C_{r-1}^{p-1} + 16(r+2-2p)C_r^{p-1}.$$

The complete value of C_r^p (assuming $C_r^0 = 0$) is given by an equation of the form

$$C_r^p = {}^0C_r^p + {}^1C_r^p\, 2^r + {}^2C_r^p\, 3^r \ldots + {}^{p-1}C_r^p\, p^r,$$

where ${}^0C_r^p$, ${}^1C_r^p$, ... are algebraical functions of r of the degrees $2p-2$, $2p-4$, &c. respectively; but as I am not able completely to effect the integration, and my only object being to give an idea of the law of the successive terms, it will be sufficient to consider the first or algebraical term ${}^0C_r^p$, which is determined by the same equation as C_r^p, and is moreover completely determined by this equation and the single additional

[1] Writing $(\beta+2)$ for α, and putting $z = e^{\frac{1}{2}x^2}\rho$, this becomes

$$\frac{d^2\rho}{dx^2} - \rho = \beta x^2 \rho - \beta x \frac{d\rho}{dx} + (8\beta + 2\beta^2)\frac{d\rho}{d\beta};$$

and if $\rho = \Sigma Z_n \beta^n$,

$$\frac{d^2 Z_n}{dx^2} - (8n+1) Z_n = \left(x^2 + 2n - 2 - x\frac{d}{dx}\right) Z_{n-1};$$

from which the successive values of Z_0, Z_1, &c. might be calculated.

relation $C_r^1 = 1$, since the arbitrary constants of the integration affect only the terms multiplied by 2^r, 3^r, &c.

Assume C_r^p

$$= \frac{1}{[p-1]^{p-1}} \{2^{p-1} L^p [r-2]^{2p-2} + 2^{p-2} M^p [r-3]^{2p-3} + \dots 2^{-p+1} X^p [r-2p]^0\};$$

and substituting this value,

$$\begin{aligned}
(1-p) L^p &= (1-p) \{L^{p-1} \}, \\
(1-p) M^p - 2p(2-2p) L^p &= (1-p) \{M^{p-1} - 11L^{p-1} \}, \\
(1-p) N^p - 2p(3-2p) M^p &= (1-p) \{N^{p-1} - 7M^{p-1} + 12L^{p-1}\}, \\
(1-p) O^p - 2p(4-2p) N^p &= (1-p) \{O^{p-1} - 3N^{p-1} + 30M^{p-1}\}, \\
&\vdots
\end{aligned}$$

the law of which is obvious, the coefficients on the second side in the qth line being 1, $4q-19$, and $(2q-3)(2q-2)$ respectively. By successive integrations and substitutions

$$\begin{aligned}
L^p - L^{p-1} &= 0, & L^p &= 1, \\
M^p - M^{p-1} &= 4p - 11, & M^p &= (p-1)(2p-7), \\
N^p - N^{p-1} &= -8p^3 + 26p^2 + 49p - 114; & &\vdots \\
&\vdots
\end{aligned}$$

(the constants determined by $M^1 = 0$, $N^1 = 0$, $O^2 = 0$, $P^2 = 0$, ... so as to make C_r^p contain positive powers only of r).

The following are a few of the complete values of C_r^p, the constants determined so as to satisfy $C_{p+1}^p = 0$ (except $C_2^1 = 1$), and the factorials being partially developed in powers of r, viz.

$$\begin{aligned}
&C_r^1 = 1, \\
&C_r^2 = (r-3)(2r-7), \\
&C_r^3 = \tfrac{1}{2}(r-4)(r-5)(4r^2 - 24r + 51), \\
&C_r^4 = \tfrac{1}{6}\{(r-5)(r-6)(r-7)(8r^3 - 60r^2 + 286r + 63) + 384(9r^2 - 93r + 242 - 2 \,.\, 4^{r-5})\}, \\
&\text{\&c.}
\end{aligned}$$

(it is curious that C_5^4, C_6^4, C_7^4, all three of them vanish). It seems hopeless to continue this investigation any further.

Returning to the equation (8), and assuming for z an expression of the same form as before, we have, corresponding to the equations before found for the coefficients C_r,

$$C_{r+2} = -(2r+1)(2r+2)(n-2r)(n-2r-1) C_r - (2r+2)(n-2r-2) \alpha C_{r+1} + 2n(\alpha^2 - 4) \frac{dC_{r+1}}{d\alpha}.$$

The case corresponding to the denominator in the multiplication of elliptic functions is that of $C_0=1$, $C_1=0$. It is easy to form the table—

$$C_0 = 1,$$

$$C_1 = 0,$$

$$C_2 = -2n(n-1),$$

$$C_3 = 8n(n-1)(n-4)\alpha,$$

$$\begin{aligned} C_4 = &-4n(n-1)(n-4)[n+75] \\ &-32n(n-1)(n-4)(n-9)\alpha^2, \end{aligned}$$

$$\begin{aligned} C_5 = &\ 96n(n-1)(n-4)(n-9)\ [n+44]\alpha \\ &+128n(n-1)(n-4)(n-9)(n-16)\alpha^3, \end{aligned}$$

$$\begin{aligned} C_6 = &-24n(n-1)(n-4)(n-9)[17n^2+403n+9000] \\ &-960n(n-1)(n-4)(n-9)(n-16)\ [n+41]\alpha^2 \\ &-512n(n-1)(n-4)(n-9)(n-16)(n-25)\ \alpha^4, \end{aligned}$$

$$\begin{aligned} C_7 = &+96n(n-1)(n-4)(n-9)(n-16)[79n^2+2825n+36180]\alpha \\ &+7168n(n-1)(n-4)(n-9)(n-16)(n-25)\ [n+42]\alpha^3 \\ &+2048n(n-1)(n-4)(n-9)(n-16)(n-25)(n-36)\ \alpha^5, \end{aligned}$$

$$\begin{aligned} C_8 = &-48n(n-1)(n-4)(n-9)[283n^4-26978n^3+277827n^2-5491932n+127764000] \\ &-3840n(n-1)(n-4)(n-9)(n-16)(n-25)\ [23n^2+1069n+23436]\alpha^2 \\ &-15360n(n-1)(n-4)(n-9)(n-16)(n-25)(n-36)\ [3n+133]\alpha^4 \\ &-8192n(n-1)(n-4)(n-9)(n-16)(n-25)(n-36)(n-49)\ \alpha^6, \end{aligned}$$

&c.

in which of course the coefficient of the highest power of n, in the successive coefficients C_r, is the value of C_r obtained from the equation (8). With regard to the law of these coefficients I have found that

$$\begin{aligned} C_r = (-)^{r+1}\, &2^{2r-3}\, n(n-1^2)\ldots\{n-(r-1)^2\}\, C_r^{\,1}\, \alpha^{r-2} \\ +\, &2^{2r-6}\, n(n-1^2)\ldots\{n-(r-2)^2\}\, C_r^{\,2}\, \alpha^{r-4} \\ +\, &2^{2r-9}\, n(n-1^2)\ldots\{n-(r-3)^2\}\, C_r^{\,3}\, \alpha^{r-6} \\ +\, &\&c. \end{aligned}$$

(where however the next term does not contain, as would at first sight be supposed, the factor $n(n-1^2)\ldots\{n-(r-4)^2\}$). And then

$$C_r^{\,1} = 1,$$

$$C_r^{\,2} = (r-3)[n(2r-7)+(r-1)(8r-7)],$$

$$\begin{aligned} C_r^{\,3} = \tfrac{1}{2}(r-4)(r-5)[\ &n^2(4r^2-24r+51) \\ &+n(32r^3-220r^2+412r-255) \\ &+2(r-1)(r-2)(32r^2-88r+51)]. \end{aligned}$$

In conclusion may be given the following results, in which, recapitulating the notation

$$x = \sqrt{k} \operatorname{sn} u, \quad \alpha = k + \frac{1}{k}, \quad \Delta x = \sqrt{(1 - \alpha x^2 + x^4)},$$

$$\sqrt{k} \operatorname{sn} 2u = \frac{2x \, \Delta x}{1 - x^4},$$

$$\sqrt{k} \operatorname{sn} 3u = \frac{x (3 - 4\alpha x^2 + 6x^4 - x^8)}{1 - 6x^4 + 4\alpha x^6 - 3x^8},$$

$$\sqrt{k} \operatorname{sn} 4u = \frac{4x \, \Delta x (1 - x^4)(1 - 2\alpha x^2 + 6x^4 - 2\alpha x^6 + x^8)}{1 - 20x^4 + 32\alpha x^6 - (26 + 16\alpha^2) x^8 + 32\alpha x^{10} - 20x^{12} + x^{16}},$$

$$\sqrt{k} \operatorname{sn} 5u = x \, \frac{\{5 - 20\alpha x^2 + (62 + 16\alpha^2) x^4 - 80\alpha x^6 - 105x^8 + 360\alpha x^{10} - (300 + 240\alpha^2) x^{12} + (368\alpha + 64\alpha^3) x^{14} - (125 + 160\alpha^2) x^{16} + 140\alpha x^{18} - 50x^{20} + x^{24}\}}{\{1 - 50x^4 + 140\alpha x^6 - (125 + 160\alpha^2) x^8 + (368\alpha + 64\alpha^3) x^{10} - (300 + 240\alpha^2) x^{12} + 360\alpha x^{14} - 105x^{16} - 80\alpha x^{18} + (62 + 16\alpha^2) x^{20} - 20\alpha x^{22} + 5x^{24}\}}$$

&c.

Thus, writing $-x^2$ for x^2, $k = 1$, and therefore $\alpha = 2$,

$$\tan 3u = x \frac{(3 + 8x^2 + 6x^4 - x^8)}{1 - 6x^4 - 8x^6 - 3x^8} = \frac{x (3 - x^2)(1 + x^2)^3}{(1 - 3x^2)(1 + x^2)^3} = \frac{x (3 - x^2)}{1 - 3x^2},$$

where $x = \tan u$. (And in general in reducing $\tan nu$ the extraneous factor in the numerator and denominator is $(1 + x^2)^{\frac{1}{2}n(n-1)}$.)

46.

NOTE ON A SYSTEM OF IMAGINARIES.

[From the *Philosophical Magazine*, vol. XXX. (1847), pp. 257—258.]

THE octuple system of imaginary quantities $i_1, i_2, i_3, i_4, i_5, i_6, i_7$, which I mentioned in a former paper [21], (and the conditions for the combination of which are contained in the symbols

$$123, \quad 246, \quad 374, \quad 145, \quad 275, \quad 365, \quad 167,$$

i.e. in the formulæ

$$i_2 i_3 = i_1, \quad i_3 i_1 = i_2, \quad i_1 i_2 = i_3, \qquad i_3 i_2 = -i_1, \quad i_1 i_3 = -i_2, \quad i_2 i_1 = -i_3,$$

with corresponding formulæ for the other triplets $i_2 i_4 i_6$, &c.,) possesses the following property; namely, if i_α, i_β, i_γ be any three of the seven quantities which do *not* form a triplet, then

$$(i_\alpha i_\beta) \,.\, i_\gamma = - i_\alpha \,.\, (i_\beta i_\gamma).$$

Thus, for instance,

$$(i_3 i_4) \,.\, i_5 = - i_7 \,.\, i_5 = - i_2\,;$$

but

$$i_3 \,.\, (i_4 i_5) = \quad i_3 \,.\, i_1 = \quad i_2,$$

and similarly for any other such combination. When i_α, i_β, i_γ form a triplet, the two products are equal, and reduce themselves each to -1, or each to $+1$, according to the order of the three quantities forming the triplet. Hence in the octuple system in question neither the commutative nor the distributive law holds, which is a still wider departure from the laws of ordinary algebra than that which is presented by Sir W. Hamilton's quaternions.

I may mention, that a system of coefficients, which I have obtained for the rectangular transformation of coordinates in n dimensions (Crelle, t. XXXII. [1846] "*Sur quelques propriétés des Déterminans gauches*" [52]), does not appear to be at all connected with any system of imaginary quantities, though coinciding in the case of $n=3$ with those mentioned in my paper "On Certain Results relating to Quaternions," *Phil. Mag.* Feb. 1845, [20].

47.

SUR LA SURFACE DES ONDES.

[From the *Journal de Mathématiques Pures et Appliquées* (Liouville), tom. XI. (1846), pp. 291—296.]

La surface des ondes est un cas particulier d'une autre surface à laquelle on peut donner le nom de *tétraédroïde*, et dont voici la propriété fondamentale :

"Le tétraédroïde est une surface du quatrième ordre, qui est coupée par les plans d'un certain tétraèdre suivant des paires de coniques par rapport auxquelles les trois sommets du tétraèdre dans ce plan sont des points conjugués. De plus : les seize points d'intersection des quatre paires de coniques sont des points singuliers de la surface, c'est-à-dire des points où, au lieu d'un plan tangent, il y a un cône tangent du second ordre."

On verra plus loin que la surface eût été suffisamment définie en disant qu'un seul de ces points était un point singulier; l'existence des quinze autres points singuliers est donc une propriété assez remarquable. Mais, avant d'aller plus loin, il convient de rappeler la signification des points conjugués par rapport à une conique : cela veut dire que la polaire de chacun des trois points passe par les deux autres. Parmi les propriétés d'un tel système, on peut citer celle-ci :

"Les points d'intersection de deux coniques, par rapport auxquelles les mêmes trois points sont des points conjugués, sont situés deux à deux sur six droites, lesquelles passent deux à deux par ces trois points."

De là : "Les quatre points singuliers compris dans chaque face du tétraèdre sont situés deux à deux sur trois paires de droites, lesquelles passent par les trois sommets dans cette face."

Autrement dit, les seize points singuliers sont situés deux à deux sur vingt-quatre droites, lesquelles passent six à six par les sommets et sont situées six à six dans les plans du tétraèdre. Donc, en considérant les six droites qui passent par le même sommet, par ces droites combinées deux à deux, il passe quinze plans, y compris trois plans du tétraèdre; en excluant ceux-ci, il y a pour chaque sommet douze plans, chacun desquels passe par quatre points singuliers; donc la courbe d'intersection d'un de ces plans avec la surface a quatre points doubles, ce qui ne peut pas arriver pour une courbe du quatrième ordre, à moins qu'elle ne se réduise à deux coniques.

Donc: "Il y a, outre les plans du tétraèdre, quarante-huit plans, douze par chaque sommet, lesquels rencontrent la surface suivant des paires de coniques."

On pourrait de même chercher combien il y a de plans qui rencontrent la surface suivant des courbes avec trois points doubles, &c. Passons aux cônes circonscrits; on démontre facilement par l'analyse cette propriété:

"Les cônes circonscrits à la surface ayant pour sommets les sommets du tétraèdre, se réduisent à des paires de cônes du second ordre, lesquelles la touchent suivant les deux coniques de la face opposée."

De plus: "Les seize plans qui touchent quatre à quatre ces paires de cônes sont des plans tangents singuliers, dont chacun touche la surface suivant une conique."

Il y a de plus, entre ces plans, des relations analogues à celles qui existent entre les points singuliers, de manière que l'on déduit de même le théorème:

"Il y a quarante-huit points, douze à douze dans les quatre plans du tétraèdre, pour chacun desquels les cônes circonscrits se réduisent à des paires de cônes du second ordre."

Ajoutons que ces quarante-huit points correspondent d'une manière particulière aux quarante-huit plans, et que les cônes circonscrits touchent la surface suivant les coniques situées dans les plans correspondants.

La réciproque d'une surface du quatrième ordre est, en général, de l'ordre trente-six; mais ici, à cause des seize points singuliers, cet ordre se réduit de trente-deux, savoir, à quatre. Et les propriétés qui viennent d'être énoncées par rapport aux cônes circonscrits montrent que la réciproque étant de cet ordre est nécessairement une surface de la même espèce; c'est-à-dire:

"La réciproque d'un tétraédroïde est aussi un tétraédroïde."

Donc le tétraédroïde est surface de la quatrième classe. Puisque cette surface n'a pas de lignes doubles ou de lignes de rebroussement, il n'y a pas de réduction dans le nombre qui exprime *le rang* de la surface, lequel est ainsi égal à douze, c'est-à-dire le cône circonscrit est ordinairement de l'ordre douze. Mais nous venons de voir que ce cône est seulement de la quatrième classe (en effet, la classe du cône est la même chose que celle de la surface); donc il y a réduction de cent vingt-huit dans la classe du cône. Les seize points singuliers de la surface donnent lieu à autant de lignes doubles dans le cône, ce qui effectue une réduction de trente-deux; il y a encore une réduction à effectuer de quatre-vingt-seize, qui doit avoir lieu à cause des lignes doubles ou des lignes de rebroussement du cône. En supposant qu'il y a y de celles-ci et x de celles-là (outre les seize lignes doubles dont on a fait mention), il faut que l'on ait

$$2x + 3y = 96,$$

où $x+y$ ne doit pas être plus grand que trente-neuf; mais cela ne suffit pas pour déterminer ces deux nombres.

Nous pouvons encore ajouter que les seize cônes qui touchent la surface aux points singuliers sont circonscrits, quatre à quatre, à quatre surfaces du second ordre, et que les seize courbes de contact des plans singuliers sont situées quatre à quatre sur quatre surfaces du second ordre.

Je vais donner maintenant une idée de la théorie analytique. En représentant par

$$x=0, \quad y=0, \quad z=0, \quad w=0$$

les équations des quatre faces du tétraèdre, et par

$$U=0$$

l'équation de la surface, U sera une fonction homogène du quatrième degré de x, y, z, w, laquelle, en faisant évanouir une quelconque des variables, doit se diviser en deux facteurs du second degré, *fonctions paires* des trois autres variables; de plus, à cause de la condition par rapport aux points singuliers, U ne peut pas contenir de terme $xyzw$. On doit donc avoir

$$U = Ax^4 + By^4 + Cz^4 + 2Fy^2z^2 + 2Gz^2x^2 + 2Hx^2y^2 + 2Lx^2w^2 + 2My^2w^2 + 2Nz^2w^2 + Pw^4,$$

où il faut que les coefficients A, B, C, P, F, G, H, L, M, N aient un système inverse[1] de la forme 0, 0, 0, 0, f, g, h, l, m, n. Donc, en formant l'inverse de ce système, on obtient, toute réduction accomplie,

$$\begin{aligned} U = {} & mnfx^4 + nlgy^4 + lmhz^4 + fghw^4 \\ & + (\ lf - mg - nh)(\ ly^2z^2 + fx^2w^2) \\ & + (-lf + mg - nh)(mz^2x^2 + gy^2w^2) \\ & + (-lf - mg + nh)(\ nx^2y^2 + hz^2w^2) = 0. \end{aligned}$$

En effet, en écrivant

$$\begin{aligned} \lambda &= \ \ lf - mg - nh, \quad \mu = -lf + mg - nh, \\ \nu &= -lf - mg + nh, \quad \omega = \ \ lf + mg + nh, \\ \nabla &= l^2f^2 + m^2g^2 + n^2h^2 - 2mngh - 2nlhf - 2lmfg, \end{aligned}$$

l'équation de la courbe pourra s'écrire sous les quatre formes

$$\begin{aligned} (2mnfx^2 + \nu ny^2 + m\mu z^2 + f\lambda w^2)^2 &= \nabla\,(n^2y^4 + m^2z^4 + f^2w^4 - 2mfz^2w^2 - 2fnw^2y^2 - 2nmy^2z^2), \\ (n\nu x^2 + 2nlgy^2 + l\lambda z^2 + g\mu w^2)^2 &= \nabla\,(l^2z^4 + g^2w^4 + n^2x^4 - 2gnw^2x^2 - 2nl\,x^2z^2 - 2lg\,z^2w^2), \\ (m\mu x^2 + l\lambda y^2 + 2lmhz^2 + h\nu w^2)^2 &= \nabla\,(h^2w^2 + m^2x^4 + l^2y^4 - 2ml\,x^2y^2 - 2lh\,y^2w^2 - 2hmw^2x^2), \\ (f\lambda x^2 + g\mu y^2 + h\nu z^2 + 2fghw^2)^2 &= \nabla\,(f^2x^4 + g^2y^4 + h^2z^4 - 2gh\,y^2z^2 - 2hf\,z^2x^2 - 2fg\,x^2y^2); \end{aligned}$$

[1] En général, quand deux systèmes de quantités sont exprimés linéairement les uns au moyen des autres, on dit que les deux systèmes de coefficients sont des systèmes inverses; on passe facilement de là à l'idée du système inverse des coefficients d'une fonction du second ordre, et de tels systèmes se rencontrent si souvent, que l'on doit avoir un terme pour exprimer sans circonlocution cette relation.

ce qui met en évidence les équations des sections de la surface par les quatre plans du tétraèdre, et aussi celles des points singuliers, lesquelles sont, en effet,

$$ny \pm mz \pm fw = 0, \quad lz \pm gw \pm nx = 0,$$
$$hw \pm mx \pm ly = 0, \quad fx \pm gy \pm hz = 0;$$

et ces plans touchent la surface suivant les courbes d'intersection avec les surfaces

$$2mnfx^2 + n\nu y^2 + m\mu z^2 + f\lambda w^2 = 0, \quad \&c., \quad \&c.,$$

ce qui démontre le théorème énoncé par rapport aux seize courbes de contact des plans singuliers, et de là celui pour les seize cônes tangents aux points singuliers.

Pour déduire de là la forme ordinaire de l'équation de la surface des ondes, écrivons

$$l = \alpha\beta\gamma\,(b\gamma - c\beta), \quad m = \alpha\beta\gamma\,(c\alpha - a\gamma), \quad n = \alpha\beta\gamma\,(a\beta - b\alpha),$$
$$f = ka\alpha\,(b\gamma - c\beta), \quad g = kb\beta\,(c\alpha - a\gamma), \quad h = kc\gamma\,(a\beta - b\alpha),$$

équations qui suffisent pour déterminer les rapports

$$a : b : c : \alpha : \beta : \gamma : k$$

au moyen de l, m, n, f, g, h. De cette manière, l'équation de la surface se réduit à

$$\alpha\beta\gamma\,(ax^2 + by^2 + cz^2)(\alpha x^2 + \beta y^2 + \gamma z^2)$$
$$- ka\alpha\,(b\gamma + c\beta)\,x^2w^2 - kb\beta\,(c\alpha + a\gamma)\,y^2w^2 - kc\gamma\,(a\beta + b\alpha)\,z^2w^2 + k^2abcw^4 = 0,$$

laquelle se réduit à la surface des ondes en écrivant

$$\frac{x}{w}\sqrt{\frac{\alpha}{k}} = \xi, \quad \frac{y}{w}\sqrt{\frac{\beta}{k}} = \eta, \quad \frac{z}{w}\sqrt{\frac{\gamma}{k}} = \zeta,$$

ξ, η, ζ étant des coordonnées rectangulaires, c'est-à-dire en faisant la transformation homographique du tétraédroïde, de manière que l'un des plans du tétraèdre passe à l'infini, et que les trois autres deviennent rectangulaires. De plus, en particularisant la transformation de manière que trois des coniques d'intersection se réduisent à des cercles, cette surface rentre dans la surface des ondes. Il va sans dire que, dans le cas général, plusieurs des points ou des plans dont nous avons parlé sont nécessairement imaginaires; l'énumération de tous les cas différents aurait été d'une longueur effrayante. Il y aurait beaucoup à dire sur les cas particuliers où quelques-unes des coniques se réduisent à des paires de droites (réelles ou imaginaires). Je me contente d'énoncer cette propriété, très-facile à démontrer, de la surface ordinaire des ondes: au cas où $c = 0$, cette surface peut être engendrée par un cercle (ayant le centre de la surface pour centre, et dans un plan passant par l'axe des z), lequel se meut de manière à passer par la conique

$$a^2x^2 + c^2y^2 = a^2b^2.$$

On trouve, dans le *Cambridge and Dublin Mathematical Journal*, t. I. [1846], p. 208, [38] la démonstration d'une autre propriété de la surface des ondes par rapport aux lignes de courbure des surfaces du second ordre.

48.

NOTE SUR LES FONCTIONS DE M. STURM.

[From the *Journal de Mathématiques Pures et Appliquées* (Liouville), tom. XI. (1846), pp. 297—299.]

On sait que la suite des fonctions

$$\begin{aligned} fx &= (x-a)(x-b)(x-c)(x-d)\ldots, \\ f_1x &= \Sigma\,(x-b)(x-c)(x-d)\ldots, \\ f_2x &= \Sigma\,(a-b)^2(x-c)(x-d)\ldots, \\ f_3x &= \Sigma\,(a-b)^2(b-c)^2(c-a)^2(x-d)\ldots, \end{aligned}$$

est de la plus grande utilité dans la théorie de la résolution numérique des équations. En effet, on obtient tout de suite à leur moyen le nombre de racines réelles comprises entre deux limites quelconques. Il était donc intéressant de chercher la manière d'exprimer ces fonctions par les coefficients de fx.

Soit, pour cela, m un nombre quelconque, pas plus grand que le degré n de cette fonction. En prenant k pour $m^{ième}$ racine de la suite a, b, c, ..., et mettant, pour abréger,

$$P = (-)^{\frac{1}{2}m(m-1)}(a-b)(a-c)\ldots(a-k)(b-c)\ldots(b-k)\ldots(j-k),$$

cela donne

$$f_mx : fx = \Sigma\frac{P^2}{(x-a)(x-b)\ldots(x-k)},$$

dans laquelle expression

$$P = \begin{vmatrix} 1, & a, \ldots, & a^{m-1}, \\ 1, & b, \ldots, & b^{m-1}, \\ \ldots & \ldots & \ldots \\ 1, & k, \ldots, & k^{m-1}, \end{vmatrix}$$

et, de plus,

$$\frac{P}{(x-a)(x-b)\dots(x-k)} = (-)^{\frac{1}{2}m(m-1)}\left[\frac{(b-c)\dots(b-k)\dots(j-k)}{(x-a)} + \dots\right],$$

dans laquelle le coefficient de x^{-r} est égal à

$$(-)^{\frac{1}{2}m(m-1)}\left[a^{r-1}(b-c)\dots(b-k)\dots(j-k)+\dots\right],$$

c'est-à-dire à

$$\begin{vmatrix} 1, & a, \dots, & a^{m-2}, & a^{r-1} \\ 1, & b, \dots, & b^{m-2}, & b^{r-1} \\ \dots & & & \\ 1, & k, \dots, & k^{m-2}, & k^{r-1} \end{vmatrix}.$$

Donc enfin le coefficient de x^{-r} dans $f_m x : fx$ est égal à

$$\Sigma \begin{vmatrix} 1, & a, \dots, & a^{m-1} \\ \dots & & \\ 1, & k, \dots, & k^{m-1} \end{vmatrix} \times \begin{vmatrix} 1, & a, \dots, & a^{m-2}, & a^{r-1} \\ \dots & & & \\ 1, & k, \dots, & k^{m-2}, & k^{r-1} \end{vmatrix}$$

où, au moyen d'une propriété connue des déterminants, en représentant comme à l'ordinaire par S_q la somme des $q^{\text{ièmes}}$ puissances de toutes les racines, ce coefficient devient égal à

$$\begin{vmatrix} S_0, & S_1, \dots, & S_{m-2}, & S_{r-1} \\ S_1, & S_2, \dots, & S_{m-1}, & S_r \\ \dots & & & \\ S_{m-1}, & S_m, \dots, & S_{2m-3}, & S_{r+m-2} \end{vmatrix}.$$

De là, en mettant

$$T_q = \Sigma \frac{a^q}{x-a},$$

$$f_m x : fx = \begin{vmatrix} S_0, & S_1, \dots, & S_{m-2}, & T_0 \\ S_1, & S_2, \dots, & S_{m-1}, & T_1 \\ \dots & & & \\ S_{m-1}, & S_m, \dots, & S_{2m-3}, & T_{m-1} \end{vmatrix}$$

en multipliant par fx, et mettant

$$Q_{m,r} = S_{m+r-1} - p_1 S_{m+r-2} \dots + (-)^r p_r S_{m-1},$$

où l'on suppose

$$fx = x^n - p_1 x^{n-1} \dots + (-)^n p_n,$$

on obtient

$$f_m x = \Sigma_r x^{n-m-r} \begin{vmatrix} S_0, & S_1, \dots, & S_{m-2}, & Q_{m,r} \\ S_1, & S_2, \dots, & S_{m-1}, & Q_{m+1,r} \\ \dots & & & \\ S_{m-1}, & S_m, \dots, & S_{2m-3}, & Q_{2m-1,r} \end{vmatrix}$$

où r peut ne s'étendre que depuis 0 jusqu'à $n-m$, puisque $f_m x$ est fonction entière. Au moyen des relations connues qui existent entre les quantités S_q, on a

$$Q_{m+s,r} = (-)^r p_{r+1} S_{m+s-2} \dots + (-)^{n-1} p_n S_{r+m+s-n-1} \dots \quad [r+m+s>n],$$

$$Q_{m+s,r} = (-)^r p_{r+1} S_{m+s-2} \dots + (-)^{r+m+s-3} p_{r+m+s-2} S_1$$
$$+ (-)^{r+m+s-2} p_{r+m+s-1} (r+m+s-1) \dots\dots\dots\dots \left[r+m+s \overset{=}{<} n\right],$$

et de là, en posant

$$Q'_{m+s,r} = (-)^{r+m-1} p_{r+m} S_{s-1} \dots + (-)^{n-1} p_n S_{r+m+s-n-1} \dots [r+m+s>n],$$

$$Q'_{m+s,r} = (-)^{r+m-1} p_{r+m} S_{s-1} \dots + (-)^{r+m+s-3} p_{r+m+s-2} S_1$$
$$+ (-)^{r+m+s-2} p_{r+m+s-1} (r+m+s-n-1) \dots \left[r+m+s \overset{=}{<} n\right],$$

on peut, par les propriétés des déterminants, réduire $Q_{m+s,r}$ à $Q'_{m+s,r}$ dans l'expression de $f_m x$. Nous avons donc exprimé cette fonction au moyen des coefficients p_1, p_2, ... et des sommes S_1, S_2, ..., S_{2m-3}, lesquelles s'expriment facilement par ces mêmes coefficients (et pour calculer $f_1 x$, $f_2 x$, ..., $f_n x$, on aurait seulement besoin de calculer ces sommes une fois pour toutes jusqu'à S_{2n-3}); il serait donc facile de former des tables de ces fonctions, pour les équations d'un degré quelconque, ce qui pourrait à peine s'effectuer d'aucune autre manière.

49.

SUR QUELQUES FORMULES DU CALCUL INTÉGRAL.

[From the *Journal de Mathématiques Pures et Appliquées* (Liouville), tom. XII. (1847), pp. 231—240.]

SOIT $x+y\sqrt{-1}$ ou $x+iy$ une quantité imaginaire quelconque; faisons

$$\rho=\sqrt{x^2+y^2},\qquad \theta=\text{arc}\tan\frac{y}{x}\ \ldots\ldots\ldots\ldots\ldots\ldots\ldots\ldots\ldots\ (1),$$

ρ étant une quantité positive, et θ un arc compris entre les limites $\frac{1}{2}\pi$, $-\frac{1}{2}\pi$. Cela posé, écrivons

$$\left.\begin{aligned}(x+iy)^m&=\rho^m e^{im\theta} &&(x\text{ positif}),\\ (x+iy)^m&=\rho^m e^{im(\theta\pm\pi)} &&(x\text{ négatif});\end{aligned}\right\}\ldots\ldots\ldots\ldots\ldots\ldots\ldots\ldots(2),$$

(dans la seconde de ces formules, il faut prendre le signe supérieur ou inférieur, selon que y est positif ou négatif.) Au cas de x positif, la valeur du second membre sera ce que M. Cauchy a appelé *valeur principale* de $(x+iy)^m$. Au cas de x négatif, on peut aussi, à ce qu'il me semble, nommer cette valeur *valeur principale*. Cela paraît contraire à la théorie de M. Cauchy (*Exercices de Mathématiques*, t. I. [1826] p. 2); mais la démonstration que l'on y trouve de l'impossibilité d'une valeur principale pour x négatif ne s'applique qu'au cas où l'on suppose que le signe $\pm$ est toujours le même sans avoir égard au signe de y. Seulement, selon nos définitions, il importe de remarquer qu'il n'y a pas de valeur principale pour x négatif, au cas particulier où $y=0$; ou plutôt dans ce cas, et dans ce cas seulement, la valeur principale devient indéterminée.

Soit, en particulier, $x=0$; les deux formules conduisent au même résultat, savoir

$$(iy)^m=(\pm y)^m e^{\pm\frac{1}{2}m\pi i}\ldots\ldots\ldots\ldots\ldots\ldots\ldots\ldots\ldots\ldots\ldots(3),$$

le signe comme auparavant. Car en considérant x comme infiniment petit positif, on obtient

$$\theta = \pm \tfrac{1}{2}\pi,$$

et, en considérant x comme infiniment petit négatif,

$$\theta = \mp \tfrac{1}{2}\pi,$$

et de là

$$\theta \pm \pi = \pm \tfrac{1}{2}\pi.$$

Ainsi cette formule est toujours vraie, sans qu'il soit nécessaire de considérer iy comme limite de $x + iy$, x positif ou x négatif.

Remarquons encore que cette fonction $(iy)^m$ reste continue quand y passe par zéro, ce qui a lieu aussi pour $(x + iy)^m$, x positif, mais non pas pour $(x + iy)^m$, x négatif.

Les mêmes remarques s'appliquent aux valeurs principales des logarithmes, lesquelles doivent se définir d'une manière analogue par les équations

$$\left.\begin{array}{ll} \log(x+iy) = \log\rho + i\theta & (x \text{ positif}), \\ \log(x+iy) = \log\rho + i(\theta \pm \pi) & (x \text{ négatif}), \end{array}\right\} \quad \ldots\ldots\ldots\ldots\ldots\ldots \quad (4),$$

le signe ambigu, comme auparavant.

On démontre sans difficulté que ces valeurs principales satisfont en tout cas aux équations

$$\left.\begin{array}{rl} (x+iy)^m (x'+iy')^m = & [(x+iy)(x'+iy')]^m, \\ \log(x+iy)\log(x'+iy') = & \log[(x+iy)(x'+iy')] \end{array}\right\} \ldots\ldots\ldots\ldots\ldots (5);$$

seulement ces équations deviennent indéterminées au cas où, l'une des quantités x, x', $xx' - yy'$ étant négative, la quantité correspondante y, y', $xy' + x'y$ s'évanouit.

Au moyen de cette définition de la valeur principale d'un logarithme, on obtient

$$\int_\beta^\alpha \frac{dx}{A + Bx} = \log\left(\frac{A + B\alpha}{A + B\beta}\right) \ldots\ldots\ldots\ldots\ldots\ldots\ldots\ldots\ldots\ldots (6),$$

où α, β sont réels, et A, B sont assujettis à la seule condition qu'au cas où α, β seraient de signes contraires, la partie imaginaire de $\dfrac{A}{B}$ ne s'évanouisse pas. En effet, dans ce cas, l'intégrale et la valeur principale du logarithme deviennent toutes les deux indéterminées. Sans doute il y a une valeur que l'on peut appeler *principale* de l'intégrale, mais cette valeur n'est égale à *aucun* des logarithmes de $\dfrac{A + B\alpha}{A + B\beta}$, et les notions des valeurs principales d'une intégrale et d'un logarithme n'ont pas de rapport ensemble. Ce résultat s'accorde avec celui que j'ai trouvé dans mon Mémoire "*Sur les fonctions doublement périodiques*," [25].

Je passe à quelques autres applications de ces principes, qui ont rapport à la théorie des fonctions Γ. Soit d'abord r un nombre positif plus petit que l'unité, et écrivons

$$U=\int_{-\infty}^{\infty}(ix)^{r-1}\,e^{ix}\,dx\,;$$

on obtient

$$U=\int_0^{\infty}(ix)^{r-1}\,e^{ix}\,dx+\int_0^{\infty}(-ix)^{r-1}\,e^{-ix}\,dx,$$

ou enfin, puisque x est positif dans ces deux intégrales,

$$U=e^{\frac{1}{2}(r-1)\pi i}\int_0^{\infty}x^{r-1}\,e^{ix}\,dx+e^{-\frac{1}{2}(r-1)\pi i}\int_0^{\infty}x^{r-1}\,e^{-ix}\,dx\,;$$

au moyen de la formule connue,

$$\int_0^{\infty}x^{r-1}\,e^{\pm ix}\,dx=e^{\pm\frac{1}{2}r\pi i}\,\Gamma r,$$

on en conclut

$$U=2\cos\left(r-\tfrac{1}{2}\right)\pi\,.\,\Gamma r=2\sin r\pi\,.\,\Gamma r,$$

savoir

$$\sin r\pi\,.\,\Gamma r=\tfrac{1}{2}\int_{-\infty}^{\infty}(ix)^{r-1}\,e^{ix}\,dx\,.$$

On obtiendrait de même

$$0=\tfrac{1}{2}\int_{-\infty}^{\infty}(-ix)^{r-1}\,e^{ix}\,dx\,.$$

Au premier coup d'œil, ces équations pourraient paraître en contradiction l'une avec l'autre: mais cela n'est pas ainsi, parce qu'il n'est pas vrai que $(-ix)^{r-1}$ soit égal à $(-1)^{r-1}(ix)^{r-1}$, $(-1)^{r-1}$ étant facteur invariable; mais au contraire $(-ix)^{r-1}=e^{\pm(r-1)\pi i}(ix)^{r-1}$, selon que x est positif ou négatif.

Dans la première de ces équations, on peut remplacer ix par $c+ix$, et dans la seconde, $-ix$ par $c-ix$, c étant positif; mais cela n'est pas permis pour c négatif, à cause de la discontinuité de valeur de $(c+ix)^{r-1}$ ou $(c-ix)^{r-1}$ dans ce cas, quand x passe par zéro. En multipliant la première équation par e^{-c}, et différentiant un nombre quelconque de fois, en admettant, pour les valeurs négatives de r, l'équation

$$\Gamma(r+1)=r\Gamma r,$$

la formule ne change pas de forme, et l'on obtient

$$\sin r\pi\Gamma r\,.\,e^{-c}=\tfrac{1}{2}\int_{-\infty}^{\infty}(c+ix)^{r-1}\,e^{ix}\,dx\;\ldots\ldots\ldots\ldots\ldots\ldots\ldots\ldots\;(7)\,;$$

et de même, au moyen de la seconde équation,

$$0=\tfrac{1}{2}\int_{-\infty}^{\infty}(c-ix)^{r-1}\,e^{ix}\,dx\ldots\ldots\ldots\ldots\ldots\ldots\ldots\ldots(8),$$

dans lesquelles formules c est positif, et r est un nombre quelconque entre 1, $-\infty$, sans exclure la limite supérieure; seulement, pour $r=0$, le facteur $\sin r\pi\Gamma r$ se réduit à π. Au cas où r est plus grand que zéro, on peut, si l'on veut, écrire aussi $c=0$. Dans tous les cas, on peut remplacer $\sin r\pi\Gamma r$ par $\frac{\pi}{\Gamma(1-r)}$. Il importe de remarquer que ces mêmes intégrales sont absolument inexprimables au cas où c est négatif: en effet, en écrivant $-c$ au lieu de c (c positif), on obtiendrait

$$\int_{-\infty}^{\infty}(-c+ix)^{r-1}e^{ix}\,dx=e^{(r-1)\pi i}\int_{0}^{\infty}(c-ix)^{r-1}e^{ix}\,dx+e^{-(r-1)\pi i}\int_{-\infty}^{0}(c-ix)^{r-1}e^{ix}\,dx.$$

Mais, par la seconde des équations dont il s'agit,

$$0=\int_{0}^{\infty}(c-ix)^{r-1}e^{ix}\,dx+\int_{-\infty}^{0}(c-ix)^{r-1}e^{ix}\,dx\,;$$

donc

$$\int_{-\infty}^{\infty}(-c+ix)^{r-1}e^{ix}\,dx=-2i\sin r\pi\int_{0}^{\infty}(c-ix)^{r-1}e^{ix}\,dx.$$

Or l'intégrale au second membre ne peut pas s'exprimer par les transcendantes connues, à moins que r ne soit entier. Écrivons encore, c étant toujours positif,

$$\left.\begin{aligned} I&=\int_{0}^{\infty}(c+ix)^{r-1}e^{ix}\,dx,\\ I_1&=\int_{0}^{\infty}(c-ix)^{r-1}e^{ix}\,dx,\\ I_2&=\int_{0}^{\infty}(c+ix)^{r-1}e^{-ix}\,dx,\\ I_3&=\int_{0}^{\infty}(c-ix)^{r-i}e^{-ix}\,dx.\end{aligned}\right\}\quad\ldots\ldots\ldots\ldots\ldots\ldots\ldots\ldots\ldots\ (9).$$

Toutes les fonctions

$$\int(\pm\,c\,\pm ix)^{r-1}e^{\pm ix}\,dx,$$

entre les limites 0, ∞, ou $-\infty$, 0, ou $-\infty$, ∞, s'expriment facilement au moyen de I, I_1, I_2, I_3. Mais ces quatre fonctions ne sont pas connues; seulement, au moyen des équations qui viennent d'être trouvées, on obtient

$$\left.\begin{aligned} 2\sin r\pi\,\Gamma r\,.\,e^{-c}&=I+I_3,\\ 0&=I_1+I_2.\end{aligned}\right\}\ldots\ldots\ldots\ldots\ldots\ldots\ldots\ldots\ldots\ (10).$$

On déduit encore de ces mêmes formules:

$$\left.\begin{aligned} &\sin r\pi\,\Gamma r\,.\,e^{-c}\\ &=\int_{-\infty}^{\infty}(c+ix)^{r-1}\cos x\,dx=\ \ i\int_{-\infty}^{\infty}(c+ix)^{r-1}\sin x\,dx,\\ &=\int_{-\infty}^{\infty}(c-ix)^{r-1}\cos x\,dx=-i\int_{-\infty}^{\infty}(c-ix)^{r-1}\sin x\,dx.\end{aligned}\right\}\ldots\ldots\ldots\ (11).$$

En supposant que $r-1$ est entier négatif, l'intégrale

$$\int_0^\infty (c^2+x^2)^{r-1}\cos x dx$$

se décompose facilement dans une suite d'intégrales de la forme

$$\int_{-\infty}^\infty (c\pm ix)^{p-1}\cos x dx;$$

et, en prenant la somme de celles-ci, on obtient la formule

$$\int_0^\infty (c^2+x^2)^{r-1}\cos x dx=\frac{\pi e^{-c}(2c)^{2r-1}}{\Gamma^2(1-r)}\int_0^\infty \theta^{-r}(\theta+2c)^{-r}e^{-\theta}d\theta \text{ .... (12)},$$

laquelle est due à M. Catalan. Cependant, ni cette démonstration ni celle de M. Catalan ne s'appliquent au cas où r n'est pas entier; la formule subsiste encore dans ce cas, comme M. Serret l'a démontré rigoureusement. En essayant de la vérifier, je suis tombé sur cette autre formule:

$$\left.\begin{aligned}&\int_0^\infty (c^2+x^2)^{r-1}\cos x dx\\ &=\frac{\sqrt{\pi}e^{-c}(2c)^{2r-1}}{2\Gamma(1-r)}\int_0^\infty \frac{(\sqrt{\theta+2c}+\sqrt{\theta})^{1-2r}+(\sqrt{\theta+2c}-\sqrt{\theta})^{1-2r}}{\sqrt{\theta}\sqrt{\theta+2c}}e^{-\theta}d\theta\end{aligned}\right\}\text{ ...... (13)},$$

ce qui suppose, comme auparavant, que $\overline{r-1}$ soit négatif; en comparant les deux valeurs, on est conduit au résultat singulier (en écrivant α au lieu de $2c$, et $\frac{1}{2}(1-p)$ pour r):

$$\left.\begin{aligned}&\int_0^\infty \frac{(\sqrt{\theta+\alpha}+\sqrt{\theta})^p+(\sqrt{\theta+\alpha}-\sqrt{\theta})^p}{\sqrt{\theta}\sqrt{\theta+\alpha}}e^{-\theta}d\theta\\ &=\frac{2\sqrt{\pi}}{\Gamma\frac{1}{2}(p+1)}\int_0^\infty \theta^{\frac{1}{2}(p-1)}(\theta+\alpha)^{\frac{1}{2}(p-1)}e^{-\theta}d\theta\end{aligned}\right\}\text{ .... (14)},$$

lequel peut se démontrer sans difficulté, quand p est entier positif impair, en développant les deux membres suivant les puissances de α; cela se fait au moyen de

$$\frac{(\sqrt{1+x}+1)^p+(\sqrt{1+x}-1)^p}{\sqrt{1+x}}=\mathbf{S}_r\left[x^r\frac{2^{p-2r}\Gamma(p-r)}{\Gamma(p-2r)\Gamma(r+1)}\right]\text{ ....(15)},$$

où r s'étend depuis 0 jusqu'à $\frac{1}{2}(p-1)$. J'ai déduit de cette formule (14) des formules assez remarquables qui se rapportent aux attractions, lesquelles paraîtront dans un numéro prochain du *Cambridge and Dublin Mathematical Journal*, [41]. On peut encore démontrer cette formule singulière:

$$\Gamma(r+a-1)=\frac{1}{2\sin r\pi}\int_{-\infty}^\infty (ix)^{r-1}(-ix)^{a-1}e^{ix}dx \text{ .... (16)};$$

en effet, pour la vérifier, il suffit de réduire l'intégrale, d'abord à

$$\int_0^\infty (ix)^{r-1}(-ix)^{a-1}e^{ix}dx+\int_0^\infty(-ix)^{r-1}(ix)^{a-1}e^{-ix}dx,$$

puis à

$$e^{\frac{1}{2}(r-a)\pi i}\int_0^\infty x^{r+a-2}e^{ix}dx+e^{\frac{1}{2}(a-r)\pi i}\int_0^\infty x^{r+a-2}e^{-ix}dx,$$

dont les deux parties s'obtiennent au moyen d'une formule donnée ci-dessus. Si, au lieu de $(-ix)^{a-1}$, l'on avait $(ix)^{a-1}$, l'intégrale se réduirait à zéro; mais cela rentre dans une formule plus simple.

J'ajouterai encore ces deux formules-ci,

$$\left.\begin{aligned}2\pi c^a e^{-c}&=\int_{-\infty}^{\infty}\frac{(-ix)^a}{c+ix}e^{ix}dx\;,\\ 0&=\int_{-\infty}^{\infty}\frac{(ix)^a}{c+ix}e^{-ix}dx,\end{aligned}\right\}\quad\text{.......................... (17)},$$

dont je supprime la démonstration. En écrivant dans la dernière $-x$ au lieu de x, puis ajoutant, on a

$$\pi c^{a-1}e^{-c}=\int_{-\infty}^{\infty}\frac{(-ix)^a}{c^2+x^2}e^{ix}dx\quad\text{..............................(18)};$$

d'où l'on déduit tout de suite cette formule de M. Cauchy

$$\tfrac{1}{2}\pi c^{a-1}e^{-c}=\int_0^\infty\frac{x^a}{c^2+x^2}\cos\left(\tfrac{1}{2}a\pi-x\right)dx\quad\text{.......................(19)}.$$

La formule (7) peut être considérée comme une définition de la fonction Γr au cas de r négatif; et, à ce point de vue, elle a, ce me semble, quelques avantages sur celle que M. Cauchy a donnée au moyen des intégrales extraordinaires. Je passe à la définition, au moyen d'une intégrale définie, de la seconde intégrale eulérienne,

$$\mathfrak{B}(m,\ n)=\frac{\Gamma m\,\Gamma n}{\Gamma(m+n)}\quad\text{................................ (20)},$$

quand m ou n, ou tous les deux, sont négatifs. Soit d'abord m et n tous les deux positifs, mais n plus petit que l'unité; on obtient au moyen de la formule (7) et de la définition ordinaire des fonctions Γ,

$$\Gamma m\,\Gamma n=\frac{1}{2\sin n\pi}\int_0^\infty dx\int_{-\infty}^{\infty}dy\,x^{m-1}(iy)^{n-1}e^{-x+iy},$$

et de là, en mettant

$$y=\alpha x,\quad dy=xd\alpha,$$

et intégrant par rapport à x,

$$\mathfrak{B}(m,\ n)=\frac{1}{2\sin n\pi}\int_{-\infty}^{\infty}\frac{(i\alpha)^{n-1}d\alpha}{(1-i\alpha)^{m+n}},$$

ou en écrivant $k+\alpha i$ au lieu de αi, k étant positif,

$$\mathfrak{F}(m,\ n) = \frac{1}{2 \sin n\pi} \int_{-\infty}^{\infty} \frac{(k+\alpha i)^{n-1}\, d\alpha}{(1-k-\alpha i)^{m+n}} \quad \ldots\ldots\ldots\ldots\ldots\ldots \quad (21),$$

formule qui est vraie, même pour les valeurs négatives de n. En effet, si cette formule est vraie pour une valeur quelconque particulière de n, elle sera aussi vraie pour la valeur $n+p$, p étant entier positif quelconque; ce qui se démontre au moyen des formules de réduction: donc il ne s'agit que de la démontrer au cas où n est positif et plus petit que l'unité, et dans ce cas, puisque la fonction à intégrer est toujours finie, on peut écrire sans crainte $k+\alpha i$ au lieu de αi, ce qui la réduit à la formule qui vient d'être démontrée; donc cette formule (21) est la définition cherchée, au cas où n est négatif, ou positif et plus petit que l'unité.

Si, de plus, m est négatif, ou positif et plus petit que l'unité, $1-m$ sera positif, et l'on déduira

$$\mathfrak{F}(m,\ n) = \frac{\Gamma m\, \Gamma n}{\Gamma(m+n)} = \frac{\Gamma(1-m-n)\, \Gamma n}{\Gamma(1-m)} \frac{\sin(m+n)\pi}{\sin m\pi},$$

c'est-à-dire

$$\mathfrak{F}(m,\ n) = \frac{\sin(m+n)\pi}{\sin m\pi} \mathfrak{F}(1-m-n,\ n);$$

ou enfin, par l'équation (21),

$$\mathfrak{F}(m,\ n) = \frac{\sin(m+n)\pi}{2 \sin m\pi \sin n\pi} \int_{-\infty}^{\infty} (k+i\alpha)^{n-1} (1-k-i\alpha)^{m-1}\, d\alpha,$$

laquelle suppose seulement que $m+n$ soit négatif, ou positif et plus petit que l'unité. Elle présente une analogie assez frappante avec l'équation ordinaire

$$\mathfrak{F}(m,\ n) = \int_0^1 \alpha^{n-1} (1-\alpha)^{m-1}\, d\alpha$$

qui correspond aux valeurs positives de m, n; de même que l'équation (21) est analogue à cette autre forme

$$\mathfrak{F}(m,\ n) = \int_0^{\infty} \frac{\alpha^{n-1}\, d\alpha}{(1+\alpha)^{m-n}},$$

qui correspond aussi aux valeurs positives de m et n.

On peut se proposer de vérifier l'équation (m et n positifs et plus petits que l'unité)

$$\Gamma m\, \Gamma n = \frac{1}{4 \sin m\pi \sin n\pi} \int_{-\infty}^{\infty} dx \int_{-\infty}^{\infty} dy\, (ix)^{m-1} (iy)^{n-1} e^{i(x+y)},$$

en transformant le second membre au moyen de $x=\alpha y$. Pour cela, on distinguera quatre cas, selon que x et y sont tous les deux positifs ou négatifs, ou l'un positif et l'autre négatif. En mettant dans les deux premiers

$$x = \alpha y, \quad dx = y\, d\alpha,$$

et en intégrant par rapport à y, on trouvera que les deux portions correspondantes de l'intégrale double se réuniront en

$$-\Gamma(m+n) \,.\, 2\cos(m+n)\pi \int_0^\infty \frac{\alpha^{m-1}\,d\alpha}{(1+\alpha)^{m+n}},$$

c'est-à-dire à

$$-2\cos(m+n)\pi\,\Gamma m\,\Gamma n.$$

Pour les deux autres portions de l'intégrale double, en écrivant

$$x = -\alpha y,$$

on verra qu'il faut encore distinguer les deux cas $\alpha < 1$ et $\alpha > 1$. Les quatre intégrales ainsi obtenues se réuniront cependant dans les deux

$$2\cos n\pi\,\Gamma(m+n) \int_0^1 \frac{\alpha^{m-1}\,d\alpha}{(1-\alpha)^{m+n}}, \qquad 2\cos m\pi \int_1^\infty \frac{\alpha^{m-1}\,d\alpha}{(\alpha-1)^{m+n}},$$

savoir, après quelques réductions faciles, dans celles-ci,

$$\frac{2\cos n\pi \sin m\pi}{\sin(m+n)\pi}\,\Gamma m\,\Gamma n, \qquad \frac{2\cos m\pi \sin m\pi}{\sin(m+n)}\,\Gamma m\,\Gamma n,$$

lesquelles se réuniront en

$$\cos(m-n)\pi\,\Gamma m\,\Gamma n.$$

On obtient donc enfin l'équation identique

$$4\sin m\pi \sin n\pi = 2\cos(m-n)\pi - 2\cos(m+n)\pi;$$

ce qui suffit pour la vérification dont il s'agit.

50.

SUR QUELQUES THÉORÈMES DE LA GÉOMÉTRIE DE POSITION.

[From the *Journal für die reine und angewandte Mathematik* (Crelle), tome XXXI. (1846), pp. 213—227.]

§ I.

En prenant pour donné un système quelconque de points et de droites, on peut mener par deux points donnés des nouvelles droites, ou trouver des points nouveaux, savoir les points d'intersection de deux des droites données; et ainsi de suite. On obtient de cette manière un nouveau système de points et de droites, qui peut avoir la propriété que plusieurs des points sont situés dans une même droite, ou que plusieurs des droites passent par le même point; ce qui donne lieu à autant de théorèmes de géométrie de position. On a déjà étudié la théorie de plusieurs de ces systèmes; par exemple de celui de quatre points; de six points, situés deux à deux sur trois droites qui se rencontrent dans un même point; de six points trois à trois sur deux droites, ou plus généralement, de six points sur une conique (ce dernier cas, celui de l'hexagramme mystique de Pascal, n'est pas encore épuisé; nous y reviendrons dans la suite), et même de quelques systèmes dans l'espace. Cependant il existe des systèmes plus généraux que ceux qui ont été examinés, et dont les propriétés peuvent être aperçues d'une manière presque intuitive, et qui, à ce que je crois, sont nouveaux. Commençons par le cas le plus simple. Imaginons un nombre n de points situés d'une manière quelconque dans l'espace, et que nous désignerons par $1, 2, 3, \ldots n$. Qu'on fasse passer par toutes les combinaisons de deux points des droites, et par toutes les combinaisons de trois points des plans; puis coupons ces droites et ces plans par un plan quelconque, les droites selon des points, et les plans selon des droites. Soit $\alpha\beta$ le point qui correspond à la droite menée par les deux points α, β; soit de même $\beta\gamma$ le point qui correspond à celle menée par les points β, γ, et ainsi de suite. Soit de plus $\alpha\beta\gamma$ la droite qui correspond au plan passant par les trois points α, β, γ, etc. Il est clair que les trois points $\alpha\beta$, $\alpha\gamma$, $\beta\gamma$ seront situés dans la droite $\alpha\beta\gamma$. Donc en représentant par

N_2, N_3, ... les nombres des combinaisons de n lettres prises deux à deux, trois à trois etc. à la fois, on a le théorème suivant:

THÉORÈME I. *On peut former un système de N_2 points situés trois à trois sur N_3 droites: savoir en représentant les points par* 12, 13, 23, *etc. et les droites par* 123, *etc., les points* 12, 13, 23 *seront situés sur la droite* 123, *et ainsi de suite.*

Pour $n=3$, ou $n=4$, cela est tout simple; on aura trois points sur une droite, ou six points trois à trois sur quatre droites; il n'en résulte aucune propriété géométrique. Pour $n=5$ on a dix points, trois à trois sur autant de droites, savoir les points

12, 13, 14, 15, 23, 24, 25, 34, 35, 45

et les droites

123, 124, 125, 134, 135, 145, 234, 235, 245, 345.

Les points 12, 13, 14, 23, 24, 34 sont les angles d'un quadrilatère quelconque[1], le point 15 est tout à fait arbitraire, le point 25 est situé sur la droite passant par les points 12 et 15, mais sa position sur cette droite est arbitraire. On déterminera depuis les points 35, 45; 35 comme point d'intersection des droites passant par 13 et 15 et par 23 et 25, c'est-à-dire des droites 135 et 235, et de même 45 comme point d'intersection des lignes 145 et 245. Les points 35 et 45 auront la propriété géométrique d'être en ligne droite avec 34, ou bien tous les trois seront dans une même droite 345.

Étudions de plus près la figure que nous venons de former. En prenant le cinq numéros dans un ordre déterminé, par exemple dans l'ordre naturel 1, 2, 3, 4, 5, les cinq points 12, 23, 34, 45, 51 pourront être considérés comme formant un pentagone que nous représenterons par la notation (12345). Les côtés de ce pentagone sont évidemment 123, 234, 345, 451, 512. De même les points 13, 35, 52, 24, 41 peuvent être considérés comme formant le pentagone (13524) dont les côtés sont 135, 352, 524, 241, 413. Ce pentagone est circonscrit au premier, car ses côtés passent évidemment par les angles 15, 23, 45, 12, 34 du premier: mais il est de même inscrit à celui-ci, car ses angles sont situés respectivement dans les côtés 123, 345, 512, 234, 451 de ce même pentagone. Donc les pentagones

(12345), (13524)

sont à la fois circonscrits et inscrits l'un à l'autre, donc:

THÉORÈME II. *La figure composée de dix points, trois à trois dans dix droites, peut être considérée (même de six manières différentes) sous la forme de deux pentagones, inscrits et circonscrits l'un à l'autre.*

Ou encore

THÉORÈME III. *Étant donné un pentagone quelconque, on peut toujours trouver un autre pentagone qui y est à la fois circonscrit et inscrit. Ce second pentagone peut satisfaire à une seule condition donnée quelconque.*

[1] Il faut avoir égard toujours à la différence entre *quadrilatère* et *quadrangle;* chaque quadrilatère a quatre côtés et six angles, chaque quadrangle a quatre angles et six côtés.

Si par exemple le second pentagone doit avoir un de ses angles sur un point donné d'un côté du premier, la construction se déduit tout de suite de ce qui précède.

Ces paires correspondantes de pentagones forment une figure connue. On en trouve la construction dans une note de M. [J. T.] Graves dans le *Philosophical Magazine* [vol. XV. 1839], mais la même figure est encore mieux connue sous un autre point de vue. En effet, considérons le point 12, et les droites 123, 124, 125 qui passent par ce point; puis les triangles dont les angles sont 13, 14, 15 et 23, 24, 25. Les côtés de ces mêmes triangles sont 134, 135, 145 et 234, 235, 245, et les côtés correspondants se rencontrent dans les points 34, 35, 45 qui sont en ligne droite. Donc le théorème sur les pentagones est le suivant:

"*Si les angles de deux triangles sont situés deux à deux dans trois droites qui se rencontrent dans un point, leurs côtés homologues se coupent dans trois points en ligne droite.*"

Remarquons aussi que ce théorème particulier (en n'empruntant rien des trois dimensions de l'espace) reproduit le théorème général relatif au nombre n. Il n'y a pour cela qu'à considérer n droites passant par le même point, et qui peuvent être désignées par 1, 2, 3, ... n. En choisissant d'abord les points 12, 13, tout triangle dont les trois angles sont situés dans les droites 1, 2, 3, pendant que deux de ses côtés passent par 12, 13, a la propriété que le troisième côté passe par un point déterminé 23 situé dans la droite passant par 12, 13. En prenant arbitrairement le point 14, on obtient avec les droites 1, 3, 4 ou 1, 2, 4 les nouveaux points 34, 24 qui sont en ligne droite avec 23, et ainsi de suite.

Passons au cas $n = 6$. Il existe ici quinze points situés trois à trois sur vingt droites, ou bien vingt droites qui se coupent quatre à quatre en quinze points. Il n'y a point ici des systèmes d'hexagones, mais il existe un système de neuf points qui est assez remarquable. Divisons d'une manière quelconque les numéros 1, 2, 3, 4, 5, 6 en deux suites par trois, par exemple en 1, 3, 5 et 2, 4, 6, et considérons les neuf points

12, 14, 16,
32, 34, 36,
52, 54, 56.

Les droites qui passent par 12 et 32, 14 et 34, 16 et 36, savoir 132, 134, 136, se rencontrent dans le même point 13. De même les droites qui passent par 32 et 52, 34 et 54, 36 et 56 se rencontrent dans 35, et les droites qui passent par 12 et 52, 14 et 54, 16 et 56 se rencontrent dans 15. Les points 13, 15 et 35 sont sur la même droite 135. En considérant les points 12, 14, 16 comme formant un triangle, et de même les points 32, 34, 36 et 52, 54, 56, cela revient à dire que les droites menées par les angles homologues des triangles prises deux à deux, se rencontrent trois à trois dans trois points situés dans la même droite. · Ou bien, ce que l'on savait déjà par le théorème 3: les côtés homologues des triangles se rencontrent trois à trois dans trois points situés en ligne droite. En effet, les côtés des triangles sont 124, 126, 146 pour la première, et 324, 326, 346 et 524, 526, 546 pour les deux autres. Les trois premiers côtés se rencontrent dans 24, les autres dans 26 et 46, et ces trois points sont dans la droite 246. Maintenant tout cela arrive également en combinant les colonnes verticales, ou en considérant les neuf points comme formant les trois autres triangles dont les angles sont 12, 32, 52; 14, 34, 54; 16, 36, 56. Cela donne lieu au théorème suivant:

THÉORÈME IV. *Le système de quinze points, situés trois à trois sur vingt droites, contient (et cela même de dix manières différentes) un système de neuf points qui ont la propriété de former de deux manières différentes trois triangles, tels, que les droites qui passent par leurs angles homologues, prises deux à deux, se rencontrent dans trois points qui sont en ligne droite, tandis que les côtés homologues des triangles se coupent trois à trois en trois autres points qui sont aussi en ligne droite. Dans la seconde manière de former les triangles, ces deux systèmes de trois points en ligne droite sont seulement échangés.*

Il ne reste qu'à savoir combien il y en a d'arbitraires dans le système de quinze points situés trois à trois sur vingt droites. En supposant le système formé pour le nombre cinq, on peut prendre arbitrairement 16 et 26 sur la droite 126 qui est déterminée par les points 12 et 16. Donc 12, 13, 14, 15 et 16 sont arbitraires et 23, 24, 25, 26 sont arbitrairement situés sur des droites données. L'existence des droites 345, 346, 356, 456 constitue autant de théorèmes géométriques; c'est-à-dire, chacune de ces droites est déterminée par *trois points.*

En essayant d'approfondir la théorie de six points sur la même conique, on rencontrera un système de neuf points, tel que ceux que nous venons d'examiner; mais il est moins général. Il existe des relations entre les points qui n'ont pas lieu dans le système général. Je renvoie cette discussion à une section séparée de ce mémoire, et je passe au cas de $n=7$.

Pour ce cas on a tout de suite le théorème suivant:

THÉORÈME V. *Le système de vingt et un points situés trois à trois sur trente-cinq droites, peut être considéré (même de cent vingt manières différentes) comme composé de trois heptagones, le premier circonscrit au second, le second au troisième et le troisième au premier. Les heptagones par exemple peuvent être* (1234567), (1357246), (1526374).

Dans ce système 12, 13, 14, 15, 16, 17 sont arbitraires, et 23, 24, 25, 26, 27 le sont sur des droites données; les droites 345, 346, 347, 356, 357, 367, 456, 457, 467, 567 sont déterminées chacune par trois points. Dans le cas général 12, 13 ... $1n$ sont arbitraires, et 23 ... $2n$ le sont sur des droites données. Il existe $\frac{1}{6}(n-2)(n-3)(n-4)$ droites dont chacune est déterminée par trois points. Un théorème analogue à celui-ci a lieu quand n est un nombre premier: savoir le suivant:

THÉORÈME VI. *Le système de N_2 points, situés trois à trois sur N_3 droites, peut être considéré $\left(\text{même de } \frac{1.2\ldots(n-2)}{n-1} \text{ manières}\right)$ comme composé de $\frac{1}{2}(n-1)$ n-gones, le premier circonscrit au second, le second au troisième, etc., et le dernier au premier.*

Je ne connais pas d'autres cas où l'idée des nombres premiers se présente dans la géométrie. Il sera peut-être possible de trouver des théorèmes analogues à III, IV, V, pour toutes les formes du nombre n, mais je n'ai pas encore examiné cela.

Le théorème général I, peut être considéré comme l'expression d'un fait analytique, qui doit également avoir lieu en considérant quatre coordonnées au lieu de trois. Ici une interprétation géométrique a lieu, qui s'applique aux points *dans l'espace.* On peut en effet, *sans recourir à aucune notion métaphysique à l'égard de la possibilité de l'espace à quatre dimensions,* raisonner comme suit (tout cela pourra aussi être traduit facilement en langue purement analytique): En supposant quatre dimensions de l'espace, il faudra considérer des *lignes* déterminées par deux points, des *demi-plans* déterminés par trois points, et des *plans* déterminés par quatre points; (deux plans se coupent alors suivant un demi-plan, etc.). L'espace ordinaire doit être considéré comme plan, et il coupera un plan selon un plan ordinaire, un demi-plan selon une ligne ordinaire, et une ligne selon un point ordinaire. Tout cela posé: en considérant un nombre n de points, et les combinant deux à deux, trois à trois, et quatre à quatre par des lignes, des demi-plans et des plans, puis coupant le système par l'espace considéré comme plan, on obtient le théorème suivant de géométrie à trois dimensions:

THÉORÈME VII. *On peut former un système de N_2 points, situés trois à trois dans N_3 droites qui elles-mêmes sont situées quatre à quatre dans N_4 plans. En représentant les points par* 12, 13, *etc., les points situés dans la même droite sont* 12, 13, 23; *et les droites étant représentées par* 123 *etc. comme auparavant, les droites* 123, 124, 134, 234 *sont situées dans le même plan* 1234.

En coupant cette figure par un plan, on obtient le théorème suivant de géométrie plane:

THÉORÈME VIII. *On peut former un système de N_3 points situés quatre à quatre dans N_4 droites. Les points doivent être représentées par la notation* 123, *etc. et les droites par* 1234, *etc. Alors* 123, 124, 134, 234 *sont dans la même droite désignée par* 1234.

De même, en considérant un espace à $p+2$ dimensions, on obtient la proposition suivante, encore plus générale:

THÉORÈME IX. *On peut former dans l'espace un système de N_p points, qui passent $p+1$ à $p+1$ par N_{p+1} droites, situées $p+2$ à $p+2$ dans N_{p+2} plans, ou bien pour la géométrie plane, un système de N_p points, situés $p+1$ à $p+1$ dans N_{p+1} droites.*

Des théorèmes analogues à IV et V seraient probablement très nombreux et très compliqués.

Les réciproques polaires auront évidemment lieu pour tous ces théorèmes; on pourrait aussi les démontrer directement d'une manière analogue.

§ II.

SUR LE THÉORÈME DE PASCAL.

En considérant six points sur la même conique, et les prenant dans un ordre déterminé, pour en former un hexagone, on sait que les côtés opposés se rencontrent dans trois points situés en ligne droite. En prenant les points dans un ordre quelconque, on en peut former soixante hexagones, à chacun desquels correspond une droite; il s'agit maintenant de trouver les relations entre ces droites.

M. Steiner a prouvé dans son ouvrage *Systematische Entwickelungen u. s. w.* [1832], que ces soixante droites passent trois à trois par vingt points, et il ajoute que ces vingt points sont situés quatre à quatre sur quinze droites. La première partie de ce théorème peut être démontrée assez facilement, comme nous le verrons: mais pour la seconde partie, je n'ai pas réussi à trouver les combinaisons de quatre points qui doivent être situés en ligne droite, et il me paraît même qu'il est impossible de les trouver[1].

Cherchons les combinaisons des droites qui doivent passer trois à trois par le même point.

Soient 1, 2, 3, 4, 5, 6 les six points situés sur la même conique. Considérons d'abord l'hexagone 123456 que l'on obtient en prenant les points dans un ordre déterminé. Suivant le théorème de Pascal les trois points

12.45, 23.56, 34.61

(où 12.45 désigne le point d'intersection des lignes passant par les points 1, 2 et 4, 5) sont situés en ligne droite. Considérons les six hexagones

1 2 3 4 5 6
1 4 3 6 5 2
1 6 3 2 5 4

---

1 4 3 2 5 6
1 2 3 6 5 4
1 6 3 4 5 2

qu'on tire du premier en permutant les nombres 2, 4, 6 correspondants aux sommets alternés de l'hexagone. Pour les trois premiers on fait les permutations cycliques de ces nombres (savoir 246, 462, 624), pour les trois autres on fait d'abord une inversion 426,

[1] Je ne sais pas s'il existe une démonstration de la seconde partie du théorème; je n'ai pu la trouver nulle part. Au cas que cette partie du théorème n'était pas correcte, il paraît que l'on devra peut-être lui substituer la proposition suivante: "Les vingt points déterminent deux à deux dix lignes qui passent trois à trois par dix points." On verra dans ce qui suit, de quelle manière il faudrait combiner ces points. [Voir 55.]

puis les permutations cycliques (426, 264, 642). En écrivant les combinaisons des points qui doivent être situés en ligne droite, on a

12.45	23.56	34.61
36.12	56.14	52.34
45.36	14.23	16.52
14.25	43.56	32.61
36.14	56.12	54.42
25.36	12.34	61.54

Suivant cette table les points sur la même horizontale sont en ligne droite.

On remarquera d'abord que les trois premières droites passent par les angles des triangles dont les côtés sont 36, 45, 12 et 14, 23, 56. Les côtés homologues de ces triangles se rencontrent en 36.14, 45.23, 12.56 qui sont en ligne droite, c'est-à-dire, par un théorème déjà cité: les trois lignes passent par un même point. On aurait été conduit au même résultat en observant que les trois premières droites passent par les triangles dont les côtés sont 14, 23, 56 et 52, 16, 34 ou enfin 52, 16, 34 et 36, 45, 12. De même les trois dernières droites passent par le même point. Donc il a été démontré ce qui suit:

THÉORÈME X. *En considérant les trois hexagones qu'on obtient en permutant cycliquement les angles alternés du premier, les trois droites qui y correspondent se rencontrent dans un même point. Les soixante lignes passent donc trois à trois par vingt points.*

Ajoutons qu'aux trois hexagones de ce théorème correspondent d'une manière particulière trois autres hexagones, ou que les vingt points doivent se combiner deux à deux d'une manière particulière.

Mais on se formera une idée plus claire du système en remarquant que les neuf droites

36, 45, 12
14, 23, 56
25, 61, 34

ont entre elles une relation qui est polaire réciproque de celle entre les neuf points du théorème IV. Pour faciliter cette comparaison, je prendrai d'abord le théorème analogue pour les tangentes d'une conique.

THÉORÈME XI. Soient 1, 3, 5 et 2, 4, 6 des tangentes à une même conique et 12, etc. les points d'intersection de ces droites: les neuf points

36, 45, 12
14, 23, 56
25, 61, 34

peuvent être déterminés au moyen de six points de l'espace A, B, C, α, β, γ, de manière que $A\alpha$, etc. représente le point d'intersection de la droite passant par A, α avec le plan de la figure. Les points sont correspondants entre eux de cette manière:

$$\begin{array}{ccc} A\alpha, & A\beta, & A\gamma \\ B\alpha, & B\beta, & B\gamma \\ C\alpha, & C\beta, & C\gamma \end{array}$$

seulement les points 36, 23, 34 etc. sont en ligne droite, ce qui n'aurait pas lieu pour les points $A\alpha$, $B\beta$, $C\gamma$, si la position de A, B, C, α, β, γ était arbitraire. On est donc conduit à ce problème:

Trouver six points A, B, C, α, β, γ dans l'espace, tels, qu'en représentant par $A\alpha$, etc. l'intersection de la droite menée par $A\alpha$ avec un plan donné, les combinaisons des points

$$\begin{array}{l} (A\alpha,\ B\beta,\ C\gamma) \\ (A\beta,\ B\gamma,\ C\alpha) \\ (A\gamma,\ B\alpha,\ C\beta) \\ (A\alpha,\ B\gamma,\ C\beta) \\ (A\beta,\ B\alpha,\ C\gamma) \\ (A\gamma,\ B\beta,\ C\alpha) \end{array}$$

soient en ligne droite.

Pour le théorème de Pascal, cela donne:

THÉORÈME XII. *Soient* 1, 3, 5 *et* 2, 4, 6 *des points d'une conique, les neuf lignes*

$$\begin{array}{ccc} 36, & 45, & 12 \\ 14, & 23, & 56 \\ 25, & 61, & 34 \end{array}$$

peuvent être considérées comme les projections des lignes

$$\begin{array}{ccc} A\alpha, & A\beta, & A\gamma \\ B\alpha, & B\beta, & B\gamma \\ C\alpha, & C\beta, & C\gamma \end{array}$$

sur le plan de la figure, où A, B, C, α, β, γ *sont six plans, dont la relation reste encore à déterminer.*

En effectuant la solution du problème que j'ai indiquée on aurait, à ce qu'il me semble, un point de vue tout à fait nouveau d'envisager les coniques.

Je vais ajouter encore quelques réflexions sur la manière de chercher les relations qui existent entre les vingt points. En écrivant seulement les angles alternés des hexagones, on a cette table:

1.2.3
1.2.4
1.2.5
1.2.6
1.3.4
1.3.5
1.3.6
1.4.5
1.4.6
1.5.6

A chaque symbole correspondent six hexagones, qui, à ce que nous avons vu, se partagent en deux paires de trois hexagones, et à chaque combinaison de trois, il correspond un point. Il y a donc deux points qui correspondent au symbole 1.3.5, deux qui correspondent au symbole 1.3.6, deux au symbole 1.5.6 etc. En représentant donc par $\overline{35}$, $\overline{36}$, $\overline{56}$, les droites passant par ces paires de points, il me paraît probable que ces droites aient ensemble les relations du théorème I, (savoir que $\overline{35}$, $\overline{36}$, $\overline{56}$ se rencontrent dans un point etc.), ce qui donnerait lieu au théorème hypothétique que j'ai énoncé dans une note. Voilà, à ce que je puis apercevoir, la seule manière symétrique de combiner les droites. Mais au moins les symboles

1.3.5
1.3.6
1.5.6

ont entre eux des rapports singuliers. En effet, écrivons pour chacun les neuf points du théorème XII, on a ce tableau:

36,	45,	12
14,	23,	56
25,	61,	34
35,	64,	12
14,	23,	56
26,	15,	34
35,	46,	12
14,	25,	36
26,	13,	54

qui ne contient que quatorze points. Cela mérite des recherches ultérieures.

Démonstration analytique du théorème de Pascal, et de la première partie de celui de M. Steiner. Formules relatives au même sujet.

Soient $P=0$, $Q=0$, $R=0$ les équations des lignes **12**, 34, 56. On démontrera assez facilement que les équations des lignes 45, 61, **23** peuvent être représentées par

$$\begin{aligned} P+\nu Q+\mu R&=0,\\ \nu P+\ \ Q+\lambda R&=0,\\ \mu P+\lambda Q+\ \ R&=0. \end{aligned}$$

En effet les six points 1, 2, 3, 4, 5, 6 seront situés dans la conique

$$P^2+Q^2+R^2+\lambda+\frac{1}{\lambda}QR+\mu+\frac{1}{\mu}PR+\nu+\frac{1}{\nu}PQ=0\,;$$

car en faisant dans cette équation $P=0$, l'équation se réduit à

$$\frac{1}{\lambda}(Q+\lambda R)(\lambda Q+R)=0\,;$$

c'est-à-dire, la conique contient les points déterminés par

$$\begin{aligned} (\,P=0, &\qquad \nu P+Q+\lambda R=0\,),\\ (\,P=0, &\qquad \mu P+\lambda Q+R=0\,), \end{aligned}$$

ou bien les points 1, 2; et de même elle contient les autres points 3, 4, 5, 6. Les fonctions P, Q, R sont censées contenir chacune deux constantes arbitraires; donc on a neuf constantes arbitraires dans ce système, qui par conséquent est tout-à-fait général. On peut former le système suivant d'équations:

$$\begin{array}{rrrrl}
12. & P & & & =0,\\
13. & \lambda\mu P+ & Q+ & \lambda R & =0,\\
14. & \lambda P+ & \mu Q+ & \lambda\mu R & =0,\\
15. & P+ & \nu Q+ & \nu\lambda R & =0,\\
16. & \nu P+ & Q+ & \lambda R & =0,\\
23. & \mu P+ & \lambda Q+ & R & =0,\\
24. & P+ & \mu\lambda Q+ & \mu R & =0,\\
25. & \lambda P+ & \nu\lambda Q+ & \nu R & =0,\\
26. & \nu\lambda P+ & \lambda Q+ & R & =0,\\
34. & & Q & & =0,\\
35. & \mu P+ & \mu\nu Q+ & R & =0,\\
36. & \mu\nu P+ & \mu Q+ & \nu R & =0,\\
45. & P+ & \nu Q+ & \mu R & =0,\\
46. & \nu P+ & Q+ & \nu\mu R & =0,\\
56. & & & R & =0.
\end{array}$$

Écrivons les équations des lignes comprises dans la table de neuf points ci-dessus donnés. On a d'abord

$$\begin{array}{lll} \mu\nu P + \mu Q + \nu R = 0, & P + \nu Q + \mu R = 0, & P = 0, \\ \lambda P + \mu Q + \lambda\mu R = 0, & \mu P + \lambda Q + R = 0, & R = 0, \\ \lambda P + \nu\lambda Q + \nu R = 0, & \nu P + Q + \lambda R = 0, & Q = 0. \end{array}$$

En combinant la seconde et la troisième colonne verticale du tableau, on obtient pour les trois points d'intersection des côtés opposés de l'hexagone 123456, les équations

$$(P = 0, \quad \mu Q + \nu R = 0),$$
$$(R = 0, \quad \lambda P + \mu Q = 0),$$
$$(Q = 0, \quad \lambda P + \nu R = 0),$$

qui appartiennent à trois points situés sur la droite

$$\lambda P + \mu Q + \nu R = 0,$$

ce qui suffit pour démontrer le théorème de Pascal.

On obtient de même, en combinant les autres paires de colonnes verticales, deux systèmes de trois points, respectivement situés dans les droites

$$\frac{P}{\lambda} + \frac{Q}{\mu} + \frac{R}{\nu} = 0 \quad \text{et}$$
$$\left(\frac{P}{\lambda} + \frac{Q}{\mu} + \frac{R}{\nu}\right) \lambda\mu\nu + \lambda P + \mu Q + \nu R = 0,$$

lesquelles, avec la droite qu'on vient de trouver,

$$\lambda P + \mu Q + \nu R = 0,$$

se rencontrent évidemment dans un même point, déterminé par les deux équations

$$\lambda P + \mu Q + \nu R = 0 \quad \text{et}$$
$$\frac{P}{\lambda} + \frac{Q}{\mu} + \frac{R}{\nu} = 0.$$

Voilà une démonstration de la première partie du théorème de M. Steiner. Les équations que nous venons de trouver appartiennent au point d'intersection des trois droites qui correspondent au premier des trois hexagones du symbole 1.3.5. Pour trouver l'autre point correspondant de la même manière à ce symbole, il faut combiner les colonnes horizontales, ce qui donne pour les coordonnées de ce point :

$$(\lambda - \mu\nu) P = (\mu - \nu\lambda) Q = (\nu - \lambda\mu) R.$$

En cherchant de même les expressions des points qui correspondent aux symboles 1.3.6 et 1.5.6, on obtient des résultats moins élégants, mais qui valent peut-être la peine d'être énoncés ici.

Je forme cette table complète :

Systèmes de trois lignes, qui se rencontrent dans un point.

Pour 1.3.5

I. $$\begin{cases} \lambda P + \mu Q + \nu R = 0, \\ \dfrac{P}{\lambda} + \dfrac{Q}{\mu} + \dfrac{R}{\nu} = 0, \\ \lambda P + \mu Q + \nu R + \lambda\mu\nu\left(\dfrac{P}{\lambda} + \dfrac{Q}{\mu} + \dfrac{R}{\nu}\right) = 0; \end{cases}$$

II. $$\begin{cases} (\lambda - \mu\nu)\,P = (\nu - \lambda\mu)\,R, \\ (\nu - \lambda\mu)\,R = (\mu - \nu\lambda)\,Q, \\ (\mu - \nu\lambda)\,Q = (\lambda - \mu\nu)\,P; \end{cases}$$

Pour 1.3.6

I. $$\begin{cases} \mu P + \lambda Q + \lambda\mu\nu R = 0, \\ \nu\lambda P + \mu\nu Q + R = 0, \\ (\mu P + \lambda Q + \lambda\mu\nu R) + (\nu\lambda P + \mu\nu Q + R) = 0; \end{cases}$$

II. $$\begin{cases} (\mu - \nu\lambda)\,P + (1 - \mu\nu\lambda)\,R = 0, \\ (1 - \nu\lambda\mu)\,R + (\lambda - \nu\mu)\,Q = 0, \\ (\lambda - \mu\nu)\,Q - (\mu - \nu\lambda)\,P = 0; \end{cases}$$

Pour 1.5.6

I. $$\begin{cases} (\mu\nu - \lambda\mu^2\nu^2 - \lambda + \lambda\mu^2)\,P + (\mu - \nu\lambda)\,Q + (\mu^2\nu - \lambda\mu\nu^2)\,R = 0, \\ (\mu\nu - \lambda\mu^2\nu^2 - \lambda + \lambda\nu^2)\,P + (\mu\nu^2 - \lambda\mu^2\nu)\,Q + (\nu - \lambda\mu)\,R = 0, \\ (\lambda\mu^2 - \lambda\nu^2)\,P + (\mu - \nu\lambda - \mu\nu^2 + \lambda\mu^2\nu)\,Q + (\mu^2\nu - \lambda\mu\nu^2 - \nu + \lambda\mu)\,R = 0; \end{cases}$$

II. $$\begin{cases} (\lambda - \lambda\nu^2 + \mu\nu - \lambda\mu^2)\,P + (\mu - \lambda\mu^2\nu)\,Q + (\nu - \lambda\mu\nu^2)\,R = 0, \\ (\lambda\nu^2 - \lambda\mu^2\nu^2 - \mu\nu + \lambda\mu^2)\,P + (\nu\lambda - \mu\nu^2)\,Q + (\lambda\mu - \mu^2\nu)\,R = 0, \\ (\lambda - \lambda\mu^2\nu^2)\,P + (\nu\lambda - \mu\nu^2 + \mu - \lambda\mu^2\nu)\,Q + (\lambda\mu - \mu^2\nu + \nu - \lambda\mu\nu^2)\,R = 0. \end{cases}$$

Note sur le théorème de M. Brianchon.

On peut donner une démonstration semblable de ce théorème, en prenant pour les équations des six tangentes celles-ci :

$$\begin{aligned} &1. \quad P = 0, \\ &3. \quad Q = 0, \\ &5. \quad R = 0, \\ &4. \quad \alpha P + \beta Q + \gamma R = 0, \\ &6. \quad \alpha' P + \beta' Q + \gamma' R = 0, \\ &2. \quad \alpha'' P + \beta'' Q + \gamma'' R = 0, \end{aligned}$$

et en cherchant la relation entre les coefficients qui est nécessaire pour que ces six équations appartiennent aux tangentes d'une même conique. On obtient facilement

$$(\alpha\gamma' - \alpha'\gamma)(\beta'\alpha'' - \beta''\alpha')(\gamma''\beta - \gamma\beta'') = (\alpha'\gamma'' - \alpha''\gamma')(\beta''\alpha - \beta\alpha'')(\gamma\beta' - \gamma'\beta),$$

ce qui exprime aussi la condition pour que les trois diagonales se rencontrent dans un même point.

51.

PROBLÈME DE GÉOMÉTRIE ANALYTIQUE.

[From the *Journal für die reine und angewandte Mathematik* (Crelle), tom. XXXI. (1846), pp. 227—230.]

Trouver explicitement les coordonnées des centres de similitude de deux surfaces du second ordre, dont chacune est circonscrite à une même surface de cet ordre.

LEMME.

Soit

$$U = Ax^2 + By^2 + Cz^2 + 2Fyz + 2Gzx + 2Hxy + 2Lxw + 2Myw + 2Nzw + Pw^2 \ldots\ldots(1),$$

l'expression générale d'une fonction homogène du second ordre à quatre variables, et considérons les dérivées

$$KU = \begin{vmatrix} A, & H, & G, & L \\ H, & B, & F, & M \\ G, & F, & C, & N \\ L, & M, & N, & P \end{vmatrix}, \qquad F_{po}\,U = \begin{vmatrix} & \xi, & \eta, & \rho, & \omega \\ \alpha, & A, & H, & G, & L \\ \beta, & H, & B, & F, & M \\ \gamma, & G, & F, & C, & N \\ \delta, & L, & M, & N, & P \end{vmatrix} \ldots\ldots\ldots\ldots (2),$$

où dans $F_{po}\,U$ les lettres o, p écrites en bas servent à indiquer les variables α, β, γ, δ et ξ, η, ρ, ω qui doivent entrer dans l'expression de cette fonction; par exemple $F_{pp}U$ est ce que devient $F_{po}U$, en écrivant ξ, η, ρ, ω au lieu de α, β, γ, δ.

Cela posé, soit

$$U = Ax^2 + By^2 + Cz^2 + 2Fyz + 2Gzx + 2Hxy + 2Lxw + 2Myw + 2Nzw + 2Pw^2, \ldots (3).$$

$$V = \alpha x + \beta y + \gamma z + \delta w, \ldots\ldots\ldots\ldots\ldots\ldots\ldots\ldots\ldots\ldots\ldots\ldots\ldots\ldots (4).$$

on aura cette équation identique:

$$F_{pp}(U+V^2)\,KU - F_{pp}(U)\,K(U+V^2) = (F_{op}U)^2 \qquad (5),$$

qui subsiste même pour un nombre quelconque de variables.

L'expression analytique du théorème consiste en effet en ce que les réciproques de deux surfaces du second ordre, circonscrites l'une à l'autre, sont deux surfaces du second ordre qui ont cette même relation. Car en prenant $\frac{x}{w}$, $\frac{y}{w}$, $\frac{z}{w}$ pour les coordonnées d'un point, les équations de deux surfaces circonscrites l'une à l'autre sont

$$\left.\begin{array}{l} U=0, \\ U+V^2=0 \end{array}\right\} \qquad (6).$$

Les équations de leurs réciproques polaires (par rapport à $x^2+y^2+z^2+w^2=0$) sont

$$F_{pp}U=0, \quad F_{pp}(U+V^2)=0,$$

c'est-à-dire, et vertu du théorème qui vient d'être posé:

$$F_{pp}U=0, \qquad K(U+V^2)\,F_{pp}U+(F_{op}U)^2=0 \qquad (7),$$

où $K(U+V^2)$ est constant; c'est-à-dire les premières parties des équations ne diffèrent entre elles que par le carré de la fonction linéaire $(F_{op}U)$; ce qui prouve le théorème en question.

SOLUTION.

Soient

$$U+V_1^2=0 \text{ et } U+V_2^2=0 \qquad (8),$$

les équations des deux surfaces, dont chacune est circonscrite à $U=0$. Les expressions de V_1 et V_2 sont

$$\left.\begin{array}{l} V_1=\alpha_1x+\beta_1y+\gamma_1z+\delta_1w, \\ V_2=\alpha_2x+\beta_2y+\gamma_2z+\delta_2w, \end{array}\right\} \qquad (9),$$

et

et les lettres o_1, o_2 écrites en bas se rapporteront à α_1, β_1, γ_1, δ_1 et à α_2, β_2, γ_2, δ_2 respectivement. Mettons de plus, pour abréger,

$$\left.\begin{array}{l} K(U+V_1^2)=K_1, \\ K(U+V_2^2)=K_2, \end{array}\right\} \qquad (10).$$

Les polaires des deux surfaces ont pour équations

$$\left.\begin{array}{l} K_1F_{pp}(U)+(F_{o_1p}U)^2=0, \\ K_2F_{pp}(U)+(F_{o_2p}U)^2=0, \end{array}\right\} \qquad (11),$$

et ces surfaces polaires se rencontrent évidemment selon les courbes situées dans les plans exprimés par les équations

$$\sqrt{K_2}F_{o_1p}U \pm \sqrt{K_1}F_{o_2p}U=0 \qquad (12);$$

équations qu'on peut écrire sous cette forme très simple:

$$F_{o'p}(U)=0 \qquad (13),$$

en mettant

$$\left.\begin{aligned} \alpha' &= \sqrt{K_2}\,\alpha_1 \pm \sqrt{K_1}\,\alpha_2,\\ \beta' &= \sqrt{K_2}\,\beta_1 \pm \sqrt{K_1}\,\beta_2,\\ \gamma' &= \sqrt{K_2}\,\gamma_1 \pm \sqrt{K_1}\,\gamma_2,\\ \delta' &= \sqrt{K_2}\,\delta_1 \pm \sqrt{K_1}\,\delta_2, \end{aligned}\right\} \qquad \text{(14)},$$

et supposant que o' se rapporte à α', β', γ', δ'. On a enfin, en se servant de la notation complète des déterminants,

$$\begin{vmatrix} \xi, & \eta, & \rho, & \omega\\ \alpha', & A, & H, & G, & L\\ \beta', & H, & B, & F, & M\\ \gamma', & G, & F, & C, & N\\ \delta', & L, & M, & N, & P \end{vmatrix} = 0 \qquad \text{(15)},$$

équation qui est double, à cause des doubles valeurs de α', β', γ', δ'. Les pôles de ces plans sont les centres de similitude des deux surfaces données. Soit donc identiquement

$$\begin{vmatrix} \xi, & \eta, & \rho, & \omega\\ \alpha', & A, & H, & G, & L\\ \beta', & H, & B, & F, & M\\ \gamma', & G, & F, & C, & N\\ \delta', & L, & M, & N, & P \end{vmatrix} = \mathfrak{A}\xi + \mathfrak{B}\eta + \mathfrak{C}\rho + \mathfrak{D}\omega \ \ldots\text{(16)},$$

on a

$$x : y : z : w = \mathfrak{A} : \mathfrak{B} : \mathfrak{C} : \mathfrak{D} \qquad \text{(17)},$$

pour les coordonnées $\frac{x}{w}$, $\frac{y}{w}$, $\frac{z}{w}$ des deux centres de similitude. $\mathfrak{A}$, $\mathfrak{B}$, $\mathfrak{C}$, $\mathfrak{D}$ sont données par l'équation (16), savoir par

$$\mathfrak{A} = \begin{vmatrix} . & 1 & . & . & .\\ \alpha', & A, & H, & G, & L\\ \beta', & H, & B, & F, & M\\ \gamma', & G, & F, & C, & N\\ \delta', & L, & M, & N, & P \end{vmatrix} \quad \text{ou} \quad \mathfrak{A} = -\begin{vmatrix} \alpha', & H, & G, & L\\ \beta', & B, & F, & M\\ \gamma', & F, & C, & N\\ \delta', & M, & N, & P \end{vmatrix}, \ \&\text{c.} \quad \text{(18)},$$

α', β', γ', δ' sont donnés par (14), et K_1, K_2 représentent ce que devient le déterminant

$$\begin{vmatrix} A, & H, & G, & L\\ H, & B, & F, & M\\ G, & F, & C, & N\\ L, & M, & N, & P \end{vmatrix} \qquad \text{(19)},$$

en écrivant $A+\alpha_1^2$, $B+\beta_1^2$, $C+\gamma_1^2$, $P+\delta_1^2$, $F+\beta_1\gamma_1$, $G+\gamma_1\alpha_1$, $H+\alpha_1\beta_1$, $L+\alpha_1\delta_1$, $M+\beta_1\delta_1$, $N+\gamma_1\delta_1$, ou $A+\alpha_2^2$, &c. au lieu de A, B, C, P, F, G, H, L, M, N.

52.

SUR QUELQUES PROPRIÉTÉS DES DÉTERMINANTS GAUCHES.

[From the *Journal für die reine und angewandte Mathematik* (Crelle), tom. XXXII. (1846), pp. 119—123.]

I.

Je donne le nom de *déterminant gauche,* à un déterminant formé par un système de quantités $\lambda_{r.s}$ qui satisfont aux conditions

$$\lambda_{r.s} = -\lambda_{s.r} \; [r \neq s] \quad \ldots\ldots (1).$$

J'appelle aussi un tel système, *système gauche.* On obtiendra des formules plus simples (quoique cette supposition ne soit pas essentielle), en considérant seulement les systèmes pour lesquels on a aussi

$$\lambda_{r.r} = 1 \quad \ldots\ldots (2).$$

Je suppose dans tout ce qui va suivre, que le déterminant est de l'ordre n, et que par conséquent les suffixes variables r, s, &c., s'étendent toujours depuis l'unité jusqu'à n.

En posant les équations

$$\Sigma_r \lambda_{r.s} x_r = P_s, \quad \Sigma_s \lambda_{r.s} x_s = Q_r \quad \ldots\ldots (3),$$

j'exprime les systèmes inverses qui déterminent les P, Q par les x, de la manière suivante:

$$Kx_r = \Sigma_s \Lambda_{r.s} P_s, \;\text{ et }\; Kx_s = \Sigma_r \Lambda_{r.s} Q_r \quad \ldots\ldots (4),$$

où K désigne le déterminant formé avec les quantités $\lambda_{r.s}$, et $\Lambda_{r.s}$ le coefficient différentiel de K par rapport à $\lambda_{r.s}$; bien entendu, que la différentiation doit être effectuée avant d'avoir particularisé ces quantités par les équations (1), (2). On sait que ces fonctions Λ satisfont aux conditions

$$\left.\begin{array}{l}\Sigma_r \lambda_{r.s} \Lambda_{r.s'} = 0\ ,\ s \neq s', \\ \Sigma_r \lambda_{r.s} \Lambda_{r.s} = K, \\ \Sigma_r \lambda_{r.s} \Lambda_{r'.s} = 0\ ,\ r \neq r', \\ \Sigma_s \lambda_{r.s} \Lambda_{r.s} = K.\end{array}\right\} \quad \ldots\ldots\ldots\ldots \quad (5).$$

Je tire des équations (4), en échangeant r et s dans la dernière de ces équations:

$$\Sigma_s \Lambda_{r.s} P_s = \Sigma_s \Lambda_{s.r} Q_s \quad \ldots\ldots\ldots\ldots \quad (6),$$

et de là, en multipliant respectivement les différentes équations de ce système par $\lambda_{s'.r}$, et prenant la somme de ces produits:

$$\Sigma_s (\Sigma_r \lambda_{s'.r} \Lambda_{r.s}) P_s = \Sigma_s (\Sigma_r \lambda_{s'.r} \Lambda_{s.r}) Q_s \ldots\ldots\ldots\ldots (7).$$

On a d'abord par les équations (5)

$$\Sigma_s (\Sigma_r \lambda_{s'.r} \Lambda_{s.r}) Q_s = K Q_{s'} \ldots\ldots\ldots\ldots (8);$$

puis par les équations (1) et (2)

$$\Sigma_r \lambda_{s'.r} \Lambda_{r.s} = 2\Lambda_{s'.s} - \Sigma_r \lambda_{r.s'} \Lambda_{r.s} \ldots\ldots\ldots\ldots (9),$$

c'est-à-dire par les équations (5):

$$\left.\begin{array}{ll} & \Sigma_r \lambda_{s'.r} \Lambda_{r.s} = 2\Lambda_{s'.s};\ s' \neq s \\ \text{et} & \Sigma_r \lambda_{s'.r} \Lambda_{r.s'} = 2\Lambda_{s'.s'} - K\end{array}\right\} \ldots\ldots\ldots\ldots (10);$$

ce qui donne

$$\Sigma_s (\Sigma_r \lambda_{s'.r} \Lambda_{r.s}) P_s = 2 (\Sigma_s \Lambda_{s'.s} P_s) - K P_{s'} \ldots\ldots\ldots\ldots (11).$$

Substituant les équations (8) et (11) dans la formule (7), on obtiendra, en écrivant r au lieu de s':

$$K Q_r = 2 (\Sigma_s \Lambda_{r.s} P_s) - K P_r \quad \ldots\ldots\ldots\ldots \quad (12),$$

et également

$$K P_s = 2 (\Sigma_r \Lambda_{r.s} Q_r) - K Q_s \quad \ldots\ldots\ldots\ldots \quad (13).$$

Posant maintenant

$$\left.\begin{array}{ll} & K \alpha_{r.s} = 2\Lambda_{r.s};\ \ r \neq s \\ \text{et} & K \alpha_{r.r} = 2\Lambda_{r.r} - K\end{array}\right\} \ldots\ldots\ldots\ldots (14):$$

les formules (12), (13) se changeront en

$$Q_r = \Sigma_s \alpha_{r.s} P_s \ \text{ et } \ P_s = \Sigma_r \alpha_{r.s} Q_r \quad \ldots\ldots\ldots\ldots \quad (15):$$

équations qui sont nécessairement équivalentes. On a donc identiquement

$$\left.\begin{array}{ll}(1) & \left\{\begin{array}{l}\Sigma_r\, \alpha_{r.s}\alpha_{r.s'} = 0\,;\ s \neq s', \\ \Sigma_r\, \alpha_{r.s}\alpha_{r.s} = 1,\end{array}\right. \\ (2) & \left\{\begin{array}{l}\Sigma_s\, \alpha_{r.s}\alpha_{r'.s} = 0\,;\ r \neq r', \\ \Sigma_s\, \alpha_{r.s}\alpha_{r.s} = 1:\end{array}\right.\end{array}\right\} \ldots\ldots\ldots\ldots(16),$$

c'est-à-dire, on a trouvé un système de n^2 quantités $\alpha_{r.s}$, fonctions explicites et rationnelles d'un nombre $\frac{1}{2}n(n-1)$ de variables indépendantes, qui satisfont identiquement aux formules (16, 1) et (16, 2). On sait qu'en géométrie cela veut dire que pour $n=2$ ou $n=3$ de tels systèmes donnent les coefficients propres à effectuer la transformation de deux systèmes de coordonnées rectangulaires; nous dirons par analogie, que des systèmes qui satisfont aux équations (16) pour une valeur quelconque de n, sont propres à effectuer la transformation entre deux systèmes de coordonnées rectangulaires. On a donc le théorème suivant:

Les coefficients propres à la transformation de coordonnées rectangulaires, peuvent être exprimés rationnellement au moyen de quantités arbitraires $\lambda_{r.s}$, soumises aux conditions

$$\lambda_{s.r} = -\lambda_{s.r}\ [r \neq s]\,;\quad \lambda_{r.r} = 1.$$

Pour développer les formules, il faut d'abord former le déterminant K de ce système, puis le système inverse $\Lambda_{r.s}, \ldots$ et écrire

$$K\alpha_{r.s} = 2\Lambda_{r.s}\,[r \neq s]\,;\quad K\alpha_{r.r} = 2\Lambda_{r.r} - K\,;$$

ce qui donne le système cherché.

Soit par exemple $n=3$. Écrivons pour le système des quantités $\lambda_{r.s}$:

$$\left.\begin{array}{ccc} 1, & \nu, & -\mu, \\ -\nu, & 1, & \lambda, \\ \mu, & -\lambda, & 1,\end{array}\right\} \ldots\ldots\ldots\ldots (17),$$

ce qui donne $K = 1 + \lambda^2 + \mu^2 + \nu^2$, et pour le système des fonctions $\Lambda_{r.s}$

$$\left.\begin{array}{ccc} 1+\lambda^2, & \lambda\mu + \nu\ , & \nu\lambda - \mu, \\ \lambda\mu - \nu\ , & 1 + \mu^2, & \mu\nu + \lambda, \\ \nu\lambda + \mu, & \mu\nu - \lambda\ , & 1 + \nu^2.\end{array}\right\} \ldots\ldots\ldots\ldots (18).$$

De là on obtient pour le système de coefficients α, β, γ; α', β', γ'; $\alpha'', \beta'', \gamma''$:

$$\left.\begin{array}{lll} K\alpha = 1 + \lambda^2 - \mu^2 - \nu^2, & K\alpha' = 2(\lambda\mu + \nu)\ , & K\alpha'' = 2(\nu\lambda - \mu)\ , \\ K\beta = 2(\lambda\mu - \nu)\ , & K\beta' = (1 + \mu^2 - \nu^2 - \lambda^2), & K\beta'' = 2(\mu\nu + \lambda)\ , \\ K\gamma = 2(\nu\lambda + \mu)\ , & K\gamma' = 2(\mu\nu - \lambda)\ , & K\gamma'' = (1 + \nu^2 - \lambda^2 - \mu^2)\,;\end{array}\right\} \ldots (19),$$

ce qui se rapporte à la transformation

$$\left.\begin{array}{ll} x = \alpha\, x_1 + \beta\, y_1 + \gamma\, z_1, & x_1 = \alpha x + \alpha' y + \alpha'' z\,, \\ y = \alpha' x_1 + \beta' y_1 + \gamma' z_1, & y_1 = \beta x + \beta' y + \beta'' z, \\ z = \alpha'' x_1 + \beta'' y_1 + \gamma'' z_1, & z_1 = \gamma x + \gamma' y + \gamma'' z,\end{array}\right\} \ldots\ldots\ldots (20),$$

de deux systèmes de coordonnées rectangulaires. En effet, les coefficients λ, μ, ν ont une signification géométrique: Les axes Ax_1, Ay_1, Az_1 vont coïncider avec les axes Ax, Ay, Az, par la rotation θ autour d'un certain axe AR (qu'on peut nommer "Axe résultant"). En prenant f, g, h pour les inclinaisons de cet axe à Ax, Ay, Az, on a $\lambda = \tan \frac{1}{2}\theta \cos f$, $\mu = \tan \frac{1}{2}\theta \cos g$, $\nu = \tan \frac{1}{2}\theta \cos h$. Cette expression de l'axe AR est due à Euler; les quantités λ, μ, ν ont été introduites pour la première fois, par M. Olinde Rodrigues, dans un mémoire "Sur les lois géométriques qui régissent les déplacements d'un système solide" (Liouville, tom. V. [1840]), où il donne [des expressions semblables à celles] qu'on vient de trouver ici, pour les coefficients de la transformation, en termes de λ, μ, ν. Ces mêmes quantités λ, μ, ν (il y a à remarquer cela en passant) sont liées de la manière la plus étroite avec celles de la belle théorie de Sir W. Hamilton sur les Quaternions. Je les ai appliquées à la théorie de la rotation d'un corps solide. Avant de donner une idée des résultats auxquels je suis parvenu, je passe aux formules de transformation qui se rapportent au cas de $n = 4$. Je prends ici pour le système des quantités λ:

$$\left.\begin{matrix} 1, & a, & b, & c, \\ -a, & 1, & -h, & g, \\ -b, & h, & 1, & -f, \\ -c, & -g, & f, & 1, \end{matrix}\right\} \quad \ldots\ldots\ldots\ldots (21),$$

ce qui donne, en mettant pour abréger, $af + bg + ch = \theta$,

$$K = 1 + a^2 + b^2 + c^2 + f^2 + g^2 + h^2 + \theta^2 \ldots\ldots\ldots\ldots (22),$$

et puis pour les quantités $\Lambda_{r.s}$ le système

$$\left.\begin{matrix} 1+f^2+g^2+h^2, & f\theta+a+bh-cg, & g\theta+b+cf-ah, & h\theta+c+ag-bf, \\ -f\theta-a+bh-cg, & 1+f^2+b^2+c^2, & -c\theta-h+fg-ab, & b\theta+g+hf-ca, \\ -g\theta-b+cf-ah, & c\theta+h+fg-ab, & 1+g^2+c^2+a^2, & -a\theta-f+gh-bc, \\ -h\theta-c+ag-bf, & -b\theta-g+hf-ca, & a\theta+f+gh-bc, & 1+h^2+a^2+b^2, \end{matrix}\right\} \ldots (23),$$

de manière que pour

$$\left.\begin{matrix} K\alpha, & K\alpha', & K\alpha'', & K\alpha''', \\ K\beta, & K\beta', & K\beta'', & K\beta''', \\ K\gamma, & K\gamma', & K\gamma'', & K\gamma''', \\ K\delta, & K\delta', & K\delta'', & K\delta''', \end{matrix}\right\} \quad \ldots\ldots\ldots\ldots (24),$$

on obtient le système suivant:

$$\begin{matrix} f^2+g^2+h^2-a^2-b^2-c^2, & 2(f\theta+a+bh-cg), & 2(g\theta+b+cf-ah), & 2(h\theta+c+ag-bf), \\ f\theta-a+bh-cg), & 1+f^2+b^2+c^2-g^2-h^2-a^2, & 2(-c\theta-h+fg-ab), & 2(b\theta+g+hf-ca), \\ g\theta-b+cf-ah), & 2(c\theta+h+fg-ab), & 1+g^2+c^2+a^2-f^2-h^2-b^2, & 2(-a\theta-f+gh-bc), \\ h\theta-c+ag-bf), & 2(-b\theta-g+hf-ca), & 2(a\theta+f+gh-bc), & 1+h^2+a^2+b^2-f^2-g^2-c^2, \end{matrix}$$

ainsi de suite pour des valeurs quelconques de n.

II.

Maintenant je vais citer les formules que j'ai présentées dans le *Journal de Cambridge*, t. III. (1843) p. 225, [6], pour la rotation d'un corps solide autour d'un point fixe. Mettant, comme à l'ordinaire, les vitesses angulaires autour des axes principaux p, q, r, les *moments* du corps pour ces mêmes axes $=A$, B, C, et la fonction des forces $=V$: les équations citées pourront être écrites sous la forme

$$dt=\frac{dp}{P}=\frac{dq}{Q}=\frac{dr}{R}=\frac{d\lambda}{\Lambda}=\frac{d\mu}{\mathrm{M}}=\frac{d\nu}{\mathrm{N}} \quad\ldots\ldots\ldots\ldots (26),$$

$$\left.\begin{aligned}
P&=\frac{1}{A}\left((B-C)\,qr+\tfrac{1}{2}\left\{(1+\lambda^2)\frac{dV}{d\lambda}+(\lambda\mu+\nu)\frac{dV}{d\mu}+(\lambda\nu-\mu)\frac{dV}{d\nu}\right\}\right),\\
Q&=\frac{1}{B}\left((C-A)\,rp+\tfrac{1}{2}\left\{(\mu\lambda-\nu)\frac{dV}{d\lambda}+(1+\mu^2)\frac{dV}{d\mu}+(\mu\nu+\lambda)\frac{dV}{d\nu}\right\}\right),\\
R&=\frac{1}{C}\left((A-B)\,pq+\tfrac{1}{2}\left\{(\nu\lambda+\mu)\frac{dV}{d\lambda}+(\mu\nu-\lambda)\frac{dV}{d\mu}+(1+\nu^2)\frac{dV}{d\nu}\right\}\right),\\
\Lambda&=\tfrac{1}{2}\{(1+\lambda^2)\,p+(\lambda\mu-\nu)\,q+(\lambda\nu+\mu)\,r\},\\
\mathrm{M}&=\tfrac{1}{2}\{(\mu\lambda+\nu)\,p+(1+\mu^2)\,q+(\mu\nu-\lambda)\,r\},\\
\mathrm{N}&=\tfrac{1}{2}\{(\nu\lambda-\mu)\,p+(\nu\mu+\lambda)\,q+(1+\nu^2)\,r\}.
\end{aligned}\right\}\ldots (27).$$

En effet, pour obtenir ces formules, il n'y a qu'à chercher au moyen de λ, μ, ν, et de leurs dérivées par rapport aux temps λ', μ', ν', l'expression de la fonction $T=\frac{1}{2}(Ap^2+Bq^2+Cr^2)$ {qui exprime la demi-somme des forces vives} : cela fait, les formules générales que Lagrange a données pour la solution des problèmes de dynamique, conduisent immédiatement aux équations en question. Dans le mémoire cité j'ai intégré ces équations pour le cas où la fonction V est zéro, et en prenant, comme dans la théorie ordinaire, le plan invariable pour plan des deux axes. Ce n'est que dernièrement que j'ai trouvé la manière convenable de traiter ce système d'équations; je le fais au moyen de deux nouvelles variables Ω, v, entre lesquelles je trouve une équation différentielle dont les variables sont séparées, et j'exprime en termes de celles-ci les autres variables du problème, y compris le temps; et cela sans aucune supposition particulière, relative à la position des axes des coordonnées par rapport au plan invariable. Le développement de cette théorie paraîtra dans le prochain No. du *Cambridge and Dublin Mathematical Journal*, [37]. Je m'occupe aussi de la recherche des formules pour les variations des constantes arbitraires relatives aux perturbatrices. Il serait bien intéressant (comme problème d'analyse pure) d'étendre ces recherches au cas d'une valeur quelconque de n; il faudrait pour cela, chercher les valeurs des quantités analogues à p, q, r, former une fonction T, en prenant la somme des carrés de chacune de ces quantités, chaque carré multiplié par un coefficient constant, et puis former les équations $\frac{d}{dt}\cdot\frac{dT}{d\lambda'}-\frac{dT}{d\lambda}=0$, &c., analogues aux équations de Dynamique. Mais je n'ai encore rien trouvé sur ce sujet.

53.

RECHERCHES SUR L'ÉLIMINATION, ET SUR LA THÉORIE DES COURBES.

[From the *Journal für die reine und angewandte Mathematik* (Crelle), tome XXXIV. (1847), pp. 30—45.]

En désignant par U, V, W, ... des fonctions homogènes des ordres m, n, p, &c. et d'un nombre égal de variables respectivement, et en supposant que ces fonctions soient les plus générales possibles, c'est-à-dire que le coefficient de chaque terme soit une lettre indéterminée: on sait que les équations $U=0$, $V=0$, $W=0$, ... offrent une relation $\Theta=0$, dans laquelle les variables n'entrent plus, et où la fonction Θ, que l'on peut nommer *Résultant complet des équations*, est homogène et de l'ordre np ... par rapport aux coefficients de U, de l'ordre mp ... par rapport à ceux de V, et ainsi de suite, tandis qu'elle n'est pas décomposable en facteurs. Cela posé, supposons que les coefficients de U, V, ..., au lieu d'être tous indéterminés, soient des fonctions quelconques d'un certain nombre de quantités arbitraires. Substituant ces valeurs dans la fonction Θ, cette fonction sera toujours le *Résultant complet* des équations. Mais Θ peut quelquefois être décomposable en facteurs, dont quelques-uns doivent être éliminés. En effet les coefficients de U, V, ... peuvent contenir des quantités ξ, η, ... censées comme variables, et d'autres quantités α, β, ... qui sont constantes, et il peut s'agir de la relation entre ξ, η, ... qui est nécessaire pour que les équations $U=0$, $V=0$, &c. puissent subsister conjointement. Dans ce cas tout facteur Λ du résultant complet Θ, qui ne contient pas les coefficients variables ξ, η, ..., doit être rejeté. En supprimant ces facteurs, et exprimant par Φ le facteur qui reste, cette fonction est alors ce que nous nommerons *Résultant réduit*. Cependant les facteurs Λ, proprement dits, ne sont jamais des facteurs étrangers, et ce n'est qu'à cause du point de vue particulier, auquel on a envisagé la question, qu'ils ont été rejetés; d'un autre point de vue ces facteurs auraient pu devenir le *Résultant réduit*. Nous les nommerons donc, à l'exemple de M. Sylvester, *Facteurs spéciaux*.

Pour faire voir tout cela avec plus de clarté, prenons pour exemple le problème suivant de Géométrie analytique:

"Trouver les équations du système des lignes tirées par un point donné aux points d'intersection de deux courbes données."

Soient $U=0$, $V=0$ les équations des deux courbes, U, V étant des fonctions homogènes des variables x, y, z, des ordres m et n respectivement, ce qui revient à prendre $x : z$ et $y : z$ pour coordonnées d'un point. De même il faut exprimer par (α, β, γ) le point dont les coordonnées sont $\alpha : \gamma$ et $\beta : \gamma$; et ainsi pour tous cas semblables. Il s'entend, qu'on suppose partout que les coefficients de U, V, ou de U, restent absolument indéterminés. Représentons par (α, β, γ) le point donné, et par (ξ, η, ζ) un point quelconque d'une des lignes dont il s'agit. En éliminant x, y, z entre les équations

$$U=0,\ V=0,\ \text{et}\ x(\beta\zeta-\gamma\eta)+y(\gamma\xi-\alpha\zeta)+z(\alpha\eta-\beta\xi)=0 \quad \ldots\ldots\ldots\ldots(1),$$

on obtient l'équation cherchée $\Theta=0$. Ici Θ est une fonction homogène de l'ordre mn par rapport à $\beta\zeta-\gamma\eta$, $\gamma\xi-\alpha\zeta$ et $\alpha\eta-\beta\xi$; de l'ordre n par rapport aux coefficients de U; et de l'ordre n par rapport à ceux de V; de plus cette fonction est décomposable en mn facteurs linéaires par rapport à ξ, η, ζ, dont les coefficients sont des fonctions irrationnelles de α, β, γ et des coefficients de U et V (en effet toute fonction homogène de $\beta\zeta-\gamma\eta$, $\gamma\xi-\alpha\zeta$, $\alpha\eta-\beta\xi$ est douée de cette propriété, qui subsiste encore en échangeant entre eux ξ, η, ζ et α, β, γ). Chaque facteur linéaire, égalé à zéro, appartient à une des lignes en question. Voilà pourquoi $\Theta=0$ est considérée comme équation du système des lignes.

Soit maintenant proposé le problème:

"Trouver l'équation du système des tangentes tirées d'un point fixe à une courbe donnée."

Il y a ici à éliminer x, y, z entre les équations

$$\left.\begin{array}{l} U=0, \\ \alpha\dfrac{dU}{dx}+\beta\dfrac{dU}{dy}+\gamma\dfrac{dU}{dz}=0\ \text{et} \\ x(\beta\zeta-\gamma\eta)+y(\gamma\xi-\alpha\zeta)+z(\alpha\eta-\beta\xi)=0. \end{array}\right\} \ldots\ldots\ldots\ldots\ldots\ldots\ldots(2).$$

Le résultant complet est une fonction homogène de l'ordre $n(n-1)$ par rapport à $\beta\zeta-\gamma\eta$, $\gamma\xi-\alpha\zeta$ et $\alpha\eta-\beta\xi$ {n représente ici l'ordre de la fonction U}, de l'ordre n par rapport à α, β, γ, et de l'ordre $2n-1$ par rapport aux coefficients de U. Mais il existe dans ce cas un facteur spécial U_0 qui est ce que devient U en écrivant α, β, γ à la place de x, y, z. En le mettant de côté, le résultant réduit Φ est fonction de l'ordre $n(n-1)$ par rapport à $\beta\zeta-\gamma\eta$, $\gamma\xi-\alpha\zeta$ et $\alpha\eta-\beta\xi$, et de l'ordre $2(n-1)$ par rapport aux coefficients de U, et l'équation $\Phi=0$ correspond au système de tangentes.

En mettant x, y, z à la place de $\beta\zeta-\gamma\eta$, $\gamma\xi-\alpha\zeta$, $\alpha\eta-\beta\xi$, Φ devient une fonction de l'ordre $n(n-1)$ par rapport à x, y, z, et de l'ordre $2(n-1)$, comme elle l'était ci-dessus, par rapport aux coefficients de U. Nous désignerons cette nouvelle valeur de Φ par $\mathsf{F}U$; c'est-à-dire nous représenterons par $\mathsf{F}U$ le résultant réduit des équations

$$\left.\begin{aligned} &U=0,\\ &\alpha\frac{dU}{dx}+\beta\frac{dU}{dy}+\gamma\frac{dU}{dz}=0, \text{ et}\\ &x\mathrm{x}+y\mathrm{y}+z\mathrm{z}=0, \end{aligned}\right\} \quad \ldots\ldots\ldots\ldots\ldots\ldots(3),$$

$\mathsf{F}U$ étant une fonction des ordres $n(n-1)$ et $2(n-1)$ par rapport à x, y, z, et aux coefficients de U. On sait que l'équation $\mathsf{F}U=0$ est celle de la polaire réciproque de la courbe, par rapport à la conique auxiliaire $x^2+y^2+z^2=0$.

Or les équations (2) peuvent être écrites aussi sous la forme

$$\left.\begin{aligned} &U=0,\\ &\alpha\frac{dU}{dx}+\beta\frac{dU}{dy}+\gamma\frac{dU}{dz}=0,\\ &\xi\frac{dU}{dx}+\eta\frac{dU}{dy}+\zeta\frac{dU}{dz}=0. \end{aligned}\right\} \quad \ldots\ldots\ldots\ldots\ldots\ldots(4).$$

Ici le résultant complet est de l'ordre $n(n-1)$ par rapport à α, β, γ, ou à ξ, η, ζ (car ce résultant complet doit être comme ci-dessus fonction de ce même ordre de $\beta\zeta-\gamma\eta$, $\gamma\xi-\alpha\zeta$ et $\alpha\eta-\beta\xi$) et de l'ordre $(n-1)(3n-1)$ par rapport aux coefficients de U. Le facteur spécial dans ce cas est donc une fonction de l'ordre $3(n-1)^2$ des coefficients de U, et il est facile de trouver sa forme; car on satisferait aux équations (4) en posant

$$\frac{dU}{dx}=0, \quad \frac{dU}{dy}=0, \quad \frac{dU}{dz}=0 \quad \ldots\ldots\ldots\ldots\ldots\ldots(5);$$

le résultant de ces équations, que nous désignerons toujours par KU, doit donc se présenter comme facteur spécial du résultant complet du système. Mais KU étant une fonction des coefficients de l'ordre $3(n-1)^2$, elle est précisément le facteur spécial dont il s'agit. (Il est clair que l'équation $KU=0$ serait la condition nécessaire, pour que la courbe pût avoir un point multiple.)

Reprenons le premier système. On satisfait à la dernière équation en écrivant

$$x=\alpha l+\xi m, \quad y=\beta l+\eta m, \quad z=\gamma l+\zeta m \quad \ldots\ldots\ldots\ldots(6).$$

Soient $[U]$, $[V]$ ce que deviennent U, V par cette substitution. En éliminant l, m entre les deux équations

$$[U]=0, \quad [V]=0,$$

on obtiendra le même résultant Θ que ci-dessus. En effet, le résultant de ces deux équations est des ordres n et m par rapport aux coefficients de U et V, et de l'ordre $2mn$ par rapport à α, β, γ, ξ, η, ζ: donc il faut qu'il soit égal à Θ, à un facteur numérique près. On a donc le théorème suivant:

THÉORÈME I. L'équation du système des droites menées par un point donné (α, β, γ) aux points d'intersection des deux courbes $U=0$, $V=0$, se trouve en éliminant les nouvelles variables l, m entre les deux équations $[U]=0$ et $[V]=0$, où $[U]$, $[V]$ sont ce que deviennent U et V par les substitutions

$$x=\alpha l+\xi m, \quad y=\beta l+\eta m, \quad z=\gamma l+\zeta m \quad \ldots\ldots\ldots\ldots(7).$$

En opérant également sur le second système d'équations, on obtient directement le résultant réduit, sans que l'opération soit embarrassée par aucun facteur spécial. En effet, on peut remplacer le système par

$$\left.\begin{aligned} &\alpha \frac{dU}{dx}+\beta \frac{dU}{dy}+\gamma \frac{dU}{dz}=0, \\ &\xi \frac{dU}{dx}+\eta \frac{dU}{dy}+\zeta \frac{dU}{dz}=0, \\ &x(\beta\zeta-\gamma\eta)+y(\gamma\xi-\alpha\zeta)+z(\alpha\eta-\beta\xi)=0, \end{aligned}\right\} \quad \ldots\ldots\ldots\ldots(8),$$

et en faisant les substitutions (6) dans la dernière équation, les deux autres équations se changent en

$$\frac{d[U]}{dl}=0 \quad \text{et} \quad \frac{d[U]}{dm}=0,$$

où $[U]$ est ce que devient U par cette substitution. Le résultant de ces équations est de l'ordre $2n(n-1)$ par rapport à $\alpha, \beta, \gamma, \xi, \eta, \zeta$ {c'est-à-dire une fonction de $\beta\zeta-\gamma\eta$, $\gamma\xi-\alpha\zeta$, $\alpha\eta-\beta\xi$, de l'ordre $n(n-1)$}, et de l'ordre $2(n-1)$ par rapport aux coefficients de U; donc il n'y a plus de facteur spécial. De là suit:

THÉORÈME II. L'équation du système de tangentes menées du point donné (α, β, γ) à la courbe $U=0$, se trouve en éliminant l, m entre les équations

$$\frac{d[U]}{dl}=0, \quad \frac{d[V]}{dl}=0,$$

$[U]$ étant ce que devient U par les substitutions $x=\alpha l+\xi m$, $y=\beta l+\eta m$ et $z=\gamma l+\zeta m$. En représentant l'équation par $\Phi=0$, Φ est une fonction de $\beta\zeta-\gamma\eta$, $\gamma\xi-\alpha\zeta$, $\alpha\eta-\beta\xi$, et en remplaçant ces fonctions par x, y, z, on obtient l'équation de la polaire réciproque (par rapport à $x^2+y^2+z^2=0$) de la courbe donnée.

Ce beau théorème est dû à M. Joachimsthal, qui me l'a communiqué l'été passé pendant mon séjour à Berlin, avec une démonstration.

On déduit de là, comme cas très particulier, une forme du résultant des deux équations $ax^2+2bxy+cy^2=0$ et $a'x^2+2b'xy+c'y^2=0$, citée dans mon mémoire sur les hyperdéterminants (t. XXX. de ce journal, [16]). En effet, soit $x^2=\mathrm{z}$, $xy=-\mathrm{y}$, $y^2=\mathrm{x}$, et $U=\mathrm{xz}-\mathrm{y}^2$, on aura évidemment $U=0$,

$$a\frac{dU}{d\mathrm{x}}+b\frac{dU}{d\mathrm{y}}+c\frac{dU}{d\mathrm{z}}=0, \quad \text{et} \quad a'\frac{dU}{d\mathrm{x}}+b'\frac{dU}{d\mathrm{y}}+c'\frac{dU}{d\mathrm{z}}=0,$$

donc en cherchant le résultant de ces équations de la manière indiquée dans le théorème, on obtient la formule en question $4(ac-b^2)(a'c'-b'^2)-(ac'+a'c-2bb')^2=0$. Cependant la véritable généralisation de cette formule, à ce que je crois, reste encore à trouver.

Il suit des principes développés dans le mémoire cité, que le résultant des deux équations $L=0$, $M=0$ (où L, M sont des fonctions homogènes des deux variables l, m) peut toujours être présenté sous la forme $\Theta=\nabla LL\ldots MM\ldots$, où ∇ est une composition d'expressions symboliques, telles que $(12)^\alpha(13)^\beta\ldots$, dans lesquelles $(12)=\partial_{l_1}\partial_{m_2}-\partial_{m_1}\partial_{l_2}$, &c. Par exemple pour $L=al^2+2blm+cm^2$, $M=a'l^2+2b'lm+c'm^2$, on a

$$\Theta=\{(12)^2(34)^2-(13)^2(24)^2\}\,LL\,MM,$$

$\{$c'est-à-dire, comme ci-dessus, $\Theta=4(ac-b^2)(a'c'-b'^2)-(ac'+a'c-2bb')^2\}$. En appliquant cette théorie à l'élimination des inconnues entre les équations $[U]=0$, $[V]=0$, on obtient

$$\partial_l=\alpha\partial_x+\beta\partial_y+\gamma\partial_z,\quad \text{et}\quad \partial_m=\xi\partial_x+\eta\partial_y+\zeta\partial_z,$$

et en faisant

$$\beta\zeta-\gamma\eta=\mathrm{x},\quad \gamma\xi-\alpha\zeta=\mathrm{y},\quad \alpha\eta-\beta\xi=\mathrm{z},$$

et $$\partial_{y_1}\partial_{z_2}-\partial_{z_2}\partial_{y_1}=(12)',\quad \partial_{z_1}\partial_{x_2}-\partial_{x_2}\partial_{z_1}=(12)'',\quad \partial_{x_1}\partial_{y_2}-\partial_{x_2}\partial_{y_1}=(12)''',$$

cela donne

$$(12)=\mathrm{x}\,(12)'+\mathrm{y}\,(12)''+\mathrm{z}\,(12)''':$$

équation qui peut être présentée sous la forme abrégée

$$(12)=(P\,12).$$

On obtient le résultant cherché en faisant cette substitution dans tous les symboles que contient ∇, et en introduisant U, V au lieu de $[U]$, $[V]$.

Les mêmes remarques peuvent être appliquées au cas d'une élimination entre les deux équations $\dfrac{dL}{dl}=0$, $\dfrac{dL}{dm}=0$, car le résultant sera exprimé ici sous la même forme $\Theta=\nabla LL\ldots$. Cherchons par exemple l'équation de la polaire réciproque d'une courbe du second ordre: en observant que le résultant des équations $\dfrac{dL}{dl}=0$, $\dfrac{dL}{dm}=0$ (où L est une fonction du second degré) sera exprimé sous la forme $\Theta=(12)^2LL$, on obtient immédiatement pour la réciproque de la courbe du second ordre $U=0$, l'équation

$$\mathsf{F}U=(P12)^2\,U\,U=0,$$

laquelle se réduit en effet à la forme connue.

Passons au cas d'une courbe du *troisième ordre*. Comme pour une fonction de deux variables, le résultant des équations

$$\frac{dL}{dl}=0,\ \frac{dL}{dm}=0,$$

aura la forme

$$\Theta = (12)^2 (34)^2 (13) (42) LLLL,$$

on a pour la polaire de $U = 0$,

$$-2\mathsf{F} U = (P12)^2 (P34)^2 (P13) (P42)\ UUUU = 0,$$

et il serait facile de calculer par là le coefficient d'une puissance ou d'un produit quelconque des variables. Par exemple le coefficient de z^6 se réduit à

$$\{(12)'''\}^2 \{(34)'''\}^2 \{(13)'''\}^2 \{(42)'''\}\ UUUU,$$

ou, toute réduction faite, et en supposant

$$6U = ax^3 + by^3 + cz^3 + 3iy^2z + 3jz^2x + 3kx^2y + 3i_1yz^2 + 3j_1zx^2 + 3k_1xy^2 + 6lxyz$$

à: $2(6bcii_1 - 4i^3c - 4i_1^3b + 3i^2i_1^2 - b^2c^2)$. L'équation complètement développée, que j'ai donnée pour cette polaire dans le *Cambridge and Dublin Mathematical Journal*, t. I. [1846], p. 97 [35], et que j'ai obtenue par une élimination directe, pourra ainsi être vérifiée.

Nous allons passer maintenant à la théorie des *points d'inflexion* et des *tangentes doubles* de la courbe $U = 0$. Ces singularités peuvent être traitées par une analyse semblable, en remarquant que parmi les $n(n-1)$ tangentes, menées à la courbe d'un point P situé sur la courbe, la tangente en ce point se présente généralement deux fois, et trois fois, selon que le point P est un point d'inflexion, ou un point de contact d'une tangente double.

Désignons comme ci-dessus par (U) ce que devient U en écrivant $lx + m\xi$, $ly + m\eta$, $lz + m\zeta$ à la place de x, y et z. En mettant $\xi\partial_x + \eta\partial_y + \zeta\partial_z = \partial$, on a évidemment

$$[U] = l^n U + l^{n-1} m\, \partial U + \frac{1}{1\,.\,2} l^{n-2} m^2 \partial^2 U + \dots ;$$

et en éliminant l, m entre $\dfrac{d[U]}{dl} = 0$, $\dfrac{d[U]}{dm} = 0$, on obtient un résultant $\Theta = 0$, où Θ est une fonction de U, ∂U, $\partial^2 U$, ... $\partial^n U$, et qui a, comme le remarque M. Joachimsthal, la propriété dont il s'agit. En écrivant $U = 0$, Θ contient le facteur $(\partial U)^2$, et en mettant de côté ce facteur et écrivant $\partial U = 0$, Θ contient le facteur $(\partial^2 U)^2$; et ainsi de suite. Nous avons supposé que le point P, auquel appartiennent les coordonnées x, y et z, est un point de la courbe; de manière que l'on a actuellement $U = 0$. En faisant donc cette supposition et en éliminant le facteur $(\partial U)^2$, l'équation $\Theta = 0$ prend la forme $X\partial U + Y(\partial^2 U)^2 = 0$ {puisqu'en mettant $\partial U = 0$, l'équation contiendra à gauche le facteur $(\partial^2 U)^2$}. Dans le cas où P est un point d'inflexion, ou un point de contact d'une tangente double, cette équation contiendra ∂U comme facteur: donc il faut que $Y(\partial^2 U)^2$ contienne ce facteur, c'est-à-dire: ou $\partial^2 U$, ou Y, contiendra le facteur ∂U. Dans le premier cas il s'agit d'un point d'inflexion, dans le second cas d'un point de contact d'une tangente double.

Considérons d'abord les points d'inflexion. Comme $\partial^2 U$ contient ∂U comme facteur, il faut que cette fonction devienne zéro pour toutes les valeurs de ξ, η, ζ qui font évanouir ∂U. Désignons par L, M, N les coefficients différentiels du premier ordre de U, de manière que $\partial U = L\xi + M\eta + N\zeta$. Cette quantité s'évanouit identiquement en faisant $\xi = \beta N - \gamma M$, $\eta = \gamma L - \alpha N$, $\zeta = \alpha M - \beta L$: donc il faut que $\partial^2 U$ s'évanouisse par la substitution de ces valeurs, quelles que soient les quantités α, β, γ. En désignant par D ce que devient le symbole D par cette substitution, c'est-à-dire en faisant

$$D = \alpha(M\partial_x - N\partial_y) + \beta(N\partial_x - L\partial_z) + \gamma(L\partial_y - M\partial_x),$$

la condition d'un point d'inflexion se réduit tout simplement à

$$D^2 U = 0:$$

équation dans laquelle les symboles ∂_x, ∂_y, ∂_z ne doivent pas contenir L, M, N. Nous reviendrons sur cette équation dans une note; pour le moment il suffit de remarquer, qu'en vertu de relations qui existent entre L, M, N et les dérivées a, b, c, f, g, h du second ordre, on a identiquement

$$(n-1)^2 D^2 U = n(n-1)\Psi . U - (\alpha x + \beta y + \gamma z)^2 (\nabla U),$$

où Ψ est une fonction de α, β, γ et des dérivées a, b, c, f, g, h, dont la forme sera donnée dans la suite, et

$$\nabla U = abc - af^2 - bg^2 - ch^2 + 2fgh.$$

Donc à cause de $U = 0$, la seule condition pour déterminer les points d'inflexion est l'équation

$$\nabla U = 0,$$

qui est de l'ordre $3(n-2)$ par rapport aux variables, et de l'ordre 3 par rapport aux coefficients de U. Cela est déjà connu par les recherches de M. Hesse. J'ai donné ici cette équation pour faire voir la liaison qui existe entre cette question et celle de trouver les tangentes doubles, à laquelle je vais passer maintenant.

On obtient l'équation qui détermine les points de contact de ces tangentes, en faisant les mêmes substitutions $\xi = \beta N - \gamma M$, &c. dans la fonction Y, et en égalant à zéro les coefficients des différentes puissances ou produits de α, β, γ. Remarquons que cette fonction Y s'obtient par une fonction de l'ordre $n^2 - n$ par rapport à x, y, z, ou à ξ, η, ζ, et de l'ordre $2(n-1)$ par rapport aux coefficients, en éliminant les facteurs $(\partial U)^2$ et $(\partial^2 U)^2$, qui ensemble montent au degré $4n-6$ par rapport à x, y, z, à 6 par rapport à ξ, η, ζ, et à 4 par rapport aux coefficients. Donc Y est des degrés $n^2 - n - 6$, $n^2 - 5n - 6$ et $2n - 6$ par rapport à x, y, z, à ξ, η, ζ, et aux coefficients de U, respectivement. En substituant donc ξ, η, ζ, cette fonction devient de l'ordre $n^2 - n - 6$ par rapport à α, β, γ, de l'ordre $n^3 - 2n^2 - 10n + 12$ par rapport à x, y, z {savoir $(n^2 - 5n + 6) + (n-1)(n^2 - n - 6)$}, et de l'ordre $n^2 + n - 12$ par rapport aux coefficients {savoir $(2n-6) + (n^2 - n - 6)$}. Mais on sait que les conditions de l'évanouissement de Y doivent se réduire à une seule équation; et cela ne peut arriver que de la même manière dont la réduction analogue a eu lieu pour les points d'inflexion. En écrivant donc $[Y]$ à la place de ce que devient Y après la substitution, il faut que l'on ait identiquement:

$$[Y] = \Lambda\, U + N(\Pi U),$$

N étant fonction de α, β, γ du degré $n^2 - n - 6$. Il paraît de plus probable que cette fonction aura la même forme que celle pour les points d'inflexion, savoir $N = (\alpha x + \beta y + \gamma z)^{n^2-n-6}$. ($\Pi U$, comme ci-dessus ∇U, est censé exprimer non pas un produit, mais une dérivée de U: de même plus bas PU et QU.) Cela étant, ΠU sera du degré $(n-2)(n^2-9)$ par rapport à x, y, z {c'est-à-dire $n^3 - 2n^2 - 10n + 12 - (n^2 - n - 6)$}, et du degré $n^2 + n - 12$ par rapport aux coefficients. On a donc le théorème suivant:

THÉORÈME. On trouve les points de contact de tangentes doubles, en combinant avec l'équation de la courbe une équation $\Pi U = 0$ de l'ordre $(n-2)(n^2-9)$ par rapport aux variables, et de l'ordre $n^2 + n - 12$ par rapport aux coefficients; c'est-à-dire, puisqu'il correspond deux points de contact à une tangente double, le nombre de ces tangentes est égal à $\frac{1}{2}n(n-2)(n^2-9)$: théorème démontré indirectement par M. Plücker.

Cherchons maintenant les équations du système des tangentes aux points d'inflexion et du système des tangentes doubles.

Pour trouver la première équation, il faut éliminer x, y, z entre les trois équations

$$U = 0, \quad \nabla U = 0, \quad \mathrm{x}\frac{dU}{dx} + \mathrm{y}\frac{dU}{dy} + \mathrm{z}\frac{dU}{dz} = 0.$$

Le résultant complet sera du degré $3n(n-2)$ par rapport à x, y et z, et de l'ordre $9n^2 - 18n + 6$ {savoir $3(n-2)(n-1) + 3n(n-1) + 3n(n-2)$} par rapport aux coefficients; mais il existe ici le facteur spécial $(KU)^2$, et ce facteur étant éliminé, on obtient un résultant réduit $QU = 0$, du degré $3n(n-2)$ par rapport aux variables, et du même degré $3n(n-2)$ par rapport aux coefficients.

De même on aura l'équation du système des tangentes doubles, en éliminant x, y, z, entre les trois équations

$$U = 0, \quad \Pi U = 0, \quad \text{et} \quad \mathrm{x}\frac{dU}{dx} + \mathrm{y}\frac{dU}{dy} + \mathrm{z}\frac{dU}{dz} = 0.$$

Le résultant complet est ici du degré $n(n-2)(n^2-9)$ par rapport à x, y, z, et du degré $3n^4 - 5n^3 - 29n^2 + 57n - 18$ {savoir $(n-1)(n-2)(n^2-9) + n(n-2)(n^2-9) + n(n-1)(n^2+n-12)$} par rapport aux coefficients. Mais il existe de même dans ce cas un facteur spécial $(KU)^{n^2-n-6}$ {du degré $3(n-1)^2(n^2-n-6) = 3n^4 - 9n^3 - 9n^2 + 33n - 18$}, et en l'éliminant, le résultant réduit sera du degré $4n(n-2)(n-3)$ par rapport aux coefficients. Mais le terme de cette équation à gauche sera évidemment un *carré;* on aura donc pour l'équation du système des tangentes doubles, la relation $PU = 0$, où PU est une fonction du degré $\frac{1}{2}n(n-2)(n^2-9)$ par rapport aux variables, et du degré $2n(n-2)(n-3)$ par rapport aux coefficients. Donc on pourra former le tableau suivant des degrés des différentes équations obtenues:

	Degrés par rapport aux variables.	Degrés par rapport aux coefficients.
Equation de la courbe $U=0$, ...	n	1
Condition d'un point multiple $KU=0$, ...	0	$3(n-1)^2$
Polaire réciproque $\mathsf{F}U=0$, ...	$n(n-1)$	$2(n-1)$
Courbe des inflexions $\nabla U=0$, ...	$3n(n-2)$	3
Courbes des contacts des tangentes doubles $\Pi U=0$,	$(n-2)(n^2-9)$	$(n+4)(n-3)$
Systèmes des tangentes aux points d'inflexion $QU=0$,	$3n(n-2)$	$3n(n-2)$
Système des tangentes doubles $PU=0$, ...	$\frac{1}{2}n(n-2)(n^2-9)$	$2n(n-2)(n-3)$.

La polaire de la polaire réciproque $\mathsf{FF}U$ sera évidemment du degré $(n^2-n)(n^2-n-1)$ par rapport aux variables, et du degré $4(n-1)(n^2-n-1)$ par rapport aux coefficients. Cette polaire de la polaire contiendra, comme on le sait par la théorie géométrique développée par M. Plücker, les facteurs U, $(PU)^2$ et $(QU)^3$; il faut y ajouter encore le facteur constant KU, et l'on aura définitivement l'équation

$$\mathsf{FF}U=(KU)(PU)^2(QU)^3 \,.\, U,$$

dans laquelle il sera facile à vérifier que les deux côtés sont des mêmes degrés par rapport aux variables et aux constantes. En effet on a

$$4(n-1)(n^2-n-1)=3(n-1)^2+4n(n-2)(n-3)+9n(n-2)+1,$$

et

$$n(n-1)(n^2-n-1)=n(n-2)(n^2-9)+9n(n-2)+n.$$

Mon but a été ici de donner une idée précise des théorèmes à démontrer, pour former une théorie toute analytique des polaires réciproques; je n'ai fait qu'avancer ces théorèmes (sans chercher à les démontrer), pour faire voir leur liaison avec la théorie de l'élimination et avec celle des hyperdéterminants; c'est à cette dernière en particulier qu'il faut, je crois, recourir pour démontrer la formule donnée ci-dessus $[Y]=\Lambda \,.\, U+(\alpha x+\beta y+\gamma z)^{n^2-n-6}(\Pi U)$, et pour trouver définitivement la forme de la dérivée ΠU, au moyen de laquelle on déterminera les points de contact des tangentes doubles. Je serais bien aise que ces recherches puissent de quelque manière faciliter la solution du problème des réciproques des *surfaces:* objet, qui est resté encore dans une complète obscurité.

Note sur les points d'inflexion.

Je vais d'abord rassembler plusieurs formules qui se rapportent au système des coefficients dans le développement de D^2U. On a

$$D^2U=\mathrm{a}\alpha^2+\mathrm{b}\beta^2+\mathrm{c}\gamma^2+2\mathrm{f}\beta\gamma+2\mathrm{g}\gamma\alpha+2\mathrm{h}\alpha\beta,$$

où

$$\mathrm{a}=M^2c-2MNf+N^2b\,,$$
$$\mathrm{b}=N^2a-2NLg+L^2c\,,$$
$$\mathrm{c}=L^2b-2LMh+M^2a,$$

$$\mathrm{f} = -MNa + NLh + LMg - L^2 f,$$
$$\mathrm{g} = +MNh - NLb + LMf - M^2 g,$$
$$\mathrm{h} = +MNg + NLf - LMc - N^2 h.$$

Nous écrirons de plus, pour abréger,

$$bc - f^2 = \mathfrak{A}, \quad gh - af = \mathfrak{F}, \quad \nabla = abc - af^2 - bg^2 - ch^2 + 2fgh,$$
$$ca - g^2 = \mathfrak{B}, \quad hf - bg = \mathfrak{G},$$
$$ab - h^2 = \mathfrak{C}, \quad fg - ch = \mathfrak{H},$$
$$\mathrm{bc} - \mathrm{f}^2 = (\mathfrak{A}), \quad \&c.$$
&c.

et enfin

$$\Phi = \mathfrak{A}L^2 + \mathfrak{B}M^2 + \mathfrak{C}N^2 + 2\mathfrak{F}MN + 2\mathfrak{G}NL + 2\mathfrak{H}MN.$$

On obtient identiquement

$$L\mathrm{a} + M\mathrm{h} + N\mathrm{g} = 0,$$
$$L\mathrm{h} + M\mathrm{b} + N\mathrm{f} = 0,$$
$$L\mathrm{g} + M\mathrm{f} + N\mathrm{c} = 0;$$

$$(\mathfrak{A}) = L^2\Phi, \quad (\mathfrak{B}) = M^2\Phi, \quad (\mathfrak{C}) = N^2\Phi,$$
$$(\mathfrak{F}) = MN\Phi, \quad (\mathfrak{G}) = NL\Phi, \quad (\mathfrak{H}) = LM\Phi;$$

$$f\mathrm{f} - MN\mathfrak{F} = (Nh - Lf)(Lf - Mg),$$
$$g\mathrm{g} - NL\mathfrak{G} = (Lf - Mg)(Mg - Nh),$$
$$h\mathrm{h} - LM\mathfrak{H} = (Mg - Nh)(Nh - Lf);$$

$$4L^2MN\mathfrak{F} + \mathrm{b}(L^2 b + M^2 a) + \mathrm{c}(N^2 a + L^2 c) - \mathrm{bc} - 2L^2 a\mathrm{a} = (L^2 b - M^2 a)(N^2 a + L^2 c),$$
$$4M^2NL\mathfrak{G} + \mathrm{c}(M^2 c + N^2 b) + \mathrm{a}(L^2 b + M^2 a) - \mathrm{ca} - 2M^2 b\mathrm{b} = (M^2 c - N^2 b)(L^2 b - M^2 a),$$
$$4N^2LM\mathfrak{H} + \mathrm{a}(N^2 a + L^2 c) + \mathrm{b}(M^2 c + N^2 b) - \mathrm{ab} - 2N^2 c\mathrm{c} = (N^2 a - L^2 c)(M^2 c - N^2 b);$$

$$4L^2MN\mathfrak{A} - \mathrm{a}^2 + 2\mathrm{a}(M^2 c + N^2 b) = -(M^2 c - N^2 b)^2,$$
$$4M^2NL\mathfrak{B} - \mathrm{b}^2 + 2\mathrm{b}(N^2 a + L^2 c) = -(N^2 a - L^2 c)^2,$$
$$4N^2LM\mathfrak{C} - \mathrm{c}^2 + 2\mathrm{c}(L^2 b + M^2 a) = -(L^2 b - M^2 a)^2;$$

$$L^2MN\mathfrak{A} + N\mathrm{g}(+Lf + Mg - Nh) + M\mathrm{h}(Lf - Mg + Nh) - \mathrm{gh} = -MN(Nh - Mg)^2,$$
$$M^2NL\mathfrak{B} + L\mathrm{h}(-Lf + Mg + Nh) + N\mathrm{f}(Lf + Mg - Nh) - \mathrm{hf} = -NL(Lf - Nh)^2,$$
$$N^2LM\mathfrak{C} + M\mathrm{f}(+Lf - Mg + Nh) + L\mathrm{g}(Lf + Mg + Nh) - \mathrm{fg} = -LM(Mg - Lf)^2;$$

$$2\nabla L^2 = -a^2\mathrm{a} + (ab - 2h^2)\mathrm{b} + (ca - 2g^2)\mathrm{c} + 2(af - 2gh)\mathrm{f} - 2ag\,\mathrm{g} - 2ah\,\mathrm{h},$$
$$2\nabla M^2 = +(ab - 2h^2)\mathrm{a} - b^2\mathrm{b} + (bc - 2f^2)\mathrm{c} - 2bf\,\mathrm{f} + 2(bg - 2hf)\mathrm{g} - 2bh\,\mathrm{h},$$
$$2\nabla N^2 = +(ca - 2g^2)\mathrm{a} + (bc - 2f^2)\mathrm{b} - c^2\mathrm{c} - 2cf\,\mathrm{f} - 2cg\,\mathrm{g} + 2(ch - 2fg)\mathrm{h};$$

$$2\nabla MN = +(af-2gh)\,\mathrm{a} - bf\mathrm{b} - cf\mathrm{c} - 2bc\mathrm{f} - 2ch\mathrm{g} - 2bg\mathrm{h},$$
$$2\nabla NL = -ag\mathrm{a} + (bg-2hf)\,\mathrm{b} - cg\mathrm{c} - 2ch\mathrm{f} - 2ca\mathrm{g} - 2af\mathrm{h},$$
$$2\nabla LM = -ah\mathrm{a} - bh\mathrm{b} + (ch-2fg)\,\mathrm{c} - 2bg\mathrm{f} - 2af\mathrm{g} - 2ab\mathrm{h}.$$

Dans ces équations, si $Lx+My+Nz$ était facteur de $ax^2+by^2+cz^2+2fyz+2gzx+2hxy$, on aurait évidemment $\mathrm{a}=0$, $\mathrm{b}=0$, $\mathrm{c}=0$, $\mathrm{f}=0$, $\mathrm{g}=0$, $\mathrm{h}=0$, et de là $\nabla=0$, $\Phi=$ un carré, ce qui donnerait des formules plus simples, et auxquelles on pourrait encore donner plusieurs autres formes. Par exemple on tire d'un de ces systèmes, $-1 \div MN\mathfrak{F} = (Mg-Nh) \div \Theta$, &c., où $\Theta = (Mg-Nh)(Nh-Lf)(Lf-Mg)$, et de là les expressions

$$\frac{L}{\mathfrak{F}} + \frac{M}{\mathfrak{G}} + \frac{N}{\mathfrak{H}} = 0,$$

$$\frac{L^2f}{\mathfrak{F}} + \frac{M^2g}{\mathfrak{G}} + \frac{N^2h}{\mathfrak{H}} = 0,$$

$$\frac{L^3f^2}{\mathfrak{F}} + \frac{M^3g^2}{\mathfrak{G}} + \frac{N^3h^3}{\mathfrak{H}} = LMN,$$

auxquelles on pourrait ajouter encore plusieurs systèmes analogues, ce qui se ferait sans la moindre difficulté.

Jusqu'ici L, M, N, a, b, c, f, g, h ont été des quantités quelconques. En supposant qu'elles soient les dérivées du premier et du second ordre d'une fonction U, on a

$$(n-1)\,L = ax+hy+gz,$$
$$(n-1)\,M = hx+by+fz,$$
$$(n-1)\,N = gx+fy+cz,$$
$$n\,(n-1)\,U = ax^2+by^2+cz^2+2fyz+2gxz+2hxy,$$

et la substitution de ces valeurs donne la formule du texte:

$$(n-1)^2\,D^2U = n\,(n-1)\,\Psi\,.\,U - (\alpha x+\beta y+\gamma z)^2\,(\nabla U),$$

(∇U) étant ici $=\nabla$, et Ψ étant donnée par l'équation

$$\Psi = \mathfrak{A}\alpha^2 + \mathfrak{B}\beta^2 + \mathfrak{C}\gamma^2 + 2\mathfrak{F}\beta\gamma + 2\mathfrak{G}\gamma\alpha + 2\mathfrak{H}\alpha\beta.$$

Cela peut être vérifié facilement par la substitution actuelle: mais nous allons le démontrer par la théorie des hyperdéterminants. En effet, soit (123) le déterminant symbolique formé avec $\partial_{x_1}, \partial_{y_1}, \partial_{z_1}$; $\partial_{x_2}, \partial_{y_2}, \partial_{z_2}$; $\partial_{x_3}, \partial_{y_3}, \partial_{z_3}$, $(A23)$ le déterminant symbolique formé avec α, β, γ; $\partial_{x_2}, \partial_{y_2}, \partial_{z_2}$; $\partial_{x_3}, \partial_{y_3}, \partial_{z_3}$, et ainsi de suite; puis soient T_1, T_2, ... les fonctions $x\partial_{x_1}+y\partial_{y_1}+z\partial_{z_1}$, $x\partial_{x_2}+y\partial_{y_2}+z\partial_{z_2}$, &c. et $\rho=\alpha x+\beta y+\gamma z$, on aura identiquement:

$$\rho\,(123) = -T_1\,(A23) - T_2\,(A31) - T_3\,(A12),$$

et en élevant au carré les deux membres de cette équation, ajoutant le facteur $U_1U_2U_3$, et réduisant les variables après avoir différentié suivant x, y, z, puis en tenant compte des équations

$$(123)^2\, U_1U_2U_3 = 6\nabla U,$$
$$T_1^2\, A\, (23)^2\, U_1U_2U_3 = 2n\,(n-1)\, U\,.\,\Psi,$$
$$T_1T_2A\,(23)\,A\,(31)\,U_1U_2U_3 = -(n-1)^2\, D^2U,$$

on obtiendra le résultat dont il s'agit.

Il sera aussi facile de déduire de cette formule quelques propriétés de la courbe $\nabla U=0$, dans les cas d'un point double ou d'un point de rebroussement de la courbe $U=0$. En effet, pour un point double, les dérivées L, M, N de U, et par suite D^2U, et ses dérivées du premier ordre, s'évanouissent. De plus, pour un point de rebroussement on a $\Psi=0$. Donc, pour un point double on a $\nabla U=0$, où la courbe exprimée par cette équation passe par chaque point double (y compris les points de rebroussement) de la courbe $U=0$. Prenons la dérivée de l'équation en question, en affectant du symbole $\partial=\xi\partial_x+\eta\partial_y+\zeta\partial_z$ ses deux membres. Supprimant les termes qui se réduisent à zéro aux points doubles, cela donne

$$0=(\alpha x+\beta y+\gamma z)^2\,\partial\nabla U;$$

c'est-à-dire qu'il y aura aussi à chacun de ces points un point double de la courbe $\nabla U=0$. Passant aux dérivées du second ordre, on aura

$$(n-1)^2\,\partial^2D^2U=n\,(n-1)\,\Psi\partial^2U-(\alpha x+\beta y+\gamma z)^2\,\partial^2\nabla U.$$

Or ici

$$\partial^2D^2U=\alpha^2\,\partial^2\mathrm{a}+\beta^2\partial^2\mathrm{b}+\gamma^2\,\partial^2\mathrm{c}+2\beta\gamma\partial^2\mathrm{f}+2\gamma\alpha\partial^2\mathrm{g}+2\alpha\beta\partial^2\mathrm{h},$$

et $\partial^2\mathrm{a}$, &c. se réduisent à

$$2\,(c\partial M^2-2f\partial M\partial N+b\partial N^2),\ \text{\&c.}$$

savoir (en mettant $a\xi+h\eta+g\zeta$, $h\xi+b\eta+f\zeta$, $g\xi+f\eta+c\zeta$ à la place de ∂L, ∂M, ∂N, et en ayant égard à la condition $\nabla U=0$) à $2\partial^2U\,.\,\mathfrak{A}$, &c.; on aura donc $\partial^2Du=2\partial^2U\,.\,\Psi$, ou enfin

$$(n-1)\,(n-2)\,\partial^2U\,.\,\Psi=-(\alpha x+\beta y+\gamma z)^2\,\partial^2\nabla U,$$

c'est-à-dire: les deux courbes $U=0$, $\nabla U=0$, se toucheront dans les points doubles. Enfin pour un point de rebroussement $\partial^2\nabla U$ s'évanouit, c'est-à-dire, il existe un point triple dans la courbe $\nabla U=0$. Mais il peut être démontré que deux branches de la courbe se touchent en ce point, et qu'elles touchent aussi la courbe $U=0$; c'est-à-dire qu'il y aura dans la courbe $\nabla U=0$ un point de rebroussement, et une autre branche de la courbe qui passe par ce point. Pour cela il faudra passer aux dérivées du troisième ordre. Cela donne, en supprimant les termes qui s'évanouissent:

$$(n-1)^2\,\partial^3\,D^2U=3n\,(n-1)\,\partial\Psi\,.\,\partial^2U-(\alpha x+\beta y+\gamma z)^2\,\partial^3\nabla U.$$

Ici on a

$$\partial^3 D^2 U = \alpha^2 \partial^3 \mathrm{a} + \&c.,$$

$$\partial^3 \mathrm{a} = 6\,(\partial c \partial M^2 - 2\partial f \partial M \partial N + \partial b \partial N^2) + 6\,\{c \partial M \partial^2 M - f(\partial M \partial^2 N + \partial N \partial^2 M) + b \partial N \partial^2 N\},\ \&c.,$$

et les deux lignes de cette expression se réduisent à $6\,\partial^2 U . \partial\mathfrak{A}$, et à zéro, respectivement. En effet, en remplaçant ∂M, ∂N par leurs valeurs, le coefficient de ξ^2 dans la première ligne devient $6\,(h^2 \partial c - 2gh\partial f + g^2 \partial b) = 6a\,(b\partial c + c\partial b - 2f\partial f)$ (et à cause de $\mathfrak{B} = 0$, $\mathfrak{C} = 0$, $\mathfrak{F} = 0) = 6a\partial\mathfrak{A}$; et également pour les autres termes. De même, le coefficient de ξ^2 dans la seconde ligne devient $6\,\{ch\partial h - f(g\partial h + h\partial g) + bg\partial g\}$. En y ajoutant $3\,(h^2\partial c - 2gh\partial f + g^2 \partial b)$, savoir $3a\partial\mathfrak{A}$, la somme se réduit à $3\partial\,(ch^2 - 2fgh + bg^2) = 3\partial\,(a\mathfrak{A} - \nabla) = 3a\partial\mathfrak{A}$; donc le coefficient en question s'évanouit. En cherchant de la même manière les autres coefficients, on trouvera les valeurs dont il s'agit, et ainsi $\partial^3 \mathrm{a} = 6\partial^2 U . \partial\mathfrak{A}$, &c. et de là $\partial^3 D^2 U = 6\,\partial^2 U \partial\Psi$; donc enfin

$$3\,(n-1)\,(n-2)\,\partial^2 U \partial\Psi = -\,(\alpha x + \beta y + \gamma z)^2\,\partial^3 \nabla U\,;$$

ce qui suffit pour démontrer le théorème, qui peut être énoncé comme suit:

THÉORÈME. "Il existe un point double de la courbe $\nabla U = 0$, pour chaque point double de la courbe $U = 0$, et les deux courbes ont des tangentes communes dans ces points. De plus, il existe un point triple de la courbe $\nabla U = 0$, pour chaque point de rebroussement de la courbe $U = 0$, savoir un point de rebroussement dont les deux branches touchent la tangente de la courbe $U = 0$, et encore une troisième branche, qui passe par le point de rebroussement."

Il suit de là que dans le cas d'un point double, ce point doit être considéré comme la réunion de six points d'intersection, et dans celui d'un point de rebroussement, de huit points d'intersection. C'est de cette manière que l'on se rend compte du théorème de M. Plücker qui dit que l'effet de ces deux singularités est de faire disparaître respectivement six ou huit points d'inflexion de la courbe donnée.

Examinons, en finissant, la théorie des points *d'osculation*. Il est facile de voir que la condition d'un tel point (savoir d'un point dans lequel la tangente coupe la courbe en quatre points consécutifs) consiste en ce que $\partial^3 U$ contient ∂U comme facteur, ou, autrement dit, que l'équation $D^3 U = 0$ est *identiquement* vraie. On obtient ainsi *dix* conditions, qui se réduisent assez facilement à *six*; mais pour les réduire à *une seule* condition, il faut prendre la dérivée, avec le symbole D de la valeur donnée ci-dessus de $D^2 U$. On doit cependant ne pas oublier que $D . D^2 U$, outre le terme $D^3 U$, contient aussi des termes que l'on obtient en faisant opérer les symboles ∂_x, ∂_y, ∂_z sur les quantités L, M, N qui entrent dans $D^2 U$. Car il est convenu que les symboles ∂_x, ∂_y, ∂_z dans $D^3 U$, ne doivent pas affecter les lettres L, M, N. Cependant il est remarquable que dans le cas actuel, ces termes se détruisent entièrement. En effet, en les désignant par Ω, on obtient

$$\Omega = 2D\,[(M\gamma - N\beta)\,(a\partial_L + h\partial_M + g\partial_N) + (N\alpha - L\beta)\,(h\partial_L + b\partial_M + f\partial_N)$$
$$+ (L\beta - M\gamma)\,(g\partial_L + f\partial_M + c\partial_N)]\,.\,DU,$$

ou il faut d'abord effectuer les opérations ∂_L, ∂_M, ∂_N qui se rapportent à D, et ensuite rapporter les ∂_x, ∂_y, ∂_z à la fonction U. Cela donne

$$\Omega = 2\left[\alpha(N\partial_y - M\partial_z) + \beta(L\partial_z - N\partial_y) + \gamma(M\partial_x - L\partial_y)\right] \times$$

$$\left\{\begin{array}{l} \alpha^2\,[N(f\partial_y - b\partial_z) - M(c\partial_y - f\partial_z)] \\ +\beta^2\,[L(g\partial_z - c\partial_x) - N(a\partial_z - g\partial_x)] \\ +\gamma^2\,[M(h\partial_x - a\partial_y) - L(b\partial_x - h\partial_y)] \\ +\beta\gamma\,[-L(h\partial_z + g\partial_y - 2f\partial_x) + M(a\partial_z - g\partial_x) + N(a\partial_y - h\partial_x)] \\ +\gamma\alpha\,[-M(f\partial_x + h\partial_z - 2g\partial_y) + N(b\partial_x - h\partial_y) + L(b\partial_z - f\partial_y)] \\ +\alpha\beta\,[-N(g\partial_y + f\partial_x - 2h\partial_z) + L(c\partial_y - f\partial_z) + M(c\partial_x - g\partial_z)] \end{array}\right\} U,$$

où ∂_x, ∂_y, ∂_z se rapportent seulement à U. Or tous les termes de cette expression *s'évanouissent.* Par exemple le coefficient de α^3 devient

$$(N\partial_y - M\partial_z)\,[N(f\partial_y - b\partial_z) - M(c\partial_y - f\partial_z)]\,U$$
$$= [N^2(f\partial_y^2 - b\partial_z\partial_y) + M^2(c\partial_y\partial_z - f\partial_z^2) - MN(c\partial_y^2 - f\partial_y\partial_z + f\partial_y\partial_z - b\partial_z^2)\,U = 0,$$

et ainsi pour les autres termes; donc on a $\Omega = 0$.

Donc, en transportant à l'autre côté de l'équation les termes qui contiennent U, DU ou ∇U, on obtient cette formule très simple:

$$(n-1)^2 D^3 U = -(\alpha x + \beta y + \gamma z)^2 D\nabla U,$$

ou la seule condition d'un point d'osculation (en ayant égard à l'équation $\nabla U = 0$) se réduit à $D\nabla U = 0$. Savoir les dérivées $\partial_x \nabla U$, $\partial_y \nabla U$, $\partial_z \nabla U$ doivent être proportionnelles à $\partial_x U$, $\partial_y U$, $\partial_z U$ (ce qui équivaut à une seule condition, en vertu de $U = 0$, $\nabla U = 0$; comme on le voit facilement). Cela donne le théorème suivant.

THÉORÈME. "Dans le cas d'un point *d'osculation*, les deux courbes $U = 0$, $\nabla U = 0$ se touchent."

Il n'y a presque pas de doute que la dérivée $\nabla\nabla U$ ne se réduise *toujours* à la forme $R \,.\, U + S \,.\, \nabla U$. En effet, M. Hesse l'a démontré pour les fonctions de trois variables et du troisième degré, et moi, je l'ai vérifié pour les fonctions de deux variables d'un degré quelconque. Cela étant, les points d'inflexion de la courbe $\nabla U = 0$ sont situés aux points d'intersection avec $U = 0$, et au cas où les deux courbes se touchent, ce point de contact doit être considéré comme la réunion de trois points d'intersection: donc,

"Tout point *d'osculation* peut être envisagé comme point de réunion de *trois* points *d'inflexion.*"

Nous démontrerons encore, d'une manière conforme à celle dont nous avons trouvé l'expression de D^2U, l'expression qui vient d'être donnée pour D^3U à moyen de la formule

$$\rho^2(123)^2 = \{T_1(A23) + T_2(A31) + T_3(A12)\}^2.$$

En multipliant les deux membres par $(A14)+(A24)+(A34)$, le terme à gauche peut être présenté sous la forme $\rho^2(A\theta 4)(123)^2$, où ∂_{x_θ}, ∂_{y_θ}, ∂_{z_θ} se rapportent à tous les systèmes de variables. En y appliquant le produit $U_1U_2U_3U_4$ (les variables identiques après les différentiations), on obtiendra $6\rho^2(A\theta 4)U_4\nabla U=-6\,D\nabla U.\rho^2$. Pour la droite de l'équation on a d'abord trois termes comme $T_1^2(A14)(A23)^2\,U_1U_2U_3U_4$, lesquels se détruisent évidemment, à cause de $(A14)\,U_1U_4=0$; puis six termes de la forme $(A14)(A23)(A31)\,U_1U_2U_3U_4$, qui se détruisent aussi, puisqu'en échangeant 1,3 et 2,4, le terme change de signe; puis trois termes comme $T_1^2\{(A24)+(A34)\}(A23)^2\,U_1U_2U_3U_4$, &c. savoir $T_1^2\,U_1\{(A24)+(A34)\}(A23)^2\,U_2U_3U_4=-2n(n-1)\,U.D\Psi$; c'est-à-dire tous ces termes sont $6n(n-1)\,U\,D\Psi$; enfin trois termes $2T_1T_2(A23)(A31)(A34)\,U_1U_2U_3U_4$ $=2(n-1)^2\,D^3U$, ou, tous pris ensemble, $6(n-1)^2D^3U$. Donc, en supprimant le terme $-6n(n-1)\,U\,D\Psi$, à cause de $U=0$, on obtient la même équation que ci-dessus, savoir:

$$(n-1)^2D^3U=-(\alpha x+\beta y+\gamma z)^2D\nabla U.$$

On pourrait croire qu'il y a une équation analogue $(n-1)^2D^4U=-(\alpha x+\beta y+\gamma z)^2D^2\nabla U$ pour les dérivées du *quatrième* degré, mais cela n'est pas. En effet, il est facile de voir que pour un point d'osculation, $D^2\nabla U$ se réduit à $\nabla\nabla U$, à un facteur près, c'est-à-dire l'on aurait $D^2\nabla U=0$, puisque $\nabla\nabla U$ s'évanouit aux points d'osculation: donc on aurait généralement pour un point d'osculation $D^4U=0$; mais cela est seulement le caractère des points d'osculation d'un plus haut degré, savoir de ceux où la tangente rencontre la courbe en cinq points consécutifs. Pour le quatrième degré le problème devient trop compliqué pour être traité de cette manière.

54.

NOTE SUR LES HYPERDÉTERMINANTS.

[From the *Journal für die reine und angewandte Mathematik* (Crelle), tome XXXIV. (1847), pp. 148—152.]

I. SOIT V une fonction homogène de x, y de $2p^{\text{ième}}$ ordre. En égalant à zéro les coefficients différentiels du $p^{\text{ième}}$ ordre de cette fonction, pris par rapport à x, y, et en éliminant ces variables, on obtiendra entre les coefficients de la fonction un nombre p d'équations. Or parmi ces équations il y aura toujours une seule du second ordre, savoir

$$B(U,\ U) = 0$$

(suivant la notation dans mon mémoire sur les hyperdéterminants, t. XXX. [(1846), 16]. Par exemple, en écrivant t au lieu de $x : y$ on a identiquement

$$\begin{aligned} &c\,(at + b) \\ -\,&b\,(bt + c) \\ =\,&(ac - b^2)\,t, \end{aligned}$$

$$\begin{aligned} &(\ et \qquad\quad)\,(at^2 + 2bt + c) \\ -\,&(4dt + 2e)\,(bt^2 + 2ct + d) \\ +\,&(3ct + 2d)\,(ct^2 + 2dt + e) \\ =\,&(ae - 4bd + 3c^2)\,t^3, \end{aligned}$$

$$\begin{aligned} &(\ gt^2 \qquad\qquad\qquad\quad)\,(at^3 + 3bt^2 + 3ct + d\,) \\ -\,&(\ 6ft^2 + \ 3gt \qquad\ \ )\,(bt^3 + 3ct^2 + 3dt + e\,) \\ +\,&(15et^2 + 18ft + 6g\,)\,(ct^3 + 3dt^2 + 3et + f\,) \\ -\,&(10dt^2 + 15et + 6f\,)\,(dt^3 + 3et^2 + 3ft + g\,) \\ =\,&(ag - 6bf + 15ec - 10d^2)\,t^5, \end{aligned}$$

et ainsi de suite: il ne reste qu'à déterminer la loi des coefficients numériques des facteurs à gauche. Pour cela, représentons par A, B, C, ... les coefficients de $(1+s)^{2p}$, et par A', B', C', ... les coefficients de $(1-s)^{-p}$. On aura pour ces nombres le système suivant:

$$\begin{array}{lll} +A'A, & & \\ -A'B, & -B'A, & \\ +A'C, & +B'B, & +C'A, \\ \cdots\cdots & \cdots\cdots & \cdots\cdots \\ \pm I, & \pm K, & \pm L \ldots. \end{array}$$

Les nombres I, K, L ... de la dernière ligne seront à déterminer au moyen d'une autre règle: il faut faire évanouir les sommes des nombres dans la même ligne verticale, ces nombres étant pris avec leurs signes actuels. Je suis parvenu par *induction* à ces formules, mais il ne serait pas, je crois, très difficile de les démontrer directement.

Il me paraît possible que tous les hyperdéterminants qui se rapportent à la fonction V, puissent être trouvés en éliminant entre les équations (en nombre de p) dont il s'agit, et cela dans le cas où U est de l'ordre $2p$ ou $2p+1$; au moins cette règle se vérifie pour les fonctions de deuxième, troisième et quatrième ordres, et cela paraissait (à priori) moins probable pour les dérivées d'un degré plus élevé que pour celles du second degré, pour lesquelles, comme on vient de le voir, il est effectivement vrai.

Cela étant, il y aura seulement un nombre p de dérivées indépendantes pour les fonctions du $2p^{\text{ième}}$ et du $(2p+1)^{\text{ième}}$ ordre: conclusion que je ne puis pas démontrer.

II. Soit $\nabla = 6abcd + 3b^2c^2 - a^2d^2 - 4ac^3 - 4b^3d$, et représentons par ∇_1 le déterminant formé avec les coefficients différentiels du second ordre de ∇ par rapport à a, b, c, d, on aura

$$\nabla_1 = 3\nabla^2$$

(propriété qui a un rapport singulier avec celle qu'a démontrée M. Eisenstein par rapport aux coefficients du premier ordre de la même fonction ∇). La démonstration que je puis donner de ce théorème est à la vérité assez compliquée, mais je ne vois pas d'autre. En mettant

$$p = \tfrac{2}{3}(bd - c^2), \quad q = \tfrac{1}{3}(bc - ad), \quad r = \tfrac{2}{3}(ac - b^2),$$

on obtient

$$\nabla_1 = 81 \begin{vmatrix} a^2, & ab, & ac-3r, & ad+9q \\ ba, & b^2+2r, & bc-q, & bd-3p \\ ca-3r, & cb-q, & c^2+2p, & cd \\ da+9q, & db-3p, & dc, & d^2 \end{vmatrix}$$

où le déterminant ne contient que les termes du quatrième ou troisième degré en p, q, r, et en développant on a

$$\begin{aligned}\nabla_1 = 81\,\{&9(pr-q^2)^2 - 2a^2p^3 - 2d^2r^3 - 12q(abp^2+cdr^2) - 18q^2(b^2p+c^2r) \\ &- 6(pr+q^2)(acp+bdr) - 2adq(3pr-q^2) - 18bcq(pr+q^2)\},\end{aligned}$$

d'où, en réduisant au moyen des expressions

$$ap = -(2bq + cr), \quad dr = -(2cq + bp), \quad ad = bc - 3q,$$

on tire $\nabla_1 = 81 \{9(pr - q^2)^2 + 4(b^2p + 2bcq + c^2r)(pr + q^2) + 6q^2(3pr - q^2)\}$,

ou enfin, au moyen de $2(b^2p + 2bcq + c^2r) = -3pr$:

$$\nabla_1 = 243(pr - q^2)^2 = 3\nabla^2;$$

voilà l'équation qu'il s'agissait à démontrer.

III. En considérant $a:d$, $b:d$, $c:d$ comme représentants les trois coordonnées d'un point, ou, si l'on veut, des fonctions linéaires de ces coordonnées, l'équation $\nabla = 0$ appartient évidemment à une surface développable (de quatrième ordre). Mais la condition pour que l'équation $U = 0$ (V étant une fonction homogène de quatre variables) appartienne à une surface développable, est, que le déterminant formé avec les coefficients différentiels du second ordre de la fonction, s'évanouisse (théorème de M. Hesse, t. XXVIII. [(1844) pp. 97—107, "Ueber die Wendepunkte der Curven dritter Ordnung"]). Donc il faut que ∇_1 s'évanouisse au moyen de $\nabla = 0$, c'est-à-dire, il faut que ∇_1 contienne le facteur ∇; ce qui s'accorde parfaitement avec l'équation qui vient d'être présentée. Mais il ne peut être prouvé de cette manière que l'autre facteur doit aussi se réduire à ∇, et même cela n'est pas vrai si ∇_1 vient d'une fonction d'un plus haut degré que le quatrième.

Il suit de cela qu'en supposant toujours que les coefficients soient des fonctions linéaires des coordonnées, le résultat $\Theta = 0$ de l'élimination de x, y entre $\frac{dU}{dx} = 0$, $\frac{dU}{dy} = 0$ appartient toujours à une surface développable. De même l'élimination de x, y entre les trois équations $\frac{d^2U}{dx^2} = 0$, $\frac{d^2U}{dx\,dy} = 0$, $\frac{d^2U}{dy^2} = 0$ conduit aux équations de l'arête de rebroussement de la surface, et de plus, en éliminant entre les équations $\frac{d^3U}{dx^3} = 0$, $\frac{d^3U}{dx^2\,dy} = 0$, $\frac{d^3U}{dx\,dy^2} = 0$, $\frac{d^3U}{dy^3} = 0$, on obtient les points de rebroussement de l'arête de rebroussement. Cela conduit à quelques résultats remarquables.

Par exemple, la surface développable dont l'équation est

$$\nabla = 6abcd + 3b^2c^2 - 4ac^3 - 4b^3d - a^2d^2,$$

a pour arête de rebroussement la courbe dont les équations (équivalentes à deux équations seulement) sont $bd - c^2 = 0$, $ad - bc = 0$, $ac - b^2 = 0$, ce qui est une *courbe du troisième ordre* seulement. Car en considérant deux quelconques de ces trois équations, par exemple celles-ci: $bd - c^2 = 0$, $ad - bc = 0$, ces équations appartiennent à deux surfaces du second ordre qui ont en commun la droite $d = 0$, $c = 0$: cela s'accorde avec un résultat que j'ai donné dans mon mémoire sur les surfaces développables dans le journal de M. Liouville [t. X. (1845), 30].

Egalement, en considérant une équation de quatrième degré en t, on obtient une surface développable.

$$(ae - 4bd + 3c^2)^3 - 27(ace + 2bcd - ad^2 - b^2e - c^3)^2 = 0,$$

qui a pour arête de rebroussement la courbe du sixième ordre exprimée par les équations $ae - 4bd + 3c^2 = 0$, $ace + 2bcd - ad^2 - b^2e - c^3 = 0$. Je n'ai pas complètement réussi à expliquer pourquoi cette courbe du sixième ordre a une osculatrice développable, seulement du sixième ordre, mais cette réduction s'opère en partie au moyen des points de rebroussement de la courbe, qui se trouvent au moyen des équations $at+b=0$, $bt+c=0$, $ct+d=0$, $dt+e=0$. Pour savoir à combien de points ces équations correspondent, il faut remarquer que, a, b, c, d, e étant des fonctions linéaires des coordonnées, on aura toujours entre ces quantités une équation linéaire telle que

$$Aa+Bb+Cc+Dd+Ee=0,$$

où A, B, ... sont des constantes. Donc, en éliminant a, b, c, d, e, on obtient $A-Bt+Ct^2-Dt^3+Et^4=0$, équation du quatrième ordre, et à chaque valeur de t il correspond un des points dont il s'agit; donc la courbe du sixième ordre a quatre points de rebroussement.

Également la surface développable qui correspond à une équation du $m^{\text{ième}}$ ordre, est de l'ordre $2(m-1)$; l'arête de rebroussement est de l'ordre $3(m-2)$, et il y a dans cette courbe un nombre $4(m-3)$ de points de rebroussement. Il faut toujours se rappeler que ces surfaces développables ne sont pas les surfaces développables les plus générales qui existent de l'ordre $2(m-1)$, excepté dans le cas des surfaces développables du quatrième ordre.

IV. Il vaut peut-être la peine de donner en passant une démonstration de ce théorème de M. Chasles: "Le plan qui passe par trois points qui se meuvent avec des vitesses uniformes dans trois droites quelconques, enveloppe une surface développable du quatrième degré." En effet, en supposant que $\alpha:\delta$, $\beta:\delta$, $\gamma:\delta$; $\alpha':\delta'$, $\beta':\delta'$, $\gamma':\delta'$; $\alpha'':\delta''$, $\beta'':\delta''$, $\gamma'':\delta''$, soient des fonctions linéaires du temps ((δ, δ', δ'') peuvent être constants, ou, si l'on veut, des fonctions linéaires du temps, ce qui correspond à un cas un peu plus général que celui de M. Chasles), on peut prendre ces valeurs pour coordonnées des trois points mobiles. Donc, en prenant $x:w$, $y:w$, $z:w$ pour coordonnées d'un point quelconque du plan, on obtient l'équation de ce plan, en égalant à zéro le déterminant formé avec les valeurs x, y, z, w; α, β, γ, δ; α', β', γ', δ'; α'', β'', γ'', δ''; ce qui donne une équation de la forme $a+3bt+3ct^2+dt^3=0$, a, b, c, d étant des fonctions linéaires de x, y, z, w; et cela suffit pour démontrer le théorème dont il s'agit.

V. En finissant j'indiquerai un principe de classification des courbes à double courbure qui me paraît être de quelque importance; savoir, on pourra distinguer les courbes qui ne peuvent pas être l'intersection *complète* de deux surfaces, de celles qui peuvent l'être. Par exemple, en faisant passer par une courbe donnée du troisième ordre deux surfaces du second ordre, la courbe n'est pas l'intersection complète des deux surfaces; celles-ci se coupent dans cette courbe et dans une certaine droite. Quel est le théorème analogue pour les courbes du $n^{\text{ième}}$ ordre? Peut-on, par exemple, toujours combiner avec une courbe donnée d'un ordre quelconque, une autre courbe qui est l'intersection complète de deux surfaces, de manière que l'ensemble des deux courbes soit une intersection complète de deux surfaces? Et si non: de quelle manière trouvera-t-on les équations générales d'une courbe de $n^{\text{ième}}$ ordre? Quel est le degré de généralité de ces équations? Il y a une foule d'autres questions qu'on pourrait ici proposer. J'ai proposé une question analogue dans le point de vue analytique, mais elle est restée sans réponse.

55.

SUR QUELQUES THÉORÈMES DE LA GÉOMÉTRIE DE POSITION.

[From the *Journal für die reine und angewandte Mathematik* (Crelle), tome XXXIV. (1847), pp. 270—275. Continued from t. XXXI. p. 227, **50**.]

§ III.

LORSQUE j'avais sous la plume la première partie de ce mémoire, je ne savais pas que la dernière partie du théorème de M. Steiner sur l'hexagramme de Pascal (savoir que les vingt points d'intersection des soixante droites sont situés, par quatre, dans quinze droites) avait déjà été démontrée d'une manière aussi simple qu'élégante par M. Plücker dans son mémoire, "*Über ein neues Princip der Geometrie*" (t. V. [1828] p. 269). En supposant maintenant cette démonstration connue, je veux examiner de plus près la corrélation de ces vingt points, en adoptant une notation plus commode.

Soit $\alpha\beta\gamma\delta\epsilon\zeta$ une permutation quelconque des nombres 1, 2, 3, 4, 5, 6 : cette permutation peut être nommée *directe* ou *inverse*, selon qu'elle est formée par un nombre pair ou impair d'inversions. Des six permutations $\alpha\beta\gamma\delta\epsilon\zeta$, $\alpha\delta\gamma\zeta\epsilon\beta$, $\alpha\zeta\gamma\beta\epsilon\delta$; $\alpha\delta\gamma\beta\epsilon\zeta$, $\alpha\beta\gamma\zeta\epsilon\delta$, $\alpha\zeta\gamma\delta\epsilon\beta$, les trois premières ou les trois dernières sont *directes*. Nous représenterons les trois permutations directes par $(\alpha\gamma\epsilon)$. Les trois droites que donne le théorème de Pascal, appliqué aux hexagones correspondants à ce symbole, se coupent dans un des vingt points dont il s'agit: point qui peut être représenté par la même notation $(\alpha\gamma\epsilon)$. En supposant que $\alpha\beta\gamma\delta\epsilon\zeta$ est une permutation directe, le point $\alpha\gamma\epsilon$ correspond au point $\begin{pmatrix}\alpha\gamma\epsilon\\ \beta\delta\zeta\end{pmatrix}$ de M. Plücker. Partout dans cette section on pourra changer les mots "directe" et "inverse."

Les six permutations des lettres α, γ, ϵ ne donnent qu'un seul point ; de manière que les vingt points, dont il s'agit, sont

$$\begin{matrix} 123, & 124, & 125, & 126, & 134, & 135, & 136, & 145, & 146, & 156, \\ 234, & 235, & 236, & 245, & 246, & 256, & 345, & 346, & 356, & 456. \end{matrix}$$

Or, pour trouver comment il faudra combiner ces points, j'écris

$$\begin{matrix} \alpha\gamma\epsilon \\ \beta\delta\zeta \end{matrix}$$

et je tire de là le système

$$\begin{matrix} \alpha\gamma\epsilon & \alpha\delta\zeta & \beta\gamma\zeta & \beta\delta\epsilon \\ \beta\delta\zeta\ ' & \beta\gamma\epsilon\ ' & \alpha\delta\epsilon\ ' & \alpha\gamma\zeta\ \cdot \end{matrix}$$

Les quatre points

$$\alpha\gamma\epsilon\,,\quad \alpha\delta\zeta\,,\quad \beta\gamma\zeta,\quad \beta\delta\epsilon$$

seront situés sur la même droite, que l'on peut représenter par $\alpha\beta\,.\,\gamma\delta\,.\,\epsilon\zeta$. Les quinze combinaisons, quatre à quatre, des vingt points, seront

	135,	146,	236,	245	dans la droite 12 . 34 . 56
	136,	154,	234,	256	„ „ „ 12 . 35 . 64
	134,	165,	235,	264	„ „ „ 12 . 36 . 45
	146,	152,	342,	356	„ „ „ 13 . 45 . 62
	142,	165,	345,	362	„ „ „ 13 . 46 . 25
	145,	126,	346,	325	„ „ „ 13 . 42 . 56
	152,	163,	453,	462	„ „ „ 14 . 56 . 23
A.	153,	126,	456,	423	„ „ „ 14 . 52 . 36
	156,	132,	452,	436	„ „ „ 14 . 53 . 62
	163,	124,	564,	523	„ „ „ 15 . 62 . 34
	164,	132,	562,	534	„ „ „ 15 . 63 . 42
	162,	143,	563,	542	„ „ „ 15 . 64 . 23
	124,	135,	625,	634	„ „ „ 16 . 23 . 45
	125,	143,	623,	645	„ „ „ 16 . 24 . 53
	123,	154,	624,	653	„ „ „ 16 . 25 . 34,

où les droites s'obtiennent en permutant dans 12 . 34 . 56 d'abord les derniers trois numéros et puis dans ces trois permutations les derniers cinq numéros. Par là la manière de trouver les droites est claire.

Cette énonciation des points et des droites, dont il s'agit, en même temps qu'elle est parfaitement symétrique, est la seule qui se présente naturellement. Cependant la symétrie en est si compliquée et si peu manifeste qu'il sera bon d'adopter une autre notation. Pour cela, je forme le tableau auxiliaire suivant, dont l'arrangement est assez clair:

135	*ace*	351	*cea*	513	*eac*
246	*bdf*	246	*fbd*	246	*dfb*
146	*acb*	346	*cef*	546	*ead*
235	*edf*	251	*abd*	213	*cfb*
236	*acd*	256	*ceb*	216	*eaf*
145	*bef*	341	*fad*	543	*dcb*
245	*acf*	241	*ced*	243	*eab*
136	*bde*	356	*fba*	516	*dfc*

De là, en écrivant

$$B. \quad \left\{ \begin{array}{ll} 123 = bcf\ , & 234 = abe\ , \\ 124 = cde\ , & 235 = def\ , \\ 125 = abd\ , & 236 = acd\ , \\ 126 = aef\ , & 245 = acf\ , \\ 134 = adf\ , & 246 = bdf\ , \\ 135 = ace\ , & 256 = bce\ , \\ 136 = bde\ , & 345 = bcd\ , \\ 145 = bef\ , & 346 = cef\ , \\ 146 = abc\ , & 356 = abf\ , \\ 156 = cdf\ , & 456 = ade\ ; \end{array} \right.$$

et de plus

$$C. \quad \left\{ \begin{array}{l} \begin{array}{l|l|l} 12.34.56 = ac\ , & 13.45.62 = ab\ , & 14.56.23 = bd\ , \\ 12.35.64 = be\ , & 13.46.25 = cd\ , & 14.52.36 = ae\ , \\ 12.36.45 = df\ , & 13.42.56 = ef\ , & 14.53.62 = cf\ , \end{array} \\ \begin{array}{l|l} 15.62.34 = de\ , & 16.23.45 = ce\ , \\ 15.63.42 = bc\ , & 16.24.53 = ad\ , \\ 15.64.23 = ef\ , & 16.25.34 = bf\ ; \end{array} \end{array} \right.$$

savoir (en représentant les points 123, 124, &c., par *bcf*, *cde*, &c., et les droites 12.34.56, 12.35.64, &c., par *ac*, *be*, &c.) on verra dans le tableau (A) que les points situés dans la droite *ac* sont *ace*, *acb*, *acd*, *acf*, que les points dans la droite *be* sont *bed*, *bef*, *bea*, *bec*, et ainsi de suite, de manière que ce système des vingt points et des quinze droites est précisément le système *réciproque* de celui des quinze points et des vingt droites que nous avons considéré dans la première section de ce mémoire; ou, autrement dit, que les vingt points et les quinze droites sont les projections sur un plan, des points et des droites d'intersection de six plans dans l'espace. Seulement la figure plane, ainsi formée, contient quatorze quantités arbitraires, tandis que le système de six points sur une conique n'en contient que onze, de sorte qu'il doit y avoir des relations entre ces six plans. On obtient donc la forme suivante plus complète du théorème de M. Steiner (théorème qui en même temps est le complément du théorème XII. § 1 de ce mémoire).

THÉORÈME XIV. "Les soixante droites correspondantes aux hexagones formés par six points d'une conique se coupent trois à trois dans vingt points qui peuvent être considérés comme les projections des points d'intersection d'un système de six plans (dont d'ailleurs la liaison reste encore à chercher)."

Également

THÉORÈME XIII. "Les soixante points correspondants aux hexagones formés par six tangentes d'une conique sont situés trois à trois sur vingt droites déterminées par des plans qui passent par trois points quelconques d'un système de six points dans l'espace (la liaison de ce système de six points étant encore à chercher)."

§ IV.

Soient a, f; b, g; c, h les points correspondants d'un système de points situés sur la même droite, et en involution. Nommons "*faisceau*" les trois côtés d'un quadrilatère qui se coupent dans un même point et "*triangle*" les trois côtés qui ne se coupent pas dans un même point (de sorte que dans tout quadrilatère il y aura quatre *faisceaux* et quatre *triangles*). Les quadrilatères dont les côtés passent par les six points en involution, peuvent être classés en deux systèmes: Dans le premier les faisceaux passeront par f, g, h; f, b, c; g, c, a; h, a, b, et les triangles par a, b, c; a, g, h; b, h, f; c, f, g; dans l'autre il en sera le contraire. Deux quadrilatères qui appartiennent à ces deux systèmes respectivement, peuvent être dits "en rapport inverse l'un à l'autre." Soient $ABCD$, $A'B'C'D'$ deux quadrilatères non situés dans le même plan, et soumis à la condition que les côtés

$$DA,\ B'C';\ DB,\ C'A';\ DC,\ A'B';\ BC,\ D'A';\ CA,\ D'B';\ AB,\ D'C'$$

coupent la droite dans les points

$$f,\ g,\ h,\ a,\ b,\ c$$

respectivement. Les deux tétraèdres $A'BCD$, $AB'C'D'$, et également les tétraèdres $AB'CD$ et $A'BC'D'$; $ABC'D$ et $A'B'CD'$; $ABCD'$ et $A'B'C'D$ seront respectivement inscrits et circonscrits l'un à l'autre. Car en considérant par exemple ces deux-ci: $A'BCD$, $AB'C'D'$: A' est dans le plan $B'C'D'$, B est dans le plan $C'D'A$ (parce que les droites AB, $C'D'$ se rencontrent); C est dans le plan $D'AB'$ (parce que AC et $B'D'$ se rencontrent), et D est dans le plan $AB'C'$ (parce que AD et $B'C'$ se rencontrent). Également A, B', C', D'

sont situés dans les plans *BCD*, *CDA'*, *DA'B* et *A'BC* respectivement; et cela vérifie la relation dont il s'agit. Ce théorème est dû à M. Möbius qui l'a obtenu par son Calcul Barycentrique (ce Journal, t. III. [1828], p. 273), et en considérant un système polaire dans lequel le plan réciproque d'un point quelconque passe toujours par le point même (Statik, c. VI. § 86, et ce Journal t. X. [1833], p. 317). On trouve aussi quelques remarques sur ce sujet dans l'ouvrage "*Systematische Entwickelungen u. s. w.*" de M. Steiner, p. 247. Je ne croyais pas inutile d'en faire voir la relation avec la théorie de l'involution. Remarquons aussi que non seulement les quadrilatères *ABCD* et *A'B'C'D'*, mais aussi ceux-ci *ABC'D'* et *A'B'CD*, *ACB'D'* et *A'C'BD*, *ADB'C'* et *A'D'BC* sont en rapport inverse entre eux. Par cela la symétrie de la figure est complétée; mais on n'en tire pas de nouveaux systèmes de tétraèdres inscrits et circonscrits.

M. Möbius a démontré qu'il n'existe pas des quadrilatères réels simples, à quatre côtés et quatre angles, inscrits et circonscrits. Mais en considérant les points imaginaires, l'existence en est possible, et on trouve des systèmes de cette sorte parmi les neuf points d'inflexion d'une courbe de troisième ordre. Je renvoie cette discussion à une autre occasion, § V.

Je me bornerais, sans examiner de plus près la figure qui en résulte, à démontrer le théorème suivant: "Si un point et n droites sont donnés, les points d'intersection de chaque droite avec la polaire du point, relative aux autres $n-1$ droites, sont situés sur une même droite polaire du point, relative au système des droites." Je prends un système de droites, considéré comme formant une courbe, et j'entends par polaire ou droite polaire, la dernière des polaires successives du point, relative à la courbe. En représentant analytiquement la courbe par $V=0$, V est une fonction homogène d'un ordre quelconque en x, y, z; et si $\alpha:\beta:\gamma$ sont les valeurs de $x:y:z$ relatives au point, l'équation

$$(\alpha\partial_x+\beta\partial_y+\gamma\partial_z)^{p-1}U=0$$

est celle de l'une quelconque des polaires successives.

Soit $p=0$, $q=0$, $r=0$, ... les équations des droites, p, q, r, ... seront des fonctions linéaires de x, y, z. Soient comme plus haut $x:y:z=\alpha:\beta:\gamma$ les équations qui déterminent le point: l'équation de la droite polaire du point, relative aux n droites, est

$$(\alpha\partial_x+\beta\partial_y+\gamma\partial_z)^{n-1}pqr\,\ldots=0.$$

Soient a, b, c, ... ce que deviennent p, q, r, ... en écrivant α, β, γ, ... au lieu de x, y, z, ..., on obtient aisément

$$\alpha\partial_x+\beta\partial_y+\gamma\partial_z=a\partial_p+b\partial_q+\ldots;$$

et de là on tire

$$(a\partial_p+b\partial_q+c\partial_r+\ldots)^{n-1}pqr\,\ldots=0,$$

pour la polaire cherchée. Les différentiations étant effectuées selon p, q, r, ..., comme variables indépendantes, on obtient

$$pbc\ldots+aqc\ldots+\ldots=0$$

ou, ce qui est la même chose:

$$\frac{p}{a}+\frac{q}{b}+\frac{r}{c}+\ldots=0.$$

De même la polaire du point, relative aux droites $q=0$, $r=0$, ..., est

$$\frac{q}{b}+\frac{r}{c}+\ldots=0,$$

et par cette raison l'intersection de cette droite avec $p=0$ est évidemment située sur la droite polaire du point, relative à toutes les droites; ce qui prouve la proposition dont il s'agit.

Par exemple, en considérant les droites BC, CA, AB qui passent par trois points A, B, C: la polaire d'un point O, relative aux droites AB, AC, est une droite $A\alpha$ telle que AB, AC; AO, $A\alpha$ forment un faisceau harmonique. Soit α le point d'intersection de $A\alpha$ et BC, et supposons le même pour les points β, γ: les points α, β, γ seront situés (comme on le sait) sur une même droite, qui est celle que je nomme *polaire* de O, relative aux côtés du triangle, et que M. Plücker a nommé "*harmonicale*." On sait de plus que les droites $A\alpha$, $B\beta$, $C\gamma$ peuvent être construites comme suit: en prenant a, b, c pour les points d'intersection de OA, OB, OC avec BC, CA, AB respectivement, les droites bc, BC; ca, CA; ab, AB se rencontreront dans les points α, β, γ; ce qui offre une règle facile pour construire la polaire d'un point, relative à un nombre quelconque de droites.

Remarquons en finissant que la conique qui passe par les points A, B, C, et qui touche deux des trois droites $A\alpha$, $B\beta$, $C\gamma$, touche aussi la troisième et est effectivement la polaire conique du point, relative aux trois côtés du triangle. En combinant cela avec la propriété connue que la $r^{\text{ième}}$ polaire de O', relative à la même courbe, passe par O, si la $(n-r)^{\text{ième}}$ polaire d'un point O, relative à une courbe du $n^{\text{ième}}$ ordre, passe par un point O', on obtiendra le théorème 22 de M. Steiner (ce Journal, t. IV. [1828] p. 209).

56.

DEMONSTRATION OF A GEOMETRICAL THEOREM OF JACOBI'S.

[From the *Cambridge and Dublin Mathematical Journal*, vol. III. (1848), pp. 48—49.]

THE theorem in question (*Crelle*, t. XII. [1834] p. 137) may be thus stated:

"If a cone be circumscribed about a surface of the second order, the focal lines of the cone are generating lines of a surface of the second order confocal to the given surface and which passes through the vertex of the cone."

Let (α, β, γ) be the coordinates of the given point,

$$\frac{x^2}{a^2}+\frac{y^2}{b^2}+\frac{z^2}{c^2}=1$$

the equation to the given surface. The equation of the circumscribed cone referred to its vertex is

$$\left(\frac{x^2}{a^2}+\frac{y^2}{b^2}+\frac{z^2}{c^2}\right)\left(\frac{\alpha^2}{a^2}+\frac{\beta^2}{b^2}+\frac{\gamma^2}{c^2}-1\right)-\left(\frac{\alpha x}{a^2}+\frac{\beta y}{b^2}+\frac{\gamma z}{c^2}\right)^2=0,$$

whence it is easily seen that the equation of the supplementary cone (i.e. the cone generated by lines through the vertex at right angles to the tangent planes of the cone in question) is

$$(\alpha x+\beta y+\gamma z)^2-a^2x^2-b^2y^2-c^2z^2=0. \text{ [1]}$$

Suppose we have identically

$$(\alpha x+\beta y+\gamma z)^2-a^2x^2-b^2y^2-c^2z^2-h\,(x^2+y^2+z^2)=(lx+my+nz)\,(l'x+m'y+n'z);$$

$lx+my+nz=0$ will determine the direction of one of the cyclic planes of the supplementary cone, and hence taking the centre for the origin the equations of the focal lines of the circumscribed cone are

$$\frac{x-\alpha}{l}=\frac{y-\beta}{m}=\frac{z-\gamma}{n}.$$

[1] $Ax^2+By^2+Cz^2+2Fyz+2Gxz+2Hxy=0$ being the equation of the first cone, that of the supplementary cone is $\mathfrak{A}x^2+\mathfrak{B}y^2+\mathfrak{C}z^2+2\mathfrak{F}yz+2\mathfrak{G}zx+2\mathfrak{H}xy=0$, these letters [denoting the inverse coefficients $BC-F^2$, &c.].

It only remains therefore to determine the values of l, m, n from the last equation but one. The condition which expresses that the first side of this equation divides itself into factors is easily reduced to

$$\frac{\alpha^2}{a^2+h}+\frac{\beta^2}{b^2+h}+\frac{\gamma^2}{c^2+h}=1 \quad\text{.................................(A).}$$

Next, since the equation is identical, write

$$x=\frac{\alpha}{a^2+h},\quad y=\frac{\beta}{b^2+h},\quad z=\frac{\gamma}{c^2+h},$$

we deduce

$$0=\frac{l\alpha}{a^2+h}+\frac{m\beta}{b^2+h}+\frac{n\gamma}{c^2+h} \quad\text{.............................. (B).}$$

Again, putting

$$x=\frac{l}{a^2+h},\quad y=\frac{m}{b^2+h},\quad z=\frac{n}{c^2+h},$$

the whole equation divides by

$$\left(\frac{ll'}{a^2+h}+\frac{mm'}{b^2+h}+\frac{nn'}{c^2+h}+1\right),$$

a factor whose value is easily seen to be $=-1$. And rejecting this, we have

$$\frac{l^2}{a^2+h}+\frac{m^2}{b^2+h}+\frac{n^2}{c^2+h}=0 \quad\text{.............................. (C).}$$

Thus of the three equations (A), (B), and (C), the first determines h, and the remaining two give the ratios $l : m : n$. It is obvious that

$$\frac{x^2}{a^2+h}+\frac{y^2}{b^2+h}+\frac{z^2}{c^2+h}=1$$

is the equation of the surface confocal with the given surface which passes through the point (α, β, γ). The generating lines at this point are found by combining this equation with that of the tangent plane at the same point, viz.

$$\frac{\alpha x}{a^2+h}+\frac{\beta y}{b^2+h}+\frac{\gamma z}{c^2+h}=1;$$

and since these two equations are satisfied by $x=\alpha+lr$, $y=\beta+mr$, $z=\gamma+nr$, if l, m, n are determined by the equations above, it follows that the focal lines of the cone are the generating lines of the surface, the theorem which was to be demonstrated. It is needless to remark that of the three confocal surfaces, the hyperboloid of one sheet has alone real generating lines; this is as it should be, since a cone has six focal lines, of which four are always imaginary.

57.

ON THE THEORY OF ELLIPTIC FUNCTIONS.

[From the *Cambridge and Dublin Mathematical Journal*, vol. III. (1848), pp. 50—51.]

WE have seen [45] that the equation

$$n(n-1)x^2z+(n-1)(\alpha x-2x^3)\frac{dz}{dx}+(1-\alpha x^2+x^4)\frac{d^2z}{dx^2}-2n(\alpha^2-4)\frac{dz}{d\alpha}=0$$

is integrable, in the case of n an odd number, in the form $z=B_0+B_1x^2\ldots+B_{\frac{1}{2}(n-1)}x^{n-1}$; and the coefficients at the beginning of the series have already been determined; to find those at the end of it, the most convenient mode of writing the series will be

$$z=\mu\Sigma\frac{(-)^r D_r}{1.2\ldots(2r+1)}x^{n-1-2r},$$

and then the coefficients D_r are determined by

$$D_{r+2}=(2r+3)(n-2r-3)D_{r+1}\alpha-2n(\alpha^2-4)\frac{dD_{r+1}}{d\alpha}$$
$$-(2r+3)(2r+2)(n-2r-2)(n-2r-1)D_r.$$

The first coefficients then are

$$D_0=1,$$

$$D_1=(n-1)\alpha,$$

$$D_2=2(n-1)(n+6)+(n-1)(n-9)\alpha^2,$$

$$D_3=6(n-1)(n-9)(n+10)\alpha+(n-1)(n-9)(n-25)\alpha^3,$$

$$\begin{aligned}D_4=&-36(n-1)(n^3-13n^2+36n+420)\\&+12(n-1)(n-9)(n-25)(n+14)\alpha^2\\&+(n-1)(n-9)(n-25)(n-49)\alpha^4,\end{aligned}$$

$$D_5 = -12(n-1)(n-9)(47n^3 - 355n^2 + 3188n + 31500)\alpha$$
$$+ 20(n-1)(n-9)(n-25)(n-49)(n+18)\alpha^3$$
$$+ (n-1)(n-9)(n-25)(n-49)(n-81)\alpha^5,$$

$$D_6 = -24(n-1)(n-9)(\ 23n^4 + 2375n^3 - 14638n^2 + 116100n + \ 693000)$$
$$-12(n-1)(n-9)(493n^4 - 8882n^3 + 70317n^2 - 361641n - 7276500)\alpha^2$$
$$+ 30(n-1)(n-9)(n-25)(n-49)(n-81)(n+22)\alpha^4$$
$$+ (n-1)(n-9)(n-25)(n-36)(n-49)(n-81)(n-121)\alpha^6,$$
&c.

And, in general,

$$D_r = \qquad (n-1)(n-9)\ldots\{n-(2r-1)^2\}\alpha^r$$
$$+ r(r-1)(n-1)(n-9)\ldots\{n-(2r-3)^2\}(n+4r-2)\alpha^{r-2},$$
&c.

(where however the next term does not contain the factor $(n-1)(n-9)\ldots\{n-(2r-5)^2\}$).

In the case when $n=\nu^2$, then in order that the constant term may reduce itself to unity, we must assume

$$\mu = (-)^{\frac{1}{2}(\nu-1)}\,\nu\,;$$

this is evident from what has preceded.

58.

NOTES ON THE ABELIAN INTEGRALS.—JACOBI'S SYSTEM OF DIFFERENTIAL EQUATIONS.

[From the *Cambridge and Dublin Mathematical Journal*, vol. III. (1848), pp. 51—54.]

THE theory of elliptic functions depends, it is well known, on the differential equation $\frac{dx}{\sqrt{(fx)}}+\frac{dy}{\sqrt{(fy)}}=0$, ($fx$ denoting a rational and integral function of the fourth order), the integral of which was discovered by Euler, though first regularly derived from the differential equation by Lagrange. The theory of the Abelian integrals depends in like manner, as is proved by Jacobi, in the memoir "Considerationes generales de transcendentibus Abelianis" (*Crelle*, t. IX. [1832] p. 394) to depend, upon the system of equations

$$\Sigma\frac{dx}{\sqrt{(fx)}}=0,\quad \Sigma\frac{xdx}{\sqrt{(fx)}}=0,\ \ldots\ldots\ \Sigma\frac{x^{n-2}\,dx}{\sqrt{(fx)}}=0 \ldots\ldots\ldots\ldots\ldots\ldots(1),$$

where fx is a rational and integral function of the order $2n-1$ or $2n$, and the sums Σ contain n terms.

The integration of this system of equations is of course virtually comprehended in Abel's theorem; the problem was to obtain $(n-1)$ integrals each of them containing a single independent arbitrary constant. One such integral was first obtained by Richelot (*Crelle*, t. XXIII. [1842] p. 354), "Ueber die Integration eines merkwürdigen Systems Differentialgleichungen," by a method founded on that of Lagrange for the solution of Euler's equation; and a second integral very ingeniously deduced from it. A complete system of integrals in the required form is afterwards obtained, not by direct integration, but by means of Abel's theorem: there is this objection to them, however, that any one of them contains two roots of the equation $fx=0$. The next paper on the subject is one by Jacobi, "Demonstratio Nova theorematis Abeliani" (*Crelle*, t. XXIV. [1842] p. 28), in which a complete system of equations is deduced by direct integration, each of which contains only a single root of the equation $fx = 0$. But in Richelot's second memoir "Einige neue Integralgleichungen des Jacobischen Systems Differentialgleichungen" (*Crelle*, t. XXV. [1843] p. 97), the equations are obtained by direct integration in a form not involving any of the roots of this equation; the method employed in obtaining them being in a great measure founded upon the memoir just quoted of Jacobi's. The following is the process of integration.

Denoting the variables by $x_1, x_2 \ldots x_n$, and writing

$$F\alpha = (\alpha - x_1)(\alpha - x_2) \ldots (\alpha - x_n),$$

so that

$$F'x_1 = (x_1 - x_2) \ldots (x_1 - x_n),$$
$$\&c.$$

then the system of differential equations is satisfied by assuming that $x_1, x_2 \ldots x_n$ are functions of a new variable t, determined by the equations

$$\frac{dx_1}{dt} = \frac{\sqrt{(fx_1)}}{F'x_1}, \quad \&c.$$

(In fact these equations give $\Sigma \dfrac{dx}{\sqrt{(fx)}} = dt\, \Sigma \dfrac{1}{F'x} = 0$, &c.)

From these we deduce, by differentiation,

$$\frac{d^2x_1}{dt^2} = \frac{1}{2} \frac{d}{dx_1} \frac{fx_1}{(F'x_1)^2} + \frac{\sqrt{(fx_1)}}{F'x_1} \Sigma' \frac{\sqrt{(fx)}}{(x_1 - x) F'x}$$

(where Σ' refers to all the roots except x_1) and a set of analogous equations for $x_2, x_3 \ldots x_n$.

Dividing this by $\alpha - x_1$, where α is arbitrary, and reducing by

$$\frac{1}{(\alpha - x_1)(x_1 - x)} = \frac{1}{2(\alpha - x)(\alpha - x_1)} \left(1 - \frac{x + x_1 - 2\alpha}{x_1 - x}\right),$$

we have

$$\frac{1}{\alpha - x_1} \frac{d^2x_1}{dt^2} = \frac{1}{2(\alpha - x_1)} \frac{d}{dx_1} \frac{fx_1}{(F'x_1)^2}$$
$$+ \frac{1}{2} \frac{\sqrt{(fx_1)}}{(\alpha - x_1) F'x_1} \Sigma' \frac{\sqrt{(fx)}}{(\alpha - x) F'x} - \frac{1}{2}\Sigma' \frac{\sqrt{(fx)}\sqrt{(fx_1)}}{F'x\, F'x_1} \frac{(x_1 + x - 2\alpha)}{(\alpha - x)(\alpha - x_1)(x_1 - x)},$$

that is

$$\frac{1}{\alpha - x_1} \frac{d^2x_1}{dt^2} = \frac{1}{2(\alpha - x_1)} \frac{d}{dx_1} \frac{fx_1}{(F'x_1)^2}$$
$$+ \frac{1}{2} \frac{\sqrt{(fx_1)}}{(\alpha - x_1) F'x_1} \Sigma \frac{\sqrt{(fx)}}{(\alpha - x) F'x} - \frac{1}{2} \frac{fx_1}{(\alpha - x_1)^2 (F'x_1)^2}$$
$$- \frac{1}{2}\Sigma' \frac{\sqrt{(fx)}\sqrt{(fx_1)}}{F'x\, F'x_1} \frac{(x_1 + x - 2\alpha)}{(\alpha - x)(\alpha - x_1)(x_1 - x)};$$

and taking the sum of all the equations of this form, the last term disappears on account of the factor $x_1 - x$ in the denominator, and the result is

$$\Sigma \frac{1}{\alpha - x} \frac{d^2x}{dt^2} = \tfrac{1}{2}\Sigma \frac{1}{\alpha - x} \frac{d}{dx} \frac{fx}{(F'x)^2} + \tfrac{1}{2}\left\{\Sigma \frac{\sqrt{(fx)}}{(\alpha - x) F'x}\right\}^2 - \tfrac{1}{2}\Sigma \frac{fx}{(\alpha - x)^2 (F'x)^2}.$$

This being premised, assume

$$y = \sqrt{(F\alpha)},$$

which, by differentiation, gives

$$\frac{dy}{dt} = -\tfrac{1}{2}y\Sigma \frac{1}{\alpha - x} \frac{dx}{dt} = -\tfrac{1}{2}y\Sigma \frac{\sqrt{(fx)}}{(\alpha - x) F'x},$$

and thence

$$\frac{d^2y}{dt^2} = -\tfrac{1}{2} \frac{dy}{dt} \Sigma \frac{1}{\alpha - x} \frac{dx}{dt} + \tfrac{1}{2}y\Sigma \frac{1}{(\alpha - x)^2} \left(\frac{dx}{dt}\right)^2 - \tfrac{1}{2}y\Sigma \frac{1}{(\alpha - x)} \frac{d^2x}{dt^2},$$

that is
$$\frac{d^2y}{dt^2} = \tfrac{1}{4}y\left(\Sigma \frac{\sqrt{(fx)}}{(\alpha - x) F'x}\right)^2 - \tfrac{1}{2}y\Sigma \frac{fx}{(\alpha - x)^2 (F'x)^2} - \tfrac{1}{2}y\Sigma \frac{1}{\alpha - x} \frac{d^2x}{dt^2}.$$

Substituting the preceding value of

$$\Sigma \frac{1}{\alpha - x} \frac{d^2x}{dt^2},$$

we have
$$\frac{d^2y}{dt^2} = -\tfrac{1}{4}y\Sigma \frac{fx}{(\alpha - x)^2 (F'x)^2} - \tfrac{1}{4}y\Sigma \frac{1}{\alpha - x} \frac{d}{dx} \frac{fx}{(F'x)^2};$$

that is
$$4\frac{d^2y}{dt^2} + y\left\{\Sigma \frac{fx}{(\alpha - x)^2 (F'x)^2} + \Sigma \frac{1}{(\alpha - x)} \frac{d}{dx} \frac{fx}{(F'x)^2}\right\} = 0.$$

Now the fractional part of $\dfrac{f\alpha}{(F\alpha)^2}$ is equal to

$$\Sigma \frac{fx}{(\alpha - x)^2 (F'x)^2} + \Sigma \frac{1}{\alpha - x} \frac{d}{dx} \frac{fx}{(F'x)^2}.$$

Also if L be the coefficient of x^{2n} in $f\alpha$, the integral part is simply equal to L, (since $(F\alpha)^2$ is a function of the order $2n$, in which the coefficient of α^{2n} is unity). Hence the coefficient of y in the last equation is simply

$$\frac{f\alpha}{(F\alpha)^2} - L, \quad = \frac{f\alpha}{y^4} - L;$$

or we have

$$4\frac{d^2y}{dt^2} + y\left(\frac{f\alpha}{y^4} - L\right) = 0,$$

viz. multiplying by the factor $2\frac{dy}{dt}$, and integrating,

$$4\left(\frac{dy}{dt}\right)^2 - \frac{f\alpha}{y^2} - Ly^2 = C.$$

Hence replacing y and $\frac{dy}{dt}$ by their values

$$\sqrt{(F\alpha)} \text{ and } -\tfrac{1}{2}\sqrt{(F\alpha)}\,\Sigma\frac{\sqrt{(fx)}}{(\alpha - x)\,F'x},$$

we have

$$F\alpha\left\{\Sigma\frac{\sqrt{(fx)}}{(\alpha - x)\,F'x}\right\}^2 - \frac{fa}{F\alpha} - LF\alpha = C,$$

for one of the integrals of the proposed system of equations: and since α is arbitrary, the complete system is obtained by giving any $(n-1)$ particular values to α, and changing the value of the constant of integration C; or by expanding the first side of the equation in terms of α, and equating the different coefficients to arbitrary constants. The *à posteriori* demonstration that all the results so obtained are equivalent to $(n-1)$ independent equations would probably be of considerable interest.

59.

ON THE THEORY OF ELIMINATION.

[From the *Cambridge and Dublin Mathematical Journal*, vol. III. (1848), pp. 116—120.]

SUPPOSE the variables X_1, $X_2 \ldots$, g in number, are connected by the h linear equations

$$\begin{aligned}\Theta_1 &= \alpha_1 X_1 + \alpha_2 X_2 \ldots = 0,\\ \Theta_2 &= \beta_1 X_1 + \beta_2 X_2 \ldots = 0,\\ &\vdots\end{aligned}$$

these equations not being all independent, but connected by the k linear equations

$$\begin{aligned}\Phi_1 &= \alpha_1' \Theta_1 + \alpha_2' \Theta_2 + \ldots = 0,\\ \Phi_2 &= \beta_1' \Theta_1 + \beta_2' \Theta_2 + \ldots = 0,\\ &\vdots\end{aligned}$$

these last equations not being independent, but connected by the l linear equations

$$\begin{aligned}\Psi_1 &= \alpha_1'' \Phi_1 + \alpha_2'' \Phi_2 + \ldots = 0,\\ \Psi_2 &= \beta_1'' \Phi_1 + \beta_2'' \Phi_2 + \ldots = 0,\\ &\vdots\end{aligned}$$

and so on for any number of systems of equations.

Suppose also that $g - h + k - l + \ldots = 0$; in which case the number of quantities X_1, $X_2, \ldots$ will be equal to the number of really independent equations connecting them, and we may obtain by the elimination of these quantities a result $\nabla = 0$.

To explain the formation of this final result, write

$$\nabla = \begin{array}{|cc|cc|}
\hline
\alpha_1, & \beta_1 \ldots & & \\
\alpha_2, & \beta_2 & & \\
\vdots & & & \\
\hline
\alpha_1', & \alpha_2' \ldots & \alpha_1'', & \beta_1'' \ldots \\
\beta_1', & \beta_2' & \alpha_2'', & \beta_2'' \\
\vdots & & \vdots & \\
\hline
 & & \alpha_1''', & \alpha_2''' \ldots \\
 & & \beta_1''', & \beta_2''' \\
 & & \vdots & \\
\hline
\end{array}$$

which for shortness may be thus represented,

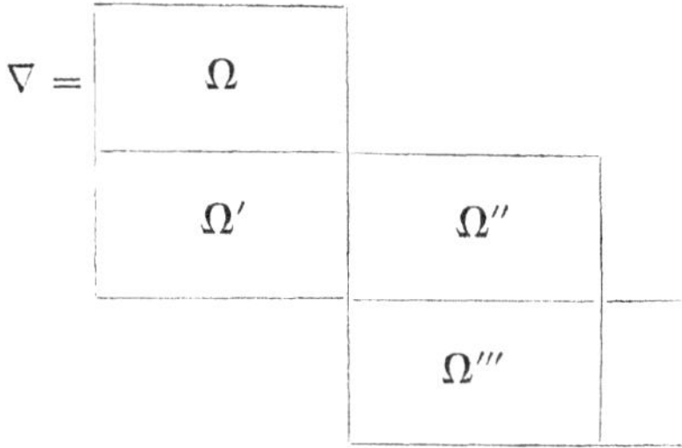

where Ω, Ω', Ω'', Ω''', Ω'''', ... contain respectively h, h, l, l, n, n, ... vertical rows, and g, k, k, m, m, p, ... horizontal rows.

It is obvious, from the form in which these systems have been arranged, what is meant by speaking of a certain number of the vertical rows of Ω' and the *supplementary* vertical rows of Ω; or of a certain number of the horizontal rows of Ω'' and the *supplementary* horizontal rows of Ω', &c.

Suppose that there is only one set of equations, or $g=h$: we have here only a single system Ω, which contains h vertical and h horizontal rows, and ∇ is simply the determinant formed with the system of quantities Ω. We may write in this case $\nabla = Q$.

Suppose that there are two sets of equations, or $g=h-k$: we have here two systems Ω, Ω', of which Ω contains h vertical and $h-k$ horizontal rows, Ω' contains h vertical and k horizontal rows. From any k of the h vertical rows of Ω' form a determinant, and call this Q'; from the supplementary $h-k$ vertical rows of Ω form a determinant, and call this Q: then Q' divides Q, and we have $\nabla = Q \div Q'$.

Suppose that there are three sets of equations, or $g = h - k + l$: we have here three systems, Ω, Ω', Ω'', of which Ω contains h vertical and $h - k + l$ horizontal rows, Ω' contains h vertical and k horizontal rows, Ω'' contains l vertical and k horizontal rows. From any l of the k horizontal rows of Ω'' form a determinant, and call this Q''; from the $k - l$ supplementary horizontal rows of Ω', choosing the vertical rows at pleasure, form a determinant, and call this Q'; from the $h - k + l$ supplementary vertical rows of Ω form a determinant, and call this Q: then Q'' divides Q', this quotient divides Q, and we have $\nabla = Q \div (Q' \div Q'')$.

Suppose that there are four sets of equations, or $g = h - k + l - m$: we have here four systems, Ω, Ω', Ω'', and Ω''', of which Ω contains h vertical and $h - k + l - m$ horizontal rows, Ω' contains h vertical and k horizontal rows, Ω'' contains l vertical and k horizontal rows, and Ω''' contains l vertical and m horizontal rows. From any m of the l vertical rows of Ω''' form a determinant, and call this Q'''; from the $l - m$ supplementary vertical rows of Ω'', choosing the horizontal rows at pleasure, form a determinant, and call this Q''; from the $k - l + m$ supplementary horizontal rows of Ω', choosing the vertical rows at pleasure, form a determinant, and call this Q'; from the $h - k + l - m$ supplementary vertical rows of Ω form a determinant, and call this Q: then Q''' divides Q'', this quotient divides Q', this quotient divides Q, and $\nabla = Q \div \{Q' \div (Q'' \div Q''')\}$. The mode of proceeding is obvious.

It is clear, that if all the coefficients α, β, ... be considered of the order unity, ∇ is of the order $h - 2k + 3l -$ &c.

What has preceded constitutes the theory of elimination alluded to in my memoir "On the Theory of Involution in Geometry," *Journal*, vol. II. p. 52—61, [40]. And thus the problem of eliminating any number of variables x, y ... from the same number of equations $U = 0$, $V = 0$, ... (where U, V, ... are homogeneous functions of any orders whatever) is completely solved; though, as before remarked, I am not in possession of any method of arriving *at once* at the final result in its most simplified form; my process, on the contrary, leads me to a result encumbered by an extraneous factor, which is only got rid of by a number of successive divisions less by two than the number of variables to be eliminated.

To illustrate the preceding method, consider the three equations of the second order,

$$U = a\ x^2 + b\,y^2 + c\ z^2 + l\ yz + m\ zx + n\ xy = 0,$$
$$V = a'\,x^2 + b'y^2 + c'\,z^2 + l'\,yz + m'\,zx + n'\,xy = 0,$$
$$W = a''x^2 + b''y^2 + c''z^2 + l''yz + m''zx + n''xy = 0.$$

Here, to eliminate the fifteen quantities x^4, y^4, z^4, y^3z, z^3x, x^3y, yz^3, zx^3, xy^3, y^2z^2, z^2x^2, x^2y^2, x^2yz, y^2zx, z^2xy, we have the eighteen equations

$$x^2U = 0,\quad y^2U = 0,\quad z^2U = 0,\quad yzU = 0,\quad zxU = 0,\quad xyU = 0,$$
$$x^2V = 0,\quad y^2V = 0,\quad z^2V = 0,\quad yzV = 0,\quad zxV = 0,\quad xyV = 0,$$
$$x^2W = 0,\quad y^2W = 0,\quad z^2W = 0,\quad yzW = 0,\quad zxW = 0,\quad xyW = 0,$$

equations, however, which are not independent, but are connected by

$$a''x^2V+b''y^2V+c''z^2W+l''yzV+m''zxV+n''xyV$$
$$-(a'x^2W+b'y^2W+c'z^2W+l'yzW+m'zxW+n'xyW)=0,$$

$$ax^2W+by^2W+cz^2W+lyzW+mzxW+nxyW$$
$$-(a''x^2U+b''y^2U+c''z^2U+l''yzU+m''zxU+n''xyU)=0,$$

$$a'x^2U+b'y^2U+c'z^2U+l'yzU+m'zxU+n'xyU$$
$$-(ax^2V+by^2V+cz^2V+lyzV+mzxV+nxyV)=0.$$

Arranging these coefficients in the required form, we have the following value of ∇.

$$\left|\begin{array}{cccccccccccccccccc}
a & & & & & & a' & & & & & & a'' & & & & & \\
 & b & & & & & & b' & & & & & & b'' & & & & \\
 & & c & & & & & & c' & & & & & & c'' & & & \\
 & l & & b & & & & l' & & b' & & & & l'' & & b'' & & \\
 & & m & & c & & & & m' & & c' & & & & m'' & & c'' & \\
n & & & & & a & n' & & & & & a' & n'' & & & & & a'' \\
 & & l & c & & & & & l' & c' & & & & & l'' & c'' & & \\
m & & & & a & & m' & & & & a' & & m'' & & & & a'' & \\
 & n & & & & b & & n' & & & & b' & & n'' & & & & b'' \\
 & c & b & l & & & & c' & b' & l' & & & & c'' & b'' & l'' & & \\
c & & a & & m & & c' & & a' & & m' & & c'' & & a'' & & m'' & \\
b & a & & & & n & b' & a' & & & & n' & b'' & a'' & & & & n'' \\
l & & & a & n & m & l' & & & a' & n' & m' & l'' & & & a'' & n'' & m'' \\
 & m & & n & b & l & & m' & & n' & b' & l' & & m'' & & n'' & b'' & l'' \\
 & & n & m & l & c & & & n' & m' & l' & c' & & & n'' & m'' & l'' & c'' \\
\hline
 & & & & & & a'' & b'' & c'' & l'' & m'' & n'' & -a' & -b' & -c' & -l' & -m' & -n' \\
-a'' & -b'' & -c'' & -l'' & -m'' & -n'' & & & & & & & a & b & c & l & m & n \\
a' & b' & c' & l' & m' & n' & -a & -b & -c & -l & -m & -n & & & & & &
\end{array}\right|$$

which may be represented as before by

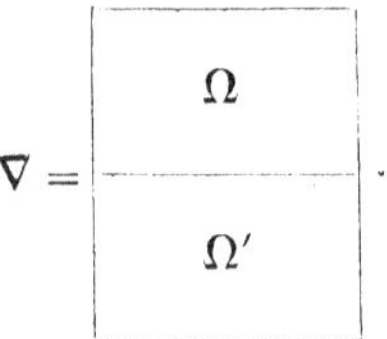

Thus, for instance, selecting the first, second, and sixth lines of Ω' to form the determinant Q', we have $Q' = a''(a'b'' - a''b')$; and then Q must be formed from the third, fourth, fifth, seventh, &c.... eighteenth lines of Ω. (It is obvious that if Q' had been formed from the first, second, and third lines of Ω', we should have had $Q' = 0$; the corresponding value of Q would also have vanished, and an illusory result be obtained; and similarly for several other combinations of lines.)

60.

ON THE EXPANSION OF INTEGRAL FUNCTIONS IN A SERIES OF LAPLACE'S COEFFICIENTS.

[From the *Cambridge and Dublin Mathematical Journal*, vol. III. (1848), pp. 120, 121.]

SUPPOSE

$$S = A\mu^s + A_1\mu^{s-1} + \dots,$$
$$= \alpha Q_s + \alpha_1 Q_{s-1} + \dots \quad \dots\dots(1),$$

where, as usual, Q_0, Q_1, &c. are the coefficients of the successive powers of p in $(1 - 2\mu p + p^2)^{-\frac{1}{2}}$. Assume

$$S = \frac{\sqrt{(1 + \Delta^2)}}{\sqrt{(1 - 2\mu\Delta + \Delta^2)}}\, C \quad \dots\dots(2),$$

where Δ refers to C; then expanding this expression, first in powers of μ, and then in a series of terms of the form $\sqrt{(1 + \Delta^2)}\,.\,Q$, and comparing these with the preceding values of S,

$$A_{s-q} = \frac{1\,.\,3 \dots 2q - 1}{2\,.\,4 \dots \quad 2q}\left(\frac{2\Delta}{1 + \Delta^2}\right)^q.\, C,$$

$$\alpha_{s-r} = \sqrt{(1 + \Delta^2)}\,.\,\Delta^r.\, C \quad \dots\dots(3).$$

Assume $\dfrac{2\Delta}{1 + \Delta^2} = \delta$, that is,

$$\Delta = \frac{1}{\delta}\{1 - \sqrt{(1 - \delta^2)}\}, \quad \sqrt{(1 + \Delta)^2} = \frac{\sqrt{2}}{\delta}\{1 - \sqrt{(1 - \delta^2)}\}^{\frac{1}{2}} \quad \dots\dots(4).$$

Then

$$\delta^q\, C = \frac{2\,.\,4 \dots \quad 2q}{1\,.\,3 \dots 2q - 1} A_{s-q},$$

$$\alpha_{s-r} = \frac{\sqrt{2}}{\delta^{r+1}}\{1 - \sqrt{(1 - \delta^2)}\}^{r+\frac{1}{2}}.\, C \quad \dots\dots(5);$$

or, expanding this last equation in powers of δ,

$$a_{s-r} = \frac{(2r+1)}{2^r}\,\delta^r\left(\frac{1}{2r+1} + \frac{1}{2}\frac{\delta^2}{2^2} + \frac{(2r+7)}{2\,.\,4}\frac{\delta^4}{2^4} + \frac{(2r+9)(2r+11)}{2\,.\,4\,.\,6}\frac{\delta^4}{2^6} + \ldots\right)C \ldots (6);$$

or, replacing the successive terms of the form $\delta^q\,.\,C$ by their respective values,

$$a_{s-r} = (2r+1)\left\{\frac{2\,.\,4 \ldots \quad 2r}{3\,.\,5 \ldots (2r+1)\,2^r}A_{s-r} + \frac{4\,.\,6 \ldots (2r+4)}{3\,.\,5 \ldots (2r+3)\,2^{r+2}}A_{s-r-2} \ldots\right.$$

$$\left. + \frac{(2k+2)(2k+4) \ldots \quad (2r+4k)}{3 \quad . \quad 5 \quad \ldots (2r+2k+1)\,2^{r+2k}}A_{s-r-2k} \ldots\right\} \ldots\ldots\ldots\ldots (7).$$

Thus, if $S = \mu^s$, so that $A_{s-r} = 0$, except in the particular case $A = 1$,

$$a_{2k-1} = 0,$$

$$a_{2k} = \frac{2s-4k+1}{2^s}\,\frac{(2k+2)(2k+4) \ldots \quad 2s}{3 \quad . \quad 5 \quad \ldots (2s-2k+1)} \ldots\ldots\ldots\ldots\ldots\ldots\ldots\ldots (8),$$

or

$$\mu^s = \frac{1}{2^s}\Sigma\left\{(2s-4k+1)\frac{(2k+2)(2k+4) \ldots \quad 2s}{3 \quad . \quad 5 \quad \ldots (2s-2k+1)}Q_{s-2k}\right\} \ldots\ldots\ldots\ldots (9),$$

which of course includes the preceding case. By substituting the expanded values of the coefficients Q, or again, by determining the value of $(1-\mu)^s$ in terms of these coefficients, and equating it with that given in Murphy's *Electricity*, [8°. Cambridge, 1833], p. 10, or in a variety of other ways, a series of identical equations involving sums of factorials may readily be obtained. The mode of employing the general theory of the separation of symbols made use of in the preceding example, may easily be applied to the solution of analogous questions.

61.

ON GEOMETRICAL RECIPROCITY.

[From the *Cambridge and Dublin Mathematical Journal*, vol. III. (1848), pp. 173—179.]

THE fundamental theorem of reciprocity in plane geometry may be thus stated.

"The points and lines of a plane P may be considered as corresponding to the lines and points of a plane P' in such a manner that to a set of points in a line in the first figure, there corresponds a set of lines through a point in the second figure, (namely through the point corresponding to the line); and to a set of lines through a point in the first figure, there corresponds a set of points in a line in the second figure, (namely in the line corresponding to the point)."

And from this theorem, without its being in any respect necessary further to particularize the nature of the correspondence, or to consider in any manner the relative position of the two planes, an endless variety of propositions and theories may be deduced, as, for instance, the duality of all theorems which relate to the purely descriptive properties of figures, the theory of the singular points and tangents of curves, &c.

Suppose, however, that the two planes coincide, so that a point may be considered indifferently as belonging to the first or to the second figure: an entirely independent series of propositions (which, properly speaking, form no part of the general theory of reciprocity) result from this particularization. In general, the line in the second figure which corresponds to a point considered as belonging to the first figure, and the line in the first figure which corresponds to the same point considered as belonging to the second figure, will not be identical; neither will the point in the second figure which corresponds with a line considered as belonging to the first figure, and the point in the first figure which corresponds to the same line considered as belonging to the second figure, be identical.

In the particular case where these lines and points are respectively identical (the identity of the lines implies that of the points and *vice versâ*) we have the theory of "reciprocal polars." Here, where it is unnecessary to define whether the points or lines belong to the first or second figures, the line corresponding to a point and the point corresponding to a line are spoken of as the polar of the point and the pole of the line, or as reciprocal polars.

"The points which lie in their respective polars are situated in a conic, to which the polars are tangents." Or, stating the *same* theorem conversely,

"The lines which pass through their respective poles are tangents to a conic, the points of contact being the poles."

To determine the polar of a point, let two tangents be drawn through this point to the conic, the points of contact are the poles of the tangents; hence the line joining them is the polar of the point of intersection of the tangents, that is, "The polar of a point is the line joining the points of contact of the tangents which pass through the point."

Conversely, and by the same reasoning,

"The pole of a line is the intersection of the tangents at the points where the line meets the conic."

The actual geometrical constructions in the several cases where the point is within or without the conic, or the line does or does not intersect the conic, do not enter into the plan of the present memoir.

Passing to the general case where the lines and points in question are not identical, which I should propose to term the theory of "Skew Polars" (Polaires Gauches), we have the theorem,

"Considering the points in the first figure which are situated in their respective corresponding lines in the second figure, or the points in the second figure which are situated in their respective corresponding lines in the first figure, in either case the points are situated in the same conic (which will be spoken of as the 'pole conic'), and the lines are tangents to the same conic (which will be spoken of as the 'polar conic'), and these two conics have a double contact." This theorem is evidently identical with the converse theorem.

The corresponding lines to a point in the pole conic are the tangents through this point to the polar conic; viz. one of these tangents is the corresponding line when the point is considered as belonging to the first figure, and the other tangent is the corresponding line when the point is considered as belonging to the second figure.

The corresponding points to a tangent of the polar conic are the points where this line intersects the pole conic; viz. one of these points is the corresponding point when the line is considered as belonging to the first figure, and the other is the corresponding point when the line is considered as belonging to the second figure.

Let i be a point in the pole conic, and when i is considered as belonging to the first figure, let iI_1 be considered as the corresponding line in the second figure (I_1 being the point of contact on the polar conic).

Then if j be another point in the pole conic, in order to determine which of the tangents is the line in the second figure which corresponds to j considered as a point of the first figure, let iI_2 be the other tangent through I: the points of contact of the tangents through j may be marked with the letters J_1, J_2, in such order that I_1J_2, I_2J_1 meet in the line of contact of the two conics, and then jJ_1 is the required corresponding line. Again, I and i, as before, if B be a tangent to the polar conic, then, marking the point of contact as J_1, let J_2 be so determined that I_1J_2, I_2J_1 meet in the line of contact of the conics: the tangent to the polar conic at J_2 will meet the pole conic in one of the points where it is met by the line B, and calling this point j, B considered as belonging to the second figure will have j for its corresponding point in the first figure. Similarly, if the point of contact had been marked J_2, J_1 would be determined by an analogous construction, and the tangent at J_1 would meet the pole conic in one of the points where it is met by the line B (viz. the other point of intersection); and representing this by j', B considered as belonging to the first figure would have j' for its corresponding point in the second figure, that is, considered as belonging to the second figure, it would have j for its corresponding point in the first figure (the same as before).

Similar considerations apply in the case where a tangent A of the polar conic, considered as belonging to one of the figures, has for its corresponding point in the other figure one of its points of intersection with the polar conic; in fact, if A represents the line iI_1, then A, considered as belonging to the second figure has i for its corresponding point in the first figure, which shows that this question is identical with the former one.

To appreciate these constructions it is necessary to bear in mind the following system of theorems, the third and fourth of which are the polar reciprocals of the first and second.

If there be two conics having a double contact, such that K is the line joining the points of contact, and k the point of intersection of the tangents at the points of contact:

1. If two tangents to one of the conics meet the other in i, i_1 and j, j_1 respectively, then, properly selecting the points j, j_1, the lines ij_1, i_1j meet in K. And

2. The line joining the points of intersection of the tangents at i, j_1, and of the tangents at i_1, j passes through k. Also

3. If from two points of one of the conics, tangents be drawn touching the other in the points I, I_1 and J, J_1, then, properly selecting the points J, J_1, the lines IJ_1, I_1J meet in K. And

4. The line joining the points of intersection of the tangents at I_1J_1 and of the tangents at I_1, J passes through k.

These theorems are in fact particular cases of two theorems relating to two conics having a double contact with a given conic.

It may be remarked also that the corresponding points to a tangent of the pole conic are the points of contact of the tangents to the polar conic which pass through the point of contact of the given tangent, and the corresponding lines to a point of the polar conic are the tangents to the pole conic at the points where it is intersected by the tangent at the point in question.

We have now to determine the corresponding lines to a given point and the corresponding points to a given line, which is immediately effected by means of the preceding results.

Thus, if the point be given,

"Through the point draw tangents to the polar conic, meeting the pole conic in A_1, A_2 and B_1, B_2 (so that A_1B_2 and A_2B_1, intersect on the line joining the points of contact of the conics), then A_2B_2 and A_1B_1 are the required lines."

In fact A_1, B_1 and A_2, B_2 are pairs of points corresponding to the two tangents, so that A_1B_1 and A_2B_2 are the lines which correspond to their point of intersection, that is, to the given point, and similarly for the remaining constructions. Again,

"Through the point draw tangents to the pole conic, and from the points of contact draw tangents to the polar conic, touching it in α_1, α_2 and β_1, β_2 (so that $\alpha_1\beta_2$ and $\alpha_2\beta_1$ intersect on the line joining the points of contact of the conic), then $\alpha_1\beta_1$ and $\alpha_2\beta_2$ are the required lines."

So that A_1, B_1, α_1, β_1 are situated in the same line, and also A_2, B_2, α_2, β_2.

Again, if the line be given,

"Through the points where the line meets the pole conic draw tangents to the polar conic C_1, C_2 and D_1, D_2 (so that the points C_1D_2 and C_2D_1 lie on a line passing through the intersection of the tangents at the points of contact of the tangents), then C_1D_1 and C_2D_2 are the required points."

Again,

"At the points where the line meets the polar conic draw tangents meeting the pole conic, and let γ_1, γ_2 and δ_1, δ_2 be the tangents to the pole conic at these points (so that the points $\gamma_1\delta_2$ and $\gamma_2\delta_1$ lie on a line through the intersection of the tangents at the points of contact of the conics), then γ_1, δ_1 and γ_2, δ_2 are the required points"; so that C_1, D_1, γ_1, δ_1 pass through the same point and also C_2, D_2, γ_2, δ_2.

"The preceding constructions have been almost entirely taken from Plücker's "System der Analytischen Geometrie," § 3, Allgemeine Betrachtungen über Coordinaten-bestimmung. I subjoin analytical demonstrations of some of the theorems in question.

Using x, y, z to determine the position of a variable point, and putting for shortness

$$\xi = ax + a'y + a''z,$$
$$\eta = bx + b'y + b''z,$$
$$\zeta = cx + c'y + c''z.$$

then if the position of a point be determined by the coordinates α, β, γ, the equation of one of the corresponding lines is

$$\alpha\xi + \beta\eta + \gamma\zeta = 0,$$

(that of the other is obtainable from this by writing a, b, c; a', b', c'; a'', b'', c'', for a, a', a''; b, b', b''; c, c', c''). Hence if the point lies in the corresponding line, this equation must be satisfied by putting α, β, γ for x, y, z; or, substituting x, y, z in the place of α, β, γ, the point must lie in the conic

$$U = ax^2 + b'y^2 + c''z^2 + (b'' + c')\, yz + (c + a'')\, zx + (a' + b)\, xy = 0,$$

(which equation is evidently not altered by interchanging the coefficients, as above). Again, determining the curve traced out by the line $\alpha\xi + \beta\eta + \gamma\zeta = 0$, when α, β, γ are connected by the equation into which $U = 0$ is transformed by the substitution of these letters for x, y, z; we obtain

$$V = -\begin{vmatrix} & \xi\ , & \eta\ , & \zeta \\ \xi, & 2a\ , & a' + b, & a'' + c \\ \eta, & a' + b, & 2b'\ , & b'' + c' \\ \zeta, & a'' + c, & b'' + c', & 2c'' \end{vmatrix} = 0,$$

which is also a conic. It only remains to be seen that the conics $U = 0$, $V = 0$ have a double contact. Writing for shortness

$$\nabla = \begin{vmatrix} a\ , & b\ , & c \\ a'\ , & b'\ , & c' \\ a'', & b'', & c'' \end{vmatrix}$$

it may be seen by expansion that the following equation is identically true,

$$V = 4\nabla U - [x\,(ab'' - a''b + a'c - ac') + y\,(b'c - bc' + b''a' - b'a'') + z\,(c''a' - c'a'' + cb'' - c''b)]^2,$$

which proves the property in question.

Suppose the equations of the two conics to be given, and let it be required to determine the corresponding lines to the point defined by the coordinates α, β, γ.

Writing, to abbreviate,

$$\left\{\begin{aligned} U &= Ax^2 + By^2 + Cz^2 + 2Fyz + 2Gzx + 2Hxy, \\ U_0 &= A\alpha^2 + B\beta^2 + C\gamma^2 + 2F\beta\gamma + 2G\gamma\alpha + 2H\alpha\beta, \\ W &= A\alpha x + B\beta y + C\gamma z + F(\beta z + \gamma y) + G(\gamma x + \alpha z) + H(\alpha y + \beta x), \\ P &= lx + my + nz, \\ P_0 &= l\alpha + m\beta + n\gamma, \\ K &= ABC - AF^2 - BG^2 - CH^2 + 2FGH, \end{aligned}\right.$$

$$\mathfrak{A} = BC - F^2, \quad \mathfrak{B} = CA - G^2, \quad \mathfrak{C} = AB - H^2,$$
$$\mathfrak{F} = GH - AF, \quad \mathfrak{G} = HF - BG, \quad \mathfrak{H} = FG - CH,$$

$$\Theta = \mathfrak{A}l^2 + \mathfrak{B}m^2 + \mathfrak{C}n^2 + 2\mathfrak{F}mn + 2\mathfrak{G}nl + 2\mathfrak{H}lm,$$

$$\begin{aligned}\square = &(\mathfrak{A}l + \mathfrak{H}m + \mathfrak{G}n)(\gamma y - \beta z)\\ &+ (\mathfrak{H}l + \mathfrak{B}m + \mathfrak{F}n)(\alpha z - \gamma x)\\ &+ (\mathfrak{G}l + \mathfrak{F}m + \mathfrak{C}n)(\beta x - \alpha y);\end{aligned}$$

suppose $U = 0$ represents the equation of the polar conic, $U - P^2 = 0$ that of the pole conic. The two tangents drawn to the polar conic are represented by $UU_0 - W^2 = 0$, and by determining k in such a way that $UU_0 - W^2 - k(U - P^2)$ may divide into factors the equation

$$UU_0 - W^2 - k(U - P^2) = 0,$$

represents the lines passing through the points of intersection of the tangents with the pole conic. Thus if $k = U_0$, the equation reduces itself to $U_0P^2 - W^2 = 0$, or $W = \pm\sqrt{(U_0)}\,P$, the equation of two straight lines each of which passes through the point of intersection of the lines $P = 0$, $W = 0$, (that is, of the line of contact of the conics, and the ordinary polar of the point with respect to the polar conic); these are in fact the lines A_1B_2, A_2B_1 intersecting in the line of contact. The remaining value of k is not easily determined, but by a somewhat tedious process I have found it to be

$$= K(U_0 - P_0^2) \div (\Theta - K).$$

In fact, substituting the value, it may be shown that

$$\square^2 + K(PP_0 - W)^2 = K(U_0 - P_0^2)(U - P^2) + (\Theta - K)(UU_0 - W^2),$$

which is an equation of the required form. To verify this, we have, by a simple reduction,

$$\Theta(UU_0 - W^2) - \square^2 = K(UP_0^2 - 2WPP_0 + U_0P^2);$$

or, writing for shortness

$$\gamma y - \beta z = \xi, \quad \alpha z - \gamma x = \eta, \quad \beta x - \alpha y = \zeta,$$

$$\begin{aligned}&(\mathfrak{A}l^2 + \mathfrak{B}m^2 + \mathfrak{C}n^2 + 2\mathfrak{F}mn + 2\mathfrak{G}nl + \mathfrak{H}lm)(\mathfrak{A}\xi^2 + \mathfrak{B}\eta^2 + \mathfrak{C}\zeta^2 + 2\mathfrak{F}\eta\zeta + 2\mathfrak{G}\zeta\xi + 2\mathfrak{H}\xi\eta)\\ &\qquad - [\mathfrak{A}l\xi + \mathfrak{B}m\eta + \mathfrak{C}n\zeta + \mathfrak{F}(n\eta + m\zeta) + \mathfrak{G}(l\zeta + n\xi) + \mathfrak{H}(m\xi + l\eta)]^2\\ &= K\{A(m\zeta - n\eta)^2 + B(n\xi - l\zeta)^2 + C(l\eta - m\xi)^2\\ &\qquad + 2F(n\xi - l\zeta)(l\eta - m\xi) + 2G(l\eta - m\xi)(m\zeta - n\eta) + 2H(m\zeta - n\eta)(n\xi - l\zeta)\},\end{aligned}$$

which is easily seen to be identically true.

62.

ON AN INTEGRAL TRANSFORMATION.

[From the *Cambridge and Dublin Mathematical Journal*, vol. III. (1848), pp. 286—287.]

THE following transformation, given for elliptic functions by Gudermann (*Crelle*, t. XXIII. [1842], p. 330) is useful for some other integrals.

If
$$y = \frac{dbc - dba - dca + abc - (bc - ad)\, z}{(bc - ad) + (d - b - c + a)\, z},$$

then, putting
$$K = (bc - ad) + (d - b - c + a)\, z,$$

we have, supposing $a < b < c < d$, so that $(b-a)$, $(c-a)$, $(d-b)$, $(d-c)$ are positive,

$$\begin{aligned} K(y-a) &= (b-a)(c-a)(d-z), \\ K(y-b) &= (b-a)(d-b)(c-z), \\ K(y-c) &= (c-a)(d-c)(b-z), \\ K(y-d) &= (d-b)(d-c)(a-z), \\ K^2\, dy &= -(b-a)(c-a)(d-b)(d-c)\, dz. \end{aligned}$$

In particular, if $\alpha + \beta + \gamma + \delta = -2$,

$$(y-a)^\alpha (y-b)^\beta (y-c)^\gamma (y-d)^\delta\, dy = -M (z-a)^\delta (z-b)^\gamma (z-c)^\beta (z-d)^\alpha\, dz,$$

where
$$M = (b-a)^{\alpha+\beta+1} (c-a)^{\alpha+\gamma+1} (d-b)^{\beta+\delta+1} (d-c)^{\gamma+\delta+1}.$$

Thus, if $\alpha = \beta = \gamma = \delta = -\frac{1}{2}$,

$$\frac{dy}{\{-(y-a)(y-b)(y-c)(y-d)\}^{\frac{1}{2}}} = \frac{-dz}{\{-(z-a)(z-b)(z-c)(z-d)\}^{\frac{1}{2}}}.$$

In any case when $y = a$, $y = b$, the corresponding values of z are $z = d$, $z = c$; the last formula becomes by this means

$$\int_a^b \frac{dy}{\{-(y-a)(y-b)(y-c)(y-d)\}^{\frac{1}{2}}} = \int_c^d \frac{dy}{\{-(z-a)(z-b)(z-c)(z-d)\}^{\frac{1}{2}}}.$$

63.

DÉMONSTRATION D'UN THÉORÈME DE M. BOOLE CONCERNANT DES INTÉGRALES MULTIPLES.

[From the *Journal de Mathématiques Pures et Appliquées* (Liouville), tom. XIII. (1848), pp. 245—248.]

THÉORÈME. "Soient P, Q des fonctions de n variables $x, y, \ldots$; lesquelles fonctions satisfont à la condition

$$\int_{-\infty}^{\infty} \ldots dxdy \ldots e^{-(Pv+Qw)\,i} = \frac{\pi^{\frac{1}{2}n} e^{-\frac{1}{4}n\pi i}\, e^{-G(v,\,w)\,i}}{\sqrt{H(v,\, w)}} \quad \ldots\ldots\ldots\ldots\ldots\ldots\ldots (1),$$

où, comme à l'ordinaire, $i = \sqrt{-1}$; $G(v, w)$, $H(v, w)$ sont des fonctions homogènes de v, w des ordres 1 et n respectivement (on verra, dans la suite, qu'il y a plusieurs fonctions P, Q qui satisfont à une équation de cette forme).

Cela étant, posons

$$V = \int \ldots dxdy \ldots \frac{f(P)}{Q^{\frac{1}{2}n+q}} \quad \ldots\ldots\ldots\ldots\ldots\ldots\ldots (2),$$

les limites de l'intégration étant données par la condition $P = 1$; et soient

$$G\left(1, \frac{1}{s}\right) = \sigma, \quad H\left(1, \frac{1}{s}\right) = s^{-n}\phi \quad \ldots\ldots\ldots\ldots\ldots\ldots\ldots (3).$$

On aura pour l'intégrale V cette formule,

$$V = \frac{\pi^{\frac{1}{2}n}}{\Gamma(\frac{1}{2}n + q)} \int_0^{\infty} \frac{Ss^{-q-1}\, ds}{\sqrt{\phi}} \quad \ldots\ldots\ldots\ldots\ldots\ldots\ldots (4),$$

dans laquelle

$$S = \frac{(1-\sigma)^{-q}}{\Gamma(-q)} \int_0^1 t^{-q-1} f[\sigma + t(1-\sigma)]\, dt \quad \ldots\ldots\ldots\ldots\ldots\ldots\ldots (5)."$$

Ce théorème remarquable est dû à M. Boole, qui me l'a communiqué sous une forme un peu différente[1], en me priant d'y suppléer la démonstration et d'en faire part aux géomètres; il ne m'a fallu, pour le prouver, que modifier un peu le procédé dont s'est servi M. Boole même, dans son Mémoire: "On a certain Multiple Definite Integral," *Irish Transactions*, t. XXI. [1848].

Je vais donc reproduire cette démonstration en l'appliquant au problème dont il s'agit.

On démontre par une analyse semblable à peu près à celle par laquelle se démontre le théorème de Fourier, que l'expression

$$\frac{1}{\pi\Gamma\left(\frac{1}{2}n+q\right)}\int_0^1 d\alpha\int_0^\infty dv\int_0^\infty dw\, e^{[(\alpha-P)v-Qw+\frac{1}{2}(\frac{1}{2}n+q)\pi]i}\ldots w^{\frac{1}{2}n+q-1}f\alpha \text{ .... } (6),$$

se réduit (en n'y faisant attention qu'à la partie réelle) à $\dfrac{fP}{Q^{\frac{1}{2}n+q}}$ ou à zéro, selon que la quantité P se trouve ou ne se trouve pas comprise entre les limites 0, 1. Donc, en substituant cette intégrale triple dans l'expression de V, on peut étendre depuis $-\infty$ jusqu'à ∞ les intégrations par rapport aux variables $x, y, \ldots$; de cette manière, et en réduisant par l'équation (1), on obtient tout de suite

$$V=\frac{\pi^{\frac{1}{2}n-1}\, e^{\frac{1}{2}q\pi i}}{\Gamma\left(\frac{1}{2}n+q\right)}\int_0^1 d\alpha\int_0^\infty dv\int_0^\infty dw\,\frac{e^{[\alpha v-G(v,\,w)]i}\ldots w^{\frac{1}{2}n+q-1}}{\sqrt{H(v,\,w)}}f\alpha \text{ .... } (7).$$

Donc, en écrivant

$$w=\frac{v}{s}, \qquad dw=-\frac{v\,ds}{s^2}$$

{ce qui donne, par les équations (3),

$$G(v,\,w)=v\sigma, \quad H(v,\,w)=v^n s^{-n}\phi\},$$

les limites par rapport à la nouvelle variable s seront ∞, 0, et l'on obtiendra, en changeant l'ordre des intégrations,

$$V=\frac{\pi^{\frac{1}{2}n}}{\Gamma\left(\frac{1}{2}n+q\right)}\int_0^\infty\frac{ds\, Ss^{-q-1}}{\sqrt{\phi}} \text{ .... } (8),$$

dans laquelle expression

$$S=\frac{1}{\pi}e^{\frac{1}{2}q\pi i}\int_0^1 d\alpha\int_0^\infty dv\, v^q\, e^{(\alpha-\sigma)vi}f\alpha \text{ .... } (9);$$

[1] M. Boole écrit

$$S=\left(-\frac{d}{d\sigma}\right)^q f\sigma,$$

expression à la vérité plus simple, mais qui donne lieu, ce me semble, à quelques difficultés.

et il ne s'agit plus que de faire voir l'identité de cette valeur avec celle qui est donnée par l'équation (5). Pour cela, je remarque que l'on aura

$$\left.\begin{aligned}\int_0^\infty dv\, v^q\, e^{(\alpha-\sigma)\, vi} &= \Gamma(q+1)\, e^{\frac{1}{2}(q+1)\pi i} \quad (\alpha-\sigma)^{-q-1}, \\ \text{ou} \qquad &= \Gamma(q+1)\, e^{-\frac{1}{2}(q+1)\pi i}\, (\sigma-\alpha)^{-q-1},\end{aligned}\right\} \dots\dots\dots\dots\dots (10),$$

selon que $(\alpha-\sigma)$ est positif ou négatif; les valeurs correspondantes de $\dfrac{1}{\pi} e^{\frac{1}{2}q\pi i}\displaystyle\int_0^\infty dv\, v^q\, e^{(\alpha-\sigma) vi}$ sont

$$\frac{1}{\pi}\, e^{(q+\frac{1}{2})\pi i}\, \Gamma(q+1)\,(\alpha-\sigma)^{-q-1} \quad \text{et} \quad \frac{1}{\pi} e^{-\frac{1}{2}\pi i}\, \Gamma(q+1)\,(\sigma-\alpha)^{-q-1} \dots\dots\dots (11);$$

or, en ne faisant attention qu'aux parties réelles, et en réduisant par une propriété connue des fonctions Γ, ces valeurs se réduisent à $\dfrac{1}{\Gamma(-q)}(\alpha-\sigma)^{-q-1}$ et zéro respectivement; d'après cela, l'équation (9) se réduit à

$$S = \frac{1}{\Gamma(-q)} \int_\sigma^1 (\alpha-\sigma)^{-q-1} f\alpha\, d\alpha \dots\dots\dots\dots\dots\dots\dots\dots\dots (12),$$

et enfin, en écrivant

$$\alpha = \sigma + t\,(1-\sigma), \quad d\alpha = (1-\sigma)\, dt,$$

on obtient pour S la valeur donnée par l'équation (5), de manière que la formule dont il s'agit se trouve complétement démontrée.

Il paraît difficile de trouver les formes générales de P, Q {rien n'étant, je crois, connu sur la solution des équations telles que l'équation (1)}; mais des formes particulières se présentent assez facilement. Ainsi, en ne considérant que les exemples que m'a donnés M. Boole (lesquels j'ai depuis vérifiés), soit

$$P = 2\,(lx + my + \dots), \quad Q = v^2 + x^2 + y^2 + \dots \quad \dots\dots\dots\dots\dots\dots (13).$$

En substituant ces valeurs dans l'équation (1), les intégrations s'effectuent sans difficulté et sous la forme nécessaire, et l'on obtient

$$G(v,\ w) = wv^2 - \frac{(l^2 + m^2 + \dots)\, v^2}{w}, \quad H(v,\ w) = w^n,$$

ou enfin

$$\sigma = \frac{v^2}{s} - (l^2 + m^2 + \dots)\, s, \quad \phi = 1 \dots\dots\dots\dots\dots\dots\dots\dots\dots (14).$$

Soit encore (ce qui comprend, comme cas particulier, le problème des attractions)

$$P = \frac{x^2}{f^2} + \frac{y^2}{g^2} + \dots, \quad Q = v^2 + (a-x)^2 + (b-y)^2 + \dots \quad \dots\dots\dots\dots (15);$$

on obtient sans plus de difficulté

$$\phi = \frac{(s+f^2)(s+g^2)\ldots}{f^2g^2}, \quad \sigma = \frac{v}{s} + \frac{a^2}{f^2+s} + \frac{b^2}{g^2+s} \ldots \quad \ldots\ldots\ldots\ldots (16).$$

Soit encore, pour dernier exemple,

$$P = l^2x^2 + \frac{L^2}{x^2} + \ldots, \quad Q = v^2 + \lambda^2x^2 + \frac{\Lambda^2}{x^2} + \ldots \quad \ldots\ldots\ldots\ldots (17);$$

on aura, pour ce cas,

$$\left.\begin{aligned} &\phi = (l^2s+\lambda^2)(L^2s+\Lambda^2)\ldots, \\ &\sigma = \frac{1}{s}\left\{v^2 + 2\sqrt{(l^2s+\lambda^2)(L^2s+\Lambda^2)} + \ldots\right\} \end{aligned}\right\} \ldots\ldots\ldots\ldots (18);$$

les formules qui se rapportent à cet exemple aussi bien qu'au premier sont, je crois, entièrement nouvelles.

64.

SUR LA GÉNÉRALISATION D'UN THÉORÈME DE M. JELLETT, QUI SE RAPPORTE AUX ATTRACTIONS.

[From the *Journal de Mathématiques Pures et Appliquées* (Liouville), tom. XIII. (1848), pp. 264—268.]

LES formules qu'a données M. Jellett pour exprimer les attractions d'un ellipsoïde au moyen de l'expression de la surface de l'ellipsoïde réciproque (t. XI. de ce Journal, [1846], page 92) peuvent s'étendre au cas d'un nombre quelconque de variables.

Pour démontrer cela, je pars de cette formule

$$\left.\begin{aligned}&\int \frac{\phi\left(\dfrac{x^2}{f^2}+\dfrac{y^2}{g^2}\ldots\right)dx\,dy\ldots}{[(a-x)^2+(b-y)^2\ldots+u^2]^{\frac{1}{2}n+q}}\ \left(\text{limites } \frac{x^2}{f^2}+\frac{y^2}{g^2}\ldots=1\right)\\ &\qquad=\frac{fg\ldots\pi^{\frac{1}{2}n}}{\Gamma\left(\frac{1}{2}n+q\right)}\int_\eta^\infty \frac{Ss^{-q-1}\,ds}{\sqrt{(s+f^2)(s+g^2)\ldots}},\end{aligned}\right\}\ldots\ldots\ldots\ldots(1),$$

dans laquelle n est le nombre des variables $x, y, \ldots$, et où

$$S=\frac{(1-\sigma)^{-q}}{\Gamma(-q)}\int_0^1 t^{-q-1}\,\phi\,[\sigma+t(1-\sigma)]\,dt\ldots\ldots\ldots\ldots\ldots\ldots\ldots\ldots(2),$$

$$\sigma=\frac{a^2}{s+f^2}+\frac{b^2}{s+g^2}\ldots+\frac{u^2}{s},$$

$$1=\frac{a^2}{\eta+f^2}+\frac{b^2}{\eta+g^2}\ldots+\frac{u^2}{\eta},$$

formule due à M. Boole qui l'a démontrée sous une forme un peu différente (*Irish Transactions*, t. XXI.). La modification que j'y ai introduite se trouve démontrée dans le *Cambridge and Dublin Mathematical Journal*, t. II. p. 223([1]) [44].

[1] Cette formule peut d'ailleurs se déduire comme cas particulier de la formule très-générale de M. Boole que M. Cayley a démontrée dans le cahier précédent. (J. L.); [63].

On déduit de là, en écrivant fx, gy, ... au lieu de x, y, ..., en réduisant à zéro les quantités a, b, ..., u (ce qui donne aussi $\eta = 0$), et en donnant une forme convenable à la fonction ϕ,

$$\int \frac{(x^2+y^2\ldots)^{\frac{5}{2}-\frac{1}{2}n}\,dx\,dy\ldots}{(f^2x^2+g^2y^2\ldots)^2} = \frac{2\pi^{\frac{1}{2}n}}{\Gamma(\frac{1}{2}n-2)}\int_0^\infty \frac{s^{\frac{1}{2}n-3}\,ds}{\sqrt{(s+f^2)(s+g^2)\ldots}} \quad \ldots\ldots\ldots\ldots(3)$$

(les limites de l'intégrale au premier membre de cette équation étant données par $x^2+y^2\ldots = 1$).

Donc, en écrivant

$$\Sigma = f^2g^2\ldots\int \frac{(x^2+y^2\ldots)^{\frac{5}{2}-\frac{1}{2}n}\,dx\,dy\ldots}{(f^2x^2+g^2y^2\ldots)^2},$$

on aura

$$\Sigma = \frac{2\pi^{\frac{1}{2}n}f^2g^2\ldots}{\Gamma(\frac{1}{2}n-2)}\int_0^\infty \frac{s^{\frac{1}{2}n-3}\,ds}{\sqrt{(s+f^2)(s+g^2)\ldots}} \quad \ldots\ldots\ldots\ldots(4).$$

Soit Σ' ce que devient Σ en écrivant $\frac{1}{f}$, $\frac{1}{g}$, ... au lieu de f, g, ... : en écrivant en même temps $\frac{1}{s}$ au lieu de s, on obtient

$$\Sigma' = \frac{2\pi^{\frac{1}{2}n}}{\Gamma(\frac{1}{2}n-2)}\frac{1}{fg\ldots}\int_0^\infty \frac{s\,ds}{\sqrt{(s+f^2)(s+g^2)\ldots}} \quad \ldots\ldots\ldots\ldots(5);$$

et de là, en écrivant

$$\frac{1}{f^2}\frac{d}{df}\Sigma' f = F,\quad \frac{1}{g^2}\frac{d}{dg}\Sigma' g = G,\ldots \quad \ldots\ldots\ldots\ldots(6),$$

on déduit

$$F = \frac{-2\pi^{\frac{1}{2}n}}{\Gamma(\frac{1}{2}n-2)}\frac{1}{fg\ldots}\int_0^\infty \frac{s}{s+f^2}\frac{ds}{\sqrt{(s+f^2)(s+g^2)\ldots}},\ \&\text{c.} \quad \ldots\ldots\ldots(7).$$

Cela étant, remarquons que l'intégrale

$$\int_0^\infty \frac{ds}{\sqrt{(s+f^2)(s+g^2)\ldots}} \quad \ldots\ldots\ldots\ldots(8)$$

est fonction homogène de l'ordre $(2-n)$ par rapport aux quantités f, g, ... (en effet, cela se voit tout de suite en faisant $s = f^2\theta$). Donc

$$\int_0^\infty \left(-\frac{f^2}{s+f^2}-\frac{g^2}{s+g^2}\ldots\right)\frac{ds}{\sqrt{(s+f^2)(s+g^2)\ldots}} = (2-n)\int_0^\infty \frac{ds}{\sqrt{(s+f^2)(s+g^2)\ldots}},$$

ou, en écrivant $\frac{s}{s+f^2}-1$, &c., au lieu de $-\frac{f^2}{s+f^2}$, &c.,

$$\int_0^\infty \left(\frac{s}{s+f^2}+\frac{s}{s+g^2}+\ldots\right)\frac{ds}{\sqrt{(s+f^2)(s+g^2)\ldots}} = 2\int_0^\infty \frac{ds}{\sqrt{(s+f^2)(s+g^2)\ldots}} \quad \ldots\ldots\ldots(9),$$

et de là aussi

$$\int_0^\infty \left(\frac{-s}{s+f^2}+\frac{s}{s+g^2}+\dots\right)\frac{ds}{\sqrt{(s+f^2)(s+g^2)\dots}}=2f^2\int_0^\infty \frac{1}{s+f^2}\frac{ds}{\sqrt{(s+f^2)(s+g^2)\dots}}, \text{ \&c. } (10).$$

Donc enfin

$$\left.\begin{array}{l}\displaystyle\int_0^\infty \left(1-\frac{a^2}{s+f^2}\dots\right)\frac{ds}{\sqrt{(s+f^2)(s+g^2)\dots}}\\ \displaystyle=\tfrac{1}{2}\int_0^\infty \left\{\begin{array}{l}\displaystyle\left(\frac{s}{s+f^2}+\frac{s}{s+g^2}+\dots\right)\\ \displaystyle-\frac{a^2}{f^2}\left(\frac{-s}{s+f^2}+\frac{s}{s+g^2}+\dots\right)-\frac{b^2}{g^2}\left(\frac{s}{s+f^2}+\frac{-s}{s+g^2}\dots\right)-\dots\end{array}\right\}\frac{ds}{\sqrt{(s+f^2)(s+g^2)\dots}}\end{array}\right\} (11),$$

c'est-à-dire

$$\left.\begin{array}{l}\displaystyle\int_0^\infty \left(1-\frac{a^2}{s+f^2}-\frac{b^2}{s+g^2}\dots\right)\frac{ds}{\sqrt{(s+f^2)(s+g^2)\dots}}\\ \displaystyle=-\frac{\Gamma(\frac{1}{2}n-2)\,.\,fg\dots}{4\pi^{\frac{1}{2}n}}\left\{\begin{array}{l}(F+G+\dots)\\ \displaystyle-\frac{a^2}{f^2}(-F+G+\dots)-\frac{b^2}{g^2}(F-G+\dots)-\dots\end{array}\right\}\end{array}\right\}\dots\dots(12).$$

En particularisant d'une manière convenable la formule (1), on obtient, pour le cas de $\dfrac{a^2}{f^2}+\dfrac{b^2}{g^2}\dots>1$, cette formule connue

$$\left.\begin{array}{l}\displaystyle V=\int\frac{dx\,dy\dots}{[(a-x)^2+(b-y)^2\dots]^{\frac{1}{2}n-1}}\\ \displaystyle\quad=\frac{\pi^{\frac{1}{2}n}fg\dots}{\Gamma(\frac{1}{2}n-1)}\int_0^\infty\left(1-\frac{a^2}{s+f^2}-\dots\right)\frac{ds}{\sqrt{(s+f^2)\dots}}\end{array}\right\}\dots\dots\dots\dots\dots(13)$$

(l'équation des limites étant, comme auparavant, $\dfrac{x^2}{f^2}+\dfrac{y^2}{g^2}+\dots=1$);

et de là, vu la formule (12), résulte

$$V=-\frac{f^2g^2\dots}{2(n-2)}\left[(F+G\dots)-\frac{a^2}{f^2}(-F+G+\dots)-\frac{b^2}{g^2}(F-G+\dots)\dots\right]\dots\dots(14).$$

L'expression de Σ, en écrivant $r\cos\alpha$, $r\cos\beta,\dots$ au lieu de x, $y,\dots$, remplaçant $dx\,dy\dots$ par $r^{n-1}\,dr\,dS$, et intégrant depuis $r=0$ jusqu'à $r=1$, se réduit à

$$\Sigma=f^2g^2\dots\int\frac{dS}{(f^2\cos^2\alpha+g^2\cos^2\beta+\dots)^2}\dots\dots\dots\dots\dots\dots\dots(15),$$

de sorte qu'au cas de $n=3$ cette fonction se réduit à l'expression qu'a donnée M. Jellett pour la surface d'un ellipsoïde. Donc, en se rappelant que les attractions sont représentées par $\frac{dV}{da}, \frac{dV}{db}, \frac{dV}{dc}$, on voit que l'équation (14) équivaut, pour ce cas, aux formules de M. Jellett.

Remarquons, qu'en transformant l'intégrale (4) de la même manière dont nous avons transformé l'intégrale (8), on obtient

$$\Sigma = \frac{\pi^{\frac{1}{2}n} f^2 g^2 \ldots}{\Gamma(\frac{1}{2}n - 1)} \int_0^\infty \left(\frac{1}{s+f^2} + \frac{1}{s+g^2} \ldots\right) \frac{s^{\frac{1}{2}n-2}\, ds}{\sqrt{(s+f^2)(s+g^2)\ldots}} \quad \ldots\ldots\ldots\ldots(16),$$

ce qui donne pour $n=3$ cette expression très-simple de la surface de l'ellipsoïde aux demi-axes f, g, h,

$$\Sigma = \pi f^2 g^2 h^2 \int_0^\infty \left(\frac{1}{s+f^2} + \frac{1}{s+g^2} + \frac{1}{s+h^2}\right) \frac{s^{-\frac{1}{2}}\, ds}{\sqrt{(s+f^2)(s+g^2)(s+h^2)}} \quad \ldots\ldots\ldots\ldots(17),$$

formule qui se vérifie tout de suite au cas de $f=g=h$.

L'expression encore plus simple que donne l'équation (4), savoir,

$$\Sigma = -\pi f^2 g^2 h^2 \int_0^\infty \frac{s^{-\frac{3}{2}}\, ds}{\sqrt{(s+f^2)(s+g^2)(s+h^2)}} \quad \ldots\ldots\ldots\ldots\ldots\ldots\ldots\ldots(18),$$

n'est pas exempte de difficulté à cause de la valeur apparemment infinie du second membre de l'équation.

Au cas d'une sphère, cela se réduit à

$$\Sigma = -\pi f^2 \int_0^\infty \frac{s^{-\frac{3}{2}}\, ds}{(1+s)^{\frac{3}{2}}},$$

ce qui serait, en effet, exact si la formule

$$\int_0^\infty \frac{s^{m-1}\, ds}{(1+s)^{m+n}} = \frac{\Gamma m\, \Gamma n}{\Gamma(m+n)}$$

subsistait pour les valeurs négatives de m. Cela nous apprend que les intégrales de la forme

$$\int_0^\infty \frac{s^{-q-1}\, ds}{\sqrt{(s+f^2)(s+g^2)\ldots}} \quad \ldots\ldots\ldots\ldots\ldots\ldots\ldots\ldots\ldots\ldots\ldots(19)$$

ne sont pas à rejeter au cas des valeurs positives de q; il est même facile, en répétant continuellement le procédé de réduction que nous venons d'employer, de présenter ces intégrales sous une forme où il n'y a plus de terme infini. En m'aidant de l'analogie de quelques formules qui se trouvent dans mon Mémoire *Sur quelques formules du calcul intégral* (t. XI. de ce Journal, p. 231 [49]), je crois même pouvoir avancer que cette intégrale doit se remplacer par

$$\frac{-1}{2\sin q\pi} \int_{-\infty}^{\infty} \frac{(k+si)^{-q-1}\, ds}{\sqrt{(k+si+f^2)(k+si+g^2)\ldots}} \quad \ldots\ldots\ldots\ldots\ldots\ldots\ldots(20)$$

où, comme à l'ordinaire, $i=\sqrt{-1}$, et où k dénote une quantité quelconque dont la partie réelle ne s'évanouit pas. Mais je renvoie cette discussion à une autre occasion.

65.

NOUVELLES RECHERCHES SUR LES FONCTIONS DE M. STURM.

[From the *Journal de Mathématiques Pures et Appliquées* (Liouville), tom. XIII. (1848), pp. 269—274.]

EN développant une remarque faite par M. Sylvester dans un Mémoire publié il y a huit ou neuf ans dans le *Philosophical Magazine*, j'ai trouvé des expressions assez simples des fonctions de M. Sturm, composées au moyen des coefficients mêmes; ce qui convient mieux que d'exprimer ces fonctions, comme je l'ai déjà fait dans le t. XI. de ce Journal, p. 297, [48], par les sommes des puissances. D'ailleurs il ne m'est plus nécessaire de parler des expressions de M. Sylvester, ou même des divisions successives de M. Sturm; mais ma méthode fait voir directement que les fonctions que je vais définir sont douées de la propriété fondamentale sur laquelle se repose la théorie de M. Sturm, à savoir que, en considérant trois fonctions successives, la première et la dernière fonction sont de signe contraire pour toute valeur de la variable qui fait évanouir la fonction intermédiaire; cependant je n'ai pas encore réussi à démontrer dans toute sa généralité l'équation identique d'où dépend cette propriété.

Soient d'abord V, V' des fonctions du même degré n,

$$V = ax^n + bx^{n-1} + \dots,$$
$$V' = a'x^n + b'x^{n-1} + \dots,$$

et écrivons

$$-F_1 = \begin{vmatrix} V, & V' \\ a & a' \end{vmatrix},$$

$$F_2 = \begin{vmatrix} xV, & V, & xV', & V' \\ a & . & a' & . \\ b & a & b' & a' \\ c & b & c' & b' \end{vmatrix},$$

$$-F_3 = \begin{vmatrix} x^2V, & xV, & V, & x^2V', & xV', & V' \\ a & . & . & a' & . & . \\ b & a & . & b' & a' & . \\ c & b & a & c' & b' & a' \\ d & c & b & d' & c' & b' \\ e & d & c & e' & d' & c' \end{vmatrix}, \text{ \&c.}$$

(ce qui suffit pour faire voir la loi de ces fonctions successives F_1, F_2, ...). Il résulte des propriétés élémentaires des déterminants que ces fonctions sont des ordres $\overline{n-1}$, $\overline{n-2}$, $\overline{n-3}$, &c., respectivement, par rapport à la variable x. En effet, dans F_1 le coefficient de x^n se réduit à $\begin{vmatrix} a, & a' \\ a & a' \end{vmatrix}$, savoir, à zéro; de même, dans F_2, les coefficients de x^n et x^{n-1} se réduisent chacun à zéro; et ainsi de suite.

Soient encore

$$P_1 = \begin{vmatrix} a, & a' \\ b & b' \end{vmatrix}, \qquad P_1' = \begin{vmatrix} a, & a' \\ c & c' \end{vmatrix},$$

$$P_2 = \begin{vmatrix} a, & . & a', & . \\ b & a & b' & a' \\ c & b & c' & b' \\ d & c & d' & c' \end{vmatrix} \qquad P_2' = \begin{vmatrix} a, & . & a', & . \\ b & a & b' & a' \\ c & b & c' & b' \\ e & d & e' & d' \end{vmatrix},$$

$$P_3 = \begin{vmatrix} a, & . & . & a', & . & . \\ b & a & . & b' & a' & . \\ c & b & a & c' & b' & a' \\ d & c & b & d' & c' & b' \\ e & d & c & e' & d' & c' \\ f & e & d & f' & e' & d' \end{vmatrix}, \text{ \&c.}, \qquad P_3' = \begin{vmatrix} a, & . & . & a', & . & . \\ b & a & . & b' & a' & . \\ c & b & a & c' & b' & a' \\ d & c & b & d' & c' & b' \\ e & d & c & e' & d' & c' \\ g & f & e & g' & f' & e' \end{vmatrix}, \text{ \&c.}$$

(ce qui suffit pour indiquer la loi). On aura entre ces différentes fonctions F, P, P' cette suite remarquable d'équations identiques,

$$P_1^2 F_3 + (xP_1P_2 + P_1P_2' + P_1'P_2)\, F_2 + P_2^2\, F_1 = 0,$$
$$P_2^2 F_4 + (xP_2P_3 + P_2P_3' + P_2'P_3)\, F_3 + P_3^2\, F_2 = 0,$$
&c.,

lesquelles équations, dans ce Mémoire, seront prises pour vraies. Cela étant, il est évident que F_1 et F_3 seront de signe contraire pour toute valeur de x qui fait évanouir F_2; F_2 et F_4 seront de signe contraire pour toute valeur de x qui fait évanouir F_3; et ainsi de suite.

Ces formules renferment le cas où les deux fonctions V, V' ne sont pas du même degré (en effet, pour les y adapter, on n'a besoin que de faire évanouir quelques-uns des premiers coefficients de V ou de V'). Il est donc permis de supposer que V' soit la dérivée de V. Dans ce cas, $F_1 = aV'$, et on verra dans un moment que les fonctions F_2, F_3, ... contiennent chacune le facteur a^2, de manière qu'il convient d'écrire $F_2 = a^2V_2$, $F_3 = a^2V_3$, ... Ce facteur a^2 peut être évidemment écarté, et ce sera de même avec le facteur a de F, pourvu, ce que je supposerai dans la suite, que a soit positif. On aura de cette manière les fonctions V, V', V_2, V_3, ... douées des propriétés des fonctions de M. Sturm. En effet, elles seront précisément les fonctions fx, f_1x, f_2x, ... du Mémoire déjà cité, ce qui cependant pourrait être difficile à démontrer à priori.

On déduit tout de suite des expressions de F_2, F_3, ... ,

$$aV_2 = -\begin{vmatrix} V, & xV', & V' \\ a & na & . \\ b & \overline{n-1}b & na \end{vmatrix}, \quad aV_3 = \begin{vmatrix} xV, & V, & x^2V', & xV', & V' \\ a & . & na & . & . \\ b & a & \overline{n-1}b & na & . \\ c & b & \overline{n-2}c & \overline{n-1}b & na \\ d & c & \overline{n-3}d & \overline{n-2}c & \overline{n-1}b \end{vmatrix}, \text{ \&c.}$$

Ces formules se simplifient au moyen des propriétés connues des déterminants, et en écrivant

$$xV' - nV = -U,$$

cela donne

$$aV_2 = \begin{vmatrix} V, & U, & V' \\ a & . & . \\ b & b & na \end{vmatrix}, \quad aV_3 = \begin{vmatrix} xV, & V, & xU, & U, & V' \\ a & . & . & . & . \\ b & a & b & . & . \\ c & b & 2c & b & na \\ d & c & 3d & 3c & \overline{n-1}b \end{vmatrix}, \text{ \&c.};$$

ou enfin, et en écrivant un autre terme de la suite, afin de mieux faire voir la loi,

$$V_2 = -\begin{vmatrix} U, & V' \\ b & na \end{vmatrix}, \quad V_3 = -\begin{vmatrix} V, & xU, & U, & V' \\ a & b & . & . \\ b & 2c & b & na \\ c & 3d & 2c & \overline{n-1}b \end{vmatrix},$$

$$V_4 = -\begin{vmatrix} xV, & V, & x^2U, & xU, & V' \\ a & . & b & . & . \\ b & a & 2c & b & . \\ c & b & 3d & 2c & na \\ d & c & 4e & 3d & \overline{n-1}b \\ e & d & 5f & 4e & \overline{n-2}c \end{vmatrix}, \text{ \&c.},$$

formules dans lesquelles

$$\begin{cases} V = ax^n + bx^{n-1} + cx^{n-2} + \dots, \\ V' = nax^{n-1} + \overline{n-1}\,bx^{n-2} + \dots, \\ U = bx^{n-1} + 2cx^{n-2} + \dots, \end{cases}$$

et où, en substituant ces valeurs, on peut commencer pour V_2 avec le terme qui contient x^{n-2}, pour V_3 avec le terme qui contient x^{n-3}, et ainsi de suite, puisque les termes des ordres plus hauts s'évanouissent identiquement. Voilà, je crois, les expressions les plus simples des fonctions de M. Sturm.

Je donnerai en conclusion ces formes développées des fonctions jusqu'à V_4.

$$V = ax^n + bx^{n-1} + cx^{n-2} + \dots;$$

$$V' = nax^{n-1} + \overline{n-1}\,bx^{n-2} + \overline{n-2}\,cx^{n-3} + \dots;$$

$$\begin{aligned} V_2 = &- na\,\{2cx^{n-2} + 3dx^{n-3} + \dots \\ &+ b\,\{\overline{n-1}\,bx^{n-2} + \overline{n-2}\,cx^{n-3} + \dots; \end{aligned}$$

$$\begin{aligned} V_3 = &\ [2nabc - \overline{n-1}\,b^3] \,\{dx^{n-3} + ex^{n-4} + \dots \\ &+ [-2n^2ac + \overline{n-1}\,ab^2] \,\{4ex^{n-3} + 5fx^{n-4} + \dots \\ &+ [3na^2d - (3n-2)\,abc + (n-1)\,b^3] \,\{3dx^{n-3} + 4ex^{n-4} + \dots \\ &+ [-3abd + 4ac^2 - b^2c] \,\{(n-2)\,cx^{n-3} + (n-3)\,dx^{n-4} + \dots; \end{aligned}$$

$$\begin{aligned} V_4 = &\ A\,\{fx^{n-4} + gx^{n-5} + \dots \\ &+ B\,\{ex^{n-4} + fx^{n-5} + \dots \\ &+ C\,\{6gx^{n-4} + 7hx^{n-5} + \dots \\ &+ D\,\{5fx^{n-4} + 6gx^{n-5} + \dots \\ &+ E\,\{4ex^{n-4} + 5fx^{n-5} + \dots \\ &+ F\,\{\overline{n-3}\,dx^{n-4} + \overline{n-4}\,ex^{n-5} + \dots; \end{aligned}$$

dans cette dernière expression j'ai mis, pour abréger,

$$\begin{aligned} A = &\ 9na^2bd^2 - 8na^2bce + (4n-4)\,ab^3e - (10n-12)\,ab^2cd \\ &+ (4n-8)\,abc^3 - (n-2)\,b^3c^2 + (2n-2)\,b^4d, \end{aligned}$$

$$\begin{aligned} B = &\ 10na^2bcf - 12na^2bde - 16na^2c^2e + 18na^2cd^2 - (5n-5)\,ab^3f \\ &+ (18n-16)\,ab^2ce + (3n-9)\,ab^2d^2 - (24n-36)\,abc^2d + (8n-16)\,ac^4 \\ &- (3n-3)\,b^4e + (5n-7)\,b^3cd - (2n-4)\,b^2c^3, \end{aligned}$$

$$\begin{aligned} C = &\ 8na^3ce - 9na^3d^2 - (4n-4)\,a^2b^2e + (10n-12)\,a^2bcd \\ &- (4n-8)\,a^2c^3 - (2n-2)\,ab^3d + (n-2)\,ab^2e^2, \end{aligned}$$

$$\begin{aligned} D = &- 10na^3cf + 12na^3de + (5n-5)\,a^2b^2f - (2n-8)\,a^2bce \\ &- (12n-9)\,a^2bd^2 + (4n-12)\,a^2c^2d - (n-1)\,ab^3e + (9n-9)\,ab^2cd \\ &- (4n-8)\,abe^3 - (2n-2)\,b^4d + (n-2)\,b^3c^2, \end{aligned}$$

$$\begin{aligned}E = {} & 15na^3df - 16na^3e^2 - (15n-10)\,a^2bcf + (17n-12)\,a^2bde \\ & + (16n-16)\,a^2c^2e - (15n-18)\,a^2cd^2 + (5n-5)\,ab^3f \\ & - (17n-18)\,ab^2ce + (14n-24)\,abc^2d - (n-3)\,ab^2d^2 \\ & - (4n-8)\,ac^4 + (3n-3)\,b^4e - (3n-5)\,b^3cd + (n-2)\,b^2c^3,\end{aligned}$$

$$\begin{aligned}F = {} & -15a^2bdf + 16a^2be^2 + 20a^2c^2f - 27a^2d^3 - 48a^2cde \\ & - 5ab^2cf + 7ab^2de - 4abc^2e - 18abcd^2 + 4ac^3d \\ & - 3b^3ce + 4b^3d^2 - b^2c^2d^2.\end{aligned}$$

Il serait évidemment inutile de vouloir pousser plus loin ces calculs.

[MS. addition in my copy of Liouville.

Quoique ces expressions soient ce qu'il y a de plus simple pour le calcul numérique des fonctions de M. Sturm, cependant sous le point de vue analytique il convient de modifier un peu la forme de ces expressions. En effet en écrivant

$$V = ax^n + \frac{n}{1}\,bx^{n-2} + \ldots,$$

mettons

$$P = ax^{n-1} + \frac{n-1}{1}\,bx^{n-2} + \ldots, \ = ax^{n-1} + \theta_1\,bx^{n-2} + \ldots,$$

$$Q = bx^{n-1} + \frac{n-1}{1}\,cx^{n-2} + \ldots, \ = bx^{n-1} + \theta_1\,cx^{n-2} + \ldots,$$

l'on aura

$$\begin{aligned}V' &= nP,\\ U' &= nQ,\\ V &= Px + Q\,;\end{aligned}$$

et cela donne après une réduction légère

$$V_2 = \begin{vmatrix} P, & Q \\ a & b \end{vmatrix}, \qquad V_3 = -\begin{vmatrix} xP, & P, & xQ, & Q \\ a & . & b & . \\ \theta_1 b & a & \theta_1 c & b \\ \theta_2 c & \theta_1 b & \theta_2 d & \theta_1 c \end{vmatrix},$$

$$V_4 = \begin{vmatrix} x^2P, & xP, & P, & x^2Q, & xQ, & Q \\ a & . & . & b & . & . \\ \theta_1 b & a & . & \theta_1 c & b & . \\ \theta_2 c & \theta_1 b & a & \theta_2 d & \theta_1 c & b \\ \theta_3 d & \theta_2 c & \theta_1 b & \theta_3 e & \theta_2 d & \theta_1 c \\ \theta_4 e & \theta_3 d & \theta_2 c & \theta_4 f & \theta_3 e & \theta_2 d \end{vmatrix}$$

et ainsi de suite.]

66.

SUR LES FONCTIONS DE LAPLACE.

[From the *Journal de Mathématiques Pures et Appliquées* (Liouville), tom. XIII. (1848), pp. 275—280].

J'AI réussi à étendre à un nombre quelconque de variables la théorie des fonctions de Laplace, en me fondant sur le théorème que voici:

"Les coefficients $l, m, \ldots, l', m', \ldots$ étant assujettis aux conditions

$$\left.\begin{array}{l} l^2 + m^2 + \ldots = 0, \\ l'^2 + m'^2 + \ldots = 0, \end{array}\right\} \quad \ldots\ldots(1),$$

et les limites de l'intégration étant données par

$$x^2 + y^2 + \ldots = 1 \quad \ldots\ldots(2),$$

l'on aura pour toutes valeurs entières et positives de s, s', excepté pour $s = s'$,

$$\int (lx + my + \ldots)^s (l'x + m'y + \ldots)^{s'} dxdy \ldots = 0 \quad \ldots\ldots(3),$$

et pour $s = s'$,

$$\int (lx + my + \ldots)^s (l'x + m'y + \ldots)^s dxdy \ldots = N_s (ll' + mm' + \ldots)^s \ldots\ldots(4),$$

en faisant, pour abréger,

$$N_s = \frac{\pi^{\frac{1}{2}n}\,\Gamma(s+1)}{2^s\,\Gamma(\frac{1}{2}n + s + 1)} \quad \ldots\ldots(5),$$

(où n dénote le nombre des variables)."

En admettant d'abord comme vrai ce théorème qui sera démontré plus bas, je remarque qu'il est permis d'écrire $\frac{d}{da}, \frac{d}{db}, \ldots, \frac{d}{da'}, \frac{d}{db'}, \ldots$, au lieu de $l, m, \ldots, l', m', \ldots$, où ces nouveaux symboles se rapportent à de certaines fonctions f, f' de $a, b, \ldots$ et de $a', b', \ldots$ respectivement. De là ce nouveau théorème:

"Les fonctions f, f' étant assujetties aux conditions

$$\left.\begin{aligned} &\frac{d^2f}{da^2}+\frac{d^2f'}{db^2}+\ldots=0,\\ &\frac{d^2f'}{da'^2}+\frac{d^2f'}{db^2}+\ldots=0, \end{aligned}\right\} \quad\ldots\ldots\ldots\ldots\ldots\ldots(6),$$

on aura (entre les mêmes limites qu'auparavant), excepté pour $s=s'$,

$$\int\left(x\frac{d}{da}+y\frac{d}{db}+\ldots\right)^s f\,.\left(x\frac{d}{da'}+y\frac{d}{db'}+\ldots\right)^{s'}f'\,.\,dxdy\ldots=0\ldots\ldots\ldots\ldots(7),$$

et pour $s=s'$,

$$\left.\begin{aligned} &\int\left(x\frac{d}{da}+y\frac{d}{db}+\ldots\right)^s f\,.\left(x\frac{d}{da'}+y\frac{d}{db'}+\ldots\right)^{s}f'\,.\,dxdy\ldots\\ &=N_s\left(\frac{d}{da}\frac{d}{da'}+\frac{d}{db}\frac{d}{db'}+\ldots\right)^s ff'. \end{aligned}\right\}\quad\ldots\ldots\ldots\ldots(8)."$$

Il est facile de voir que les expressions

$$\left(x\frac{d}{da}+y\frac{d}{db}+\ldots\right)^s f,\quad \left(x\frac{d}{da'}+y\frac{d}{db'}+\ldots\right)^{s'} f'$$

satisfont à l'équation

$$\frac{d^2u}{dx^2}+\frac{d^2u}{dy^2}+\ldots=0\quad\ldots\ldots\ldots\ldots\ldots\ldots\ldots\ldots(9),$$

et, de plus, qu'elles sont les fonctions entières et homogènes, des degrés s et s' respectivement, les plus générales qui puissent satisfaire à cette équation. On a donc ce théorème :

"Soient V_s, $W_{s'}$ les fonctions entières et homogènes des degrés s et s' respectivement, les plus générales qui satisfassent à l'équation (9) ; on aura toujours, excepté au cas de $s=s'$,

$$\int V_s\, W_{s'}\, dxdy\ldots=0\quad\ldots\ldots\ldots\ldots\ldots\ldots\ldots\ldots(10)$$

(les limites étant les mêmes qu'auparavant)."

Écrivons à présent

$$f=(a^2+b^2+\ldots)^{-\frac{1}{2}n+1},$$

valeur qui satisfait à la première des équations (8), et nous obtiendrons par la différentiation successive, en faisant attention à la seconde de ces mêmes équations,

$$\left.\begin{aligned} &\left(\frac{d}{da}\frac{d}{da'}+\frac{d}{db}\frac{d}{db'}+\ldots\right)^s(a^2+b^2+\ldots)^{-\frac{1}{2}n+1}f'\\ &=(-)^s\,2^s\,\frac{\Gamma(\frac{1}{2}n+s-1)}{\Gamma(\frac{1}{2}n-1)}\left(a\frac{d}{da'}+b\frac{d}{db'}+\ldots\right)^s f'\,(a^2+b^2+\ldots)^{-\frac{1}{2}n-s+1} \end{aligned}\right\}\quad\ldots(11).$$

En représentant, comme auparavant, par W_s la fonction

$$\left(x\frac{d}{da'}+y\frac{d}{db'}+\dots\right)^s f',$$

soit W_s' ce que devient W_s en écrivant a, b, ... au lieu de x, y, ... , c'est-à-dire écrivons

$$W_s'=\left(a\frac{d}{da'}+b\frac{d}{db'}+\dots\right)^s f'.$$

On déduit de là, et au moyen de l'équation (11), en substituant dans l'équation (8), la formule

$$\left.\begin{aligned}\frac{(-)^s}{\Gamma(s+1)}\int\left(x\frac{d}{da}+y\frac{d}{db}+\dots\right)^s(a^2+b^2+\dots)^{-\frac{1}{2}n+1}\,W_s\,dxdy\dots\\ =M_s(a^2+b^2+\dots)^{-\frac{1}{2}n-s+1}\,W_s'\end{aligned}\right\}\qquad\dots\dots\dots\dots(12),$$

en faisant, pour abréger,

$$\Gamma(s+1)\,M_s=\frac{\Gamma\,2^s(\frac{1}{2}n+s-1)}{\Gamma(\frac{1}{2}n-1)}\,N_s,$$

ou bien

$$M_s=\frac{4\pi^{\frac{1}{2}n}}{\Gamma(\frac{1}{2}n-1)\,(n+2s)\,(n+2s-2)}\qquad\dots\dots\dots\dots(13).$$

Soit maintenant

$$\frac{1}{[(a-x)^2+\dots]^{\frac{1}{2}n-1}}=\frac{Q_0}{(a^2+b^2+\dots)^{\frac{1}{2}n-1}}\dots+\frac{Q_s}{(a^2+b^2+\dots)^{\frac{1}{2}n+s-1}}\qquad\dots\dots\dots\dots(14),$$

ou, autrement dit, soit

$$Q_s=\frac{(-)^s}{\Gamma(s+1)}\left(x\frac{d}{da}+y\frac{d}{db}+\dots\right)^s(a^2+b^2+\dots)^{-\frac{1}{2}n+1}\dots\dots\dots\dots(15),$$

l'équation (12) devient

$$\int Q_s W_s\,dxdy\dots=M_s W_s'\qquad\dots\dots\dots\dots(16)$$

{où la valeur de M_s est donnée par l'équation (13)}. Les deux équations (10) et (16) contiennent la théorie des fonctions W_s, Q_s, lesquelles comprennent évidemment, comme cas particulier, les fonctions de Laplace.

Pour démontrer le théorème exprimé par les équations (1), (2), (3), (4), (5), je fais d'abord abstraction des équations (1), et j'écris

$$\begin{aligned}x&=\ l\xi+\ l'\eta+\ l''\zeta\dots,\\ y&=m\xi+m'\eta+m''\zeta\dots,\end{aligned}$$

où les coefficients sont tels que l'équation

$$x^2+y^2+\dots=p\xi^2+2q\xi\eta+p'\eta^2+p''\zeta^2+\dots$$

soit identiquement vraie. Cela suppose que les valeurs de p, p', q soient respectivement $l^2+m^2+\dots$, $l'^2+m'^2+\dots$, et $ll'+mm'+\dots$, et que les sommes de produits telles que $ll''+mm''+\dots$, $l'l''+m'm''+\dots$ se réduisent chacune à zéro. De là

$$\begin{aligned} dxdy\dots &= \sqrt{pp'-q^2}\,\sqrt{p''}\dots d\xi\,d\eta\,d\zeta\dots, \\ lx+my+\dots &= p\xi+q\eta, \\ l'x+m'y+\dots &= q\xi+p'\eta. \end{aligned}$$

En représentant par I l'intégrale au premier membre de l'équation (3), cela donne

$$I=\sqrt{pp'-q^2}\,\sqrt{p''}\dots\int(p\xi+q\eta)^s(q\xi+p'\eta)^{s'}\,d\xi\,d\eta\,d\zeta\dots,$$

l'équation des limites étant

$$p\xi^2+2q\xi\eta+p'\eta^2+p''\zeta^2+\dots=1.$$

Cette intégrale se simplifie en écrivant

$$\xi\sqrt{p}+\frac{q\eta}{\sqrt{p}}=\xi_{,},\quad \eta\sqrt{p'-\frac{q^2}{p}}=\eta_{,},\quad \zeta\sqrt{p''}=\zeta_{,},\dots;$$

car alors

$$\begin{aligned} \sqrt{pp'-q^2}\,\sqrt{p''}\dots d\xi\,d\eta\,d\zeta\dots &= d\xi_{,}\,d\eta_{,}\,d\zeta_{,}\dots, \\ p\xi+q\eta &= \sqrt{p}\,\xi_{,}, \\ q\xi+p'\eta &= \frac{1}{\sqrt{p}}(q\xi_{,}+\sqrt{pp'-q^2}\,\eta_{,}), \end{aligned}$$

et de là

$$I=p^{\frac{1}{2}(s-s')}\int\xi_{,}^s(q\xi_{,}+\sqrt{pp'-q^2}\,\eta_{,})^{s'}\,d\xi_{,}\,d\eta_{,}\,d\zeta_{,}\dots,$$

l'équation des limites étant

$$\xi_{,}^2+\eta_{,}^2+\dots=1.$$

En supposant $s>s'$, l'intégrale s'évanouit pour $p=0$, et de même quand $s'>s$, elle s'évanouit pour $p'=0$. Donc, en écrivant $p=0$, $p'=0$, on aura toujours, excepté pour $s=s'$, l'équation $I=0$; ce qui revient à l'équation (3). Au cas de $s=s'$, en écrivant de même $p=0$, $p'=0$, on trouve

$$I=q^s\int\xi_{,}^s(\xi_{,}+i\eta_{,})^s\,d\xi_{,}\,d\eta_{,}\,d\zeta_{,}\dots$$

(où, comme à l'ordinaire, $i=\sqrt{-1}$). En faisant attention à la valeur de q, et en comparant avec l'équation (4), cette dernière équation sera démontrée en vérifiant la formule

$$N_s=\int\xi_{,}^s(\xi_{,}+i\eta_{,})^s\,d\xi_{,}\,d\eta_{,}\,d\zeta_{,}\dots.$$

Soient pour cela

$$\xi_{,}=\rho\cos\theta,\qquad \eta_{,}=\rho\sin\theta.$$

Cela donne (en omettant la partie imaginaire qui s'évanouit évidemment)

$$N_s = \int \rho^{2s+1} \cos^s \theta \cos s\theta \, d\rho \, d\theta \, d\zeta_, \ldots .$$

En effectuant d'abord l'intégration par rapport à $\zeta_,, \ldots$, les limites de ces variables sont données par

$$\zeta_,^2 + \ldots = 1 - \rho^2,$$

et l'on trouve tout de suite

$$N_s = \frac{\pi^{\frac{1}{2}n-1}}{\Gamma(\frac{1}{2}n)} \int \rho^{2s+1} (1-\rho^2)^{\frac{1}{2}n-1} \cos^s \theta \cos s\theta \, d\rho \, d\theta.$$

Cette formule doit être intégrée depuis $\rho = 0$ à $\rho = 1$, et depuis $\theta = 0$ à $\theta = 2\pi$; mais en multipliant par quatre, on peut n'étendre l'intégration par rapport à θ que depuis $\theta = 0$ jusqu'à $\theta = \frac{1}{2}\pi$. De là

$$N_s = \frac{4\pi^{\frac{1}{2}n-1}}{\Gamma(\frac{1}{2}n)} \int_0^1 \rho^{2s+1} (1-\rho^2)^{\frac{1}{2}n-1} \, d\rho \int_0^{\frac{1}{2}\pi} \cos^s \theta \cos s\theta \, d\theta;$$

et enfin, au moyen des formules connues

$$\int_0^1 \rho^{2s+1} (1-\rho^2)^{\frac{1}{2}n-1} \, d\rho = \frac{\Gamma(s+1)\,\Gamma(\frac{1}{2}n)}{2\,\Gamma(\frac{1}{2}n+s+1)},$$

$$\int_0^{\frac{1}{2}\pi} \cos^s \theta \cos s\theta \, d\theta = \frac{\pi}{2^{s+1}},$$

on retrouve la formule (5), laquelle il s'agissait de démontrer. Ainsi le théorème fondamental est complétement établi.

67.

NOTE SUR LES FONCTIONS ELLIPTIQUES.

[From the *Journal für die reine und angewandte Mathematik* (Crelle), tom. XXXVII. (1848), pp. 58—60.]

SOIT $x=\sqrt{k}\sin\operatorname{am}u$ et $\alpha=k+\dfrac{1}{k}$: la fonction $\sqrt{k}\sin\operatorname{am}nu$ (où n est un entier), peut être exprimée sous forme d'une fraction dont le dénominateur est une fonction rationnelle et entière par rapport à x et α. En exprimant ce dénominateur par z, on aura

$$n^2(n^2-1)x^2z+(n^2-1)(\alpha x-2x^3)\frac{dz}{dx}+(1-\alpha x^2+x^4)\frac{d^2z}{dx^2}-2n^2(\alpha^2-4)\frac{dz}{d\alpha}=0\ldots\ (1).$$

Cette équation est due à M. Jacobi, (voyez les deux mémoires "Suite des notices sur les fonctions elliptiques," t. III. [1828] p. 306 et t. IV. [1829] p. 185.)

En essayant d'intégrer cette équation à moyen d'une suite ordonnée suivant les puissances de x, et en considérant en particulier les cas $n=2$, 3, 4 et 5, j'ai trouvé que les différentes puissances de α se présentent et disparaissent d'une manière assez bizarre: (voyez mon mémoire "On the theory of elliptic functions," *Cambridge and Dublin Math. Journal,* t. II. [1847], p. 256, [45].) J'ai reconnu depuis que cela vient de ce que la valeur de z est composée de plusieurs séries indépendantes; une quelconque de ces séries ordonnée selon les puissances descendantes de α va à l'infini; mais, en combinant les différentes séries, les termes qui contiennent les puissances négatives de α se détruisent. Par rapport à x chacune de ces séries ne contient que des puissances paires et positives, car les puissances négatives qui y entrent apparemment, se réduisent toujours à zéro. En effet, on satisfait à l'équation (1) en supposant pour z une expression de la forme

$$z_s=2^{2s(n-2s)}\alpha^{s(n-2s)}Z_0\ldots+(-1)^q\,2^{2s(n-2s)-2q}\alpha^{s(n-2s)-q}\,.\,Z_q\ldots\qquad\ldots\ldots\ldots\ldots\ (2)$$

(où s est arbitraire). Cela donne pour Z_q l'équation aux différences mêlées:

$$\left[x^2 \frac{d^2}{dx^2} - (n^2-1)\, x\frac{d}{dx} + 2n^2 s\,(n-2s) - 2n^2 q\right] Z_q$$
$$+ \left[\left\{n^2(n^2-1)\,x^2 + x^4\frac{d^2}{dx^2} - (2n^2-2)\,x^3\frac{d}{dx}\right\} + \frac{d^2}{dx^2}\right] 4Z_{q-1}$$
$$- 128n^2\,[s\,(n-2s) - q + 2]\, Z_{q-2} = 0.$$

Pour intégrer cette équation, supposons

$$Z_q = \Sigma \frac{Z_q^{\sigma}}{\Gamma(\sigma+1)\,\Gamma(q-\sigma+1)}\, x^{2ns+2q-4\sigma},$$

où la sommation se rapporte à σ et s'étend depuis $\sigma = 0$ jusqu'à $\sigma = q$. Toute réduction faite, et ayant mis pour plus de simplicité $n^2 - 2ns = \lambda$, $2ns = \mu$, on obtient pour Z_q^{σ} l'expression

4.
$$[-(q-\sigma)\,\lambda - \sigma\mu + (q-2\sigma)^2]\, Z_q^{\sigma}$$
$$+ (q-\sigma)(\lambda - 2q + 4\sigma + 2)(\lambda - 2q + 4\sigma + 1)\, Z_{q-1}{}^{\sigma}$$
$$+ \sigma\,(\mu + 2q - 4\sigma + 2)(\mu + 2q - 4\sigma + 1)\, Z_{q-1}{}^{\sigma-1}$$
$$+ 16\sigma\,(q-\sigma)\,[\lambda\mu - 2\,(q-2)(\lambda+\mu)]\, Z_{q-2}{}^{\sigma-1} = 0.$$

En supposant la valeur de Z_0^0 égale à l'unité, les valeurs de Z_q^{σ} se trouvent complétement déterminées; malheureusement la loi des coefficients n'est pas en évidence, excepté dans le cas de $\sigma = 0$, ou $\sigma = q$. En calculant les termes successifs, on obtient

$$z_s = (4\alpha)^{s(n-2s)} \,.\, x^{2ns}$$
$$- (4\alpha)^{s(n-2s)-1} \left\{ \begin{array}{l} \frac{1}{1}\lambda \,.\, x^{2ns+2} \\ + \frac{1}{1}\mu \,.\, x^{2ns-2} \end{array} \right.$$
$$+ (4\alpha)^{s(n-2s)-2} \left\{ \begin{array}{l} \frac{1}{1\,.\,2}\, \lambda\,(\lambda-3)\, x^{2ns+4} \\ + \frac{1}{1\,.\,1}\left(\lambda\mu + 2 - \frac{10\lambda\mu}{\lambda+\mu}\right) x^{2ns} \\ + \frac{1}{1\,.\,2}\, \mu\,(\mu-3)\, x^{2ns-4} \end{array} \right.$$
$$- (4\alpha)^{s(n-2s)-3} \left\{ \begin{array}{l} \frac{1}{1\,.\,2\,.\,3}\, \lambda\,(\lambda-4)(\lambda-5)\, x^{2ns+6} \\ + \frac{1}{1\,.\,2\,.\,1}\,\lambda\left(\mu\,(\lambda-3) + 40 - \frac{20\lambda\mu}{\lambda+\mu}\right) x^{2ns+2} \\ + \frac{1}{1\,.\,1\,.\,2}\,\mu\left(\lambda\,(\mu-3) + 40 - \frac{20\lambda\mu}{\lambda+\mu}\right) x^{2ns-2} \\ + \frac{1}{1\,.\,2\,.\,3}\, \mu\,(\mu-4)(\mu-5)\, x^{2ns-6} \end{array} \right.$$
$$+ \&c.$$

On aurait une valeur assez générale de z en multipliant les différentes fonctions z_s chacune par une constante arbitraire, et en sommant les produits; mais dans le cas actuel où z dénote le dénominateur de $\sqrt{k}\sin\operatorname{am} nu$, la valeur convenable de z se réduit à

$$z = z_0 \pm z_1 \ldots \pm z_s \ldots,$$

où s est un nombre entier et positif, entre zéro et $\frac{1}{2}n$ ou $\frac{1}{2}(n-1)$. On aura par exemple dans le cas $n=5$ (les signes étant tous positifs si n est impair, et alternativement positifs et négatifs si n est pair), en supprimant les puissances négatives de α (lesquelles s'entredétruisent):

$$z_0 = 1,$$

$$z_1 = 64\alpha^3x^{10} - \alpha^2(160x^8 + 240x^{12}) + \alpha(140x^6 + 368x^{10} + 360x^{14}) - (50x^4 + 125x^8 + 300x^{12} + 275x^{16}),$$

$$z_2 = 16\alpha^2x^{20} - \alpha(80x^{18} + 20x^{22}) + (170x^{16} + 62x^{20} + 5x^{24}),$$

et de là:

$$z = 1 - 50x^4 + 140\alpha x^6 - (125 + 160\alpha^2)x^8 + (368\alpha + 64\alpha^3)x^{10} - (300 + 240\alpha^2)x^{12} + 360\alpha x^{14} - 105x^{16} - 80\alpha x^{18} + (62 + 16\alpha^2)x^{20} - 20\alpha x^{22} + 5x^{24};$$

ce qui est effectivement la valeur de z que j'ai trouvée dans le mémoire cité pour ce cas particulier.

68.

ON THE APPLICATION OF QUATERNIONS TO THE THEORY OF ROTATION.

[From the *Philosophical Magazine*, vol. XXXIII. (1848), pp. 196—200.]

IN a paper published in the *Philosophical Magazine*, February 1845, [20], I showed how some formulæ of M. Olinde Rodrigues relating to the rotation of a solid body might be expressed in a very simple form by means of Sir W. Hamilton's theory of quaternions. The property in question may be thus stated. Suppose a solid body which revolves through an angle θ round an axis passing through the origin and inclined to the axes of coordinates at angles a, b, c. Let

$$\lambda = \tan \tfrac{1}{2}\theta \cos a, \qquad \mu = \tan \tfrac{1}{2}\theta \cos b, \qquad \nu = \tan \tfrac{1}{2}\theta \cos c,$$

and write

$$\Lambda = 1 + i\lambda + j\mu + k\nu\,;$$

let x, y, z be the coordinates of a point in the body previous to the rotation, x_1, y_1, z_1 those of the same point after the rotation, and suppose

$$\begin{aligned} \Pi &= ix + jy + kz\,, \\ \Pi_1 &= ix_1 + jy_1 + kz_1; \end{aligned}$$

then the coordinates after the rotation may be determined by the formula

$$\Pi_1 = \Lambda \Pi \Lambda^{-1}\,;$$

viz., developing the second side of this equation,

$$\begin{aligned} \Pi_1 = {} & i\,(\alpha\, x + \beta\, y + \gamma z\ ) \\ & + j\,(\alpha' x + \beta' y + \gamma' z) \\ & + k\,(\alpha'' x + \beta'' y + \gamma'' z), \end{aligned}$$

where, putting to abbreviate $\kappa = 1 + \lambda^2 + \mu^2 + \nu^2$, we have

$$\begin{array}{lll} \kappa\alpha = 1 + \lambda^2 - \mu^2 - \nu^2, & \kappa\alpha' = 2(\lambda\mu + \nu), & \kappa\alpha'' = 2(\lambda\nu - \mu), \\ \kappa\beta = 2(\lambda\mu - \nu), & \kappa\beta' = 1 - \lambda^2 + \mu^2 - \nu^2, & \kappa\beta'' = 2(\mu\nu + \lambda), \\ \kappa\gamma = 2(\lambda\nu + \mu), & \kappa\gamma' = 2(\mu\nu - \lambda), & \kappa\gamma'' = 1 - \lambda^2 - \mu^2 + \nu^2; \end{array}$$

these values satisfying identically the well-known system of equations connecting the quantities $\alpha, \beta, \gamma, \alpha', \beta', \gamma', \alpha'', \beta'', \gamma''$.

The quantities a, b, c, θ being immediately known when λ, μ, ν are known, these last quantities completely determine the direction and magnitude of the rotation, and may therefore be termed the coordinates of the rotation; Λ will be the quaternion of the rotation. I propose here to develope a few of the consequences which may be deduced from the preceding formulæ.

Suppose, in the first place, $\Pi = \Lambda - 1$, then $\Pi_1 = \Lambda - 1$, which evidently implies that the point is on the axis of rotation. The equation $\Pi_1 = \Pi$ gives the identical equations

$$\begin{array}{l} \lambda(\alpha - 1) + \mu\,\beta + \nu\,\gamma = 0, \\ \lambda\,\alpha' + \mu(\beta' - 1) + \nu\,\gamma' = 0, \\ \lambda\,\alpha'' + \mu\,\beta'' + \nu(\gamma'' - 1) = 0; \end{array}$$

from which, by changing the signs of λ, μ, ν, we derive

$$\begin{array}{l} \lambda(\alpha - 1) + \mu\,\alpha' + \nu\,\alpha'' = 0, \\ \lambda\,\beta + \mu(\beta' - 1) + \nu\,\beta'' = 0, \\ \lambda\,\gamma + \mu\,\gamma' + \nu(\gamma'' - 1) = 0. \end{array}$$

Hence evidently, whatever be the value of Π,

$$\Lambda\Pi\Lambda^{-1} - \Pi = 0,$$

if after the multiplication i, j, k are changed into λ, μ, ν, a property which will be required in the sequel.

By changing the signs of λ, μ, ν, we also deduce

$$\begin{aligned} \Lambda^{-1}\Pi\Lambda = {} & i\,(\alpha x + \alpha' y + \alpha'' z) \\ & + j\,(\beta x + \beta' y + \beta'' z) \\ & + k\,(\gamma x + \gamma' y + \gamma'' z), \end{aligned}$$

where $\alpha, \beta, \gamma, \alpha', \beta', \gamma', \alpha'', \beta'', \gamma''$ are the same as before.

Let the question be proposed to compound two rotations (both axes of rotation being supposed to pass through the origin). Let L be the first axis, Λ the quaternion of rotation, L' the second axis, which is supposed to be fixed in space, so as not to alter its direction by reason of the first rotation, Λ' the corresponding quaternion of rotation. The combined effect is given at once by

$$\Pi_1 = \Lambda'(\Lambda\Pi\Lambda^{-1})\Lambda'^{-1},$$

that is,

$$\Pi_1 = \Lambda'\Lambda\Pi(\Lambda'\Lambda)^{-1};$$

or since (if Λ_1 be the quaternion for the combined rotation) $\Pi_1 = \Lambda_1\Pi\Lambda_1^{-1}$, we have clearly

$$\Lambda_1 = M_1\Lambda'\Lambda,$$

M_1 denoting the reciprocal of the real part of $\Lambda'\Lambda$, so that

$$M_1^{-1} = 1 - \lambda\lambda' - \mu\mu' - \nu\nu'.$$

Retaining this value, the coefficients of the combined rotation are given by

$$\lambda_1 = M_1(\lambda + \lambda' + \mu'\nu - \mu\nu'),$$
$$\mu_1 = M_1(\mu + \mu' + \nu'\lambda - \nu\lambda'),$$
$$\nu_1 = M_1(\nu + \nu' + \lambda'\mu - \lambda\mu');$$

to which may be joined [if $\kappa_1 = 1 + \lambda_1^2 + \mu_1^2 + \nu_1^2$],

$$\kappa_1 = M_1^2\,\kappa\kappa',$$

κ, κ', κ_1 as before. Λ or Λ' may be determined with equal facility in terms of Λ', Λ_1, or Λ, Λ_1. These formulæ are given in my paper on the rotation of a solid body (*Cambridge Mathematical Journal*, vol. III. p. 226, [6]).

If the axis L' be fixed in the body and moveable with it, its position after the first rotation is obtained from the formula $\Pi_1 = \Lambda\Pi\Lambda^{-1}$ by writing $\Pi = \Lambda' - 1$. Representing by $\Lambda'' - 1$ the corresponding value of Π_1, we have $\Lambda'' = \Lambda\Lambda'\Lambda^{-1}$, which is the value to be used instead of Λ' in the preceding formula for the combined rotation, thus the quaternion of rotation is proportional to $\Lambda\Lambda'\Lambda^{-1}\Lambda$, that is to $\Lambda\Lambda'$. Hence here

$$\Lambda_1 = M_1\Lambda\Lambda',$$

which only differs from the preceding in the order of the quaternion factors. If the fixed and moveable axes be mixed together in any order whatever, the fixed axes taken in order being L, L', ... and the moveable axes taken in order being L_0, L_0' ... then the combined effect of the rotations is given by

$$\Lambda_1 = M \ldots \Lambda''\Lambda'\Lambda\Lambda_0\Lambda_0' \ldots,$$

M being the reciprocal of the real term of the product of all the quaternions.

Suppose next the axes do not pass through the same point. If α, β, γ be the coordinates of a point in L, and

$$\Gamma = \alpha i + \beta j + \gamma k,$$

then the formula for the rotation is

$$\Pi_1 - \Gamma = \Lambda(\Pi - \Gamma)\Lambda^{-1},$$

or

$$\Pi_1 = \Lambda\Pi\Lambda^{-1} - (\Lambda\Gamma\Lambda^{-1} - \Gamma),$$

where the first term indicates a rotation round a parallel axis through the origin, and the second term a translation.

For two axes L, L' fixed in space,

$$\Pi_1 = \Lambda'\Lambda\Pi(\Lambda'\Lambda)^{-1} - (\Lambda'\Gamma'\Lambda'^{-1} - \Gamma') - \Lambda'(\Lambda\Gamma\Lambda^{-1} - \Gamma)\Lambda'^{-1};$$

and so on for any number, the last terms being always a translation. If the two axes are parallel, and the rotations equal and opposite,

$$\Lambda = \Lambda'^{-1},$$

whence

$$\Pi_1 = \Pi + \Lambda'(\Gamma - \Gamma')\Lambda'^{-1}(\Gamma - \Gamma');$$

or there is only a translation. The constant term vanishes if i, j, k are changed into λ', μ', ν', which proves that the translation is in a plane perpendicular to the axes.

Any motion of a solid body being represented by a rotation and a translation, it may be required to resolve this into two rotations. We have

$$\Pi_1 = \Lambda_1\Pi\Lambda_1^{-1} + T,$$

where T is a given quaternion whose constant term vanishes. Hence, comparing this with the general formula just given for the combination of two rotations,

$$\Lambda_1 = M_1\Lambda'\Lambda,$$
$$T = -(\Lambda'\Gamma'\Lambda'^{-1} - \Gamma') - \Lambda'(\Lambda\Gamma\Lambda^{-1} - \Gamma)\Lambda'^{-1},$$

the second of which equations may be simplified by putting $\Lambda'^{-1}T\Lambda' = S$, by which it may be reduced to

$$S = (\Lambda'^{-1}\Gamma'\Lambda' - \Gamma') - (\Lambda\Gamma\Lambda^{-1} - \Gamma),$$

which, with the preceding equation $\Lambda_1 = M_1\Lambda'\Lambda$, contains the solution of the problem. Thus if Λ or Λ' be given, the other is immediately known; hence also S is known. If in the last equation, after the multiplication is completely effected, we change i, j, k into λ, μ, ν, or λ', μ', ν', we have respectively,

$$S = \Lambda'^{-1}\Gamma'\Lambda' - \Gamma', \quad S = -(\Lambda\Gamma\Lambda^{-1} - \Gamma),$$

which are equations which must be satisfied by the coefficients of Γ' and Γ respectively. Thus if the direction of one axis is given, that of the other is known, and the axes must lie in certain known planes. If the position of one of the axes in its plane be assumed, the equation containing S divides itself into three others (equivalent to two independent equations) for the determination of the position in its plane of the other axis. If the axes are parallel, λ, μ, ν are proportional to λ', μ', ν'; or changing i, j, k into λ, μ, ν, or λ', μ', ν', we have $S = 0$; or what is the same thing, $T = 0$, which shows that the translation must be perpendicular to the plane of the two axes.

If p, q, r have their ordinary signification in the theory of rotation, then from the values in the paper in the *Cambridge Mathematical Journal* already quoted,

$$\kappa(ip + jq + kr) = 2\frac{d\Lambda}{dt}\Lambda + \frac{d\kappa}{dt};$$

but I have not ascertained whether this formula leads to any results of importance. It may, however, be made use of to deduce the following property of quaternions, viz. if $\Lambda_1 = M_1\Lambda'\Lambda$, M_1 as before, then

$$\frac{1}{\kappa_1}\left(2\frac{d\Lambda_1}{dt}\Lambda_1 + \frac{d\kappa_1}{dt}\right) = \frac{1}{\kappa}\left(2\frac{d\Lambda}{dt}\Lambda + \frac{d\kappa}{dt}\right),$$

in which the coefficients of Λ' are considered constant.

To verify this *à posteriori*, if in the first place we substitute for κ_1 its value $M_1^2\kappa\kappa'$, we have

$$\frac{d\kappa_1}{dt} = M_1^2\kappa'\frac{d\kappa}{dt} + \frac{2}{M_1}\frac{dM_1}{dt}\kappa_1,$$

and thence

$$\frac{d\Lambda_1}{dt}\Lambda_1 + \frac{1}{M_1}\frac{dM_1}{dt}\kappa_1 = M^2\kappa'\frac{d\Lambda}{dt}\Lambda.$$

Also

$$\frac{d\Lambda_1}{dt}\Lambda_1 = \left(\frac{1}{M_1}\frac{dM_1}{dt}\Lambda_1 + M_1\Lambda'\frac{d\Lambda}{dt}\right)\Lambda_1 = \frac{1}{M_1}\frac{dM_1}{dt}\Lambda_1^2 + M_1^2\Lambda'\frac{d\Lambda}{dt}\Lambda'\Lambda,$$

which reduces the equation to

$$\frac{1}{M_1}\frac{dM_1}{dt}(\Lambda_1^2 + \kappa_1) + M_1^2\Lambda'\frac{d\Lambda}{dt}\Lambda'\Lambda = M_1^2\kappa_1\frac{d\Lambda}{dt}\Lambda.$$

Hence observing that

$$\Lambda_1^2 + \kappa_1 = 2\Lambda_1 = 2M_1\Lambda'\Lambda,$$

and omitting the factor Λ from the resulting equation,

$$\frac{2}{M_1^2}\frac{dM_1}{dt}\Lambda' + \Lambda'\frac{d\Lambda}{dt}\Lambda' = \kappa'\frac{d\Lambda}{dt};$$

or since

$$\frac{1}{M_1} = 1 - \lambda\lambda' - \mu\mu' - \nu\nu',$$

substituting and dividing by Λ', we obtain

$$2\left(\lambda'\frac{d\lambda}{dt} + \mu'\frac{d\mu}{dt} + \nu'\frac{d\nu}{dt}\right) = \kappa'\Lambda'^{-1}\frac{d\Lambda}{dt} - \frac{d\Lambda}{dt}\Lambda',$$

or finally,

$$2\left(\lambda'\frac{d\lambda}{dt} + \mu'\frac{d\mu}{dt} + \nu'\frac{d\nu}{dt}\right) = (1 - i\lambda' - j\mu' - k\nu')\left(i\frac{d\lambda}{dt} + j\frac{d\mu}{dt} + k\frac{d\nu}{dt}\right)$$

$$- \left(i\frac{d\lambda}{dt} + j\frac{d\mu}{dt} + k\frac{d\nu}{dt}\right)(1 + i\lambda' + j\mu' + k\nu')$$

$$= -(i\lambda' + j\mu' + k\nu')\left(i\frac{d\lambda}{dt} + j\frac{d\mu}{dt} + k\frac{d\nu}{dt}\right) - \left(i\frac{d\lambda}{dt} + j\frac{d\mu}{dt} + k\frac{d\nu}{dt}\right)(i\lambda' + j\mu' + k\nu'),$$

which is obviously true.

69.

SUR LES DÉTERMINANTS GAUCHES.

[From the *Journal für die reine und angewandte Mathematik* (Crelle), tom. XXXVIII. (1848), pp. 93—96: continued from t. XXXII. p. 123, **52**.]

J'AI nommé "*Déterminant gauche*" un déterminant formé par un système de quantités $\lambda_{r.s}$, qui satisfont aux conditions

$$\lambda_{s.r} = -\lambda_{r.s}\,(r \neq s) \quad\ldots\ldots\ldots\ldots\ldots\ldots\ldots\ldots\ldots\ldots\ldots (1),$$

(où les valeurs de r, s s'étendent depuis l'unité jusqu'à n).

Or ces déterminants peuvent facilement être exprimés par un système de déterminants pareils, dont les termes satisfont à ces conditions même dans le cas où les valeurs de r et s deviennent égales, ou pour lesquels on a

$$\lambda_{r.s} = -\lambda_{s.r}\,(r \neq s)\;;\quad \lambda_{r.r} = 0 \quad\ldots\ldots\ldots\ldots\ldots\ldots\ldots\ldots\ldots (2)\;;$$

ces déterminants peuvent être nommés "*gauches* et *symétriques*."

En effet, soit Ω le déterminant gauche dont il s'agit, cette fonction peut être présentée sous la forme

$$\Omega = \Omega_0 + \Omega_1\lambda_{1.1} + \Omega_2\lambda_{2.2}\ldots + \Omega_{12}\lambda_{1.1}\lambda_{2.2}\ldots \quad\ldots\ldots\ldots\ldots\ldots\ldots (3),$$

où Ω_0 est ce que devient Ω si $\lambda_{1.1}$, $\lambda_{2.2}$, &c. sont réduits à zéro; Ω_1 est ce que devient le coefficient de $\lambda_{1.1}$ sous la même condition, et ainsi de suite; c'est-à-dire: Ω_0 est le déterminant formé par les quantités $\lambda_{r.s}$ en supposant que ces quantités satisfassent aux conditions (2), et en donnant à r les valeurs $1, 2, 3, \ldots n$; Ω_1 est le determinant formé pareillement en donnant à r, s les valeurs $2, 3, \ldots n$; Ω_2 s'obtient en donnant à r, s les valeurs $1, 3, \ldots n$, et ainsi de suite; cela est aisé de voir si l'on range les quantités $\lambda_{r.s}$ en forme de carré.

Or les déterminants gauches et symétriques (savoir les déterminants dont les termes satisfont aux conditions (2)) se réduisent à zéro pour un n *impair*, et pour un n *pair* aux carrés des fonctions que M. Jacobi a traitées dans son mémoire "Ueber die Pfaff'sche Integrations-Methode" (t. II. [1827], p. 354, de ce journal) et dans le mémoire "Theoria novi multiplicatoris æquationum differentialium" t. XXIX. [1845], p. 236, &c.

En effet, on voit d'abord (par ce que dit M. Jacobi) que le déterminant s'évanouit pour un n *impair*, et que pour un n pair il aura pour *facteur* la fonction dont il s'agit; mais je ne sais pas si l'on a déjà remarqué que l'autre facteur se réduit à la même fonction.

On obtient ces fonctions (dont je reprends ici la théorie), par les propriétés générales d'un déterminant, défini de la manière que voici: en exprimant par $(12 \ldots n)$ une fonction quelconque dans laquelle entrent les nombres symboliques $1, 2, \ldots n$, et par $\pm$, le signe correspondant à une permutation quelconque de ces nombres, la fonction

$$\Sigma \pm (12 \ldots n)$$

(où Σ désigne la somme de tous les termes qu'on obtient en permutant ces nombres d'une manière quelconque) est ce qu'on nomme *Déterminant*. On pourrait encore généraliser cette définition en admettant plusieurs systèmes de nombres $1, 2, \ldots n$; $1', 2', \ldots n'$; ... qui alors devraient être permutés indépendamment les uns des autres; on obtiendrait de cette manière une infinité d'autres fonctions, mentionnées (t. XXX. [1846] p. 7). Dans le cas des déterminants ordinaires, auquel je ne m'arrêterai pas ici, on aura $(12 \ldots n) = \lambda_{\alpha.1} \lambda_{\beta.2} \ldots \lambda_{\kappa.n}$. Pour les cas des fonctions dont il s'agit (les fonctions de M. Jacobi), on supposera n *pair*, et l'on écrira

$$(12 \ldots n) = \lambda_{1.2} \lambda_{3.4} \ldots \lambda_{n-1,\, n},$$

où $\lambda_{r.s}$ sont des quantités quelconques qui satisfont aux équations (1). La fonction sera composée d'un nombre $1.2 \ldots n$ de termes; mais parmi eux il n'y aura que $1.3 \ldots (n-1)$ termes différents qui se trouveront répétés $2^{\frac{1}{2}n}(1.2 \ldots \frac{1}{2}n)$ fois, et qu'on obtiendra en permutant cycliquement d'abord les $n-1$ derniers nombres, puis les $n-3$ derniers nombres de chaque permutation, et ainsi de suite; le signe étant toujours $+$. Il pourra être démontré, comme pour les déterminants, que ces fonctions changent de signe en permutant deux quelconques des nombres symboliques, et qu'elles s'évanouissent si deux de ces nombres deviennent identiques. De plus, en exprimant par $[12 \ldots n]$ la fonction dont il s'agit, la règle qui vient d'être énoncée, donnera pour la formation de ces fonctions:

$$[12 \ldots n] = \lambda_{1.2} [34 \ldots n] + \lambda_{1.3} [4 \ldots n-2] \ldots + \lambda_{1.n} [23 \ldots (n-1)].$$

Cela posé, revenons aux déterminants gauches et symétriques; et soit d'abord n un nombre impair. Alors le déterminant sera composé de plusieurs termes, chacun multiplié par le produit de l'une des quantités $\lambda_{1.2}, \lambda_{1.3}, \ldots \lambda_{1.n}$ par une des quantités $\lambda_{2.1}, \lambda_{3.1}, \ldots \lambda_{n.1}$. Il est facile de voir que pour chaque terme, multiplié par $\lambda_{1.\alpha} \lambda_{\beta.1} (\alpha \neq \beta)$, il existera un terme *égal* et de signe contraire, multiplié par $\lambda_{1.\beta} \lambda_{\alpha.1}$. Or $\lambda_{1.\alpha} \lambda_{\beta.1} = \lambda_{\alpha.1} \lambda_{1.\beta}$: donc ces deux termes se *détruiront*. Restent les termes multipliés par $\lambda_{1.\alpha}\lambda_{\alpha.1}$: le coefficient d'un terme de cette forme sera un déterminant gauche de l'ordre $n-2$, mais précisément de la forme du déterminant même de l'ordre n dont il s'agit. Or n étant impair, $n-2$ le sera également: donc tout déterminant gauche et symétrique s'évanouira, si cela a lieu pour les déterminants pareils d'un ordre inférieur de deux unités. Or pour $n=3$ on a évidemment $\lambda_{1.2}\lambda_{2.3}\lambda_{3.1}+\lambda_{2.1}\lambda_{3.2}\lambda_{1.3}=0$, donc:

"Tout déterminant gauche et symétrique d'un ordre *impair* est zéro."

Soit maintenant n un nombre *pair*. Considérons le déterminant gauche et symétrique plus général qu'on obtient en donnant à r les valeurs α, 2, 3, ... n, et à s les valeurs β, 2, 3, ... n. En développant cette fonction comme on vient de le faire dans le cas d'un n *impair*, il se présentera d'abord un terme multiplié par $\lambda_{\alpha.\beta}$. Mais ce terme sera un déterminant gauche et symétrique de l'ordre $n-1$ (savoir celui que l'on obtiendrait en donnant à r, s les valeurs 2, 3, ... n), et comme $n-1$ est impair, ce terme s'évanouira. Puis le coefficient de $-\lambda_{\alpha.\alpha'}\lambda_{\beta'.\beta}$ (ou de $\lambda_{\alpha.\alpha'}\lambda_{\beta.\beta'}$) sera le déterminant gauche et symétrique qu'on obtient en donnant à r les valeurs 2, 3, ... n (α excepté), et à s les valeurs 2, 3, ... n (β' excepté). En admettant que ce déterminant gauche se réduit à $[(\alpha'+1)\ldots n2\ldots(\alpha'-1)]\,[(\beta'+1)\ldots n2\ldots(\beta'-1)]$, on aura

$$\lambda_{\alpha.\alpha'}[(\alpha'+1)\ldots(\alpha'-1)]\,.\,\lambda_{\beta.\beta'}[(\beta'+1)\ldots(\beta'-1)]$$

pour le terme général, et la somme de tous ces termes se réduira à

$$\{\lambda_{\alpha.2}[34\ldots n]+\lambda_{\alpha.3}[4\ldots n2]+\ldots\}\,.\,\{\lambda_{\beta.2}[34\ldots n]+\lambda_{\beta.3}[4\ldots n2]+\ldots\},$$

ou enfin à $[\alpha 23\ldots n]\,[\beta 23\ldots n]$. Or si le théorème en question a lieu pour $n-2$, il aura lieu aussi pour n. Pour $n=2$ le déterminant est $\lambda_{\alpha.\beta}\lambda_{2.2}-\lambda_{2.\beta}\lambda_{\alpha.2}$, c'est-à-dire $\lambda_{\alpha.2}\lambda_{\beta.2}$ ou $[\alpha 2]\,.\,[\beta 2]$: donc la même chose a toujours lieu et on obtient le théorème que voici:

"Le déterminant gauche et symétrique qu'on obtient en donnant à r les valeurs α, 2, 3, ... n, et à s les valeurs β, 2, 3, ... n (où n est *pair*), se réduit à

$$[\alpha 23\ldots n]\,.\,[\beta 23\ldots n];$$

et en particulier, en donnant à r, s les valeurs 1, 2, ... n, ce déterminant se réduit à $[12\ldots n]^2$."

Appliquons ces théorèmes à la réduction de l'équation (3). En supposant pour plus de simplicité $\lambda_{r.r}=1$, on aura pour un n *pair*:

$$\Omega=[123\ldots n]^2+[34\ldots n]^2+[56\ldots n]^2+\ldots+1;$$
$$\vdots$$
$$+[24\ldots n]^2$$
$$\vdots$$

et pour un *n impair:*

$$\begin{aligned}\Omega = \quad & [23 \ldots n]^2 + [45 \ldots n]^2 + \ldots + 1.\\ & \qquad\qquad\quad \vdots \\ + & [13 \ldots n]^2 \\ & \vdots\end{aligned}$$

Particulièrement pour $n=4$ on obtient:

$$\Omega = [1234]^2 + [12]^2 + [13]^2 + [14]^2 + [23]^2 + [34]^2 + [42]^2 + 1,$$

$$= (\lambda_{1.2}\lambda_{3.4} + \lambda_{1.3}\lambda_{2.4} + \lambda_{1.4}\lambda_{2.3})^2 + \lambda_{1.2}^2 + \lambda_{1.3}^2 + \lambda_{1.4}^2 + \lambda_{3.4}^2 + \lambda_{4.2}^2 + \lambda_{2.3}^2 + 1,$$

et pour $n=3$:

$$\Omega = [23]^2 + [31]^2 + [12]^2 + 1,$$

$$= \lambda_{2.3}^2 + \lambda_{3.1}^2 + \lambda_{1.2}^2 + 1:$$

résultats qui s'accordent parfaitement avec ceux que j'ai donnés dans mon premier mémoire sur ce sujet.

Il serait possible de trouver pour les quantités $\Lambda_{r,s}$ (mémoire cité) des expressions analogues à celles que nous venons de donner pour Ω: mais elles seraient beaucoup plus compliquées.

70.

SUR QUELQUES THÉORÈMES DE LA GÉOMÉTRIE DE POSITION.

[From the *Journal für die reine und angewandte Mathematik* (Crelle), tom. XXXVIII. (1848), pp. 97—104: continued from t. XXXIV. p. 275, **55**.]

§ V.

Le théorème de Pascal appliqué au cas où une conique se réduit à deux droites, donne lieu à un système de neuf points situés trois à trois dans neuf droites qui passent réciproquement trois à trois par les neuf points. Je représenterai ces points par

1, 4, 7; 2, 5, 8; 3, 6, 9,

et les droites par

135, 426, 789; 129, 483, 756; 186, 459, 723.

On peut prendre pour la conique dont il s'agit, deux quelconques de ces droites qui ne se rencontrent pas dans un des neuf points, ou ce qui revient au même, deux quelconques des droites qui appartiennent à un des trois systèmes dans lesquels les droites viennent d'être divisées; alors la troisième droite sera celle qui contient les points de rencontre des côtés opposés de l'hexagone. Par exemple, en prenant pour conique les deux droites 135, 246, l'hexagone sera 123456, et les points de rencontre des côtés opposés, savoir de 12 et 45, de 23 et 56, et de 34 et 61, seront respectivement 9, 7, 8; c'est-à-dire des points situés dans la droite 789; de manière que:

Théorème XV. "La figure formée par neuf points situés 3 à 3 dans 9 droites peut être considérée, de neuf manières différentes, comme résultante du théorème de Pascal appliqué au cas où la conique se réduit à deux droites."

Il y a, comme l'a remarqué M. Graves dans le mémoire cité du *Philosophical Magazine,* une autre manière assez singulière d'envisager la figure. Considérons pour cela les trois triangles 126, 489, 735, qui peuvent être dérivés assez simplement de l'arrangement 135, 426, 789. Le premier de ces triangles est circonscrit au second; car les côtés 26, 61, 12 contiennent respectivement les points 4, 8, 9 ; également le second est circonscrit au troisième ; et le troisième au premier. On tire encore du même arrangement 135, 426, 789 un autre système pareil de triangles; savoir 189, 435, 726. Pareillement les arrangements 129, 483, 756 et 186, 459, 723 donnent lieu chacun à deux systèmes analogues. Donc:

THÉORÈME XVI. La figure formée par neuf points situés 3 à 3 dans 9 droites peut être considérée, de six manières différentes, comme composée de trois triangles circonscrits cycliquement l'un à l'autre.

Représentons maintenant par 1, 2, 3, 4, 5, 6, 7, 8, 9 les neuf points d'inflexion d'une courbe du troisième ordre (six de ces points seront nécessairement imaginaires). Ces points sont, comme on sait, situés 3 à 3 dans douze droites qui passent réciproquement quatre à quatre par les neuf points, et qui peuvent être représentées par

135, 426, 789 ; 129, 483, 756 ; 186, 459, 723 ; 147, 258, 369 ;

de sorte que ce système est un cas particulier de celui que nous venons de considérer. En considérant deux quelconques des droites qui ont un point en commun, par exemple 147, 186, et les deux autres droites qui passent par ce même point 135, 129, on obtient par là deux quadrilatères 4876 et 3952 qui ont pour centre commun le point 7 et dont chacun est circonscrit à l'autre. On aurait pu obtenir par ces mêmes droites 147, 186, 135, 129 des autres systèmes pareils ; donc

THÉORÈME XVII. Huit points quelconques, parmi neuf points d'inflexion d'une courbe du troisième ordre, peuvent être considérés de trois manières différentes comme formant deux quadrilatères circonscrits l'un à l'autre. Les diagonales de ces six quadrilatères se réduisent à quatre droites qui passent par le neuvième point d'inflexion.

§ VI. *Sur les figures réciproques.*

Soient $\xi : \omega$, $\eta : \omega$, $\rho : \omega$ les coordonnées d'un point. Les deux plans définis par les équations

$$\left.\begin{array}{r}(a\ \xi+b\ \eta+c\ \rho+d\ \omega)\ x\\+(a'\ \xi+b'\ \eta+c'\ \rho+d'\ \omega)\ y\\+(a''\ \xi+b''\ \eta+c''\ \rho+d''\ \omega)\ z\\+(a'''\xi+b'''\eta+c'''\rho+d'''\omega)\ w\end{array}\right\}=0,\quad\text{et}\quad\left.\begin{array}{r}(a\xi+a'\eta+a''\rho+a'''\omega)\ x\\+(b\xi+b'\eta+b''\rho+b'''\omega)\ y\\+(c\xi+c'\eta+c''\rho+c'''\omega)\ z\\+(d\xi+d'\eta+d''\rho+d'''\omega)\ w\end{array}\right\}=0,$$

seront les *plans réciproques* du point donné.

Il peut arriver que le plan réciproque d'un point quelconque passe par ce point même; ce qui implique aussi l'identité des deux plans réciproques. Ce cas particulier a été l'objet des recherches de M. Möbius. Je parlerai dans la suite des réciproques de cette espèce en les appelant *Réciproques gauches.*

Mais généralement, pour que les plans réciproques d'un point passent par ce point même, il faut que le point soit situé sur une certaine surface du second ordre; et alors tous les réciproques des points situés sur cette surface seront des plans tangents d'une autre surface du second ordre. Ces deux surfaces pourront être identiques ; ce qui implique aussi l'identité des deux plans réciproques. Ce cas particulier constitue en effet la théorie connue des *Polaires réciproques.*

Je me propose ici d'examiner la théorie du cas général, où les deux surfaces ne sont point identiques. On pourra démontrer que dans ce cas les deux surfaces ont nécessairement en commun quatre droites: ces droites se rencontrent de manière à former un quadrilatère gauche que je représenterai par $ABCD$. Il est évident que les deux surfaces se touchent aux points A, B, C, D. En effet, le plan DAB est le plan tangent de l'une et de l'autre de ces surfaces au point A ; et de même ABC, BCD et CDA sont respectivement les plans tangents aux points B, C et D. Les deux réciproques du point A se réduisent à ce même plan DAB, et il en est également pour les points B, C, D: il suit de là que les droites AC et BD sont réciproques l'une à l'autre, tandis que les droites AB, BC, CD, DA sont respectivement réciproques chacune à elle-même.

Les réciproques d'un point quelconque passent par la droite qui est l'intersection du plan polaire du point par rapport à la première surface, et de la réciproque gauche du point déterminé d'une manière qui sera expliquée tout à l'heure. Donc les réciproques d'un point quelconque de la première surface passent par la droite qui est l'intersection du plan tangent de la première surface au point dont il s'agit, et de la réciproque gauche du même point; de manière que si cette droite d'intersection était connue, les réciproques d'un point de la première surface seraient les plans tangents de la seconde surface menés par cette droite d'intersection ; ou, pour trouver les réciproques d'un point de la première surface, on n'a qu'à chercher la réciproque gauche dont je viens de parler.

Dans ce système de réciproques gauches, les réciproques gauches des points A, B, C, D sont (comme dans le système des réciproques que nous considérons) les plans DAB, ABC, BCD et CAD, et de là les droites AC et DB sont réciproques gauches l'une à l'autre, tandis que les droites AB, BC, CD, DA sont réciproques gauches chacune à elle-même. Mais cela ne suffit pas pour déterminer le système des réciproques gauches. En effet, le système des réciproques que nous considérons n'est pas complètement déterminé au moyen des deux surfaces ; il contient encore une quantité arbitraire (on peut facilement se satisfaire de cela). Or remarquons que la réciproque gauche d'un plan quelconque, passant par la droite AB (ou par l'une quelconque des droites AB, BC, CD, DA), est située dans cette même droite. Considérons un plan donné quelconque pAB passant par cette droite AB; la réciproque gauche de ce plan sera un point P de la droite AB, dont la position pourra être prise à volonté. Au moyen de ce plan pAB et de sa réciproque gauche P, sur la droite AB, on pourra facilement construire le système complète des réciproques gauches. Car soit q un point quelconque, et représentons par Q la réciproque gauche du plan qAB (Q sera aussi un point de la droite AB) ; considérons les quatre plans DAB, CAB, pAB, qAB qui se rencontrent selon la droite AB, et qui ont respectivement pour réciproques gauches les points A, B, P, Q (situés sur cette même droite AB) : le rapport anharmonique des quatre plans sera égal au rapport anharmonique des quatre points; ce qui suffit pour déterminer le point Q, puisque les quatres plans et les trois autres points sont donnés, et la construction graphique pour déterminer ce point Q est parfaitement connue. Il est d'ailleurs évident que la droite menée par un point donné, de sorte qu'elle rencontre des droites réciproques gauches l'une à l'autre, est située dans la réciproque gauche de ce point. Donc, en menant par le point q la droite qui rencontre les deux droites AC et BD, le plan passant par cette droite et par le point Q, est la réciproque gauche du point q qu'il s'agissait de trouver[1]. Également, on pourrait construire la réciproque gauche d'un plan donné.

Donc enfin, pour trouver les réciproques d'un point de la première surface:

(A) "Construisez le plan tangent à ce point de la première surface, et construisez de la manière expliquée ci-dessus, la réciproque gauche du point. Par la droite d'intersection de ces deux plans menez deux plans tangents à la seconde surface: ces deux plans seront les réciproques qu'il s'agissait de trouver."

On pourra construire d'une manière analogue les réciproques d'un plan tangent de la première surface. En effet:

(B) "En construisant les réciproques du point de contact avec la première surface du plan dont il s'agit, les points de contact avec la seconde surface, de ces deux plans, seront les réciproques dont il s'agissait."

Également, pour trouver les réciproques d'un plan tangent donné de la seconde surface:

(C) "Construisez le point de contact de ce plan avec la seconde surface, et construisez la réciproque gauche de ce même plan: la droite menée par ces deux points rencontrera la première surface dans deux points qui seront les réciproques que l'on désirait."

Et pour trouver les réciproques d'un point de la seconde surface:

(D) "Construisez les réciproques du plan tangent passant par le point donné de la seconde surface: les plans tangents à la première surface passant par ces deux points, seront les réciproques qu'il s'agissait de trouver."

En effet les théorèmes (C, D) ne sont que des transformations des théorèmes (A, B) au moyen de la théorie des polaires réciproques.

Enfin, pour trouver les réciproques d'un point quelconque, menez par ce point trois plans tangents ou à la première ou à la seconde surface, et construisez les réciproques de ces plans au moyen du théorème (B ou C): les deux plans menés par ces points réciproques, pris trois et trois ensemble, et combinés de manière que l'intersection des deux plans coïncide avec l'intersection de la polaire par rapport à la première surface et de la réciproque gauche du point donné, selon la remarque ci-dessus, seront les réciproques cherchées; et de la même manière pour les réciproques d'un plan quelconque.

[1] Il y a un cas très simple qui mérite d'être considéré; savoir celui où le point P coïncide avec A. Dans ce cas Q coïncide aussi avec A, et la réciproque gauche de q se réduit au plan qAC. De même, si les points P, B coïncident, la réciproque gauche de q se réduit au plan qBD.

Je n'essaierai pas d'énumérer ici le grand nombre de relations descriptives qui pourraient être tirées des constructions qu'on vient d'expliquer. L'on remarquera sans peine l'analogie parfaite qui existe pour toute cette théorie et la théorie correspondante de la géométrie à deux dimensions, telle que M. Plücker l'a exposée, "System der analytischen Geometrie," [4°, Berlin, 1835], pp. 78—83. On pourra aussi consulter sur ce point un mémoire [61] que je viens de composer pour le *Cambridge and Dublin Mathematical Journal,* et qui paraîtra prochainement.

La vérification analytique de ces théorèmes donne lieu à des développements assez intéressants. Pour obtenir l'équation de la première des surfaces du second ordre, il suffit d'écrire ξ, η, ρ, ω au lieu de x, y, z, w, dans l'équation de l'un ou de l'autre des plans réciproques. En supposant que $x=0$, $y=0$, $z=0$, $w=0$ soient des plans conjugués par rapport à la surface, et en remarquant que chacune de ces coordonnées peut être censée contenir un facteur constant, l'équation de la première surface peut être écrite sous la forme

$$x^2+y^2+z^2+w^2=0.$$

On aura alors pour a, b, c, d; a', b', c', d'; a'', b'', c'', d''; a''', b''', c''', d''' un système de la forme $l, -h, g, -a$; $h, l, -f, -b$; $-g, f, l, -c$; a, b, c, l, et les équations des plans réciproques deviendront

$$(\xi x+\eta y+\rho z+\omega w) \pm \left\{\begin{array}{l} x(\quad . \quad -h\eta+g\rho-a\omega) \\ +y(\ \ h\xi \quad . \quad -f\rho-b\omega) \\ +z(-g\xi+f\eta \quad . \quad -cw) \\ +w(\ \ a\xi+b\eta+c\rho \quad . \quad) \end{array}\right\}=0;$$

ce qui prouve le théorème énoncé ci-dessus; savoir que les deux réciproques passent par la droite d'intersection de la polaire et de la réciproque gauche du point. On obtient sans difficulté, pour l'équation de la seconde surface :

$$(x-hy+gz-aw)^2+(hx+y-fz-bw)^2+(-gx+fy+z-cw)^2+(ax+by+cz+w)^2=0,$$

ou sous une autre forme plus commode:

$$(x^2+y^2+z^2+w^2)$$
$$+(\,.-hy+gz-aw)^2+(hx\,.-fz-bw)^2+(-gx+fy\,.-cw)^2+(ax+by+cz\,.\,)^2=0.$$

Les coordonnées des points A, B, C, D seront déterminées au moyen des expressions

$$\frac{.-hy+gz-aw}{x}=\frac{hx\,.-fz-bw}{y}=\frac{-gx+fy\,.-cw}{z}=\frac{ax+by+cz\,.}{w},$$

ou, en introduisant la quantité indéterminée s:

$$\begin{aligned} sx-hy+gz-aw&=0,\\ hx+sy-fz-bw&=0,\\ -gx+fy+sz-cw&=0,\\ ax+by+cz+sw&=0. \end{aligned}$$

En effet, on démontrera aisément, non seulement que les points définis par ces équations sont situés dans l'intersection des deux surfaces, mais aussi que les deux surfaces se touchent dans ces points-ci; ce qui fait voir qu'en combinant convenablement les quatre points, les deux surfaces doivent se couper (comme nous l'avons déjà avancé), suivant les droites AB, BC, CD, DA.

Je vais examiner encore de plus près le système d'équations qui déterminent ces points. On en tire tout de suite, pour déterminer s, l'équation

$$s^4 + \mu s^2 + \Theta^2 = 0,$$

où l'on a fait pour abréger:

$$\mu = a^2 + b^2 + c^2 + f^2 + g^2 + h^2,$$
$$\Theta = af + bg + ch.$$

Supposons encore

$$s^2 f + a\Theta = A, \quad s^2 g + b\Theta = B, \quad s^2 h + c\Theta = C$$

et

$$s^2 a + f\Theta = F, \quad s^2 b + g\Theta = G, \quad s^2 c + h\Theta = H:$$

l'équation qui détermine s, pourra être écrite sous cette autre forme:

$$AF + BG + CH = 0.$$

J'ai trouvé que les équations des droites AC et BD peuvent être présentées sous les formes assez élégantes

$$\left.\begin{array}{rrrr} & -Hy & +Gz & -Aw = 0, \\ Hx & . & -Fz & -Bw = 0, \\ -Gx & +Fy & . & -Cw = 0, \\ Ax & +By & +Cz & . \quad = 0, \end{array}\right\} \qquad \left.\begin{array}{rrrr} & -Cy & +Bz & -Fw = 0, \\ Cx & . & -Az & -Gw = 0, \\ -Bx & +Ay & . & -Hw = 0, \\ Fx & +Gy & +Hz & . \quad = 0, \end{array}\right\}$$

où chacun de ces systèmes d'équations n'est équivalent qu'à un système de deux équations. En entrechangeant les racines de l'équation en s^2, c'est-à-dire en écrivant $\dfrac{\Theta^2}{s^2}$ au lieu de s^2, les deux systèmes ne font que s'entrechanger; comme en effet cela doit être.

Enfin, les équations des plans ACD, BAC, ou des plans ABD, BDC peuvent être exprimées sous les formes

$$PQ = (\,. - Hy + Gz - Aw)^2 + (Hx . - Fz - Bw)^2 + (-Gx + Fy . - Cw)^2 + (Ax + By + Cz\,.)^2 = 0,$$
$$RS = (\,. - Cy + Bz - Fw)^2 + (Cx . - Az - Gw)^2 + (-Bx + Ay . - Hw)^2 + (Fx + Gy + Hz\,.)^2 = 0\,;$$

équations desquelles on tire les relations identiques

$$PQ + \;\; RS = -s^2(\mu^2 - 4\Theta^2)(x^2 + y^2 + z^2 + w^2),$$
$$\Theta^2 PQ + s^4 RS = s^4(\mu^2 - 4\Theta^2)$$
$$\times\,[(\,. - hy + gz - aw)^2 + (hx \ldots fz - bw)^2 + (-gx - fy . - cw)^2 + (ax + by + cz\,.)^2],$$

qui mettent en évidence ce que l'on savait déjà, savoir, que les deux surfaces contiennent les droites AB, BC, CD, DA.

Mettons pour un moment, pour abréger:

$$\begin{array}{llll} . & -h\eta & +g\rho - a\omega = l\,, \\ h\xi & . & -f\rho - b\omega = m, \\ -g\xi - f\eta & & . \quad -c\omega = n\,, \\ a\xi + b\eta & +c\rho & . \quad = p\,, \end{array}$$

de sorte que l'équation de la réciproque gauche du point (ξ, η, ρ, ω) devient

$$lx + my + nz + pw = 0\,;$$

supposons de plus que, en combinant cette équation avec les équations des droites AC et BD respectivement, on obtient $x : y : z : w = X : Y : Z : W$ et $x : y : z : w = X_, : Y_, : Z_, : W_,$ respectivement. De là on tire

$$\left.\begin{array}{l} X = \quad . \quad - Cm + Bn - Fp\,, \\ Y = \quad Cl \quad . \quad - An - Gp\,, \\ Z = - Bl + Am \quad . \quad - Hp, \\ W = \quad Fl + Gm + Hn \quad . \end{array}\right\} \qquad \left.\begin{array}{l} X_, = \quad . \quad - Hm + Gn - Ap, \\ Y_, = \quad Hl \quad . \quad + Fn - Bp, \\ Z_, = - Gl + Fm \quad . \quad - Cp\,, \\ W_, = \quad Al + Bm + Cn \quad . \end{array}\right\}$$

et de là, identiquement:

$$\begin{aligned} \Theta\,(\Theta^2 - s^4)\;\xi &= \Theta X - s^2 X_,\,, \\ \Theta\,(\Theta^2 - s^4)\;\eta &= \Theta Y - s^2 Y_,\,, \\ \Theta\,(\Theta^2 - s^4)\,\rho &= \Theta Z - s^2 Z_,\,, \\ \Theta\,(\Theta^2 - s^4)\;\omega &= \Theta W - s^2 W_,\,; \end{aligned}$$

ce qui correspond en effet au théorème énoncé ci-dessus, savoir que la réciproque gauche d'un point rencontre les droites AC, BD en deux points tels que la droite passant par ces deux points passe par le point même. Les démonstrations des différents théorèmes d'analyse dont je me sers, n'ont point de difficultés.

71.

NOTE SUR LES FONCTIONS DU SECOND ORDRE.

[From the *Journal für die reine und angewandte Mathematik* (Crelle), tom. XXXVIII. (1848), pp. 105—106.]

SOIENT $x, y, \ldots$ des variables dont le nombre est $2n$ ou $2n+1$; représentons par $\xi, \eta, \ldots$ un nombre égal de fonctions linéaires des variables $x, y, \ldots$ et soit

$$\begin{aligned} U &= \xi x \;\; + \eta y \;\; + \ldots \\ &= Ax^2 + By^2 + \ldots + 2Hxy + \ldots \end{aligned}$$

et

$$V = \begin{vmatrix} . & \xi, & \eta \ldots \\ \xi, & A, & H \\ \eta, & H, & B \\ \vdots & & \end{vmatrix}.$$

J'ai trouvé que les deux fonctions U et V peuvent être réduites à la forme

$$\begin{aligned} U &= \lambda\,\Omega + \mu\,\Theta, \\ V &= \lambda'\Omega + \mu'\Theta, \end{aligned}$$

où λ, μ, λ', μ' sont des coefficients constants, Ω est une fonction du second ordre des n variables (fonctions linéaires de $x, y, \ldots$), et Θ est une fonction de n ou de $n+1$ variables (fonctions linèaires de $x, y, \ldots$); selon que le nombre des variables $x, y, \ldots$ est $2n$ ou $2n+1$.

Par exemple pour trois variables x, y, z, on a

$$\begin{aligned} U &= \lambda\,x^2 + \mu\,yz, \\ V &= \lambda' x^2 + \mu' yz; \end{aligned}$$

et les coniques représentées par les équations $U=0$, $V=0$ ont entre elles un contact double. Cela se rapporte à la théorie des réciproques dans la géométrie plane, telle que M. Plücker l'a présentée dans son "System der analytischen Geometrie" [4°, Berlin, 1835].

De même pour quatre variables x, y, z, w on a

$$U = \lambda\, xy + \mu\, zw,$$
$$V = \lambda' xy + \mu' zw,$$

et les surfaces représentées par les équations $U = 0$, $V = 0$ se coupent dans quatre droites, ou bien se touchent aux quatre sommets d'un quadrilatère gauche. Cela se rapporte également à la théorie des réciproques dans l'espace: théorie dont j'ai parlé dans mon Mémoire "Sur quelques théorèmes de la géométrie de position," § VI, [70].

Il y a à remarquer que dans le cas où les coefficients de $x, y, \ldots$ dans les fonctions $\xi, \eta, \ldots$ forment un système symétrique, c'est-à-dire, où

$$\begin{aligned} \xi &= Ax + Hy + \ldots, \\ \eta &= Hx + By + \ldots, \\ &\vdots \end{aligned}$$

on a $\lambda : \mu = \lambda' : \mu'$, et de là $V = KU$, K étant donné par l'équation

$$K = \begin{vmatrix} A, & H, \ldots \\ H, & B, \\ \vdots & \end{vmatrix}$$

ce qui est connu.

Remarquons enfin que dans le cas général où

$$\begin{aligned} \xi &= a\,x + b\,y + \ldots, \\ \eta &= a'x + b'y + \ldots, \\ &\vdots \end{aligned}$$

la fonction V ne change pas de valeur en écrivant au lieu de ces expressions:

$$\begin{aligned} \xi &= ax + a'y + \ldots, \\ \eta &= bx + b'y + \ldots, \\ &\vdots \end{aligned}$$

propriété qui se rapporte aussi à la théorie des réciproques.

72.

NOTE ON THE THEORY OF PERMUTATIONS.

[From the *Philosophical Magazine*, vol. XXXIV. (1849), pp. 527—529.]

It seems worth inquiring whether the distinction made use of in the theory of determinants, of the permutations of a series of things all of them different, into positive and negative permutations, can be made in the case of a series of things not all of them different. The ordinary rule is well known, viz. permutations are considered as positive or negative according as they are derived from the primitive arrangement by an even or an odd number of inversions (that is, interchanges of two things); and it is obvious that this rule fails when two or more of the series of things become identical, since in this case any given permutation can be derived indifferently by means of an even or an odd number of inversions. To state the rule in a different form, it will be convenient to enter into some preliminary explanations. Consider a series of n things, all of them different, and let $abc\ldots$ be the primitive arrangement; imagine a symbol such as (xyz) (u) $(vw)\ldots$ where x, y, &c., are the entire series of n things, and which symbol is to be considered as furnishing a rule by which a permutation is to be derived from the primitive arrangement $abc\ldots$ as follows, viz. the (xyz) of the symbol denotes that the letters x, y, z in the primitive arrangement $abc\ldots$ are to be interchanged x into y, y into z, z into x. The (u) of the symbol denotes that the letter u in the primitive arrangement $abc\ldots$ is to remain unaltered. The (vw) of the symbol denotes that the letters v, w in the primitive arrangement are to be interchanged v into w and w into v, and so on. It is easily seen that any permutation whatever can be derived (and derived in one manner only) from the primitive arrangement by means of a rule such as is furnished by the symbol in question[1]; and moreover that the number of inversions requisite in order to obtain the permutation by means of the rule in question, is always the smallest number of

[1] See on this subject Cauchy's "Mémoire sur les Arrangemens &c.", *Exercises d'Analyse et de Physique Mathématique*, t. III. [1844], p. 151.

inversions by which the permutation can be derived. Let α, β ... be the number of letters in the components (xyz), (u) (vw), &c., λ the number of these components. The number of inversions in question is evidently $\overline{\alpha-1}+\overline{\beta-1}+$&c., or what comes to the same thing, this number is $(n-\lambda)$. It will be convenient to term this number λ the exponent of irregularity of the permutation, and then $(n-\lambda)$ may be termed the supplement of the exponent of irregularity. The rule in the case of a series of things, all of them different, may consequently be stated as follows: "a permutation is positive or negative according as the supplement of the exponent of irregularity is even or odd." Consider now a series of things, not all of them different, and suppose that this is derived from the system of the same number of things abc ... all of them originally different, by supposing for instance $a=b=$ &c., $f=g=$ &c. A given permutation of the system of things not all of them different, is of course derivable under the supposition in question from several different permutations of the series abc .... Considering the supplements of the exponents of irregularity of these last-mentioned several permutations, we may consider the given permutation as positive or negative according as the *least of these numbers* is even or odd. Hence we obtain the rule, "a permutation of a series of things not all of them different, is positive or negative according as the minimum supplement of irregularity of the permutation is even or odd, the system being considered as a particular case of a system of the same number of things all of them different, and the given permutation being successively considered as derived from the different permutations which upon this supposition reduce themselves to the given permutation." This only differs from the rule, "a permutation of a series of things, not all of them different, is positive or negative according as the minimum number of inversions by which it can be obtained is even or odd, the system being considered &c.," inasmuch as the former enunciation is based upon and indicates a direct method of determining the minimum number of inversions requisite in order to obtain a given permutation; but the latter is, in simple cases, of the easier application. As a very simple example, treated by the former rule, we may consider the permutation 1212 derived from the primitive arrangement 1122. Considering this primitive arrangement as a particular case of $abcd$, there are four permutations which, on the suppositions $a=b=1$, $c=d=2$, reduce themselves to 1212, viz. $acbd$, $bcad$, $adbc$, $bdac$, which are obtained by means of the respective symbols $(a)(bc)(d)$; $(abc)(d)$; $(a)(bdc)$; $(abdc)$, the supplements of the exponents of irregularity being therefore 1, 2, 2, 3, or the permutation being negative; in fact it is obviously derivable by means of an inversion of the two mean terms.

73.

ABSTRACT OF A MEMOIR BY DR HESSE ON THE CONSTRUCTION OF THE SURFACE OF THE SECOND ORDER WHICH PASSES THROUGH NINE GIVEN POINTS.

[From the *Cambridge and Dublin Mathematical Journal*, vol. IV. (1849), pp. 44—46.]

THE construction to be presently given of the surface of the second order which passes through nine given points, is taken from a memoir by Dr Hesse (*Crelle*, t. XXIV. [1842], p. 36). It depends upon the following lemma, which is there demonstrated.

LEMMA. The polar plane of a fixed point P with respect to any surface of the second order passing through seven given points, passes through a fixed point Q (which may be termed the harmonic pole of the point P with respect to the system of surfaces of the second order).

PROBLEM. Given the seven points 1, 2, 3, 4, 5, 6, 7, and a point P, to construct the harmonic pole Q of the point P with respect to the system of surfaces of the second order passing through the seven points.

The required point Q may be considered as the intersection of the polar planes of the point P with respect to any three hyperboloids, each of which passes through the seven given points; any such hyperboloid may be considered as determined by means of three of its generating lines. These considerations lead to the construction following.

1. Connecting the points 1 and 2, and also the points 3 and 4, by two straight lines, and determining the three lines, each of which passes through one of the points 5, 6, 7, and intersects both of the first-mentioned lines, the three lines so determined are generating lines of a hyperboloid passing through the seven points.

Two other systems of generating lines (belonging to two new hyperboloids) are determined by the like construction, interchanging the points 1, 2, 3, 4. And by interchanging all the seven points we obtain 105 systems of generating lines (belonging to as many different hyperboloids, unless some of these hyperboloids are identical).

2. It remains to be shown how the polar plane of the point P with respect to one of the 105 hyperboloids may be constructed. Drawing through the point P three lines, each of which passes through two of the three given generating lines of the hyperboloid in question, the points of intersection of the lines so determined with the generating lines which they respectively intersect, are points of the hyperboloid. Hence, constructing upon each of the three lines in question the harmonic pole of the point P with respect to the two points of intersection, the plane passing through the three harmonic poles is the polar plane of P with respect to the hyperboloid. Hence, constructing the polar planes of P with respect to any three of the 105 hyperboloids, the point of intersection of these three polar planes is the required point Q.

PROBLEM. To construct the polar plane of a point P with respect to the surface of the second order which passes through nine given points 1, 2, 3, 4, 5, 6, 7, 8, 9.

Consider any seven of the nine points, e.g. the points 1, 2, 3, 4, 5, 6, 7, and construct the harmonic pole of the point P with respect to the system of surfaces of the second order passing through these seven points. By permuting the different points we obtain 36 different points Q, all of which lie in the same plane. This plane (which is of course determined by any three of the thirty-six points) is the required polar plane. Hence we obtain the solution of

PROBLEM. To construct the surface of the second order which passes through nine given points 1, 2, 3, 4, 5, 6, 7, 8, 9.

Assuming the point P arbitrarily, construct the polar plane of this point with respect to the surface of the second order passing through the nine points. Join the point P with any one of the nine points, e.g. the point 1, and on the line so formed determine the harmonic pole R of the point 1 with respect to the point P, and the point where the line $P1$ is intersected by the polar plane. R is a point of the required surface of the second order, which surface is therefore determined by giving every possible position to the point P.

This construction is the complete analogue of Pascal's theorem *considered as a construction for describing the conic section which passes through five given points.* And it would appear that the principles by means of which the construction is obtained ought to lead to the analogue of Pascal's theorem considered in its ordinary form, that is, *as a relation between six points of a conic,* or in other words to the solution of the problem to determine the relation between ten points of a surface of the second order; but this problem, one of the most interesting in the theory of surfaces of the second order, remains as yet unsolved. The problem last mentioned was proposed as a prize question by the Brussels Academy, which subsequently proposed the more general question to determine the analogue of Pascal's theorem for surfaces of the second order. This of course admitted of being answered in a variety of different ways, according to the different ways of viewing the theorem of Pascal. Thus, M. Chasles, considering Pascal's theorem as a property of a conic intersected by the three sides of a triangle, discovered the following very elegant analogous theorem for surfaces of the second order.

"The six edges of a tetrahedron may be considered as intersecting a surface of the second order in twelve points lying three and three upon four planes, each one of which contains three points lying on edges which pass through the same angle of the tetrahedron; these planes meet the faces opposite to these angles in four straight lines which are generating lines (of the same species) of a certain hyperboloid."

It is hardly necessary to remark that all the properties involved in the present memoir are such as to admit of being transformed by the theory of reciprocal polars.

74.

ON THE SIMULTANEOUS TRANSFORMATION OF TWO HOMOGENEOUS FUNCTIONS OF THE SECOND ORDER.

[From the *Cambridge and Dublin Mathematical Journal*, vol. IV. (1849), pp. 47—50.]

THE theory of the simultaneous transformation by linear substitutions of two homogeneous functions of the second order has been developed by Jacobi in the memoir "De binis quibuslibet functionibus &c., *Crelle*, t. XII. [1834], p. 1; but the simplest method of treating the problem is the one derived from Mr Boole's Theory of Linear Transformations, combined with the remark in his "Notes on Linear Transformations," in the *Cambridge Mathematical Journal*, vol. IV. [1845], p. 167. As I shall have occasion to refer to the results of this theory in the second part of my paper "On the Attraction of Ellipsoids," in the present number of the *Journal* [75], I take this opportunity of developing the formula in question; considering for greater convenience the case of three variables only.

Suppose that by a linear transformation,

$$\begin{aligned} x &= \alpha\ x_1 + \beta\ y_1 + \gamma\ z_1, \\ y &= \alpha'\, x_1 + \beta'\, y_1 + \gamma'\, z_1, \\ z &= \alpha'' x_1 + \beta'' y_1 + \gamma'' z_1, \end{aligned}$$

we have identically,

$$ax^2 + by^2 + cz^2 + 2fyz + 2gxz + 2hxy = a_1x_1^2 + b_1y_1^2 + c_1z_1^2 + 2f_1y_1z_1 + 2g_1z_1x_1 + 2h_1x_1y_1,$$

$$Ax^2 + By^2 + Cz^2 + 2Fyz + 2Gzx + 2Hxy = A_1x_1^2 + B_1y_1^2 + C_1z_1^2 + 2F_1y_1z_1 + 2G_1z_1x_1 + 2H_1x_1y_1.$$

Of course, whatever be the values of a, b, c, f, g, h, the same transformation gives

$$\mathrm{a}x^2 + \mathrm{b}y^2 + \mathrm{c}z^2 + 2\mathrm{f}yz + 2\mathrm{g}zx + 2\mathrm{h}xy = \mathrm{a}_1x_1^2 + \mathrm{b}_1y_1^2 + \mathrm{c}_1z_1^2 + 2\mathrm{f}_1y_1z_1 + 2\mathrm{g}_1z_1x_1 + 2\mathrm{h}_1x_1y_1,$$

provided that we have

$$\begin{cases} \mathrm{a}_1 = \mathrm{a}\alpha^2 + \mathrm{b}\alpha'^2 + \mathrm{c}\alpha''^2 + 2\mathrm{f}\alpha'\alpha'' + 2\mathrm{g}\alpha''\alpha + 2\mathrm{h}\alpha\alpha', \\ \mathrm{b}_1 = \mathrm{a}\beta^2 + \mathrm{b}\beta'^2 + \mathrm{c}\beta''^2 + 2\mathrm{f}\beta'\beta'' + 2\mathrm{g}\beta''\beta + 2\mathrm{h}\beta\beta', \\ \mathrm{c}_1 = \mathrm{a}\gamma^2 + \mathrm{b}\gamma'^2 + \mathrm{c}\gamma''^2 + 2\mathrm{f}\gamma'\gamma'' + 2\mathrm{g}\gamma''\gamma + 2\mathrm{h}\gamma\gamma', \\ \mathrm{f}_1 = \mathrm{a}\beta\gamma + \mathrm{b}\beta'\gamma' + \mathrm{c}\beta''\gamma'' + \mathrm{f}(\beta'\gamma'' + \beta''\gamma') + \mathrm{g}(\beta''\gamma + \beta\gamma'') + \mathrm{h}(\beta\gamma' + \beta'\gamma), \\ \mathrm{g}_1 = \mathrm{a}\gamma\alpha + \mathrm{b}\gamma'\alpha' + \mathrm{c}\gamma''\alpha'' + \mathrm{f}(\gamma'\alpha'' + \gamma''\alpha') + \mathrm{g}(\gamma''\alpha + \gamma\alpha'') + \mathrm{h}(\gamma\alpha' + \gamma'\alpha), \\ \mathrm{h}_1 = \mathrm{a}\alpha\beta + \mathrm{b}\alpha'\beta' + \mathrm{c}\alpha''\beta'' + \mathrm{f}(\alpha'\beta'' + \alpha''\beta') + \mathrm{g}(\alpha''\beta + \alpha\beta'') + \mathrm{h}(\alpha\beta' + \alpha'\beta). \end{cases}$$

Representing for a moment the equations between the pairs of functions of the second order by

$$u = u_1, \quad U = U_1, \quad v = v_1,$$

we have, whatever be the value of λ,

$$\lambda u + U + v = \lambda u_1 + U_1 + v_1.$$

Hence, if

$$\begin{vmatrix} \alpha, & \beta, & \gamma \\ \alpha', & \beta', & \gamma' \\ \alpha'', & \beta'', & \gamma'' \end{vmatrix} = \Pi; \text{ then}$$

$$\begin{vmatrix} \lambda a_1 + A_1 + \mathrm{a}_1, & \lambda h_1 + H_1 + \mathrm{h}_1, & \lambda g_1 + G_1 + \mathrm{g}_1 \\ \lambda h_1 + H_1 + \mathrm{h}_1, & \lambda b_1 + B_1 + \mathrm{b}_1, & \lambda f_1 + F_1 + \mathrm{f}_1 \\ \lambda g_1 + G_1 + \mathrm{g}_1, & \lambda f_1 + F_1 + \mathrm{f}_1, & \lambda c_1 + C_1 + \mathrm{c}_1 \end{vmatrix} = \Pi^2 \begin{vmatrix} \lambda a + A + \mathrm{a}, & \lambda h + H + \mathrm{h}, & \lambda g + G + \mathrm{g} \\ \lambda h + H + \mathrm{h}, & \lambda b + B + \mathrm{b}, & \lambda f + F + \mathrm{f} \\ \lambda g + G + \mathrm{g}, & \lambda f + F + \mathrm{f}, & \lambda c + C + \mathrm{c} \end{vmatrix}.$$

Hence, since a, b, c, f, g, h, are arbitrary,

$$\begin{vmatrix} \lambda a_1 + A_1, & \lambda h_1 + H_1, & \lambda g_1 + G_1 \\ \lambda h_1 + H_1, & \lambda b_1 + B_1, & \lambda f_1 + F_1 \\ \lambda g_1 + G_1, & \lambda f_1 + F_1, & \lambda c_1 + C_1 \end{vmatrix} = \Pi^2 \begin{vmatrix} \lambda a + A, & \lambda h + H, & \lambda g + G \\ \lambda h + H, & \lambda b + B, & \lambda f + F \\ \lambda g + G, & \lambda f + F, & \lambda c + C \end{vmatrix}$$

which determine the relations which must subsist between the coefficients of the functions of the second order. We derive

$$\begin{vmatrix} a_1, & h_1, & g_1 \\ h_1, & b_1, & f_1 \\ g_1, & f_1, & c_1 \end{vmatrix} = \Pi^2 \begin{vmatrix} a, & h, & g \\ h, & b, & f \\ g, & f, & c \end{vmatrix}$$

and by comparing the coefficients of a, &c., if we write for shortness,

$$\mathfrak{A} = \begin{vmatrix} 1, & . & . \\ ., & \lambda b + B, & \lambda f + F \\ ., & \lambda f + F, & \lambda c + C \end{vmatrix} \text{ \&c., then}$$

$$\mathfrak{A}_1\alpha^2 + \mathfrak{B}_1\beta^2 + \mathfrak{C}_1\gamma^2 + 2\mathfrak{F}_1\beta\gamma + 2\mathfrak{G}_1\gamma\alpha + 2\mathfrak{H}_1\alpha\beta = \Pi^2\mathfrak{A},$$
$$\mathfrak{A}_1\alpha'^2 + \mathfrak{B}_1\beta'^2 + \mathfrak{C}_1\gamma'^2 + 2\mathfrak{F}_1\beta'\gamma' + 2\mathfrak{G}_1\gamma'\alpha' + 2\mathfrak{H}_1\alpha'\beta' = \Pi^2\mathfrak{B},$$
$$\mathfrak{A}_1\alpha''^2 + \mathfrak{B}_1\beta''^2 + \mathfrak{C}_1\gamma''^2 + 2\mathfrak{F}_1\beta''\gamma'' + 2\mathfrak{G}_1\gamma''\alpha'' + 2\mathfrak{H}_1\alpha''\beta'' = \Pi^2\mathfrak{C},$$

$$\mathfrak{A}_1\alpha'\alpha'' + \mathfrak{B}_1\beta'\beta'' + \mathfrak{C}_1\gamma'\gamma'' + \mathfrak{F}_1(\beta'\gamma'' + \beta''\gamma') + \mathfrak{G}_1(\gamma'\alpha'' + \gamma''\alpha') + \mathfrak{H}_1(\alpha'\beta'' + \alpha''\beta') = \Pi^2\mathfrak{F},$$
$$\mathfrak{A}_1\alpha''\alpha + \mathfrak{B}_1\beta''\beta + \mathfrak{C}_1\gamma''\gamma + \mathfrak{F}_1(\beta''\gamma + \beta\gamma'') + \mathfrak{G}_1(\gamma''\alpha + \gamma\alpha'') + \mathfrak{H}_1(\alpha''\beta + \alpha\beta'') = \Pi^2\mathfrak{G},$$
$$\mathfrak{A}_1\alpha\alpha' + \mathfrak{B}_1\beta\beta' + \mathfrak{C}_1\gamma\gamma' + \mathfrak{F}_1(\beta\gamma' + \beta'\gamma) + \mathfrak{G}_1(\gamma\alpha' + \gamma'\alpha) + \mathfrak{H}_1(\alpha\beta' + \alpha'\beta) = \Pi^2\mathfrak{H},$$

each of which virtually contains three equations on account of the indeterminate quantity λ. A somewhat more elegant form may be given to these equations; thus the first of them is

$$\begin{vmatrix} & \alpha, & \beta, & \gamma, \\ \alpha, & \lambda a_1 + A_1, & \lambda h_1 + H_1, & \lambda g_1 + G_1 \\ \beta, & \lambda h_1 + H_1, & \lambda b_1 + B_1, & \lambda f_1 + F_1 \\ \gamma, & \lambda g_1 + G_1, & \lambda f_1 + F_1, & \lambda c_1 + C_1 \end{vmatrix} = \Pi^2 \begin{vmatrix} 1, & . & . \\ ., & \lambda b + B, & \lambda f + F \\ ., & \lambda f + F, & \lambda c + C \end{vmatrix}$$

from which the form of the whole system is sufficiently obvious. The actual values of the coefficients α, β, &c. can only be obtained in the particular case where $f_1 = g_1 = h_1 = F_1 = G_1 = H_1 = 0$. If we suppose besides (which is no additional loss of generality) that $a_1 = b_1 = c_1 = 1$, then the whole system of formulæ becomes

$$(A_1 + \lambda)(B_1 + \lambda)(C_1 + \lambda) = \Pi^2 \begin{vmatrix} \lambda a + A, & \lambda h + H, & \lambda g + G \\ \lambda h + H, & \lambda b + B, & \lambda f + F \\ \lambda g + G, & \lambda f + F, & \lambda c + C \end{vmatrix};$$

$$1 = \Pi^2 \begin{vmatrix} a, & h, & g \\ h, & b, & f \\ g, & f, & c \end{vmatrix} \text{ or } \Pi^2 = \kappa^{-1} \text{ suppose; and then}$$

$$(B_1 + \lambda)(C_1 + \lambda)\alpha^2 + (C_1 + \lambda)(A_1 + \lambda)\beta^2 + (A_1 + \lambda)(B_1 + \lambda)\gamma^2 = \frac{1}{\kappa}\mathfrak{A},$$

$$(B_1 + \lambda)(C_1 + \lambda)\alpha'^2 + (C_1 + \lambda)(A_1 + \lambda)\beta'^2 + (A_1 + \lambda)(B_1 + \lambda)\gamma'^2 = \frac{1}{\kappa}\mathfrak{B},$$

$$(B_1 + \lambda)(C_1 + \lambda)\alpha''^2 + (C_1 + \lambda)(A_1 + \lambda)\beta''^2 + (A_1 + \lambda)(B_1 + \lambda)\gamma''^2 = \frac{1}{\kappa}\mathfrak{C},$$

$$(B_1 + \lambda)(C_1 + \lambda)\alpha'\alpha'' + (C_1 + \lambda)(A_1 + \lambda)\beta'\beta'' + (A_1 + \lambda)(B_1 + \lambda)\gamma'\gamma'' = \frac{1}{\kappa}\mathfrak{F},$$

$$(B_1 + \lambda)(C_1 + \lambda)\alpha''\alpha + (C_1 + \lambda)(A_1 + \lambda)\beta''\beta + (A_1 + \lambda)(B_1 + \lambda)\gamma''\gamma = \frac{1}{\kappa}\mathfrak{G},$$

$$(B_1 + \lambda)(C_1 + \lambda)\alpha\alpha' + (C_1 + \lambda)(A_1 + \lambda)\beta\beta' + (A_1 + \lambda)(B_1 + \lambda)\gamma\gamma' = \frac{1}{\kappa}\mathfrak{H},$$

where, writing down the expanded values of $\mathfrak{A}$, $\mathfrak{B}$, $\mathfrak{C}$, $\mathfrak{F}$, $\mathfrak{G}$, $\mathfrak{H}$,

$$(\lambda b + B)(\lambda c + C) - (\lambda f + F)^2 = \mathfrak{A},$$
$$(\lambda c + C)(\lambda a + A) - (\lambda g + G)^2 = \mathfrak{B},$$
$$(\lambda a + A)(\lambda b + B) - (\lambda h + H)^2 = \mathfrak{C},$$

$$(\lambda g + G)(\lambda h + H) - (\lambda a + A)(\lambda f + F) = \mathfrak{F},$$
$$(\lambda h + H)(\lambda f + F) - (\lambda b + B)(\lambda g + G) = \mathfrak{G},$$
$$(\lambda f + F)(\lambda g + G) - (\lambda c + C)(\lambda h + H) = \mathfrak{H}.$$

By writing successively $\lambda = -A_1$, $\lambda = -B_1$, $\lambda = -C_1$, we see in the first place that A_1, B_1, C_1 are the roots of the same cubic equation, and we obtain next the values of α^2, β^2, γ^2, &c. in terms of these quantities A_1, B_1, C_1, and of the coefficients a, b, &c., A, B, &c. It is easy to see how the above formulæ would have been modified if a_1, b_1, c_1, instead of being equal to unity, had one or more of them been equal to unity with a negative sign. It is obvious that every step of the preceding process is equally applicable whatever be the number of variables.

75.

ON THE ATTRACTION OF AN ELLIPSOID.

[From the *Cambridge and Dublin Mathematical Journal*, vol. IV. (1849), pp. 50—65.]

Part I.—On Legendre's Solution of the Problem of the Attraction of an Ellipsoid on an External Point.

I propose in the following paper to give an outline of Legendre's investigation of the attraction of an ellipsoid upon an exterior point, ["Mémoire sur les Intégrales Doubles," Paris, Mem. Acad. Sc. for 1788, published 1791, pp. 454—486], one of the earliest and (notwithstanding its complexity) most elegant solutions of the problem. It will be convenient to begin by considering some of the geometrical properties of a system of cones made use of in the investigation.

§ 1. The equation of the ellipsoid referred to axes parallel to the principal axes, and passing through the attracted point, may be written under the form

$$l(x-a)^2 + m(y-b)^2 + n(z-c)^2 - k = 0,$$

(where $\sqrt{\frac{k}{l}}$, $\sqrt{\frac{k}{m}}$, $\sqrt{\frac{k}{n}}$ are the semiaxes, and a, b, c are the coordinates of the attracted point referred to the principal axes). Or putting $la^2 + mb^2 + nc^2 - k = \delta$, this equation becomes

$$lx^2 + my^2 + nz^2 - 2(lax + mby + ncz) + \delta = 0.$$

The cones in question are those which have the same axes and directions of circular section as the cone having its vertex in the attracted point and circumscribed about the ellipsoid. The equation of the system of cones (containing the arbitrary parameter ω) is

$$(lx^2 + my^2 + nz^2)\,\delta - (lax + mby + ncz)^2 + \omega^2(x^2 + y^2 + z^2) = 0\,;$$

or as it may also be written,

$$(\omega^2 + l\delta - l^2a^2)\,x^2 + (\omega^2 + m\delta - m^2b^2)\,y^2 + (\omega^2 + n\delta - n^2c^2)\,z^2 - 2mnbcyz - 2nlcazx - 2lmabxy = 0.$$

For $\omega = 0$, the cone coincides with the circumscribed cone; as ω increases, the aperture of the cone gradually diminishes, until for a certain value, $\omega = \Omega$, the cone reduces itself to a straight line (the normal of the confocal ellipsoid through the attracted point). It is easily seen that Ω^2 is the positive root of the equation

$$\frac{l^2a^2}{\Omega^2 + l\delta} + \frac{m^2b^2}{\Omega^2 + m\delta} + \frac{n^2c^2}{\Omega^2 + n\delta} = 1,$$

a different form of which may be obtained by writing $\Omega^2 = \frac{k\delta}{\xi}$, ξ being then determined by means of the equation

$$\frac{la^2}{k + l\xi} + \frac{mb^2}{k + m\xi} + \frac{nc^2}{k + n\xi} = 1;$$

that is,

$$\sqrt{\left(\frac{k}{l} + \xi\right)}, \quad \sqrt{\left(\frac{k}{m} + \xi\right)}, \quad \sqrt{\left(\frac{k}{n} + \xi\right)}$$

are the semiaxes of the confocal ellipsoid through the attracted point.

In the case where ω remains indeterminate, it is obvious that the cone intersects the ellipsoid in the curve in which the ellipsoid is intersected by a certain hyperboloid of revolution of two sheets, having the attracted point for a focus, and the plane of contact of the ellipsoid with the circumscribed cone (that is the polar plane of the attracted point) for the corresponding directrix plane: also the excentricity of the hyperboloid is $\frac{1}{\omega}\sqrt{(l^2a^2 + m^2b^2 + n^2c^2)}$, which suffices for its complete determination. For $\omega = 0$, the hyperboloid reduces itself to the plane of contact of the ellipsoid with the circumscribed cone, and for $\omega = \Omega$, the hyperboloid and the ellipsoid have a double contact, viz. at the points where the ellipsoid is intersected by the normal to the confocal ellipsoid through the attracted point.

If ω remains constant while k is supposed to vary, that is, if the ellipsoid vary in magnitude (the position and proportion of its axes remaining unaltered), the locus of the intersection of the cone and the ellipsoid is a surface of the fourth order defined by the equation

$$(lx^2 + my^2 + nz^2 - lax - mby - ncz)^2 = \omega^2 (x^2 + y^2 + z^2),$$

and consisting of an exterior and an interior sheet meeting at the attracted point, which is a conical point on the surface, viz. a point where the tangent plane is

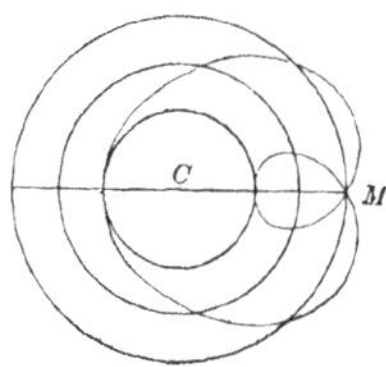

replaced by a tangent cone. The general form of this surface is easily seen from the figure, in which the ellipsoid has been replaced by a sphere, and the surface in question is that generated by the revolution of the curve round the line CM. The surface of the fourth order being once described for any particular value of ω, the cone corresponding to any one of the series of similar, similarly situated, and concentric ellipsoids is at once determined by means of the intersection of the ellipsoid in question with the surface of the fourth order. It is clear too that there is always one of these ellipsoids which has a double contact with the surface of the fourth order, viz. at the points where this ellipsoid is intersected by the normal to the confocal ellipsoid through the attracted point; thus there is always an ellipsoid for which the cone corresponding to a given value of ω reduces itself to a straight line.

Consider the attracting ellipsoid, which for distinction may be termed the ellipsoid S, and the two cones C, C', which correspond to the values ω, $\omega - d\omega$ of the variable parameter. Legendre shows that the attraction of the portion of the ellipsoid S included between the two cones C, C' is independent of the quantity k, which determines the magnitude of the ellipsoid: that is, if there be any other ellipsoid T similarly situated and concentric to and with the ellipsoid S, and two cones D, D', which for the ellipsoid T correspond to the same values ω, $\omega - d\omega$ of the variable parameter; then the attraction of the portion of the ellipsoid S, included between the two cones C, C', is equal to the attraction of the portion of the ellipsoid T included between the two cones D and D'. By taking for the ellipsoid T the ellipsoid for which the cone D reduces itself to a straight line, the aperture of the cone D' is indefinitely small, and the attraction of the portion of the ellipsoid T included within the cone D' is at once determined; and thus the attraction of the portion of the ellipsoid S included between the cones C, C' is obtained in a finite form. Hence the attraction of the portion of the ellipsoid S included between any two cones $C_{,}$, $C_{,,}$ corresponding to the values $\omega_{,}$, $\omega_{,,}$ of the variable parameter, is expressed by means of a single integral, and by extending the integration from $\omega = 0$ to $\omega = \sqrt{\left(\frac{k\delta}{\xi}\right)}$, the attraction of the whole ellipsoid is obtained in the form of a single integral readily reducible to that given by the ordinary solutions. It is clear too that the attraction of the portion of the ellipsoid S included between any two cones $C_{,}$, $C_{,,}$, is equal to that of the portion of the ellipsoid T included between the corresponding cones $D_{,}$ and $D_{,,}$. Hence also, assuming for the ellipsoid T, that for which the cone $D_{,,}$ reduces itself to a straight line, and supposing that the cones $C_{,}$ and $D_{,}$ coincide with the circumscribing cones, the attraction of the portion of the ellipsoid S exterior to the cone $C_{,,}$ is equal to the attraction of the entire ellipsoid T. More generally, the attraction of the portion of the ellipsoid S included between the cones $C_{,}$ and $C_{,,}$ is equal to the attraction of the shell included between the surfaces of the two ellipsoids, for which the cones $D_{,}$ and $D_{,,}$ respectively reduce themselves to straight lines.

§ 2. Proceeding to the analytical solution, and resuming the equation of the ellipsoid

$$lx^2 + my^2 + nz^2 - 2(lax + mby + ncz) + \delta = 0,$$

and that of the cone

$$(lx^2 + my^2 + nz^2)\,\delta - (lax + mby + ncz)^2 + \omega^2(x^2 + y^2 + z^2) = 0;$$

consider a radius vector on the conical surface such that the cosines of its inclinations to the axes are

$$\frac{P}{\Theta}, \frac{Q}{\Theta}, \frac{R}{\Theta}, \{\Theta = \sqrt{(P^2 + Q^2 + R^2)}\},$$

P, Q, R and Θ being functions of the parameter ω, and of another variable ϕ, which determines the position of the radius vector upon the conical surface. Also let ρ be the length of the portion of the radius vector which lies within the ellipsoid; then representing by dS the spherical angle corresponding to the variations of ω and ϕ, the attraction in the direction of the axis of x is given by the formula

$$A = \iint \rho \frac{P}{\Theta} dS.$$

Also by a known formula

$$dS = \frac{1}{\Theta^3}\left\{P\left(\frac{dQ}{d\phi}\frac{dR}{d\omega} - \frac{dR}{d\phi}\frac{dQ}{d\omega}\right) + Q\left(\frac{dR}{d\phi}\frac{dP}{d\omega} - \frac{dR}{d\omega}\frac{dP}{d\phi}\right) + R\left(\frac{dP}{d\phi}\frac{dQ}{d\omega} - \frac{dP}{\delta\omega}\frac{dQ}{d\phi}\right)\right\},$$

and it is easy to obtain

$$\rho = \frac{2\omega\Theta^2}{lP^2 + mQ^2 + nR^2}.$$

The quantities P, Q, R have now to be expressed as functions of ω, ϕ, so that their values substituted for x, y, z, may satisfy identically the equation of the cone. This may be done by assuming

$$P = p,$$

$$Q = mb\,(\omega^2 + n\delta)\left(la + \frac{D}{U}\cos\phi\right) - \frac{D\sqrt{p}}{U} nc \sin\phi,$$

$$R = nc\,(\omega^2 + m\delta)\left(la + \frac{D}{U}\cos\phi\right) + \frac{D\sqrt{p}}{U} mb \sin\phi,$$

where

$$p = (\omega^2 + m\delta)(\omega^2 + n\delta) - m^2b^2(\omega^2 + n\delta) - n^2c^2(\omega^2 + m\delta),$$

$$U^2 = m^2b^2(\omega^2 + n\delta) + n^2c^2(\omega^2 + m\delta),$$

$$D^2 = (\omega^2 + l\delta)(\omega^2 + m\delta)(\omega^2 + n\delta)\left\{\frac{l^2a^2}{\omega^2 + l\delta} + \frac{m^2b^2}{\omega^2 + m\delta} + \frac{n^2c^2}{\omega^2 + n\delta} - 1\right\};$$

a system of values which, in point of fact, depend upon the following geometrical considerations: by treating x as a constant in the equation of the cone, that is, in effect by considering the sections of the cone by planes parallel to that of yz, the equation of the cone becomes that of an ellipse; transforming first to a set of axes through the centre and then to a set of conjugate axes, one of which passes through the point where the plane of the ellipse is intersected by the axis of x, then the equation takes the form $\frac{\xi^2}{A^2} + \frac{\eta^2}{B^2} = 1$, and is satisfied by $\xi = A\cos\phi$, $\eta = B\sin\phi$, and $\frac{y}{x}$, $\frac{z}{x}$ being of course linear functions of these values, the preceding expressions may be obtained.

The substitution of the above values of P, Q, R (a somewhat tedious one which does not occur in the process actually made use of by Legendre) gives the very simple result,

$$dS = \frac{P^{\frac{1}{2}}\omega\, d\omega\, d\phi}{\Theta};$$

and the formula for the attraction becomes

$$A = 2\iint \frac{P^{\frac{3}{2}}\omega^2\, d\omega\, d\phi}{lP^2 + mQ^2 + nR^2},$$

which is of the form $A = 2\int I\omega^2 d\omega$, where

$$I = \int \frac{P^{\frac{3}{2}} d\phi}{lP^2 + mQ^2 + nR^2},$$

which last integral, taken between the limits $\phi = 0$ and $\phi = 2\pi$, and multiplied by $2\omega^2 d\omega$, expresses the attraction of the portion of the ellipsoid included between two consecutive cones. The integration is evidently possible, but the actual performance of it is the great difficulty of Legendre's process. The result, as before mentioned, is independent of the quantity k, or, what comes to the same thing, of the quantity δ: assuming this property (an assumption which in fact resolves itself into the consideration of the ellipsoid for which the cone reduces itself to a straight line, as before explained), the integral is at once obtained by writing $\delta = \Delta$ where Δ represents the positive root of the equation

$$\frac{l^2a^2}{\omega^2 + l\Delta} + \frac{m^2b^2}{\omega^2 + m\Delta} + \frac{n^2c^2}{\omega^2 + n\Delta} - 1 = 0.$$

This gives

$$P = \frac{l^2a^2(\omega^2 + m\Delta)(\omega^2 + n\Delta)}{\omega^2 + l\Delta},$$

$$Q = lmab\,(\omega^2 + n\Delta),$$

$$R = lnac\,(\omega^2 + m\Delta),$$

values independent of ϕ, or the value of I is found by multiplying the quantity under the integral sign by 2π: and hence we have

$$A = 4\pi la \int \frac{\omega^2(\omega^2 + l\Delta)^{\frac{1}{2}}(\omega^2 + m\Delta)^{\frac{3}{2}}(\omega^2 + n\Delta)^{\frac{3}{2}}\, d\omega}{l^3a^2(\omega^2 + m\Delta)^2(\omega^2 + n\Delta)^2 + m^3b^2(\omega^2 + n\Delta)^2(\omega^2 + l\Delta)^2 + n^3c^2(\omega^2 + n\Delta)^2(\omega^2 + l\Delta)^2},$$

where of course Δ is to be considered as a function of ω. By integrating from $\omega = \omega_{\prime}$ to $\omega = \omega_{\prime\prime}$, we have the attraction of the portion of the ellipsoid included between any two of the series of cones, and to obtain the attraction of the whole ellipsoid we must integrate from $\omega = 0$ to $\omega = \sqrt{\left(\frac{k\delta}{\xi}\right)}$, where ξ is determined as before by the equation

$$\frac{la^2}{k + l\xi} + \frac{mb^2}{k + m\xi} + \frac{nc^2}{k + n\xi} = 1;$$

and it is obvious that for this value of ω we have $\Delta = \delta$. The expression for the attraction is easily reduced to a known form by writing $y = \frac{k\Delta}{\omega^2}$; this gives

$$A = 4\pi la \int \frac{k^{\frac{1}{2}} (k+ly)^{\frac{1}{2}} (k+my)^{\frac{3}{2}} (k+ny)^{\frac{3}{2}} \omega d\omega}{l^3a^2 (k+my)^2 (k+ny)^2 + m^3b^2 (k+ny)^2 (k+ly)^2 + n^3c^2 (k+ly)^2 (k+my)^2}.$$

Also

$$\omega^2 = k\left(\frac{l^2a^2}{k+ly} + \frac{m^2b^2}{k+my} + \frac{n^2c^2}{k+ny}\right);$$

whence

$$\omega d\omega = -\frac{k\,[l^3a^2 (k+my)^2 (k+ny)^2 + m^3b^2 (k+ny)^2 (k+ly)^2 + n^3c^2 (k+ly)^2 (k+my)^2]}{2 (k+ly)^2 (k+my)^2 (k+ny)^2},$$

and thus

$$A = 2\pi k^{\frac{3}{2}} la \int \frac{dy}{(k+ly)^{\frac{3}{2}} (k+my)^{\frac{1}{2}} (k+ny)^{\frac{1}{2}}},$$

where for the entire ellipsoid the integral is to be taken from $y = \xi$ to $y = \infty$. A better known form is readily obtained by writing $x^2 = \frac{k+l\xi}{k+ly}$, in which case the limits for the entire ellipsoid are $x = 0$, $x = 1$.

It may be as well to indicate the first step of the reduction of the integral I, viz. the method of resolving the denominator into two factors. We have identically,

$$(\Delta - \delta)(lP^2 + mQ^2 + nR^2) = \omega^2 (P^2 + Q^2 + R^2) + \Delta (lP^2 + mQ^2 + nR^2) - (laP + mbQ + ncR)^2,$$

and the second side of this equation is resolvable into two factors independently of the particular values of P, Q, R. Representing this second side for a moment in the notation of a general quadratic function, or under the form

$$AP^2 + BQ^2 + CR^2 + 2FQR + 2GRP + 2HPQ,$$

we have the required solution,

$$lP^2 + mQ^2 + nR^2 =$$

$$\frac{1}{A}\,[AP + \{H + \sqrt{(-\mathfrak{C})}\}\,Q + \{G + \sqrt{(-\mathfrak{B})}\}\,R]\,[AP + \{H - \sqrt{(-\mathfrak{C})}\}\,Q + \{G - \sqrt{(-\mathfrak{B})}\}\,R];$$

where, as usual, $\mathfrak{B} = CA - G^2$, $\mathfrak{C} = AB - H^2$, and the roots must be so taken that $\sqrt{(-\mathfrak{B})}\sqrt{(-\mathfrak{C})} = \mathfrak{F}$ $\{\mathfrak{F} = (GH - AF)\}$.

I have purposely restricted myself so far to the problem considered by Legendre: the general transformation, of which the preceding is a particular case, and also a simpler mode of effecting the integration, are given in the next part of this paper.

PART II.—ON A FORMULA FOR THE TRANSFORMATION OF CERTAIN MULTIPLE INTEGRALS.

Consider the integral

$$V = \int F(x, y, \ldots)\, dx\, dy \ldots,$$

where the number of variables $x, y, \ldots$ is equal to n, and $F(x, y, \ldots)$ is a homogeneous function of the order μ.

Suppose that $x, y, \ldots$ are connected by a homogeneous equation $\psi(x, y, \ldots) = 0$ containing a variable parameter ω (so that ω is a homogeneous function of the order zero in the variables $x, y, \ldots$). Then, writing

$$r^2 = x^2 + y^2 + \ldots, \quad x = r\alpha, \quad y = r\beta, \ldots$$

the quantities $\alpha, \beta, \ldots$ are connected by the equations

$$\alpha^2 + \beta^2 + \ldots = 1, \quad \psi(\alpha, \beta, \ldots) = 0,$$

and we may therefore consider them as functions of ω and of $(n-2)$ independent variables θ, ϕ, &c.; whence

$$dx\, dy \ldots = r^{n-1} \nabla\, dr\, d\omega\, d\theta \ldots,$$

where

$$\nabla = \begin{vmatrix} \alpha, & \beta, & \ldots \\ \dfrac{d\alpha}{d\omega}, & \dfrac{d\beta}{d\omega}, & \\ \dfrac{d\alpha}{d\theta}, & \dfrac{d\beta}{d\theta}, & \\ \vdots & & \end{vmatrix}$$

Also

$$F(x, y, \ldots) = r^{\mu} F(\alpha, \beta \ldots),$$

and therefore

$$V = \int r^{\mu+n-1} F(\alpha, \beta, \ldots) \nabla\, dr\, d\omega\, d\theta \ldots,$$

or, integrating with respect to r,

$$\int r^{\mu+n-1}\, dr = \frac{1}{\mu+n} r^{\mu+n},$$

which, taken between the proper limits, is a function of $\alpha, \beta, \ldots$, equal $f(\alpha, \beta, \ldots)$ suppose; this gives

$$V = \int f(\alpha, \beta, \ldots) F(\alpha, \beta, \ldots) \nabla\, d\omega\, d\theta \ldots,$$

in which I shall assume that the limits of ω are constant. If, in order to get rid of the condition $\alpha^2 + \beta^2 \ldots = 1$, we assume

$$\alpha = \frac{p}{\rho}, \quad \beta = \frac{q}{\rho}, \ldots, \quad \rho^2 = p^2 + q^2 + \ldots$$

the preceding expression for V becomes

$$V=\int f\left(\frac{p}{\rho},\ \frac{q}{\rho},\ \ldots\right) F(p,\ q,\ \ldots)\,\frac{1}{\rho^{\mu+n}}\,D\,d\omega\,d\theta\,\ldots,$$

in which

$$D=\begin{vmatrix} p\,, & q\,, & \ldots \\ \dfrac{dp}{d\omega}, & \dfrac{dq}{d\omega}, & \\ \dfrac{dp}{d\theta}, & \dfrac{dq}{d\theta}, & \\ \vdots & & \end{vmatrix}$$

Assume

$$\begin{aligned} p &= P\xi + P'\eta + P''\zeta \ldots \\ q &= Q\xi + Q'\eta + Q''\zeta \ldots \\ &\vdots \end{aligned}$$

where the number of variables $\xi,\ \eta,\ \zeta\ldots$ (functions in general of $\omega,\ \theta$, &c.) is n, and where the coefficients $P,\ Q$, &c. are supposed to be functions of ω only. We have

$$\frac{dp}{d\theta}=P\frac{d\xi}{d\theta}+P'\frac{d\eta}{d\theta}+P''\frac{d\zeta}{d\theta}+\ldots,$$

$$\frac{dq}{d\theta}=Q\frac{d\xi}{d\theta}+Q'\frac{d\eta}{d\theta}+Q''\frac{d\zeta}{d\theta}+\ldots,$$

and, substituting these values as well as those of $p,\ q$, &c., but retaining the terms $\dfrac{dp}{d\omega},\ \dfrac{dq}{d\omega}$, &c. in their original form, the determinant D resolves itself into the sum of a series of products,

$$\begin{vmatrix} \dfrac{dp}{d\omega}, & \dfrac{dq}{d\omega}, \ldots \\ P'\,, & Q'\,, \\ P''\,, & Q''\,, \\ \vdots & \end{vmatrix}\ \begin{vmatrix} 1 & . & . & \ldots \\ . & \eta\,, & \zeta\,, & \\ . & \dfrac{d\eta}{d\theta}, & \dfrac{d\zeta}{d\theta}, & \\ \vdots & & & \end{vmatrix}.$$

Let Ψ be the function to which $\psi\,(p,\ q,\ \ldots)$ is changed by the substitution of the above values of $p,\ q,\ldots$ so that Ψ is a homogeneous function of $\xi,\ \eta,\ \zeta,\ldots$ and we have the relation $\Psi=0$. (It will be convenient to consider $\xi,\ \eta,\ \zeta,\ldots$ as functions of $\omega,\ \theta$, &c., such as to satisfy identically this last equation.) We deduce

$$\frac{1}{X}\begin{vmatrix} 1, & . & . & \ldots \\ . & \eta\,, & \zeta\,, & \\ . & \dfrac{d\eta}{d\theta}, & \dfrac{d\zeta}{d\theta}, & \\ \vdots & & & \end{vmatrix}=\frac{1}{Y}\begin{vmatrix} . & 1, & . & \ldots \\ \xi\,, & . & \zeta\,, & \\ \dfrac{d\xi}{d\theta}, & . & \dfrac{d\zeta}{d\theta}, & \\ \vdots & & & \end{vmatrix}=\text{\&c.}=S\text{ suppose},$$

where for shortness $X=\frac{d\Psi}{d\xi}$, $Y=\frac{d\Psi}{d\eta}$, &c. The substitution of these values gives

$$D=\begin{vmatrix} & \frac{dp}{d\omega}, & \frac{dq}{d\omega}, \dots \\ X, & P\ , & Q\ , \\ Y, & P', & Q', \\ Z, & P'', & Q'', \\ \vdots & & \end{vmatrix} S,$$

where it will be remarked that the successive horizontal lines (after the first) of the determinant are the differential coefficients of Ψ, p, q, ... with respect to ξ, with respect to η, &c. In general, if ψ denote any function of p, q, ... these quantities being themselves functions of ω, ξ, η, ..., and ξ, η, ... containing ω; also if Ψ be what ψ becomes when for p, q, ... we substitute their values in ω, ξ, η, ...; then we have identically

$$\begin{vmatrix} \frac{d\psi}{d\omega}, & \frac{dp}{d\omega}, & \frac{dq}{d\omega}, \dots \\ X\ , & P\ , & Q\ , \\ Y\ , & P', & Q', \\ Z\ , & P'', & Q'', \\ \vdots & & \end{vmatrix} = 0. \ [1]$$

In the present case however, writing for shortness $\psi(p, q, \dots)=\psi$, this function ψ contains ω explicitly as well as implicitly through p, q, &c. The formula is still true if for $\frac{d\psi}{d\omega}$ we substitute $\frac{d(\psi)}{d\omega}-\frac{d\psi}{d\omega}$, $\frac{d\psi}{d\omega}$ on the second side denoting a partial differential coefficient taken only so far as ω is explicitly contained in ψ. And considering p, q, ... as functions of ω, ξ, η, ... (ξ, η, ... themselves functions of ω and of other variables which need not here be considered), ψ or Ψ vanishes identically, and we have $\frac{d(\psi)}{d\omega}=0$. Hence, in the last formula, we have to write $-\frac{d\psi}{d\omega}$ instead of $\frac{d\psi}{d\omega}$, and we thus derive

$$\frac{d\psi}{d\omega}\begin{vmatrix} P, & Q, \dots \\ P', & Q', \\ \vdots & \end{vmatrix} = \begin{vmatrix} . & \frac{dp}{d\omega}, & \frac{dq}{d\omega}, \dots \\ X, & P\ , & Q\ , \\ Y, & P', & Q', \\ Z, & P'', & Q'', \\ \vdots & & \end{vmatrix},$$

[1] This formula, or one equivalent to it, is given in Jacobi's memoir "De Determinantibus Functionalibus," *Crelle*, t. XXII. [1841] p. 319.

whence D becomes

$$D = \begin{vmatrix} P, & Q, \dots \\ P', & Q', \\ \vdots & \end{vmatrix} \frac{d\psi}{d\omega} S,$$

which is the value to be made use of in the equation

$$V = \int f\left(\frac{p}{\rho}, \frac{q}{\rho} \dots\right) F(p, q, \dots) \frac{1}{\rho^{\mu+n}} D \, d\omega \, d\theta \dots .$$

The principal use of the formula is where ψ is a homogeneous function of the second order of $p, q, \dots$. Thus, suppose

$$\psi = \tfrac{1}{2}(Ap^2 + Bq^2 \dots + 2Hpq + \dots),$$

also, for the sake of conformity to the usual notation in the theory of transformation of quadratic functions, writing $\alpha, \alpha', \dots \beta, \beta', \dots$ instead of $P, P', \dots Q, Q', \dots$ and putting after the differentiations $\xi = 1$, we have

$$\begin{aligned} p &= \alpha + \alpha'\eta + \alpha''\zeta + \dots, \\ q &= \beta + \beta'\eta + \beta''\zeta + \dots, \\ &\vdots \end{aligned}$$

values which we may assume to give rise to the equation

$$(Ap^2 + Bq^2 \dots + 2Hpq \dots) = (1 - \eta^2 - \zeta^2 - \dots),$$

(where $\eta, \zeta, \dots$ are taken to be functions of θ, &c. such as to satisfy identically the equation $1 = \eta^2 + \zeta^2 + \dots$).

Hence, by a well-known property, if

$$\begin{vmatrix} A, & H, \dots \\ H, & B, \\ \vdots & \end{vmatrix} = \kappa,$$

we have

$$\begin{vmatrix} \alpha, & \beta, \dots \\ \alpha', & \beta', \\ \vdots & \end{vmatrix} = \sqrt{\left\{\frac{(-)^{n-1}}{\kappa}\right\}};$$

so that, observing that in the present case $X = 1$, and therefore

$$S = \begin{vmatrix} \eta, & \zeta, \dots \\ \dfrac{d\eta}{d\theta}, & \dfrac{d\zeta}{d\theta}, \\ \vdots & \end{vmatrix}$$

we have

$$V = \int \frac{(-)^{\frac{1}{2}(n-1)}}{\kappa^{\frac{1}{2}}} f\left(\frac{p}{\rho}, \frac{q}{\rho}, \dots\right) F(p, q, \dots) \frac{1}{\rho^{\mu+n}} \frac{d\psi}{d\omega} S \, d\omega \, d\theta \dots .$$

The remainder of the process of integration may in many cases be effected by the method made use of by Jacobi in the memoir "De binis quibuslibet functionibus &c." *Crelle*, t. XII. [1834] p. 1, viz. the coefficients α, α', &c., β, &c., may in addition to the conditions which they are already supposed to satisfy, be so determined as to reduce any homogeneous function of p, q, r, ... entering into the integral to a form containing the squares only of the variables. This method is applied in the memoir in question to the integrals of n variables, analogous to those which give the attraction of an ellipsoid; and that directly without effecting an integration with respect to the radius vector. I proceed to show how the preceding investigations lead to Legendre's integral, and how the method in question effects with the utmost simplicity the integration which Legendre accomplished by means of what Poisson has spoken of as inextricable calculations.

Consider in particular the formula

$$V=\int\frac{(x,\ y\ \ldots)^h\,dx\,dy\ \ldots}{(x^2+y^2\ \ldots)^{\frac{3}{2}-i}},$$

the number of variables being as before n, and $(x,\ y,\ldots)^h$ denoting a homogeneous function of the order h. The equation for the limits is assumed to be

$$l\,(x-a)^2+m\,(y-b)^2+\ldots=k.$$

Assume

$$\psi\,(x,\ y\,,\ \ldots)=\omega^2\,(x^2+y^2+\ldots)+\delta\,(lx^2+my^2+\ldots)-(lax+mby+\ldots)^2,$$

(where δ, $=la^2+mb^2\ldots-k$, is taken to be positive); or more simply,

$$\psi\,(x,\ y\ \ldots)=(\omega^2+l\delta-l^2a^2)\,x^2+(\omega^2+m\delta-m^2b^2)\,y^2+\ldots-2lm\,ab\,xy-\ldots$$

Here $\mu=h+2i-3$. Also, putting for shortness

$$lap+mbq+\ldots=\Lambda,\quad lp^2+mq^2+\ldots=\Phi,$$

it is easy to obtain

$$f\left(\frac{p}{\rho},\ \frac{q}{\rho},\ldots\right)=\frac{1}{h+2i+n-3}\ \frac{\rho^{h+2i+n-3}\left[(\Lambda+\omega\rho)^{h+2i+n-3}-(\Lambda-\omega\rho)^{h+2i+n-3}\right]}{\Phi^{h+2i+n-3}}.$$

Also
$$F(p,\ q,\ldots)=\frac{(p,\ q,\ldots)^h}{\rho^{3-2i}},\qquad \frac{d\psi}{d\omega}=2\omega\rho^2,$$

$$\kappa=(\omega^2+l\delta)\,(\omega^2+m\delta)\ldots\left(1-\frac{l^2a^2}{\omega^2+l\delta}-\frac{m^2b^2}{\omega^2+m\delta}-\text{\&c.}\right),$$

values which give

$$V=\frac{2}{h+2i+n-3}\int\frac{(-)^{\frac{1}{2}(n-1)}}{\kappa^{\frac{1}{2}}}\ \frac{\omega\rho^{2i-1}\left[(\Lambda+\omega\rho)^{h+2i+n-3}-(\Lambda-\omega\rho)^{h+2i+n-3}\right](p,\ q,\ldots)^h\,Sd\omega\,d\theta\ \ldots}{\Phi^{h+2i+n-3}},$$

where it will be remembered that p, q,... are linear functions (with constant terms) of $(n-1)$ variables η, ζ,..., these last mentioned quantities being themselves functions of $(n-2)$ variables θ, &c. such that $1-\eta^2-\zeta^2-\ldots=0$ identically. If besides we suppose that $\Phi=lp^2+mq^2+\ldots$ reduces itself to the form $\frac{1}{P}-\frac{\eta^2}{Q}-$ &c., we have, by the formula of the paper "On the Simultaneous Transformation of two Homogeneous Equations of the Second Order," [74],

$$\left(1-\frac{\lambda}{P}\right)\left(1-\frac{\lambda}{Q}\right)\ldots=$$

$$\frac{1}{\kappa}(\omega^2+l\delta-l\lambda)(\omega^2+m\delta-m\lambda)\ldots\left(1-\frac{l^2a^2}{\omega^2+l\delta-l\lambda}-\frac{m^2b^2}{\omega^2+m\delta-m\lambda}-\ldots\right),$$

which is true, whatever be the value of λ.

It seems difficult to proceed further with the general formula, and I shall suppose $n=3$, $i=0$, $h=1$, $(x, y\ldots)^h=x$, or write

$$V=\int\frac{x\,dx\,dy\,dz}{(x^2+y^2+z^2)^{\frac{3}{2}}},$$

the equation of the limits being

$$l(x-a)^2+m(y-b)^2+n(z-c)^2=k.$$

Here we may assume $\eta=\cos\theta$, $\zeta=\sin\theta$, (values which give $S=1$). And we have

$$V=2\int\frac{\omega^2d\omega}{\kappa^{\frac{1}{2}}}\int\frac{(\alpha+\alpha'\cos\theta+\alpha''\sin\theta)\,d\theta}{\frac{1}{P}-\frac{\cos^2\theta}{Q}-\frac{\sin^2\theta}{R}},$$

from $\theta=0$ to $\theta=2\pi$; or, what comes to the same thing,

$$V=8\int\frac{\omega^2\,d\omega}{\kappa^{\frac{1}{2}}}\int\frac{\alpha\,d\theta}{\frac{1}{P}-\frac{\cos^2\theta}{Q}-\frac{\sin^2\theta}{R}},$$

from $\theta=0$ to $\theta=\frac{1}{2}\pi$. Hence

$$V=4\pi\int\frac{\alpha\omega^2P\sqrt{(QR)}\,d\omega}{\kappa^{\frac{1}{2}}\sqrt{\{(P-Q)(P-R)\}}};$$

we have from the formulæ of the paper before quoted,

$$\alpha^2=\frac{QR}{\kappa}\,\frac{(B-mP)(C-nP)-F^2}{(P-Q)(P-R)},$$

B, C, F, being the coefficients of y^2, z^2, yz in $\psi(x, y, z)$, viz.

$$B=\omega^2+m\delta-m^2b^2,\quad C=\omega^2+n\delta-n^2c^2,\quad F=mnbc;$$

and consequently

$$V=4\pi\int\omega^2\,d\omega\,\frac{PQR}{\kappa}\,\frac{\{(B-mP)(C-nP)-F^2\}^{\frac{1}{2}}}{(P-Q)(P-R)}.$$

Also from the equation

$$\left(1-\frac{\lambda}{P}\right)\left(1-\frac{\lambda}{Q}\right)\left(1-\frac{\lambda}{R}\right)=$$

$$\frac{1}{\kappa}(\omega^2+l\delta-l\lambda)(\omega^2+m\delta-m\lambda)(\omega^2+n\delta-n\lambda)\left(1-\frac{l^2a^2}{\omega^2+l\delta-l\lambda}-\frac{m^2b^2}{\omega^2+m\delta-m\lambda}-\frac{n^2c^2}{\omega^2+n\delta-n\lambda}\right);$$

differentiating with respect to λ, and writing $\lambda=P$,

$$\frac{-\frac{1}{P}\left(1-\frac{P}{Q}\right)\left(1-\frac{P}{R}\right)}{(\omega^2+l\delta-lP)(\omega^2+m\delta-mP)(\omega^2+n\delta-nP)}$$

$$=-\frac{1}{\kappa}\left\{\frac{l^3a^2}{(\omega^2+l\delta-lP)^2}+\frac{m^3b^2}{(\omega^2+m\delta-mP)^2}+\frac{n^3c^2}{(\omega^2+n\delta-nP)^2}\right\},$$

or, as this may be written,

$$\frac{PQR}{\kappa(P-Q)(P-R)}$$

$$=\frac{1}{(\omega^2+l\delta-lP)(\omega^2+m\delta-mP)(\omega^2+n\delta-nP)\left\{\frac{l^3a^2}{(\omega^2+l\delta-lP)^2}+\frac{m^3b^2}{(\omega^2+m\delta-mP)^2}+\frac{n^3c^2}{(\omega^2+n\delta-nP)^2}\right\}},$$

and from the values first written down, for B, C, F, we obtain $(B-mP)(C-nP)-F^2$

$$=(\omega^2+m\delta)(\omega^2+n\delta)-m^2b^2(\omega^2+n\delta)-n^2c^2(\omega^2+m\delta)-mP(\omega^2+n\delta-n^2c^2)-nP(\omega^2+m\delta-m^2b^2)+mnP^2$$

$$=(\omega^2+m\delta-mP)(\omega^2+n\delta-nP)-m^2b^2(\omega^2+n\delta-nP)-n^2c^2(\omega^2+m\delta-mP)$$

$=\dfrac{l^2a^2(\omega^2+m\delta-mP)(\omega^2+n\delta-nP)}{\omega^2+l\delta-lP}$, the last reduction being effected by means of the equation

$$1-\frac{l^2a^2}{\omega^2+l\delta-lP}-\frac{m^2b^2}{\omega^2+m\delta-mP}-\frac{n^2c^2}{\omega^2+n\delta-nP}=0.$$

Hence

$$\frac{PQR}{\kappa}\frac{\{(B-mP)(C-nP)-F^2\}^{\frac{1}{2}}}{(P-Q)(P-R)}$$

$$=\frac{la}{(\omega^2+l\delta-lP)^{\frac{3}{2}}(\omega^2+m\delta-mP)^{\frac{1}{2}}(\omega^2+n\delta-nP)^{\frac{1}{2}}\left\{\frac{l^3a^2}{(\omega^2+l\delta-lP)^2}+\frac{m^3b^2}{(\omega^2+m\delta-mP)^2}+\frac{n^3c^2}{(\omega^2+n\delta-nP)^2}\right\}}.$$

Substituting this value, and multiplying out the fractions in the denominator,

$$V=4\pi la\times$$

$$\int\frac{\omega^2(\omega^2+l\delta-lP)^{\frac{1}{2}}(\omega^2+m\delta-mP)^{\frac{3}{2}}(\omega^2+n\delta-nP)^{\frac{3}{2}}\,d\omega}{l^3a^2(\omega^2+m\delta-mP)^2(\omega^2+n\delta-nP)^2+m^3b^2(\omega^2+n\delta-nP)^2(\omega^2+l\delta-lP)^2+n^3c^2(\omega^2+l\delta-lP)^2(\omega^2+m\delta-mP)^2},$$

the reduction of which integral has been already treated of in the former part of this present memoir.

76.

ON THE TRIPLE TANGENT PLANES OF SURFACES OF THE THIRD ORDER.

[From the *Cambridge and Dublin Mathematical Journal*, vol. IV. (1849), pp. 118—132.]

A SURFACE of the third order contains in general a certain number of straight lines. Any plane through one of these lines intersects the surface in the line and in a conic, that is in a curve or system of the third order having two double points. Such a plane is therefore a double tangent plane of the surface, the double points (or points where the line and conic intersect) being the points of contact. By properly determining the plane, the conic will reduce itself to a pair of straight lines. Here the plane intersects the surface in three straight lines, that is in a curve or system of the third order having three double points, and the plane is therefore a triple tangent plane, the three double points or points of intersection of the lines taken two and two together being the points of contact. The number of lines and triple tangent planes is determined by means of a theorem very easily demonstrated, viz. that through each line there may be drawn five (and only five) triple tangent planes. Thus, considering any triple tangent plane, through each of the three lines in this plane there may be drawn (in addition to the plane in question) four triple tangent planes: these twelve new planes give rise to twenty-four new lines upon the surface, making up with the former three lines, twenty-seven lines upon the surface. It is clear that there can be no lines upon the surface besides these twenty-seven; for since the three lines upon the triple tangent plane are the complete intersection of this plane with the surface, every other line upon the surface must meet the triple tangent plane in a point upon one of the three lines, and must therefore lie in a plane passing through one of these lines, such plane (since it meets the surface in two lines and therefore in a third line) being obviously a triple tangent plane. Hence the whole number of lines upon the surface is twenty-seven; and it immediately follows that the number of triple tangent planes is forty-five. The number of lines upon the surface may also be obtained by the following method, which has the advantage of not assuming *à priori* the existence of a line upon the surface. Imagine the cone having for its vertex a given point not upon the surface and circumscribed about the surface, every double tangent plane of the cone is also a double tangent plane of the surface, and therefore intersects the surface in a straight line (and a conic). And, conversely, if there be any line upon the surface, the plane through this line and the vertex of the cone will be a double tangent plane of the cone. Hence the number of double tangent planes of the cones is precisely that of the lines upon the surface. By the theorems in Mr [Dr] Salmon's paper "On the degree of a surface reciprocal to a given one," *Journal*, vol. II. [1847] p. 65, the cone is of the sixth order and has no double lines and six cuspidal lines: hence by the formula in Plücker's "Theorie der algebraischen Curven," [1839] p. 211, stated so as to apply to cones instead of plane curves, viz. n being the order, x the number of double lines, y that of the cuspidal lines, u that of the double tangent planes, then

$$u = \tfrac{1}{2}n(n-2)(n^2-9) - (2x+3y)(n^2-n-6) + 2x(x-1) + 6xy + \tfrac{9}{2}y(y-1),$$

the number of double tangent planes is twenty-seven, which is therefore also the number of lines upon the surface.

Suppose the equation of one of the triple tangent planes to be $w=0$, and let $x=0$, $y=0$, be the equation of *any* two triple tangent planes intersecting the plane $w=0$ in two of the lines in which it meets the surface. Let $z=0$ be the equation of a triple tangent plane meeting $w=0$ in the remaining line in which it intersects the surface. The equation of the surface of the third order is in every case of the form $wP+kxyz=0$, P being a function of the second order, but of the four different planes which the equation $z=0$ may be supposed to represent, one of them such that the function P resolves itself into the product of a pair of factors, and for the remaining three this resolution into factors does not take place. This will be obvious from the sequel: at present I shall suppose that the plane $z=0$ is of the latter class, or that $P=0$ represents a proper surface of the second order. Since $x=0$, $y=0$, $z=0$, are treble tangent planes of the surface, each of these planes must be a tangent plane of the surface of the second order $P=0$, and this will be the case if we assume

$$P = x^2+y^2+z^2+w^2$$
$$+ yz\left(mn+\frac{1}{mn}\right) + zx\left(nl+\frac{1}{nl}\right) + xy\left(lm+\frac{1}{lm}\right) + xw\left(l+\frac{1}{l}\right) + yw\left(m+\frac{1}{m}\right) + zw\left(n+\frac{1}{n}\right);$$

and considering x, y, z and w as each of them implicitly containing an arbitrary constant, this is the most general function which satisfies the conditions in question.

We are thus led to the equation of the surface of the third order:

$$U = w\left\{x^2+y^2+z^2+w^2+\right.$$
$$\left. yz\left(mn+\frac{1}{mn}\right) + zx\left(nl+\frac{1}{nl}\right) + xy\left(lm+\frac{1}{lm}\right) + xw\left(l+\frac{1}{l}\right) + yw\left(m+\frac{1}{m}\right) + zw\left(n+\frac{1}{n}\right)\right\} + kxyz = 0.$$

I have found that by expressing the parameter k in the particular form

$$k = \frac{p^2 - \left(lmn - \dfrac{1}{lmn}\right)^2}{2\left(p - lmn - \dfrac{1}{lmn}\right)},$$

or, as this equation may be more conveniently written,

$$k = \frac{p^2 - \beta^2}{2(p-\alpha)}; \qquad \alpha = lmn + \frac{1}{lmn}, \qquad \beta = lmn - \frac{1}{lmn}, \text{[1]}$$

the equations of all the planes are expressible in a rational form. These equations are in fact the following: [I have added, here and in the table p. 450, the reference numbers 12′, 23′, &c. constituting a different notation for the lines and planes.]

(w)	$w = 0.$	12′
(θ)	$lx + my + nz + w\left[1 + \frac{1}{k}\left(l - \frac{1}{l}\right)\left(m - \frac{1}{m}\right)\left(n - \frac{1}{n}\right)\right] = 0,$	23′
$(\bar\theta)$	$\frac{x}{l} + \frac{y}{m} + \frac{z}{n} + w\left[1 + \frac{1}{k}\left(l - \frac{1}{l}\right)\left(m - \frac{1}{m}\right)\left(n - \frac{1}{n}\right)\right] = 0,$	31′
(x)	$x = 0,$	12 . 34 . 56
(y)	$y = 0,$	42′
(z)	$z = 0,$	14′
(ξ)	$x + \frac{1}{k}\left(m - \frac{1}{m}\right)\left(n - \frac{1}{n}\right) w = 0,$	1′2
(η)	$y + \frac{1}{k}\left(n - \frac{1}{n}\right)\left(l - \frac{1}{l}\right) w = 0,$	2′3
(ζ)	$z + \frac{1}{k}\left(l - \frac{1}{l}\right)\left(m - \frac{1}{m}\right) w = 0,$	3′1
(f)	$lx + \frac{y}{m} + \frac{z}{n} + w = 0,$	41′
(g)	$\frac{x}{l} + my + \frac{z}{n} + w = 0,$	34′
(h)	$\frac{x}{l} + \frac{y}{m} + nz + w = 0,$	13 . 24 . 56

[1] A somewhat more elegant form is obtained by writing $p = 2q + \alpha$; this gives

$$k = \frac{2}{q} \cdot (q + lmn)\left(q + \frac{1}{lmn}\right), \text{ \&c.}$$

$(\bar{\text{f}})\quad \frac{x}{l}+my+nz+w=0,$ ....24′

$(\bar{\text{g}})\quad lx+\frac{y}{m}+nz+w=0,$ 14.23.56

$(\bar{\text{h}})\quad lx+my+\frac{z}{n}+w=0,$ ....43′

$(\text{x})\quad x+\frac{l(p-\alpha)+2mn}{p+\beta}w=0,$ 12.35.46

$(\text{y})\quad y+\frac{m(p-\alpha)+2nl}{p+\beta}w=0,$ .... 52′

$(\text{z})\quad z+\frac{n(p-\alpha)+2lm}{p+\beta}w=0,$ .... 15′

$(\bar{\text{x}})\quad x+\frac{\frac{1}{l}(p-\alpha)+\frac{2}{mn}}{p-\beta}w=0,$ 12.36.45

$(\bar{\text{y}})\quad y+\frac{\frac{1}{m}(p-\alpha)+\frac{2}{nl}}{p-\beta}w=0,$ .... 62′

$(\bar{\text{z}})\quad z+\frac{\frac{1}{n}(p-\alpha)+\frac{2}{lm}}{p-\beta}w=0,$ .... 16′

$(\text{l})\quad -\frac{2n}{m(p-\alpha)}x+\frac{1}{m}y+nz+w=0,$ .... 56′

$(\text{m})\quad lx-\frac{2l}{n(p-\alpha)}y+\frac{1}{n}z+w=0,$ .... 45′

$(\text{n})\quad \frac{1}{l}x+my-\frac{2m}{l(p-\alpha)}z+w=0,$ .... 4′6

$(\bar{\text{l}})\quad -\frac{2m}{n(p-\alpha)}x+my+\frac{1}{n}z+w=0,$ 15.26.34

$(\bar{\text{m}})\quad \frac{1}{l}x-\frac{2l}{l(p-\alpha)}y+nz+w=0,$ 16.24.35

$(\bar{\text{n}})\quad lx+\frac{1}{m}y-\frac{2l}{m(p-\alpha)}z+w=0,$ 14.25.36

$(\text{l}_,)\quad -\frac{n(p-\alpha)}{2m}x+\frac{y}{m}+nz+w=0,$ .... 65′

$(\text{m}_,)\quad lx-\frac{l(p-\alpha)}{2n}y+\frac{1}{n}z+w=0,$ .... 46′

$(\text{n}_,)\quad \frac{1}{l}x+my-\frac{m(p-\alpha)}{2l}z+w=0,$ .... 4′5

$(\bar{\text{l}}_{,})\quad -\frac{m(p-\alpha)}{2n}x+my+\frac{1}{n}z+w=0,$ 16.25.34

$(\bar{\text{m}}_{,})\quad \frac{1}{l}x-\frac{n(p-\alpha)}{2l}y+nz+w=0,$ 15.24.36

$(\bar{\text{n}}_{,})\quad lx+\frac{1}{m}y-\frac{l(p-\alpha)}{2m}z+w=0,$ 14.26.35

$(\text{p})\quad -\frac{2x}{p-\alpha}+ny+mz+[mn(p-\alpha)-2l(1-m^2-n^2)]\frac{w}{p+\beta}=0,$ .... 51′

$(\text{q})\quad nx-\frac{2y}{p-\alpha}+lz+[nl(p-\alpha)-2m(1-n^2-l^2)]\frac{w}{p+\beta}=0,$ .... 35′

$(\text{r})\quad mx+ly-\frac{2z}{p-\alpha}+[lm(p-\alpha)-2n(1-l^2-m^2)]\frac{w}{p+\beta}=0,$ ... 13.25.46

$(\bar{\text{p}})\quad -\frac{2x}{p-\alpha}+\frac{1}{n}y+\frac{1}{m}z+\left[\frac{1}{mn}(p-\alpha)-\frac{2}{l}\left(1-\frac{1}{m^2}-\frac{1}{n^2}\right)\right]\frac{w}{p-\beta}=0,$ ...... 26′

$(\bar{\text{q}})\quad \frac{1}{n}x-\frac{2y}{p-\alpha}+\frac{1}{l}z+\left[\frac{1}{nl}(p-\alpha)-\frac{2}{m}\left(1-\frac{1}{n^2}-\frac{1}{l^2}\right)\right]\frac{w}{p-\beta}=0,$ ...16.23.45

$(\bar{\text{r}})\quad \frac{1}{m}x+\frac{1}{l}y-\frac{2z}{p-\alpha}+\left[\frac{1}{lm}(p-\alpha)-\frac{2}{n}\left(1-\frac{1}{l^2}-\frac{1}{m^2}\right)\right]\frac{w}{p-\beta}=0,$ ....63′

$(\text{p}_{,})\quad -\frac{p-\alpha}{2}x+\frac{y}{n}+\frac{z}{m}-lmn\left[\frac{1}{l}\left(1-\frac{1}{m^2}-\frac{1}{n^2}\right)(p-\alpha)-\frac{2}{mn}\right]\frac{w}{p+\beta}=0,$ ....25′

$(\text{q}_{,})\quad \frac{x}{n}-\frac{p-\alpha}{2}y+\frac{z}{l}-lmn\left[\frac{1}{m}\left(1-\frac{1}{n^2}-\frac{1}{l^2}\right)(p-\alpha)-\frac{2}{nl}\right]\frac{w}{p+\beta}=0,$ ...15.23.46

$(\text{r}_{,})\quad \frac{x}{m}+\frac{y}{l}-\frac{p-\alpha}{2}z-lmn\left[\frac{1}{n}\left(1-\frac{1}{l^2}-\frac{1}{m^2}\right)(p-\alpha)-\frac{2}{lm}\right]\frac{w}{p+\beta}=0,$ .... 53′

$(\bar{\text{p}}_{,})\quad -\frac{p-\alpha}{2}x+ny+mz-\frac{1}{lmn}[l(1-m^2-n^2)(p-\alpha)-2mn]\frac{w}{p-\beta}=0,$ .... 61′

$(\bar{\text{q}}_{,})\quad nx-\frac{p-\alpha}{2}y+lz-\frac{1}{lmn}[m(1-n^2-l^2)(p-\alpha)-2nl]\frac{w}{p-\beta}=0,$ .... 36′

$(\bar{\text{r}}_{,})\quad mx+ly-\frac{p-\alpha}{2}z-\frac{1}{lmn}[n(1-l^2-m^2)(p-\alpha)-2lm]\frac{w}{p-\beta}=0.$ ...13.26.45

In fact, representing the several functions on the left-hand side of these equations respectively by the letters placed opposite to them respectively, the function U is expressible in the sixteen forms following

$$\begin{aligned} U &= w\text{f}\bar{\text{f}} + k\xi yz, \\ &= w\text{g}\bar{\text{g}} + k\eta zx, \\ &= w\text{h}\bar{\text{h}} + k\zeta xy, \\ &= w\theta\bar{\theta} + k\xi\eta\zeta, \end{aligned}$$

$$
\begin{aligned}
&= w\,\mathrm{l}\bar{\mathrm{l}}_{,} + k\mathrm{y}\bar{\mathrm{z}}x,\\
&= w\mathrm{m}\bar{\mathrm{m}}_{,} + k\mathrm{z}\bar{\mathrm{x}}y,\\
&= w\,\mathrm{n}\bar{\mathrm{n}}_{,} + k\mathrm{x}\bar{\mathrm{y}}z,\\
&= w\,\mathrm{l}_{,}\mathrm{l} + k\bar{\mathrm{y}}\mathrm{z}x,\\
&= w\mathrm{m}_{,}\bar{\mathrm{m}} + k\bar{\mathrm{z}}\mathrm{x}y,\\
&= w\,\mathrm{n}_{,}\bar{\mathrm{n}} + k\bar{\mathrm{x}}\mathrm{y}z,\\
&= w\,\mathrm{pp}_{,} + k\xi\mathrm{yz},\\
&= w\,\mathrm{qq}_{,} + k\eta\mathrm{zx},\\
&= w\,\mathrm{rr}_{,} + k\zeta\mathrm{xy},\\
&= w\,\mathrm{p}\bar{\mathrm{p}}_{,} + k\xi\bar{\mathrm{y}}\bar{\mathrm{z}},\\
&= w\,\mathrm{q}\bar{\mathrm{q}}_{,} + k\eta\bar{\mathrm{z}}\bar{\mathrm{x}},\\
&= w\,\mathrm{r}\bar{\mathrm{r}}_{,} + k\zeta\bar{\mathrm{x}}\bar{\mathrm{y}},
\end{aligned}
$$

(being the forms containing w, out of a complete system of one hundred and twenty different forms).

The forty-five planes pass five and five through the twenty-seven lines in the following manner:

(a_1) $(w, x, \xi, \mathrm{x}, \bar{\mathrm{x}})$ ... 12,	(a_4) $(x, \mathrm{g}, \bar{\mathrm{h}}, \mathrm{l}, \bar{\mathrm{l}}_{,})$ ... 34,	(a_7) $(\mathrm{x}, \mathrm{m}_{,}, \mathrm{n}, \mathrm{q}_{,}, \mathrm{r})$ ... 46,
(b_1) $(w, y, \eta, \mathrm{y}, \bar{\mathrm{y}})$ ... 2′,	(b_4) $(y, \mathrm{h}, \bar{\mathrm{f}}, \bar{\mathrm{m}}, \bar{\mathrm{m}}_{,})$ ... 24,	(b_7) $(\mathrm{y}, \mathrm{n}_{,}, \mathrm{l}, \mathrm{r}_{,}, \mathrm{p})$ ... 5,
(c_1) $(w, z, \zeta, \mathrm{z}, \bar{\mathrm{z}})$ ... 1,	(c_4) $(z, \mathrm{f}, \bar{\mathrm{g}}, \bar{\mathrm{n}}, \bar{\mathrm{n}}_{,})$ ... 14,	(c_7) $(\mathrm{z}, \mathrm{l}_{,}, \mathrm{m}, \mathrm{p}_{,}, \mathrm{q})$ ... 5′,
(a_2) $(\xi, \bar{\mathrm{f}}, \theta, \bar{\mathrm{p}}, \mathrm{p}_{,})$ ... 2,	(a_5) $(x, \bar{\mathrm{g}}, \mathrm{h}, \mathrm{l}, \mathrm{l}_{,})$ ... 56,	(a_8) $(\bar{\mathrm{x}}, \bar{\mathrm{m}}_{,}, \bar{\mathrm{n}}, \bar{\mathrm{q}}_{,}, \bar{\mathrm{r}})$ ... 36,
(b_2) $(\eta, \bar{\mathrm{g}}, \theta, \bar{\mathrm{q}}, \mathrm{q}_{,})$ ... 23,	(b_5) $(y, \bar{\mathrm{h}}, \mathrm{f}, \mathrm{m}, \mathrm{m}_{,})$ ... 4,	(b_8) $(\bar{\mathrm{y}}, \bar{\mathrm{n}}_{,}, \bar{\mathrm{l}}, \bar{\mathrm{r}}_{,}, \bar{\mathrm{p}})$ ... 26,
(c_2) $(\zeta, \bar{\mathrm{h}}, \theta, \bar{\mathrm{r}}, \mathrm{r}_{,})$ ... 3′,	(c_5) $(z, \bar{\mathrm{f}}, \mathrm{g}, \mathrm{n}, \mathrm{n}_{,})$ ... 4′,	(c_8) $(\bar{\mathrm{z}}, \bar{\mathrm{l}}_{,}, \bar{\mathrm{m}}, \bar{\mathrm{p}}_{,}, \bar{\mathrm{q}})$ ... 16,
(a_3) $(\xi, \mathrm{f}, \bar{\theta}, \mathrm{p}, \bar{\mathrm{p}}_{,})$ ... 1′,	(a_6) $(\mathrm{x}, \mathrm{m}, \mathrm{n}_{,}, \mathrm{q}, \mathrm{r}_{,})$ ... 35,	(a_9) $(\bar{\mathrm{x}}, \mathrm{m}, \mathrm{n}_{,}, \bar{\mathrm{q}}, \bar{\mathrm{r}}_{,})$ ... 45,
(b_3) $(\eta, \mathrm{g}, \bar{\theta}, \mathrm{q}, \bar{\mathrm{q}}_{,})$ ... 3,	(b_6) $(\mathrm{y}, \mathrm{n}, \mathrm{l}_{,}, \mathrm{r}, \mathrm{p}_{,})$ ... 25,	(b_9) $(\bar{\mathrm{y}}, \mathrm{n}, \mathrm{l}_{,}, \bar{\mathrm{r}}, \bar{\mathrm{p}}_{,})$ ... 6,
(c_3) $(\zeta, \mathrm{h}, \bar{\theta}, \mathrm{r}, \bar{\mathrm{r}}_{,})$ ... 13,	(c_6) $(\mathrm{z}, \mathrm{l}, \mathrm{m}_{,}, \mathrm{p}, \mathrm{q}_{,})$ ... 15,	(c_9) $(\bar{\mathrm{z}}, \mathrm{l}, \mathrm{m}_{,}, \bar{\mathrm{p}}, \bar{\mathrm{q}}_{,})$ ... 6′,

where each line may be represented by the letter placed oppositive to the system of planes passing through it. The twenty-seven lines lie three and three upon the forty-five planes in the following manner:

(w) $a_1b_1c_1$,	(f) $a_3b_5c_4$,	(l) $a_5b_7c_9$,	(p) $a_3b_7c_6$,
(θ) $a_2b_2c_2$,	(g) $b_3c_5a_4$,	(m) $b_5c_7a_9$,	(q) $b_3c_7a_6$,
(θ) $a_3b_3c_3$,	(h) $c_3a_5b_4$,	(n) $c_5a_7b_9$,	(r) $c_3a_7b_6$,

(x)	$a_1a_4a_5$,	$(\bar{\text{f}})$	$a_2b_4c_5$,	$(\bar{\text{l}})$	$a_4b_8c_6$,	$(\bar{\text{p}})$	$a_2b_8c_9$,
(y)	$b_1b_4b_5$,	$(\bar{\text{g}})$	$b_2c_4a_5$,	$(\bar{\text{m}})$	$b_4c_8a_6$,	$(\bar{\text{q}})$	$b_2c_8a_9$,
(z)	$c_1c_4c_5$,	$(\bar{\text{h}})$	$c_2a_4b_5$,	$(\bar{\text{n}})$	$c_4a_8b_6$,	$(\bar{\text{r}})$	$c_2a_8b_9$,
(ξ)	$a_1a_2a_3$,	(x)	$a_1a_6a_7$,	$(\text{l}_{,})$	$a_5b_9c_7$,	$(\text{p}_{,})$	$a_2b_6c_7$,
(η)	$b_1b_2b_3$,	(y)	$b_1b_6b_7$,	$(\text{m}_{,})$	$b_5c_9a_7$,	$(\text{q}_{,})$	$b_2c_6a_7$,
(ζ)	$c_1c_2c_3$,	(z)	$c_1c_6c_7$,	$(\text{n}_{,})$	$c_5a_9b_7$,	$(\text{r}_{,})$	$c_2a_6b_7$,
		$(\bar{\text{x}})$	$a_1a_8a_9$,	$(\bar{\text{l}}_{,})$	$a_4b_6c_8$,	$(\bar{\text{p}}_{,})$	$a_3b_9c_8$,
		$(\bar{\text{y}})$	$b_1b_8b_9$,	$(\bar{\text{m}}_{,})$	$b_4c_6a_8$,	$(\bar{\text{q}}_{,})$	$b_3c_9a_8$,
		$(\bar{\text{z}})$	$c_1c_8c_9$,	$(\bar{\text{n}}_{,})$	$c_4a_6b_8$,	$(\bar{\text{r}}_{,})$	$c_3a_9b_8$.

The preceding method was the one that first occurred to me, and which appears to conduct most simply to the actual analytical expressions for the forty-five planes; but it is worth noticing that the relations between the lines and planes might have been obtained almost without algebraical developments, if we had supposed that P, instead of representing a proper surface of the second order, had represented a pair of planes. This would have conducted at once to one of the one hundred and twenty forms U, e.g. $U = w\theta\bar{\theta} + k\xi\eta\zeta$. Or changing the notation so as to include k in one of the linear functions, $U = ace - bdf$, and it is indeed obvious *à priori*, by merely reckoning the number of arbitrary constants, that any function of the third order can be put under this form. If we suppose $a = \mu b$ to be the equation of one of the triple tangent planes through the intersection of the planes a and b, the plane $a = \mu b$ meets the surface in the same lines in which it meets the hyperboloid $\mu ce - df = 0$, that is, the two lines in the plane are generating lines of different species, and consequently one of them meets the pair of lines cd and ef, and the other of them meets the pair of lines cf and de (where cd represents the line of intersection of the planes $c = 0$, $d = 0$, &c.). This suggests a notation for the lines in question, viz. each line may be represented by the three lines which it meets, or by the symbols $ab.cd.ef$ and $ab.cf.de$. Or observing that μ has three values, and that the same considerations apply *mutatis mutandis* to the planes through bc and ca, the whole system of lines may be represented by the notation,

ab,	ad,	af,
cb,	cd,	cf,
eb,	ed,	ef,
$(ab\,.\,cd\,.\,ef)_1$,	$(ab\,.\,cd\,.\,ef)_2$,	$(ab\,.\,cd\,.\,ef)_3$,
$(ad\,.\,cf\,.\,eb)_1$,	$(ad\,.\,cf\,.\,eb)_2$,	$(ad\,.\,cf\,.\,eb)_3$,
$(af\,.\,cb\,.\,ed)_1$,	$(af\,.\,cb\,.\,ed)_2$,	$(af\,.\,cb\,.\,ed)_3$,
$(ab\,.\,cf\,.\,ed)_1$,	$(ab\,.\,cf\,.\,ed)_2$,	$(ab\,.\,cf\,.\,ed)_3$,
$(ad\,.\,cb\,.\,ef)_1$,	$(ad\,.\,cb\,.\,ef)_2$,	$(ad\,.\,cb\,.\,ef)_3$,
$(af\,.\,cd\,.\,eb)_1$,	$(af\,.\,cd\,.\,eb)_2$,	$(af\,.\,cd\,.\,eb)_3$,

where the last eighteen lines have been divided into two systems of nine each. The five planes through $(ab.cd.ef)_1$ may be considered as cutting the surface in

$$ab;\ (ab.cf.ed)_1,$$
$$cd;\ (af.cd.eb)_1,$$
$$ef;\ (ad.cb.ef)_1,$$
$$(ad.cf.eb)_2;\ (af.cb.ed)_3,$$
$$(ad.cf.eb)_3;\ (af.cb.ed)_2,$$

(which supposes however that the distinguishing suffixes 1, 2, 3, are added to the different planes according to a certain rule). And similarly for the lines in the planes through the other lines represented by symbols of the like form. The five planes through ab intersect the surface in the lines

$$\begin{array}{ll} cb, & eb, \\ ad, & af, \\ (ab.cd.ef)_1, & (ab.cf.ed)_1, \\ (ab.cd.ef)_2, & (ab.cf.ed)_2, \\ (ab.cd.ef)_3, & (ab.cf.ed)_3, \end{array}$$

and similarly for the planes through the other lines represented by symbols of a like form.

Observing that ξ, η, ζ, correspond to b, d, f, and w, θ, $\bar{\theta}$, to a, c, e respectively, ab corresponds to the intersection of w and ξ, i.e. to a_1, &c.; also $(ab.cd.ef)_1$, $(ab.cd.ef)_2$, $(ab.cd.ef)_3$ correspond to three lines meeting a_1, b_2, and c_3, i.e. to a_5, a_7, a_9, &c.; and the system of the twenty-seven lines as last written down corresponds to the system,

$$\begin{array}{lll} a_1, & b_1, & c_1, \\ a_2, & b_2, & c_2, \\ a_3, & b_3, & c_3, \\ \\ a_5, & a_7, & a_9, \\ b_5, & b_7, & b_9, \\ c_5, & c_7, & c_9, \\ \\ a_4, & a_6, & a_8, \\ b_4, & b_6, & b_8, \\ c_4, & c_6, & c_8. \end{array}$$

The investigations last given are almost complete in themselves as the geometrical theory of the subject: there is however some difficulty in seeing *à priori* the nature of the correspondence between the planes which determines which are the planes which ought to be distinguished with the same one of the symbolic numbers, 1, 2, 3.

There is great difficulty in conceiving the complete figure formed by the twenty-seven lines, indeed this can hardly I think be accomplished until a more perfect notation is discovered. In the mean time it is easy to find theorems which partially exhibit the properties of the system. For instance, any two lines, a_1, b_2, which do not meet are intersected by five other lines, a_2, b_1, a_5, a_7, a_9, (no two of which meet). Any four of these last-mentioned lines are intersected by the lines a_1, b_2 and no other lines, but any three of them, e.g. a_5, a_7, a_9, are intersected by the lines a_1, b_2, and by some third line (in the case in question the line c_3). Or generally any three lines, no two of which meet, are intersected by three other lines, no two of which meet. Again, the lines which do not meet any one of the lines a_5, a_7, a_9, are a_2, a_3, b_3, b_1, c_1, c_2: these lines form a hexagon, the pairs of the opposite sides of which, a_2, b_1; a_3, c_1; b_3, c_2, are met by the pairs a_1, b_2; c_3, a_1 and a_1, b_2, respectively, viz. by pairs out of the system of three lines intersecting the system a_5, a_7, a_9. And the lines a_5, a_7, a_9 may be considered as representing any three lines no two of which meet. Again, consider three lines in the same triple tangent plane, e.g. a_1, b_1, c_1, and the hexahedron formed by any six triple tangent planes passing two and two through these lines, e.g. the planes x, y, z, ξ, η, ζ. These planes contain (independently of the lines a_1, b_1, c_1) the twelve lines a_2, a_3, a_4, a_5, b_2, b_3, b_4, b_5, c_2, c_3, c_4, c_5. Consider three contiguous faces of the hexahedron, e.g. x, y, z, the lines in these planes, viz. a_4, b_5, c_4, a_5, b_4, c_5, form a hexagon the opposite sides of which intersect in a point, or in other words these six lines are generating lines of a hyperboloid. The same property holds for the systems x, η, ζ; ξ, y, ζ; ξ, η, z. But for the system ξ, η, ζ, the six lines are a_2, b_2, c_2, and a_3, b_3, c_3, which form two triangles, and similarly for the systems ξ, y, z; x, η, z; and x, y, ζ; so that the twelve lines form four hexagons (the opposite sides of which intersect) circumscribed round four of the angles of the hexahedron, and four pairs of triangles about the opposite four angles of the hexahedron. The number of such theorems might be multiplied indefinitely, and the number of different combinations of lines or planes to which each theorem applies is also very considerable.

Consider the four planes x, ξ, x, $\bar{\mathrm{x}}$, and represent for a moment the equations of these planes by $x + Aw = 0$, $x + Bw = 0$, $x + Cw = 0$, $x + Dw = 0$, so that

$$A = 0,\quad B = \frac{1}{k}\left(m - \frac{1}{m}\right)\left(n - \frac{1}{n}\right),\quad C = \frac{l(p-\alpha) + 2mn}{p+\beta},\quad D = \frac{\frac{1}{l}(p-\alpha) + \frac{2}{mn}}{p-\beta}.$$

By the assistance of

$$B - C = \frac{-1}{lmn(p^2-\beta^2)}[ln(p-\alpha) + 2m][lm(p-\alpha) + 2n],$$

it is easy to obtain

$$\frac{(A-C)(D-B)}{(A-D)(B-C)} = lmn\frac{p-\beta}{p+\beta}\frac{[l(p-\alpha)+2mn][m(p-\alpha)+2nl][n(p-\alpha)+2lm]}{[mn(p-\alpha)+2l][nl(p-\alpha)+2m][lm(p-\alpha)+2n]},$$

which remains unaltered for cyclical permutations of l, m, n, i.e. the anharmonic ratio of x, ξ, x, $\bar{\mathrm{x}}$ is the same as that of y, η, y, $\bar{\mathrm{y}}$, or z, ζ, z, $\bar{\mathrm{z}}$; there is of course no correspondence of x to y or ξ to η, &c., the correspondence is by the general properties of anharmonic ratios, a correspondence of the system x, ξ, x, $\bar{\mathrm{x}}$, to any one of the systems (y, η, y, $\bar{\mathrm{y}}$), or (η, y, $\bar{\mathrm{y}}$, y), or (y, $\bar{\mathrm{y}}$, y, η), or ($\bar{\mathrm{y}}$, y, η, y), indifferently. The theorems may be stated generally as follows: "Considering two lines in the same triple tangent plane, the remaining triple tangent planes through these two lines respectively are homologous systems."

Suppose the surface of the third order intersected by an arbitrary plane. The curve of intersection is of course one of the third order, and the positions upon this curve of six of the points in which it is intersected may be arbitrarily assumed. Let these points be the points in which the plane is intersected by the lines a_1, b_1, a_6, b_6, c_6, a_8; or as we may term them, the points a_1, b_1, a_6, b_6, c_6, a_8.(1) The point c_1 is of course the point in which the line a_1b_1 intersects the curve. The straight lines $a_4b_6c_8$, $b_4c_6a_8$, $c_4a_6b_8$, and $a_4b_8c_6$, $b_4c_8a_6$, $c_4a_8b_6$, show that c_4 and b_4 are the points in which a_8b_6, and a_8c_6 intersect the curve, and then b_8 and c_8 are determined as the intersections of a_6c_4, a_6b_4 with the curve. The intersection of the lines b_6c_8 and b_8c_6 (which is known to be a point upon the curve by the theorem, every curve of the third order passing through eight of the points of intersection of two curves of the third order passes through the ninth point of intersection) is the point a_4. The systems a_4, b_4, c_4; a_6, b_6, c_6; a_8, b_8, c_8, determine the conjugate system a_5, b_5, c_5; a_7, b_7, c_7; a_9, b_9, c_9; by reason of the straight lines $a_1a_4a_5$, $b_1b_4b_5$, $c_1c_4c_5$; $a_1a_6a_7$, $b_1b_6b_7$, $c_1c_6c_7$; $a_1a_8a_9$, $b_1b_8b_9$, $c_1c_8c_9$, viz. a_5 is the point where a_1a_4 intersects the curve, &c. The relations of the systems (a_4, b_4, c_4; a_5, b_5, c_5), (a_6, b_6, c_6; a_7, b_7, c_7), (a_8, b_8, c_8; a_9, b_9, c_9) to the system a_1, b_1, c_1; a_2, b_2, c_2; a_3, b_3, c_3 are precisely identical. It is only necessary to show how the points a_2, b_2, c_2; a_3, b_3, c_3 of the latter system are determined by means of one of the former systems, suppose the system a_6, b_6, c_6; a_7, b_7, c_7; and to discover a compendious statement of the relation between the two systems. The points a_1, b_1, c_1; a_2, b_2, c_2; a_3, b_3, c_3; a_6, b_6, c_6; a_7, b_7, c_7, are a system of fifteen points lying on the fifteen straight lines $a_1b_1c_1$, $a_2b_2c_2$, $a_3b_3c_3$, $a_1a_2a_3$, $b_1b_2b_3$, $c_1c_2c_3$, $a_1a_6a_7$, $b_1b_6b_7$, $c_1c_6c_7$, $a_3b_7c_6$, $b_3c_7a_6$, $c_3a_7b_6$, $a_2b_6c_7$, $b_2c_6a_7$, $c_2a_6b_7$, viz. the nine points a_1, b_1, c_1; a_2, b_2, c_2; a_3, b_3, c_3 are the points of intersection of the three lines $a_1b_1c_1$, $a_2b_2c_2$, $a_3b_3c_3$ with the three lines $a_1a_2a_3$, $b_1b_2b_3$, $c_1c_2c_3$, and the remaining six points form a hexagon $a_6b_7c_6a_7b_6c_7$, of which the diagonals a_6a_7, b_6b_7, c_6c_7 pass through the points a_1, b_1, c_1, respectively, the alternate sides a_6b_7, c_6a_7, and b_6c_7 pass through the points c_2, b_2, a_2 respectively, and the remaining alternate sides b_7c_6, a_7b_6, and c_7a_6 pass through the three points a_3, b_3, c_3 respectively. The fifteen points of such a system do not necessarily lie upon a curve of the third order, as will presently be seen: in the actual case however where all the points lie upon a given curve of the third order, and the points a_1, b_1, c_1; a_6, b_6, c_6; a_7, b_7, c_7 are known, a_2, b_2, c_2; a_3, b_3, c_3 are the intersections of the curve with b_6c_7, c_6a_7, a_6b_7, b_7c_6, c_7a_6, a_7b_6 respectively, and the fact

1 In general, the point in which any line upon the surface intersects the plane in question may be represented by the symbol of the line, and the line in which any triple tangent plane intersects the plane in question may be represented by the symbol of the triple tangent plane: thus, a_1, b_1, c_1 are points in the line $a_1b_1c_1$, or in the line w, &c.

of the existence of the lines $a_1b_1c_1$, $a_2b_2c_2$, $a_3b_3c_3$, $a_1a_2a_3$, $b_1b_2b_3$, $c_1c_2c_3$ is an immediate consequence of the theorem quoted above with respect to curves of the third order—a theorem from which the entire system of relations between the twenty-seven points on the curve might have been deduced *à priori*. But returning to the system of fifteen points, suppose the lines $a_1b_1c_1$, $a_2b_2c_2$, $a_3b_3c_3$, and $a_1a_2a_3$, $b_1b_2b_3$, and also the point a_6 to be given arbitrarily. The point a_7 lies on the line a_1a_6, suppose its position upon this line to be arbitrarily assumed (in which case, since the ten points a_1, a_2, a_3, b_1, b_2, b_3, c_1, c_2, c_3, are sufficient to determine a curve of the third order, there is no curve of the third order through these points and the point a_7). If the points

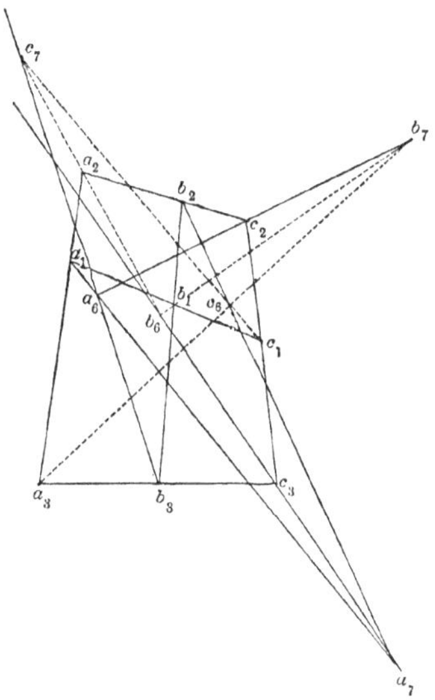

b_6, c_6, b_7, c_7 can be so determined that the sides of the quadrilateral $b_6b_7c_6c_7$, viz. b_6b_7, b_7c_6, c_6c_7, c_7b_6 pass through the points b_1, a_3, c_1, a_2 respectively, while the angles b_6, b_7, c_6, c_7 lie upon the lines a_7c_3, a_6c_2, a_7b_2 and a_6b_3 respectively, the required conditions will be satisfied by the fifteen points in question; and the solution of this problem is known. I have not ascertained whether in the case of an arbitrary position as above of the point a_7, it is possible to determine a complete system of twenty-seven points lying three and three upon forty-five lines in the same manner as the twenty-seven points upon the curve of the third order; but it appears probable that this is the case, and to determine whether it be so or not, presents itself as an interesting problem for investigation.

Suppose that the intersecting plane coincides with one of the triple tangent planes. Here we have a system of twenty-four points, lying eight and eight in three lines; the twenty-four points lie also three and three in thirty-two lines, which last-mentioned lines therefore pass four and four through the twenty-four points. If we represent by a, b, c, d, a', b', c', d' and a, b, c, d, a', b', c', d', the eight points, and eight points which lie upon two of the three lines (the order being determinate), the systems of four lines which intersect in the eight points of the third line are

$$\begin{array}{llllllll}
(a\mathrm{a}, & b\mathrm{b}, & c\mathrm{c}, & d\mathrm{d}), & (a'\mathrm{a}', & b'\mathrm{b}', & c'\mathrm{c}', & d'\mathrm{d}'); \\
(a\mathrm{b}', & b\mathrm{a}', & c'\mathrm{d}, & d'\mathrm{c}), & (a'\mathrm{b}, & b'\mathrm{a}, & c\mathrm{d}', & d\mathrm{c}'); \\
(a\mathrm{c}', & c\mathrm{a}', & d'\mathrm{b}, & b'\mathrm{d}), & (a'\mathrm{c}, & c'\mathrm{a}, & b\mathrm{d}', & d\mathrm{b}'); \\
(a\mathrm{d}', & d\mathrm{a}', & b'\mathrm{c}, & c'\mathrm{b}), & (a'\mathrm{d}, & d'\mathrm{a}, & b\mathrm{c}', & c\mathrm{b}'):
\end{array}$$

the principle of symmetry made use of in this notation (which however represents the actual symmetry of the system very imperfectly) being obviously entirely different from that of the case of an arbitrary intersecting plane. The transition case where the

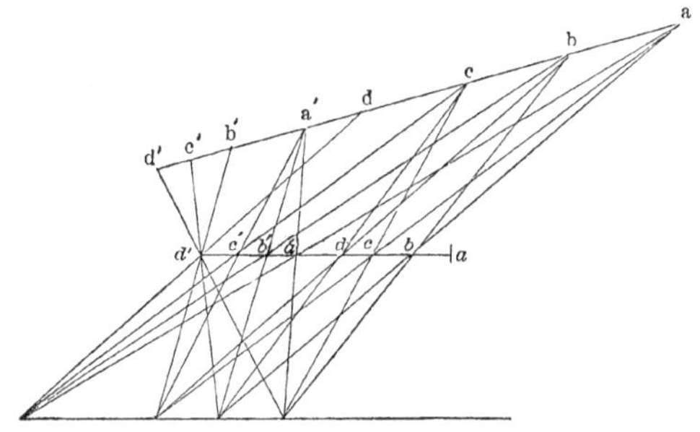

intersecting plane passes through one of the lines upon the surface (and is thus a double tangent plane) would be worth examining. It should be remarked that the preceding theory is very materially modified when the surface of the third order has one or more conical points; and in the case of a double line (for which the surface becomes a ruled surface) the theory entirely ceases to be applicable. I may mention in conclusion that the whole subject of this memoir was developed in a correspondence with Mr Salmon, and in particular, that I am indebted to him for the determination of the number of lines upon the surface and for the investigations connected with the representation of the twenty-seven lines by means of the letters a, c, e, b, d, f, as developed above.

77.

ON THE ORDER OF CERTAIN SYSTEMS OF ALGEBRAICAL EQUATIONS.

[From the *Cambridge and Dublin Mathematical Journal*, vol. IV. (1849), pp. 132—137.([1])]

SUPPOSE the variables $x, y \ldots$ so connected that any one of the ratios $x : y : z, \ldots$ or, more generally, any determinate function of these ratios, depends on an equation of the μth order. The variables $x, y, z \ldots$ are said to form a system of the μth order.

In the case of two variables x, y, supposing that these are connected by an equation $U = 0$ (U being a homogeneous function of the order μ) the variables form a system of the μth order; and, conversely, whenever the variables form a system of the μth order, they are connected by an equation of the above form.

In the case of a greater number of variables, the question is one of much greater difficulty. Thus with three variables x, y, z; if μ be resolvable into the factors μ', μ'', then, supposing the variables to be connected by the equations $U = 0$, $V = 0$, U and V being homogeneous functions of the orders μ', μ'', respectively, they will it is true form a system of the μth order, but the converse proposition does not hold: for instance, if μ is a prime number, the only mode of forming a system of the μth order would on the above principle be to assume $\mu' = \mu$, $\mu'' = 1$, that is to suppose the variables connected by an equation of the μth order and a linear equation; but this is far from being the most general method of obtaining such a system. In fact, systems not belonging to the class in question may be obtained by the introduction of subsidiary

[1] This memoir was intended to appear at the same time with Mr Salmon's "Note on a Result of Elimination," (*Journal*, vol. III. p. 169) with which it is very much connected.

variables to be eliminated: the simplest example is the following: suppose a, b, a', b', a'', b'' to be linear functions (without constant terms) of x, y, z, and write

$$\begin{cases} a\xi + b\eta = 0, \\ a'\xi + b'\eta = 0, \\ a''\xi + b''\eta = 0; \end{cases}$$

equations from which, by the elimination of ξ, η, two relations may be obtained between the variables x, y, z.

Suppose, however, from these three equations x, y, z are first eliminated: the ratio $\xi : \eta$ will evidently be determined by a cubic equation; and assuming $\xi : \eta$ to be equal to one of the roots of this, any two of the three equations may be considered as implying the third; and will likewise determine linearly the ratios $x : y : z$. Hence any determinate function of these ratios depends on a cubic equation only, or the system is one of the third order. But the order of the system may be obtained by means of the equations resulting from the elimination of ξ, η; and since this will explain the following more general example (in which the corresponding process is the only one which readily offers itself), it will be convenient to deduce the preceding result in this manner. Thus, performing the elimination, we have

$$L = (a'b'' - a''b') = 0, \quad L' = (a''b - ab'') = 0, \quad L'' = (ab' - a'b) = 0.$$

Here the equations $L = 0$, $L' = 0$, $L'' = 0$, are each of them of the second order, and any two of them may be considered as implying the third. For we have identically,

$$aL + a'L' + a''L'' = 0, \text{ [1]}$$

so that $L = 0$, $L' = 0$, gives $a''L'' = 0$, or $L'' = 0$. Nevertheless the system is imperfectly represented by means of two equations only. For instance, $L = 0$, $L' = 0$ do, of themselves, represent a system which is really of the fourth order. In fact, these equations are satisfied by $a'' = 0$, $b'' = 0$, (which is to be considered as forming a system of the first order), but these values do not satisfy the remaining equation $L'' = 0$. In other words, the equations $L = 0$, $L' = 0$ contain an extraneous system of the first order, and which is seen to be extraneous by means of the last equation L'': the system required is the system of the third order which is common to the three equations $L = 0$, $L' = 0$, $L'' = 0$.

Suppose, more generally, that x, y, z are connected by $\overline{p+1}$ equations, involving p variables ξ, η, $\zeta \ldots$,

$$\begin{aligned} &a\xi + b\eta + c\zeta + \ldots = 0, \\ &a'\xi + b'\eta + c'\zeta + \ldots = 0; \\ &\vdots \end{aligned}$$

[1] Also $bL + b'L' + b''L'' = 0$: but since by the elimination of L'', taking into account the actual values of L and L', we obtain an identical equation, these two relations may be considered as equivalent to a single one.

or what comes to the same by the equations (equivalent to two independent relations)

$$\left\|\begin{matrix} a, & a', & a'', \ldots & a^{(p)} \\ b, & b', & b'', \ldots & b^{(p)} \\ \vdots & & & \end{matrix}\right\| = 0\,;$$

(where the number of horizontal rows is p). Consider x, y, z, as connected by the two equations

$$\left|\begin{matrix} a, \ldots & a^{(p-2)}, & a^{(p-1)} \\ b, \ldots & b^{(p-2)}, & b^{(p-1)} \\ \vdots & & \end{matrix}\right| = 0, \qquad \left|\begin{matrix} a, \ldots & a^{(p-2)}, & a^{p} \\ b, \ldots & b^{(p-2)}, & b^{p} \\ \vdots & & \end{matrix}\right| = 0:$$

these form a system of the order p^2, but they involve the extraneous system

$$\left\|\begin{matrix} a, & b, & \ldots\ldots \\ \vdots & & \\ a^{(p-2)}, & b^{(p-2)}, & \ldots \end{matrix}\right\| = 0.$$

Suppose $\phi(p)$ is the order of the system in question, then the order of this last system is $\phi(p-1)$ and hence $\phi(p) = p^2 - \phi(p-1)$: observing that $\phi(2) = 3$, this gives directly $\phi(p) = \frac{1}{2}p(p+1)$. Hence the order of the system is $\frac{1}{2}p(p+1)$.

Suppose x, y, z, connected by equations of the form $U=0$, $V=0$, $W=0$; U, V, W being linear in x, y, z, and homogeneous functions of the orders m, n, p respectively in ξ, η. By eliminating x, y, z, the ratio $\xi : \eta$ will be determined by an equation of the order $m+n+p$; and since when this is known the ratios $x : y : z$ are linearly determinable, we have $m+n+p$ for the order μ of the system.

Thus, if $m = n = p = 2$, selecting the particular system

$$a\xi^2 + 2b\xi\eta + c\eta^2 = 0,$$
$$b\xi^2 + 2c\xi\eta + d\eta^2 = 0,$$
$$c\xi^2 + 2d\xi\eta + e\eta^2 = 0,$$

it is possible in this case to obtain two resulting equations of the orders two and three respectively, and which consequently constitute the system of the sixth order, without containing any extraneous system. In fact, from the identical equation

$$\begin{aligned} & e\xi\,.\,(a\xi^2 + 2b\xi\eta + c\eta^2) \\ & - (4d\xi + 2e\eta)(b\xi^2 + 2c\xi\eta + d\eta^2) \\ & + (3c\xi + 2d\eta)(c\xi^2 + 2d\xi\eta + e\eta^2) \\ & \qquad = (ae - 4bd + 3c^2)\,\xi^3, \end{aligned}$$

and

$$\begin{aligned}&(ce-d^2)(a\xi^2+2b\xi\eta+c\eta^2)\\&+(cd-be)(b\xi^2+2c\xi\eta+d\eta^2)\\&+(bd-c^2)(c\xi^2+2d\xi\eta+e\eta^2)\\&\qquad=(ace-ad^2-b^2e-c^3+2bcd)\,\xi^2,\end{aligned}$$

we deduce

$$\begin{cases}ae-4bd+3c^2=0,\\ ace-ad^2-b^2e-c^3+2bcd=0;\end{cases}$$

which form the system in question, and may for shortness be represented by $I=0$, $J=0$.

The three equations in ξ, η may be considered as expressing that

$$\begin{aligned}a\xi^3+3b\xi^2\eta+3c\xi\eta^2+d\eta^3&=0,\\ b\xi^3+3c\xi^2\eta+3d\xi\eta^2+e\eta^3&=0,\end{aligned}$$

have a pair of equal roots in common; in other words, that it is possible to satisfy identically

$$(A\xi+B\eta)(a\xi^3+3b\xi^2\eta+3c\xi\eta^2+d\eta^3)+(A'\xi+B'\eta)(b\xi^3+3c\xi^2\eta+3d\xi\eta^2+e\eta^3)=0.$$

Equating to zero the separate terms of this equation, and eliminating A, B, A', B', we obtain

$$\left\|\begin{matrix}. & a, & 3b, & 3c, & d\\ . & b, & 3c, & 3d, & e\\ a, & 3b, & 3c, & d, & .\\ b, & 3c, & 3d, & e, & .\end{matrix}\right\|=0.$$

It is not at first sight obvious what connection these equations have with the two, $I=0$, $J=0$, but by actual expansion they reduce themselves to the following five,

$$\begin{aligned}3\,[2(ce-d^2)\,I-3eJ]&=0,\\ 3\,[(be-cd)\,I-3dJ]&=0,\\ [-(ae+2bd-3c^2)\,I+9cJ]&=0,\\ 3\,[(ad-bc)\,I-3bJ]&=0,\\ 3\,[2(ac-b^2)\,I-3aJ]&=0;\end{aligned}$$

which are satisfied by $I=0$, $J=0$. By the theorem above given, the equations are to be considered as forming a system of the tenth order; the system must therefore be considered as composed of the system $I=0$, $J=0$, and of a system of the fourth order. The system of the fourth order may be written in the form

$$\begin{aligned}&2(ac-b^2):ad-bc:ae+2bd-3c^2:be-cd:2(ce-d^2):3J\\&=\quad a\quad:\quad b\quad:\quad 3c\quad:\quad d\quad:\quad e\quad:\quad I:\end{aligned}$$

but to justify this, it must be shown first that these equations reduce themselves to two independent equations; and next that system is really one of the fourth order. We may remark in the first place, that if

$$u, \ = a\xi^4 + 4b\xi^3\eta + 6c\xi^2\eta^2 + 4d\xi\eta^3 + e\eta^4,$$

is a perfect square, the coefficients will be proportional to those of $\dfrac{d^2u}{d\xi^2}\dfrac{d^2u}{d\eta^2} - \left(\dfrac{d^2u}{d\xi d\eta}\right)^2$. [1] Thus the conditions requisite in order that u may be a perfect square, are given by the system

$$\begin{array}{ccccccccc} 2(ac-b^2) & : & (ad-bc) & : & ae+2bd-3c^2 & : & be-cd & : & 2(ce-d^2) \\ = \quad a & : & b & : & 3c & : & d & : & e, \end{array}$$

or these equations are equivalent to two independent equations only (this may be easily verified *à posteriori*); and by writing $3J$ in the form

$$e(ac-b^2) - 2d(ad-bc) + c(ae+2bd-3c^2) - 2b(be-cd) + a(ce-d^2),$$

the remaining equations of the complete system (3) are immediately deduced; thus the latter system contains only two independent equations. (The preceding reasoning shows that the system (3) expresses the conditions in order that the equations

$$a\xi^3 + 3b\xi^2\eta + 3c\xi\eta^2 + d\eta^3 = 0, \quad b\xi^3 + 3c\xi^2\eta + 3d\xi\eta^2 + e\eta^2 = 0,$$

may have a pair of unequal roots in common: we have already seen that the equations $I=0$, $J=0$ represent the conditions in order that these two equations may have a pair of equal roots in common.) Finally, to verify *à posteriori* the fact of the system (3) being one only of the fourth order, me may, as Mr Salmon has done in the memoir above referred to, represent the system by the two equations

$$a(ce-d^2) - e(ac-b^2) = 0, \quad e(ad-bc) - 2b(ce-d^2) = 0,$$

that is, by $$ad^2 - eb^2 = 0, \quad 2bd^2 - 3bce + ade = 0.$$

These equations contain the extraneous system $(a=0,\ b=0)$ and the extraneous systems $(b=0,\ d=0)$ and $(d=0,\ e=0)$, each of which last, as Mr Salmon has remarked from geometrical considerations, counts double, or the system is one of the 4th order only.

[1] More generally whatever be the order of u, if u contain a square factor, this square factor may easily be shown to occur in $\dfrac{d^2u}{d\xi^2}\dfrac{d^2u}{d\eta^2} - \left(\dfrac{d^2u}{d\xi d\eta}\right)^2$.

78.

NOTE ON THE MOTION OF ROTATION OF A SOLID OF REVOLUTION.

[From the *Cambridge and Dublin Mathematical Journal*, vol. IV. (1849), pp. 268—270.]

USING the notation employed in my former papers on the subject of rotation (*Cambridge Math. Journal*, vol. III. pp. 224—232, [6]; and *Cambridge and Dublin Math. Journal*, vol. I. pp. 167—264, [37]), suppose $B = A$, then r is constant, equal to n suppose; and writing

$$\frac{(A-C)\,n}{A} = \nu, \quad \text{or } C = \left(1 - \frac{\nu}{n}\right) A\,;$$

also putting

$$\theta = \nu t + \gamma,$$

(where γ is an arbitrary constant) the values of p, q, r are easily seen to be given by the equations

$$\begin{cases} p = M \sin\theta, \\ q = M\cos\theta, \\ r = n, \end{cases}$$

(where M is arbitrary). And consequently

$$\begin{aligned} h &= A\ \{M^2 + n\,(n-\nu)\}, \\ k^2 &= A^2\,\{M^2 + \ \ (n-\nu)^2\}. \end{aligned}$$

Also, since $\mathrm{a}^2 + \mathrm{b}^2 + \mathrm{c}^2 = k^2$, we may write

$$\begin{aligned} \mathrm{a} &= -\,k \sin i \cos j, \\ \mathrm{b} &= \quad k \cos i \cos j, \\ \mathrm{c} &= \quad k \sin j\,; \end{aligned}$$

k having the value above given, and the angles i, j being arbitrary.

From the equations (12) and (15) in the second of the papers quoted, we deduce

$$\begin{aligned} 2v &= k\{k + A(n-\nu)\sin j + MA\cos j\cos(\theta+i)\},\\ \Phi &= k\{n\sin j + M\cos j\cos(\theta+i)\},\\ \nabla &= -\nu kMA\cos j\sin(\theta+i), \end{aligned}$$

(values which verify as they should do the equation (19)). Hence, from the equation (27), writing $\dfrac{2dv}{\nabla} = dt = \dfrac{1}{\nu}\,d\theta$, we have

$$2\tan^{-1}\frac{\Omega}{k} = \delta + \frac{1}{\nu}\int d\theta\,\frac{h + kn\sin j + kM\cos j\cos(\theta+i)}{k + A(n-\nu)\sin j + MA\cos j\cos(\theta+i)}.$$

This is easily integrated; but the only case which appears likely to give a simple result is when the quantity under the integral sign is constant, or

$$A(h + kn\sin j) = k\{k + A(n-\nu)\sin j\},$$

or

$$Ah - k^2 + Ak\nu\sin j = 0;$$

that is,

$$A(n-\nu) + k\sin j = 0:$$

whence

$$\sin j = -\frac{A(n-\nu)}{k},\quad \cos j = \frac{AM}{k},\quad \text{or}\ \tan j = \frac{-(n-\nu)}{M} = -\frac{C}{A}\frac{n}{M}.$$

Observing that $\frac{1}{2}\pi - j$ is the inclination of the axis of z to the normal to the invariable plane, this equation shows that the supposition above is not any restriction upon the generality of the motion, but amounts only to supposing that the axis of z (which is a line fixed in space) is taken upon the surface of a certain right cone having for its axis the perpendicular to the invariable plane. Resuming the solution of the problem, we have

$$2\tan^{-1}\frac{\Omega}{k} = \delta + \frac{k}{\nu A}\,\theta,$$

which may also be written under the form

$$2\tan^{-1}\frac{\Omega}{k} = \delta_1 + \frac{kt}{A},$$

$\left(\text{where } \delta_1 = \delta + \dfrac{k\gamma}{\nu A}\right)$. And hence

$$\Omega = k\tan\tfrac{1}{2}\left(\delta_1 + \frac{kt}{A}\right) = k\tan\psi,$$

where

$$\psi = \tfrac{1}{2}\left(\delta_1 + \frac{kt}{A}\right).$$

Substituting these values,

$$\text{a} = -MA\sin i,\quad \text{b} = MA\cos i,\quad \text{c} = -A(n-\nu),$$

$$2v = A^2M^2\{1+\cos(\theta+i)\} = 2M^2A^2\cos^2\tfrac{1}{2}(\theta+i);$$

and substituting in the equations (14) the values of λ, μ, ν reduce themselves to

$$\begin{cases} \lambda = \dfrac{1}{MA\cos\frac{1}{2}(\theta+i)}\{k\tan\psi\sin\tfrac{1}{2}(\theta-i) - A(n-\nu)\cos\tfrac{1}{2}(\theta-i)\}, \\ \mu = \dfrac{1}{MA\cos\frac{1}{2}(\theta+i)}\{k\tan\psi\cos\tfrac{1}{2}(\theta-i) + A(n-\nu)\sin\tfrac{1}{2}(\theta-i)\}, \\ \nu = \tan\tfrac{1}{2}(\theta+i); \end{cases}$$

where, recapitulating, $\theta = \nu t + \gamma$, $2\psi = \dfrac{kt}{A} + \delta_1$.

I may notice, in connexion with the problem of rotation, a memoir, "Specimen Inaugurale de motu gyratorio corporis rigidi &c.," by A. S. Rueb (Utrecht, 1834), which contains some very interesting developments of the ordinary solution of the problem, by means of the theory of elliptic functions.

79.

ON A SYSTEM OF EQUATIONS CONNECTED WITH MALFATTI'S PROBLEM, AND ON ANOTHER ALGEBRAICAL SYSTEM.

[From the *Cambridge and Dublin Mathematical Journal*, vol. IV. (1849), pp. 270—275.]

CONSIDER the equations

$$\begin{aligned} by^2 + cz^2 + 2fyz &= \theta^2 a\,(bc - f^2), \\ cz^2 + ax^2 + 2gzx &= \theta^2 b\,(ca - g^2), \\ ax^2 + by^2 + 2hxy &= \theta^2 c\,(ab - h^2); \end{aligned}$$

or, as they may be more conveniently written,

$$\begin{aligned} by^2 + cz^2 + 2fyz &= \theta^2 a\mathfrak{A}, \\ cz^2 + ax^2 + 2gzx &= \theta^2 b\mathfrak{B}, \\ ax^2 + by^2 + 2hxy &= \theta^2 c\mathfrak{C}. \end{aligned}$$

The second and third equations give

$$(g^2\mathfrak{C} - h^2\mathfrak{B})\,x^2 - b^2\mathfrak{B}y^2 + c^2\mathfrak{C}z^2 + 2cg\mathfrak{C}zx - 2bh\mathfrak{B}xy = 0,$$

hence $\{(g^2\mathfrak{C} - h^2\mathfrak{B})\,x - bh\mathfrak{B}y + cg\mathfrak{C}z\}^2 - \mathfrak{B}\mathfrak{C}\,(-bgy + chz)^2 = 0$, and consequently

$$(g^2\mathfrak{C} - h^2\mathfrak{B})\,x - b\sqrt{\mathfrak{B}}\,(g\sqrt{\mathfrak{C}} + h\sqrt{\mathfrak{B}})\,y + c\sqrt{\mathfrak{C}}\,(g\sqrt{\mathfrak{C}} + h\sqrt{\mathfrak{B}})\,z = 0:$$

dividing this by $g\sqrt{\mathfrak{C}} + h\sqrt{\mathfrak{B}}$, and writing down the system of equations to which the equation thus obtained belongs,

$$\begin{aligned} (g\sqrt{\mathfrak{C}} - h\sqrt{\mathfrak{B}})\,x - {} & b\sqrt{\mathfrak{B}}\ y + c\sqrt{\mathfrak{C}}\ z = 0, \\ a\sqrt{\mathfrak{A}}\ x + {} & (h\sqrt{\mathfrak{A}} - f\sqrt{\mathfrak{C}})\,y - c\sqrt{\mathfrak{C}}\ z = 0, \\ -a\sqrt{\mathfrak{A}}\ x + {} & b\sqrt{\mathfrak{B}}\ y + (f\sqrt{\mathfrak{B}} - g\sqrt{\mathfrak{A}})\,z = 0. \end{aligned}$$

Hence also

$$\begin{aligned}
&(\ \ h\sqrt{\mathfrak{A}}+b\sqrt{\mathfrak{B}}-f\sqrt{\mathfrak{C}})\,y-(\ \ g\sqrt{\mathfrak{A}}-f\sqrt{\mathfrak{B}}+c\sqrt{\mathfrak{C}})\,z=0,\\
&(-g\sqrt{\mathfrak{A}}+f\sqrt{\mathfrak{B}}+c\sqrt{\mathfrak{C}})\,z-(\ \ a\sqrt{\mathfrak{A}}+h\sqrt{\mathfrak{B}}-g\sqrt{\mathfrak{C}})\,x=0,\\
&(\ \ a\sqrt{\mathfrak{A}}-h\sqrt{\mathfrak{B}}+g\sqrt{\mathfrak{C}})\,x-(-h\sqrt{\mathfrak{A}}+b\sqrt{\mathfrak{B}}+f\sqrt{\mathfrak{C}})\,y=0\,;
\end{aligned}$$

these equations may be written

$$\frac{1}{K}\{\ \ \mathfrak{F}\sqrt{\mathfrak{A}}-\mathfrak{G}\sqrt{\mathfrak{B}}-\mathfrak{H}\sqrt{\mathfrak{C}}+\sqrt{(\mathfrak{A}\mathfrak{B}\mathfrak{C})}\}\,[\{\mathfrak{G}+\sqrt{(\mathfrak{C}\mathfrak{A})}\}\,y-\{\mathfrak{H}+\sqrt{(\mathfrak{A}\mathfrak{B})}\}\,z]=0,$$

$$\frac{1}{K}\{-\mathfrak{F}\sqrt{\mathfrak{A}}+\mathfrak{G}\sqrt{\mathfrak{B}}-\mathfrak{H}\sqrt{\mathfrak{C}}+\sqrt{(\mathfrak{A}\mathfrak{B}\mathfrak{C})}\}\,[\{\mathfrak{H}+\sqrt{(\mathfrak{A}\mathfrak{B})}\}\,z-\{\mathfrak{F}+\sqrt{(\mathfrak{B}\mathfrak{C})}\}\,x]=0,$$

$$\frac{1}{K}\{-\mathfrak{F}\sqrt{\mathfrak{A}}-\mathfrak{G}\sqrt{\mathfrak{B}}+\mathfrak{H}\sqrt{\mathfrak{C}}+\sqrt{(\mathfrak{A}\mathfrak{B}\mathfrak{C})}\}\,[\{\mathfrak{F}+\sqrt{(\mathfrak{B}\mathfrak{C})}\}\,x-\{\mathfrak{G}+\sqrt{(\mathfrak{C}\mathfrak{A})}\}\,y]=0\,;$$

where, as usual,

$$\begin{aligned}
&\mathfrak{F}=gh-af,\quad \mathfrak{G}=hf-bg,\quad \mathfrak{H}=fg-ch,\\
&K=abc-af^2-bg^2-ch^2+2fgh\,;
\end{aligned}$$

(in fact the coefficient of y in the first equation is

$$\frac{1}{K}\{(\mathfrak{F}\mathfrak{G}-\mathfrak{C}\mathfrak{H})\sqrt{\mathfrak{A}}+(\mathfrak{A}\mathfrak{C}-\mathfrak{G}^2)\sqrt{\mathfrak{B}}-(\mathfrak{G}\mathfrak{H}-\mathfrak{A}\mathfrak{F})\sqrt{\mathfrak{C}}\},=h\sqrt{\mathfrak{A}}+b\sqrt{\mathfrak{B}}-f\sqrt{\mathfrak{C}},$$

as it should be, and similarly for the coefficients of the remaining terms). We have therefore

$$\{\mathfrak{F}+\sqrt{(\mathfrak{B}\mathfrak{C})}\}\,x=\{\mathfrak{G}+\sqrt{(\mathfrak{C}\mathfrak{A})}\}\,y=\{\mathfrak{H}+\sqrt{(\mathfrak{A}\mathfrak{B})}\}\,z\,;$$

or, what comes to the same thing,

$$\begin{aligned}
yz&=\tfrac{1}{2}s\,\{\mathfrak{F}+\sqrt{(\mathfrak{B}\mathfrak{C})}\},\\
zx&=\tfrac{1}{2}s\,\{\mathfrak{G}+\sqrt{(\mathfrak{C}\mathfrak{A})}\},\\
xy&=\tfrac{1}{2}s\,\{\mathfrak{H}+\sqrt{(\mathfrak{A}\mathfrak{B})}\}.
\end{aligned}$$

Now

$$\begin{aligned}
&a\,\{\mathfrak{G}+\sqrt{(\mathfrak{A}\mathfrak{C})}\}\,\{\mathfrak{H}+\sqrt{(\mathfrak{A}\mathfrak{B})}\}=\{\mathfrak{F}+\sqrt{(\mathfrak{B}\mathfrak{C})}\}\,\{abc-fgh+f\sqrt{(\mathfrak{B}\mathfrak{C})}-g\sqrt{(\mathfrak{C}\mathfrak{A})}-h\sqrt{(\mathfrak{A}\mathfrak{B})}\},\\
&b\,\{\mathfrak{H}+\sqrt{(\mathfrak{A}\mathfrak{B})}\}\,\{\mathfrak{F}+\sqrt{(\mathfrak{B}\mathfrak{C})}\}=\{\mathfrak{G}+\sqrt{(\mathfrak{C}\mathfrak{A})}\}\,\{abc-fgh-f\sqrt{(\mathfrak{B}\mathfrak{C})}+g\sqrt{(\mathfrak{C}\mathfrak{A})}-h\sqrt{(\mathfrak{A}\mathfrak{B})}\},\\
&c\,\{\mathfrak{F}+\sqrt{(\mathfrak{B}\mathfrak{C})}\}\,\{\mathfrak{G}+\sqrt{(\mathfrak{C}\mathfrak{A})}\}=\{\mathfrak{H}+\sqrt{(\mathfrak{A}\mathfrak{B})}\}\,\{abc-fgh-f\sqrt{(\mathfrak{B}\mathfrak{C})}-g\sqrt{(\mathfrak{C}\mathfrak{A})}+h\sqrt{(\mathfrak{A}\mathfrak{B})}\},
\end{aligned}$$

{as readily appears by writing the first of these equations under the form

$$a\,\{\mathfrak{G}+\sqrt{(\mathfrak{A}\mathfrak{C})}\}\,\{\mathfrak{H}+\sqrt{(\mathfrak{A}\mathfrak{B})}\}=\{\mathfrak{F}+\sqrt{(\mathfrak{B}\mathfrak{C})}\}\,\{a\mathfrak{A}-f\mathfrak{F}+f\sqrt{(\mathfrak{B}\mathfrak{C})}-g\sqrt{(\mathfrak{C}\mathfrak{A})}-h\sqrt{(\mathfrak{A}\mathfrak{B})}\},$$

and comparing the rational term and the coefficients of $\sqrt{(\mathfrak{B}\mathfrak{C})}$, $\sqrt{(\mathfrak{C}\mathfrak{A})}$, $\sqrt{(\mathfrak{A}\mathfrak{B})}$}.

Hence, observing the values of yz, zx, xy, we find

$$x^2 = \frac{s}{2a}\{abc - fgh + f\surd(\mathfrak{BC}) - g\surd(\mathfrak{CA}) - h\surd(\mathfrak{AB})\},$$

$$y^2 = \frac{s}{2b}\{abc - fgh - f\surd(\mathfrak{BC}) + g\surd(\mathfrak{CA}) - h\surd(\mathfrak{AB})\},$$

$$z^2 = \frac{s}{2c}\{abc - fgh - f\surd(\mathfrak{BC}) - g\surd(\mathfrak{CA}) - h\surd(\mathfrak{AB})\}.$$

Hence, forming the value of any one of the functions $by^2 + cz^2 + 2fyz$, $cz^2 + ax^2 + 2gxy$, $ax^2 + by^2 + 2hxy$, we obtain $s = \theta^2$; or we have

$$\left\{\begin{array}{l} x^2 = \dfrac{\theta^2}{2a}\{abc - fgh + f\surd(\mathfrak{BC}) - g\surd(\mathfrak{CA}) - h\surd(\mathfrak{AB})\}, \\ y^2 = \dfrac{\theta^2}{2b}\{abc - fgh - f\surd(\mathfrak{BC}) + g\surd(\mathfrak{CA}) - h\surd(\mathfrak{AB})\}, \\ z^2 = \dfrac{\theta^2}{2c}\{abc - fgh - f\surd(\mathfrak{BC}) - g\surd(\mathfrak{CA}) + h\surd(\mathfrak{AB})\}, \\ yz = \frac{1}{2}\theta^2\{\mathfrak{F} + \surd(\mathfrak{BC})\}, \\ zx = \frac{1}{2}\theta^2\{\mathfrak{G} + \surd(\mathfrak{CA})\}, \\ xy = \frac{1}{2}\theta^2\{\mathfrak{H} + \surd(\mathfrak{AB})\}. \end{array}\right.$$

It may be remarked that the equations

$$\begin{aligned} b\ y^2 + c\ z^2 + 2f\ yz &= L, \\ c'\ z^2 + a'x^2 + 2g'\ yz &= M, \\ a''x^2 + b''y^2 + 2h''xy &= N, \end{aligned}$$

in which the coefficients are supposed to be such that the functions

$$\begin{aligned} &M(a''x^2 + b''y^2 + 2h''xy) - N(c'\ z^2 + a'\ x^2 + 2g'\ yz), \\ &N(b\ y^2 + c\ z^2 + 2f\ yz) - L\ (a''x^2 + b''y^2 + 2h''yz), \\ &L\ (c'\ z^2 + a'x^2 + 2g'yz) - M(b\ y^2 + c\ z^2 + 2f\ yz), \end{aligned}$$

are each of them decomposable into linear factors, may always be reduced to a system of equations similar to those which have just been solved.

Suppose

$$f = g = h = \frac{1}{\theta^2} = \sqrt{\left(\frac{abc}{a+b+c}\right)} = r,$$

and write $\surd X$, $\surd Y$, $\surd Z$ instead of x, y, z. The equations to be solved become

$$\begin{aligned} bY + cZ + 2r\surd(YZ) &= (b+c)\,r, \\ cZ + aX + 2r\surd(ZX) &= (c+a)\,r, \\ aX + bY + 2r\surd(XY) &= (a+b)\,r, \end{aligned}$$

where $r^2 = \dfrac{abc}{a+b+c}$, and the solution is

$$\text{x} = \frac{r}{2a}\{a+b+c-r+\surd(r^2+a^2)-\surd(r^2+b^2)-\surd(r^2+c^2)\},$$

$$\text{y} = \frac{r}{2b}\{a+b+c-r+\surd(r^2+a^2)+\surd(r^2+b^2)-\surd(r^2+c^2)\},$$

$$\text{z} = \frac{r}{2c}\{a+b+c-r-\surd(r^2+a^2)-\surd(r^2+b^2)+\surd(r^2+c^2)\},$$

$$\surd(\text{yz}) = \tfrac{1}{2}\{r-a+\surd(r^2+a^2)\},$$
$$\surd(\text{zx}) = \tfrac{1}{2}\{r-b+\surd(r^2+b^2)\},$$
$$\surd(\text{xy}) = \tfrac{1}{2}\{r-c+\surd(r^2+c^2)\},$$

a system of formulæ which contain the solution of the problem "In a given triangle to inscribe three circles such that each circle touches the remaining two circles and also two sides of the triangle." In fact, if r denote the radius of the inscribed circle, and a, b, c the distances of the angles of the triangle from the points where the sides are touched by the inscribed circle (quantities which it is well known satisfy the condition $r^2 = \dfrac{abc}{a+b+c}$), also if x, y, z denote the radii of the required circles, there is no difficulty whatever in obtaining for the determination of x, y, z, the above system of equations. The problem in question was first proposed and solved by an Italian geometer named Malfatti, and has been called after him Malfatti's problem. His solution, dated 1803, and published in the 10th volume of the Transactions of the Italian Academy of Sciences, appears to have consisted in showing that the values first found for the radii of the three circles satisfy the equations given above, without any indication of the process of obtaining the expressions for these radii. Further information as to the history of the problem may be found in the memoir "Das Malfattische Problem neu gelöst von C. Adams," Winterthur, 1846.

In connexion with the preceding investigations may be considered the problem of determining l and m from the equations

$$B(l+\theta)^2 - 2H(l+\theta)m + (A+1)m^2 = 0,$$
$$A(m+\theta)^2 - 2H(m+\theta)l + (B+1)l^2 = 0;$$

which express that the function

$$\theta^2 U + (lx+my)^2, \qquad (U = Ax^2 + 2Hxy + By^2),$$

has for one of its factors a factor of $U+x^2$, and for the other of its factors a factor of $U+y^2$. There is no difficulty in solving these equations; and if we write

$$K = AB - H^2, \quad \varpi_1 = \surd(-K-B), \quad \varpi_2 = \surd(-K-A),$$

the result is easily shown to be

$$l : m : \theta = A(B+H+\varpi_1) : B(A+H+\varpi_2) : (H+\varpi_1)(H+\varpi_2) - AB.$$

But the problem may be considered as the problem for two variables, analogous to that of determining the conic having a double contact with a given conic, and touching three conics each of them having a double contact with the given conic; and in this point of view I was led to the following solution. If we assume

$$Bl - Hm = u, \quad -Hl + Am = v,$$

or, what is the same thing,

$$Kl = Au + Hv, \quad Km = Hu + Bv,$$

then putting

$$\nabla = Au^2 + 2Huv + Bv^2,$$

the two equations become after some reduction

$$(u - K\theta)^2 = -\varpi_1^2\left(K\theta^2 + \frac{1}{K}\nabla\right),$$

$$(v - K\theta)^2 = -\varpi_2^2\left(K\theta^2 + \frac{1}{K}\nabla\right).$$

Hence, writing $K\theta^2 + \frac{1}{K}\nabla = -s^2$, we have

$$u = K\theta + \varpi_1 s, \quad v = K\theta + \varpi_2 s, \quad \nabla + K^2\theta^2 + Ks^2 = 0;$$

and substituting these values of u, v in the last equation,

$$A(K\theta + \varpi_1 s)^2 + 2H(K\theta + \varpi_1 s)(K\theta + \varpi_2 s) + B(K\theta + \varpi_2 s)^2 + K^2\theta^2 + Ks^2 = 0,$$

or reducing,

$$K^2\theta^2(A + 2H + B + 1) + 2K\theta s\{(A+H)\varpi_1 + (H+B)\varpi_2\} + s^2(A\varpi_1^2 + 2H\varpi_1\varpi_2 + B\varpi_2^2 + K) = 0;$$

whence

$$[K\theta(A+2H+B+1) + s\{(A+H)\varpi_1 + (H+B)\varpi_2\}]^2$$

$$= s^2[\{(A+H)\varpi_1 + (H+B)\varpi_2\}^2 - (A+2H+B+1)(A\varpi_1^2 + 2H\varpi_1\varpi_2 + B\varpi_2^2 + K)],$$

$$= s^2\{-K(\varpi_1 - \varpi_2)^2 - (A\varpi_1^2 + 2H\varpi_1\varpi_2 + B\varpi_2^2) - K(A+2H+B+1)\},$$

$$= s^2\{-(A+K)\varpi_1^2 - (B+K)\varpi_2^2 - K(A+2H+B+1) + 2(K-H)\varpi_1\varpi_2\},$$

$$= s^2\{2\varpi_1^2\varpi_2^2 - K(A+2H+B+1) + 2(K-H)\varpi_1\varpi_2\},$$

$$= s^2(\varpi_1\varpi_2 + K - H)^2:$$

and therefore

$$K\theta\,(A+2H+B+1)+s\,\{(A+H)\,\varpi_1+(H+B)\,\varpi_2\}=s\,(\varpi_1\varpi_2+K-H),$$

giving

$$s=\frac{K\theta\,(A+2H+B+1)}{\varpi_1\varpi_2-(A+H)\,\varpi_1-(H+B)\,\varpi_2+K-H}.$$

But

$$Kl=Au+Hv=(A+H)\,K\theta+(A\varpi_1+H\varpi_2)\,s,\quad Km=(H+B)\,K\theta+(H\varpi_1+B\varpi_2)\,s,$$

and substituting the above value of s, we obtain, after some simple reductions,

$$l:m:\theta=(\varpi_1\varpi_2+K-H)\,(A+H-\varpi_2):(\varpi_1\varpi_2+K-H)\,(B+H-\varpi_2)$$
$$:\{\varpi_1\varpi_2-(A+H)\,\varpi_1-(B+H)\,\varpi_2+K-H\},$$

a result which presents itself in a very different form from the one previously obtained: if, however, the terms of this proportion be multiplied by the factor

$$\frac{H\varpi_1\varpi_2-K\varpi_1-K\varpi_2+AB-KH}{(A+2H+B+1)\,K},$$

they become (as they ought to do) identical with those of the former proportion, and the identical equations to which this process gives rise are not without interest.

80.

SUR QUELQUES TRANSMUTATIONS DES LIGNES COURBES.

[From the *Journal de Mathématiques Pures et Appliquées* (Liouville), tom. XIV. (1849) pp. 40—46.]

JE suppose qu'une courbe quelconque soit représentée par une équation homogène entre trois variables x, y, z, et je me propose d'examiner (en ne faisant attention qu'aux droites et aux coniques) ce que signifient les transmutations

I. $$\xi = \sqrt{x}, \quad \eta = \sqrt{y}, \quad \zeta = \sqrt{z},$$

II. $$\xi = x^2, \quad \eta = y^2, \quad \zeta = z^2,$$

III. $$\xi = \frac{1}{x}, \quad \eta = \frac{1}{y}, \quad \zeta = \frac{1}{z},$$

ces quantités étant prises dans chaque cas pour de nouvelles coordonnées. Les théorèmes auxquels on se trouve conduit comprennent les théorèmes qu'obtient M. William Roberts par sa méthode générale de transmutation en écrivant $n = \frac{1}{2}$, $n = 2$. (*Voir* sa Note sur ce sujet, t. XIII. [1848] de ce Journal, page 209).[1]

I. Soit

$$\xi = \sqrt{x}, \quad \eta = \sqrt{y}, \quad \zeta = \sqrt{z}.$$

Je supposerai ici et partout dans ce mémoire que PQR soit le triangle formé par les droites

$$x = 0, \quad y = 0, \quad z = 0.$$

[1] M. William Roberts a bien voulu me parler l'été passé à Dublin, avant que cette Note eût paru, des théorèmes qui devaient s'y trouver: cela me suggéra, et je lui communiquai, peu de jours après, la deuxième et la troisième de mes méthodes de transmutation. On verra dans la suite que ce sont la première et la deuxième de mes méthodes qui correspondent aux cas $n = \frac{1}{2}$, $n = 2$, respectivement, mais que la troisième est aussi très étroitement liée avec le cas $n = \frac{1}{2}$.

Cela posé, en considérant les nouvelles coordonnées ξ, η, ζ, on voit tout de suite qu'un système d'équations telles que

$$\xi : \eta : \zeta = \alpha : \beta : \gamma$$

correspond à un point; qu'une équation linéaire quelconque

$$A\xi + B\eta + C\zeta = 0$$

correspond à une conique qui touche les trois côtés du triangle PQR (ou, si l'on veut, inscrite à ce triangle); et qu'une équation quelconque du second ordre

$$a\xi^2 + b\eta^2 + c\zeta^2 + 2f\zeta\eta + 2g\xi\zeta + 2h\eta\xi = 0$$

correspond à une courbe du quatrième ordre qui touche les trois côtés du triangle PQR, chaque côté deux fois. Et réciproquement, de telles coniques et de telles courbes du quatrième ordre peuvent toujours se représenter par une équation linéaire ou par une équation du second ordre entre les coordonnées ξ, η, ζ.

On déduit de là cette propriété générale:

"Tout théorème descriptif qui se rapporte à des points, à des droites, et à des coniques, conduit à un théorème qui se rapporte d'une manière analogue à des points, à des coniques qui touchent chacune trois droites fixes, et à des courbes du quatrième ordre qui touchent chacune ces trois droites fixes, deux fois chaque droite."

En supposant que les coefficients g, h se réduisent à zéro, l'équation

$$a\xi^2 + b\eta^2 + c\zeta^2 + 2f\eta\zeta = 0$$

est celle d'une conique tangente aux deux droites PQ, PR, et la courbe du quatrième ordre se réduit à deux coniques égales et superposées l'une à l'autre. De là:

"Tout théorème descriptif qui se rapporte à des points, à des droites, et à une conique, conduit à un théorème qui se rapporte d'une manière analogue à des points, à des coniques qui touchent chacune trois droites fixes, et à une conique qui touche deux de ces droites fixes."

En particulier, en supposant que les deux droites fixes que la conique touche soient les deux droites tangentes à cette conique qui passent par un des foyers, et que la troisième droite fixe soit à l'infini:

"Tout théorème descriptif qui se rapporte à des points, à des droites, et à une conique, conduit à un théorème qui se rapporte d'une manière analogue à des points, à des paraboles qui ont pour foyer commun un point fixe, et à une conique qui a aussi ce point fixe pour un de ses foyers."

II. Soit à présent

$$\xi = x^2, \quad \eta = y^2, \quad \zeta = z^2.$$

Un système d'équations telles que

$$\xi : \eta : \zeta = \alpha : \beta : \gamma$$

correspond à quatre points qui formeront ce que l'on peut nommer *un système symétrique par rapport aux points* P, Q, R, ou tout simplement *un système symétrique.* On voit sans peine que les quatre points d'un même système symétrique sont tels, que le quadrilatère formé par ces quatre points a pour points de concours des diagonales et des côtés opposés les points P, Q, R. Pour ne pas interrompre la suite des raisonnements, il convient d'entrer dans quelques détails relatifs à ce sujet. Nous nommerons *courbe symétrique par rapport aux points* P, Q, R, ou simplement *courbe symétrique,* toute courbe lieu d'un système symétrique. Cela posé, il y a une infinité de coniques symétriques; savoir, toute conique par rapport à laquelle les points P, Q, R sont des points conjugués (cela veut dire par rapport à laquelle l'un quelconque de ces points a pour polaire la droite menée par les deux autres) est une conique symétrique. Remarquons qu'une courbe symétrique du quatrième ordre peut se réduire à un système de deux coniques telles, que les points de chaque système symétrique de la courbe soient partagés deux à deux sur les deux coniques. En effet, considérons une conique telle, que, par rapport à cette conique, le point P ait pour polaire la droite QR. Il est facile de construire une autre conique telle, que l'ensemble des deux coniques soit une courbe symétrique. Pour cela, menons une transversale quelconque PMN qui rencontre la conique donnée aux points M, N: soient M', N' les points de rencontre des droites QM, NR et des droites QN, MR; le lieu des deux points M', N' (lesquels seront en ligne droite avec le point P) sera la conique dont il s'agit. Nous dirons que les deux coniques sont des coniques supplémentaires.

En revenant à notre but actuel, une équation linéaire quelconque

$$A\xi + B\eta + C\zeta = 0$$

correspond à une conique symétrique. Une équation du second ordre quelconque

$$a\xi^2 + b\eta^2 + c\zeta^2 + 2f\eta\zeta + 2g\zeta\xi + 2h\xi\eta = 0$$

correspond à une courbe symétrique du quatrième ordre. Et réciproquement, toute conique symétrique, ou toute courbe symétrique du quatrième ordre, peut se représenter par une équation linéaire, ou du second ordre, avec les coordonnées ξ, η, ζ. De là cette propriété générale:

"Tout théorème descriptif qui se rapporte à des points, à des droites, et à des coniques, conduit à un théorème qui se rapporte d'une manière analogue à des systèmes symétriques (par rapport à trois points fixes), à des coniques symétriques, et à des courbes symétriques du quatrième ordre."

Dans ce théorème on peut, si l'on veut, considérer l'un des points d'un système symétrique comme représentant le système, et substituer aux mots *systèmes symétriques* (par rapport à trois points fixes), le mot *points.*

En supposant que dans la courbe du quatrième ordre on ait à la fois

$$ac - g^2 = 0, \quad ab - h^2 = 0,$$

il est facile de voir que la courbe du quatrième ordre se réduit à deux coniques supplémentaires.

En effet, ces conditions étant remplies, en rétablissant les valeurs de ξ, η, ζ, on obtient une équation qui se divise en deux équations, telles que

$$\alpha x^2 + \beta y^2 + \gamma z^2 - \delta yz = 0, \quad \alpha x^2 + \beta y^2 + \gamma z^2 + \delta yz = 0,$$

qui appartiennent, comme on le voit sans peine, à une paire de coniques supplémentaires.

De là, en remarquant qu'il est permis de faire abstraction de l'une de ces coniques:

"Tout théorème qui se rapporte à des points, à des droites, et à une conique, conduit à un théorème qui se rapporte d'une manière analogue à des points, à des coniques symétriques (par rapport à trois points fixes), et à une conique telle, que, par rapport à cette conique, l'un des trois points fixes a pour polaire la droite menée par les deux autres points fixes."

Supposons en particulier que les deux points fixes dont nous venons de parler soient les points où la droite à l'infini est rencontrée par un cercle qui a pour centre le troisième point fixe; ce troisième point fixe sera le centre tant des coniques symétriques par rapport à ces trois points fixes, que de la conique par rapport à laquelle le troisième point fixe a pour polaire la droite menée par les deux autres points fixes. De plus, les asymptotes de l'une quelconque des coniques symétriques formeront avec les droites imaginaires asymptotes du cercle un faisceau harmonique, et de là les asymptotes de la conique dont il s'agit seront à angle droit, ou, autrement dit, cette conique sera une hyperbole équilatère. De là enfin:

"Tout théorème descriptif qui se rapporte à des points, à des droites, et à une conique, conduit à un théorème qui se rapporte d'une manière analogue à des points, à des hyperboles équilatères et concentriques, et à une conique concentrique avec les hyperboles."

III. Soit enfin

$$\xi = \frac{1}{x}, \quad \eta = \frac{1}{y}, \quad \zeta = \frac{1}{z}.$$

Cette supposition conduit à l'une des méthodes de transmutation données par M. Steiner parmi les observations générales qui forment la conclusion de son ouvrage intitulé: *Systematische Entwickelung* &c., [Berlin, 1832], méthode qu'obtient M. Steiner au moyen de la théorie de l'hyperboloïde gauche. Je la reproduis ici tant pour en faire voir la théorie analytique qu'à cause de son analogie avec les méthodes I. et II.

Il est évident d'abord qu'un système d'équations telles que

$$\xi : \eta : \zeta = \alpha : \beta : \gamma$$

correspond à un point; de même, une équation linéaire quelconque, telle que

$$A\xi + B\eta + C\zeta = 0,$$

correspond à une conique qui passe par les trois points P, Q, R, et une équation quelconque du second ordre, telle que

$$a\xi^2 + b\eta^2 + c\zeta^2 + 2f\eta\zeta + 2g\zeta\xi + 2h\xi\eta = 0,$$

correspond à une courbe du quatrième ordre qui a les trois points P, Q, R pour points doubles. Réciproquement, de telles coniques et de telles courbes du quatrième ordre peuvent toujours se représenter par des équations linéaires et par des équations du second ordre. De là :

"Tout théorème descriptif qui se rapporte à des points, à des droites, et à des coniques, conduit à un théorème qui se rapporte d'une manière analogue à des points, à des coniques qui passent par trois points fixes, et à des courbes du quatrième ordre qui ont ces trois points fixes pour points doubles."

En supposant que l'on ait à la fois $a = 0$, $b = 0$, la courbe du quatrième ordre se réduit à une conique qui passe par les deux points Q, R (ou, si l'on veut, au système formé de cette conique et des droites PQ, PR). De là :

"Tout théorème descriptif qui se rapporte à des points, à des droites, et à une conique, conduit à un théorème qui se rapporte d'une manière analogue à des points, à des coniques qui passent par trois points fixes, et à une conique qui passe par deux de ces points fixes."

On ne peut pas particulariser ce théorème de manière à obtenir des théorèmes intéressants. Mais, en prenant les polaires réciproques des deux théorèmes qui viennent d'être obtenus, on a ces propriétés nouvelles :

"Tout théorème descriptif qui se rapporte à des droites, à des points, et à des coniques, conduit à un théorème qui se rapporte d'une manière analogue à des droites, à des coniques qui touchent chacune trois droites fixes, et à des courbes de la *quatrième classe* qui touchent chacune ces mêmes droites fixes, deux fois chaque droite ;"

et :

"Tout théorème qui se rapporte à des droites, à des points, et à une conique, conduit à un théorème qui se rapporte d'une manière analogue à des droites, à des coniques qui touchent chacune trois droites fixes, et à une conique qui touche deux de ces droites fixes."

De là, comme dans la méthode I. :

"Tout théorème qui se rapporte à des droites, à des points, et à une conique, conduit à un théorème qui se rapporte d'une manière analogue à des droites, à des paraboles qui ont pour foyer commun un point fixe, et à une conique qui a ce même point fixe pour un de ses foyers ;"

théorème que l'on doit comparer avec le troisième théorème que donne la méthode I.

Pour compléter cette théorie, nous aurions dû ajouter les deux théorèmes polaires réciproques du premier et du deuxième théorème que donne la méthode I. ; mais cela se fait sans la moindre peine. Quant au premier et au deuxième théorème que donne la méthode II., les polaires réciproques de ces deux théorèmes ne conduisent à rien de nouveau.

81.

ADDITION AU MÉMOIRE SUR QUELQUES TRANSMUTATIONS DES LIGNES COURBES.

[From the *Journal de Mathématiques Pures et Appliquées* (Liouville), tom. XV. (1850), pp. 351—356: continued from t. XIV. p. 46, **80**.]

Je me propose de résumer ici la théorie des courbes du quatrième ordre, auxquelles donne lieu la première de mes méthodes de transmutation appliquée à une conique quelconque.

L'équation d'une telle courbe est de la forme

$$ax + by + cz + 2f\sqrt{yz} + 2g\sqrt{zx} + 2h\sqrt{xy} = 0,$$

et nous avons déjà vu que cette courbe a pour tangentes doubles les trois droites $x = 0$, $y = 0$, $z = 0$ (droites que nous avons représentées par QR, RP, PQ). Pour trouver les autres propriétés de la courbe, mettons l'équation sous la forme

$$\left(a - \frac{gh}{f}\right)x + \left(b - \frac{hf}{g}\right)y + \left(c - \frac{fg}{h}\right)z + \left(\sqrt{\frac{gh}{f}}\sqrt{x} + \sqrt{\frac{hf}{g}}\sqrt{y} + \sqrt{\frac{fg}{h}}\sqrt{z}\right)^2 = 0\,;$$

puis, en écrivant, pour plus de simplicité, $\frac{fx}{gh}$, $\frac{gy}{hf}$, $\frac{fz}{gh}$ au lieu de x, y, z, cette équation devient

$$\left(\frac{af}{gh} - 1\right)x + \left(\frac{bg}{hf} - 1\right)y + \left(\frac{ch}{fg} - 1\right)z + (\sqrt{x} + \sqrt{y} + \sqrt{z})^2 = 0\,;$$

et de là, en mettant

$$\left(\frac{af}{gh} - 1\right)x + \left(\frac{bg}{hf} - 1\right)y + \left(\frac{ch}{fg} - 1\right)z = -w,$$

l'équation de la courbe prend cette forme très simple

$$\sqrt{x}+\sqrt{y}+\sqrt{z}+\sqrt{w}=0,$$

en se souvenant toujours que les quantités x, y, z, w satisfont à l'équation linéaire qui vient d'être donnée. Je représenterai dans la suite cette équation linéaire par

$$\alpha x+\beta y+\gamma z+\delta w=0.$$

Il est évident que la droite $w=0$, de même que les droites QR, RP, PQ, est tangente double de la courbe. De plus, ces quatre droites sont le système complet des tangentes doubles, car la courbe a, comme nous allons le voir, trois points doubles: en effet, la forme rationnelle de l'équation est

$$(x^2+y^2+z^2+w^2-2yz-2zx-2xy-2xw-2yw-2zw)^2-64xyzw=0,$$

et, au moyen de l'identité

$$(x+y)^2(z+w)^2-16xyzw=(x-y)^2(z+w)^2+(x+y)^2(z-w)^2-(x-y)^2(z-w)^2,$$

cette équation rationnelle se transforme en

$$[(x-y)^2-(z-w)^2]^2-4(x+y-z-w)[(x+y)(z-w)^2-(z+w)(x-y)^2]=0,$$

laquelle fait voir que le point $(x=y,\ z=w)$ est un point double; de là aussi les points $(x=z,\ w=y)$, $(x=w,\ y=z)$ sont des points doubles. Remarquons en passant qu'en supposant que les coefficients α, β, γ, δ restent indéterminés, les droites $x=y$, $x=z$, $x=w$ seront des droites quelconques par les points $(x=0,\ y=0)$, $(x=0,\ z=0)$, $(x=0,\ w=0)$ respectivement, et ces droites une fois connues, les droites $y=z$, $z=w$, $w=y$ seront déterminées, la première au moyen des points $(y=0,\ z=0)$, $(y=x,\ z=x)$, la deuxième au moyen des points $(z=0,\ w=0)$, $(z=x,\ w=x)$, et la troisième au moyen des points $(w=0,\ y=0)$, $(w=x,\ y=x)$; et les trois droites ainsi déterminées se couperont nécessairement dans un même point. Cela revient au théorème suivant:

"Les trois points doubles d'une courbe du quatrième ordre avec trois points doubles sont les centres d'un quadrangle dont les côtés passent par les angles du quadrilatère formé par les tangentes doubles de la courbe."

Cette propriété des courbes du quatrième ordre dont il s'agit (je veux dire celle d'avoir trois points doubles) aurait dû faire partie du théorème général donné auparavant pour cette première méthode de transmutation.

En supposant que la conique à transmuter passe par le point P, on aura

$$\alpha+\delta=0,$$

et il suit de là que le point double $(x=w,\ y=w)$, identique dans ce cas avec le point P, se change en point de rebroussement, et en même temps que les droites PQ, PR ne sont plus des tangentes doubles proprement dites, mais se réduisent à des tangentes simples qui passent par le point de rebroussement. Ajoutons que la tangente au point de rebroussement est la droite $x=w$.

En supposant que la conique à transmuter passe à la fois par les deux points P et Q, nous aurons

$$\alpha+\delta=0,\quad \beta+\delta=0.$$

Ici les deux points doubles ($x=w$, $y=z$), ($y=w$, $x=z$), identiques dans ce cas avec les points P, Q, deviennent des points de rebroussement, les droites QR, RP, PQ ne sont plus des tangentes doubles proprement dites, mais les droites RP, PQ sont les tangentes simples qui passent par les deux points de rebroussement respectivement, et la droite PQ est la droite menée par les deux points de rebroussement. Ajoutons que les tangentes aux deux points de rebroussement respectivement sont les droites $x=w$ et $y=w$.

On sait qu'un cercle quelconque peut s'envisager comme conique qui passe par deux points fixes, savoir [les points circulaires à l'infini, c'est-à-dire] les points où l'infini, considéré comme droite, est rencontré par les deux droites imaginaires auxquelles se réduit un cercle évanouissant quelconque. En nommant ces droites les *axes imaginaires* de leur point d'intersection, prenons pour les droites PQ, PR les axes imaginaires d'un point quelconque P, et pour la droite QR l'infini. Cela étant, un cercle quelconque sera transmuté dans une courbe du quatrième ordre ayant deux points de rebroussement aux points où l'infini est rencontré par les axes imaginaires du point P, ou, ce qui est la même chose, d'un point quelconque, et ayant de plus un point double. Et le point P, comme point d'intersection de deux axes imaginaires tangents de la courbe, est un foyer de la courbe (voyez le Mémoire de M. Plücker: *Ueber solche Punkte die bei Curven höhern Ordnung als der zweiten den Brennpunkten der Kegelschnitte entsprechen*, Journal de M. Crelle, t. X. [1833] pp. 84—91). Cela suffit pour faire voir que la courbe est un limaçon de Pascal ayant le point P pour le foyer qui n'est pas le point double. En effet, prenant pour vrai le théorème: "Les ovales de Descartes ont deux points de rebroussement aux points où l'infini est rencontré par les axes imaginaires d'un point quelconque[1]," comme cela revient à huit conditions, et qu'un ovale de Descartes peut être déterminé de manière à satisfaire à six conditions (ce qui fait en tout quatorze conditions, nombre des conditions qui déterminent une courbe du quatrième ordre), toute courbe du quatrième ordre avec deux points de rebroussement, tels que nous venons de les mentionner, sera un ovale de Descartes, et si, de plus, la courbe du quatrième ordre a un point double, elle se réduira en limaçon (cas particulier, comme on sait, des ovales de Descartes). Donc, en résumé, tout cercle est transmuté dans un limaçon ayant un point fixe pour le foyer qui n'est pas le point double, théorème qui se rapporte à la méthode de M. Roberts pour le cas $n=\frac{1}{2}$. L'on doit cependant remarquer que cette méthode est due à M. Chasles. En effet, on trouve dans la Note citée de l'*Aperçu historique*, non-seulement la propriété des droites de se transmuter en des paraboles, mais aussi

[1] M. Chasles a remarqué en passant (Note XXI. de l'*Aperçu historique* [Bruxelles, 1839]) que les ovales de Descartes ont deux points conjugués imaginaires à l'infini, théorème moins complet que celui que je viens d'énoncer. Pour la démonstration du théorème complet, voyez la Note à la suite de ce mémoire.

celle des cercles de se transmuter en des ovales de Descartes (seulement M. Chasles paraît ne pas avoir remarqué que ces ovales étaient nécessairement des limaçons), et c'est la lecture de cette Note qui m'a appris cette théorie de la transmutation des cercles.

En supposant que la conique à transmuter passe par les trois points P, Q, R, nous aurons

$$\alpha+\delta=0, \quad \beta+\delta=0, \quad \gamma+\delta=0;$$

les points doubles, identiques (dans ce cas) avec les points P, Q, R, deviennent des points de rebroussement, et les droites QR, RP, PQ, au lieu d'être des tangentes doubles, sont tout simplement les droites qui passent chacune par deux points de rebroussement. Ajoutons que les tangentes de la courbe, aux trois points de rebroussement respectivement, sont les droites $x-w=0$, $y-w=0$, $z-w=0$.

Il y a encore un cas particulier à considérer, savoir celui où la conique à transmuter est telle que, par rapport à cette conique, les points Q et R sont situés chacun dans la polaire de l'autre; on a alors $f=0$, cas qui échappe à l'analyse ci-devant employée. On voit sans peine que les deux points doubles $(y=w,\ x=z)$, $(z=w,\ y=z)$ deviennent ici identiques, ce qui donne lieu à un point d'osculation. La droite QR et la droite $w=0$ ne sont plus des tangentes doubles proprement dites, mais ces droites deviennent l'une et l'autre identiques avec la tangente au point d'osculation.

Note sur les ovales de Descartes.

De l'équation de ces ovales,

$$\sqrt{(x-a)^2+y^2}=m\sqrt{x^2+y^2}+n,$$

on tire d'abord

$$(1-m^2)(x^2+y^2)-2ax+a^2-n^2=2mn\sqrt{x^2+y^2},$$

puis

$$\begin{aligned}(1-m^2)^2(x^2+y^2)^2-4a(1-m^2)\,x\,(x^2+y^2)\\ +2\left[a^2(1-m^2)-n^2(1+m^2)\right](x^2+y^2)+4a^2x^2-4a(a^2-n^2)\,x+(a^2-n^2)^2=0.\end{aligned}$$

Pour trouver la nature de la courbe à l'infini, mettons $x+yi=\xi$, $x-yi=\eta$, $i=\sqrt{-1}$, et introduisons la quantité ζ de manière à rendre l'équation homogène. Cela donne

$$\begin{aligned}(1-m^2)^2\xi^2\eta^2-2a(1-m^2)(\xi+\eta)\,\xi\eta\zeta\\ +2\left[a^2(1-m^2)-n^2(1+m^2)\right]\xi\eta\zeta^2+a^2(\xi+\eta)^2\zeta^2-2a(a^2-n^2)(\xi+\eta)\,\zeta^3+(a^2-n^2)^2\zeta^4=0;\end{aligned}$$

ce qui fait voir, sans la moindre peine, qu'il y a des points de rebroussement aux points $(\xi=0,\ \zeta=0)$, $(\eta=0,\ \zeta=0)$, savoir, aux points où l'infini, considéré comme droite, est rencontré par les droites $x+yi=0$, $x-yi=0$, qui sont les axes imaginaires du point $x=0$, $y=0$.

Nous pouvons remarquer, en passant, que l'équation

$$(1-m^2)(x^2+y^2)-2ax+(a^2-n^2)=2mn\sqrt{x^2+y^2},$$

conduit, avec beaucoup de facilité, à une autre propriété, donnée par M. Chasles dans la Note déjà citée. En effet, en mettant

$$a'=\frac{n^2-a^2}{a(m^2-1)},\qquad m'=\frac{m^2(n^2-1)}{a(m^2-1)},\qquad n'=\frac{n}{a},$$

cette équation se transforme en

$$(1-m'^2)(x^2+y^2)-2a'x+(a'^2-n'^2)=2m'n'\sqrt{x^2+y^2},$$

et par là on voit que l'équation primitive peut se transformer en

$$\sqrt{(x-a')^2+y^2}=m'\sqrt{x^2+y^2}+n',$$

c'est-à-dire, il y a toujours un troisième foyer de la courbe.

Il ne reste qu'à démontrer que la transformée (selon la méthode de M. Chasles) d'un cercle est toujours un limaçon. Soit, pour cela,

$$r^2-2ar\cos\theta+\delta=0$$

l'équation du cercle; en mettant $\sqrt{mr}$ au lieu de r, et $\frac{1}{2}\theta$ au lieu de θ, cette équation devient $mr-2a\sqrt{mr}\cos\frac{1}{2}\theta+\delta=0$, ce qui donne $(mr+\delta)^2=2a^2mr(1+\cos\theta)$, ou, en mettant $r\cos\theta=x$ et en réduisant à une forme rationnelle,

$$[m^2(x^2+y^2)+\delta^2-2a^2mx]^2-4m^2(\delta-a^2)(x^2+y^2)=0,$$

ce qui appartient évidemment à un ovale de Descartes. En mettant $y=0$, l'équation devient

$$[m^2x^2+2m(\delta-2a^2)x+\delta^4](mx-\delta)^2=0,$$

c'est-à-dire le point $mx-\delta=0$, $y=0$ est point double, ou la courbe est le limaçon de Pascal.

82.

ON THE TRIADIC ARRANGEMENTS OF SEVEN AND FIFTEEN THINGS.

[From the *Philosophical Magazine*, vol. XXXVII. (1850), pp. 50—53.]

THERE is no difficulty in forming with seven letters, a, b, c, d, e, f, g, a system of seven triads containing every possible duad; or, in other words, such that no two triads of the system contain the same duad. One such system, for instance, is

$$abc,\quad ade,\quad afg,\quad bdf,\quad beg,\quad cdg,\quad cef;$$

and this is obviously one of six different systems obtained by permuting the letters a, b, c. We have therefore six different systems containing the triad *abc*; and there being the same number of systems containing the triads *abd, abe, abf* and *abg* respectively, there are in all thirty-five different systems, each of them containing every possible duad. It is deserving of notice, that it is impossible to arrange the thirty-five triads formed with the seven letters into five systems, each of them possessing the property in question. In fact, if this could be done, the system just given might be taken for one of the systems of seven triads. With this system we might (of the systems of seven triads which contain the triad *abd*) combine either the system

$$abd,\quad acg,\quad aef,\quad bce,\quad bfg,\quad dcf,\quad deg,$$

or the system

$$abd,\quad acf,\quad aeg,\quad bcg,\quad bef,\quad dce,\quad dfg;$$

(but any one of the other *abd* systems would be found to contain a triad in common with the given *abc* system, and therefore cannot be made use of: for instance, the system *abd, acg, aef, bcf, beg, dce, dfg* contains the triad *beg* in common with the given *abc* system): and whichever of the two proper *abd* systems we select to combine with the given *abc* system, it will be found that there is no *abe* system which does not contain some triad in common, either with the *abc* system or with the *abd* system.

The order of the letters in a triad has been thus far disregarded. There are some properties which depend upon considering the triads obtained by cyclical permutations of the three letters as identical, but distinct from the triads obtained by a permutation of two letters, or inversion. Thus *abc, bca, cab* are to be considered as identical *inter se*, but distinct from the triads *acb, cba, bac*, which are also identical *inter se*. I write down the system (equivalent, as far as the mere combination of the letters is concerned, to the system at the commencement of this paper)

ade, afg, bdf, bge, cdg, cef, cba,

derived, it is to be observed, from a pair of triads, *ade, afg*, by a cyclical permutation of the *e, f, g*, and by successively changing the *a* into *b* and into *c*, the remaining triad of the system being the letters *a, b, c* taken in an inverse order. Let it be proposed to derive the system in the same manner from any other two triads of the system; for instance, from the triads *acb, ade*. The process of derivation gives

acb, ade, gcd, geb, fce, fbd, fga, [1]

which is, in fact, the original system. But attempt to derive the system from the two triads *ebg, efc*, the process of derivation gives

ebg, efc, dbf, dcg, abc, agf, ade,

which is not the original system, inasmuch as the triads *dbf, dcg, abc, agf* are inversions of the triads *bdf, cdg, cba, afg* of the original system. The point to be attended to, however, is, that *both* triads of the pair *dbf, dcg*, or of the pair *abc, agf*, are inversions of the triads of the corresponding pair in the original system; the pair is either reproduced (as the pair *efc, dbf*), or there is an inversion of both triads. Where there is no such inversion of the triads of a pair, the system may be said to be properly reproduced; and where there is inversion of the triads of one or more pairs, to be improperly reproduced. There is no difficulty in seeing that the system is properly reproduced from a pair of triads containing in common any one of the letters *a, b, c* or *d*, and improperly reproduced from pairs of triads containing in common any one of the letters *d, e* or *f*. It is owing to the reproduction, proper or improper, of the system from any pair of duads that it is possible to form a system of "octaves" analogous to the quaternions of Sir William Hamilton; the impossibility of a corresponding system of fifteen imaginary quantities arises from the circumstance of there being always, in whatever manner the system of triads is formed, an inversion of a single triad of some one or more pairs of triads containing a letter in common. When the system is considered as successively derived from different pairs, the system is not according to the previous definition reproduced either properly or improperly. A system of triads having the necessary properties with respect to the mere combination of the letters (viz. that $\alpha\beta\gamma$ and $\alpha\delta\epsilon$ being any two

[1] The order of the letters *f, g* is selected so as to reproduce the original system so far as the mere combination of the letters is concerned.

triads having a letter in common, there shall be triads such as $\zeta\beta\delta$, $\zeta\gamma\epsilon$, and $\eta\beta\epsilon$, $\eta\gamma\delta$) may easily be found; the system to be presently given of the triads of fifteen things would answer the purpose. And so would many other systems.

Dropping the consideration of the order of the letters which form a triad, I pass to the case of a system of fifteen letters, *a*, *b*, *c*, *d*, *e*, *f*, *g*, *h*, *i*, *j*, *k*, *l*, *m*, *n*, *o*. It is possible in this case, not only to form systems of thirty-five triads containing every possible duad, but this can be done in such manner that the system of thirty-five triads can be arranged in seven systems of five triads, each of these systems containing the fifteen letters[1]. My solution is obtained by a process of derivation from the arrangements *ab . cf . dg . eh* and *ab . cd . ef . gh* as follows; viz. the triads are

iab	*jac*	*kaf*	*lad*	*mag*	*nae*	*oah*
icf	*jfb*	*kbc*	*lce*	*mch*	*ncd*	*ocg*
idg	*jde*	*kdh*	*lgb*	*mbd*	*ngf*	*ofd*
ieh	*jhg*	*kge*	*lhf*	*mfe*	*nhb*	*obe*

and a system formed with *i*, *j*, *k*, *l*, *m*, *n* *o*, which are then arranged in the form

klo	*ino*	*jmo*	*ilm*	*jln*	*ijk*	*kmn*
iab	*jac*	*lad*	*nae*	*kaf*	*mag*	*oah*
ncd	*mdb*	*kbc*	*ocg*	*mch*	*lce*	*icf*
mef	*keg*	*ieh*	*jfb*	*obe*	*ofd*	*jde*
jgh	*lhf*	*nfg*	*khd*	*idg*	*nhb*	*lbg*

an arrangement, which, it may be remarked, contains eight different systems (such as have been considered in the former part of this paper) of seven letters such as *i*, *j*, *k*, *l*, *m*, *n*, *o*; and seven of other seven letters, such as *i*, *j*, *k*, *a*, *b*, *c*, *f*[2]. The theory of the arrangement seems to be worth further investigation.

Assuming that the four hundred and fifty-five triads of fifteen things can be arranged in thirteen systems of thirty-five triads, each system of thirty-five triads containing every possible duad, it seems natural to inquire whether the thirteen systems can be obtained from any one of them by cyclical permutations of thirteen letters. This is, I think, impossible. For let the cyclical permutation be of the letters *a*, *b*, *c*, *d*, *e*, *f*, *g*, *h*, *i*, *j*, *k*, *l*, *m*. Consider separately the triads which contain the letter *n* and the letter *o*; neither of these systems of triads contains the letter, whatever it is, which forms a triad with *n* and *o*. Hence, omitting the letters *n*, *o*, we have two different sets, each of them of six duads, and composed of the same twelve letters. And each of these systems of duads ought, by the cyclical permutation

[1] The problem was proposed by Mr Kirkman, and has, to my knowledge, excited some attention in the form "To make a school of fifteen young ladies walk together in threes every day for a week so that each two may walk together." It will be seen from the text that I am uncertain as to the existence of a solution to the further problem suggested by Mr Sylvester, "to make the school walk every week in the quarter so that each three may walk together."

[2] [I have somewhat altered this sentence so as to express more clearly what appeared to be the meaning of it.]

in question, to produce the whole system of the seventy-eight duads of the thirteen letters. Hence arranging the duads of the thirteen letters in the form

$$\begin{array}{l} ab\,.\,bc\,.\,cd\,.\,de\,.\,ef\,.\,fg\,.\,gh\,.\,hi\,.\,ij\,.\,jk\,.\,kl\,.\,lm\,.\,ma \\ ac\,.\,bd\,.\,ce\,.\,df\,.\,eg\,.\,fh\,.\,gi\,.\,hj\,.\,ik\,.\,jl\,.\,km\,.\,la\,.\,mb \\ ad\,.\,be\,.\,cf\,.\,dg\,.\,eh\,.\,fi\,.\,gj\,.\,hk\,.\,il\,.\,jm\,.\,ka\,.\,lb\,.\,mc \\ ae\,.\,bf\,.\,cg\,.\,dh\,.\,ei\,.\,fj\,.\,gk\,.\,hl\,.\,im\,.\,ja\,.\,kb\,.\,lc\,.\,md \\ af\,.\,bg\,.\,ch\,.\,di\,.\,ej\,.\,fk\,.\,gl\,.\,hm\,.\,ia\,.\,jb\,.\,kc\,.\,ld\,.\,me \\ ag\,.\,bh\,.\,ci\,.\,dj\,.\,ek\,.\,fl\,.\,gm\,.\,ha\,.\,ib\,.\,jc\,.\,kd\,.\,le\,.\,mf \end{array}$$

and consequently the duads of each set ought to be situated one duad in each line. Suppose the sets of duads are composed of the letters a, b, c, d, e, f, g, h, i, j, k, l, it does not appear that there is any set of six duads composed of these letters, and situated one duad in each line, other than the single set al, bk, cj, di, eh, fg; and the same being the case for any twelve letters out of the thirteen, the derivation of the thirteen systems of thirty-five triads by means of the cyclical permutations of thirteen letters is impossible. And there does not seem to be any obvious rule for the derivation of the thirteen systems from any one of them, or any *prima facie* reason for believing that the thirteen systems do really exist, it having been already shown that such systems do not exist in the case of seven things.

83.

ON CURVES OF DOUBLE CURVATURE AND DEVELOPABLE SURFACES.

[From the *Cambridge and Dublin Mathematical Journal*, vol. v. (1850), pp. 18—22.]

This is translated with some slight alterations from the Memoir in Liouville, t. x. (1845) pp. 245—250, 30.

84.

ON THE DEVELOPABLE SURFACES WHICH ARISE FROM TWO SURFACES OF THE SECOND ORDER.

[From the *Cambridge and Dublin Mathematical Journal,* vol. v. (1850), pp. 46—57.]

Any two surfaces considered in relation to each other give rise to a curve of intersection, or, as I shall term it, an Intersect and a circumscribed Developable[1] or Envelope. The Intersect is of course the edge of regression of a certain Developable which may be termed the Intersect-Developable, the Envelope has an edge of regression which may be termed the Envelope-Curve. The order of the Intersect is the product of the orders of the two surfaces, the class of the Envelope is the product of the classes of the two surfaces. When neither the Intersect breaks up into curves of lower order, nor the Envelope into Developables of lower class, the two surfaces are said to form a proper system. In the case of two surfaces of the second order (and class) the Intersect is of the fourth order and the Envelope of the fourth class. Every proper system of two surfaces of the second order belongs to one of the following three classes:—*A*. There is no contact between the surfaces; *B*. There is an ordinary contact; *C*. There is a singular contact. Or the three classes may be distinguished by reference to the conjugates (conjugate points or planes) of the system. *A*. The four conjugates are all distinct; *B*. Two conjugates coincide; *C*. Three conjugates coincide.

To explain this it is necessary to remark that in the general case of two surfaces of the second order not in contact (that is for systems of the class *A*) there is a certain tetrahedron such that with respect to either of the surfaces (or more generally with respect to any surface of the second order passing through the Intersect of the system

[1] The term 'Developable' is used as a substantive, as the reciprocal to 'Curve,' which means curve of double curvature. The same remark applies to the use of the term in the compound Intersect-Developable. For the signification of the term 'singular contact,' employed lower down, see Mr Salmon's memoir 'On the Classification of Curves of Double Curvature,' [same volume] p. 32.

or inscribed in the Envelope) the angles and faces of the tetrahedron are reciprocals of each other, each angle of its opposite face, and *vice versâ*. The angles of the tetrahedron are termed the conjugate points of the system, and the faces of the tetrahedron are termed the conjugate planes of the system, and the term conjugates may be used to denote indifferently either the conjugate planes or the conjugate points. A conjugate plane and the conjugate point reciprocal to it are said to correspond to each other. Each conjugate point is evidently the point of intersection of the three conjugate planes to which it does not correspond, and in like manner each conjugate plane is the plane through the three conjugate points to which it does not correspond.

In the case of a system belonging to the class B, two conjugate points coincide together in the point of contact forming what may be termed a double conjugate point, and in like manner two conjugate planes coincide in the plane of contact (that is the tangent plane through the point of contact) forming what may be termed a double conjugate plane. The remaining conjugate points and planes may be distinguished as single conjugate points and single conjugate planes. It is clear that the double conjugate plane passes through the three conjugate points, and that the double conjugate point is the point of intersection of the three conjugate planes: moreover each single conjugate plane passes through the single conjugate point to which it does not correspond and the double conjugate point; and each single conjugate point lies on the line of intersection of the single conjugate plane to which it does not correspond and the double conjugate plane.

In the case of a system belonging to the class (C), three conjugate points coincide together in the point of contact forming what may be termed a triple conjugate point, and three conjugate planes coincide together in the plane of contact forming a triple conjugate plane. The remaining conjugate point and conjugate plane may be distinguished as the single conjugate point and single conjugate plane. The triple conjugate plane passes through the two conjugate points and the triple conjugate point lies on the line of intersection of the two conjugate planes; the single conjugate plane passes through the triple conjugate point and the single conjugate point lies on the triple conjugate plane.

Suppose now that it is required to find the Intersect-Developable of two surfaces of the second order. If the equations of the surfaces be $\Upsilon = 0$, $\Upsilon' = 0$ (Υ, Υ' being homogeneous functions of the second order of the coordinates ξ, η, ζ, ω), and x, y, z, w represent the coordinates of a point upon the required developable surface: if moreover U, U' are the same functions of x, y, z, w that Υ, Υ' are of ξ, η, ζ, ω and X, Y, Z, W; X', Y', Z', W' denote the differential coefficients of U, U' with respect to x, y, z, w; then it is easy to see that the equation of the Intersect-Developable is obtained by eliminating ξ, η, ζ, ω between the equations

$$\Upsilon = 0, \quad \Upsilon' = 0,$$

$$X\xi + Y\eta + Z\zeta + W\omega = 0,$$

$$X'\xi + Y'\eta + Z'\zeta + W'\omega = 0.$$

If, for shortness, we suppose

$$\begin{aligned} \overline{F} &= YZ' - Y'Z, & \overline{L} &= XW' - X'W, \\ \overline{G} &= ZX' - Z'X, & \overline{M} &= YW' - Y'W, \\ \overline{H} &= XY' - X'Y, & \overline{N} &= ZW' - Z'W, \end{aligned}$$

(values which give rise to the identical equation

$$\overline{L}\overline{F} + \overline{M}\overline{G} + \overline{N}\overline{H} = 0),$$

then, λ, μ, ν, ρ denoting any indeterminate quantities, the two linear equations in ξ, η, ζ, ω are identically satisfied by assuming

$$\begin{aligned} \xi &= \quad . \quad \overline{N}\mu - \overline{M}\nu + \overline{F}\rho, \\ \eta &= -\overline{N}\lambda \quad . \quad + \overline{L}\nu + \overline{G}\rho, \\ \zeta &= \overline{M}\lambda - \overline{L}\mu \quad . \quad + \overline{H}\rho, \\ \omega &= -\overline{F}\lambda - \overline{G}\mu - \overline{H}\nu \quad . \end{aligned}$$

and, substituting these values in the equations $\Upsilon = 0$, $\Upsilon' = 0$, we have two equations:

$$A\lambda^2 + B\mu^2 + C\nu^2 + 2F\mu\nu + 2G\nu\lambda + 2H\lambda\mu + 2L\lambda\rho + 2M\mu\rho + 2N\nu\rho = 0,$$
$$A'\lambda^2 + B'\mu^2 + C'\nu^2 + 2F'\mu\nu + 2G'\nu\lambda + 2H'\lambda\mu + 2L'\lambda\rho + 2M'\mu\rho + 2N'\nu\rho = 0,$$

which are of course such as to permit the four quantities λ, μ, ν, ρ to be simultaneously eliminated. The coefficients of these equations are obviously of the fourth order in x, y, z, w.

Suppose for a moment that these coefficients (instead of being such as to permit this simultaneous elimination of λ, μ, ν, ρ) denoted any arbitrary quantities, and suppose that the indeterminates λ, μ, ν, ρ were besides connected by two linear equations,

$$a\lambda + b\mu + c\nu + d\rho = 0,$$
$$a'\lambda + b'\mu + c'\nu + d'\rho = 0;$$

then, putting

$$\begin{aligned} bc' - b'c &= f, & ad' - a'd &= l, \\ ca' - c'a &= g, & bd' - b'd &= m, \\ ab' - a'b &= h, & cd' - c'd &= n, \end{aligned}$$

(values which give rise to the identical equation $lf + mg + nh = 0$), and effecting the elimination of λ, μ, ν, ρ between the four equations, we should obtain a final equation $\square = 0$, in which $\square$ is a homogeneous function of the second order in each of the systems of coefficients A, B, &c., and A', B', &c., and a homogeneous function of the fourth order (indeterminate to a certain extent in its form on account of the identical equation $lf+mg+nh=0$) in the coefficients f, g, h, l, m, n [1]. But re-establishing the actual values of the coefficients A, B, &c., A', B', &c. (by which means the function $\square$ becomes a function of the sixteenth order in x, y, z, w) the quantities f, g, h, l, m, n ought, it is clear, to disappear of themselves; and the way this happens is that the function $\square$ resolves itself into the product of two factors M and Ψ, the latter of which is independent of f, g, h, l, m, n. The factor M is a function of the fourth order in these quantities, and it is also of the eighth order in the variables x, y, z, w: the factor Ψ is consequently of the eighth order in x, y, z, w. And the result of the elimination being represented by the equation $\Psi=0$, the Intersect-Developable in the general case, or (what is the same thing) for systems of the class (A), is of the eighth order. In the case of a system of the class (B) the equation obtained as above contains as a factor the square, and in the case of a system of the class (C) the cube, of the linear function which equated to zero is the equation of the plane of contact. The Intersect-Developable of a system of the class (B) is therefore a Developable of the sixth order, and that of a system of the class (C) a Developable of the fifth order. The elimination is in every case most simply effected by supposing two of the quantities λ, μ, ν, ρ to vanish (e.g. ν and ρ): the equations between which the elimination has to be effected then are

$$A\lambda^2+B\mu^2+2H\lambda\mu=0,$$
$$A'\lambda^2+B'\mu^2+2H'\lambda\mu=0;$$

and the result may be presented under the equivalent forms

$$(AB'+A'B-2HH')^2-4(AB-H^2)(A'B'-H'^2)=0,$$

and

$$(AB'-A'B)^2+4(AH'-A'H)(BH'-B'H)=0,$$

the latter of which is the most convenient. These two forms still contain an extraneous factor of the eighth order in x, y, z, w, of which they can only be divested by substituting the actual values of A, B, H, A', B', H'.

A. Two surfaces forming a system belonging to this class may be represented by equations of the form

$$a\,x^2+b\,y^2+c\,z^2+d\,w^2=0,$$
$$a'x^2+b'y^2+c'z^2+d'w^2=0,$$

[1] I believe the result of the elimination is

$$\square=4(PR-Q^2)=0,$$

where, if we write $uA+u'A'=A$, &c., the quantities P, Q, R are given by the equation (identical with respect to u, u')

$$Pu^2+2Quu'+Ru'^2=(Aa^2+\dots)(Aa'^2+\dots)-(Aaa'+\dots)^2$$
$$=u^2\{(BC-F^2)f^2+\dots\}+uu'\{(BC'+B'C-2FF')f^2+\dots\}+u'^2\{(B'C'-F'^2)f^2+\dots\}$$

a theorem connected with that given in the second part of my memoir 'On Linear Transformations' (*Journal*, vol. I. p. 109) [see 14, p. 100]. I am not in possession of any verification *à posteriori* of what is subsequently stated as to the resolution into factors of the function $\square$ and the forms of these factors.

where $x=0$, $y=0$, $z=0$, and $w=0$, are the equations of the four conjugate planes. There is no particular difficulty in performing the operations indicated by the general process given above; and if we write, in order to abbreviate,

$$\begin{aligned} bc'-b'c&=f, & ad'-a'd&=l, \\ ca'-c'a&=g, & bd'-b'd&=m, \\ ab'-a'b&=h, & cd'-c'd&=n, \end{aligned}$$

(values which satisfy the identical equation $lf+mg+nh=0$), the result after all reductions is

$$\begin{aligned} &l^2f^4y^4z^4+m^2g^4z^4x^4+n^2h^4x^4y^4+l^4f^2x^4w^4+m^4g^2y^4w^4+n^4h^2z^4w^4 \\ &+2mng^2h^2x^4y^2z^2+2nlh^2f^2y^4z^2x^2+2lmf^2g^2z^4x^2y^2 \\ &-2m^2n^2ghw^4y^2z^2-2n^2l^2hfw^4z^2x^2-2l^2m^2fgw^4x^2y^2 \\ &+2fmg^2l^2x^4z^2w^2+2gnh^2m^2y^4x^2w^2+2hlf^2n^2z^4y^2w^2 \\ &-2fnh^2l^2x^4y^2w^2-2glf^2m^2y^4z^2w^2-2hmg^2n^2z^4x^2w^2 \\ &+2(mg-nh)(nh-lf)(lf-mg)x^2y^2z^2w^2=0, \end{aligned}$$

which is therefore the equation of the Intersect-Developable for this case. The discussion of the geometrical properties of the surface will be very much facilitated by presenting the equation under the following form, which is evidently one of a system of six different forms,

$$\begin{aligned} &\{m(gx^2+nw^2)(hy^2-gz^2+lw^2)-l(-fy^2+nw^2)(-hx^2+fz^2+mw^2)\}^2 \\ &\qquad -4fglmx^2y^2(hy^2-gz^2+lw^2)(-hx^2+fz^2+mw^2)=0. \end{aligned}$$

B. Two surfaces forming a system belonging to this class may be represented by equations such as

$$\begin{aligned} a\,x^2+b\,y^2+c\,z^2+2n\,zw&=0, \\ a'x^2+b'y^2+c'z^2+2n'zw&=0, \end{aligned}$$

in which $x=0$, $y=0$ are the equations of the single conjugate planes, $z=0$ that of the double conjugate plane or plane of contact, $w=0$ that of an indeterminate plane through the two single conjugate points. If we write

$$\begin{aligned} bc'-b'c&=f, & an'-a'n&=p, \\ ca'-c'a&=g, & bn'-b'n&=q, \\ ab'-a'b&=h, & cn'-c'n&=r, \end{aligned}$$

(values which satisfy the identical equation $pf+qg+rh=0$), the result after all reductions is

$$\begin{aligned} &r^4h^2z^6+2pr^2h(rh-qg)z^4x^2-2qr^2h(pf-rh)z^4y^2 \\ &+4p^2qr^2hz^3x^2w-4pq^2r^2hz^3y^2w \\ &+p^2(rh-qg)^2z^2x^4+q^2(pf-rh)^2z^2y^4+2pq(4r^2h^2-fgpq)z^2x^2y^2 \\ &+4p^3q(rh-qg)zx^4w+4pq^3zy^4w-4p^2q^2(qg-pf)zx^2y^2w \\ &+4p^4q^2x^4w^2+4p^2q^4y^4w^2+8p^3q^3x^2y^2w^2+4p^2qrh^2x^4y^2+4pq^2rh^2x^2y^4=0, \end{aligned}$$

which is therefore the equation of the Intersect-Developable for systems of the case in question. The equation may also be presented under the form

$$\{q(px^2+rz^2)(hy^2-gz^2+2pzw)-p(qy^2+rz^2)(-hx^2+fz^2+2qzw)\}^2$$
$$+4p^2q^3x^2y^2(hy^2-gz^2+2pzw)(-hx^2+fz^2+2qzw)=0,$$

which it is to be remarked contains the extraneous factor z^2. The following is also a form of the same equation,

$$\{r(qg-pf)z^3-fp^2zx^2+gq^2zy^2+2pq(px^2+qy^2)w\}^2$$
$$-4pq(px^2+qy^2+rz^2)\{r(hy^2-gz^2)(fz^2-hx^2)+2pq(gx^2-fy^2)zw\}=0.$$

C. Two surfaces forming a system belonging to this class may be represented by equations of the form

$$a\,x^2+b\,y^2+2f\,yz+2n\,zw=0,$$
$$a'x^2+b'y^2+2f'yz+2n'zw=0,$$

in which $bn'-b'n=0$, $af'-a'f=0$. In these equations $x=0$ is the equation of a properly chosen plane passing through the two conjugate points, $y=0$ is the equation of the single conjugate plane, $z=0$ that of the triple conjugate plane, and $w=0$ is the equation of a properly chosen plane passing through the single conjugate point. Or without loss of generality, we may write

$$\alpha\,(x^2-2yz)+\beta\,(y^2-2zw)=0,$$
$$\alpha'(x^2-2yz)+\beta'(y^2-2zw)=0,$$

where x, y, z and w have the same signification as before[1]. The result after all reductions is

$$4z^3w^2+12y^2z^2w+9y^4z-24x^2yzw-4x^2y^3+8x^4w=0,$$

which may also be presented under the forms

$$z(y^2-2zw)^2-4y(y^2-2zw)(x^2-2yz)+8w(x^2-2yz)^2=0,$$

and
$$z(3y^2+2zw)^2-4x^2(y^3-2x^2w+6yzw)=0.$$

[In these three equations and in the last two equations of p. 495 as originally printed, there was by mistake, an interchange of the letters x and y.]

[1] Of course in working out the equation of the Intersect-Developable, it is simpler to employ the equations $x^2-2yz=0$, $y^2-2zw=0$. These equations belong to two cones which pass through the Intersect and have their vertices in the triple conjugate point and single conjugate point respectively. I have not alluded to these cones in the text, as the theory of them does not come within the plan of the present memoir, the immediate object of which is to exhibit the equations of certain developable surfaces—but these cones are convenient in the present case as furnishing the easiest means of defining the planes $x=0$, $w=0$. If we represent for a moment the single conjugate point by S and the triple conjugate point by T (and the cones through these points by the same letters), then the point T is a point upon the cone S, and the triple conjugate plane which touches the cone S along the line TS touches the cone T along some generating line TM. Let the other tangent plane through the line TS to the cone T be TM', where M' may represent the point where the generating line in question meets the cone S; and we may consider M as the point of intersection of the line TM with the tangent plane through the line SM' to the cone S: then the plane TMM' is the plane represented by the equation $x=0$, and the plane SMM' is that represented by the equation $w=0$. We may add that $y=0$ is the equation of the plane TSM', and $z=0$ that of the plane TSM.

Proceeding next to the problem of finding the envelope of two surfaces of the second order, this is most readily effected by the following method communicated to me by Mr Salmon. Retaining the preceding notation, the equation $U + kU' = 0$ belongs to a surface of the second order passing through the Intersect of the two surfaces $U = 0$, $U' = 0$. The polar reciprocal of this surface $U + kU' = 0$ is therefore a surface inscribed in the envelope of the reciprocals of the two surfaces $U = 0$, $U' = 0$, and consequently this envelope is the envelope (in the ordinary sense of the word) of the reciprocal of the surface $U + kU' = 0$, k being considered as a variable parameter. It is easily seen that the reciprocal of the surface $U + kU' = 0$ is given by an equation of the form

$$\mathrm{A} + 3\mathrm{B}k + 3\mathrm{C}k^2 + \mathrm{D}k^3 = 0,$$

in which A, B, C, D are homogeneous functions of the second order in the coordinates x, y, z, w. Differentiating with respect to k, and performing the elimination, we have for the equation of the envelope in question,

$$(\mathrm{AD} - \mathrm{BC})^2 - 4(\mathrm{AC} - \mathrm{B}^2)(\mathrm{BD} - \mathrm{C}^2) = 0\,;$$

or the envelope is, in general or (what is the same thing) for a system of the class (A), a developable of the eighth order. For a system of the class (B) the equation contains as a factor, the square of the linear function which equated to zero is the equation of the plane of contact; or the envelope is in this case a Developable of the sixth order. And in the case of a system of the class (C) the equation contains as a factor the cube of this linear function; or the envelope is a developable of the fifth order only.

A. We may take for the two surfaces the reciprocals (with respect to $x^2 + y^2 + z^2 + w^2 = 0$) of the equations made use of in determining the Intersect-Developable. The equations of these reciprocals are

$$b\,c\,d\,x^2 + c\,d\,a\,y^2 + d\,a\,b\,z^2 + a\,b\,c\,w^2 = 0\,,$$

$$b'c'd'x^2 + c'd'a'y^2 + d'a'b'z^2 + a'b'c'w^2 = 0\,;$$

and it is clear from the form of them (as compared with the equations of the surfaces of which they are the reciprocals) that $x = 0$, $y = 0$, $z = 0$, $w = 0$, are still the equations of the conjugate planes. We have, introducing the numerical factor 3 to avoid fractions,

$$3\,\{(b + kb')(c + kc')(d + kd')\,x^2 + (c + kc')(d + kd')(a + ka')\,y^2$$
$$+ (d + kd')(a + ka')(b + kb')\,z^2 + (a + ka')(b + kb')(c + kc')\,w^2\}$$

$= \mathrm{A} + 3\mathrm{B}k + 3\mathrm{C}k^2 + \mathrm{D}k^3$, which determine the values of A, B, C, D.

We have in fact

$$\mathrm{A} = 3\,(bcdx^2 + cday^2 + dabz^2 + abcw^2)$$
$$\mathrm{B} = (b'cd + bc'd + bcd')\,x^2 + \ldots\ldots$$
$$\mathrm{C} = (bc'd' + b'cd' + b'c'd)\,x^2 + \ldots\ldots$$
$$\mathrm{D} = 3\,(b'c'd'x^2 + c'd'a'y^2 + d'a'b'z^2 + a'b'c'w^2)$$

and these values give (with the same signification as before of f, g, h, l, m, n)

$$2(\mathrm{AC}-\mathrm{B}^2) = Aa^2 + Bb^2 + Cc^2 + 2Fbc + 2Gca + 2Hab + 2Lad + 2Mbd + 2Ncd + Dd^2,$$
$$2(\mathrm{BD}-\mathrm{C}^2) = Aa'^2 + Bb'^2 + Cc'^2 + 2Fb'c' + 2Gc'a' + 2Ha'b' + 2La'd' + 2Mb'd' + 2Nc'd' + Dd'^2,$$
$$\mathrm{AD}-\mathrm{BC} = Aaa' + Bbb' + Ccc' + F(bc'+b'c) + G(ca'+c'a) + H(ab'+a'b)$$
$$+ L(ad'+a'd) + M(bd'+b'd) + N(cd'+c'd) + Ddd',$$

where

$$\begin{aligned}
\mathrm{A} &= n^2y^4 + m^2x^4 + f^2w^4 + 2fmz^2w^2 - 2fny^2w^2 + 2nmy^2z^2,\\
\mathrm{B} &= l^2z^4 + n^2x^4 + g^2w^4 + 2gnx^2w^2 - 2glz^2w^2 + 2lnz^2x^2,\\
\mathrm{C} &= m^2x^4 + l^2y^4 + h^2w^4 + 2hly^2w^2 - 2hmx^2w^2 + 2mlx^2y^2,\\
\mathrm{D} &= f^2x^4 + g^2y^4 + h^2z^4 - 2ghy^2z^2 - 2hfz^2x^2 - 2fgy^2z^2,\\
\mathrm{F} &= l^2y^2z^2,\\
\mathrm{G} &= m^2z^2x^2,\\
\mathrm{H} &= n^2x^2y^2,\\
\mathrm{L} &= f^2x^2w^2,\\
\mathrm{M} &= g^2y^2w^2,\\
\mathrm{N} &= h^2z^2w^2;
\end{aligned}$$

and then $4(\mathrm{AC}-\mathrm{B}^2)(\mathrm{BD}-\mathrm{C}^2)-(\mathrm{AD}-\mathrm{BC})^2 =$

$$\begin{aligned}
&(BC-F^2)f^2 + (CA-G^2)g^2 + (AB-H^2)h^2 + (AD-L^2)l^2 + (BD-M^2)m^2 + (CD-N^2)n^2\\
&\quad + 2(GH-AF)gh + 2(HF-BG)hf + 2(FG-CH)fg\\
&\quad - 2(MN-DF)mn - 2(NL-DG)nl - 2(LM-DH)lm\\
&\quad + 2(AM-LH)lh + 2(BN-NF)mf + 2(CL-NG)ng\\
&\quad - 2(AN-LG)lg - 2(BL-MH)mh - 2(CM-NF)nf\\
&\quad + 2(NH-MG)lf + 2(LF-NH)mg + 2(MG-LF)nh.
\end{aligned}$$

Substituting the values of A, B, &c., in this expression, the result after all reductions is

$$\begin{aligned}
&f^2m^2n^2x^8 + g^2n^2l^2y^8 + h^2l^2m^2z^8 + f^2g^2h^2w^8\\
&+ 2gl^2n(mg-nh)y^6z^2 + 2hm^2l(nh-lf)z^6x^2 + 2fn^2m(lf-mg)x^6y^2\\
&- 2hl^2m(mg-nh)y^2z^6 - 2fm^2n(nh-lf)z^2x^6 - 2gn^2l(lf-mg)x^2y^6\\
&+ 2f^2mn(mg-nh)x^6w^2 + 2g^2nl(nh-lf)y^6w^2 + 2h^2lm(lf-mg)z^6w^2\\
&- 2f^2gh(mg-nh)x^2w^6 - 2fg^2h(nh-lf)y^2w^6 - 2fgh^2(lf-mg)z^2w^6\\
&+ f^2(l^2f^2-6ghmn)w^4x^4 + g^2(m^2g^2-6hfnl)w^4y^4 + h^2(n^2h^2-6lmfg)w^4z^4\\
&+ l^2(l^2f^2-6ghmn)y^4z^4 + m^2(m^2g^2-6hfnl)z^4x^4 + n^2(n^2h^2-6lmfg)x^4y^4\\
&+ 2gh(ghmn-3f^2l^2)w^4y^2z^2 + 2hf(hfnl-3g^2m^2)w^4z^2x^2 + 2fg(fglm-3h^2n^2)w^4x^2y^2\\
&+ 2hm(ghmn-3f^2l^2)z^4x^2w^2 + 2fn(hfnl-3g^2m^2)x^4y^2w^2 + 2gl(fglm-3h^2n^2)y^4z^2w^2\\
&- 2gn(ghmn-3f^2l^2)y^4x^2w^2 - 2hl(hfnl-3g^2m^2)z^4y^2w^2 - 2fm(fglm-3h^2n^2)x^4z^2w^2\\
&- 2mn(ghmn-3f^2l^2)x^4y^2z^2 - 2nl(hfnl-3g^2m^2)y^4z^2x^2 - 2lm(fglm-3h^2n^2)z^4x^2y^2\\
&- 2(mg-nh)(nh-lf)(lf-mg)x^2y^2z^2w^2 = 0,
\end{aligned}$$

which is therefore the equation of the envelope for this case. The equation may also be presented under the form

$$w^2\Theta + (mnx^2 + nly^2 + lmz^2)^2 (f^2x^4 + g^2y^4 + h^2x^4 - 2ghy^2z^2 - 2hfz^2x^2 - 2fgx^2y^2) = 0;$$

and there are probably other forms proper to exhibit the different geometrical properties of the surface, but with which I am not yet acquainted.

B. Here taking for the two surfaces the reciprocals of the equations made use of in determining the Intersect-Developable, the equations of these reciprocals are

$$n^2\, b\, x^2 + n^2\, a\, y^2 - a\, b\, c\, w^2 + 2n\, a\, b\, zw = 0,$$

$$n'^2b'x^2 + n'^2a'y^2 - a'b'c'w^2 + 2n'a'b'zw = 0,$$

which are similar to the equations of the surfaces of which they are reciprocal, only z and w are interchanged, so that here $x = 0$, $y = 0$ are the single conjugate planes, $z = 0$ is an indeterminate plane passing through the single conjugate points, and $w = 0$ is the equation of the double conjugate plane or plane of contact.

The values of A, B, C, D are

$$\begin{aligned}
\mathrm{A} &= 3\,(n^2bx^2 + n^2ay^2 - abcw^2 + 2nabzw),\\
\mathrm{B} &= \,(2nn'b + n^2b')\,x^2 + \ldots\ldots\\
\mathrm{C} &= \,(2nn'b' + n'^2b)\,x^2 + \ldots\ldots\\
\mathrm{D} &= 3\,(n'^2b'x^2 + n'^2a'y^2 - a'b'c'w^2 + 2n'a'b'zw).
\end{aligned}$$

Hence, using f, g, h, p, q, r in the same sense as before, we have for $2(\mathrm{AC}-\mathrm{B}^2)$, $2(\mathrm{BD}-\mathrm{C}^2)$, $(\mathrm{AD}-\mathrm{BC})$ expressions of the same form as in the last case (p, q, r being written for l, m, n), but in which

$$\begin{aligned}
\mathrm{A} &= f^2w^4 + 4q^2z^2w^2 + 8qry^2w^2 - 4fqzw^3,\\
\mathrm{B} &= g^2w^4 + 4p^2z^2w^2 + 8prx^2w^2 + gpzw^3,\\
\mathrm{C} &= h^2w^4,\\
\mathrm{D} &= 2q^2x^4 + 2p^2y^4 + 4h^2z^2w^2 + 4pqx^2y^2 - 8qhx^2zw + 8phy^2zw + 2ghy^2w^2 + 2fhx^2w^2,\\
\mathrm{F} &= -\,p^2y^2w^2,\\
\mathrm{G} &= -\,q^2x^2w^2,\\
\mathrm{H} &= 0,\\
\mathrm{L} &= 2pqy^2zw,\\
\mathrm{M} &= 2pqx^2zw,\\
\mathrm{N} &= -\,2h^2zw^3\ .
\end{aligned}$$

The substitution of these values gives after all reductions the result

$$\begin{aligned}&f^2g^2h^2w^6 + 4(pf - qg)fgh^2zw^5\\&+ 4(r^2h^2 - 6pqgf)h^2z^2w^4 + 2(q^2g^2 + 2prfh)fhx^2w^4 + 2(p^2f^2 + 2qrgh)ghy^2w^4\\&- 16(pf - qg)z^3w^3 - 4(q^2g^2 - 4p^2f^2 - 6pqfg)qhx^2zw^3 - 4(p^2f^2 - 4q^2g^2 - 6pqfg)phy^2zw^3\\&+ 16p^2q^2h^2z^4w^2 - 8(pf + 4qg)q^2phx^2z^2w^2 - 8(qg + 4pf)pq^2hy^2z^2w^2\\&+ (q^2g^2 + 8prfh)q^2x^4w^2 + (p^2f^2 + 8qrgh)p^2y^4w^2 + 2(10r^2h^2 - pqfg)pqx^2y^2w^2\\&- 16p^2q^3hx^2z^3w + 16p^3q^2hy^2z^3w\\&+ 4(4pf + 5qg)pq^3x^4zw - 4(4qg + 5pf)p^3qy^4zw - 4(pf - qg)p^2q^2x^2y^2zw\\&+ 4p^2q^4x^4z^2 + 4p^4q^2y^4z^2 + 8p^3q^3x^2y^2z^2 + 4pq^4rx^6 + 4p^4qry^6 + 12p^2q^3rx^4y^2 + 12p^3q^2rx^2y^4 = 0;\end{aligned}$$

which is therefore the equation of the envelope for this case. This equation may be presented under the form

$$w\Psi + 4pq(qx^2 + py^2)^2(qrx^2 + rpy^2 + pqz^2) = 0,$$

and there are probably other forms which I am not yet acquainted with.

C. The reciprocals of the two surfaces made use of in determining the Intersect-Developable, although in reality a system of the same nature with the surfaces of which they are reciprocals, are represented by equations of a somewhat different form. There is no real loss of generality in replacing the two surfaces by the reciprocals of the cones $x^2 = 2yz$, $y^2 = 2zw$; or we may take the two conics

$$(x^2 - 2yz = 0,\ w = 0) \text{ and } (y^2 - 2zw = 0,\ x = 0),$$

for the surfaces of which the envelope has to be found, these conics being, it is evident, the sections by the planes $w = 0$ and $x = 0$ respectively of the cones the Intersect-Developable of which was before determined. The process of determining the envelope is however essentially different: supposing the plane $\xi x + \eta y + \zeta z + \omega w = 0$ to be the equation of a tangent plane to the two conics (that is, of a plane passing through a tangent of each of the conics) then the condition of touching the first conic gives $\xi^2 - 2\eta\zeta = 0$, and that of touching the second conic gives $\eta^2 - 2\zeta\omega = 0$. We have therefore to find the envelope (in the ordinary sense of the word) of the plane $\xi x + \eta y + \zeta z + \omega w = 0$, in which the coefficients ξ, η, ζ, ω are variable quantities subject to the conditions

$$\xi^2 - 2\eta\zeta = 0, \quad \eta^2 - 2\zeta\omega = 0.$$

The result which is obtained without difficulty by the method of indeterminate multipliers, [or more easily by writing $\xi : \eta : \zeta : w = 2\theta^3 : 2\theta^2 : \theta^4 : 2$] is

$$8y^4z - 32y^2z^2w + 32z^3w^2 - 27x^4w + 27x^2yzw - 4x^2y^3 = 0,$$

which may also be written under the form

$$8z(y^2 - 2zw)^2 - x^2\{4y^3 + 9(3x^2 - 8yz)w\} = 0.$$

[Another form, containing the factor w, is $4(y^2 + 2zw)^3 - (2y^3 + 27x^2w - 36yzw)^2 = 0$.]

85.

NOTE ON A FAMILY OF CURVES OF THE FOURTH ORDER.

[From the *Cambridge and Dublin Mathematical Journal*, vol. v. (1850), pp. 148—152.]

THE following theorem, in a slightly different and somewhat less general form, is demonstrated in Mr Hearn's "Researches on Curves of the Second Order, &c.", Lond. 1846: "The locus of the pole of a line, $u+v+w=0$, with respect to the conics passing through the angles of the triangle ($u=0$, $v=0$, $w=0$), and touching a fixed line $\alpha u+\beta v+\gamma w=0$, is the curve of the fourth order,

$$\sqrt{\{\alpha u(v+w-u)\}}+\sqrt{\{\beta v(w+u-v)\}}+\sqrt{\{\gamma w(u+v-w)\}}=0\,;\text{"}$$

the difference in fact being, that with Mr Hearn the indeterminate line $u+v+w=0$ is replaced by the line ∞, so that the poles in question become the centres of the conics.

Previous to discussing the curve of the fourth order, it will be convenient to notice a property of curves of the fourth order with three double points. Such curves contain eleven arbitrary constants: or if we consider the double points as given, then five arbitrary constants. From each double point may be drawn two tangents to the curve; any five of the points of contact of these tangents determine the curve, and consequently determine the sixth point of contact. The nature of this relation will be subsequently explained; at present it may be remarked that it is such that, if three of the points of contact (each one of such points of contact corresponding to a different double point) lie in a straight line, the remaining three points of contact also lie in a straight line. A curve of the fourth order having three given double points and besides such that the points of contact of the tangents from the double points lie three and three in two straight lines, contains therefore four arbitrary constants. Now it is easily seen that the curve in question has three double points, viz. the points given by the equations

$$(u=0,\ v-w=0),\quad (v=0,\ w-u=0),\quad (w=0,\ u-v=0),$$

points which may be geometrically defined as the projections from the angles of the triangle ($u=0$, $v=0$, $w=0$) upon the opposite sides, of the point ($u=v=w$) which is the harmonic with respect to the triangle of the given line $u+v+w=0$. Moreover, the lines $u=0$, $v=0$, $w=0$ (being lines which, as we have seen, pass through the double points) touch the curve in three points lying in a line, viz. the given line $\alpha u+\beta v+\gamma w=0$. Hence the curve in question is a curve with three double points, such that the points of contact of the tangents from the double points lie three and three in two straight lines. Considering the double points as given, the functions u, v, w contain two arbitrary ratios, and the ratios of the quantities α, β, γ being arbitrary, the equation of the curve contains four arbitrary constants, or it represents the general curve of the class to which it has been stated to belong.

As to the investigation of the above-mentioned theorem with respect to curves of the fourth order with three double points, the general form of the equation of such a curve is

$$\frac{a}{x^2}+\frac{b}{y^2}+\frac{c}{z^2}+\frac{2f}{yz}+\frac{2g}{zx}+\frac{2h}{xy}=0,$$

where the double points are the angles of the triangle ($x=0$, $y=0$, $z=0$). It may be remarked in passing, that the six tangents at the double point touch the conic

$$ax^2+by^2+cz^2-2fyz-2gxz-2hxy=0.$$

To determine the tangents through ($y=0$, $z=0$), we have only to write the equation to the curve under the form

$$\left(\frac{a}{x}+\frac{h}{y}+\frac{g}{z}\right)^2+\frac{C}{y^2}+\frac{B}{z^2}-\frac{2F}{yz}=0:$$

the points of contact are given by the system

$$\frac{a}{x}+\frac{h}{y}+\frac{g}{z}=0,$$

$$\frac{C}{y^2}+\frac{B}{z^2}-\frac{2F}{yz}=0;$$

the latter equation (which evidently belongs to a pair of lines) determining the tangents. The former equation is that of a conic passing through the angles of the triangle $x=0$, $y=0$, $z=0$: since the tangents pass through the point ($y=0$, $z=0$) they evidently each intersect the conic in one other point only. The equation of the tangents shows that these lines are the tangents through the point $y=0$, $z=0$ to the conic whose equation is

$$aA^2x^2+bB^2y^2+cC^2z^2+2fBCyz+2gCAzx+2hABxy=0.$$

To complete the construction of the points of contact it may be remarked, that the equations which determine the coordinates of these points may be presented under the form

$$Ax=\{A \qquad\qquad\qquad \}\,x,$$

$$By=\left\{H-g\sqrt{\left(\frac{-K}{a}\right)}\right\}x,$$

$$Cz=\left\{G+h\sqrt{\left(\frac{-K}{a}\right)}\right\}x,$$

whence also

$$\begin{aligned} aAx + hBy + gCz &= Kx, \\ hAx + bBy + fCz &= G\sqrt{\left(\frac{-K}{a}\right)}\,x, \\ gAx + fBy + cCz &= -H\sqrt{\left(\frac{-K}{a}\right)}\,x : \end{aligned}$$

or writing for shortness ξ, η, ζ instead of the linear functions forming the first sides of these equations respectively, we have

$$\frac{\xi}{\sqrt{(-aK)}} = \frac{\eta}{G} = -\frac{\zeta}{H};$$

from which it follows at once that the equation to the line forming the two points of contact is

$$\frac{\eta}{G} + \frac{\zeta}{H} = 0;$$

and this may again be considered as the line joining the points $\left(\xi = 0,\ \frac{\xi}{F} + \frac{\eta}{G} + \frac{\zeta}{H} = 0\right)$ and $(\eta = 0,\ \zeta = 0)$.

Now $\xi = 0$, $\eta = 0$, $\zeta = 0$, are the polars of the double points (or angles of the triangle $x = 0$, $y = 0$, $z = 0$) with respect to the last-mentioned conic. The equation of the line joining the point $(y = 0,\ z = 0)$ with the point $(\eta = 0,\ \zeta = 0)$, is easily seen to be $GBy - CHz = 0$; from which it follows, that the lines forming each double point with the intersection of the polars of the other two double points meet in a point, the coordinates of which are given by

$$x : y : z = \frac{1}{AF} : \frac{1}{BG} : \frac{1}{CH},$$

and the polar of this point is the line $\frac{\xi}{F} + \frac{\eta}{G} + \frac{\zeta}{H} = 0$. The construction of this line is thus determined; and we have already seen how this leads to the construction of

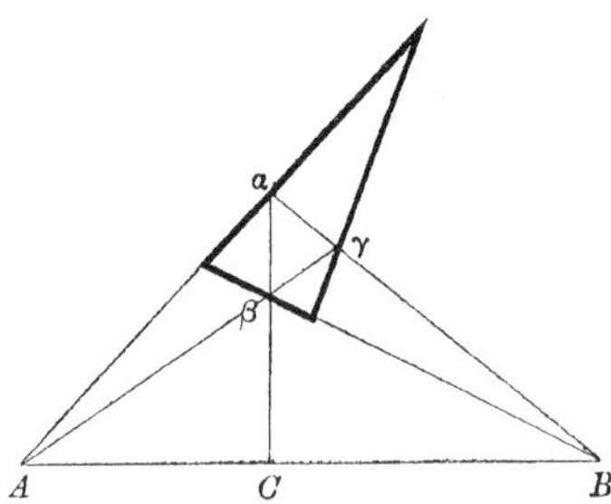

the lines joining the points of contact of the tangents from the double points. In fact, in the figure, if $\alpha\beta\gamma$ be the triangle whose sides are $\xi = 0$, $\eta = 0$, $\zeta = 0$, and A, B, C the points of intersection of the sides of this triangle with the line

$$\frac{\xi}{F} + \frac{\eta}{G} + \frac{\zeta}{H} = 0,$$

the lines in question are $A\alpha$, $B\beta$, $C\gamma$. Moreover, the points of contact upon $A\alpha$ are harmonics with respect to A, α, and in like manner the points of contact on $B\beta$, $C\gamma$ are respectively harmonics with respect to B, β and to C, γ. This proves that if three of the points of contact are in the same straight line, the remaining three are also in the same straight line; in fact, we may consider three of the points of contact as given by

$$\xi : \eta : \zeta = \sqrt{(-aK)} : G : -H,$$
$$\xi : \eta : \zeta = -F : \sqrt{(-bK)} : H,$$
$$\xi : \eta : \zeta = F : -G : \sqrt{(-cK)};$$

and the condition that these may be in the same line is

$$\sqrt{(-a)}\sqrt{(-b)}\sqrt{(-c)} + GH\sqrt{(-a)} + HF\sqrt{(-b)} + FG\sqrt{(-c)} = 0,$$

which remains unaltered when the signs of all the roots are changed. The equation just obtained may be considered as the condition which must exist between the coefficients of

$$\frac{a}{x^2} + \frac{b}{y^2} + \frac{c}{z^2} + \frac{2f}{yz} + \frac{2g}{zx} + \frac{2h}{xy} = 0,$$

in order that this curve may be transformable into the form

$$\sqrt{\{\alpha u(v+w-u)\}} + \sqrt{\{\beta v(w+u-v)\}} + \sqrt{\{\gamma w(u+v-w)\}} = 0.$$

86.

ON THE DEVELOPABLE DERIVED FROM AN EQUATION OF THE FIFTH ORDER.

[From the *Cambridge and Dublin Mathematical Journal*, vol. v. (1850), pp. 152—159.]

MÖBIUS, in his "Barycentrische Calcul," [Leipzig, 1827], has considered, or rather suggested for consideration, the family of curves of double curvature given by equations such as $x : y : z : w = A : B : C : D$, where A, B, C, D are rational and integral functions of an indeterminate quantity t. Observing that the plane $Ax + By + Cz + Dw = 0$ may be considered as the polar of the point determined by the system of equations last preceding, the reciprocal of the curve of Möbius is the developable, which is the envelope of a plane the coefficients in the equation of which are rational and integral functions of an indeterminate quantity t, or what is equivalent, homogeneous functions of two variables ξ, η. Such an equation may be represented by $U = a\xi^n + n\xi^{n-1}\eta + \ldots = 0$, (where a, b, &c. are linear functions of the coordinates); and we are thus led to the developables noticed, I believe for the first time, in my "Note sur les Hyper-determinants," *Crelle*, t. XXXIV. p. 148, [54]. I there remarked, that not only the equation of the developable was to be obtained by eliminating ξ, η from the first derived equations of $U = 0$; but that the second derived equations conducted in like manner to the edge of regression, and the third derived equations to the cusps or stationary points of the edge of regression. It followed that the order of the surface was $2(n-1)$, that of the edge of regression $3(n-2)$, and the number of stationary points $4(n-3)$. These values lead at once, as Mr Salmon pointed out to me, to the table,

$$\begin{aligned}
m &= 3(n-2),\\
n &= n,\\
r &= 2(n-1),\\
\alpha &= 0,\\
\beta &= 4(n-3),\\
g &= \tfrac{1}{2}(n-1)(n-2),\\
h &= \tfrac{1}{2}(9n^2 - 53n + 80),\\
x &= 2(n-2)(n-3),\\
y &= 2(n-1)(n-3),
\end{aligned}$$

where the letters in the first column have the same signification as in my memoir in Liouville, [30], translated in the last number of the *Journal*. The order of the nodal line is of course $2(n-2)(n-3)$; Mr Salmon has ascertained that there are upon this line $6(n-3)(n-4)$ stationary points and $\frac{4}{3}(n-3)(n-4)(n-5)$ real double points, (the stationary points lying on the edge of regression, and with the stationary points of the edge of regression forming the system of intersections of the nodal line and edge of regression, and the real double points being triple points upon the surface). Also, that the number of apparent double points of the nodal line is

$$(n-3)(2n^3-18n^2+57n-65).$$

The case of U a function of the second order gives rise to the cone $ac-b^2=0$. When U is a function of the third order, we have the developable

$$4(ac-b^2)(bd-c^2)-(ad-bc)^2=0,$$

which is the general developable of the fourth order having for its edge of regression the curve of the third order,

$$ac-b^2=0,\quad bd-c^2=0,\quad ad-bc=0,$$

which is likewise the most general curve of this order: there are of course in this case no stationary points on the edge of regression. In the case where U is of the fourth order we have the developable of the sixth order,

$$(ae-4bd+3c^2)^3-27(ace+2bcd-ad^2-b^2e-c^3)^2=0;$$

having for its edge of regression the curve of the sixth order,

$$ae-4bd+3c^2=0,\quad ace+2bcd-ad^2-b^2e-c^3=0,$$

with four stationary points determined by the equations

$$\frac{a}{b}=\frac{b}{c}=\frac{c}{d}=\frac{d}{e}.$$

The form exhibiting the nodal line of the surface has been given in the *Journal* by Mr Salmon. I do not notice it here, but pass on to the principal subject of the present paper, which is to exhibit the edge of regression and the stationary points of this edge of regression for the developable obtained from the equation of the fifth order,

$$U=a\xi^5+5b\xi^4\eta+10c\xi^3\eta^2+10d\xi^2\eta^3+5e\xi\eta^4+f\eta^5=0;$$

viz. that represented by the equation

$$\square=0=a^4f^4+160a^3cef^2+\ldots-4000b^2c^3e^3,$$

[I do not reproduce here this expression for the discriminant of the binary quintic] a result for which I am indebted to Mr Salmon.

To effect the reduction of this expression, consider in the first place the equations which determine the stationary points of the edge of regression. Writing instead of $\xi : \eta$ the single letter t, these equations are

$$at^2 + 2bt + c = 0,$$
$$bt^2 + 2ct + d = 0,$$
$$ct^2 + 2dt + e = 0,$$
$$dt^2 + 2et + f = 0,$$

write for shortness

$$A = 2\,(bf - 4ce + 3d^2),$$
$$B = \quad af - 3be + 2cd,$$
$$C = 2\,(ae - 4bd + 3c^2),$$

and let α, 3β, 3γ, δ represent the terms of

$$\left\|\begin{matrix} a, & b, & c, & d \\ b, & c, & d, & e \\ c, & d, & e, & f \end{matrix}\right\|,$$

viz.

$$\alpha = bdf - be^2 + 2cde - c^2f - d^3,$$
$$3\beta = adf - ae^2 - bcf + bde + c^2e - cd^2,$$
$$3\gamma = acf - ade - b^2f + bd^2 + bce - c^2d,$$
$$\delta = ace - ad^2 - b^2e + 2bcd - c^3;$$

it is obvious at first sight that the result of the elimination of t from the four quadratic equations is the system (equivalent of course to three equations),

$$\alpha = 0, \quad \beta = 0, \quad \gamma = 0, \quad \delta = 0.$$

The system in question may however be represented under the more simple form (which shows at once that the number of stationary points is, as it ought to be, eight),

$$A = 0, \quad B = 0, \quad C = 0;$$

this appears from the identical equations,

$$\begin{aligned} &(2ct + 3d)\,(bt^2 + 2ct + d) \\ -&(2bt + 4c)\,(ct^2 + 2dt + e) \\ &\qquad + b\,(dt^2 + 2et + f) = \tfrac{1}{2}A\,; \end{aligned}$$

$$\begin{aligned} &(2ct + 3d)\,(at^2 + 2bt + c) \\ &\qquad - c\,(bt^2 + 2ct + d) \\ -&(2at + 3b)\,(ct^2 + 2dt + e) \\ &\qquad + a\,(dt^2 + 2et + f) = B\,; \end{aligned}$$

$$\begin{aligned}&(2bt + 3c)(at^2 + 2bt + c)\\ &- (2at + 4b)(bt^2 + 2ct + d)\\ &\qquad + a(ct^2 + 2dt + e) = \tfrac{1}{2}C,\end{aligned}$$

(formulæ the first and third of which are readily deduced from an equation given in the Note on Hyperdeterminants above quoted). The connexion between the quantities A, B, C and α, β, γ, δ, is given by

$$\begin{aligned}Aa - 2Bb + Cc &= -6\delta,\\ Ab - 2Bc + Cd &= -6\gamma,\\ Ac - 2Bd + Ce &= -6\beta,\\ Ad - 2Be + Cf &= -6\alpha.\end{aligned}$$

The theory of the stationary points being thus obtained, the next question is that of finding the equations of the edge of regression. We have for this to eliminate t from the three cubic equations,

$$\begin{aligned}at^3 + 3bt^2 + 3ct + d &= 0,\\ bt^3 + 3ct^2 + 3dt + e &= 0,\\ ct^3 + 3dt^2 + 3et + f &= 0:\end{aligned}$$

treating the quantities t^3, t^2, t^1, t^0 as if they were independent, we at once obtain

$$\beta t + \alpha = 0,\quad \delta t + \gamma = 0,\quad \gamma t^2 - \alpha = 0,\quad \delta t^2 - \beta = 0;$$

or as this system may be more conveniently written,

$$\beta t + \alpha = 0,\quad \gamma t + \beta = 0,\quad \delta t + \gamma = 0.$$

But the most simple forms are obtained from the identical equations,

$$\begin{aligned}&ft(at^3 + 3bt^2 + 3ct + d),\\ &-(3et + f)(bt^3 + 3ct^2 + 3dt + e),\\ &+(2dt + e)(ct^3 + 3dt^2 + 3et + f) = t^3(Bt + A);\end{aligned}$$

$$\begin{aligned}&(bt + c)(at^3 + 3bt^2 + 3ct + d),\\ &-(at + 3b)(bt^3 + 3ct^2 + 3dt + e),\\ &+\quad a\quad (ct^3 + 3dt^2 + 3et + f) = \quad Ct + B;\end{aligned}$$

equations which, combined with those which precede, give the complete system

$$\beta t + \alpha = 0,\quad \gamma t + \beta = 0,\quad \delta t + \gamma = 0,\quad Bt + A = 0,\quad Ct + B = 0:$$

or the equations of the edge of regression are given by the system (equivalent of course to two equations),

$$\left\|\begin{matrix} \alpha, & \beta, & \gamma, & A, & B \\ \beta, & \gamma, & \delta, & B, & C \end{matrix}\right\| = 0.$$

The simplest mode of verifying *à posteriori* that the edge of regression is only of the ninth order, appears to be to consider this curve as the common intersection of the three surfaces of the seventh order:

$$A^3a - 3A^2Bb + 3AB^2c - B^3d = 0,$$
$$A^3b - 3A^2Bc + 3AB^2d - B^3e = 0,$$
$$A^3c - 3A^2Bd + 3AB^2e - B^3f = 0,$$

(which are at once obtained by combining the equation $Bt + A = 0$ with the cubic equations in t). It is obvious from a preceding equation that if the equations first given are multiplied by fA, $-3eA + fB$, $2dA - eB$, and added, an identical result is obtained. This shows that the curve of the forty-ninth order, the intersection of the first two surfaces, is made up of the curve in question, the curve of the fourth order $A = 0$, $B = 0$ (which reckons for thirty-six, as being a triple line on each surface), and the curve which is common to the two surfaces of the seventh order and the surface $2dA - eB = 0$. The equations of this last curve may be written,

$$e(af - 3be + 2cd) - 4d(bf - 4ce + 3d^2) = 0,$$
$$e^3a - 6e^2db + 12ed^2c - 8d^4 = 0,$$
$$e^3b - 6e^2dc + 4ed^3 = 0;$$

or, observing that these equations are

$$f(ae - 4bd) - 3\ (be^2 - 6ced + 4d^3) = 0,$$
$$e^2(ae - 4bd) - 2d(be^2 - 6ced + 4d^3) = 0,$$
$$e\ (be^2 - 6ced + 4d^3) = 0;$$

the last-mentioned curve is the intersection of

$$ae - 4bd = 0,$$
$$be^2 - 6ced + 4d^3 = 0,$$

where the second surface contains the double line $e = 0$, $d = 0$, which is also a single line upon the first surface. Omitting this extraneous line, the intersection is of the *fourth* order; and we may remark that, in passing, it is determined (exclusively of the double line) as the intersection of the three surfaces

$$ae - 4bd = 0,$$
$$be^2 - 6ced + 4d^3 = 0,$$
$$a^2d - 6abc + 4b^3 = 0,$$

being in fact of the species IV. 4, of Mr Salmon's paper, "On the Classification of Curves of Double Curvature" [*Journal*, vol. V. (1850), pp. 23—46]. But returning to the question in hand; since $49 = 9 + 4 + 36$, we see that the curve common to the three surfaces of the seventh order, or the edge of regression, is, as it ought to be, of the ninth order. It only remains to express the equation of the developable surface in terms of the functions A, B, C, α, β, γ, δ, which determine the stationary points and edge of regression; I have satisfied myself that the required formula is

$$\square = (AC - B^2)^2 - 1152\,\{A\,(\beta\delta - \gamma^2) - B\,(\alpha\delta - \beta\gamma) + C\,(\alpha\gamma - \beta^2)\} = 0,$$

where the quantities α, β, γ, δ may be replaced by their values in A, B, C; and it will be noticed that when this is done, the terms of $\square$ are each of them at least of the third order in the last-mentioned functions.

I propose to term the family of developables treated of in this paper, 'planar developables.' In general, the coefficients of the generating plane of a developable being algebraical functions of a variable parameter t, the equation rationalized with respect to the parameter belongs to a system of n different planes; the developable which is the envelope of such a system may be termed a 'multiplanar developable,' and in the particular case of n being equal to unity, we have a planar developable. It would be very desirable to have some means of ascertaining from the equation of a developable what the degree of its 'planarity' is.

P.S.—At the time of writing the preceding paper I was under the impression that the only surface of the fourth order through the edge of regression was that given by the equation $AC - B^2 = 0$; but Mr Salmon has since made known to me an entirely new form of the function $\square$, the component functions of which, equated to zero, give six different surfaces of the fourth order, each of them passing through the edge of regression. The form in question is

$$3\,\square = LL' + 64MM' - 64NN',$$

where

$$\begin{aligned}
L &= a^2f^2 + 225b^2e^2 - 32ace^2 - 32b^2df + 480bd^3 + 480c^3e - 34abef + 76acdf - 12bc^2f \\
&\qquad - 12ad^2e - 820bcde - 320c^2d^2, \\
L' &= 3a^2f^2 - 45b^2e^2 + 64ace^2 + 64b^2df - 22abef - 12acdf - 36bc^2f \\
&\qquad - 36ad^2e + 20bcde, \\
M &= 10bcd^2 - 18ad^3 - 15bc^2e + 32acde + 6b^2cf - 9ac^2f + 2abdf + a^2ef - 9abe^2, \\
M' &= 10c^2de - 18c^3f - 15bd^2e + 32bcdf + 6ade^2 - 9b^2ef + 2acef + abf^2 - 9ad^2f, \\
N &= 10b^2d^2 - 15b^2ce - 12acd^2 + 18ac^2e + abde - 2a^2e^2 + 6b^3f - 9abcf + 3a^2df, \\
N' &= 10c^2e^2 - 15bde^2 - 12c^2df + 18bd^2f + bcef - 2b^2f^2 + 6ae^3 - 9adef + 3acf^2,
\end{aligned}$$

where it should be noticed that

$$L + 3L' = -10\,(AC - B^2).$$

The expressions of L, L', M, M', N, N' as linear functions of A, B, C (also due to Mr Salmon) are

$$L = (11\,ae + 28bd - 39c^2)\,A + (\,af - 75be + 74cd)\,B + (\,11bf + 28ce - 39d^2)\,C,$$

$$L' = (-7ae + 4bd + 3c^2)\,A + (3af + 15be - 18cd)\,B + (-7bf + 4ce + 3d^2)\,C,$$

$$M = 3\,(bc - ad)\,A + 3\,(ae - c^2)\,B + (cd - af)\,C,$$

$$M' = (cd - af)\,A + 3\,(bf - d^2)\,B + 3\,(de - cf)\,C,$$

$$N = 3\,(b^2 - ac)\,A + 3\,(ad - bc)\,B + (bd - ae)\,C,$$

$$N' = (ce - bf)\,A + 3\,(cf - de)\,B + 3\,(e^2 - df)\,C.$$

I propose resuming the subject of these forms, and the general theory, in a subsequent paper. [This was never written.]

87.

NOTES ON ELLIPTIC FUNCTIONS (FROM JACOBI).

[From the *Cambridge and Dublin Mathematical Journal,* vol. v. (1850), pp. 201—204.]

THESE Notes are mere translations from Jacobi's "Note sur une nouvelle application de l'analyse des fonctions elliptiques à l'algèbre," [*Crelle*, t. VII. (1831) pp. 41—43], and from the addition to the notice by him of the third supplement to Legendre's "Théorie des fonctions elliptiques" [*Crelle*, t. VIII. (1832) pp. 413—417].

88.

ON THE TRANSFORMATION OF AN ELLIPTIC INTEGRAL.

[From the *Cambridge and Dublin Mathematical Journal*, vol. v. (1850), pp. 204—206.]

THE following is a demonstration of a formula proved incidentally by Mr Boole (*Journal*, vol. II. [1847] p. 7), in a paper "On the Attraction of a Solid of Revolution on an External Point."

Let
$$U=\int_{-1}^{1}\frac{dx}{\sqrt{[(1-x^2)\{1-(mx+n)^2\}]}};$$

then, assuming
$$ix=\frac{\alpha+iy}{1-i\alpha y},$$

(so that $x=\pm 1$ gives $y=\pm 1$), we obtain
$$1-x^2=\frac{(1+\alpha^2)(1-y^2)}{(1-i\alpha y)^2},$$

$$mx+n=\frac{(n-im\alpha)+(m-in\alpha)\,y}{1-i\alpha y}.$$

Assume therefore
$$i\alpha+(n-im\alpha)(m-in\alpha)=0,$$

whence
$$-i\alpha=\frac{(1-m^2-n^2)+\Delta}{2mn}\qquad(\Delta^2=1+m^4+n^4-2m^2-2n^2-2m^2n^2),$$

we find

$$1-(mx+n)^2=\frac{1-(n-im\alpha)^2}{(1-i\alpha y)^2}\{1-(m-in\alpha)^2y^2\},$$

and also

$$dx=\frac{(1+\alpha^2)\,dy}{(1-i\alpha y)^2},$$

whence

$$U=\sqrt{\left\{\frac{1+\alpha^2}{1-(n-im\alpha)^2}\right\}}\int_{-1}^{1}\frac{dy}{\sqrt{[(1-y^2)\{1-(m-in\alpha)^2y^2\}]}},$$

that is

$$U=2\sqrt{\left\{\frac{1+\alpha^2}{1-(n-im\alpha)^2}\right\}}\int_0^1\frac{dy}{\sqrt{[(1-y^2)\{1-(m-in\alpha)^2y^2\}]}}.$$

But since

$$n-im\alpha=\frac{1-m^2+n^2+\Delta}{2n},$$

$$m-in\alpha=\frac{1+m^2-n^2+\Delta}{2m},$$

we have

$$1-(n-im\alpha)^2=-\frac{\Delta}{2n^2}\quad(\Delta+1-m^2+n^2),$$

$$1+\alpha^2=-\frac{\Delta}{2m^2n^2}(\Delta+1-m^2-n^2);$$

and therefore

$$\frac{1+\alpha^2}{1-(n-im\alpha)^2}=\frac{1}{m^2}\,\frac{\Delta+1-m^2-n^2}{\Delta+1-m^2+n^2}$$

$$=\frac{1}{m^2}\,\frac{(1-m^2-n^2+\Delta)(1-m^2+n^2-\Delta)}{(1-m^2+n^2+\Delta)(1-m^2+n^2-\Delta)}=\frac{2(1+m^2-n^2+\Delta)}{4m^2};$$

consequently

$$U=\frac{1}{m}\sqrt{\{2(1+m^2-n^2+\Delta)\}}\int_0^1\frac{dy}{\sqrt{\left[(1-y^2)\left\{1-\left(\frac{1+m^2-n^2+\Delta}{2m}\right)^2y^2\right\}\right]}}.$$

Write

$$k=\frac{1+m^2-n^2+\Delta}{2m},\qquad \lambda^2=\frac{4m}{(1+m)^2-n^2};$$

then

$$U=\frac{4\sqrt{k}}{\lambda}\,\frac{1}{\sqrt{\{(1+m)^2-n^2\}}}\int_0^1\frac{dy}{\sqrt{\{(1-y^2)(1-k^2y^2)\}}};$$

where λ and k are connected by the relation that exists for the transformation of the second order, viz.

$$\lambda=\frac{2\sqrt{k}}{1+k},$$

as may be immediately verified; hence, assuming

$$y = \frac{\lambda z}{\sqrt{k}} \sqrt{\left(\frac{1 - z^2}{1 - \lambda^2 z^2}\right)},$$

which gives

$$\int_0^1 \frac{dy}{\sqrt{\{(1 - y^2)(1 - k^2 y^2)\}}} = \frac{\lambda}{\sqrt{k}} \int_0^1 \frac{dz}{\sqrt{\{(1 - z^2)(1 - \lambda^2 z^2)\}}},$$

we find

$$U = \frac{4}{\sqrt{\{(1+m)^2 - n^2\}}} \int_0^1 \frac{dz}{\sqrt{\left\{(1 - z^2)\left(1 - \frac{4m}{(1+m)^2 - n^2} z^2\right)\right\}}};$$

that is

$$\int_{-1}^{1} \frac{dx}{\sqrt{[(1 - x^2)\{1 - (mx + n)\}^2]}} = 4 \int_0^1 \frac{dz}{\sqrt{(1 - z^2)[\{(1+m)^2 - n^2\} - 4mz^2]}}.$$

Writing here

$$x = \cos\theta, \quad z = \cos\tfrac{1}{2}\phi,$$

then

$$\int_0^\pi \frac{d\theta}{\sqrt{\{1 - (m\cos\theta + n)^2\}}} = \int_0^\pi \frac{d\phi}{\sqrt{(1 + m^2 - n^2 - 2m\cos\phi)}};$$

or if

$$m = \frac{r}{a}, \qquad n = -\frac{iz}{a},$$

then finally

$$\int_0^\pi \frac{d\theta}{\sqrt{\{a^2 + (z + ir\cos\theta)^2\}}} = \int_0^\pi \frac{d\phi}{\sqrt{(a^2 + r^2 + z^2 - 2ar\cos\phi)}},$$

the formula in question.

89.

ON THE ATTRACTION OF ELLIPSOIDS (JACOBI'S METHOD).

[From the *Cambridge and Dublin Mathematical Journal*, vol. v. (1850), pp. 217—226.]

IN a letter published in 1846 in *Liouville's Journal* (t. XI. p. 341) Jacobi says, "Il y a quatorze ans, je me suis posé le problème de chercher l'attraction d'un ellipsoïde homogène exercée sur un point extérieur quelconque par une méthode analogue à celle employée par Maclaurin par rapport aux points situés dans les axes principaux. J'y suis parvenu par trois substitutions consécutives. La première est une transformation de coordonnées; par la seconde le radical $\sqrt{(1 - m^2 \sin^2\beta \cos^2\psi - n^2 \sin^2\beta \sin^2\psi)}$ qui entre dans la double intégrale transformée est rendu rationnel au moyen de la double substitution

$$m \sin\beta \cos\psi = \sin\eta \cos\theta, \quad m \sin\beta \sin\psi = \sin\eta \sin\theta;$$

la troisième est encore une transformation de coordonnées. La recherche du sens géométrique de ces trois substitutions m'a conduit à approfondir la théorie des surfaces confocales par rapport auxquelles je découvris quantité de beaux théorèmes dont je communiquai quelques-uns des principaux à M. Steiner. Considérons l'ellipsoïde confocal mené par le point attiré P et le point p de l'ellipsoïde proposé, conjugué à P. Soient Q et q deux autres points conjugués quelconques situés respectivement sur l'ellipsoïde extérieur et intérieur. Menons de P un premier cône tangent à l'ellipsoïde intérieur, de p un second cône tangent à l'ellipsoïde extérieur. Ce dernier, tout imaginaire qu'il est, a ses trois axes réels (ainsi que ses deux droites focales). La première substitution ramène les axes de l'ellipsoïde à ceux du premier cône (c'est la substitution employée par Poisson, mais que j'avais antérieurement traitée et même étendue à un nombre quelconque de variables dans le mémoire *De binis Functionibus homogeneis &c.* [*Crelle*, t. XII. (1834) pp. 1—69]). Par la seconde substitution les angles que la droite Pq forme avec les axes du premier cône sont ramenés aux angles que la droite pQ forme avec les axes du second. Par la dernière substitution, on retourne de ces axes aux axes de l'ellipsoïde. La seconde substitution répond à un théorème de géométrie remarquable, savoir que 'Les cosinus des angles que la droite Pq forme avec deux des axes du premier cône sont en raison constante avec les cosinus des angles que la droite

pQ forme avec deux des axes du second cône; ces deux axes sont les tangents situés respectivement dans les sections de plus grande et de moindre courbure de chaque ellipsoïde, le troisième axe étant la normale à l'ellipsoïde.' Tout cela semble difficile à établir par la synthèse."

The object of this paper is to develope the above method of finding the attraction of an ellipsoid.

Consider an exterior ellipsoid, the squared semiaxes of which are $f+u$, $g+u$, $h+u$; and an interior ellipsoid, the squared semiaxes of which are $f+\bar{u}$, $g+\bar{u}$, $h+\bar{u}$. Let u, p, q be the elliptic coordinates of a point P on the exterior ellipsoid, the elliptic coordinates of the corresponding point $\bar{P}$ on the interior ellipsoid will be $\bar{u}$, p, q, and if a, b, c and $\bar{a}$, $\bar{b}$, $\bar{c}$ represent the ordinary coordinates of these points (the principal axes being the axes of coordinates), we have

$$a^2=\frac{(f+u)(f+q)(f+r)}{(f-g)(f-h)},\qquad \bar{a}^2=\frac{(f+\bar{u})(f+q)(f+r)}{(f-g)(f-h)},$$

$$b^2=\frac{(g+u)(g+q)(g+r)}{(g-h)(g-f)},\qquad \bar{b}^2=\frac{(g+\bar{u})(g+q)(g+r)}{(g-h)(g-f)},$$

$$c^2=\frac{(h+u)(h+q)(h+r)}{(h-g)(h-f)},\qquad \bar{c}^2=\frac{(h+\bar{u})(h+q)(h+r)}{(h-f)(h-g)}.$$

I form the systems of equations

$$a_1^2=\frac{(u+f)(u+g)(u+h)}{(u-q)(u-r)},\qquad \bar{a}_1^2=\frac{(\bar{u}+f)(\bar{u}+g)(\bar{u}+h)}{(\bar{u}-q)(\bar{u}-r)},$$

$$b_1^2=\frac{(q+f)(q+g)(q+h)}{(q-r)(q-u)},\qquad \bar{b}_1^2=\frac{(q+f)(q+g)(q+h)}{(\bar{q}-r)(q-\bar{u})},$$

$$c_1^2=\frac{(r+f)(r+g)(r+h)}{(r-u)(r-q)},\qquad \bar{c}_1^2=\frac{(r+f)(r+g)(r+h)}{(r-\bar{u})(r-q)}.$$

$$\alpha=\frac{a_1a}{f+u},\quad \beta=\frac{a_1b}{g+u},\quad \gamma=\frac{a_1c}{h+u},$$

$$\alpha'=\frac{b_1a}{f+q},\quad \beta'=\frac{b_1b}{g+q},\quad \gamma'=\frac{b_1c}{h+q},$$

$$\alpha''=\frac{c_1a}{f+r},\quad \beta''=\frac{c_1b}{g+r},\quad \gamma''=\frac{c_1c}{h+r},$$

$$\bar{\alpha}=\frac{\bar{a}_1\bar{a}}{f+\bar{u}},\quad \bar{\beta}=\frac{\bar{a}_1\bar{b}}{g+\bar{u}},\quad \bar{\gamma}=\frac{\bar{a}_1\bar{c}}{h+\bar{u}},$$

$$\bar{\alpha}'=\frac{\bar{b}_1\bar{a}}{f+q},\quad \bar{\beta}'=\frac{\bar{b}_1\bar{b}}{g+q},\quad \bar{\gamma}'=\frac{\bar{b}_1\bar{c}}{h+q},$$

$$\bar{\alpha}''=\frac{\bar{c}_1\bar{a}}{f+r},\quad \bar{\beta}''=\frac{\bar{c}_1\bar{b}}{g+r},\quad \bar{\gamma}''=\frac{\bar{c}_1\bar{c}}{h+r}.$$

And then writing

$$\begin{aligned} X &= \alpha X_1 + \alpha' Y_1 + \alpha'' Z_1, \\ Y &= \beta X_1 + \beta' Y_1 + \beta'' Z_1, \\ Z &= \gamma X_1 + \gamma' Y_1 + \gamma'' Z_1, \end{aligned} \qquad \begin{aligned} \bar{X} &= \alpha \bar{X}_1 + \alpha' \bar{Y}_1 + \alpha'' \bar{Z}_1, \\ \bar{Y} &= \beta \bar{X}_1 + \beta' \bar{Y}_1 + \beta'' \bar{Z}_1, \\ \bar{Z} &= \gamma \bar{X}_1 + \gamma' \bar{Y}_1 + \gamma'' \bar{Z}_1, \end{aligned}$$

if X, Y, Z are the cosines of the inclinations of a line $P\bar{Q}$ to the principal axes of the ellipsoids, X_1, Y_1, Z_1 will be the cosines of the inclinations of this line to the principal axes of the cone having P for its vertex, and circumscribed about the interior ellipsoid. In like manner, $\bar{X}$, $\bar{Y}$, $\bar{Z}$ being the cosines of the inclinations of a line $\bar{P}Q$ to the principal axes of the ellipsoids, $\bar{X}_1$, $\bar{Y}_1$, $\bar{Z}_1$ will be the cosines of the inclinations of this line to the principal axes of the cone having $\bar{P}$ for its vertex and circumscribed about the exterior ellipsoid. Assuming that the points Q, $\bar{Q}$ are situated upon the exterior and interior ellipsoids respectively, suppose that X_1, Y_1, Z_1 and $\bar{X}_1$, $\bar{Y}_1$, $\bar{Z}_1$ are connected by the equivalent systems of equations,

$$X_1 = \sqrt{(u - \bar{u})} \sqrt{\left(\frac{\bar{X}_1^2}{u - \bar{u}} + \frac{\bar{Y}_1^2}{u - q} + \frac{\bar{Z}_1^2}{u - r}\right)},$$

$$Y_1 = \sqrt{\left(\frac{\bar{u} - q}{u - q}\right)} \bar{Y}_1,$$

$$Z_1 = \sqrt{\left(\frac{\bar{u} - r}{u - r}\right)} \bar{Z}_1,$$

$$\bar{X}_1 = \sqrt{(\bar{u} - u)} \sqrt{\left(\frac{X_1^2}{\bar{u} - u} + \frac{Y_1^2}{\bar{u} - q} + \frac{Z_1^2}{\bar{u} - r}\right)},$$

$$\bar{Y}_1 = \sqrt{\left(\frac{u - q}{\bar{u} - q}\right)} Y_1,$$

$$\bar{Z}_1 = \sqrt{\left(\frac{u - r}{\bar{u} - r}\right)} Z_1;$$

then it will presently be shown that the points Q, $\bar{Q}$ are corresponding points, which will prove the geometrical theorem of Jacobi. Before proceeding further it will be convenient to notice the formulæ

$$1 - \frac{a^2}{f + \bar{u}} - \frac{b^2}{g + \bar{u}} - \frac{c^2}{h + \bar{u}} = \frac{\bar{u} - u}{\bar{a}_1^2},$$

$$\frac{Xa}{f + \bar{u}} + \frac{Yb}{g + \bar{u}} + \frac{Zc}{h + \bar{u}} = \frac{\bar{u} - u}{\bar{a}_1^2} \left(\frac{X_1 a_1}{\bar{u} - u} + \frac{Y_1 b_1}{\bar{u} - q} + \frac{Z_1 c_1}{\bar{u} - r}\right),$$

$$\left(\frac{Xa}{f + \bar{u}} + \frac{Yb}{g + \bar{u}} + \frac{Zc}{h + \bar{u}}\right)^2 + \frac{\bar{u} - u}{\bar{a}_1^2} \left(\frac{X^2}{f + \bar{u}} + \frac{Y^2}{g + \bar{u}} + \frac{Z^2}{h + \bar{u}}\right)$$

$$= \frac{\bar{u} - u}{\bar{a}_1^2} \left(\frac{X_1^2}{\bar{u} - u} + \frac{Y_1^2}{\bar{u} - q} + \frac{Z_1^2}{\bar{u} - r}\right) = \frac{\bar{X}_1^2}{\bar{a}_1^2};$$

and the corresponding ones

$$1-\frac{\bar{a}^2}{f+u}-\frac{\bar{b}^2}{g+u}-\frac{\bar{c}^2}{h+u}=\frac{u-\bar{u}}{a_1^2},$$

$$\frac{\bar{X}\bar{a}}{f+u}+\frac{\bar{Y}\bar{b}}{g+u}+\frac{\bar{Z}\bar{c}}{h+u}=\frac{u-\bar{u}}{a_1^2}\left(\frac{\bar{X}_1\bar{a}_1}{u-\bar{u}}+\frac{\bar{Y}_1\bar{b}_1}{u-q}+\frac{\bar{Z}_1\bar{c}_1}{u-r}\right),$$

$$\left(\frac{\bar{X}\bar{a}}{f+u}+\frac{\bar{Y}\bar{b}}{g+u}+\frac{\bar{Z}\bar{c}}{h+u}\right)^2+\frac{u-\bar{u}}{a_1^2}\left(\frac{\bar{X}^2}{f+u}+\frac{\bar{Y}^2}{g+u}+\frac{\bar{Z}^2}{h+u}\right)$$

$$=\frac{u-\bar{u}}{a_1^2}\left(\frac{\bar{X}_1^2}{u-\bar{u}}+\frac{\bar{Y}_1^2}{u-q}+\frac{\bar{Z}_1^2}{u-r}\right)=\frac{X_1^2}{a_1^2}.$$

The coordinates of the point $\bar{Q}$ are obviously $a+\rho X$, $b+\rho Y$, $c+\rho Z$ (where $\rho=P\bar{Q}$); substituting these values in the equation of the interior ellipsoid, we obtain

$$\rho^2\left(\frac{X^2}{f+\bar{u}}+\frac{Y^2}{g+\bar{u}}+\frac{Z^2}{h+\bar{u}}\right)+2\rho\left(\frac{Xa}{f+\bar{u}}+\frac{Yb}{g+\bar{u}}+\frac{Zc}{h+\bar{u}}\right)+\left(\frac{a^2}{f+u}+\frac{b^2}{g+\bar{u}}+\frac{c^2}{h+\bar{u}}-1\right)=0:$$

reducing the coefficients of this equation by the formulæ first given, and omitting a factor $\frac{\bar{u}-u}{\bar{a}_1^2}$, we obtain

$$\left\{\frac{\bar{a}_1^2\bar{X}^2}{(\bar{u}-u)^2}-\left(\frac{X_1a_1}{\bar{u}-u}+\frac{Y_1b_1}{\bar{u}_1-q}+\frac{Z_1c_1}{\bar{u}-r}\right)^2\right\}\rho^2+2\rho\left(\frac{X_1a_1}{u-u}+\frac{Y_1b_1}{u-q}+\frac{Z_1c_1}{\bar{u}-r}\right)-1=0,$$

that is,
$$\frac{a_1^2\bar{X}_1^2}{(\bar{u}-u)^2}\rho^2=\left\{\rho\left(\frac{X_1a_1}{\bar{u}-u}+\frac{Y_1b_1}{\bar{u}-q}+\frac{Z_1c_1}{\bar{u}-r}\right)-1\right\}^2,$$

or
$$\rho=\frac{1}{\frac{X_1a_1-\bar{X}_1\bar{a}_1}{\bar{u}-u}+\frac{Y_1a_1}{\bar{u}-q}+\frac{Z_1c_1}{\bar{u}-r}};$$

which is easily transformed into

$$\rho=\frac{\bar{u}-u}{a_1X_1-\bar{a}_1X_1+\frac{(f+u)\,b_1Y_1-(f+\bar{u})\,\bar{b}_1\bar{Y}_1}{f+q}+\frac{(f+u)\,c_1Z_1-(f+\bar{u})\,\bar{c}_1\bar{Z}_1}{f+r}}:$$

and this form remaining unaltered when u and $\bar{u}$ are interchanged, it follows that if $\bar{P}Q=\bar{\rho}$, then $\rho=\bar{\rho}$, which is a known theorem. The value of ρ or $\bar{\rho}$ may however be expressed in a yet simpler form; for, considering the expression

$$\frac{X}{\sqrt{(f+\bar{u})}}-\frac{\bar{X}}{\sqrt{(f+u)}}=\frac{a}{\sqrt{(f+\bar{u})}}\left\{\frac{a_1X_1}{f+u}+\frac{b_1Y_1}{f+q}+\frac{c_1Z_1}{f+r}\right\}-\frac{\bar{a}}{\sqrt{(f+u)}}\left\{\frac{\bar{a}_1\bar{X}_1}{f+\bar{u}}+\frac{\bar{b}_1\bar{Y}_1}{f+q}+\frac{\bar{c}_1\bar{Z}_1}{f+r}\right\}$$

$$=\frac{-1}{\bar{u}-u}\left\{\frac{a}{\sqrt{(f+u)}}-\frac{\bar{a}}{\sqrt{(f+u)}}\right\}$$

$$\times\left\{a_1X_1-\bar{a}_1\bar{X}_1+\frac{(f+u)\,b_1Y_1-(f+\bar{u})\,\bar{b}_1\bar{Y}_1}{f+q}+\frac{(f+u)\,c_1Z_1-(f+\bar{u})\,\bar{c}_1\bar{Z}_1}{f+r}\right\},$$

we see that

$$\frac{X}{\sqrt{(f+\bar{u})}} - \frac{\bar{X}}{\sqrt{(f+u)}} = -\frac{1}{\rho}\left(\frac{a}{\sqrt{(f+\bar{u})}} - \frac{\bar{a}}{\sqrt{(f+u)}}\right),$$

and similarly

$$\frac{Y}{\sqrt{(f+\bar{u})}} - \frac{\bar{Y}}{\sqrt{(g+u)}} = -\frac{1}{\rho}\left(\frac{b}{\sqrt{(g+\bar{u})}} - \frac{\bar{b}}{\sqrt{(g+u)}}\right),$$

$$\frac{Z}{\sqrt{(h+\bar{u})}} - \frac{\bar{Z}}{\sqrt{(h+u)}} = -\frac{1}{\rho}\left(\frac{c}{\sqrt{(h+\bar{u})}} - \frac{\bar{c}}{\sqrt{(h+u)}}\right);$$

which are in fact the equations which express that Q and $\bar{Q}$ are corresponding points.

It is proper to remark that supposing, as we are at liberty to do, that P, $\bar{P}$ are situate in corresponding octants of the two ellipsoids, then if the curve of contact of the circumscribed cone having P for its vertex divide the surface of the interior ellipsoid into two parts $\bar{M}$, $\bar{N}$, of which the former lies contiguous to $\bar{P}$: also if the curve of intersection of the tangent plane at $\bar{P}$ divide the surface of the exterior ellipsoid into two parts M, N, of which M lies contiguous to the point P; then the different points of M, $\bar{M}$ correspond to each other, as do also the different points of N, $\bar{N}$.

We may now pass to the integral calculus problem. The Attraction parallel to the axis of x is

$$A = \int \frac{x dx\, dy\, dz}{(x^2+y^2+z^2)^{\frac{3}{2}}},$$

the limits of the integration being given by

$$\frac{(x+a)^2}{f+\bar{u}} + \frac{(y+b)^2}{g+\bar{u}} + \frac{(z+c)^2}{h+\bar{u}} = 1;$$

or putting

$$x = rX, \quad y = rY, \quad z = rZ,$$

where X, Y, Z have the same signification, as before, we have

$$dx\, dy\, dz = r^2 dr\, dS,$$

and then

$$A = \int X dr\, dS = \int \rho X\, dS,$$

where ρ has the same signification as before: it will be convenient to leave the formula in this form, rather than to take at once the difference of the two values of ρ, but of course the integration is as in the ordinary methods to be performed so as to extend to the whole volume of the ellipsoid. The expression dS denotes the differential of a spherical surface radius unity, and if θ, ϕ are the parameters by which the position of ρ is determined, we have

$$dS = \begin{vmatrix} X, & Y, & Z \\ \frac{dX}{d\theta}, & \frac{dY}{d\theta}, & \frac{dZ}{d\theta} \\ \frac{dX}{d\phi}, & \frac{dY}{d\phi}, & \frac{dZ}{d\phi} \end{vmatrix} d\theta\, d\phi.$$

In the present case

$$dS = dS_1 = \begin{vmatrix} X_1, & Y_1, & Z_1 \\ \frac{dX_1}{d\bar{Y}_1}, & \frac{dY_1}{d\bar{Y}_1}, & \frac{dZ_1}{d\bar{Y}_1} \\ \frac{dX_1}{d\bar{Z}_1}, & \frac{dY_1}{d\bar{Z}_1}, & \frac{dZ_1}{d\bar{Z}_1} \end{vmatrix} d\bar{Y}_1 d\bar{Z}_1,$$

or from the values of X_1, Y_1, Z_1 in terms of $\bar{X}_1$, $\bar{Y}_1$, $\bar{Z}_1$ {observing that $\bar{X}_1$ must be replaced by its value $\sqrt{(1-\bar{Y}_1^2-\bar{Z}_1^2)}$} we deduce

$$dS = \sqrt{\left\{\frac{(\bar{u}-q)(\bar{u}-r)}{(u-q)(u-r)}\right\}}\,\frac{1}{X_1}\,d\bar{Y}_1 d\bar{Z}_1.$$

But $d\bar{S} = d\bar{S}_1 = \dfrac{1}{\bar{X}_1}\, d\bar{Y}_1 d\bar{Z}_1$,

whence

$$dS = \sqrt{\left\{\frac{(\bar{u}-q)(\bar{u}-r)}{(u-q)(u-r)}\right\}}\,\frac{\bar{X}_1 d\bar{S}}{X_1},$$

which shows that the corresponding elements of the spheres whose centres are P, $\bar{P}$, projected upon the tangent planes at P and $\bar{P}$ respectively, are in a constant ratio. It may be noticed also that if μ, $\bar{\mu}$ are the masses of the ellipsoids, the ratio in question

$$= \sqrt{\left\{\frac{(\bar{u}-q)(\bar{u}-r)}{(u-q)(u-r)}\right\}} = \frac{\bar{\mu}a_1}{\mu\bar{a}_1}.$$

We have thus

$$A = \sqrt{\left\{\frac{(\bar{u}-q)(\bar{u}-r)}{(u-q)(u-r)}\right\}}\int\frac{\bar{\rho}\bar{X}_1 X d\bar{S}}{X_1},$$

that is

$$A = \sqrt{\left\{\frac{(\bar{u}-q)(\bar{u}-r)}{(u-q)(u-r)}\right\}}\int\frac{\bar{\rho}\,(\alpha X_1 + \alpha' Y_1 + \alpha'' Z_1)\,\bar{X}_1 d\bar{S}}{X_1}.$$

The value which it will be convenient to use for $\bar{\rho}$ is that derived from the equation

$$\bar{\rho}^2\left(\frac{\bar{X}^2}{f+u}+\frac{\bar{Y}^2}{g+u}+\frac{\bar{Z}^2}{h+u}\right)+2\bar{\rho}\left(\frac{\bar{X}\bar{a}}{f+u}+\frac{\bar{Y}\bar{b}}{g+u}+\frac{\bar{Z}\bar{c}}{h+u}\right)+\frac{\bar{u}-u}{a_1^2}=0,$$

with only the transformation of expressing the radical in terms of X_1, viz.

$$\rho = \frac{\dfrac{\bar{X}\bar{a}}{f+u} + \dfrac{\bar{Y}\bar{b}}{g+u} + \dfrac{\bar{Z}\bar{c}}{h+u} + \dfrac{1}{a_1}X_1}{\dfrac{\bar{X}^2}{f+u} + \dfrac{\bar{Y}^2}{g+u} + \dfrac{\bar{Z}^2}{h+u}};$$

substituting these values and observing that Y_1 and Z_1 are rational functions of $\bar{X}$, $\bar{Y}$, and $\bar{Z}$, but that $\bar{X}_1$ is a radical, and that in order to extend the integration to the whole ellipsoid, the values corresponding to the opposite signs of X_1 will require to be added, the quantity to be integrated (omitting for the moment the exterior constant factor) is

$$= \frac{\left\{\alpha\left(\dfrac{\bar{X}\bar{a}}{f+u} + \dfrac{\bar{Y}\bar{b}}{g+u} + \dfrac{\bar{Z}\bar{c}}{h+u}\right) + \dfrac{1}{a_1}(\alpha' Y_1 + \alpha'' Z_1)\right\}\bar{X}_1 d\bar{S}}{\dfrac{\bar{X}^2}{f+u} + \dfrac{\bar{Y}^2}{g+u} + \dfrac{\bar{Z}^2}{h+u}},$$

the integration to be extended over the spherical area $\bar{S}$. Consider the quantity within { }, this is

$$\alpha\left(\frac{\bar{X}\bar{a}}{f+u} + \frac{\bar{Y}\bar{b}}{g+u} + \frac{\bar{Z}\bar{c}}{h+u}\right) + \frac{\alpha'}{a_1}\sqrt{\left(\frac{q-\bar{u}}{q-u}\right)}(\bar{\alpha}'\bar{X} + \bar{\beta}'\bar{Y} + \bar{\gamma}'\bar{Z}) + \frac{\alpha''}{a_,}\sqrt{\left(\frac{r-\bar{u}}{r-u}\right)}(\bar{\alpha}''\bar{X} + \bar{\beta}''\bar{Y} + \bar{\gamma}''\bar{Z}).$$

The coefficients of $\bar{Y}$ and $\bar{Z}$ vanish, in fact that of $\bar{Y}$ is

$$\frac{a_1 a}{f+u}\frac{\bar{b}}{g+u} + \frac{b_1 a}{a_1(f+q)}\sqrt{\left(\frac{q-\bar{u}}{q-u}\right)}\frac{\bar{b}_1\bar{b}}{g+q} + \frac{c_1 a}{a_1(f+r)}\sqrt{\left(\frac{r-\bar{u}}{r-u}\right)}\frac{\bar{c}_1\bar{b}}{g+r}$$

$$= \frac{a\bar{b}}{a_1}\left\{\frac{a_1^2}{(f+u)(g+u)} + \frac{b_1\bar{b}_1}{(f+q)(g+q)}\sqrt{\left(\frac{q-\bar{u}}{q-u}\right)} + \frac{c_1\bar{c}_1}{(f+r)(g+r)}\sqrt{\left(\frac{r-\bar{u}}{r-u}\right)}\right\}$$

$$= \frac{ab_1}{a_1}\left\{\frac{a_1^2}{(f+u)(g+u)} + \frac{b_1^2}{(f+q)(g+q)} + \frac{c_1^2}{(f+r)(g+r)}\right\} = 0;$$

and similarly for the coefficient of $\bar{Z}$.

The coefficient of $\bar{X}$ is in like manner shown to be

$$\frac{a\bar{a}}{a_1}\left\{\frac{a_1^2}{(f+u)^2} + \frac{b_1^2}{(f+q)^2} + \frac{c_1^2}{(f+r)^2}\right\} = \frac{a\bar{a}}{a_1}\frac{(f-g)(f-h)}{(f+u)(f+q)(f+r)} = \frac{a\bar{a}}{a^2 a_1} = \frac{\bar{a}}{aa_1};$$

or the quantity in question is simply

$$\frac{\bar{a}}{aa_1}\bar{X}.$$

Multiplying this by $\bar{X}_1, = \bar{\alpha} X + \bar{\alpha}' Y + \bar{\alpha}'' Z$, the terms containing $\bar{X}\bar{Y}$, $\bar{X}\bar{Z}$ vanish after the integration, and we need only consider the term $\frac{\bar{a}\,\bar{\alpha}}{a a_1} \bar{X}^2$, or what is the same $\frac{\bar{a}^2 \bar{a}_1}{a a_1 (f+\bar{u})} \bar{X}^2$. Hence

$$A = \sqrt{\left\{\frac{(\bar{u}-q)(\bar{u}-r)}{(u+q)(u-r)}\right\}} \frac{\bar{a}^2\bar{a}_1}{a a_1 (f+\bar{u})} \int \frac{\bar{X}^2 d\bar{S}}{\frac{\bar{X}^2}{f+u} + \frac{\bar{Y}^2}{g+u} + \frac{\bar{Z}^2}{h+u}}.$$

The value of the corresponding function $\bar{A}$ (that is, the attraction of the exterior ellipsoid upon $\bar{P}$) is

$$\bar{A} = \frac{\bar{a}}{f+u} \int \frac{\bar{X}^2 d\bar{S}}{\frac{\bar{X}^2}{f+u} + \frac{\bar{Y}^2}{g+u} + \frac{\bar{Z}^2}{h+u}},$$

the limits being the same, whence

$$A \div \bar{A} = \sqrt{\left\{\frac{(\bar{u}-q)(\bar{u}-r)}{(u-q)(u-r)}\right\}} \frac{f+u}{f+\bar{u}} \frac{\bar{a}\bar{a}_1}{a a_1} = \frac{\sqrt{(\bar{u}+g)}\sqrt{(\bar{u}+h)}}{\sqrt{(u+g)}\sqrt{(u+h)}};$$

or we have

$$A = \frac{\sqrt{(\bar{u}+g)}\sqrt{(\bar{u}+h)}}{\sqrt{(u+g)}\sqrt{(u+h)}} \bar{A}, \quad B = \frac{\sqrt{(\bar{u}+h)}\sqrt{(\bar{u}+f)}}{\sqrt{(u+h)}\sqrt{(\bar{u}+f)}} \bar{B}, \quad C = \frac{\sqrt{(\bar{u}+f)}\sqrt{(\bar{u}+g)}}{\sqrt{(u+f)}\sqrt{(u+g)}} \bar{C},$$

formulæ which constitute in fact Ivory's theorem.

Let K, $\bar{K}$ denote the attractions in the directions of the normals at P, $\bar{P}$, we have

$$K = \frac{\bar{\mu} a_1}{\mu \bar{a}_1} \int \bar{X}_1 d\bar{S}, \quad \bar{K} = \int \bar{X}_1 d\bar{S},$$

or

$$K = \frac{\bar{\mu} a_1}{\mu \bar{a}_1} K;$$

and it is important to remark that this is true not only for the entire ellipsoids; but if $\mathfrak{M}$, $\mathfrak{N}$ denote the attractions of the cones standing on the portions $\bar{M}$, $\bar{N}$ of the surface of the interior ellipsoid, and $\bar{\mathfrak{M}}$, $\bar{\mathfrak{N}}$ the attractions of the portions of the exterior ellipsoid bounded by the tangent plane at $\bar{P}$, and the portions M, N of the surface of the exterior ellipsoid, then

$$\mathfrak{M} = -\frac{\bar{\mu} a_1}{\mu a_1} \bar{\mathfrak{M}}, \quad \mathfrak{N} = \frac{\bar{\mu} a_1}{\mu a_1} \bar{\mathfrak{N}},$$

where obviously

$$K = \mathfrak{N} - \mathfrak{M}, \quad \bar{K} = \bar{\mathfrak{N}} + \bar{\mathfrak{M}};$$

this theorem is so far as I am aware new.

90.

NOTE SUR QUELQUES FORMULES RELATIVES AUX CONIQUES.

[From the *Journal für die reine und angewandte Mathematik* (Crelle), tom. XXXIX. (1850), pp. 1—3.]

SOIT, comme à l'ordinaire:

$$\left.\begin{aligned}
\mathfrak{A} &= BC - F^2, \\
\mathfrak{B} &= CA - G^2, \\
\mathfrak{C} &= AB - H^2, \\
\mathfrak{F} &= GH - AF, \\
\mathfrak{G} &= HF - BG, \\
\mathfrak{H} &= FG - CH, \\
K &= ABC - AF^2 - BG^2 - CH^2 + 2FGH,
\end{aligned}\right\} \quad \ldots\ldots\ldots\ldots\ldots\ldots(1)$$

et désignons, pour abréger, les fonctions

$$Ax^2 + By^2 + Cz^2 + 2Fyz + 2Gxz + 2Hxy,$$

$$A\alpha x + B\beta y + C\gamma z + F(\beta z + \gamma y) + G(\gamma x + \alpha z) + H(\alpha y + \beta x),$$

$$A(\beta z - \gamma y)^2 + B(\gamma x - \alpha z)^2 + C(\alpha y - \beta x)^2 + 2F(\gamma x - \alpha z)(\alpha y - \beta x)$$
$$+ 2G(\alpha y - \beta x)(\beta z - \gamma y) + 2H(\beta z - \gamma y)(\gamma x - \alpha z);$$

&c.

par $\qquad Ax^2 + \ldots\,; \quad A\alpha x + \ldots\,; \quad A(\beta z - \gamma y)^2 + \ldots\,;$ &c.

Cela posé, soient $\mathfrak{A} + \mathrm{a}$, $\mathfrak{B} + \mathrm{b}$, $\mathfrak{C} + \mathrm{c}$, $\mathfrak{F} + \mathrm{f}$, $\mathfrak{G} + \mathrm{g}$, $\mathfrak{H} + \mathrm{h}$, $K + \mathrm{k}$ ce que deviennent $\mathfrak{A}$, $\mathfrak{B}$, $\mathfrak{C}$, $\mathfrak{F}$, $\mathfrak{G}$, $\mathfrak{H}$, K, en écrivant $A + \alpha^2$, $B + \beta^2$, $C + \gamma^2$, $F + \beta\gamma$, $G + \gamma\alpha$, $H + \alpha\beta$ au lieu de A, B, C, F, G, H. Alors on aura d'abord

$$\mathrm{k} = \mathfrak{A}\alpha^2 + \ldots \quad \ldots\ldots\ldots\ldots\ldots\ldots\ldots\ldots\ldots\ldots(2)$$

et les quantités a, b, c, f, g, h seront données par l'équation

$$\mathrm{a}x^2 + \ldots = A\,(\beta z - \gamma y)^2 + \ldots \quad \ldots\ldots\ldots\ldots\ldots\ldots\ldots\ldots(3),$$

ou, si l'on veut, par celle-ci:

$$\mathrm{a}xx_1 + \ldots = A\,(\beta z - \gamma y)\,(\beta z_1 - \gamma y_1) + \ldots \quad \ldots\ldots\ldots\ldots\ldots\ldots(4),$$

(savoir, en considérant ces équations comme identiques par rapport à x, y, z et x_1, y_1, z_1 respectivement).

On obtient sans difficulté les équations identiques:

$$(A\alpha\alpha_1 + \ldots)\,(Axx_1 + \ldots) - (A\alpha x + \ldots)\,(A\alpha_1 x_1 + \ldots) = \mathfrak{A}\,(\beta z_1 - \gamma y_1)\,(\beta_1 z - \gamma_1 y) + \ldots, \quad (5),$$

$$(\mathfrak{A}\alpha\alpha_1 + \ldots)\,(\mathfrak{A}xx_1 + \ldots) - (\mathfrak{A}\alpha x + \ldots)\,(\mathfrak{A}\alpha_1 x_1 + \ldots) = K\,[A\,(\beta z_1 - \gamma y_1)\,(\beta_1 z - \gamma_1 y) + \ldots] \quad (6).$$

Comme on sait, la condition sous laquelle la droite $lx + my + nz = 0$ touche la conique $U = Ax^2 + \ldots = 0$ peut être présentée sous la forme $\mathfrak{A}l^2 + \ldots = 0$. Donc la condition pour que cette droite touche la conique $U + (\alpha x + \beta y + \gamma z)^2 = 0$, est

$$\mathfrak{A}l^2 + \ldots + A\,(\gamma m - \beta n)^2 + \ldots = 0 \quad \ldots\ldots\ldots\ldots\ldots\ldots\ldots\ldots \quad (7).$$

En réduisant au moyen de l'équation $(\mathfrak{A}\alpha^2 + \ldots)\,(\mathfrak{A}l^2 + \ldots) = K\,[A\,(\gamma m - \beta n)^2 + \ldots]$ (laquelle n'est qu'un cas particulier de l'équation (6)), cette condition devient:

$$(K + \mathfrak{A}\alpha^2 + \ldots)\,(\mathfrak{A}l^2 + \ldots) - (\mathfrak{A}\alpha l + \ldots)^2 = 0 \ldots\ldots\ldots\ldots\ldots\ldots(8).$$

Pour trouver la condition sous laquelle les coniques

$$U + (\alpha x + \beta y + \gamma z)^2 = 0, \quad U + (\alpha_1 x + \beta_1 y + \gamma_1 z)^2 = 0$$

se touchent, on n'a qu'à remarquer que l'équation de la tangente commune est, ou

$$(\alpha - \alpha_1)\,x + (\beta - \beta_1)\,y + (\gamma - \gamma_1)\,z = 0, \quad \text{ou} \quad (\alpha + \alpha_1)\,x + (\beta + \beta_1)\,y + (\gamma + \gamma_1)\,z = 0.$$

En ne considérant que la première de ces droites, on a pour la condition cherchée:

$$\mathfrak{A}\,(\alpha - \alpha_1)^2 + \ldots + A\,(\beta\gamma_1 - \beta_1\gamma)^2 + \ldots = 0 \ldots\ldots\ldots\ldots\ldots\ldots\ldots\ldots\ldots(9);$$

ou, en réduisant au moyen du même cas particulier de l'équation (6), on obtient cette condition sous la forme

$$(K + \mathfrak{A}\alpha^2 + \ldots)\,(K + \mathfrak{A}\alpha_1^2 + \ldots) - (K + \mathfrak{A}\alpha\alpha_1 + \ldots)^2 = 0 \ldots\ldots\ldots\ldots(10).$$

On sait que les coordonnées X, Y, Z du pôle de la droite $lx + my + nz = 0$, par rapport à la conique $U = 0$, peuvent être trouvées par l'équation

$$\lambda X + \mu Y + \nu Z = \mathfrak{A}l\lambda + \ldots,$$

(considérée comme identique par rapport à λ, μ, ν). De là les coordonnées X, Y, Z du pôle de cette même droite par rapport à la conique

$$U + (\alpha x + \beta y + \gamma z)^2 = 0$$

se trouveront par l'équation

$$\lambda X + \mu Y + \nu Z = \mathfrak{A} l\lambda + \ldots + A(\gamma m - \beta n)(\gamma\mu - \beta\nu) \ldots \quad \ldots\ldots\ldots \ (11)$$

(considérée comme identique par rapport à λ, μ, ν). Cette équation peut aussi être présentée sous la forme

$$K(\lambda X + \mu Y + \nu Z) = (K + \mathfrak{A}\alpha^2 + \ldots)(\mathfrak{A} l\lambda + \ldots) - (\mathfrak{A}\alpha\lambda + \ldots)(\mathfrak{A}\alpha l + \ldots) \quad \ldots \ (12),$$

ce qui peut être démontré facilement au moyen d'un cas particulier de l'équation (6).

Si les deux droites $lx + my + nz = 0$, $l'x + m'y + n'z = 0$, touchent la conique $U = 0$, l'équation de la droite qui passe par les points de contact sera $A(mn' - m'n)x + \ldots = 0$. Donc: si ces deux droites touchent la conique $U + (\alpha x + \beta y + \gamma z)^2 = 0$, l'équation de la droite qui passe par les deux points de contact sera

$$A(mn' - m'n)x + \ldots + (\alpha x + \beta y + \gamma z)[\alpha(mn' - m'n) + \beta(nl' - n'l) + \gamma(lm' - l'm)] = 0 \ \ldots \ (13).$$

Les formules obtenues seront utiles pour la solution du problème du mémoire suivant, [91]. Je les ai rapprochées ici pour ne pas interrompre cette solution.

91.

SUR LE PROBLÈME DES CONTACTS.

[From the *Journal für die reine und angewandte Mathematik* (Crelle), tom. XXXIX. (1850), pp. 4—13.]

JE me propose ici la solution analytique du problème suivant:

"Étant données trois coniques inscrites à une même conique: trouver une autre conique, aussi inscrite à cette conique, qui touche les trois coniques inscrites; et tirer de là les constructions géométriques ordinaires."

Je commence par récapituler quelques-unes des propriétés d'un système de trois coniques inscrites à la même conique.

Un système de six droites qui passent trois à trois par quatre points, s'appelle *quadrangle*. Les points de rencontre des côtés opposés sont les *centres* du quadrangle; les *côtés* du triangle formé par ces trois centres sont les *axes* du quadrangle. De même, un système de six points situés trois à trois sur quatre droites, s'appelle *quadrilatère*. Les droites qui passent par les angles opposés sont les *axes;* et les angles du triangle formé par les trois axes sont les *centres* du quadrilatère.

Deux coniques quelconques se coupent en quatre points qui forment un quadrangle inscrit aux deux coniques. Elles ont quatre tangentes communes qui forment un quadrilatère circonscrit aux deux coniques. Le quadrangle inscrit et le quadrilatère circonscrit ont les même centres et les mêmes axes.

Si deux coniques sont circonscrites ou inscrites l'une à l'autre, la droite qui passe par les deux points de contact s'appelle *chorde de contact*, et le point de rencontre des deux tangentes communes s'appelle *centre de contact.*

Cela posé: les coniques circonscrites à deux coniques données, peuvent être divisées en trois classes: une conique circonscrite appartient à une quelconque de ces trois classes, selon que les points de rencontre des chordes de contact de la conique circonscrite et de chacune des deux coniques données coïncide avec un quelconque des trois centres du quadrangle inscrit, ou du quadrilatère circonscrit; ou, si l'on veut, selon que la droite qui passe par les centres de contact de la conique circonscrite et des deux coniques données, coïncide avec un quelconque des trois axes du quadrangle inscrit, ou du quadrilatère circonscrit.

En considérant les deux coniques données, et une conique circonscrite, nous dirons que les deux côtés du quadrangle inscrit qui se coupent dans le point d'intersection des deux chordes de contact, sont les *axes de symptose* des deux coniques données, et que les deux angles du quadrilatère circonscrit, situés sur la droite qui passe par les deux centres de contact, sont les *centres d'homologie* des deux coniques données.

Soient maintenant inscrites trois coniques à la même conique. En combinant deux à deux ces trois coniques, les six axes de symptose se couperont trois à trois en quatre points que nous nommerons *centres de symptose*, et les six centres d'homologie seront situés trois à trois sur quatre droites que nous appelerons *axes d'homologie.*

Soient

$$U+V_1^2=0, \quad U+V_2^2=0, \quad U+V_3^2=0$$

les équations des trois coniques inscrites, où

$$\begin{aligned} U &= Ax^2+By^2+Cz^2+2Fyz+2Gxz+2Hxy, \\ V_1 &= \alpha_1 x + \beta_1 y + \gamma_1 z, \\ V_2 &= \alpha_2 x + \beta_2 y + \gamma_2 z, \\ V_3 &= \alpha_3 x + \beta_3 y + \gamma_3 z. \end{aligned}$$

Si

$$lx+my+nz=0$$

est l'équation d'une tangente commune aux coniques $U+V_1^2=0$, $U+V_2^2=0$, les formules de la "Note sur quelques formules &c." [90], en adoptant la notation de cette note, donneront les équations

$$\begin{aligned} (K+\mathfrak{A}\alpha_1^2+\ldots)(\mathfrak{A}l^2+\ldots)-(\mathfrak{A}\alpha_1 l+\ldots)^2=0, \\ (K+\mathfrak{A}\alpha_2^2+\ldots)(\mathfrak{A}l^2+\ldots)-(\mathfrak{A}\alpha_2 l+\ldots)^2=0, \end{aligned}$$

qui serviront à déterminer les valeurs de l, m, n; on obtient par là l'éxpression

$$\surd(K+\mathfrak{A}\alpha_2^2+\ldots)(\mathfrak{A}\alpha_1 l+\ldots)-\surd(K+\mathfrak{A}\alpha_1^2+\ldots)(\mathfrak{A}\alpha_2 l+\ldots),$$

qui fait voir que l'équation $lx+my+nz=0$ est satisfaite en écrivant

$$\begin{aligned} x:y:z = {} & \surd(K+\mathfrak{A}\alpha_1^2+\ldots)(\mathfrak{A}\,\alpha_1+\mathfrak{H}\beta_1+\mathfrak{G}\,\gamma_1)-\surd(K+\mathfrak{A}\alpha_2^2+\ldots)(\mathfrak{A}\,\alpha_2+\mathfrak{H}\beta_2+\mathfrak{G}\gamma_2)\ldots(1) \\ : {} & \surd(K+\mathfrak{A}\alpha_1^2+\ldots)(\mathfrak{H}\alpha_1+\mathfrak{B}\,\beta_1+\mathfrak{F}\,\gamma_1)-\surd(K+\mathfrak{A}\alpha_2^2+\ldots)(\mathfrak{H}\alpha_2+\mathfrak{B}\,\beta_2+\mathfrak{F}\gamma_2) \\ : {} & \surd(K+\mathfrak{A}\alpha_1^2+\ldots)(\mathfrak{G}\,\alpha_1+\mathfrak{F}\,\beta_1+\mathfrak{C}\,\gamma_1)-\surd(K+\mathfrak{A}\alpha_2^2+\ldots)(\mathfrak{G}\,\alpha_2+\mathfrak{F}\,\beta_2+\mathfrak{C}\gamma_2), \end{aligned}$$

et ces équations, qui peuvent aussi être présentées sous la forme plus simple

$$Ax+Hy+Gz : Hx+By+Fz : Gx+Fy+Cz \ldots\ldots (2),$$
$$= \sqrt{(K+\mathfrak{A}\alpha_1^2\ldots)}\,\alpha_2 - \sqrt{(K+\mathfrak{A}\alpha_2^2\ldots)}\,\alpha_1$$
$$: \sqrt{(K+\mathfrak{A}\alpha_1^2\ldots)}\,\beta_2 - \sqrt{(K+\mathfrak{A}\alpha_2^2\ldots)}\,\beta_1$$
$$: \sqrt{(K+\mathfrak{A}\alpha_1^2\ldots)}\,\gamma_2 - \sqrt{(K+\mathfrak{A}\alpha_2^2\ldots)}\,\gamma_1,$$

correspondent à un centre d'homologie des deux coniques $U+V_1^2=0$, $U+V_2^2=0$.

En mettant, pour abréger,

$$\sqrt{(K+\mathfrak{A}\alpha_1^2+\ldots)}=p_1,\quad \sqrt{(K+\mathfrak{A}\alpha_2^2+\ldots)}=p_2,\quad \sqrt{(K+\mathfrak{A}\alpha_3^2+\ldots)}=p_3 \ldots\ldots (3),$$

on obtient facilement, pour un des axes d'homologie des trois coniques, l'équation

$$\begin{vmatrix} & p_1, & p_2, & p_3 \\ Ax+Hy+Gz, & \alpha_1, & \alpha_2, & \alpha_3 \\ Hx+By+Fz, & \beta_1, & \beta_2, & \beta_3 \\ Gx+Fy+Cz, & \gamma_1, & \gamma_2, & \gamma_3 \end{vmatrix} = 0 \ldots\ldots (4),$$

et celles des trois autres axes d'homologie en peuvent être tirées en changeant les signes de p_1, p_2, p_3.

Remarquons que l'équation

$$\begin{vmatrix} & 1, & 1, & 1 \\ Ax+Hy+Gz, & \alpha_1, & \alpha_2, & \alpha_3 \\ Hx+By+Fz, & \beta_1, & \beta_2, & \beta_3 \\ Gx+Fy+Cz, & \gamma_1, & \gamma_2, & \gamma_3 \end{vmatrix} = 0 \ldots\ldots (5)$$

est celle de la polaire d'un des centres de symptose des trois coniques par rapport à la conique circonscrite $U=0$, savoir de celle qui est donnée par le système $V_1=V_2=V_3$, et que les équations des trois autres polaires correspondantes se trouveront en changeant les signes de α_1, β_1, γ_1, ou de α_2, β_2, γ_2, ou de α_3, β_3, γ_3.

Cherchons le pôle de l'axe d'homologie dont nous venons de trouver l'équation, par rapport à une quatrième conique inscrite $U+V^2=0$ $(V=\alpha x+\beta y+\gamma z)$. En exprimant cette équation par $lx+my+nz=0$, on obtiendra les coordonnées de ce pôle au moyen de l'équation

$$K(\lambda X+\mu Y+\nu Z)=p^2(\mathfrak{A}l\lambda+\ldots)-(\mathfrak{A}\alpha\lambda+\ldots)(\mathfrak{A}\alpha l+\ldots) \ldots\ldots (6),$$

où, pour abréger, on a mis $p=\sqrt{(K+\mathfrak{A}\alpha^2+\ldots)}$. (Mémoire cité; équation (12).) Mais ici on a

$$lx+my+nz=\begin{vmatrix} & p_1, & p_2, & p_3 \\ Ax+Hy+Gz, & \alpha_1, & \alpha_2, & \alpha_3 \\ Hx+By+Fz, & \beta_1, & \beta_2, & \beta_3 \\ Gx+Fy+Cz, & \gamma_1, & \gamma_2, & \gamma_3 \end{vmatrix}$$

ce qui donne immédiatement

$$\mathfrak{A}l\lambda + \ldots = K \begin{vmatrix} & p_1, & p_2, & p_3 \\ \lambda, & \alpha_1, & \alpha_2, & \alpha_3 \\ \mu, & \beta_1, & \beta_2, & \beta_3 \\ \nu, & \gamma_1, & \gamma_2, & \gamma_3 \end{vmatrix}, \quad \mathfrak{A}\alpha l + \ldots = K \begin{vmatrix} & p_1, & p_2, & p_3 \\ \alpha, & \alpha_1, & \alpha_2, & \alpha_3 \\ \beta, & \beta_1, & \beta_2, & \beta_3 \\ \gamma, & \gamma_1, & \gamma_2, & \gamma_3 \end{vmatrix};$$

et de là on obtient

$$\lambda X + \mu Y + \nu Z = p^2 \begin{vmatrix} & p_1, & p_2, & p_3 \\ \lambda, & \alpha_1, & \alpha_2, & \alpha_3 \\ \mu, & \beta_1, & \beta_2, & \beta_3 \\ \nu, & \gamma_1, & \gamma_2, & \gamma_3 \end{vmatrix} - (\mathfrak{A}\alpha\lambda + \ldots) \begin{vmatrix} & p_1, & p_2, & p_3 \\ \alpha, & \alpha_1, & \alpha_2, & \alpha_3 \\ \beta, & \beta_1, & \beta_2, & \beta_3 \\ \gamma, & \gamma_1, & \gamma_2, & \gamma_3 \end{vmatrix} \ldots\ldots(7);$$

savoir, en considérant cette équation comme identique par rapport à λ, μ, ν, on obtient les coordonnées X, Y, Z du point dont il s'agit. En prenant particulièrement ce pôle par rapport à la conique $U + V_2^2 = 0$, cette équation se réduit à

$$\frac{1}{p_1}(\lambda X + \mu Y + \nu Z) = p_1 \begin{vmatrix} & p_1, & p_2, & p_3 \\ \lambda, & \alpha_1, & \alpha_2, & \alpha_3 \\ \mu, & \beta_1, & \beta_2, & \beta_3 \\ \nu, & \gamma_1, & \gamma_2, & \gamma_3 \end{vmatrix} - (\mathfrak{A}\alpha_1\lambda + \ldots) \begin{vmatrix} \alpha_1, & \alpha_2, & \alpha_3 \\ \beta_1, & \beta_2, & \beta_3 \\ \gamma_1, & \gamma_2, & \gamma_3 \end{vmatrix} \ldots\ldots (8),$$

où, si l'on veut, X, Y, Z seront déterminés par les expressions

$$\left.\begin{aligned} \alpha_1 X + \beta_1 Y + \gamma_1 Z &= -p_1^2 \;\; + \mathfrak{A}\alpha_1^2 \;\; + \ldots, = -K \\ \alpha_2 X + \beta_2 Y + \gamma_2 Z &= -p_1 p_2 + \mathfrak{A}\alpha_1\alpha_2 + \ldots, \\ \alpha_3 X + \beta_3 Y + \gamma_3 Z &= -p_1 p_3 + \mathfrak{A}\alpha_1\alpha_3 + \ldots, \end{aligned}\right\} \ldots\ldots\ldots\ldots\ldots\ldots (9).$$

Le facteur $\frac{1}{p_1}$ a été supprimé. De là on obtient aussi l'équation de la droite menée par ce point, savoir par le pôle de l'axe d'homologie par rapport à la conique $U + V_1^2 = 0$, et par le centre de symptose $V_1 = V_2 = V_3$. En effet, cette équation est

$$\begin{vmatrix} V_1, & V_2, & V_3 \\ 1\;, & 1\;, & 1 \\ p_1^2 - \mathfrak{A}\alpha_1^2 - \ldots, & p_1p_2 - \mathfrak{A}\alpha_1\alpha_2 - \ldots, & p_1p_3 - \mathfrak{A}\alpha_1\alpha_3 - \ldots \end{vmatrix} = 0 \ldots\ldots (10).$$

On pourrait également chercher le point d'intersection de la polaire du centre de symptose $V_1 = V_2 = V_3$ par rapport à $U + V_1^2 = 0$, et de l'axe d'homologie; ce point serait évidemment le pôle de la droite exprimée par la dernière équation, par rapport à $U + V_1^2 = 0$.

Ces résultats seront utiles pour l'interprétation de la formule relative à la conique qui touche les trois coniques données et que nous irons chercher maintenant.

Représentons par $U+V^2=0$ l'équation de cette conique; l'équation (10) du mémoire cité donnera les expressions

$$\left.\begin{array}{l}\mathfrak{A}\alpha\alpha_1+\ldots=-K+pp_1,\\ \mathfrak{A}\alpha\alpha_2+\ldots=-K+pp_2,\\ \mathfrak{A}\alpha\alpha_3+\ldots=-K+pp_3,\end{array}\right\}\ldots\ldots\ldots\ldots\ldots\ldots\ldots\ldots\ldots \quad (11),$$

auxquelles nous ajouterons l'équation qui donne la valeur de p, savoir

$$\mathfrak{A}\alpha^2+\ldots=-K+p^2\ldots\ldots\ldots\ldots\ldots\ldots\ldots\ldots\ldots\ldots \quad (12).$$

Il n'y a qu'à substituer dans cette dernière équation les valeurs de α, β, γ que donnent les trois autres. Par là on obtient, pour déterminer p, une équation du second degré et de la forme

$$K^2L-2KpM+p^2N=0\ldots\ldots\ldots\ldots\ldots\ldots\ldots\ldots\ldots(13);$$

c'est-à-dire, en faisant $M^2-NL=\Omega^2$, $\Lambda=\dfrac{L}{\Omega+M}$, on aura $p=\Lambda K$ et de là

$$\left.\begin{array}{l}\mathfrak{A}\alpha\alpha_1+\ldots=-K(\Lambda p_1-1),\\ \mathfrak{A}\alpha\alpha_2+\ldots=-K(\Lambda p_2-1),\\ \mathfrak{A}\alpha\alpha_3+\ldots=-K(\Lambda p_3-1),\end{array}\right\}\ldots\ldots\ldots\ldots\ldots\ldots\ldots\ldots \quad (14),$$

où Λ est une quantité connue, dont la valeur sera donnée dans la suite. Pour le moment il suffit de remarquer qu'en changeant à la fois les signes de p_1, p_2, p_3, Ω, cette quantité Λ ne change que de signe. Au lieu de chercher les valeurs de α, β, γ, il vaut mieux éliminer ces quantités entre ces dernières équations et l'équation $V=\alpha x+\beta y+\gamma z=0$. Cela donne, pour trouver V, l'équation

$$\begin{vmatrix} V, & K(\Lambda p_1-1), & K(\Lambda p_2-1), & K(\Lambda p_3-1) \\ x, & \mathfrak{A}\alpha_1+\mathfrak{H}\beta_1+\mathfrak{G}\gamma_1, & \mathfrak{A}\alpha_2+\mathfrak{H}\beta_2+\mathfrak{G}\gamma_2, & \mathfrak{A}\alpha_3+\mathfrak{H}\beta_3+\mathfrak{G}\gamma_3 \\ y, & \mathfrak{H}\alpha_1+\mathfrak{B}\beta_1+\mathfrak{F}\gamma_1, & \mathfrak{H}\alpha_2+\mathfrak{B}\beta_2+\mathfrak{F}\gamma_2, & \mathfrak{H}\alpha_3+\mathfrak{B}\beta_3+\mathfrak{F}\gamma_3 \\ z, & \mathfrak{G}\alpha_1+\mathfrak{F}\beta_1+\mathfrak{C}\gamma_1, & \mathfrak{G}\alpha_2+\mathfrak{F}\beta_2+\mathfrak{C}\gamma_2, & \mathfrak{G}\alpha_3+\mathfrak{F}\beta_3+\mathfrak{C}\gamma_3 \end{vmatrix}=0\ \ldots\ (15),$$

qui peut aussi être écrite comme suit:

$$\begin{vmatrix} V, & \Lambda p_1-1, & \Lambda p_2-1, & \Lambda p_3-1 \\ Ax+Hy+Gz, & \alpha_1, & \alpha_2, & \alpha_3 \\ Hx+By+Fz, & \beta_1, & \beta_2, & \beta_3 \\ Gx+Fy+Cz, & \gamma_1, & \gamma_2, & \gamma_3 \end{vmatrix}=0\ldots\ldots\ldots\ldots \quad (16).$$

En mettant $V=0$, on aura l'équation de la chorde de contact de la conique cherchée et de la conique circonscrite $U=0$. En remarquant que l'équation de la conique cherchée peut être mise sous la forme $V = \sqrt{-U}$, l'équation de cétte conique se présentera sous la forme très simple:

$$\begin{vmatrix} \sqrt{-U}, & \Lambda p_1 - 1, & \Lambda p_2 - 1, & \Lambda p_3 - 1 \\ Ax + Hy + Gz, & \alpha_1, & \alpha_2, & \alpha_3 \\ Hx + By + Fz, & \beta_1, & \beta_2, & \beta_3 \\ Gx + Fy + Cz, & \gamma_1, & \gamma_2, & \gamma_3 \end{vmatrix} = 0 \text{.}\ldots\ldots\ldots\ldots \quad (17),$$

ou enfin, si l'on veut, sous la forme plus usitée:

$$\begin{vmatrix} \alpha_1, & \alpha_2, & \alpha_3 \\ \beta_1, & \beta_2, & \beta_3 \\ \gamma_1, & \gamma_2, & \gamma_3 \end{vmatrix}^2 U + \begin{vmatrix} & \Lambda p_1 - 1, & \Lambda p_2 - 1, & \Lambda p_3 - 1 \\ Ax + Hy + Gz, & \alpha_1, & \alpha_2, & \alpha_3 \\ Hx + By + Fz, & \beta_1, & \beta_2, & \beta_3 \\ Gx + Fy + Cz, & \gamma_1, & \gamma_2, & \gamma_3 \end{vmatrix}^2 = 0 \ldots (18);$$

la premiere de ces deux formes est peut-être la plus élégante.

Les propriétés géométriques sont absolument indépendantes de la valeur de la quantité Λ: mais pour compléter la solution, je vais donner l'expression de cette quantité. Pour cela, remarquons qu'en mettant pour abréger:

$$\Pi = \begin{vmatrix} \alpha_1, & \alpha_2, & \alpha_3 \\ \beta_1, & \beta_2, & \beta_3 \\ \gamma_1, & \gamma_2, & \gamma_3 \end{vmatrix} \ldots\ldots\ldots\ldots\ldots\ldots\ldots\ldots\ldots\ldots \quad (19),$$

$$\begin{aligned} l_1 &= \beta_2\gamma_3 - \beta_3\gamma_2, \quad l_2 = \beta_3\gamma_1 - \beta_1\gamma_3, \ldots \\ m_1 &= \gamma_2\alpha_3 - \gamma_3\alpha_2, \\ &\vdots \\ \lambda &= l_1(pp_1 - K) + l_2(pp_2 - K) + l_3(pp_3 - K), \\ &\vdots \end{aligned}$$

on aura d'abord l'équation identique

$$-K\Pi(\alpha x + \beta y + \gamma z) = \begin{vmatrix} & pp_1 - K, & pp_2 - K, & pp_3 - K \\ Ax + Hy + Gz, & \alpha_1, & \alpha_2, & \alpha_3 \\ Hx + By + Fz, & \beta_1, & \beta_2, & \beta_3 \\ Gx + Fy + Cz, & \gamma_1, & \gamma_2, & \gamma_3 \end{vmatrix},$$

ou, ce qui est la même chose,

$$K\Pi(\alpha x + \beta y + \gamma z) = \lambda(Ax + Hy + Gz) + \mu(Hx + By + Fz) + \nu(Gx + Fy + Cz).$$

Cela donne

$$\Pi(\mathfrak{A}\alpha^2 + \ldots) = \lambda\alpha + \mu\beta + \nu\gamma, \quad K\Pi(\lambda\alpha + \mu\beta + \nu\gamma) = (A\lambda^2 + \ldots),$$

c'est-à-dire

$$K\Pi^2(\mathfrak{A}\alpha^2+\ldots)=A\lambda^2+\ldots,$$

et, en vertu de cette expression, l'équation qui sert à déterminer p se réduit à

$$K\Pi^2+A\lambda^2+\ldots-K\Pi^2p^2=0 \quad\ldots\ldots\ldots\ldots\ldots\ldots\ldots\ldots\quad (20).$$

En la comparant avec l'équation $K^2L-2KpM+p^2N=0$, on obtient

$$\left.\begin{aligned} L &= \Pi^2+[A\,(l_1+l_2+l_3)^2+\ldots],\\ M &= [A\,(l_1+l_2+l_3)(l_1p_1+l_2p_2+l_3p_3)+\ldots],\\ N &= -K\Pi^2+[A\,(l_1p_1+l_2p_2+l_3p_3)^2+\ldots], \end{aligned}\right\}\ldots\ldots\ldots\ldots(21),$$

et de là, par une transformation déjà employée,

$$M^2-NL=\Omega^2=\Pi^2\left\{\begin{aligned} &-\ [A\,(l_1p_1+l_2p_2+l_3p_3)^2+\ldots]\\ &+K[A\,(l_1+l_2+l_3)^2+\ldots]+K\Pi^2\\ &-\ [\mathfrak{A}\,\{p_1(\alpha_2-\alpha_3)+p_2(\alpha_3-\alpha_1)+p_3(\alpha_1-\alpha_2)\}^2+\ldots] \end{aligned}\right\}\ldots(22);$$

mais l'interpretation de ce résultat paraît être difficile.

En revenant sur l'équation trouvée pour la conique qui touche les trois coniques données, remarquons que les signes de α_1, β_1, γ_1, ou de α_2, β_2, γ_2, ou de α_3, β_3, γ_3, peuvent être changés conjointement. Cela revient en effet à écrire $-V_1$, ou $-V_2$, ou $-V_3$ au lieu de $+V_1$, ou de $+V_2$, ou de $+V_3$, ce qui ne change pas les coniques inscrites. Mais il est facile de voir qu'en changeant à la fois les signes de V_1, V_2, V_3, on ne change pas l'équation de la conique dont il s'agit; cette équation ne change non plus, en changeant à la fois les signes de p_1, p_2, p_3, Ω; de manière que l'équation trouvée correspond réellement à 32 coniques différentes. En distinguant ces 32 coniques par des symboles de la forme

$$(\pm V_1,\ \pm V_2,\ \pm V_3,\ \pm p_1,\ \pm p_2,\ \pm p_3,\ \pm\Omega):$$

quatre symboles tels que

$$\begin{array}{lllllll} (\ V_1, & V_2, & V_3, & p_1, & p_2, & p_3, & \Omega),\\ (-V_1, & -V_2, & -V_3, & p_1, & p_2, & p_3, & \Omega),\\ (\ V_1, & V_2, & V_3, & -p_1, & -p_2, & -p_3, & -\Omega),\\ (-V_1, & -V_2, & -V_3, & -p_1, & -p_2, & -p_3, & -\Omega), \end{array}$$

ne se rapporteront qu'à une seule conique. Nous appelerons *paires de coniques* deux coniques quelconques exprimées par des symboles de la forme

$$(V_1,\ V_2,\ V_3,\ p_1,\ p_2,\ p_3,\ \pm\Omega);$$

donc les 32 coniques forment 16 paires, groupées quatre à quatre de deux manières différentes : savoir, pour former un groupe, quatre paires telles que

$$(\pm V_1,\ \pm V_2,\ V_3,\ p_1,\ p_2,\ p_3,\ \pm\Omega),$$

ou quatre paires telles que

$$(V_1,\ V_2,\ V_3,\ \pm p_1,\ \pm p_2,\ p_3,\ \pm\Omega)$$

peuvent être combinées. Ces deux espèces de groupes peuvent être distinguées par les noms *groupes par rapport à un axe d'homologie*, et *groupes par rapport à un centre de symptose.* En effet: considérons une paire de coniques, par exemple celle qui est représentée par les symboles $(V_1,\ V_2,\ V_3,\ p_1,\ p_2,\ p_3,\ \pm\Omega)$. Les équations des deux coniques de la paire sont les mêmes aux valeurs de Λ près; il est donc évident que les chordes de contact de ces deux coniques avec la conique circonscrite $U=0$ doivent se rencontrer au point d'intersection des droites

$$\begin{vmatrix} & 1, & 1, & 1 \\ Ax+Hy+Gz, & \alpha_1, & \alpha_2, & \alpha_3 \\ Hx+By+Fz, & \beta_1, & \beta_2, & \beta_3 \\ Gx+Fy+Cz, & \gamma_1, & \gamma_2, & \gamma_3 \end{vmatrix} = 0, \qquad \begin{vmatrix} & p_1, & p_2, & p_3 \\ Ax+Hy+Gz, & \alpha_1, & \alpha_2, & \alpha_3 \\ Hx+By+Fz, & \beta_1, & \beta_2, & \beta_3 \\ Gx+Fy+Cz, & \gamma_1, & \gamma_2, & \gamma_3 \end{vmatrix} = 0.$$

La première de ces équations se rapporte à la polaire d'un des centres de symptose des trois coniques inscrites par rapport à la conique circonscrite; la seconde se rapporte à un des axes d'homologie. On a donc le théorème suivant:

"Les points de rencontre des 16 chordes de contact des paires de coniques sont les points de rencontre des polaires des quatre centres de symptose par rapport à la conique circonscrite, avec les quatre axes d'homologie."

Cela suffit pour expliquer la manière dont les deux espèces de groupes ont été distinguées.

Cherchons l'équation de la droite menée par les points de contact d'une des coniques inscrites (par exemple celle que donne l'equation $U+V_1^2=0$) avec deux coniques de la même paire. En représentant par

$$U+(\alpha x+\beta y+\gamma z)^2=0 \quad \text{et} \quad U+(\alpha' x+\beta' y+\gamma' z)^2=0$$

les équations de ces deux coniques: les équations des tangentes communes seront

$$(\alpha-\alpha_1)\,x+(\beta-\beta_1)\,y+(\gamma-\gamma_1)\,z=0 \quad \text{et} \quad (\alpha'-\alpha_1)\,x+(\beta'-\beta_1)\,y+(\gamma'-\gamma_1)\,z=0,$$

et l'équation de la droite qui passe par les points de contact de ces deux droites avec la conique $U+V_1^2=0$ est:

$$A\,(\beta\gamma'-\beta'\gamma)\,x+\ldots+A\,(\gamma_1\,(\beta'-\beta)-\beta_1\,(\gamma'-\gamma))\,x+\ldots$$
$$+(\alpha_1 x+\beta_1 y+\gamma_1 z)\,[\alpha_1\,(\beta\gamma'-\beta'\gamma)+\beta_1\,(\gamma\alpha'-\gamma'\alpha)+\gamma_1\,(\alpha\beta'-\alpha'\beta)]=0 \;\ldots\ldots\; (23).$$

Pour réduire cette équation, en exprimant par Π, l_1, m_1 &c. les valeurs plus haut, mettons $\lambda=l_1\,(\Lambda p_1-1)+l_2\,(\Lambda p_2-1)+l_3\,(\Lambda p_3-1)$ &c., et soient λ', μ', ν' ce que deviennent les valeurs de λ, μ, ν en écrivant Λ' au lieu de Λ. On aura

$$\Pi\alpha=A\lambda+H\mu+G\nu, \text{ \&c.}$$

et des expressions pareilles des valeurs de λ', μ', ν'. Delà on tire

$$\Pi^2(\beta\gamma' - \beta'\gamma) = (\mu\nu' - \mu'\nu)\mathfrak{A} + (\nu\lambda' - \nu'\lambda)\mathfrak{H} + (\lambda\mu' - \lambda'\mu)\mathfrak{G},$$

&c.,

$$\mu\nu' - \mu'\nu = (m_2n_3 - m_3n_2)\{(\Lambda p_2 - 1)(\Lambda' p_3 - 1) - (\Lambda' p_2 - 1)(\Lambda p_3 - 1)\} + \&c.$$
$$= \Pi(\Lambda' - \Lambda)[\alpha_1(p_2 - p_3) + \alpha_2(p_3 - p_1) + \alpha_3(p_1 - p_2)],$$

&c.

c'est-à-dire, en supprimant le facteur commun $\dfrac{\Lambda' - \Lambda}{\Pi}$:

$$\begin{aligned}\beta\gamma' - \beta'\gamma = & \ (\mathfrak{A}\alpha_1 + \mathfrak{H}\beta_1 + \mathfrak{G}\gamma_1)(p_2 - p_3) \\ & + (\mathfrak{A}\alpha_2 + \mathfrak{H}\beta_2 + \mathfrak{G}\gamma_2)(p_3 - p_1) \\ & + (\mathfrak{A}\alpha_3 + \mathfrak{H}\beta_3 + \mathfrak{G}\gamma_3)(p_1 - p_2),\end{aligned}$$

&c.

Aussi on aura, en supprimant le même facteur:

$$\begin{aligned}\gamma_1(\beta' - \beta) - \beta_1(\gamma' - \gamma) = & \ (l_1p_1 + l_2p_2 + l_3p_3)(\gamma_1 H - \beta_1 G) \\ & + (m_1p_1 + m_2p_2 + m_3p_3)(\alpha_1 G - \gamma_1 A) \\ & + (n_1p_1 + n_2p_2 + n_3p_3)(\beta_1 A - \alpha_1 H).\end{aligned}$$

De là on obtient immédiatement

$$A(\beta\gamma' - \beta'\gamma)x + \ldots = K[V_1(p_2 - p_3) + V_2(p_3 - p_1) + V_3(p_1 - p_2)],$$

et, par une réduction un peu plus difficile,

$$\begin{aligned}& A(\gamma_1(\beta' - \beta) - \beta_1(\gamma' - \gamma))x + \ldots \\ & = (l_1p_1 + l_2p_2 + l_3p_3)[z(\mathfrak{H}\alpha_1 + \mathfrak{B}\beta_1 + \mathfrak{F}\gamma_1) - y(\mathfrak{G}\alpha_1 + \mathfrak{F}\beta_1 + \mathfrak{C}\gamma_1)] \\ & \quad + (m_1p_1 + m_2p_2 + m_3p_3)[x(\mathfrak{G}\alpha_1 + \mathfrak{F}\beta_1 + \mathfrak{C}\gamma_1) - z(\mathfrak{A}\alpha_1 + \mathfrak{H}\beta_1 + \mathfrak{G}\gamma_1)] \\ & \quad + (n_1p_1 + n_2p_2 + n_3p_3)[y(\mathfrak{A}\alpha_1 + \mathfrak{H}\beta_1 + \mathfrak{G}\gamma_1) - x(\mathfrak{H}\alpha_1 + \mathfrak{B}\beta_1 + \mathfrak{F}\gamma_1)] \\ & = V_1[p_2(\mathfrak{A}\alpha_1\alpha_3 + \ldots) - p_3(\mathfrak{A}\alpha_1\alpha_2 + \ldots)] + V_2[p_3(\mathfrak{A}\alpha_1^2 + \ldots) - p_1(\mathfrak{A}\alpha_1\alpha_3 + \ldots)] \\ & \qquad + V_3[p_1(\mathfrak{A}\alpha_1\alpha_2 + \ldots) - p_2(\mathfrak{A}\alpha_1^2 + \ldots)];\end{aligned}$$

et enfin

$$\begin{aligned}& (\alpha_1x + \beta_1y + \gamma_1z)[\alpha_1(\beta\gamma' - \beta'\gamma) + \beta_1(\gamma\alpha' - \gamma'\alpha) + \gamma_1(\alpha\beta' - \alpha'\beta)] \\ & = V_1[\mathfrak{A}\alpha_1^2 + \ldots)(p_2 - p_3) + (\mathfrak{A}\alpha_1\alpha_2 + \ldots)(p_3 - p_1) + (\mathfrak{A}\alpha_1\alpha_3 + \ldots)(p_1 - p_2)].\end{aligned}$$

Donc, en réunissant ces expressions des trois parties de l'équation dont il s'agit, cette équation se réduit à

$$\begin{aligned}& V_1[(K + \mathfrak{A}\alpha_1^2 + \ldots)(p_2 - p_3) - (\mathfrak{A}\alpha_1\alpha_2 + \ldots)p_1 + (\mathfrak{A}\alpha_1\alpha_3 + \ldots)p_1] \\ & + V_2[(K + \mathfrak{A}\alpha_1^2 + \ldots)\ p_3 - (K + \mathfrak{A}\alpha_1^2 + \ldots)p_1] \\ & + V_3[(K + \mathfrak{A}\alpha_1\alpha_2 + \ldots)\ p_1 - (K + \mathfrak{A}\alpha_1^2 + \ldots)p_2] = 0 \ \ldots\ldots\ldots\ldots\ldots\ldots\ldots(24).\end{aligned}$$

En mettant p_1^2 au lieu de $K + \mathfrak{A}\alpha_1^2 + \ldots$ et en supprimant le facteur commun p_1, on aura

$$V_1 [p_1 p_2 - (\mathfrak{A}\alpha_1\alpha_2 + \ldots) - p_1 p_3 + (\mathfrak{A}\alpha_1\alpha_3 + \ldots)] + V_2 [p_1 p_3 - (\mathfrak{A}\alpha_1\alpha_3 + \ldots) - K] + V_3 [K - p_1 p_2 + (\mathfrak{A}\alpha_1\alpha_2 + \ldots)] = 0,$$

et en remettant $p_1^2 - (\mathfrak{A}\alpha_1^2 + \ldots) = K$, on obtient pour l'équation dont il s'agit:

$$\begin{vmatrix} V_1, & V_2, & V_3 \\ 1\ , & 1\ , & 1 \\ p_1^2 - (\mathfrak{A}\alpha_1^2 + \ldots), & p_1 p_2 - (\mathfrak{A}\alpha_1\alpha_2 + \ldots), & p_1 p_3 - (\mathfrak{A}\alpha_1\alpha_3 + \ldots) \end{vmatrix} = 0 \ \ldots \ (25).$$

Cette équation est celle de la droite menée par les points de contact de deux coniques de la même paire avec la conique $U + V_1^2 = 0$. Elle est la même qui a été déjà obtenue pour la droite résultante d'une certaine construction géométrique; on est donc arrivé au théorème connu suivant:

"La droite menée par le pôle d'un axe d'homologie par rapport à une des trois coniques inscrites, et par un centre de symptose, rencontre cette même conique en deux points qui sont les points de contact de cette conique avec deux coniques de la même paire;"

ou, ce qui revient au même:

"Le point de rencontre de la polaire d'un centre de symptose, par rapport à une des trois coniques inscrites, et d'un axe d'homologie, est le point de rencontre des tangentes communes de cette même conique et de deux coniques de la même paire."

92.

NOTE SUR UN SYSTÈME DE CERTAINES FORMULES.

[From the *Journal für die reine und angewandte Mathematik* (Crelle), tom. XXXIX. (1850), pp. 14, 15.]

LES formules dont il s'agit se rapportent à la théorie de la composition des formes quadratiques. Je les présente ici pour faire voir la relation qui existe entre elles et quelques formules de mon mémoire sur les hyperdéterminants (t. XXX. p. 1), [16]. En adoptant la notation de ce mémoire, et en mettant

$$\left.\begin{array}{lll}
2A\ \ = \overset{\dagger}{\begin{Bmatrix}111\\122\end{Bmatrix}}, & 2B\ \ = \overset{\dagger}{\begin{Bmatrix}111\\222\end{Bmatrix}}, & 2C\ \ = \overset{\dagger}{\begin{Bmatrix}211\\222\end{Bmatrix}}, \\
2A' \ = \overset{\dagger}{\begin{Bmatrix}111\\212\end{Bmatrix}}, & 2B' \ = \overset{\dagger}{\begin{Bmatrix}111\\222\end{Bmatrix}}, & 2C' \ = \overset{\dagger}{\begin{Bmatrix}121\\222\end{Bmatrix}}, \\
2A'' = \overset{\dagger}{\begin{Bmatrix}111\\221\end{Bmatrix}}, & 2B'' = \overset{\dagger}{\begin{Bmatrix}111\\222\end{Bmatrix}}, & 2C'' = \overset{\dagger}{\begin{Bmatrix}112\\222\end{Bmatrix}},
\end{array}\right\}\ \ldots\ldots\ldots\ldots\ldots (1),$$

savoir, en mettant pour abréger,

$$\left.\begin{array}{ll}
a = 111, & e = 112, \\
b = 211, & f = 212, \\
c = 121, & g = 122, \\
d = 221, & h = 222;
\end{array}\right\}\ \ldots\ldots\ldots\ldots\ldots\ldots\ldots\ldots\ldots (2);$$

$$\left.\begin{array}{lll}
A\ \ = ag - ce, & 2B\ \ = ah + bg - de - cf, & C\ \ = bh - df, \\
A'\ = af - be, & 2B'\ = ah + cf - bg - de, & C'\ = ch - dg, \\
A'' = fg - eh, & 2B'' = ah + de - bg - cf, & C'' = bc - ad,
\end{array}\right\}\ \ldots\ldots\ldots (3),$$

on aura identiquement:

$$\left.\begin{array}{l}
AA' = A''a^2 + \quad 2B''ae \quad + C''e^2\ , \\
AB'\ = A''ac + B''(ag + ce) + C''eg\ , \\
AC'\ = A''c^2 + \quad 2B''cg \quad + C''g^2\ ;
\end{array}\right\}\ \ldots\ldots\ldots\ldots\ldots\ldots\ldots\ldots (4);$$

$$\left.\begin{aligned} BA' &= A''ab + B''(af + be) + C''ef\,, \\ BB' + \Theta &= A''ad + B''(ah + de) + C''eh\,, \\ BB' - \Theta &= A''be + B''(bg + cf) + C''fg\,, \\ BC' &= A''cd + B''(ch + dg) + C''gh\,; \end{aligned}\right\}\ldots\ldots\ldots\ldots(5);$$

$$\left.\begin{aligned} CA' &= A''b^2 + 2B''bf + C''f^2\,, \\ CB' &= A''bd + B''(bh + df) + C''fh\,, \\ CC' &= A''d^2 + 2B''dh + C''h^2\,; \end{aligned}\right\}\ldots\ldots\ldots\ldots(6);$$

$$\Theta = B^2 - AC = B'^2 - A'C' = B''^2 - A''C'' \ldots\ldots\ldots\ldots(7)$$

$$= \tfrac{1}{4}\{a^2h^2 + b^2g^2 + c^2f^2 + d^2e^2 - 2ahbg - 2ahcf - 2ahde - 2bgcf - 2bgde - 2cfde + 4adfg + 4bech\}.$$

En regardant la seconde et la troisième des équations (5) comme équivalentes avec la seule équation $2BB' = A''(ad + be) + B''(ah + de + bg + cf) + C''(eh + fg)$, on trouvera que les systèmes (4, 5, 6) répondent à la transformation

$$A''z_1^2 + 2B''z_1z_2 + C''z_2^2 = (Ax_1^2 + 2Bx_1x_2 + Cx_2^2)(A'y_1^2 + 2B'y_1y_2 + C'y_2^2)\ldots\ldots(8),$$

$$z_1 = ax_1x_2 + by_1x_2 + cx_1y_2 + dy_1y_2,$$

$$z_2 = ex_1x_2 + fy_1x_2 + gx_1y_2 + hy_1y_2,$$

qui appartient à une théorie dont celle des transformations linéaires n'est qu'un cas particulier.

Je profite de cette occasion pour donner une addition à la "Note sur les hyperdéterminants" (t. XXXIV.), [54]. J'y ai dit (§ III.) que je ne pouvais pas expliquer la raison de ce que la courbe du sixième ordre donnée par les équations $ae - 4bd + 3c^2 = 0$ et $ace + 2bcd - ad^2 - eb^2 - c^3 = 0$, ait une osculatrice développable qui n'est que du sixième ordre, mais que cette réduction s'opérait en partie au moyen des quatre points de rebroussement de la courbe. En effet, cette courbe, considérée comme l'intersection de deux surfaces, l'une du second et l'autre du troisième ordre, a six droites que dans le mémoire [30] cité dans cette note j'ai nommé *droites par deux points:* cela suffit pour compléter la réduction dont il s'agit. M. Salmon, à qui je dois cette remarque, m'a fait voir aussi que l'expression que j'ai donnée pour le nombre des points de rebroussement de la courbe, dans le cas d'une équation du $m^{\text{ième}}$ ordre, combinée avec les formules du mémoire mentionné, suffit pour former le tableau complèt des singularités de la famille de surfaces développables dont il s'agissait, savoir:

$$\begin{array}{ccccccc} m, & n, & r, & \alpha, & \beta, & g, & h, \\ 3(m-2), & m, & 2(m-1), & 0, & 4(m-3), & \tfrac{1}{2}(m-1)(m-2), & \tfrac{1}{2}(9n^2 - 53n + 80), \end{array}$$

$$\begin{array}{cc} x, & y, \\ 2(n-2)(n-3), & 2(-1)(n-3). \end{array}$$

Ici, dans la ligne supérieure, m, n, r, α, β, g, h, x, y ont les mêmes significations que dans le mémoire dont je viens de parler; et dans la ligne inférieure, m est le degré de l'équation primitive.

93.

NOTE SUR QUELQUES FORMULES QUI SE RAPPORTENT À LA MULTIPLICATION DES FONCTIONS ELLIPTIQUES.

[From the *Journal für die reine und angewandte Mathematik* (Crelle), tom. XXXIX. (1850), pp. 16—22.]

Les fonctions

$$\{\lambda,\ \mu,\ x,\ y\} = P_{0,0} - \left(\frac{1}{1} P_{1,0}\, x + \frac{1}{1} P_{0,1}\, y\right) + \left(\frac{1}{1\,.\,2} P_{2,0}\, x^2 + \frac{1}{1\,.\,1} P_{1,1}\, xy + \frac{1}{1\,.\,2} P_{0,2}\, y^2\right) - \&\text{c.},$$

où $P_{0,0} = 1$ et les autres coefficients sont donnés par l'équation à différences

$$\begin{aligned} &[-l\lambda - m\mu + (l-m)^2]\, P_{l,m} \\ &+ l\,(\lambda - 2l + 2m + 2)\,(\lambda - 2l + 2m + 1)\, P_{l-1,m} \\ &+ m\,(\mu + 2l - 2m + 2)\,(\mu + 2l - 2m + 1)\, P_{l,m-1} \\ &- 16lm\,[\lambda\mu - (2l + 2m - 4)\,(\lambda + \mu)]\ P_{l-1,m-1} = 0, \end{aligned}$$

jouent, comme je crois, un rôle important dans la théorie des fonctions elliptiques[1].

[1] La fonction $\{\lambda,\ \mu,\ x,\ y\}$ satisfait à l'équation

$$\begin{aligned} &-[\lambda\,(\lambda-1)\,x + \mu\,(\mu-1)\,y + 16\lambda\mu\, xy]\, u \\ &+[-(\lambda-1) + (4\lambda-6)\,x + (4\mu+2)\,y + 32\,(\lambda+\mu)\,xy]\, y\,\frac{du}{dx} \\ &+[-(\mu-1) + (4\lambda+2)\,x + (4\mu-6)\,y + 32\,(\lambda+\mu)\,xy]\, y\,\frac{du}{dy} \\ &+(1-4x-4y)\left(x^2\frac{d^2u}{dx^2} - 2xy\,\frac{d^2u}{dx\,dy} + y^2\frac{d^2u}{dy^2}\right) = 0, \end{aligned}$$

qui peut être tirée de l'équation

$$n\,(n-1)\,x^2u + (n-1)\,(\alpha x - 2x^3)\,\frac{du}{dx} + (1 - \alpha x^2 + x^4)\,\frac{d^2u}{dx^2} - 2n\,(\alpha^2 - 4)\,\frac{du}{d\alpha} = 0$$

(voyez le mémoire cité plus bas), mais qu'on obtient plus facilement au moyen de l'équation à différences à laquelle satisfont les coefficients $P_{l,\,m}$.

En effet, en faisant $x=\sqrt{k}\sin\operatorname{am} u$, $\alpha=k+\dfrac{1}{k}$ et en représentant par z le dénominateur de la fonction $\sqrt{k}\sin\operatorname{am} nu$ (où n est un entier positif quelconque), on aura

$$z=z_1+z_2+\ldots+z_s+\ldots,$$

cette série étant continuée jusqu'au terme $z_{\frac{1}{2}n}$ ou $z_{\frac{1}{2}(n-1)}$, selon que n est pair ou impair, et la fonction z étant donnée par l'équation

$$z_s=(-1)^{(n+1)s}(4\alpha)^{s(n-2s)}x^{2ns}\left\{n^2-2ns,\ 2ns,\ \frac{x^2}{4\alpha},\ \frac{1}{4\alpha x^2}\right\},$$

où cependant les termes qui contiennent des puissances négatives de α doivent être négligés. Ces formules reviennent à celles que j'ai présentées dans la "Note sur les fonctions elliptiques" (t. XXXVII.), [67].

En revenant aux fonctions $\{\lambda,\ \mu,\ x,\ y\}$, j'ai trouvé les deux formules

$$\begin{aligned} P_{l,0}&=\lambda\,[\lambda-l-1]^{l-1},\\ P_{l,1}&=\mu\lambda\,[\lambda-l-1]^{l-1}\\ &\quad+l\lambda\,[\lambda-l-1]^{l-3}\{(18l-16)\,\lambda-(16\lambda^2-10l-4)\}\\ &\quad-\frac{10l\lambda\mu}{\lambda+\mu}\lambda\,[\lambda-l]^{l-2}\end{aligned}$$

(où selon la notation de Vandermonde la factorielle $p\,(p-1)\ldots(p-q+1)$ est exprimée par $[p]^q$). De là, et en calculant la valeur de $P_{2,2}$ à l'aide de l'équation à différences, on obtient:

$$\begin{aligned} P_{0,0}&=1,\\ P_{1,0}&=\lambda,\\ P_{0,1}&=\mu,\\ P_{2,0}&=\lambda\,(\lambda-3),\\ P_{1,1}&=\left(\lambda\mu+2-\frac{10\lambda\mu}{\lambda+\mu}\right),\\ P_{0,2}&=\mu\,(\mu-3),\\ P_{3,0}&=\lambda\,(\lambda-4)\,(\lambda-5),\\ P_{2,1}&=\lambda\left((\lambda-3)\,\mu+40-\frac{20\lambda\mu}{\lambda+\mu}\right),\\ P_{1,2}&=\mu\left((\mu-3)\,\lambda+40-\frac{20\lambda\mu}{\lambda+\mu}\right),\\ P_{0,3}&=\mu\,(\mu-4)\,(\mu-5),\end{aligned}$$

$$P_{4,0} = \lambda(\lambda-5)(\lambda-6)(\lambda-7),$$

$$P_{3,1} = \lambda\left(\mu(\lambda-4)(\lambda-5) + 114\lambda - 330 - \frac{30\lambda\mu}{\lambda+\mu}(\lambda-3)\right),$$

$$P_{2,2} = \left(\lambda(\lambda-3)\,\mu(\mu-3) + 152\lambda\mu + 336 - \frac{40\lambda^2\mu^2 + 1156\lambda\mu}{\lambda+\mu} + \frac{200\lambda^2\mu^2}{(\lambda+\mu)^2}\right),$$

$$P_{1,3} = \mu\left(\lambda(\mu-4)(\mu-5) + 114\mu - 330 - \frac{30\lambda\mu}{\lambda+\mu}(\mu-3)\right),$$

$$P_{0,4} = \mu(\mu-5)(\mu-6)(\mu-7);$$

la première partie de cette table se trouve dans la note citée.

Nous remarquerons en passant que pour $y=0$, on a $\{\lambda, \mu, x, 0\} = (\frac{1}{2} + \sqrt{(\frac{1}{4} - x)})^\lambda$. On sait que la théorie de la multiplication des fonctions circulaires depend de la fonction $(x + \sqrt{(x^2-1)})^\lambda$ ou, en faisant $\frac{1}{4}x^{-2} = \mathrm{x}$, de la fonction $(\frac{1}{2} + \sqrt{(\frac{1}{4} - \mathrm{x})})^\lambda$. Cela fait espérer que l'on parviendra par les fonctions $\{\lambda, \mu, x, y\}$ à la théorie complète de la multiplication des fonctions elliptiques.

J'ai calculé les valeurs qui servent à trouver les dénominateurs z de sin am nu, où n est un quelconque des entiers 1, 2, 3, 4, 5, 6, 7. Les voici:

$n=1$, $\quad z_0 = 1$, $\qquad z = z_0$,

$n=2$, $\quad z_0 = 1$, $\quad -z_1 = -x^4$, $\qquad z = z_0 - z_1$,

$n=3$, $\quad z_0 = 1$, $\quad +z_1 = 4\alpha x^6 - (6x^4 + 3x^8)$, $\qquad z = z_0 + z_1$,

$n=4$, $\quad z_0 = 1$, $\quad -z_1 = -16\alpha^2 x^8 + 4\alpha(8x^{10} + 8x^6) - (20x^{12} + 26x^8 + 20x^4)$,

$\qquad z_2 = x^{16}$, $\qquad z = z_0 - z_1 + z_2$,

$n=5$, $\quad z_0 = 1$, $\qquad z = z_0 + z_1 + z_2$,

$$\begin{aligned} z_1 = &\ 64\alpha^3\, x^{10} \\ &- 16\alpha^2(15x^{12} + 10x^8) \\ &+ 4\alpha(90x^{14} + 92x^{10} + 35x^6) \\ &- (275x^{16} + 300x^{12} + 125x^8 + 50x^4), \end{aligned}$$

$$\begin{aligned} z_2 = &\ 16\alpha^2\, x^{20} \\ &- 4\alpha(5x^{22} + 20x^{18}) \\ &+ (5x^{24} + 62x^{20} + 170x^{16}), \end{aligned}$$

$n=6$, $\quad +z_0 = 1$, $\qquad z = z_0 - z_1 + z_2 - z_3$,

$$\begin{aligned} -z_1 = &- 256\alpha^4\, x^{12} \\ &+ 64\alpha^3(24x^{14} + 12x^{10}) \\ &- 16\alpha^2(252x^{16} + 210x^{12} + 54x^8) \\ &+ 4\alpha(1520x^{18} + 1584x^{14} + 576x^{10} + 112x^6) \\ &- (5814x^{20} + 7704x^{16} + 2400x^{12} + 444x^8 + 105x^4), \end{aligned}$$

$$
\begin{aligned}
+z_2 = {} & 256\alpha^4 \quad x^{24} \\
& - 64\alpha^3 (12x^{26} + 24x^{22}) \\
& + 16\alpha^2 (54x^{28} + 210x^{24} + 252x^{20}) \\
& - 4\alpha (112x^{30} + 576x^{26} + 1584x^{22} + 1520x^{18}) \\
& + (105x^{32} + 444x^{28} + 2400x^{24} + 7704x^{20} + 5814x^{16}), \\
-z_3 = {} & -x^{36},
\end{aligned}
$$

$n = 7,$ $\qquad z = z_0 + z_1 + z_2 + z_3.$

$$
\begin{aligned}
z_0 = {} & 1, \\
z_1 = {} & 1024\alpha^5 \quad x^{14} \\
& - 256\alpha^4 (35x^{16} + 14x^{12}) \\
& + 64\alpha^3 (1120x^{18} + 196x^{14} + 77x^{10}) \\
& - 16\alpha^2 (5425x^{20} + 5040x^{16} + 1575x^{12} + 210x^{8}) \\
& + 4\alpha (35525x^{22} + 41300x^{18} + 14934x^{14} + 2604x^{10} + 294x^{6}) \\
& - (166257x^{24} + 260750x^{20} + 220395x^{16} + 14756x^{12} + 1304x^{8} + 196x^{4}),
\end{aligned}
$$

$$
\begin{aligned}
z_2 = {} & 4096\alpha^6 \quad x^{28} \\
& - 1024\alpha^5 (21x^{30} + 28x^{26}) \\
& + 256\alpha^4 (189x^{32} + 470x^{28} + 350x^{24}) \\
& - 64\alpha^3 (952x^{34} + 3192x^{30} + 4550x^{26} + 2576x^{22}) \\
& + 16\alpha^2 (2940x^{36} + 11200x^{32} + 21750x^{28} + 25452x^{24} + 12397x^{20}) \\
& - 4\alpha (5733x^{38} + 22064x^{34} + 44324x^{30} + 82488x^{26} + 96761x^{22} + 40964x^{18}) \\
& + (7007x^{40} + 59388x^{36} + 35231x^{32} + 41132x^{28} + 278173x^{24} + 302918x^{20} + 94962x^{16}),
\end{aligned}
$$

$$
\begin{aligned}
z_3 = {} & 64\alpha^3 \quad x^{42} \\
& - 16\alpha^2 (7x^{44} + 42x^{40}) \\
& + 4\alpha (14x^{46} + 236x^{42} + 819x^{38}) \\
& - (7x^{48} + 308x^{44} + 4053x^{40} + 9842x^{36}).
\end{aligned}
$$

Pour rassembler tous mes résultats, je veux citer ceux que j'ai donné dans le *Cambridge and Dublin Mathematical Journal* [vol. II. (1847), 45, and vol. III. (1848), 57]. En écrivant la lettre p au lieu de n (symbole qui représente le carré du nombre n de ce mémoire), et en changeant les signes des termes alternatifs, on aura pour solution particulière de l'équation

$$p(p-1)x^2z + (p-1)(\alpha x - 2x^2)\frac{dz}{dx} + (1 - \alpha x^2 + x^4)\frac{d^2z}{dx^2} - 2p(\alpha^2 - 4)\frac{dz}{d\alpha} = 0$$

(savoir pour la solution qui pour $p = n^2$ se réduit au dénominateur de sin am u) la valeur

$$z = 1 - C_2 \frac{x^4}{1.2.3.4} + C_6 \frac{x^6}{1.2.3.4.5.6} - \&c.,$$

les coefficients étant déterminés au moyen de l'expression

$$C_{r+2} = -(2r+1)(2r+2)(p-2r)(p-2r-1)C_r + (2r+2)(p-2r-2)\alpha C_{r+1} - 2p(\alpha^2-4)\frac{dC_{r+1}}{d\alpha}.$$

Cela donne les valeurs particulières suivantes:

$$C_2 = 2p(p-1),$$
$$C_3 = 8p(p-1)(p-4)\alpha,$$
&c.

[viz. with the change referred to, these are the values of C_2, C_3, ... C_8 given *ante* p. 299].

On remarquera que dans ces formules le premier terme de C_8 ne contient pas, comme on pourrait l'attendre, le facteur $(p-25)$. Cela vient de ce que le coefficient C_8 est composé des coefficients des termes correspondants de z_0 et z_1, tandis que les coefficients C_7, &c., sont tout simplement des coefficients de z_0. La suite des coefficients C offre plusieurs discontinuités de cette sorte. Par exemple on obtient généralement

$$C_r = (-)^{r+1} \left\{ \begin{array}{l} 2^{2r-3}p(p-1^2)\dots(p-(r-1)^2)\,C_r^1\alpha^{r-2} \\ +\,2^{2r-6}p(p-1^2)\dots(p-(r-2)^2)\,C_r^2\alpha^{r-4} \\ +\,2^{2r-9}p(p-1^2)\dots(p-(r-3)^2)\,C_r^3\alpha^{r-6} \\ +\,\&c.; \end{array} \right.$$

mais le terme suivant ne contient pas le facteur $p(p-1^2)\dots(p-(r-4)^2)$. Quant à la loi des coefficients C_r^1, C_r^2, C_r^3, on a

$$C_r^1 = 1,$$
$$C_r^2 = (r-3)\{n(2r-7) + (r-1)(8r-7)\},$$
$$C_r^3 = (r-4)(r-5)\{n^2(4r^2-24r+51) + n(32r^3-220r^2+412r-255) + 2(r-1)(r-2)(32r^2-88r+51)\}.$$

Également, en ordonnant la série suivant les puissances descendantes de x, la quantité z étant la solution particulière qui pour $p = n^2$ (n impair) se réduit au dénominateur de sin am nu, on aura

$$z = (-1)^{\frac{1}{2}(\sqrt{p}-1)}\,.\,\sqrt{p}\left(x^{p-1} - D_1\frac{x^{p-3}}{1.2.3} + D_2\frac{x^{p-5}}{1.2.3.4.5} - \&c.\right),$$

où les coefficients D sont donnés par l'expression

$$D_{r+2} = -(2r+3)(2r+2)(p-2r-2)(p-2r-1)D_r$$
$$+(2r+3)(p-2r-3)\alpha D_{r+1} - 2p(\alpha^2-4)\frac{dD_{r+1}}{d\alpha},$$

ce qui donne les valeurs particulières suivantes:

$$D_1 = (p-1)\alpha,$$
$$D_2 = 2(p-1)(p+6) + (p-1)(p-9)\alpha^2,$$
&c.

[viz. with the change referred to, these are the values of D_1, $D_2, \ldots D_6$ given *ante* pp. 364, 365].

Les mêmes remarques sont applicables aux coefficients D; seulement la discontinuité a lieu ici dès le coefficient D_4. Il paraît que c'est cause de cette discontinuité que le signe négatif se présente aux premiers termes des coefficients D_4, &c. En effet, on a généralement:

$$D_r = (p-1)(p-9)\ldots(p-(2r-1)^2)\alpha^r$$
$$+(p-1)(p-9)\ldots(p-(2r-3)^2)r(r-1)(p+4r-2)\alpha^{r-2}$$
$$+ \&c.;$$

ici la discontinuité se présente déjà dans le terme suivant, qui ne contient pas le facteur $(p-1)(p-9)\ldots(p-(2r-5)^2)$. Et c'est précisément le terme suivant qui devient négatif dans les expressions de D_4, D_5 et D_6. Mais tout cela est moins important que la théorie des séries partielles z_s, sur lesquelles les recherches ultérieures seront à fonder.

PROBLÈME.

Donner la solution de l'équation à différences

$$(-l\lambda - m\mu + (l-m)^2)P_{l,m}$$
$$+ l(\lambda - 2l + 2m + 2)(\lambda - 2l + 2m + 1)P_{l-1,m}$$
$$+ m(\mu + 2l - 2m + 2)(\mu + 2l - 2m + 1)P_{l,m-1}$$
$$- 16lm(\lambda\mu - (2l + 2m - 4)(\lambda + \mu))P_{l-1,m-1} = 0,$$

dans laquelle $P_{0,0} = 1$. (Voyez la "Note sur quelques formules &c.")

94.

NOTE SUR L'ADDITION DES FONCTIONS ELLIPTIQUES.

[From the *Journal für die reine und angewandte Mathematik* (Crelle), tom. XLI. (1851), pp. 57—65.]

SOIT, pour observer autant que possible la symétrie:

$$Su = \sqrt{k} \sin \operatorname{am} \frac{u}{\sqrt{k}},$$

$$Cu = \quad \cos \operatorname{am} \frac{u}{\sqrt{k}},$$

$$Gu = \quad \Delta \operatorname{am} \frac{u}{\sqrt{k}},$$

$$\alpha = k + \frac{1}{k}, \qquad \tfrac{1}{2}\upsilon = \frac{K}{\sqrt{k}}, \qquad \tfrac{1}{2}\delta = \frac{K + iK'}{\sqrt{k}};$$

et soit pour abréger:

$$Su = x, \quad Sv = y, \text{ \&c.}$$

$$Cu = \sqrt{\left(1 - \frac{x^2}{k}\right)} = X_{,},$$

$$Gu = \quad \sqrt{(1 - kx^2)} = X_{,,},$$

$$Cu\, Gu = \sqrt{(1 - \alpha x^2 + x^4)} = X_{,}X_{,,} = X.$$

Cela posé, les méthodes d'Abel donnent les expressions suivantes de sin am d'une somme quelconque d'arcs: savoir

$$S(u + v + \ldots) = (-1)^{n-1} \frac{[\theta,\ \theta^3, \ldots \theta^{2n-1},\ \Theta\ ,\ \theta^2\Theta, \ldots \theta^{2n-4}\Theta]}{[1,\ \theta^2, \ldots \theta^{2n-2},\ \theta\Theta,\ \theta^3\Theta, \ldots \theta^{2n-3}\Theta]},$$

pour un nombre *impair* $2n-1$ d'arcs, et

$$S(u+v+\ldots) = -\frac{[1,\ \theta^2, \ldots \theta^{2n},\ \theta\Theta,\ \theta^3\Theta, \ldots \theta^{2n-3}\Theta]}{[\theta,\ \theta^3, \ldots \theta^{2n-1},\ \Theta,\ \theta^2\Theta, \ldots \theta^{2n-2}\Theta]}$$

pour un nombre pair $2n$ d'arcs. Dans ces expressions les symboles dans lesquelles entrent les lettres θ, Θ, sont censés représenter les déterminants, dont on obtient les termes en changeant successivement ces lettres en x, X; y, Y; &c.

J'ai trouvé qu'on a aussi

$$C(u+v+\ldots) = \frac{[\Theta_{,},\ \theta^2\Theta_{,}, \ldots \theta^{2n-2}\Theta_{,},\ \theta\Theta_{,,},\ \theta^3\Theta_{,,}, \ldots \theta^{2n-3}\Theta_{,,}]}{[1,\ \theta^2, \ldots \theta^{2n-2},\ \theta\Theta,\ \theta^3\Theta, \ldots \theta^{2n-3}\Theta]}$$

pour un nombre impair $2n-1$ d'arcs, et

$$C(u+v+\ldots) = \frac{[\theta\Theta_{,},\ \theta^3\Theta_{,}, \ldots \theta^{2n-1}\Theta_{,},\ \Theta_{,,},\ \theta^2\Theta_{,,}, \ldots \theta^{2n-2}\Theta_{,,}]}{[\theta,\ \theta^3, \ldots \theta^{2n-1},\ \Theta,\ \theta^2\Theta, \ldots \theta^{2n-2}\Theta]}$$

pour un nombre pair $2n$ d'arcs. Les valeurs correspondantes de $G(u+v+\ldots)$ se trouvent en échangeant les symboles $\Theta_{,}$ et $\Theta_{,,}$.

Particulièrement pour la somme de trois arcs on a :

$$S(u+v+w) = \frac{-[\theta,\ \theta^3,\ \Theta]}{[1,\ \theta^2,\ \theta\Theta]},$$

$$C(u+v+w) = \frac{[\Theta_{,},\ \theta^2\Theta_{,},\ \theta\Theta_{,,}]}{[1,\ \theta^2,\ \theta\Theta]},$$

$$G(u+v+w) = \frac{[\Theta_{,,},\ \theta^2\Theta_{,,},\ \theta\Theta_{,}]}{[1,\ \theta^2,\ \theta\Theta]}.$$

Pour réduire ces expressions à une forme qui soit encore applicable au cas où deux quelconques des quantités u, v, w sont égales, il n'y a qu'à multiplier les termes des fractions à droite par

$$\Omega = \frac{-(xY+yX)(yZ+zY)(zX+xZ)}{(x^2-y^2)(y^2-z^2)(z^2-x^2)};$$

cela donne, après une réduction un peu difficile :

$$-\Omega[\theta,\ \theta^3,\ \Theta] = (xYZ+yZX+zXY) - xyz(\alpha - x^2 - y^2 - z^2 + x^2y^2z^2),$$

$$\Omega[\Theta_{,},\ \theta^2\Theta_{,},\ \theta\Theta_{,,}] = (1-kx^2y^2z^2)\,X_{,}Y_{,}Z_{,} - \frac{1}{k}(yzX+zxY+xyZ)\,X_{,,}Y_{,,}Z_{,,},$$

$$\Omega[\Theta_{,,},\ \theta^2\Theta_{,,},\ \theta\Theta_{,}] = \left(1-\frac{1}{k}x^2y^2z^2\right)X_{,,}Y_{,,}Z_{,,} - k(yzX+zxY+xyZ)\,X_{,}Y_{,}Z_{,},$$

$$\Omega[1,\ \theta^2,\ \theta\Theta] = 1 - y^2z^2 - z^2x^2 - x^2y^2 + \alpha x^2y^2z^2 - xyz(xYZ+yZX+zXY),$$

de manière qu'en écrivant

$$M = 1 - x^2y^2 - y^2z^2 - z^2x^2 + \alpha x^2y^2z^2 - xyz\,(xYZ + yZX + zXY),$$

on a

$$S\,(u+v+w) = \frac{xYZ + yZX + zXY - xyz\,(\alpha - x^2 - y^2 - z^2 + x^2y^2z^2)}{M},$$

$$C\,(u+v+w) = \frac{(1 - kx^2y^2z^2)\,X_{,}Y_{,}Z_{,} - \frac{1}{k}(yzX + zxY + xyZ)\,X_{,,}Y_{,,}Z_{,,}}{M},$$

$$G\,(u+v+w) = \frac{\left(1 - \frac{1}{k}x^2y^2z^2\right)X_{,,}Y_{,,}Z_{,,} - k\,(yzX + zxY + xyZ)\,X_{,}Y_{,}Z_{,}}{M}.$$

Les mêmes formules peuvent être trouvées plus simplement en écrivant $u + \tfrac{1}{2}\upsilon + \tfrac{1}{2}\delta$, $v + \tfrac{1}{2}\upsilon + \tfrac{1}{2}\delta$ au lieu de u, v. La somme $u+v+w$ se change par là en $u+v+w+(\upsilon+\delta)$, et les fonctions $S\,(u+v+w)$, $C\,(u+v+w)$, $G\,(u+v+w)$ deviennent $S\,(u+v+w)$, $-C\,(u+v+w)$, $-G\,(u+v+w)$. De plus x, $X_{,}$, $X_{,,}$, X et y, $Y_{,}$, $Y_{,,}$, Y se changent en $-\frac{1}{x}$, $\frac{-i}{\sqrt{k}}\frac{X_{,,}}{x}$, $i\sqrt{k}\frac{X_{,}}{x}$, $\frac{X}{x^2}$ et $-\frac{1}{y}$, $\frac{-i}{\sqrt{k}}\frac{Y_{,,}}{y}$, $i\sqrt{k}\frac{Y_{,}}{y}$, $\frac{Y}{y^2}$, et l'on a

$$S\,(u+v+w) = -\begin{vmatrix} x^2, & 1, & -xX \\ y^2, & 1, & -yY \\ z, & z^3, & Z \end{vmatrix} \div \begin{vmatrix} x^3, & x, & -X \\ y^3, & y, & -Y \\ 1, & z^2, & zZ \end{vmatrix}$$

$$C\,(u+v+w) = \frac{1}{k}\begin{vmatrix} x^2X_{,,}, & X_{,,}, & kxX_{,} \\ y^2Y_{,,}, & Y_{,,}, & kyY_{,} \\ Z_{,}, & z^2Z_{,}, & zZ_{,,} \end{vmatrix} \div \begin{vmatrix} x^3, & x, & -X \\ y^3, & y, & -Y \\ 1, & z^2, & zZ \end{vmatrix}$$

$$G\,(u+v+w) = k\begin{vmatrix} x^2X_{,}, & X_{,}, & \frac{1}{k}xX_{,,} \\ y^2Y_{,}, & Y_{,}, & \frac{1}{k}yY_{,,} \\ Z_{,,}, & z^2Z_{,,}, & zZ_{,} \end{vmatrix} \div \begin{vmatrix} x^3, & x, & -X \\ y^3, & y, & -Y \\ 1, & z^2, & zZ \end{vmatrix}$$

Ces formules conduisent aux formes réduites que l'on obtient en multipliant par le facteur beaucoup plus simple

$$\Omega_{,} = -\frac{xY + yX}{x^2 - y^2}.$$

En passant, il y a à noter les équations identiques

$$\frac{(yZ + zY)(zX + xZ)}{(y^2 - z^2)(z^2 - x^2)}\begin{vmatrix} x, & x^3, & X \\ y, & y^3, & Y \\ z, & z^3, & Z \end{vmatrix} = \begin{vmatrix} x^3, & x, & -X \\ y^3, & y, & -Y \\ 1, & z^2, & zZ \end{vmatrix}, \text{ \&c.}$$

auxquelles conduit la méthode qui vient d'être expliquée. Aussi en multipliant les valeurs de $C(u+v+w)$, $G(u+v+w)$ on obtient l'équation

$$C(u+v+w)\,G(u+v+w)=\frac{\Psi XYZ+\Lambda yzX+\mathrm{M}zxY+\mathrm{N}xyZ}{\{1-y^2z^2-z^2x^2-x^2y^2+\alpha x^2y^2z^2-xyz\,(xYZ+yZX+zXY)\}^2},$$

dans laquelle

$$\begin{aligned}
\Psi &= 1\ +y^2z^2+z^2x^2+x^2y^2-4\alpha x^2y^2z^2+x^2y^2z^2\,(x^2+y^2+z^2)+x^4y^4z^4,\\
\Lambda &= \Pi+2x^2Y^2Z^2,\\
\mathrm{M} &= \Pi+2y^2Z^2X^2,\\
\mathrm{N} &= \Pi+2z^2X^2Y^2,\\
\Pi &= -\alpha+2\,(x^2+y^2+z^2)-\alpha\,(y^2z^2+z^2x^2+x^2y^2)+(2\alpha^2-4)\,x^2y^2z^2\\
&\quad +2x^2y^2z^2\,(y^2z^2+z^2x^2+x^2y^2)-\alpha x^2y^2z^2\,(x^2+y^2+z^2)-\alpha x^4y^4z^4.
\end{aligned}$$

Pour le cas de quatre arcs, je n'ai trouvé que le sin am de la *somme*. En effet on a

$$S(u+v+w+p)=-\begin{vmatrix}1, & x^2, & x^4, & xX\\ 1, & y^2, & y^4, & yY\\ 1, & z^2, & z^4, & zZ\\ 1, & t^2, & t^4, & tT\end{vmatrix}\div\begin{vmatrix}x, & x^3, & X, & x^2X\\ y, & y^3, & Y, & y^2Y\\ z, & z^3, & Z, & z^2Z\\ t, & t^3, & T, & t^2T\end{vmatrix}$$

où les termes de la fraction sont à multiplier par

$$\Omega=-\frac{(xY+yX)\,(xZ+zX)\,(xT+tX)\,(yZ+zY)\,(zT+tZ)\,(tY+yT)}{(x^2-y^2)\,(x^2-z^2)\,(x^2-t^2)\,(y^2-z^2)\,(z^2-t^2)\,(t^2-y^2)}.$$

Mais il est plus simple de se servir de la forme

$$S(u+v+w+p)=-\begin{vmatrix}x^4, & x^2, & 1, & -xX\\ y^4, & y^2, & 1, & -yY\\ 1, & z^2, & z^4, & zZ\\ 1, & t^2, & t^4, & tT\end{vmatrix}\div\begin{vmatrix}x^3, & x, & -x^2X, & -X\\ y^3, & y, & -y^2Y, & -Y\\ z, & z^3, & Z, & z^2Z\\ t, & t^3, & T, & t^2T\end{vmatrix}$$

que l'on obtient de la même manière que la forme analogue pour trois arcs. Ici le facteur est

$$\Omega_1=\frac{(yX+xY)\,(zT+tZ)}{(x^2-y^2)\,(z^2-t^2)},$$

et l'on obtient, toute réduction faite,

$$S(u+v+w+p)=\frac{\mathfrak{N}}{\mathfrak{D}},$$

$$\mathfrak{N} = (1 - x^2y^2z^2t^2)(xYZT + yZTX + zTXY + tXYZ)$$
$$- \{(\alpha - x^2 - y^2 - z^2 - t^2 + y^2z^2t^2 + z^2t^2x^2 + t^2x^2y^2 + x^2y^2z^2 - \alpha x^2y^2z^2t^2)$$
$$\times (Xyzt + Yztx + Ztxy + Txyz)\},$$

$$\mathfrak{D} = 1 - x^2y^2 - x^2z^2 - x^2t^2 - y^2z^2 - z^2t^2 - t^2y^2 - (x^2y^2 + x^2z^2 + x^2t^2 + y^2z^2 + z^2t^2 + t^2y^2)\,x^2y^2z^2t^2$$
$$+ x^4y^4z^4t^4 + \alpha(x^2y^2z^2 + y^2z^2t^2 + z^2t^2x^2 + t^2x^2y^2) + \alpha(x^2 + y^2 + z^2 + t^2)\,x^2y^2z^2t^2$$
$$+ (2 - 2\alpha^2)\,x^2y^2z^2t^2$$
$$- (x^2Y^2 + y^2X^2)\,ztZT - (x^2Z^2 + z^2X^2)\,ytYT - (x^2T^2 + t^2X^2)\,yzYZ$$
$$- (y^2Z^2 + z^2Y^2)\,xtXT - (z^2T^2 + t^2Z^2)\,xyXY - (t^2Y^2 + y^2T^2)\,xzXZ.$$

Il y a à remarquer qu'en employant la première valeur de $S(u+v+w+p)$ et le facteur correspondant, on aurait trouvé le même numérateur, et aussi le même dénominateur, ce qui donne lieu à des équations identiques, semblables à celles qui ont lieu pour le cas de trois arcs.

Revenons à l'expression

$$\begin{vmatrix} x, & x^3, & X \\ y, & y^3, & Y \\ z, & z^3, & Z \end{vmatrix} \frac{(yZ + zY)(zX + xZ)(xY + yX)}{(y^2 - z^2)(z^2 - x^2)(x^2 - y^2)}$$

qui donne le numérateur de $S(u+v+w)$. En mettant $x^2 = a$, $\frac{1}{k}X = A$, &c. on voit qu'il s'agit d'effectuer la division de

$$\begin{vmatrix} 1, & a, & A \\ 1, & b, & B \\ 1, & c, & C \end{vmatrix} (B + C)(C + A)(A + B)$$

par le produit $(b - c)(c - a)(a - b)$, les fonctions A, B, C denotant des racines carrées de fonctions rationnelles d'une forme particulière. Or, en supposant toujours que les carrés de A, B, C soient des fonctions rationnelles, et d'ailleurs d'une forme quelconque, cela peut se faire dans tous les cas particuliers au moyen de l'équation identique

$$\begin{vmatrix} 1, & a, & A \\ 1, & b, & B \\ 1, & c, & C \end{vmatrix} (B + C)(C + A)(A + B)$$

$$= \begin{vmatrix} 1, & a, & A^2 \\ 1, & b, & B^2 \\ 1, & c, & C^2 \end{vmatrix} (A^2 + B^2 + C^2 + BC + CA + AB) - \begin{vmatrix} 1, & a, & A^4 \\ 1, & b, & B^4 \\ 1, & c, & C^4 \end{vmatrix}.$$

De même le dénominateur de $S(u+v+w)$ depend de l'équation analogue

$$\begin{vmatrix} 1, & a, & aA \\ 1, & b, & bB \\ 1, & c, & cC \end{vmatrix} (B+C)(C+A)(A+B)$$

$$= \begin{vmatrix} 1, & a, & aA^2 \\ 1, & b, & bB^2 \\ 1, & c, & cC^2 \end{vmatrix} (A^2+B^2+C^2+BC+CA+AB) - \begin{vmatrix} 1, & a, & aA^4 \\ 1, & b, & bB^4 \\ 1, & c, & cC^4 \end{vmatrix};$$

et le numérateur et le dénominateur de $S(u+v+w+p)$ dependent de l'équation

$$\begin{vmatrix} 1, & a, & a^2, & aA \\ 1, & b, & b^2, & bB \\ 1, & c, & c^2, & cC \\ 1, & d, & d^2, & dD \end{vmatrix} (A+B)(A+C)(A+D)(B+C)(B+D)(C+D)$$

$$= M \begin{vmatrix} 1, & a, & a^2, & aA^2 \\ 1, & b, & b^2, & bB^2 \\ 1, & c, & c^2, & cC^2 \\ 1, & d, & d^2, & dD^2 \end{vmatrix} - N \begin{vmatrix} 1, & a, & a^2, & aA^4 \\ 1, & b, & b^2, & bB^4 \\ 1, & c, & c^2, & cC^4 \\ 1, & d, & d^2, & dD^4 \end{vmatrix} + P \begin{vmatrix} 1, & a, & a^2, & aA^6 \\ 1, & b, & b^2, & bB^6 \\ 1, & c, & c^2, & cC^6 \\ 1, & d, & d^2, & dD^6 \end{vmatrix},$$

dans laquelle

$$M = 2ab^2c^2 + \ldots + a^3b^2 + \ldots + a^3bc + \ldots + 3a^2bcd + \ldots,$$

$$N = (a+b+c+d)(a^2+b^2+c^2+d^2) + (abc+bcd+cda+dab),$$

$$P = a+b+c+d,$$

et de l'équation

$$\begin{vmatrix} 1, & a, & A, & aA \\ 1, & b, & B, & bB \\ 1, & c, & C, & cC \\ 1, & d, & D, & dD \end{vmatrix} (A+B)(A+C)(A+D)(B+C)(B+D)(C+D)$$

$$= (a^2b^2 + \ldots + abc^2 + \ldots + 2abcd) \begin{vmatrix} 1, & a, & A^2, & aA^2 \\ 1, & b, & B^2, & bB^2 \\ 1, & c, & C^2, & cC^2 \\ 1, & d, & D^2, & dD^2 \end{vmatrix} - \begin{vmatrix} 1, & a, & A^4, & aA^4 \\ 1, & b, & B^4, & bB^4 \\ 1, & c, & C^4, & cC^4 \\ 1, & d, & D^4, & dD^4 \end{vmatrix};$$

mais je n'ai pas encore trouvé la loi générale de ces équations.

Pour faciliter l'usage des symboles S, C, G, je veux exprimer par cette notation les propriétés les plus simples des fonctions elliptiques. Cela me donnera aussi l'opportunité d'arranger d'une manière particulière les formules qui se rapportent à la somme ou à la différence de deux arcs. On a d'abord

$$C^2u = 1 - \frac{1}{k} \cdot S^2u,$$

$$G^2u = 1 - k \cdot S^2u,$$

$$S'u = Cu \cdot Gu,$$

$$C'u = -\frac{1}{k} \cdot Su \cdot Gu,$$

$$G'u = -k \cdot Su \cdot Cu,$$

$$S(0) = 0, \qquad C(0) = 1, \qquad G(0) = 1,$$

$$S'(0) = 1, \qquad C'(0) = 0, \qquad G'(0) = 0,$$

$$S(-u) = -S(u), \qquad C(-u) = C(u), \qquad G(-u) = G(v),$$

$$S(\tfrac{1}{2}\upsilon) = \sqrt{k}, \qquad S(\tfrac{1}{2}\aleph) = \frac{1}{\sqrt{k}}, \qquad S(\tfrac{1}{2}\upsilon + \tfrac{1}{2}\aleph) = \infty,$$

$$C(\tfrac{1}{2}\upsilon) = 0, \qquad C(\tfrac{1}{2}\aleph) = \frac{ik'}{k}, \qquad C(\tfrac{1}{2}\upsilon + \tfrac{1}{2}\aleph) = \infty,$$

$$G(\tfrac{1}{2}\upsilon) = k', \qquad G(\tfrac{1}{2}\aleph) = 0, \qquad G(\tfrac{1}{2}\upsilon + \tfrac{1}{2}\aleph) = \infty,$$

$$S(u + \tfrac{1}{2}\upsilon) = \frac{\sqrt{k} \cdot Cu}{Gu}, \qquad S(u + \tfrac{1}{2}\aleph) = \frac{1}{\sqrt{k}} \cdot \frac{Gu}{Cu},$$

$$C(u + \tfrac{1}{2}\upsilon) = \frac{-k'}{\sqrt{k}} \cdot \frac{Su}{Gu}, \qquad C(u + \tfrac{1}{2}\aleph) = \frac{ik'}{k} \cdot \frac{1}{Cu},$$

$$G(u + \tfrac{1}{2}\upsilon) = \frac{k'}{Gu}, \qquad G(u + \tfrac{1}{2}\aleph) = \frac{-ik'}{\sqrt{k}} \cdot \frac{Su}{Gu},$$

$$S(u + \tfrac{1}{2}\upsilon + \tfrac{1}{2}\aleph) = -\frac{1}{Su},$$

$$C(u + \tfrac{1}{2}\upsilon + \tfrac{1}{2}\aleph) = \frac{-i}{\sqrt{k}} \cdot \frac{Gu}{Su},$$

$$G(u + \tfrac{1}{2}\upsilon + \tfrac{1}{2}\aleph) = i\sqrt{k} \cdot \frac{Cu}{Su},$$

$$S(u + m\upsilon + n\aleph) = (-1)^{m+n} \cdot Su,$$

$$C(u + m\upsilon + n\aleph) = (-1)^{m} \cdot Cu,$$

$$G(u + m\upsilon + n\aleph) = (-1)^{n} \cdot Gu.$$

Dans ces équations les symboles S, C, G, v, δ, k, k', i et S, G, C, δ, v, $\frac{1}{k}$, $\frac{ik'}{k}$, $-i$ peuvent être échangés les uns d'avec les autres. Les formules fondamentales qui se rapportent à deux arcs sont

$$S(u+v) = \frac{Su \,.\, Cv \,.\, Gv + Sv \,.\, Cu \,.\, Gu}{1 - S^2u \,.\, S^2v},$$

$$C(u+v) = \frac{Cu \,.\, Cv - \frac{1}{k} \,.\, Su \,.\, Gu \,.\, Sv \,.\, Gv}{1 - S^2u \,.\, S^2v},$$

$$G(u+v) = \frac{Gu \,.\, Gv - k \,.\, Su \,.\, Cu \,.\, Sv \,.\, Cv}{1 - S^2u \,.\, S^2v},$$

auxquelles on ajoutera :

$$C(u+v)\,G(u+v) = \frac{(1 + S^2u \,.\, S^2v)(Cu \,.\, Gu \,.\, Cv \,.\, Gv) - Su \,.\, Sv\,(\alpha - 2S^2u - 2S^2v + \alpha S^2u \,.\, S^2v)}{(1 - S^2u \,.\, S^2v)^2}.$$

Mais pour trouver toutes les formes différentes de ces équations, mettons pour abréger (en supposant comme auparavant $Su = x$, &c.):

$$A = xY, \qquad A' = yX,$$

$$B = X_{,}Y_{,}, \qquad B' = -\frac{1}{k}xyX_{,,}Y_{,,},$$

$$C = X_{,,}Y_{,,}, \qquad C' = -kxyX_{,}Y_{,},$$

$$P = x^2 - y^2,$$

$$Q = 1 - \frac{1}{k}x^2 - \frac{1}{k}y^2 + x^2y^2,$$

$$R = 1 - kx^2 - ky^2 + x^2y^2,$$

$$S = XY, \qquad S' = -\frac{k'^2}{k}xy,$$

$$T = xX_{,,}Y_{,}, \qquad T' = -yY_{,,}X_{,},$$

$$U = xX_{,}Y_{,,}, \qquad U' = yY_{,}X_{,,},$$

$$K = 1 - x^2y^2.$$

Alors on aura

$$S(u+v) = \frac{A + A'}{K} = \frac{P}{A - A'} = \frac{U + U'}{B - B'} = \frac{T - T'}{C - C'},$$

$$C(u+v) = \frac{B + B'}{K} = \frac{U - U'}{A - A'} = \frac{Q}{B - B'} = \frac{S + S'}{C - C'},$$

$$G(u+v) = \frac{C + C'}{K} = \frac{T + T'}{A - A'} = \frac{S - S'}{B - B'} = \frac{R}{C - C'},$$

et les valeurs correspondantes de $S(u-v)$, $C(u-v)$, $G(u-v)$ se trouveront en échangeant les signes de A', B', C', S', T', U'.

[Reverting to the functions sin am, cos am, Δ am, or say sn, cn, dn, instead of S, C, D, and introducing Dr Glaisher's very convenient notation s_1, c_1, d_1 for the sn, cn, dn of u, and s_2, c_2, d_2 for those of v, the formulæ just obtained may be written

$$\operatorname{sn}(u+v)=\frac{s_1c_2d_2+s_2c_1d_1}{1-k^2s_1^2s_2^2}=\frac{s_1^2-s_2^2}{s_1c_2d_2-s_2c_1d_1}=\frac{s_1c_1d_2+s_2c_2d_1}{c_1c_2+s_1d_1s_2d_2}=\frac{s_1d_1c_2+s_2d_2c_1}{d_1d_2+k^2s_1c_1s_2c_2},$$

$$\operatorname{cn}(u+v)=\frac{c_1c_2-s_1d_1s_2d_2}{1-k^2s_1^2s_2^2}=\frac{s_1c_1d_2-s_2c_2d_1}{s_1c_2d_2-s_2c_1d_1}=\frac{1-s_1^2-s_2^2+k^2s_1^2s_2^2}{c_1c_2+s_1d_1s_2d_2}=\frac{c_1d_1c_2d_2-k'^2s_1s_2}{d_1d_2+k^2s_1c_1s_2c_2},$$

$$\operatorname{dn}(u+v)=\frac{d_1d_2-k^2s_1c_1s_2c_2}{1-k^2s_1^2s_2^2}=\frac{s_1d_1c_2-s_2d_2c_1}{s_1c_2d_2-s_2c_1d_1}=\frac{c_1d_1c_2d_2+k'^2s_1s_2}{c_1c_2+s_1d_1s_2d_2}=\frac{1-k^2s_1^2-k^2s_2^2+k^2s_1^2s_2^2}{d_1d_2+k^2s_1c_1s_2c_2},$$

viz. we have thus a fourfold representation of the addition-equation for each of the three functions.]

De ces formules il peut être tiré un grand nombre d'équations identiques; par exemple celles-ci:

$$(A^2-A'^2)=KP,\ \&c.,\qquad S^2-S'^2=QR,\ \&c.,$$

$$(B+B')(C-C')=K(S+S'),\ \&c.,\qquad (B-B')(C+C')=K(S-S'),\ \&c.,$$

$$BC-B'C'=KS,\ \&c.,\qquad B'C-BC'=KS',\ \&c.,$$

$$(S+S')(T+T')(U+U')=(S-S')(T-T')(U-U')=PQR,$$

$$S'T'U'+S'TU+ST'U+STU'=0,$$

$$STU+ST'U'+S'TU'+S'T'U=PQR,$$

$$(A-A')(S+S')=(C-C')(U-U'),\ \&c.,$$

$$(A+A')(S-S')=(C+C')(U+U'),\ \&c.,$$

$$(A-A')(T-T')=P(C-C'),\ \&c.,$$

$$(A+A')(T+T')=P(C+C'),\ \&c.,$$

$$(A-A')(U+U')=P(B-B'),\ \&c.,$$

$$(A+A')(U-U')=P(B+B'),\ \&c.,$$

&c., &c.,

et ces équations donnent immédiatement et dans les formes les plus simples, les formules qui se rapportent aux sommes et aux produits des fonctions de $u+v$ et $u-v$, par exemple

$$S(u+v)\,S(u-v)=\frac{A^2-A'^2}{K^2},$$

savoir au moyen de la première de ces équations identiques:

$$S(u+v)\,S(u-v)=\frac{P}{K};$$

de manière que toutes ces formules peuvent être considérées comme comprises dans les équations fondamentales et dans ce système d'équations identiques.

95.

NOTE SUR QUELQUES THÉORÈMES DE LA GÉOMÉTRIE DE POSITION.

[From the *Journal für die reine und angewandte Mathematik* (Crelle), tom. XLI. (1851), pp. 66—72. Continued from t. XXXVIII. p. 104, **70.**]

§ VII.

En considérant les soixante droites auxquelles donne lieu le théorème de Pascal, et en appliquant ce théorème aux hexagones différents qui peuvent être formés par six points sur une même conique, M. Kirkman a trouvé que ces soixante droites se coupent trois à trois non seulement dans les vingt points de M. Steiner (points que M. Kirkman nomme les points g), mais aussi dans soixante points h. Il a trouvé aussi qu'il y a quatre-vingt-dix droites J, dont chacune contient deux des points h et un des quarante cinq points p, dans lesquels s'entrecoupent, deux à deux, les droites menées par deux quelconques des six points. Les recherches étendues que M. Kirkman a faites dans la géométrie de position, paraitront dans un numéro prochain du *Cambridge and Dublin Mathematical Journal*, [t. V. (1850), pp. 185—200]. En attendant, M. Kirkman a publié dans le *Manchester Courier* du 27ième Juin 1849, vingt cinq théorèmes qui contiennent les résultats de ses recherches.

Moi, j'ai depuis trouvé que les soixante points h sont situés trois à trois sur vingt droites X. Tous ces théorèmes peuvent être démontrés assez facilement quand on connait la manière suivant laquelle les points et les droites doivent être combinés en construisant les points et les droites h, J, &c. Cela se fait alors d'une manière très simple, au moyen d'une notation que je vais expliquer.

Représentons les six points sur la conique par 1, 2, 3, 4, 5, 6. En combinant ces points deux à deux par les droites 12, 13 &c., les systèmes tels que 12, 34, 56 peuvent être représentés par les combinaisons binaires des six symboles *a*, *b*, *c*, *d*, *e*, *f*, et au moyen de la table qui se trouve § III. de ce mémoire, savoir la table

(A)

12.34.56 = *ac*	13.45.62 = *ab*	14.56.23 = *bd*
12.35.64 = *be*	13.46.25 = *cd*	14.52.36 = *ae*
12.36.45 = *df*	13.42.56 = *ef*	14.53.62 = *cf*
15.62.34 = *de*	16.23.45 = *ce*	
15.63.42 = *bc*	16.24.53 = *ad*	
15.64.23 = *af*	16.25.34 = *bf*:	

le symbole *ac* dénote ici l'ensemble des droites 12, 34, 56; et ainsi de suite.

On voit que pour obtenir les six côtés d'un quelconque des soixante hexagones, il n'y a qu'à combiner les droites correspondantes, par paires, telles que *ab*, *ac*, qui ont une lettre en commun. Cela posé, les hexagones, ou, si l'on veut, les droites derivées de ces hexagones au moyen du théorème de Pascal (droites que je nommerai *droites de Pascal*), peuvent être représentées par les symboles *ab.ac*, &c., conformément à la table que voici:

(B)

213456 = *ab.ac*	214356 = *cf.ca*	215346 = *ed.eb*	216345 = *fb.fd*
564 = *ef.eb*	563 = *db.df*	463 = *fa.fd*	453 = *ec.eb*
645 = *dc.df*	635 = *ea.eb*	634 = *cb.ca*	534 = *ad.ac*
465 = *cd.ca*	365 = *ae.ac*	364 = *bc.be*	354 = *da.df*
546 = *ba.be*	536 = *fc.fd*	436 = *de.df*	435 = *bf.be*
654 = *fe.fd*	653 = *bd.be*	643 = *af.ac*	543 = *ce.ca*
314256 = *ea.ef*	315246 = *cb.cd*	316245 = *ad.ab*	415236 = *af.ae*
562 = *bd.ba*	462 = *af.ab*	452 = *ce.cd*	362 = *cb.cf*
625 = *cf.cd*	624 = *ed.ef*	524 = *fb.fe*	623 = *de.db*
265 = *fc.fe*	264 = *de.dc*	254 = *bf.ba*	263 = *ed.ea*
526 = *ae.ab*	426 = *bc.ba*	425 = *da.dc*	326 = *fa.fc*
652 = *db.dc*	642 = *fa.fe*	542 = *ec.ef*	632 = *bc.bd*
416235 = *ce.cf*	516234 = *ec.ed*		
352 = *ad.ae*	342 = *bf.bc*		
523 = *bf.bd*	423 = *ad.af*		
253 = *fb.fc*	243 = *da.de*		
325 = *ec.ea*	324 = *ce.cb*		
532 = *da.db*	432 = *fb.fa*		

Remarquons maintenant que les droites de Pascal qui passent par un point p, tel que 12.45, sont *ca.ce*, *ba.be*, *ac.ab*, *ec.eb*. Cela étant, le point 12.45 peut être représenté par la notation *cb.ae*, et de cette manière le système complet des points p est représenté par la table suivante :

(*C*)

12.34 = *bd.ef*	14.35 = *ab.de*	23.45 = *ad.bf*
12.35 = *af.cd*	14.36 = *bf.cd*	23.46 = *bc.de*
12.36 = *ab.ce*	14.56 = *af.ce*	23.56 = *ae.cf*
12.45 = *ae.be*	15.23 = *be.cd*	24.35 = *bf.ce*
12.46 = *ad.cf*	15.24 = *ae.df*	24.36 = *af.de*
12.56 = *bf.de*	15.26 = *ac.bf*	24.56 = *ab.ed*
13.24 = *ac.bd*	15.34 = *ab.cf*	25.34 = *ad.ce*
13.25 = *af.be*	15.36 = *ad.fe*	25.36 = *bd.cf*
13.26 = *bd.ec*	15.46 = *bd.ce*	25.46 = *ab.ef*
13.45 = *cf.de*	16.23 = *ab.df*	26.34 = *af.bc*
13.46 = *ae.bf*	16.24 = *cf.be*	26.35 = *ae.bd*
13.56 = *ad.bc*	16.25 = *ac.de*	26.45 = *cd.ef*
14.23 = *ac.ef*	16.34 = *ae.cd*	34.56 = *be.df*
14.25 = *ce.df*	16.35 = *bc.ef*	35.46 = *ac.df*
14.26 = *ad.be*	16.45 = *af.bd*	36.45 = *ac.be*.

Enfin les droites 12, &c. peuvent être représentées par des symboles tels que *ac.be.df*, &c., et au moyen de la table suivante, qui est pour ainsi dire la réciproque de la table (*A*) :

(*D*)

ac.be.df = 12	*ab.ed.fc* = 62	*ae.df.cb* = 36
ac.bd.fe = 56	*ab.ef.cd* = 13	*ae.dc.bf* = 52
ac.bf.ed = 34	*ab.ec.df* = 45	*ae.db.fc* = 14

ad.fb.ce = 16	*af.bc.ed* = 15
ad.fc.eb = 53	*af.be.dc* = 64
ad.fe.be = 24	*af.bd.ce* = 32.

Il y a à remarquer qu'une droite de Pascal *ab.ac* contient les points *bc.ad*, *bc.ae*, *bc.af*, et que par un point *ab.cd* passent les droites (les côtés opposés d'un hexagone) *ac.bd.ef*, *ab.bc.ef*, et les droites de Pascal *ca.cb*, *da.db*, *ac.ad*, *bc.bd*. Cela posé, en combinant les propriétés déjà connues d'avec celles que j'ai énoncées au commencement de cette section, en particularisant en même temps les combinaisons qui donnent lieu aux points et droites g, h, J, &c. et en adoptant une notation convenable pour ces points et droites, on trouvera ce qui suit :

(α) Les droites *ab.bc*, *bc.ca*, *ca.ab* se rencontrent dans un même point *abc* qui est un des vingt points g, et que j'ai dénoté par ce symbole, § III. de ce mémoire.

(β) Les points abc, abd, abe, abf sont situés sur une même droite ab qui est une des quinze droites de M. Steiner ou de M. Plücker, et que j'ai dénotée (§ III.). Je nommerai droites I ces droites.

(γ) Les droites $ab.ac$, $ac.ad$, $ad.ab$ se rencontrent dans un même point $a.ef$ qui est un des soixante points h de M. Kirkman.

(δ) Les points $b.cd$, $c.db$, $d.bc$ sont situés sur la même droite $\{bcd\}$ qui est une de mes vingt droites X.

(ϵ) Les points $ab.cd$, $e.ab$, $f.ab$ sont situés sur une même droite $(ab)\,cd$ qui est un des quatre-vingt-dix points J de M. Kirkman.

Quant aux théorèmes (α) et (β), je vais reproduire dans la notation de cette section les démonstrations de M. Plücker.

Voici le principe de la démonstration du théorème (α): principe qui s'applique aussi, comme nous le verrons, aux démonstrations des théorèmes (γ, δ, et ϵ).

Supposons qu'il s'agit de démontrer généralement que trois droites X, X', X'' se rencontrent dans un même point, et supposons que ces droites sont déterminées:

$$\begin{array}{llll} X & \text{au moyen des points } A\,, & B\,, & C\,, \\ X' & \quad ,, \quad ,, \quad ,, \quad A', & B', & C', \\ X'' & \quad ,, \quad ,, \quad ,, \quad A'', & B'', & C''; \end{array}$$

formons d'abord la table

$$(\odot)\quad \left\{\begin{array}{lll} A'A'', & B'B'', & C'C'', \\ A''A\,, & B''B\,, & C''C\,, \\ A\,A', & B\,B', & C\,C', \end{array}\right.$$

où $A'A''$, &c. sont les droites qui passent par les points A' et A'', &c.; et puis la table

$$(\text{☽})\quad \left\{\begin{array}{lll} B'B''.\,C'C'', & C'C''.\,A'A'', & A'A''.\,B'B'', \\ B''B\;.\,C''C\,, & C''C\;.\,A''A\,, & A''A\;.\,B''B\,, \\ B\,B'.\,C\,C', & C\,C'\;.\,A\,A', & A\,A'\;.\,B\,B', \end{array}\right.$$

où $B'B''.\,C'C''$, &c. sont les points d'intersection des droites $B'B''$ et $C'C''$, &c. On sait que si les points de l'une quelconque des colonnes verticales de cette dernière table sont situés sur la même droite, les droites X, X', X'' se couperont dans un même point; et réciproquement. Précisément de la même manière on démontrerait que trois points X, X', X'' sont situés sur une même droite; seulement A, B, &c. seraient des droites, $A'A''$, &c. des points; et ainsi de suite. Or les droites du théorème (α) sont déterminées,

$$\begin{array}{llll} ab.bc \text{ au moyen des points} & ac.be, & ac.bf, & ac.bd, \\ bc.ca \quad ,, \quad ,, \quad ,, & ba.ce, & ba.cf, & ba.cd, \\ ca.ab \quad ,, \quad ,, \quad ,, & cb.ae, & cb.af, & cb.ad; \end{array}$$

donc la table (⊙) se réduit à

$$\begin{array}{lll} ac.be.df, & ac.bf.de, & ac.bd.ef, \\ ba.ce.df, & ba.cf.de, & ba.cd.ef, \\ cb.ae.df, & cb.af.de, & cb.ad.ef, \end{array}$$

et la table (☽) à

$$\begin{array}{lll} be.df, & bf.de, & bd.ef, \\ ce.df, & cf.de, & cd.ef, \\ ae.df, & af.de, & ad.ef; \end{array}$$

et les points de la première colonne verticale de cette table sont situés sur la droite *ed.ef*, ceux de la deuxième colonne verticale sur la droite *fd.fe*, et ceux de la troisième colonne verticale sur la droite *de.df*: l'existence de l'une quelconque de ces droites fait voir la vérité du théorème dont il s'agit.

Pour démontrer le théorème (β), considérons à part un quelconque des points *abc, abd, abe, abf*; par exemple le point *abf*. On peut envisager ce point comme déterminé par les droites *ab.af, ab.bf*, et ces droites contiennent:

ab.af les points *bf.ac, bf.ad, bf.ae,*
ab.bf „ „ *af.bc, af.bd, af.be.*

Or *bf.ac* et *af.bc* sont situés sur la droite *ab.de.cf*; *bf.ad* et *af.bd* sur la droite *ab.ec.df*; et *bf.ae* et *af.be* sur la droite *ab.cd.ef*. De plus, les points *bf.ac, bf.ad, bf.ae* sont situés sur les droites *ab.bc, ab.bd, ab.be* respectivement, et les points *af.bc, af.bd, af.be* sont situés sur les droites *ab.ac, ab.ad, ab.ae* respectivement. Donc on peut représenter les points de la droite *ab.af* par les symboles

$$(ab.de.cf)(ab.bc), \quad (ab.ec.df)(ab.bd), \quad (ab.cd.ef)(ab.be),$$

et les points de la droite *ab.bf* par les symboles

$$(ab.de.cf)(ab.ac), \quad (ab.ec.df)(ab.ad), \quad (ab.cd.ef)(ab.ae).$$

Maintenant

Les droites	de même que les droites	se rencontrent sur la droite
ab.bd, ab.ae	*ab.be, ab.ad,*	*ab.de.cf,*
ab.be, ab.ac	*ab.bc, ab.ae,*	*ab.ec.df,*
ab.bc, ab.ad	*ab.bd, ab.ac,*	*ab.cd.ef,*

c'est à dire, il existe un système de trois hexagones dont les côtés sont

$$\begin{array}{llllll} (ab.de.cf, & ab.ec.df, & ab.cd.ef, & ab.bc, & ab.bd, & ab.be), \\ (ab.de.cf, & ab.ec.df, & ab.cd.ef, & ab.ac, & ab.ad, & ab.ae), \\ (ab.bc\quad, & ab.bd\quad, & ab.be\quad, & ab.ac, & ab.ad, & ab.ae). \end{array}$$

Ces hexagones ont pour angles les mêmes six points. Or l'existence de l'une ou de l'autre des droites *ab.af, ab.bf* suffit pour faire voir que ces six points sont situés sur la même conique: donc les côtés opposés du troisième hexagone se rencontrent dans trois points situés sur la même droite. De plus, on voit aisément que les hexagones sont précisément tels, qu'en vertu du théorème (α), les trois droites, auxquelles donnent lieu ces hexagones, se rencontrent dans un même point; et ce point sera évidemment le point *abf*. Mais les côtés opposés du troisième hexagone, savoir les droites *ab.bc* et *ab.ac*; *ab.bd* et *ab.ad*; *ab.be* et *ab.ae*, se rencontrent dans les points *abc, abd, abf*: donc les quatre points *abc, abd, abe, abf* sont situés sur la même droite: théorème dont il s'agissait.

Pour démontrer le théorème (γ), on n'a qu'à considérer les droites de ce théorème comme déterminées,

ab.ac	par les points	*bc.ad*,	*bc.ae*,	*bc.af*,
ac.ad	„ „	*cd.ab*,	*cd.ae*,	*cd.af*,
ad.ab	„ „	*bd.ac*,	*bd.ae*,	*bd.af*.

La table (⊙) se réduit alors à

bc.ad.ef,	*da.de*,	*da.df*,
cd.ab.ef,	*ba.be*,	*ba.bf*,
bd.ac.ef,	*ca.ce*,	*ca.cf*,

et la table (☽) à

(da.df)(da.de),	*af.de*,	*ae.df*,
(ba.bf)(ba.be),	*af.be*,	*ae.bf*,
(ca.cf)(ca.ce),	*af.ce*,	*ae.cf*.

Or les points de la deuxième colonne verticale sont situés sur la droite *ea.ef*, et les points de la troisième colonne verticale sur la droite *fa.fe*. L'existence de l'une ou de l'autre de ces droites fait voir que les droites *ab.ac, ac.ad, ad.ab* se rencontrent dans un même point *a.be*.

Pour démontrer le théorème (δ), il est évident que les points de la première colonne verticale de la table qui vient d'être présentée, sont situés sur la même droite. Mais ces points sont précisément les points *d.bc, b.cd, c.db*; le théorème est donc démontré.

[En remarquant que les trois droites sur lesquelles sont situés les neuf points de la table générale (☽) se rencontrent dans un même point, cette démonstration de l'existence des droites X fait voir que la droite {*bcd*} passe par le point d'intersection des droites *ae.af, ef.fa*, savoir par le point *afe*; ou bien que chacune des vingt droites X passe, non seulement par trois points h, mais aussi par un seul point g. Ce théorème est dû à M. Salmon, qui indépendamment de mes recherches à trouvé l'existence des vingt droites X. 8 août 1849. *Inserted here from* Crelle, t. XLI. p. 84.]

Enfin, pour démontrer le théorème (ϵ), nous pouvons considérer les points de ce théorème comme déterminés,

$ab.cd$	par les droites	$bc.bd$,	$ad.be.ef$,	$ac.bd.ef$,
$f.ab$	„ „	$fc.fd$,	$fc.fe$,	$fd.fe$,
$e.ab$	„ „	$ec.ed$,	$ec.fe$,	$ed.fe$.

La table (⊙) se réduit alors à

$$\begin{array}{lll} cd.ef, & f.ce, & f.de, \\ cd.eb, & cf.eb, & df.eb, \\ cd.bf, & ce.bf, & de.bf, \end{array}$$

et la table (☽) à

$$\begin{array}{lll} fe\ , & ce.cf, & de.df, \\ fe.fb, & cb.ce, & db.de, \\ fe.eb, & cb.cf, & db.df. \end{array}$$

Or les droites de la première colonne verticale de cette table se rencontrent dans le point $bf.be$, celles de la deuxième colonne verticale dans le point $c.ad$, et celles de la troisième colonne verticale dans le point $d.ac$; le théorème dont il s'agit est donc démontré. Dans cette démonstration on aurait aussi pu échanger les lettres a, b.

Les théorèmes (α) et (γ) peuvent être énoncés par le seul théorème suivant:

"Étant donnés six points sur la même conique, et menant par ces points neuf droites, de manière que chaque droite passe par deux points et que par chaque point il passe trois droites: on formera avec ces neuf droites trois hexagones différents dont chacun a les six points pour angles. Les droites de Pascal, auxquelles donnent lieu ces trois hexagones, se rencontreront dans un même point."

En supposant que le système de neuf droites contient toujours un même hexagone, il est possible de compléter de quatre manières différentes le système des neuf droites; savoir, on peut ajouter aux côtés de l'hexagone 1 les trois diagonales de l'hexagone 2, 3 ou 4, en menant une quelconque de ces diagonales et deux droites, chacune par deux angles alternés de l'hexagone. Ces quatre systèmes donnent lieu au point g, et aux trois points h, qui se trouvent sur la droite de Pascal, correspondante à l'hexagone dont il s'agit; savoir le premier système donne lieu au point g, et les trois derniers systèmes au point h.

96.

MÉMOIRE SUR LES CONIQUES INSCRITES DANS UNE MÊME SURFACE DU SECOND ORDRE.

[From the *Journal für die reine und angewandte Mathematik* (Crelle), tom. XLI. (1851), pp. 73—86.]

En considérant une surface quelconque du second ordre, le problème se presente: d'examiner les proprietés des coniques inscrites dans cette surface et des cônes circonscrits. La plupart de ces propriétés est peut-être connue[1]; cependant je crois qu'on ne les a pas encore développées systématiquement. Je me propose de donner ici l'analyse des propriétés les plus simples d'un tel système de coniques, et la solution du problème analogue au problème des tactions qui se présente ici, ainsi que quelques théorèmes relatifs au passage à un système de coniques situées dans un même plan et inscrites dans une même conique, en me réservant pour une seconde partie de ce mémoire les développements ultérieurs concernant ce passage, et la solution complète du problème analogue au problème de Malfatti, généralisé par M. Steiner.

Remarquons d'abord que les coniques inscrites et les cônes circonscrits, ainsi que les plans des coniques inscrites et les sommets des cônes circonscrits, sont des figures *réciproques* par rapport à la surface du second ordre. En considérant deux coniques inscrites quelconques, et les cônes circonscrits correspondants, on remarquera que les plans des coniques inscrites se rencontrent dans une droite. Je la nommerai *Droite de symptose.* Les sommets des cônes circonscrits seront situés dans une droite que je nommerai *Droite d'homologie.* Ces deux droites seront évidemment réciproques l'une à l'autre. Il se trouvera sur la droite d'homologie deux points dont chacun est le sommet d'un cône qui passe par les deux coniques inscrites. Ces deux points peuvent être nommés *Points d'homologie.* De même il passera par la droite de *symptose* deux plans,

[1] Voyez le mémoire de M. Steiner "Einige geometrische Betrachtungen" Journal t. I. [1826] pp. 161—184, et un mémoire de M. Olivier, Quetelet, Corresp. Math. t. V. [1829].

qui sont les plans des coniques dans lesquelles se coupent les deux cônes circonscrits. Ces deux plans peuvent être nommés *Plans de symptose.* Les plans de symptose et les points d'homologie ne sont pas seulement des figures réciproques : les deux plans de symptose passent aussi par les deux points d'homologie, chacun par le point réciproque de l'autre plan ; c'est à dire : les plans de symptose sont des plans *conjugués* par rapport à la surface du second ordre, et les points d'homologie sont des points *conjugués* par rapport à cette même surface. Remarquons aussi qu'en considérant le système formé par les plans des coniques inscrites et les plans tangents à la surface menés par la droite de symptose, on trouvera que les plans de symptose sont les plans doubles (ou si l'on veut les plans *auto-conjugués*) de l'involution. De même, en considérant le système formé par les sommets des cônes circonscrits et par les points de leur intersection avec la surface de la droite d'homologie, on trouvera que les points d'homologie sont les points doubles (ou *auto-conjugués*) de l'involution. Les deux cônes circonscrits qui ont pour sommets les deux points d'homologie, peuvent être nommés *Cônes d'homologie ;* de même, les deux coniques inscrites, situées dans les deux plans de symptose, peuvent être nommées *Coniques de symptose.* (En passant, nous remarquerons que ces coniques de symptose correspondent aux *Potenzkreise* de M. Steiner.) Il est évident que les cônes d'homologie et les coniques de symptose sont des figures réciproques.

En considérant trois coniques inscrites, et les cônes circonscrits correspondants, on verra que les plans des coniques inscrites se rencontrent dans un point que je nommerai *Point de symptose.* Les sommets des cônes circonscrits seront situés dans un plan que je nommerai *Plan d'homologie.* Ce point et le plan seront réciproques l'un à l'autre. En combinant deux à deux les coniques inscrites ou les cônes circonscrits, cela donne lieu à trois droites de symptose qui passent chacune par le point de symptose, et à trois droites d'homologie situées chacune dans le plan d'homologie. Il existe aussi six plans de symptose qui se coupent trois à trois dans quatre droites, arêtes d'une pyramide quadrilatère qui a pour axes les trois droites de symptose. Les quatre droites dont il s'agit, peuvent être nommées *Axes de symptose.* Il existe également six points d'homologie, situés trois à trois dans quatre droites, côtés d'un quadrilatère qui a pour axes les trois droites d'homologie. Les quatre droites dont il s'agit, peuvent être nommées *Axes d'homologie.* La pyramide et le quadrilatère sont des figures réciproques, et il convient de remarquer (quoique cela soit assez évident) qu'il y a ici trois points d'homologie, non-situés dans un des côtés du quadrilatère, mais contenus dans trois points de symptose qui se coupent dans une arête de la pyramide.

Par l'un quelconque des axes d'homologie il passe deux plans dont chacun touche les trois coniques inscrites ; de même il se trouve sur l'un quelconque des axes de symptose deux points, dont chacun est un point d'intersection des trois cônes circonscrits. Cela constitue la solution du problème : Trouver la conique inscrite, ou le cône circonscrit, qui touche trois coniques inscrites, ou trois cônes circonscrits. Il y a *huit* solutions de ce problème.

Avant d'aller plus loin, je vais indiquer quelques-unes des formules analytiques correspondantes à la théorie qui vient d'être expliquée.

Écrivons, pour abréger:

$$U = Ax^2 + By^2 + Cz^2 + Dw^2 + 2Fyz + 2Gxz + 2Hxy + 2Lxw + 2Myw + 2Nzw,$$
$$V = \alpha x \ + \beta y \ + \gamma z \ + \delta w,$$

et représentons par $\mathfrak{A}$, $\mathfrak{B}$, $\mathfrak{C}$, $\mathfrak{D}$, $\mathfrak{F}$, $\mathfrak{G}$, $\mathfrak{H}$, $\mathfrak{L}$, $\mathfrak{M}$, $\mathfrak{N}$ les coefficients du système inverse de A, B, C, D, F, G, H, L, M, N. Soit de plus

$$X = Ax + Hy + Gz + Lw,$$
$$\vdots$$
$$p^2 = \mathfrak{A}\alpha^2 + \mathfrak{B}\beta^2 + \mathfrak{C}\gamma^2 + \mathfrak{D}\delta^2 + 2\mathfrak{F}\beta\gamma + 2\mathfrak{G}\gamma\alpha + 2\mathfrak{H}\alpha\beta + 2\mathfrak{L}\alpha\delta + 2\mathfrak{M}\beta\delta + 2\mathfrak{N}\gamma\delta.$$

Cela posé, en prenant $U=0$ pour équation de la surface du second ordre, et $V_1=0$, $V_2=0$, $V_3=0$ pour les équations des plans des coniques inscrites, on obtient pour l'un des plans de symptose des coniques inscrites $(U=0,\ V_1=0)$, $(U=0,\ V_2=0)$ l'équation très simple $p_2V_1 = p_1V_2 = 0$. De là on tire pour les coordonnées du point d'homologie, qui est le réciproque de ce plan de symptose, les équations

$$X : Y : Z : W = p_2\alpha_1 - p_1\alpha_2 : p_2\beta_1 - p_1\beta_2 : p_2\gamma_1 - p_1\gamma_2 : p_2\delta_1 - p_1\delta_2.$$

En formant également les expressions des coordonnées d'un point d'homologie des deux autres paires de coniques inscrites, on obtient pour équation de l'un des axes d'homologie:

$$\begin{vmatrix} & X, & Y, & Z, & W \\ p_1, & \alpha_1, & \beta_1, & \gamma_1, & \delta_1 \\ p_2, & \alpha_2, & \beta_2, & \gamma_2, & \delta_2 \\ p_3, & \alpha_3, & \beta_3, & \gamma_3, & \delta_3 \end{vmatrix} = 0;$$

savoir, en choisissant quatre colonnes verticales quelconques de cette formule, on trouve que les déterminants que l'on obtient, sont tous égaux à zéro.

Nous ajouterons que la droite qui, par rapport à la conique $(U=0,\ V_1=0)$, est réciproque de cet axe d'homologie, est donnée par les équations

$$V_1 = 0, \quad V_2 : V_3 = p_1p_2 - \mathfrak{A}\alpha_1\alpha_2 - \ldots : p_1p_3 - \mathfrak{A}\alpha_1\alpha_3 - \ldots;$$

il est clair que cette droite rencontre la surface du second ordre en deux points situés dans les plans des coniques inscrites qui, au moyen de l'axe d'homologie dont l'équation vient être donnée, sont déterminés de manière à toucher les trois coniques inscrites données. Mais sans se servir des équations de cette droite, on peut déterminer l'équation des deux plans menés par l'axe d'homologie dont il s'agit, de manière à toucher la conique inscrite $(U=0,\ V_1=0)$; et la symétrie du résultat fera voir que ces deux plans touchent aussi deux autres coniques inscrites. La recherche de cette équation étant un peu difficile, je la donnerai en détail, en supposant cependant connu le théorème suivant:

En écrivant $v = \lambda x + \mu y + \nu z + \rho w$, $v' = \lambda' x + \mu' y + \nu' z + \rho' w$: les plans menés par la droite $(v=0,\ v'=0)$ de manière qu'ils touchent la conique inscrite $(U=0,\ V=0)$, sont donnés par l'équation

$$p^2\left[\mathfrak{A}(\lambda v' - \lambda' v)^2 + \ldots\right] - \left[\mathfrak{A}\alpha(\lambda v' - \lambda' v) + \ldots\right]^2 = 0.$$

Pour appliquer ce théorème au problème dont il s'agit, nous n'avons qu'à substituer V_1 au lieu de V, et qu'à écrire

$$v = \begin{vmatrix} & a, & b, & c, & d \\ & X, & Y, & Z, & W \\ p_1, & \alpha_1, & \beta_1, & \gamma_1, & \delta_1 \\ \vdots & & & & \end{vmatrix}, \quad v' = \begin{vmatrix} & a', & b', & c', & d' \\ & X, & Y, & Z, & W \\ p_1, & \alpha_1, & \beta_1, & \gamma_1, & \delta_1 \\ \vdots & & & & \end{vmatrix}$$

où les coefficients a, b, c, d; a', b', c', d' sont des quantités quelconques.

Réduisons d'abord l'expression $\mathfrak{A}\alpha_1(\lambda v' - \lambda' v) + \dots$. Pour cela, mettons dans les valeurs de v, v', les expressions $\mathfrak{A}\alpha_1 + \dots$, $\mathfrak{H}\alpha_1 + \dots$, &c. à la place de $x, y, \dots$: les quantités X, Y, Z, W deviennent alors $K\alpha_1$, $K\beta_1$, $K\gamma_1$, $K\delta_1$ (où comme à l'ordinaire K est le déterminant formé par les quantités $A, B, \dots$), et l'on obtient ainsi, aux signes près :

$$\mathfrak{A}\alpha_1\lambda + \dots = Kp_1 \begin{vmatrix} a, & b, & c, & d \\ \alpha_1, & \beta_1, & \gamma_1, & \delta_1 \\ \vdots & & & \end{vmatrix}, \quad \mathfrak{A}\alpha_1\lambda' + \dots + = Kp_1 \begin{vmatrix} a', & b', & c', & d' \\ \alpha_1, & \beta_1, & \gamma_1, & \delta_1 \\ \vdots & & & \end{vmatrix}$$

et de là

$$\mathfrak{A}\alpha_1(\lambda v' - \lambda' v) + \dots = Kp_1 \,\Box\,,$$

où

$$\Box = \begin{vmatrix} a, & b', & c, & d \\ \alpha_1, & \beta_1, & \gamma_1, & \delta_1 \\ \vdots & & & \end{vmatrix} \begin{vmatrix} & a', & b', & c', & d' \\ & X, & Y, & Z, & W \\ p_1, & \alpha_1, & \beta_1, & \gamma_1, & \delta_1 \\ \vdots & & & & \end{vmatrix} - \begin{vmatrix} a', & b', & c', & d' \\ \alpha_1, & \beta_1, & \gamma_1, & \delta_1 \\ \vdots & & & \end{vmatrix} \begin{vmatrix} & a, & b, & c, & d \\ & X, & Y, & Z, & W \\ p_1, & \alpha_1, & \beta_1, & \gamma_1, & \delta_1 \\ \vdots & & & & \end{vmatrix}:$$

formule qui au moyen des propriétés des déterminants se réduit à

$$\Box = \begin{vmatrix} X, & Y, & Z, & W \\ \alpha_1, & \beta_1, & \gamma_1, & \delta_1 \\ \vdots & & & \end{vmatrix} \begin{vmatrix} & a, & b, & c, & d, \\ & a', & b', & c', & d', \\ p_1, & \alpha_1, & \beta_1, & \gamma_1, & \delta_1, \\ \vdots & & & & \end{vmatrix}$$

Passons à l'expression $\mathfrak{A}(\lambda v' - \lambda' v)^2 + \dots$, que nous mettrons sous la forme

$$\frac{1}{K}\{A\,[\mathfrak{A}(\lambda v' - \lambda' v)^2 + \dots]^2 + \dots\}.$$

En prenant des quantités quelconques $\mathfrak{a}$, $\mathfrak{b}$, $\mathfrak{c}$, $\mathfrak{d}$, on obtient par une analyse semblable :

$$\mathfrak{A}\mathfrak{a}(\lambda v' - \lambda' v) + \dots = K\,\overline{\Box}\,,$$

où

$$\overline{\Box} = \left|\begin{array}{ccccc|ccccc} & \mathfrak{a}, & \mathfrak{b}, & \mathfrak{c}, & \mathfrak{d} & & a', & b', & c', & d' \\ & a, & b, & c, & d & & X, & Y, & Z, & W \\ p_1, & \alpha_1, & \beta_1, & \gamma_1, & \delta_1 & p_1, & \alpha_1, & \beta_1, & \gamma_1, & \delta_1 \\ \vdots & & & & & \vdots & & & & \end{array}\right| - \left|\begin{array}{ccccc|ccccc} & \mathfrak{a}, & \mathfrak{b}, & \mathfrak{c}, & \mathfrak{d} & & a, & b, & c, & d \\ & a', & b', & c', & d' & & X, & Y, & Z, & W \\ p_1, & \alpha_1, & \beta_1, & \gamma_1, & \delta_1 & p_1, & \alpha_1, & \beta_1, & \gamma_1, & \delta_1 \\ \vdots & & & & & \vdots & & & & \end{array}\right| :$$

formule qui se réduit à

$$\overline{\Box} = \left|\begin{array}{ccccc|ccccc} & \mathfrak{a}, & \mathfrak{b}, & \mathfrak{c}, & \mathfrak{d} & & a, & b, & c, & d \\ & X, & Y, & Z, & W & & a', & b', & c', & d' \\ p_1, & \alpha_1, & \beta_1, & \gamma_1, & \delta_1 & p_1, & \alpha_1, & \beta_1, & \gamma_1, & \delta_1 \\ \vdots & & & & & \vdots & & & & \end{array}\right| .$$

De là, en supprimant les facteurs constants de $\Box$ et de $\overline{\Box}$, on obtient

$$\mathfrak{a}\xi + \mathfrak{b}\eta + \mathfrak{c}\zeta + \mathfrak{d}\omega = \left|\begin{array}{ccccc} & \mathfrak{a}, & \mathfrak{b}, & \mathfrak{c}, & \mathfrak{d} \\ & X, & Y, & Z, & W \\ p_1, & \alpha_1, & \beta_1, & \gamma_1, & \delta_1 \\ \vdots & & & & \end{array}\right| ,$$

expression qui sert à définir les fonctions ξ, η, ζ, ω. L'équation qu'il s'agissait de trouver devient

$$A\xi^2 + \ldots - K \left|\begin{array}{cccc} X, & Y, & Z, & W \\ \alpha_1, & \beta_1, & \gamma_1, & \delta_1 \\ \vdots & & & \end{array}\right|^2 = 0,$$

où il faut avoir égard que l'on a $X = Ax + \ldots$, &c. Savoir, l'équation qu'on vient d'écrire se décompose nécessairement en facteurs linéaires qui, égalés à zéro, donnent les équations des plans des coniques inscrites qui touchent chacune les trois coniques inscrites données.

Nous avons obtenu ce résultat en traduisant en analyse une construction géométrique; mais on y peut aussi parvenir en considérant le problème d'une manière purement analytique. En effet: soient comme plus haut, $U = 0$ l'équation de la surface du second ordre, $V_1 = 0$, $V_2 = 0$, $V_3 = 0$ les équations des plans des trois coniques inscrites données, $V = 0$ l'équation du plan de la conique inscrite qui touche chacune de ces trois coniques. La condition pour que cette conique touche la conique inscrite située dans le plan $V_1 = 0$, est $\mathfrak{A}\alpha\alpha_1 + \ldots = pp_1$. On a donc les trois équations

$$\mathfrak{A}\alpha\alpha_1 + \ldots = pp_1,$$
$$\mathfrak{A}\alpha\alpha_2 + \ldots = pp_2,$$
$$\mathfrak{A}\alpha\alpha_3 + \ldots = pp_3.$$

Au lieu de tirer de ces équations les quantités $\alpha : \beta : \gamma : \delta$, nous ajouterons au système la nouvelle équation

$$\alpha x + \ldots = 0,$$

par laquelle il sera possible d'éliminer les quatre quantités α, β, γ, δ. En attribuant à $X, \dots$ la même signification qu'auparavant, nous mettrons les quatre équations sous la forme

$$(\mathfrak{A}\alpha + \dots)\,\alpha_1 + \dots = pp_1,$$
$$(\mathfrak{A}\alpha + \dots)\,\alpha_2 + \dots = pp_2,$$
$$(\mathfrak{A}\alpha + \dots)\,\alpha_3 + \dots = pp_3,$$
$$(\mathfrak{A}\alpha + \dots)\,X + \dots = 0\,.$$

Écrivons de plus

$$(\mathfrak{A}\alpha + \dots)\,\mathfrak{a} + \dots = p\Theta,$$

où les quantités $\mathfrak{a}$, $\mathfrak{b}$, $\mathfrak{c}$, $\mathfrak{d}$ sont arbitraires. En éliminant de ces équations les fonctions $(\mathfrak{A}\alpha + \dots)$, puis en mettant à la place de $p\Theta$ la quantité à gauche de l'équation, on obtient, à un facteur constant près,

$$(\mathfrak{A}\alpha + \dots)\,\mathfrak{a} + \dots = \begin{vmatrix} & \mathfrak{a}, & \mathfrak{b}, & \mathfrak{c}, & \mathfrak{d} \\ & X, & Y, & Z, & W \\ p_1, & \alpha_1, & \beta_1, & \gamma_1, & \delta_1 \\ \vdots & & & & \end{vmatrix};$$

cela donne d'abord

$$\mathfrak{A}\alpha\alpha_1 + \dots = p_1 \begin{vmatrix} X, & Y, & Z, & W \\ \alpha_1, & \beta_1, & \gamma_1, & \delta_1 \\ \vdots & & & \end{vmatrix},$$

puis, en écrivant $p^2 = \mathfrak{A}\alpha^2 + \dots = \dfrac{1}{K}[A\,(\mathfrak{A}\alpha + \dots)^2 + \dots]$, on a

$$p^2 = \frac{1}{K}(A\xi^2 + \dots),$$

et

$$\mathfrak{a}\xi + \dots = \begin{vmatrix} & \mathfrak{a}, & \mathfrak{b}, & \mathfrak{c}, & \mathfrak{d} \\ & X, & Y, & Z, & W \\ p_1, & \alpha_1, & \beta_1, & \gamma_1, & \delta_1 \\ \vdots & & & & \end{vmatrix},$$

et de là enfin, en substituant dans l'équation $\mathfrak{A}\alpha\alpha_1 + \dots = pp_1$, on obtient comme plus haut l'équation

$$A\xi^2 + \dots - K \begin{vmatrix} X, & Y, & Z, & W \\ \alpha_1, & \beta_1, & \gamma_1, & \delta_1 \\ \vdots & & & \end{vmatrix}^2 = 0.$$

Il est clair que cette analyse peut être appliquée à la solution d'un nombre quelconque d'équations de la forme $\mathfrak{A}\alpha\alpha_1 + \dots = pp_1$.

En revenant à la théorie *géométrique*, considérons un point quelconque que nous prendrons pour point de projection : le cône qui passe par une conique inscrite quelconque aura (comme on sait) un contact double avec le cône qui a pour sommet le point de projection. Le plan de contact sera le plan mené par le point de projection et par la droite d'intersection du plan de la conique inscrite et du plan réciproque au point de projection. En considérant plusieurs coniques inscrites ayant une droite de symptose commune, tous les cônes auxquels donnent lieu ces coniques inscrites, auront pour arêtes communes les deux droites menées par le point de projection aux points dans lesquels la surface est rencontrée par la droite de symptose commune. Ajoutons que les plans de contact des cônes dont il s'agit, avec le cône circonscrit, rencontrent le plan des deux arêtes communes dans une droite fixe, savoir dans l'une ou l'autre des droites doubles (ou auto-conjuguées) de l'involution formée par les deux arêtes communes et par les droites dans lesquelles le plan de ces deux arêtes communes rencontre le cône circonscrit. De plus, en considérant les plans tangents menés par l'une ou par l'autre des deux arêtes communes, ces plans tangents forment un système homologue à celui des plans des coniques inscrites. En considérant en particulier l'une ou l'autre des coniques de symptose de deux coniques inscrites quelconques : le plan tangent du cône correspondant est le plan double (ou auto-conjugué) de l'involution formée par les plans tangents des cônes qui correspondent aux deux coniques inscrites (c'est à dire par les plans tangents qui passent par l'arête commune dont il s'agit), et par les plans tangents de la surface du second ordre menés par cette même arête commune. C'est là en effet la propriété qui conduit à la construction des coniques de symptose de deux coniques situées dans le même plan et considerées comme inscrites dans une conique donnée.

97.

NOTE SUR LA SOLUTION DE L'ÉQUATION $x^{257}-1=0$.

[From the *Journal für die reine und angewandte Mathematik* (Crelle), tom. XLI. (1851), pp. 81—83.]

SOIT p_m la $m^{\text{ième}}$ puissance d'une racine quelconque (l'unité exceptée) de l'équation $x^{257}-1=0$, et représentons par α une racine quelconque (l'unité exceptée) de l'equation $\alpha^{256}-1=0$. En posant l'équation

$$(p_0+\alpha p_1+\alpha^2 p_2 \ldots + \alpha^{255} p_{255})^2 = M\,(p_0+\alpha^2 p_1+\alpha^4 p_2 \ldots + \alpha^{254} p_{255}),$$

on sait que la quantité M peut être exprimée en fonction rationnelle de α. Cette fonction une fois connue, donnera tout de suite la valeur de l'expression $(p_0+\alpha p_1+\alpha^2 p_2 \ldots + \alpha^{255} p_{255})^{256}$ en fonction rationnelle de α, et cela suffit pour résoudre l'équation dont il s'agit.

Une solution du problème a été donnée depuis longtemps par M. Richelot qui commence par supposer que α soit une racine primitive de l'équation $\alpha^{128}-1=0$. Cette solution est comprise, comme cas particulier, dans celle que je vais donner. La question est d'ailleurs intéressante, à cause de son rapport avec la théorie des nombres. En effet, quoiqu'en tant que je sache l'on n'a pas encore trouvé la règle pour former *à priori* la valeur de M, il est clair que les recherches de MM. Jacobi et Kummer doivent conduire à cette règle. Le résultat ici bas pourra servir pour la vérifier.

Voici la valeur que j'obtiens pour la fonction M:

$$\begin{aligned} M = -2 &+ 2\alpha - 2\alpha^{4} + 2\alpha^{5} + 2\alpha^{7} + 2\alpha^{9} - 2\alpha^{10} - 2\alpha^{14} - 2\alpha^{16} + 2\alpha^{21} + 2\alpha^{23} \\ &+ 2\alpha^{25} - 2\alpha^{26} - 2\alpha^{28} + 2\alpha^{29} - 2\alpha^{30} + 2\alpha^{33} - 2\alpha^{34} - 2\alpha^{36} + 2\alpha^{37} - 2\alpha^{38} + 2\alpha^{45} \\ &+ 2\alpha^{47} - 2\alpha^{48} + 2\alpha^{49} - 2\alpha^{50} + 2\alpha^{51} + 2\alpha^{53} - 2\alpha^{54} - 2\alpha^{60} + 2\alpha^{61} - 2\alpha^{64} + 2\alpha^{65} \\ &- 2\alpha^{66} + 2\alpha^{67} - 2\alpha^{68} + 2\alpha^{69} - 2\alpha^{72} - 2\alpha^{74} + 2\alpha^{75} - 2\alpha^{76} + 2\alpha^{79} - 2\alpha^{80} - 2\alpha^{82} \\ &- 2\alpha^{84} - 2\alpha^{86} + 2\alpha^{89} - 2\alpha^{92} - 2\alpha^{94} - 2\alpha^{100} + 2\alpha^{101} - 2\alpha^{104} - 2\alpha^{106} - 2\alpha^{108} + 2\alpha^{109} \\ &+ 2\alpha^{111} + 2\alpha^{113} - 2\alpha^{114} + 2\alpha^{115} + 2\alpha^{117} + 2\alpha^{119} - 2\alpha^{122} + 2\alpha^{123} - 2\alpha^{124} + 2\alpha^{127} - 2\alpha^{128} \\ &+ 2\alpha^{129} - 2\alpha^{130} - 2\alpha^{134} + 2\alpha^{135} - 2\alpha^{138} + 2\alpha^{141} + 2\alpha^{145} + 2\alpha^{147} + 2\alpha^{151} + 2\alpha^{153} - 2\alpha^{154} \\ &- 2\alpha^{156} - \alpha^{160} + 2\alpha^{161} + 2\alpha^{163} - 2\alpha^{164} + 2\alpha^{165} - 2\alpha^{166} + 2\alpha^{171} + 2\alpha^{177} + 2\alpha^{181} - 2\alpha^{182} \\ &+ 2\alpha^{183} - 2\alpha^{186} + 2\alpha^{187} + 2\alpha^{189} - 2\alpha^{190} - 2\alpha^{192} + 2\alpha^{195} - 2\alpha^{196} - 2\alpha^{198} + 2\alpha^{199} - 2\alpha^{206} \\ &- 2\alpha^{212} + 2\alpha^{213} - 2\alpha^{214} + 2\alpha^{215} - 2\alpha^{216} + 2\alpha^{217} - 2\alpha^{220} + 2\alpha^{221} + 2\alpha^{223} + 2\alpha^{225} - 2\alpha^{226} \\ &+ 2\alpha^{227} - 2\alpha^{228} + 2\alpha^{229} + 2\alpha^{233} - 2\alpha^{234} + 2\alpha^{235} - 2\alpha^{236} + 2\alpha^{237} - 2\alpha^{238} + 2\alpha^{239} - 2\alpha^{240} \\ &+ 2\alpha^{243} - 2\alpha^{244} - 2\alpha^{246} + 2\alpha^{247} - 2\alpha^{248} + 2\alpha^{249} - 2\alpha^{252} - 2\alpha^{254}. \end{aligned}$$

Représentons par M' ce que devient M, en supposant $\alpha^{128}+1=0$, nous obtiendrons

$$\begin{aligned}M' = {} & 2\alpha^2 - 2\alpha^4 + 2\alpha^5 + 2\alpha^6 + 2\alpha^9 - 2\alpha^{13} - 2\alpha^{14} - 2\alpha^{16} - 2\alpha^{17} - 2\alpha^{19} + 2\alpha^{21} \\ & + 2\alpha^{29} - 2\alpha^{30} + \alpha^{32} - 2\alpha^{34} - 2\alpha^{35} - 2\alpha^{43} + 2\alpha^{45} + 2\alpha^{47} - 2\alpha^{48} - 2\alpha^{50} + 2\alpha^{51} \\ & - 2\alpha^{55} + 2\alpha^{58} - 2\alpha^{59} - 2\alpha^{60} + 2\alpha^{62} + 2\alpha^{65} - 2\alpha^{66} + 2\alpha^{69} + 2\alpha^{70} - 2\alpha^{71} - 2\alpha^{72} \\ & - 2\alpha^{74} + 2\alpha^{75} - 2\alpha^{76} + 2\alpha^{78} + 2\alpha^{79} - 2\alpha^{80} - 2\alpha^{82} - 2\alpha^{85} - 2\alpha^{87} + 2\alpha^{88} - 2\alpha^{93} \\ & - 2\alpha^{94} - 2\alpha^{95} - 2\alpha^{97} + 2\alpha^{98} - 2\alpha^{99} - 2\alpha^{104} - 2\alpha^{105} - 2\alpha^{107} + 2\alpha^{110} + 2\alpha^{112} + 2\alpha^{113} \\ & - 2\alpha^{114} + 2\alpha^{116} + 2\alpha^{117} + 2\alpha^{118} + 2\alpha^{120} - 2\alpha^{121} - 2\alpha^{122} + 2\alpha^{123} + 2\alpha^{126} + 2\alpha^{127}.\end{aligned}$$

Soient M_1, M_1' ce que devient M en supposant successivement $\alpha^{128}-1=0$, $\alpha^{64}+1=0$, et soient M_2, M_2' ce que deviennent M ou M_1 en supposant successivement $\alpha^{64}-1=0$, $\alpha^{32}+1=0$, et ainsi de suite, jusqu'à M_7, M_7' qui seront ce que deviennent M ou M_1, &c. en supposant successivement $\alpha^2-1=0$, $\alpha+1=0$; nous aurons:

$$\begin{aligned}M_1 = {} & -4 + 4\alpha - 2\alpha^2 - 2\alpha^4 + 2\alpha^5 - 2\alpha^6 + 4\alpha^7 + 2\alpha^9 - 4\alpha^{10} + 2\alpha^{13} - 2\alpha^{14} \\ & - 2\alpha^{16} + 2\alpha^{17} + 2\alpha^{19} + 2\alpha^{21} + 4\alpha^{23} + 4\alpha^{25} - 4\alpha^{26} - 4\alpha^{28} + 2\alpha^{29} - 2\alpha^{30} - \alpha^{32} \\ & + 4\alpha^{33} - 2\alpha^{34} + 2\alpha^{35} - 4\alpha^{36} + 4\alpha^{37} - 4\alpha^{38} + 2\alpha^{43} + 2\alpha^{45} + 2\alpha^{47} - 2\alpha^{48} + 4\alpha^{49} \\ & - 2\alpha^{50} + 2\alpha^{51} + 4\alpha^{53} - 4\alpha^{54} + 2\alpha^{55} - 2\alpha^{58} + 2\alpha^{59} - 2\alpha^{60} + 4\alpha^{61} - 2\alpha^{62} - 4\alpha^{64} \\ & + 2\alpha^{65} - 2\alpha^{66} + 4\alpha^{67} - 4\alpha^{68} + 2\alpha^{69} - 2\alpha^{70} + 2\alpha^{71} - 2\alpha^{72} - 2\alpha^{74} + 2\alpha^{75} - 2\alpha^{76} \\ & - 2\alpha^{78} + 2\alpha^{79} - 2\alpha^{80} - 2\alpha^{82} - 4\alpha^{84} + 2\alpha^{85} - 4\alpha^{86} + 2\alpha^{87} - 2\alpha^{88} + 4\alpha^{89} - 4\alpha^{92} \\ & + 2\alpha^{93} - 2\alpha^{94} + 2\alpha^{95} + 2\alpha^{97} - 2\alpha^{98} + 2\alpha^{99} - 4\alpha^{100} + 4\alpha^{101} - 2\alpha^{104} + 2\alpha^{105} - 4\alpha^{106} \\ & + 2\alpha^{107} - 4\alpha^{108} + 4\alpha^{109} - 2\alpha^{110} + 4\alpha^{111} - 2\alpha^{112} + 2\alpha^{113} - 2\alpha^{114} + 4\alpha^{115} - 2\alpha^{116} + 2\alpha^{117} \\ & - 2\alpha^{118} + 4\alpha^{119} - 2\alpha^{120} + 2\alpha^{121} - 2\alpha^{122} + 2\alpha^{123} - 4\alpha^{124} - 2\alpha^{126} + 2\alpha^{127}.\end{aligned}$$

$$\begin{aligned}M_1' = {} & 2\alpha - 4\alpha^3 + 2\alpha^4 + 2\alpha^7 + 2\alpha^8 + 2\alpha^9 - 2\alpha^{10} - 2\alpha^{11} + 2\alpha^{12} + 2\alpha^{13} - 2\alpha^{15} + 2\alpha^{17} \\ & + 2\alpha^{18} + 2\alpha^{19} + 4\alpha^{20} + 4\alpha^{22} + 2\alpha^{23} + 2\alpha^{24} - 4\alpha^{26} - 2\alpha^{31} - \alpha^{32} + 2\alpha^{33} - 4\alpha^{38} + 2\alpha^{40} \\ & - 2\alpha^{41} + 4\alpha^{42} + 4\alpha^{44} - 2\alpha^{45} + 2\alpha^{46} - 2\alpha^{47} + 2\alpha^{49} - 2\alpha^{51} + 2\alpha^{52} + 2\alpha^{53} - 2\alpha^{54} - 2\alpha^{55} \\ & + 2\alpha^{56} - 2\alpha^{57} + 2\alpha^{60} + 4\alpha^{61} - 2\alpha^{63}.\end{aligned}$$

$$\begin{aligned}M_2 = {} & -8 + 6\alpha - 4\alpha^2 + 4\alpha^3 - 6\alpha^4 + 4\alpha^5 - 4\alpha^6 + 6\alpha^7 - 2\alpha^8 + 2\alpha^9 - 6\alpha^{10} + 2\alpha^{11} \\ & - 2\alpha^{12} + 2\alpha^{13} - 2\alpha^{14} + 2\alpha^{15} - 4\alpha^{16} + 2\alpha^{17} - 2\alpha^{18} + 2\alpha^{19} - 4\alpha^{20} + 4\alpha^{21} - 4\alpha^{22} + 6\alpha^{23} \\ & - 2\alpha^{24} + 8\alpha^{25} - 4\alpha^{26} - 8\alpha^{28} + 4\alpha^{29} - 4\alpha^{30} + 2\alpha^{31} - \alpha^{32} + 6\alpha^{33} - 4\alpha^{34} + 4\alpha^{35} - 8\alpha^{36} \\ & + 8\alpha^{37} - 4\alpha^{38} - 2\alpha^{40} + 2\alpha^{41} - 4\alpha^{42} + 4\alpha^{43} - 4\alpha^{44} + 6\alpha^{45} - 2\alpha^{46} + 6\alpha^{47} - 4\alpha^{48} + 6\alpha^{49} \\ & - 4\alpha^{50} + 6\alpha^{51} - 2\alpha^{52} + 6\alpha^{53} - 6\alpha^{54} + 6\alpha^{55} - 2\alpha^{56} + 2\alpha^{57} - 4\alpha^{58} + 4\alpha^{59} - 6\alpha^{60} + 4\alpha^{61} \\ & - 4\alpha^{62} + 2\alpha^{63}.\end{aligned}$$

$$\begin{aligned}M_2' = {} & -7 + 2\alpha^4 - 4\alpha^5 + 6\alpha^7 - 2\alpha^{10} - 2\alpha^{11} + 2\alpha^{12} - 4\alpha^{13} - 2\alpha^{14} - 4\alpha^{15} - 4\alpha^{17} + 2\alpha^{18} \\ & - 4\alpha^{19} - 2\alpha^{20} - 2\alpha^{21} + 2\alpha^{22} + 6\alpha^{25} - 4\alpha^{27} - 2\alpha^{28}.\end{aligned}$$

$$M_3 = -9 + 12\alpha - 8\alpha^2 + 8\alpha^3 - 14\alpha^4 + 12\alpha^5 - 8\alpha^6 + 6\alpha^7 - 4\alpha^8 + 4\alpha^9 - 10\alpha^{10} + 6\alpha^{11} - 6\alpha^{12} + 8\alpha^{13} - 6\alpha^{14} + 8\alpha^{15} - 8\alpha^{16} + 8\alpha^{17} - 6\alpha^{18} + 8\alpha^{19} - 6\alpha^{20} + 10\alpha^{21} - 10\alpha^{22} + 12\alpha^{23} - 4\alpha^{24} + 10\alpha^{25} - 8\alpha^{26} + 4\alpha^{27} - 14\alpha^{28} + 8\alpha^{29} - 8\alpha^{30} + 4\alpha^{31}.$$

$$M_3' = -1 + 4\alpha - 2\alpha^2 - 8\alpha^4 + 2\alpha^5 + 2\alpha^6 - 6\alpha^7 - 8\alpha^8 - 6\alpha^9 - 2\alpha^{10} + 2\alpha^{11} + 8\alpha^{12} + 2\alpha^{14} + 4\alpha^{15}.$$

$$M_4 = -17 + 20\alpha - 14\alpha^2 + 16\alpha^3 - 20\alpha^4 + 22\alpha^5 - 18\alpha^6 + 18\alpha^7 - 8\alpha^8 + 14\alpha^9 - 18\alpha^{10} + 10\alpha^{11} - 20\alpha^{12} + 16\alpha^{13} - 14\alpha^{14} + 12\alpha^{15}.$$

$$M_4' = -9 + 6\alpha + 4\alpha^2 + 6\alpha^3 + 6\alpha^5 - 4\alpha^6 + 6\alpha^7.$$

$$M_5 = -25 + 34\alpha - 32\alpha^2 + 26\alpha^3 - 40\alpha^4 + 38\alpha^5 - 32\alpha^6 + 30\alpha^7.$$

$$M_5' = -15 - 4\alpha - 4\alpha^3.$$

$$M_6 = -65 + 72\alpha - 64\alpha^2 + 56\alpha^3.$$

$$M_6' = -1 + 16\alpha.$$

$$M_7 = -129 + 128\alpha.$$

$$M_7' = -257.$$

Ces différentes expressions étant trouvées, supposons que α soit une racine primitive, et représentons par $F, F_1, F_2, \ldots F_7$ ce que deviennent $M', M_1', M_2' \ldots M_7'$ en substituant $\alpha, \alpha^2, \alpha^4, \ldots \alpha^{128}$ au lieu de α ($F = M'$, $F_7 = -257$); nous aurons

$$(p_0 + \alpha p_1 + \alpha^2 p_2 \ldots + \alpha^{255} p_{255})^{256} = -F^{128} . F_1^{64} . F_2^{32} . F_3^{16} . F_4^{8} . F_5^{4} . F_6^{2} . F_7;$$

cette équation constitue la solution dont il s'agit.

Ajoutons encore les formules beaucoup plus simples qui correspondent à l'équation $x^{17} - 1 = 0$. En supposant $\alpha^{16} - 1 = 0$, nous aurons

$$M = -2 - \alpha^4 + 2\alpha^5 + 2\alpha^7 - 2\alpha^8 + 2\alpha^9 - 2\alpha^{10} + 2\alpha^{13} - 2\alpha^{14}.$$

$$M' = -2\alpha + 2\alpha^2 - \alpha^4 + 2\alpha^6 + 2\alpha^7.$$

$$M_1 = -4 + 2\alpha - 2\alpha^2 - \alpha^4 + 4\alpha^5 - 2\alpha^6 + 2\alpha^7.$$

$$M_1' = -3 - 2\alpha - 2\alpha^3.$$

$$M_2 = -5 + 6\alpha - 4\alpha^2 + 2\alpha^3.$$

$$M_2' = -1 + 4\alpha.$$

$$M_3 = -9 + 8\alpha.$$

$$M_3' = -17;$$

et de là, en supposant que α soit une racine primitive:

$$(p_0 + \alpha p_1 + \ldots + \alpha^{15} p_{15})^{16} = (-2\alpha + 2\alpha^2 - \alpha^4 + 2\alpha^6 + 2\alpha^7)^8 . (-3 - 2\alpha^2 - 2\alpha^6)^4 . (-1 + 4\alpha^4)^2 . 17.$$

Pour $x^5 - 1 = 0$ on obtient sans peine

$$(p_0 + \alpha p_1 + \alpha^2 p_2 + \alpha^3 p_3)^4 = (-2 - \alpha^2 + 2\alpha^3)^2 . 5;$$

où α est une racine primitive de l'équation $\alpha^4 - 1 = 0$, savoir $\alpha = \pm i$.

98.

NOTE RELATIVE À LA SIXIÈME SECTION DU "MÉMOIRE SUR QUELQUES THÉORÈMES DE LA GÉOMÉTRIE DE POSITION."

[From the *Journal für die reine und angewandte Mathematik* (Crelle), tom. XLI. (1851), p. 84.]

This Note which is thus entitled in Crelle, but which in fact refers to the seventh section of the Memoir, is inserted in its proper place, *ante* p. 555.

99.

NOTE SUR QUELQUES FORMULES QUI SE RAPPORTENT À LA MULTIPLICATION DES FONCTIONS ELLIPTIQUES.

[From the *Journal für die reine und angewandte Mathematik* (Crelle), tom. XLI. (1851), pp. 85—92.]

EN revenant sur l'équation

$$\begin{aligned}&\{-l\lambda - m\mu + (l-m)^2\} P_{l,m}\\ &+ l(\lambda - 2l + 2m + 2)(\lambda - 2l + 2m + 1) P_{l-1,m}\\ &+ m(\mu + 2l - 2m + 2)(\mu + 2l - 2m + 1) P_{l,m-1}\\ &\quad - 16lm\{\lambda\mu - (2l + 2m - 4)(\lambda + \mu)\} P_{l-1,m-1} = 0,\end{aligned}$$

dans laquelle $P_{0,0} = 1$, les expressions que j'ai données pour $P_{l,0}$, $P_{l,1}$ [93, see p. 535] peuvent être écrites comme suit:

$$P_{l,0} = \lambda\,[\lambda - l - 1]^{l-1},$$

$$P_{l,1} = \mu\lambda\,[\lambda - l - 1]^{l-1} + l\lambda\,[\lambda - l]^{l-2}\left(18l - 16 + \frac{2(l-1)(l-2)}{\lambda - l}\right) + \frac{\lambda\mu}{\lambda + \mu}(-10l\lambda\,[\lambda - l]^{l-2}):$$

équations qui peuvent être représentées par

$$P_{l,0} = Q_{l,0},$$

$$P_{l,1} = Q_{l,1} + R_{l,1;1}\frac{\lambda\mu}{\lambda + \mu},$$

ce qui conduisent à la forme

$$P_{l,2} = Q_{l,2} + R_{l,2;1}\frac{\lambda\mu}{\lambda + \mu} + R_{l,2;2}\frac{\lambda^2\mu^2}{(\lambda + \mu)^2},$$

où les coefficients R ne contiennent que la seule quantité λ, et où $Q_{l,2}$ est une fonction intégrale du $l^{\text{ième}}$ ordre par rapport à λ, et du second ordre par rapport à μ.

Cela donne

$$\{-l\lambda - 2\mu + (l-2)^2\}\left\{Q_{l,2} \quad + R_{l,2;1}\,\frac{\lambda\mu}{\lambda+\mu} + R_{l,2;2}\,\frac{\lambda^2\mu^2}{(\lambda+\mu)^2}\right\}$$

$$+\, l(\lambda - 2l + 6)(\lambda - 2l + 5)\left\{Q_{l-1,2} + R_{l-1,2;1}\,\frac{\lambda\mu}{\lambda+\mu} + R_{l-1,2;2}\,\frac{\lambda^2\mu^2}{(\lambda+\mu)^2}\right\}$$

$$+\, 2(\mu + 2l - 2)(\mu + 2l - 3)\left\{Q_{l,1} \quad + R_{l,1;1}\,\frac{\lambda\mu}{\lambda+\mu}\right\}$$

$$-\, 32l\,\{\lambda\mu - 2l(\lambda+\mu)\}\left\{Q_{l-1,1} + R_{l-1,1;1}\frac{\lambda\mu}{\lambda+\mu}\right\} = 0,$$

ce qui se réduit à

$$\{-l\lambda - 2\mu + (l-2)^2\}\, Q_{l,2} - (l-2)(\lambda - l + 2)\left\{R_{l,2;1}\frac{\lambda\mu}{\lambda+\mu} + R_{l,2;2}\frac{\lambda^2\mu^2}{(\lambda+\mu)^2}\right\}$$

$$-\, 2\lambda\mu R_{l,2;1} - 2\lambda^2\mu R_{l,2;2} + 2\lambda^2 R_{l,2;2}\frac{\lambda\mu}{(\lambda+\mu)}$$

$$+\, l(\lambda - 2l + 6)(\lambda - 2l + 5)\left\{Q_{l-1,2} + R_{l-1,2;1}\frac{\lambda\mu}{\lambda+\mu} + R_{l-1,2;2}\frac{\lambda^2\mu^2}{(\lambda+\mu)^2}\right\}$$

$$+\, 2(\mu + 2l - 2)(\mu + 2l - 3)\, Q_{l,1} + 2(\mu - \lambda + 4l - 5)\,\lambda\mu R_{l,1;1}$$

$$+\, 2(\lambda - 2l + 2)(\lambda - 2l + 3)\, R_{l,1;1}\frac{\lambda\mu}{(\lambda+\mu)}$$

$$-\, 32l\,\{\lambda\mu - 2l(\lambda+\mu)\}\, Q_{l-1,1} - 32l\lambda\mu(\lambda - 2l)\, R_{l-1,1;1} + 32l\lambda^2 R_{l-1,1;1}\frac{\lambda\mu}{\lambda+\mu} = 0.$$

En ne faisant attention d'abord qu'aux termes qui contiennent des puissances négatives de $\lambda+\mu$, nous obtenons

$$-(l-2)(\lambda - l + 2)\, R_{l,2;2} + l(\lambda - 2l + 6)(\lambda - 2l + 5)\, R_{l-1,2;2} = 0$$

et

$$-(l-2)(\lambda - l + 2)\, R_{l,2;1} + l(\lambda - 2l + 6)(\lambda - 2l + 5)\, R_{l-1,2;1}$$

$$+\, 2(\lambda - 2l + 2)(\lambda - 2l + 3)\, R_{l,1;1} + 32l\lambda^2 R_{l-1,1;1} + 2\lambda^2 R_{l,2;2} = 0.$$

La première équation, en calculant la constante arbitraire au moyen de $R_{2,2,2} = 200$, donne

$$R_{l,2;2} = 100\, l(l-1)\,\lambda\,[\lambda - l + 1]^{l-3};$$

l'expression de $R_{2,2,2} = 200$ se trouve par celle de $P_{2,2}$ qui peut être écrite sous la forme

$$P_{2,2} = \lambda(\lambda-3)\,\mu(\mu-3) + 152\lambda\mu + 336 - 40\lambda^2\mu + (40\lambda^2 - 1156)\frac{\lambda\mu}{\lambda+\mu} + \frac{200\lambda^2\mu^2}{(\lambda+\mu)^2}.$$

En substituant la valeur de $R_{l,2;2}$ et celles de

$$R_{l,1;1} = -10l\lambda\,[\lambda - l]^{l-2}, \qquad R_{l-1,1;1} = -10(l-1)\,\lambda\,[\lambda - l + 1]^{l-3}$$

dans la seconde équation, on obtient

$$-(l-2)(\lambda-l+2)R_{l,2;1}+l(\lambda-2l+6)(\lambda-2l+5)R_{l-1,2;1}$$
$$-20l\lambda(\lambda-2l+3)[\lambda-l]^{l-1}-120l(l-1)\lambda^3[\lambda-l+1]^{l-3}=0,$$

c'est à dire

$$-(l-2)(\lambda-l+2)R_{l,2;1}+l(\lambda-2l+6)(\lambda-2l+5)R_{l-1,2;1}$$
$$-20l\lambda[\lambda-l]^{l-4}\{6(l-1)\lambda^2(\lambda-l+1)+(\lambda-2l+4)(\lambda-2l+3)^2(\lambda-2l+2)\}.$$

Mettons

$$R_{l,2;2}=l(l-1)\lambda[\lambda-l+1]^{l-3}\Psi_l;$$

cela donne

$$\Psi_l-\Psi_{l-1}=\frac{-20}{(l-1)(l-2)(\lambda-l+1)(\lambda-l+2)}\{6(l-1)\lambda^2(\lambda-l+1)$$
$$+(\lambda-2l+4)(\lambda-2l+3)^2(\lambda-2l+2)\},$$

ce qui devient, quelques réductions faites,

$$\Psi_l-\Psi_{l-1}=\frac{-20}{(l-1)(l-2)}\{(\lambda+1)(\lambda+2)+17(l-1)(l-2)\}$$
$$-\frac{20(l-2)(l-3)}{\lambda-l+1}+\frac{20(l-2)(l-4)}{\lambda-l+2},$$

et de là on tire

$$\Psi_l=C-340l+\frac{20(\lambda+1)(\lambda+2)}{l-1}-\frac{20(l-2)(l-3)}{\lambda-l+1}.$$

Or $R_{2,2;1}=2\Psi_2=40\lambda^2-1156$; donc $\Psi_2=20\lambda^2-578$, et de là $C=62-60\lambda$; donc enfin, en restituant la valeur de $R_{l,2;1}$,

$$R_{l,2;1}=l(l-1)\lambda[\lambda-l+1]^{l-3}\left\{62-340l-60\lambda+\frac{20(\lambda+1)(\lambda+2)}{l-1}-\frac{20(l-2)(l-3)}{\lambda-l+1}\right\}.$$

Passons à l'expression de $Q_{l,2}$. Elle donne

$$\{-l\lambda-2\mu+(l-2)^2\}Q_{l,2}+l(\lambda-2l+6)(\lambda-2l+5)Q_{l-1,2}$$
$$+2(\mu+2l-2)(\mu+2l-3)Q_{l,1}\qquad-32l\{\lambda\mu-2l(\lambda+\mu)\}Q_{l-1,1}$$
$$+2(\mu-\lambda+4l-5)\lambda\mu R_{l,1;1}$$
$$-2\lambda\mu R_{l,2;1}-2\lambda^2\mu R_{l,2;2}-32l\lambda\mu(\lambda-2l)R_{l-1,1;1}=0;$$

où la dernière ligne se réduit à

$$-2l(l-1)\mu\lambda^2[\lambda-l+1]^{l-3}\left\{62-20l-120\lambda-\frac{20(\lambda+1)(\lambda+2)}{l-1}-\frac{20(l-1)(l-2)}{\lambda-l+1}\right\}.$$

Donc on a

$$\{l\lambda + 2\mu - (l-2)^2\} Q_{l,2} - l(\lambda - 2l + 6)(\lambda - 2l + 5) Q_{l-1,2}$$

$$= 2(\mu + 2l - 2)(\mu + 2l - 3)\left\{\mu\lambda[\lambda - l - 1]^{l-1} + l\lambda[\lambda - l]^{l-2}\left(18l - 16 + \frac{2(l-1)(l-2)}{\lambda - l}\right)\right\}$$

$$- 32l\{\lambda\mu - 2l(\lambda + \mu)\}\left\{\mu\lambda[\lambda - l]^{l-2} + (l-1)\lambda[\lambda - l + 1]^{l-3}\left(18l - 34 + \frac{2(l-2)(l-3)}{\lambda - l + 1}\right)\right\}$$

$$- 20l(4l - 5 + \mu - \lambda)\lambda^2\mu[\lambda - l]^{l-2}$$

$$- 2l(l-1)\lambda^2\mu[\lambda - l + 1]^{l-3}\left\{62 - 20l - 120\lambda + \frac{20(\lambda+1)(\lambda+2)}{l-1} - \frac{20(\lambda-1)(\lambda-2)}{\lambda - l + 1}\right\},$$

équation qui peut être représentée par

$$\{l\lambda + 2\mu - (l-2)^2\} Q_{l,2} - l(\lambda - 2l + 6)(\lambda - 2l + 5) Q_{l-1,2} = \mathfrak{A}\mu^3 + \mathfrak{B}\mu^2 + \mathfrak{C}\mu + \mathfrak{D},$$

où les valeurs de $\mathfrak{A}$, $\mathfrak{B}$, $\mathfrak{C}$, $\mathfrak{D}$ sont données par les formules

$$\mathfrak{A} = 2\lambda[\lambda - l - 1]^{l-1},$$

$$\mathfrak{B} = 2(4l-5)\lambda[\lambda - l - 1]^{l-1} + 2l\lambda[\lambda - l]^{l-2}\left(18l - 16 + \frac{2(l-1)(l-2)}{\lambda - l}\right)$$

$$- 32l\lambda(\lambda - 2l)[\lambda - l]^{l-2} - 20l\lambda^2[\lambda - l]^{l-2},$$

$$\mathfrak{C} = 2(2l-2)(2l-3)\lambda[\lambda - l - 1]^{l-1}$$

$$+ 2(4l-5)l\lambda[\lambda - l]^{l-2}\left\{18l - 16 + \frac{2(l-1)(l-2)}{\lambda - l}\right\}$$

$$+ 64l^2\lambda^2[\lambda - l]^{l-2}$$

$$- 32l(l-1)(\lambda - 2l)\lambda[\lambda - l + 1]^{l-3}\left\{18l - 34 + \frac{2(l-2)(l-3)}{\lambda - l + 1}\right\}$$

$$- 20l(4l - 5 - \lambda)\lambda^2[\lambda - l]^{l-2}$$

$$- 2l(l-1)\lambda^2[\lambda - l + 1]^{l-3}\left\{62 - 20l - 120\lambda + \frac{20(\lambda+1)(\lambda+2)}{l-1} - \frac{20(l-1)(l-2)}{\lambda - l + 1}\right\},$$

$$\mathfrak{D} = 2(2l-2)(2l-3)l\lambda[\lambda - l]^{l-2}\left\{18l - 16 + \frac{2(l-1)(l-2)}{\lambda - l}\right\}$$

$$+ 64l^2(l-1)\lambda^2[\lambda - l + 1]^{l-3}\left\{18l - 34 + \frac{2(l-2)(l-3)}{\lambda - l + 1}\right\}.$$

Sans m'arrêter à réduire ces expressions aux formes les plus simples, j'écris

$$Q_{l,2} = \mu(\mu - 3)\lambda[\lambda - l - 1]^{l-1} + \mu I_l + J_l;$$

en substituant cette valeur, les termes qui contiennent μ^3 se détruisent, et la comparaison des autres termes donne

$$2I_l - 6\lambda[\lambda - l - 1]^{l-1} + \{l\lambda - (l-2)^2\}\lambda[\lambda - l - 1]^{l-1}$$

$$- l(\lambda - 2l + 6)(\lambda - 2l + 5)\lambda[\lambda - l]^{l-2} - \mathfrak{B} = 0,$$

$$2J_l+\{l\lambda-(l-2)^2\}\{-3\lambda[\lambda-l-1]^{l-1}+I_l\}$$
$$-l(\lambda-2l+6)(\lambda-2l+5)\{-3\lambda[\lambda-l]^{l-2}+I_{l-1}\}-\mathfrak{C}=0,$$

$$\{l\lambda-(l-2)^2\}J_l-l(\lambda-2l+6)(\lambda-2l+5)J_{l-1}-\mathfrak{D}=0.$$

Les valeurs de I_l et J_l peuvent être tirées sans intégration de la première et de la seconde de ces équations. La valeur ainsi trouvée de J_l satisfera à la troisième équation (ce qui cependant doit être vérifié a posteriori). Il m'a paru plus simple de tirer la fonction I_l de la première équation, et celle de J_l en intégrant la troisième équation; alors ce sera la seconde équation qu'il y a à vérifier. En effet on obtient

$$I_l=l\lambda[\lambda-l]^{l-2}\left\{36l+4-20\lambda+\frac{4(l-1)(l-2)}{\lambda-l}\right\}.$$

L'équation qui sert à déterminer J_l devient, en substituant la valeur de $\mathfrak{D}$,

$$\{l\lambda-(l-2)^2\}J_l-l(\lambda-2l+6)(\lambda-2l+5)J_{l-1}$$
$$=4l(l-1)(2l-3)\lambda[\lambda-l]^{l-2}\left\{18l-16+\frac{2(l-1)(l-2)}{\lambda-l}\right\}$$
$$+64l^2(l-1)\lambda^2[\lambda-l+1]^{l-3}\left\{18l-34+\frac{2(l-2)(l-3)}{\lambda-l+1}\right\}.$$

En écrivant

$$J_l=l(l-1)\lambda[\lambda-l+1]^{l-3}V_l,$$

on trouve

$$\{l\lambda-(l-2)^2\}V_l-(l-2)(\lambda-l+2)V_{l-1}$$
$$=8(2l-3)\frac{(\lambda-2l+3)(\lambda-2l+4)}{\lambda-l+1}\left\{9l-8+\frac{(l-1)(l-2)}{\lambda-l}\right\}$$
$$+128l\lambda\left\{9l-17+\frac{(l-2)(l-3)}{\lambda-l+1}\right\}.$$

En faisant

$$V_l=M_l+\frac{A_l}{\lambda-l+1}+\frac{B_l}{(\lambda-l+1)(\lambda-l)},$$

cette équation se réduit à

$$\{l\lambda-(l-2)^2\}M_l-(l-2)(\lambda-l+2)M_{l-1}$$
$$+lA_l-(l-2)A_{l-1}+\frac{(3l-4)A_l+lB_l-(l-2)B_{l-1}}{\lambda-l+1}+\frac{4(l-1)B_l}{(\lambda-l+1)(\lambda-l)}$$
$$=\quad 8(2l-3)(9l-8)\left\{(\lambda-3l+6)-\frac{(l-2)(l-3)}{\lambda-l+1}\right\}$$
$$+8(2l-3)(l-1)(l-2)\left\{1-\frac{2(l-3)}{\lambda-l+1}+\frac{(l-3)(l-4)}{(\lambda-l+1)(\lambda-l)}\right\}$$
$$+128l\lambda(9l-17)+128\frac{l(l-1)(l-2)(l-3)}{\lambda-l+1},$$

et cela donne tout de suite la valeur de B_l, et après quelques réductions celle de A_l.

On obtient ainsi

$$A_l = 2(l-2)(l-3)(36l+7),$$
$$B_l = 2(l-2)(l-3)(l-4)(2l-3).$$

En substituant ces valeurs, on a, toute réduction faite:

$$\{l\lambda-(l-2)^2\}M_l-(l-2)(\lambda-l+2)M_{l-1}=8\lambda(162l^2-315l+24)-24(l-2)^2(27l-26),$$

c'est à dire

$$lM_l-(l-2)M_{l-1}=1296l^2-2520l+192,$$
$$M_l-M_{l-1} \quad = \quad 648l-624,$$

ou enfin

$$M_l \quad = 324l^2-300l-528,$$
$$M_{l-1}=324l^2-948l+192:$$

valeurs qui (comme cela doit être) se changent l'une dans l'autre en entrechangeant les quantités l, $l-1$.

De là on obtient

$$J_l = l(l-1)\lambda[\lambda-l+1]^{l-3}\left\{324l^2-300l-528+\frac{2(l-2)(l-3)(36l+7)}{\lambda-l+1}\right.$$
$$\left.+\frac{2(2l-3)(l-2)(l-3)(l-4)}{(\lambda-l+1)(\lambda-l)}\right\},$$

valeur qu'on trouverait aussi par l'autre procédé indiqué ci-dessus. Or nous avons

$$Q_{l,2}=\mu(\mu-3)\lambda[\lambda-l-1]^{l-1}+\mu I_l+J_l,$$
$$P_{l,2}=Q_{l,2}+R_{l,2,1}\frac{\lambda\mu}{\lambda+\mu}+R_{l,2;2}\frac{\lambda^2\mu^2}{(\lambda+\mu)^2};$$

donc enfin, en réunissant les valeurs des différentes parties de $P_{l,2}$, on obtient

$$P_{l,2}=\mu(\mu-3)\lambda[\lambda-l-1]^{l-1}$$
$$+\mu l\lambda[\lambda-l]^{l-2}\left\{36l+4-20\lambda+\frac{4(l-1)(l-2)}{\lambda-l}\right\}$$
$$+l(l-1)\lambda[\lambda-l+1]^{l-3}\left\{324l^2-300l-528+\frac{2(l-2)(l-3)(36l+7)}{\lambda-l+1}\right.$$
$$\left.+\frac{2(2l-3)(l-2)(l-3)(l-4)}{(\lambda-l+1)(\lambda-l)}\right\}$$
$$+l(l-1)\lambda[\lambda-l+1]^{l-3}\left\{62-340l-60\lambda+\frac{20(\lambda+1)(\lambda+2)}{l-1}-\frac{20(l-2)(l-3)}{\lambda-l+1}\right\}\frac{\lambda\mu}{\lambda+\mu}$$
$$+100l(l-1)\lambda[\lambda-l+1]^{l-3}\frac{\lambda^2\mu^2}{(\lambda+\mu)^2},$$

équation qui fait suite aux équations

$$P_{l,1} = \mu\lambda\,[\lambda - l - 1]^{l-1} + l\lambda\,[\lambda - l]^{l-2}\left\{18l - 16 + \frac{2(l-1)(l-2)}{\lambda - l}\right\} - 10l\lambda\,[\lambda - l]^{l-2}\cdot\frac{\lambda\mu}{\lambda+\mu},$$

$$P_{l,0} = \lambda\,[\lambda - l - 1]^{l-1}.$$

Je vais essayer maintenant à chercher d'une manière plus systématique les termes de $P_{l,m}$ qui ne contiennent pas la quantité μ, ou bien à chercher la solution de l'équation

$$\begin{aligned}\{-l\lambda + (l-m)^2\}\,P_{l,m}&\\ + l(\lambda - 2l + 2m + 2)(\lambda - 2l + 2m + 1)\,P_{l-1,m}&\\ + 2m(l - m + 1)(2l - 2m + 1)\,P_{l,m-1}&\\ + 32lm(l + m - 2)\,P_{l-1,m-1} &= 0.\end{aligned}$$

En supposant que

$$P_{l,m} = [l]^m\,\lambda\,[\lambda - l + m - 1]^{l-m-1}\,\Sigma\,\frac{A_{l,m,p}}{[\lambda - l + m - 1]^p}$$

(où la sommation se rapporte à p, nombre qui doit être étendu depuis $p=0$ jusqu'à $p=m$), on obtiendra sans peine

$$\begin{aligned}\{-l\lambda + (l-m)^2\}\,\Sigma\,\frac{A_{l,m,p}}{[\lambda - l + m - 1]^p}&\\ + (l-m)\,\Sigma\,\frac{A_{l-1,m,p}}{[\lambda - l + m - 1]^{p-1}}&\\ + 2m(2l - 2m + 1)\,[\lambda - 2l + 2m]^2\,\Sigma\,\frac{A_{l,m-1,p}}{[\lambda - l + m - 1]^{p+1}}&\\ + 32lm(l + m - 2)\,\lambda\,\Sigma\,\frac{A_{l-1,m-1,p}}{[\lambda - l + m - 1]^p} &= 0\,;\end{aligned}$$

où p s'étend seulement jusqu'à $m-1$ dans la troisième et dans la quatrième ligne.

Pour réduire la première ligne, j'écris

$$-l\lambda + (l-m)^2 = -l(\lambda - l + m - p) + [m^2 - (m+p)\,l],$$

ce qui réduit le terme général à

$$\frac{-lA_{l,m,p}}{[\lambda - l + m - 1]^{p-1}} + \frac{[m^2 - (m+p)\,l]A_{l,m,p}}{[\lambda - l + m - 1]^p}:$$

expression qui, en écrivant $r+1$ au lieu de p dans le premier terme, et r dans le second terme, peut être remplacée par

$$(\alpha) \qquad \frac{-lA_{l,m,r+1}}{[\lambda-l+m-1]^r}+\frac{\{m^2-(m+r)\,l\}\,A_{l,m,r}}{[\lambda-l+m-1]^r}.$$

La seconde ligne donne tout de suite le terme général

$$(\beta) \qquad \frac{(l-m)\,A_{l-1,m,r+1}}{[\lambda-l+m-1]^r}.$$

Pour réduire la troisième ligne, je mets

$$[\lambda-2l+2m]^2=[\lambda-l+m-p]^2-2\,[l-m-p+1]^1\,[\lambda-l+m-p-1]^1+[l-m-p]^2,$$

ce qui réduit le terme général à

$$2m\,(2l-2m+1)\frac{A_{l,m-1,p}}{[\lambda-l+m-1]^{p-1}}-4m\,(2l-2m+1)\,\frac{[l-m-p+1]^1\,A_{l,m-1,p}}{[\lambda-l+m-1]^p}$$
$$+\frac{2m\,(2l-2m+1)\,[l-m-p]^2\,A_{l,m-1,p}}{[\lambda-l+m-1]^{p+1}}:$$

expression qui (en écrivant $r+1$ au lieu de p dans le premier terme, r dans le second terme et $r-1$ dans le troisième terme) peut être remplacée par

$$(\gamma) \quad 2m\,(2l-2m+1)\frac{A_{l,m-1,r+1}}{[\lambda-l+m-1]^r}-4m\,(2l-2m+1)\,\frac{[l-m-r+1]^2\,A_{l,m-1,r}}{[\lambda-l+m-1]^r}$$
$$+\frac{2m\,(2l-2m+1)\,[l-m-r+1]^2\,A_{l,m-1,r-1}}{[\lambda-l+m-1]^r}.$$

Pour réduire la quatrième ligne, je mets

$$\lambda=(\lambda-l+m-p)+(l-m+p),$$

ce qui réduit le terme général à

$$32lm\,(l+m-2)\frac{A_{l-1,m-1,p}}{[\lambda-l+m-1]^{p-1}}+32lm\,(l+m-2)\,\frac{(l-m+p)\,A_{l-1,m-1,p}}{[\lambda-l+m-1]^p},$$

expression qui (en écrivant $r+1$ au lieu de p dans le premier et r dans le second terme) peut être remplacée par

$$(\delta) \quad 32lm\,(l+m-2)\frac{A_{l-1,m-1,r+1}}{[\lambda-l+m-1]^r}+32lm\,(l+m-2)\,\frac{(l-m+p)\,A_{l-1,m-1,r}}{[\lambda-l+m-1]^r}.$$

Donc en réunissant les expressions (α), (β), (γ), (δ) et en faisant attention que le coefficient du terme qui contient $[\lambda - l + m - 1]^r$ au dénominateur, doit se réduire à zéro, on obtient, en arrangeant encore les termes d'une manière convenable:

$$\begin{aligned}
&2m(2l-2m+1)\,A_{l,m-1,r+1}\\
&+32lm(l+m-2)\,A_{l-1,m-1,r+1}\\
&-lA_{l,m,r+1}\\
&+(l-m)\,A_{l-1,m,r+1}\\
&-4m(2l-2m+1)(l-m-r+1)\,A_{l,m-1,r}\\
&+32lm(l+m-2)(l+m-r)\,A_{l-1,m-1,r}\\
&+(m^2-(m+r)\,l)\,A_{l,m,r}\\
&+2m(2l-2m+1)(l-m-r+1)(l-m-r)\,A_{l,m-1,r-1}=0,
\end{aligned}$$

où r s'étend depuis $r=0$ jusqu'à $r=m$, en réduisant à zéro les termes pour lesquels le troisième suffixe est négatif ou plus grand que le second suffixe. Par exemple dans le cas de $r=m$, on obtient l'équation très simple

$$\begin{aligned}
&(2l-m)\,A_{l,m,m}\\
&-2(2l-2m+1)(l-2m+1)(l-2m)\,A_{l,m-1,m-1}=0
\end{aligned}$$

qui (sous la condition $A_{l,0,0}=1$) donne sans peine la valeur générale de $A_{l,m,m}$, savoir:

$$A_{l,m,m}=[2l-m-1]^m\,[l-m-1]^m:$$

valeur qui peut être présentée sous d'autres formes en considérant à part les deux cas de m pair et de m impair. En supposant $m=1$, $m=2$, on obtient

$$A_{l,1,1}=2(l-1)(l-2),\quad A_{l,2,2}=2(2l-3)(l-2)(l-3)(l-4):$$

valeurs qui servent à vérifier des résultats déjà trouvés.

100.

NOTE SUR LA THÉORIE DES HYPERDÉTERMINANTS.

[From the *Journal für die reine und angewandte Mathematik* (Crelle), tom. XLII. (1851), pp. 368—371.]

DANS la théorie dont il s'agit, je suis parvenu à un théorème qui pourra, à ce qu'il me paraît, conduire à des développements intéressants.

Je ne considère ici que le cas d'une fonction homogène à *deux variables*, et en me servant des nouveaux termes de M. Sylvester, je nomme *Covariant* d'une fonction donnée, toute fonction qui ne change pas de forme en faisant subir aux variables des transformations linéaires quelconques, et *Invariant* toute fonction des seuls coefficients qui a la propriété mentionnée.

Cela posé, soit U une fonction donnée quelconque, du degré n par rapport aux variables, et, comme à l'ordinaire, contenant des coefficients arbitraires a, b, c, &c. Soit Q un *covariant* quelconque (y compris le cas particulier où Q est un *invariant*) de la fonction U, s le degré de Q par rapport aux variables, r le degré de cette même fonction Q par rapport aux coefficients. En supposant que la fonction U ait un facteur θ^ν (où $\theta = lx + my$ est une fonction linéaire des variables), ou autrement dit, en supposant l'équation $U = \theta^\nu V$, je dis que le *covariant* Q contiendra ce même facteur θ élevé à la puissance $r\nu - \frac{1}{2}(rn - s)$.

En effet, en se rappelant la méthode dont je me suis servi dans la seconde partie de mon mémoire sur les Hyperdéterminants (t. XXX. de ce Journal, [16]) (je suppose que le lecteur ait ce mémoire sous les yeux), on verra que cette fonction Q, supposée, comme plus haut, du degré r par rapport aux coefficients, sera nécessairement de la forme

$$Q = \overline{12}^\alpha\, \overline{13}^\beta\, \overline{23}^\gamma \ldots\, U_1 U_2 \ldots\, U_r,$$

puisque les coefficients n'entrent dans Q que par les fonctions U_1, U_2, &c. Or Q étant du degré s par rapport aux variables, on obtient $s = rn - 2(\alpha + \beta + \gamma + \ldots)$, c'est à dire:

$$\alpha + \beta + \gamma \ldots = \tfrac{1}{2}(rn - s).$$

Cela posé, puisque $U = \theta^\nu V$, on aura de même $U_1 = \theta_1{}^\nu V_1$, $U_2 = \theta_2{}^\nu V_2$, &c. Les expressions $\overline{12}$, &c. qui entrent dans l'expression de Q contiennent ∂_{x_1}, ∂_{y_1}, &c.: symboles qui doivent être remplacés par $\partial_{x_1} + l\partial_{\theta_1}$, $\partial_{y_1} + m\partial_{\theta_1}$, &c. en supposant (comme il est permis) que les nouveaux symboles ∂_{x_1}, ∂_{y_1}, &c. ne se rapportent plus à $\theta_1{}^\nu V_1$, &c., mais seulement à V_1, &c. Cela donne

$$\begin{aligned}\overline{12} &= (\partial_{x_1} + l\partial_{\theta_1})(\partial_{y_2} + m\partial_{\theta_2}) - (\partial_{x_2} + l\partial_{\theta_2})(\partial_{y_1} + m\partial_{\theta_1}) \\ &= \partial_{x_1}\partial_{y_2} - \partial_{x_2}\partial_{y_1} + (l\partial_{y_2} - m\partial_{x_2})\partial_{\theta_1} - (l\partial_{y_1} - m\partial_{x_1})\partial_{\theta_2};\end{aligned}$$

c'est à dire: $\overline{12}$ est une fonction linéaire par rapport à ∂_{θ_1}, ∂_{θ_2}, et il en sera de même pour les expressions analogues $\overline{13}$, $\overline{23}$, &c.: donc le nombre des différentiations par rapport aux quantités θ_1, θ_2, &c., prises ensemble, ne surpasse pas $\alpha + \beta + \gamma + \dots$, ou $\frac{1}{2}(rn - s)$. Or l'expression à différentier contient le facteur $\theta_1{}^\nu \theta_2{}^\nu \dots \theta_r{}^\nu$; donc, en remettant, après les différentiations, θ au lieu de $\theta_1, \theta_2, \dots \theta_r$, la fonction Q contiendra le facteur θ élévé à la puissance $r\nu - \frac{1}{2}(rn - s)$.

Tout cela suppose implicitement que l'on ait $r\nu - \frac{1}{2}(rn - s) \overline{\overline{<}} s$. Or le même raisonnement, modifié très peu, fait voir aussi que pour

$$r\nu - \tfrac{1}{2}(nr - s) > s,$$

ou plus simplement pour

$$r(\nu - \tfrac{1}{2}n) > \tfrac{1}{2}s,$$

la fonction Q doit s'évanouir d'elle-même, savoir en établissant entre les coefficients de U les relations qui expriment l'existence du facteur θ^ν. De là on tire le théorème suivant:

Étant donnée une fonction U du degré n, tout *covariant* du degré r par rapport aux coefficients et du degré s par rapport aux variables, s'évanouit en supposant que la fonction U ait un facteur θ pour lequel $r(\nu - \frac{1}{2}n) > \frac{1}{2}s$; et en particulier:

Un *invariant* quelconque de la fonction U s'évanouit en supposant que la fonction U ait un facteur θ^ν, pour lequel $\nu > \frac{1}{2}n$.

En mettant $n = 2m$ ou $2m + 1$, l'*invariant* s'évanouit en supposant que U ait le facteur θ^{m+1}.

Les conditions pour que la fonction U ait un tel facteur, se trouvent en égalant à zéro les coefficients différentiels de U du $m^{\text{ième}}$ ordre par rapport aux variables x, y, et en éliminant ces variables.

Mais avant d'aller plus loin il convient d'entrer dans quelques détails de la théorie d'une telle élimination. Je prends l'exemple le plus simple, et je suppose que l'on ait à éliminer x, y des équations

$$\begin{aligned} ax + by &= 0, \\ bx + cy &= 0, \\ cx + dy &= 0. \end{aligned}$$

On est habitué à dire que ce système équivaut à *deux équations* entre les seuls coefficients: mais cela n'est juste que dans un sens qui manque de precision. Le système équivaut plutôt à *deux relations* entre les coefficients, et ces deux relations sont exprimées par les *trois équations*, $bd - c^2 = 0$, $bc - ad = 0$, $ac - b^2 = 0$. Il n'est pas vrai que deux de ces équations embrassent *nécessairement* la troisième. En effet, la première et la seconde équations sont satisfaites en écrivant $c=0$, $d=0$, mais ces valeurs sont absolument étrangères à la question, et ne satisfont pas à la troisième équation, de manière que toutes les trois équations sont nécessaires pour exprimer les relations entre les coefficients. C'est pourquoi je dis que ces trois équations sont des résultats *distincts* de l'élimination. Et de même, pour un système quelconque d'équations, le nombre des résultats distincts de l'élimination n'est pas généralement à beaucoup près si faible que le nombre des relations entre les coefficients. Qu'on veuille consulter sur ce sujet mon mémoire "On the order of certain systems of algebraical equations," *Cambridge and Dublin Mathematical Journal*, t. IV. [1849] pp. 132—137 [77], et le mémoire de M. Salmon "On the Classification of curves of double curvature," t. V. [1850] pp. 23—46.

Je reviens à l'objet de cette note, et je suppose qu'en égalant à zéro les coefficients différentiels du $m^{\text{ième}}$ ordre de la fonction U, les équations $P=0$, $Q=0$, $R=0$, &c. forment le système entier des résultats distincts de l'élimination. Un *invariant* quelconque I s'évanouira en supposant $P=0$, $Q=0$, $R=0$, &c. Il doit donc exister une équation telle que $\lambda I = \alpha P + \beta Q + \gamma R + \ldots$, où λ, α, β, $\gamma\ldots$ sont des fonctions rationnelles et intégrales des coefficients. Mais de plus, la fonction λ doit être purement numérique, ou ce qui est le même, doit se réduire à *l'unité*, car autrement $I=0$ serait un résultat de l'élimination différent des résultats $P=0$, $Q=0$, $R=0$, &c., et ces équations ne seraient plus le système entier des résultats distincts. Donc enfin: un *invariant* quelconque I sera exprimé par une équation telle que

$$I = \alpha P + \beta Q + \gamma R + \ldots,$$

α, β, γ, ... étant des fonctions intégrales et rationnelles des coefficients.

Les résultats que je viens d'obtenir s'accordent parfaitement avec ceux dans ma "Note sur les hyperdéterminants," t. XXXIV. [1847] pp. 148—152 [54]. En effet, j'y ai fait voir qu'en supposant qu'une fonction $ax^4 + 4bx^3y + 6cx^2y^2 + 4dxy^3 + ey^4$ ait un facteur $(\alpha x + \beta y)^3$, l'élimination des variables entre les équations

$$\begin{aligned} ax^2 + 2bxy + cy^2 &= 0, \\ bx^2 + 2cxy + dy^2 &= 0, \\ cx^2 + 2dxy + ey^2 &= 0, \end{aligned}$$

donne lieu aux équations $ae - 4bd + 3c^2 = 0$, $ace + 2bcd - ad^2 - b^2e - c^3 = 0$; et les fonctions égalées à zero sont en effet les seuls *invariants* de la fonction du quatrième ordre. J'ajoute que la théorie actuelle fait voir aussi que dans le cas dont il s'agit, la dérivée

$$(ax^2 + 2bxy + cy^2)(cx^2 + 2dxy + ey^2) - (bx^2 + 2cxy + dy^2)^2,$$

ou son développement

$$(ac - b^2)x^4 + 2(ad - bc)x^3y + (ae + 2bd - 3c^2)x^2y^2 + 2(be - cd)xy^3 + (ce - d^2)y^4,$$

se réduit (à un coefficient constant près) à $(\alpha x + \beta y)^4$: et au cas où la fonction donnée du quatrième ordre est supposée être un carré, cette fonction et la dérivée qui vient d'être écrite sont égales à un facteur constant près; resultat dont je me suis servi ailleurs.

NOTES AND REFERENCES.

1. As to the history of Determinants, see Dr Muir's "List of Writings on Determinants," *Quart. Math. Jour.* vol. XVII. (1882), pp. 110—149; and the interesting analyses of the earlier papers in course of publication by him in the *R. S. E. Proceedings*, vol. XIII. (1885—86) *et seq.*

The (new?) theorem for the multiplication of two determinants was given by Binet in his "Mémoire sur un système de formules analytiques &c." *Jour. École Polyt.* t. X. (1815), pp. 29—112.

An expression for the relation between the distances of five points in space, but not by means of a determinant or in a developed form, is given by Lagrange in the Memoir "Solutions analytiques de quelques problèmes sur les pyramides triangulaires," *Mém. de Berlin*, 1773: the question was afterwards considered by Carnot in his work "Sur la relation qui existe entre les distances respectives de cinq points quelconques pris dans l'espace, suivi d'un essai sur la théorie des transversales," 4to Paris, 1806. Carnot projected four of the points on a spherical surface having for its centre the fifth point, and then, from the relation connecting the cosines of the sides and diagonals of the spherical quadrilateral, deduced the relation between the distances of the five points: this is given in a completely developed form, containing of course a large number of terms.

Connected with the question we have the theorem given by Staudt in the paper "Ueber die Inhalte der Polygone und Polyeder," *Crelle* t. XXIV. (1842), pp. 252—256; the product of the volumes of two polyhedra is expressible as a rational and integral function of the distances of the vertices of the one from those of the other polyhedron.

More general determinant-formulæ relating to the "powers" of circles and spheres have been subsequently obtained by Darboux, Clifford and Lachlan: see in particular Lachlan's Memoir, "On Systems of Circles and Spheres," *Phil. Trans.* vol. CLXXVII. (1886), pp. 481—625.

2 and 3. The investigation was suggested to me by a passage in the *Mécanique Analytique*, Ed. 2 (1811), t. I. p. 113 (Ed. 3, p. 106); after referring to a formula of Laplace, whereby it appeared that the attraction of an ellipsoid on an exterior point depends only on the quantities B^2-A^2 and C^2-A^2 which are the squares of

the eccentricities of the two principal sections through the major semiaxis A, Lagrange remarks that, starting from this result and making use of a theorem of his own in the Berlin Memoirs 1792—93, he was able to construct the series by means of the development of the radical $1 \div \sqrt{x^2+y^2+z^2-2by-2cz+b^2+c^2}$ in powers of b, c, preserving therein only the even powers of b and c, and transforming a term such as $Hb^{2m}c^{2n}$ into a determinate numerical multiple of $\frac{4}{3}\pi ABC \,.\, H(B^2-A^2)^m(C^2-A^2)^n$.

It occurred to me that Lagrange's series must needs be a series

$$\Sigma\, \zeta_p \left(A^2\frac{d^2}{da^2}+B^2\frac{d^2}{db^2}+C^2\frac{d^2}{dc^2}\right)^p \phi(a,\ b,\ c),$$

reducible to his form as a function of B^2-A^2, C^2-A^2, in virtue of the equation $\left(\frac{d^2}{da^2}+\frac{d^2}{db^2}+\frac{d^2}{dc^2}\right)\phi(a,\ b,\ c)=0$ satisfied by the function ϕ (I wrote this out some time before the Senate House Examination 1842, in an examination paper for my tutor, Mr Hopkins): and I was thus led to consider how the series in question could be transformed so as to identify it with the known expression for the attraction as a single definite integral.

I remark that my formulæ relate to the case of n variables: as regards ellipsoids the number of variables is of course $=3$: in the earlier solutions of the problem of the attraction of ellipsoids there is no ready method of making the extension from 3 to n. The case of n variables had however been considered in a most able manner by Green in his Memoir "On the determination of the exterior and interior attractions of Ellipsoids of variable densities," *Camb. Phil. Trans.* vol. V. 1835, pp. 395—430 (and Mathematical Papers, 8vo London, 1871, pp. 187—222); and in the Memoir by Lejeune-Dirichlet, "Sur une nouvelle méthode pour la détermination des intégrales multiples," *Liouv.* t. IV. (1839), pp. 164—168, although the case actually treated is that of three variables, the method can be at once extended to the case of any number of variables: it is to be noticed also that the methods of Green and Lejeune-Dirichlet are each applicable to the case of an integral involving an integer or fractional negative power of the distance. This is far more general than my formulæ, for in them the negative exponent for the squared distance is $=\frac{1}{2}n$, and, by differentiation in regard to the coordinates $a, b, \ldots$ of the attracted point, we can only change this into $\frac{1}{2}n+p$, where p is a positive integer. But in 28, the radical contained in the multiple integral is $\dfrac{1}{\{(a_1-x_1t)^2+\ldots\}^{\frac{1}{2}n-s}}$, where s is integer or fractional, and by a like process of expansion and summation I obtain a result depending on a single integral $\displaystyle\int_0^1 \frac{(1-u)^\sigma u^{i+\kappa-1}du}{\{(\zeta+h_1^2u)\ldots\}^{\frac{1}{2}}}$. And in 29, retaining throughout the general function $\phi(a_1-x_1, \ldots)$ and making the analogous transformation of the multiple integral itself, I express the integral

$$V=\int dx_1\ldots dx_n\ x_1^{2a_1+1}\ldots x_{f+1}^{2a_f+1}\ldots\ \phi(a_1-x_1, \ldots)$$

in terms of an integral $\displaystyle\int_0^1 T^{n+1}(1-T^2)^{k+f}W dT$, where $W=\displaystyle\int dx_1\ldots dx_n\, \phi(a_1-x_1T, \ldots)$.

I recall the fundamental idea of Lejeune-Dirichlet's investigation; starting with an integral $\iiint U dx dy dz$ over a given volume he replaces this by $\iiint \rho U dx dy dz$ where ρ is a discontinuous function, $=1$ for points inside, and $=0$ for points outside, the given volume; such a function is expressible as a definite integral (depending on the form of the bounding surface) in regard to a new variable θ: the limits for x, y, z, may now be taken to be ∞, $-\infty$ for each of the variables x, y, z, and it is in many cases possible to effect these integrations and thus to express the original multiple integral as a single definite integral in regard to θ.

I have not ascertained how far the wholly different method in Lejeune-Dirichlet's Memoir "Sur un moyen général de vérifier l'expression du potentiel relatif à une masse quelconque homogène ou hétérogène," *Crelle*, t. XXXII. (1846), pp. 80—84, admits of extension in regard to the number of variables, or the exponent of the radical.

4. As noticed p. 22, the investigation was suggested to me by Mr Greathead's paper, "Analytical Solutions of some problems in Plane Astronomy," *Camb. Math. Jour.* vol. I. (1839), pp. 182—187, giving the expression of the true anomaly in multiple sines of the mean anomaly. I am not aware that this remarkable expression has been elsewhere at all noticed except in a paper by Donkin, "On an application of the Calculus of Operations in the transformation of trigonometrical series," *Quart. Math. Journal*, vol. III. (1860), pp. 1—15; see p. 9, *et seq.*

5. In a terminology which I have since made use of:

The Postulandum or Capacity (□) of a curve of the order r is $=\frac{1}{2}r(r+3)$; and the Postulation (∇) of the condition that the curve shall pass through k given points is in general $=k$.

If however the k points are the mn intersections of two given curves of the orders m and n respectively, and if r is not less than m or n, and not greater than $m+n-3$, then the postulation for the passage through the mn points, instead of being $=mn$, is $=mn-\frac{1}{2}(m+n-r-1)(m+n-r-2)$.

Writing $\gamma=m+n-r$, and $\delta=\frac{1}{2}(\gamma-1)(\gamma-2)$, the theorem may be stated in the form, a curve of the order r passing through $mn-\delta$ of the mn points of intersection will pass through the remaining δ points. The method of proof is criticised by Bacharach in his paper, "Ueber den Cayley'schen Schnittpunktsatz," *Math. Ann.* t. 26 (1886), pp. 275—299, and he makes what he considers a correction, but which is at any rate an important addition to the theorem, viz. if the δ points lie in a curve of the order $\gamma-3$, then the curve of the order r through the $mn-\delta$ points does not of necessity nor in general pass through the δ points. See my paper "On the Intersection of Curves," *Math. Ann.* t. XXX. (1887), pp. 85—90.

6. The formulæ in Rodrigues' paper for the transformation of rectangular coordinates afterwards presented themselves to me in connexion with Quaternions, see 20; and again in connexion with the theory of skew determinants, see 52.

8. A correction to the theorem (18), p. 42, is made in my paper "Notes on Lagrange's theorem," *Camb. and Dubl. Math. Jour.* vol. VI. (1851), pp. 37—45.

10. This paper is connected with 5, but it is a particular investigation to which I attach little value. The like remark applies to 40.

12. The second part of this paper, pp. 75—80, relates to the functions obtained from n columns of symbolical numbers in such manner as a determinant is obtained from 2 columns, and which are consequently sums of determinants: they are the functions which have since been called Commutants; the term is due to Sylvester.

13. In modern language: Boole (in his paper "Exposition of a general theory of linear transformations," *Camb. Math. Jour.* vol. III. (1843), pp. 1—20 and 106—119) had previously shown that a discriminant was an invariant; and Hesse in the paper "Ueber die Wendepunkte der Curven dritter Ordnung," *Crelle,* t. XXVIII. (1844), pp. 68—96, had established certain covariantive properties of the ternary cubic function. I first proposed in this paper the general problem of invariants (that is, functions of the coefficients, invariantive for a linear transformation of the facients), treating it by what may be called the "tantipartite" theory: the idea is best seen from the example p. 89, viz. for the tripartite function

$$U = ax_1y_1z_1 + bx_2y_1z_1 + cx_1y_2z_1 + dx_2y_2z_1 + ex_1y_1z_2 + fx_2y_1z_2 + gx_1y_2z_2 + hx_2y_2z_2,$$

we have a function of the coefficients which is simultaneously of the forms

$$H\begin{vmatrix} a, & b, & c, & d \\ e, & f, & g, & h \end{vmatrix}, \quad H\begin{vmatrix} a, & b, & e, & f \\ c, & d, & g, & h \end{vmatrix}, \quad H\begin{vmatrix} a, & c, & e, & g \\ b, & d, & f, & h \end{vmatrix},$$

and as such it is invariantive for linear transformations of the (x_1, x_2), (y_1, y_2), (z_1, z_2).

Passing from the tantipartite form to a binary form, I obtained for the binary quartic the quadrinvariant $(I =)\ ae - 4bd + 3c^2$: as noticed at the end of the paper, the remark that there is also the cubinvariant $(J =)\ ace - ad^2 - b^2e - c^3 + 2bcd$ was due to Boole. The two functions present themselves, but without reference to the invariantive property and not in an explicit form, in Cauchy's Memoir "Sur la détermination du nombre des racines réelles dans les équations algébriques," *Jour. École Polyt.* t. X. (1815), pp. 457—548.

In p. 92 it is assumed that the invariant called θu is the discriminant of the function $U = ax_1y_1z_1w_1 \ldots + px_2y_2z_2w_2$: but, as mentioned in [ ], the assumption was incorrect. This was shown by Schläfli in his Memoir, "Ueber die Resultante eines Systemes mehrerer algebraischen Gleichungen," *Wiener Denks.* t. IV. Abth. 2 (1852), pp. 1—74: see pp. 35 *et seq.* The discriminant is there found by actual calculation to be a function (not of the order 6 as is θu, but) of the order 24, not breaking up into factors; in the particular case where the coefficients $a, \ldots p$ are equal, 1, 4, 6, 4, 1 of them to a, b, c, d, e respectively, in such wise that changing only the variables the function becomes $= (a, b, c, d, e \between x, y)^4$, then the discriminant in question does break up into factors, the value in this case being $J^6(I^3-27J^2)$ of the order 24 as in the general case, but containing the factor I^3-27J^2 which is the discriminant of the binary quartic.

14. In this paper I developed what (to give it a distinctive name) may be called the "hyperdeterminant" theory, viz. the expressions considered are of the form

$$\overline{12}^{\alpha}\,\overline{13}^{\beta}\,\overline{23}^{\gamma}\ldots U_1U_2U_3\ldots,$$

where after the differentiations the variables (x_1, y_1), (x_2, y_2), ... are to be or may be put equal to each other: it is to be noticed that although in the examples I chiefly consider constant derivatives, or invariants, the memoir throughout relates as well to covariants as invariants. The theory is to be distinguished from Gordan's process of Ueberschiebung, or derivational theory, viz. this may be considered as dealing exclusively, or nearly so, with the single class of derivatives $(V, W)^{\alpha}, = \overline{12}^{\alpha}V_1W_2$: the theorem that all the covariants of a binary function can be obtained successively by operating in this manner on the function itself and a covariant of the next inferior degree was a very important one.

15. Eisenstein's theorem may be stated as follows: the function $a^2d^2+4ac^3-6abcd+4b^3d-3b^2c^2$ (which is the discriminant of the binary cubic $(a, b, c, d\between x, y)^3$) is automorphic, viz. it is converted into a power of itself when for a, b, c, d we substitute the differential coefficients $\frac{d\phi}{da}, \frac{d\phi}{db}, \frac{d\phi}{dc}, \frac{d\phi}{dd}$ of the function itself. It is remarkable, see 54, that the function is automorphic in a different manner, viz. the Hessian determinant formed with the second differential coefficients $\frac{d^2\phi}{da^2}$, &c., is also equal to a power of the function itself. The first part of the paper relates to the function $a^2h^2+b^2g^2+\ldots+4bceh$ which had presented itself to me, 13, in the theory of linear transformations, and which is in like manner automorphic for the change $a, b, \ldots$ into $\frac{d\phi}{da}, \frac{d\phi}{db}$, &c. The function however occurs in connexion with the arithmetical theory of the composition of quadratic forms, Gauss, *Disquisitiones Arithmeticæ* (1801), and see 92. The second part gives for the binary quartic covariant an automorphic formula analogous to those previously obtained by Hesse for the ternary cubic, viz. the Hessian of any linear function of the quartic and its Hessian, is itself a linear function of the quartic and its Hessian, the coefficients depending on the invariants I, J of the quartic form.

16. This is a mere reproduction of 13 and 14, and requires no remark.

19 and 23. These papers contain a mere sketch of the application of the doubly infinite product expression of the elliptic function sn u to the problem of transformation. As noticed in 23, I purposely abstained from any consideration of the infinite limiting values of m and n.

20. The discovery of the formula $q(ix+jy+kz)q^{-1}=ix'+jy'+kz'$, as expressing a rotation, was made by Sir W. R. Hamilton some months previous to the date of this paper. As appears by the paper itself, I was led to it by Rodrigues' formulæ, see 6. For the further development of the theory, see 68.

21. The system of imaginaries $i_1, i_2, \dots i_7$ had presented itself to J. T. Graves about Christmas 1843, see his paper "On a Connection &c." *Phil. Mag.* vol. XXVI. (1845), pp. 315—320. They are called by him Octads, or Octaves.

24 and 25. These papers precede the researches of Eisenstein on the same subject. *Crelle*, t. XXXV. (1847). It was I think right that the theory of the doubly infinite products should have been investigated as in these papers: but the investigation is in some measure superseded by the beautiful theory of Weierstrass, viz. he takes for the element of the product (not a mere linear function of u, but) a linear function multiplied by an exponential factor,

$$\sigma u = u\Pi\left\{\left(1+\frac{u}{w}\right)e^{-\frac{u}{w}-\frac{1}{2}\frac{u^2}{w^2}}\right\},$$

$w=2m\omega+2m'\omega'$ where the ratio $\omega:\omega'$ is imaginary, and the product extends to all positive or negative integer values of m, m' (the simultaneous values 0, 0 excluded): in consequence of the introduction of the exponential factor the form of the bounding curve becomes immaterial, and the only condition is that it shall be ultimately everywhere at an infinite distance from the origin.

The general theory of Weierstrass in regard to the exponential factor is given in the Memoir, "Zur Theorie der eindeutigen analytischen Functionen," *Berlin. Abh.* 1876 (reprinted, *Abhandlungen aus der Functionenlehre*, 8° Berlin, 1886): and the application to Elliptic Functions is made in his lectures, edited by Schwarz, *Formeln und Lehrsätze* u. s. w. 4° Gött. 1883. See also Halphen, *Théorie des Fonctions Elliptiques*, Paris, 1887.

26. The geometrical results in regard to corresponding points on a cubic curve are many of them due to Maclaurin. See his "De linearum geometricarum proprietatibus generalibus tractatus," published as an Appendix to his Treatise on Algebra, 5 Ed. Lond. 1788. (See also De Jonquières' "Mélanges de Géométrie pure," 8° Paris, 1856.) But the theorem in the "Addition" was probably new: the curve of the third class touched by the line PP' is the curve called by me the Pippian (as represented by the contravariant equation $PU=0$), but which has since been called the Cayleyan of the cubic curve.

27. The theory of the conics of involution was so far as I am aware new.

28 and 29. See 2, 3.

30. As noticed in the paper, the investigation is directly founded upon that of Plücker for the singularities of a plane curve. It is to be observed that the definition of a "line through two points" *ligne menée par deux points* (*non consecutifs en général*) *du système*, does not exclude actual double points, for the line through an actual double point is a line through two points, coincident indeed, but not consecutive; but

it would have been proper to notice the distinction between actual and apparent double points (first made by Dr Salmon, to whom the term apparent double point, adp, is due): and the like in regard to the lines in two planes. In the translation of this paper, 83, there are two footnotes signed G. S. (Dr Salmon), giving for plane curves the formulæ $\iota - \kappa = 3(\nu - \mu)$ and $2(\tau - \delta) = (\nu - \mu)(\nu + \mu - 9)$: and for curves of double curvature and developable surfaces the analogous formulæ $\alpha - \beta = 2(n - m)$, $x - y = (n - m)$ and $2(g - h) = (n - m)(n + m - 7)$. Also in the second set of six equations, p. 210, the last three equations are replaced by

$$x = \tfrac{1}{2}m(m-1)(m^2 - m - 4) - \tfrac{1}{2}(2h + 3\beta)(2h + 3\beta + 1) - m(m-1)(2h + 3\beta) + 3h + 4\beta,$$

$$\alpha = 2m(3m - 7) - 3(4h + 5\beta),$$

$$g = \tfrac{1}{2}m(3m-7)(3m^2 - 5m - 7) + \tfrac{1}{2}(6h + 8\beta)(6h + 8\beta + 1) - 3m(m-2)(6h + 8\beta) + 19h + 24\beta,$$

viz. these are the equations serving to express x, α, g in terms of m, h, β.

For the discussion of some singularities not considered in the present paper see Zeuthen, "Sur les singularités ordinaires des courbes géométriques à double courbure," *Comptes Rendus*, t. LXVII. (1868), pp. 225—233, and "Sur les singularités ordinaires d'une courbe gauche et d'une surface développable," *Annali di Matem.* t. III. (1869–70), pp. 175—217.

40. See 10.

41 and 44. For demonstrations of Sir W. Thomson's theorem for the value of the definite integral $\int \frac{dx \ldots}{\{(x-a)^2 + \ldots + u^2\}^i (x^2 + \ldots + v^2)^{i+1}}$ see his papers "Démonstration d'un theorème d'Analyse" and "Extrait d'une lettre à M. Liouville," *Liouv.* t. X. (1845), pp. 137—147 and 364—367, also the paper "On certain definite integrals suggested by problems in the theory of Electricity," *Camb. Math. Jour.* t. II. (1845), pp. 109—121.

45. I am not aware that the equation $\sqrt{k}\,\mathrm{sn}\,u = \mathrm{H}(u) \div \Theta(u)$ had been previously demonstrated otherwise than by the circuitous process employed in the *Fundamenta Nova*.

The series $z = 1 + C_1 \frac{x^2}{1\,.\,2} \ldots + C_r \frac{x^r}{1\,.\,2 \ldots r} + \ldots$, which is the solution of the differential equation $x^2 z + \alpha x \frac{dz}{dx} + \frac{d^2 z}{dx^2} - 2(\alpha^2 - 4)\frac{dz}{dx} = 0$, and in which C_1, C_2, ... denote the coefficients of the highest powers of n in the expressions given p. 299, is in fact (as remarked by me in a later paper, *Liouv.* t. VII. (1862)) the Weierstrassian function Al (x).

47. The surface here considered, the Tetrahedroid, is the general homographic transformation of the wave surface. It is a special case of the 16-nodal quartic surface considered by Kummer in his Memoir, "Algebraische Strahlen-systeme," *Berl. Abh.* 1866, pp. 1—120, and in various papers in the *Berliner Monatsberichte*.

48. The expressions $f_2x = \Sigma (a-b)^2 (x-c)(x-d) \ldots$ for the Sturmian functions in terms of the roots, or (to use Sylvester's term), say the endoscopic expressions of these functions, were obtained by him, *Phil. Mag.* vol. XV. (1839).

It was interesting to express these in terms of the sums of powers S_1, S_2, &c., that is in terms of symmetrical functions of the coefficients, but for the actual expression of the Sturmian functions in terms of the coefficients the process is a very circuitous one, and the proper course is to start directly from the exoscopic expressions as linear functions of fx, $f'x$ also due to Sylvester (see his Memoir "On a theory of the syzygetic relations of two rational integral functions &c.," *Phil. Trans.* t. CXLIII. (1853), pp. 407—548), which is what is done in the subsequent paper 65.

49. I attach some value to this paper as a contribution to the theory of the Gamma function.

50, 55, 70, 95, 98. The general theorem of § I. was given in a very different and less suggestive notation, by an anonymous writer, "Théorèmes appartenant à la géométrie de la règle," *Gergonne* t. IX. (1818–19), pp. 289—291; viz. the statement is in effect as follows: considering in a plane or in space any n points 1, 2, 3 ... n; then joining these in order, take 12 any point in the line 1, 2; 23 any point in the line 2, 3, and so on to $n-1 . n$. Take then 123 the intersection of the lines 1, 23 and 12, 3; 234 the intersection of the lines 2, 34 and 23, 4; and so on to $n-2 . n-1 . n$. Take then 1234 the intersection of the lines 1, 234; 12, 34; and 123, 4 (viz. these three lines will meet in a point): 2345 the intersection of the lines 2, 345; 23, 45; 234, 5 (viz. these three lines will meet in a point)... and so on to $n-3 . n-2 . n-1 . n$. And so on for 12345, &c. up to 123 ... n; in the successive constructions we have four, five, ... and finally n lines which in each case meet in a point. A proof is given by Gergonne, t. XI. (1820–21).

A large part of this paper relates to the theory of the relations to each other of the 60 Pascalian lines derived from the hexagons which can be formed with the same six points upon a conic: the literature of the question is very extensive, and I hope to refer to it again in the Notes to another volume.

54. See for an addition which should have been printed with this paper, the last paragraph of 92.

63. Boole's theorem of integration which is here demonstrated is a very remarkable one, and it would be very interesting to investigate the general forms, or a larger number of particular forms, for the functions P, Q satisfying the condition mentioned in the theorem. It may be remarked that the demonstration is very closely connected with Lejeune-Dirichlet's method for the determination of certain definite integrals referred to in 2 and 3, we have a triple integral the real part of which is $= fP \div Q^{\frac{1}{2}n+q}$ or 0, according as P is or is not comprised between the limits 0 and 1. It includes Boole's formula mentioned in 44 and in 64.

65. See 48.

66. The method here employed of establishing the theory of Laplace's coefficients in n-dimensional space by means of rectangular coordinates, has I think some advantage

over the ordinary one, in which angular coordinates are introduced. The method is not given in Heine's *Kugelfunctionen*, Berlin 1878–81.

67, 93, and 99. These papers relate to the equation of differences, see p. 540, derived from Jacobi's equation $n(n-1)x^2z+(n-1)(\alpha x-2x^3)\frac{dz}{dx}+(1-\alpha x^2+x^4)\frac{d^2z}{dx^2}-2n(\alpha^2-4)\frac{dz}{d\alpha}=0$: the investigation seems to show that the integration cannot be effected in any tolerably simple form.

68. See 20.

69. This paper relates to the theory of the functions considered by Jacobi, and to which the name "Pfaffian" has since been given. It is shown that the definition of a determinant may be so extended as to include within it the Pfaffian: and it is proved that a symmetrical skew determinant is the square of a Pfaffian.

71. I doubt whether the theorem is true except in the cases $n=3$ and $n=4$.

76. As mentioned at the conclusion of the Memoir the whole subject was developed in a correspondence with Dr Salmon. Steiner's researches upon Cubic Surfaces are of later date, viz. we have his Memoir, "Ueber die Flächen dritten Grades" (read to the Berlin Academy 31 January 1856), *Crelle*, t. LIII. (1857), pp. 133—141 and *Werke*, t. II. pp. 651—659.

77. Contains the general definition of the 'order' of a system of equations.

81. Contains the remark that the Cartesian has a *cusp* at each of the two circular points at infinity.

82. The Problem of the fifteen school girls was proposed by Kirkman, *Lady's and Gentleman's Diary*, 1850: it is a particular case of the Prize-question proposed by him in the *Diary* for 1844. A great deal has been written on the subject.

83. See 30.

92. See 15.

94. This paper contains my fourfold formulæ for the addition of the elliptic functions sn, cn, and dn.

100. I remark that the terms covariant, invariant, here referred to as introduced by Sylvester, were first employed (together with many other valuable new terms) in the first part of his paper "On the Principles of the Calculus of Forms," *Camb. and Dubl. Math. Jour.* vol. VII. (1852), pp. 52—97. I hope to give in the next volume, in connexion with my Introductory Memoir on Quantics, a review of the earlier history of the subject.

END OF VOL. I.

CAMBRIDGE:
PRINTED BY C. J. CLAY, M.A. AND SONS,
AT THE UNIVERSITY PRESS.